This book is to be returned on or before
the last date stamped below.

# TITANIUM '80

## SCIENCE AND TECHNOLOGY

Proceedings of the Fourth International Conference on Titanium

Kyoto, Japan   May 19-22, 1980

## Edited by: H. Kimura
## O. Izumi

*"This international conference was supported by a grant from the Japan World Expositions Commemorative Fund."*

$D$
$669.732$
$TIT$

**A Publication of The Metallurgical Society of AIME**
P.O. Box 430
420 Commonwealth Drive
Warrendale, Pa. 15086
(412) 776-9000

# Table of Contents

## C. Superplasticity and Creep

## MECHANICAL PROPERTIES
### A. Mechanical Properties I

## B. Phase Transformation and Heat Treatments

# VOLUME 2

# DEFORMATION

# A. Deformation Dynamics

# STUDY OF THE PLASTIC DEFORMATION DYNAMICS OF HIGH PURITY TITANIUM AND DILUTE TITANIUM-OXYGEN SOLID SOLUTION

G. Baur[*] and P. Lehr[**]

[*]Centre d'Etudes de Chimie Métallurgique. Vitry, France

[**]Laboratoire de Matériaux à Haute Résistance
Ecole Nationale Supérieure de Techniques Avancées. Paris, France

## Introduction

Numerous research works [1 to 11] have demonstrated the importance of interstitial impurities, especially oxygen, in the plastic deformation dynamics of titanium. However these studies were generally carried out on titanium grades of different origins, containing non-negligeable levels of residual impurities.

The aim of the present investigation is to study the deformation dynamics of high purity iodide titanium with a very low interstitial content, particularly oxygen (90 at. ppm), and to analyse the influence of the later on the plastic behavior of a series of titanium-oxygen alloys (90 - 3200 at. ppm) prepared from the same high purity material. In carrying out the work the influences of strain rate $\varepsilon'_p$ , plastic strain $\varepsilon_p$, oxygen content and temperature were examined.

## Experimental Procedure

The starting material is a high purity iodid titanium elaborated at the Centre d'Etudes de Chimie Métallurgique [12] [13]. The chemical analysis is reported in table I.

The titanium-oxygen alloys were prepared by limited oxidation of titanium cylinders (about 30 gr. in weight) carried out in a thermobalance, at 1000° C, in purified oxygen atmosphere, under reduced pressure ($10^{-4}$ Torr.). They were then levitation-melted and homogenized. Their oxygen concentrations (table II) calculated from measured weight gains were verified, after melting and homogenizing, by activation analysis using the nuclear reaction $O^{16}(T,n)F^{18}$*. More information concerning the elaboration and the analysis of the alloys can be found in a previous publication [14]. Furthermore it has been verified that the preparation does not introduce changes in the concentration of metallic elements.

Tensile tests and stress relaxation experiments were conducted with an Instron TT-D type tensile machine, using small size polycristalline specimens (rectangular cross section 2 mm x 1 mm, length gage 15 mm). These specimens were cut in cold rolled sheets (95 % thickness reduction) parallely to rolling direction. They were further annealed under high vacuum ($10^{-7}$ Torr) during one hour at 650 or 800° C in order to obtain a nearly equivalent grain size (mean diameter of 15 to 16 μm) for all the alloys.

In relaxation tests the specimens were first elongated to a predetermined plastic strain at a strain rate $\varepsilon$ = 5.5 x $10^{-4}$ $sec^{-1}$, then the crosshead was stopped and the stress decrease $\Delta\sigma$ recorded as function of time t. Activation volumes were calculated from the slope of $\Delta\sigma$ versus log t plots.

Table I

Chemical analysis of iodid titanium[*]

| O | N | C | Fe | Zr | Cu | Mn | Cr |
|---|---|---|----|----|----|----|----|
| 30 | <30 | <10 | 26 | 39 | 14.3 | <1 | 0.8 |

[*] Concentrations in ppm by weight, determined by activation analysis, except nitrogen, analysed by the Kjeldahl technique.

Table II

Oxygen concentrations of Ti-O alloys
(atomic percent)

| designation | A | B | C | D | E | F |
|---|---|---|---|---|---|---|
| thermogravimetry[*] | 0.0192 | 0.043 | 0.1085 | 0.1561 | 0.2586 | 0.3200 |
| activation analysis[**] | | 0.030 | 0.1026 | 0.1336 | 0.2332 | 0.3104 |

[*] oxygen concentrations deduced from weight gains

[**] oxygen concentrations controlled by activation analysis.


The plastic deformation of the studied samples involves predominantly prismatic slip. The preferential orientation of the grains corresponding to the crystallization texture is in favour of this deformation mode. A calculation based on the recrystallization texture, as described by Brehm [15], leads to a Schmidt Factor of Fs = 0.433.

Isothermal stress relaxation experiments were conducted at four temperatures : 77 K, 151 K, 293 K and 423 K. The useful temperature range is in fact restricted on both high and low values. Tests made at 4.2 K have shown that plastic deformation takes place esentially by twinning, an athermal mechanism which does not give rise to relaxation. [16]. On the other hand above 473 K the occurence of a strain-stress aging effect leads to unworkable results.

Results

Stress relaxation at room temperature

The stress relaxation behavior of the whole set of Ti-O alloys at room
temperature is shown on a semi-log plot, for a plastic strain $\varepsilon^P = 0.1$, in
figure 1. Straight lines can be drawn through the data points for each alloy.
The relaxation amplitude $\Delta\sigma$, at a given time, increases with the oxygen concentration.

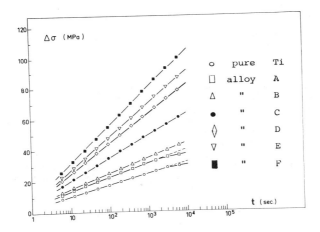

Fig. 1 - Stress relaxation curves at 295 K for high purity
titanium and Ti-O alloys (initial plastic strain
$\varepsilon^P = 0.1$)

The relaxation phenomenon is well described, at least for test durations
of the order of $10^3$ seconds, by the logarithmic function that follows :

$$\Delta\sigma = \sigma_0 - \sigma = a \, Ln \, (bt + 1) \tag{1}$$

where $\sigma_0$ is the stress from which relaxation begins, $\sigma$ the stress at time t
and a and b constants.

The logarithmic law can be interpreted in the theoretical frame of rate
controlling, thermally activated deformation process. [2] [7] [16 to 18].
According to the theory the shear strain rate $\dot{\gamma}_\rho$ is given by the relation :

$$\dot{\gamma}_\rho = \nu \exp \left(\frac{-\Delta H^*}{K \, T}\right) = \nu \exp \left(- \frac{H^* - V^* \tau^*}{K \, T}\right) \tag{2}$$

where $\Delta H^*$ is the activation enthalpy representing the thermal activation contribution to the overcoming of obstacles whose short range stress field interacts with the mobile dislocations. The activation volume $V^*$ and the energy term $H^*$ are two characteristic parameters of the nature of the obstacles, T is the absolute temperature and K Boltzman's constant.

$\tau^*$ is the thermal stress component, or effective shear stress, for the involved slip system. It can be calculated from the applied stress $\sigma$ by the relation :

$$\tau^* = F_S (\sigma - \sigma_i) \qquad (3)$$

where $\sigma_i$ is the athermal stress component (or long-range internal stress) and $F_s$ the Schmidt Factor of the slip system. The activation volume $V^*$ can be evaluated, on a semi-log plot, from the slope "a" of the relaxation straight lines using the following formula [5] [19] ;

$$\frac{1}{a} = \frac{V^* F_S}{K T} (1 + \frac{\theta S}{R L}) \qquad (4)$$

The corrective term in brackets is introduced to take into account the elastic relaxation of the tensile machine during the test. R is the machine's stiffness, L and S respectively the length and the cross section of the tensile specimen and $\theta$ ($\theta = d\sigma/d\varepsilon$) the work hardening rate of the material for the studied plastic strain level $\varepsilon p$.

According to the above relationship the activation volume, in given conditions, is inversely proportional to the slope of the relaxation straight lines. Consequently according to figure 1 the activation volume decreases when the oxygen content of the alloys increases.

On the other hand the activation volume can be expressed as follows :

$$V^* = 1^* b \ d \qquad (5)$$

where $1^*$ is the mean length of dislocation segments involved in the thermally activated process, b is their Burgers vector and d the activation length swept out per succesful thermal fluctuation.

The length $1^*$ can be identified in a first aproximation as the average spacing between obstacles. The decreasing of the activation volume when the oxygen content increases indicates that the interstitial oxygen atoms are directly concerned in the plastic deformation rate controlling mechanism.

Figure 2 shows the influence of the initial plastic strain $\varepsilon p$ on the room temperature activation volume. For a given $\varepsilon p$ value the activation volume increase when the oxygen concentration decreases. For the more concentrated alloys (D, E and F) the activation volume appears to be independent of plastic strain. This observation is in agreement with the hypothesis of a single interaction mechanism between moving dislocations and interstitial oxygen atoms.

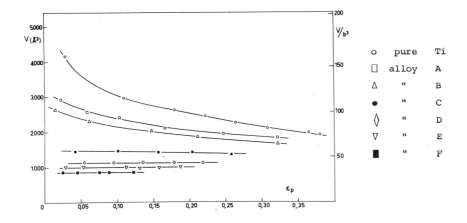

Fig. 2 – Activation volume versus plastic strain
at 295 K, for high purity titanium and Ti–O
alloys.

However, for pure titanium and the dilute alloys the activation volume $V^*$
diminishes when the plastic strain increases. This fact indicates that the density of obstacles interacting with moving dislocations increases with plastic
strain $\varepsilon_p$. It suggests, consequently, that at the lower oxygen concentration
levels a contribution to the plastic deformation dynamics is brought by the
forest dislocations cutting mechanism.

Internal and thermal stress components.

According to the concept of thermally activated plastic deformation the
flow stress $\sigma$ can be considered as the sum of two basic stress components :

$$\sigma = \sigma^* + \sigma_i \qquad (6)$$

where $\sigma^*$ is the thermal stress which is particularly sensitive to temperature,
strain rate and chemical composition, and $\sigma_i$ the long range internal stress.

In a relaxation experiment the flow stress $\sigma$, plotted versus time, on
principle tends asymptotically towards a limit value ; this value is the internal stress. In practice, however, this limit is not easily reached.

In the present investigation, the values of the internal stress were
obtained, at room temperature, using an incremental unloading procedure which
derives from the method proposed by Mc Ewen and coworkers. [20]. At the other
temperatures, the internal stress was evaluated by extrapolation from the room
temperature value, assuming that its variation with temperature is only depending on the shear modulus $\mu$ according to the following relation :

$$\frac{d\sigma_i}{dT} = \frac{\sigma_i}{\mu} = \frac{d\mu}{dT} \qquad (7)$$

    In figure 3 and 4 are reported, as an example, the variations of the
thermal and internal flow stress components with respect to plastic strain
for high purity titanium and Ti -0.26 at % O alloy. For the alloy in consideration the thermal component $\sigma^*$ appears to be fairly independent of plastic strain. This fact implies that work hardening affects only the internal
stress component. On the other hand for high purity titanium (fig.3) and more
generally for Ti-O alloys whose oxygen content is lower than approximately
0.1 at %, a very slight increase of the thermal stress with plastic strain is
observed. Such an increase is more important the lower the oxygen content is.
This point agrees with the above mentioned hypothesis concerning a rate controlling contribution of the "forest" dislocations cutting mechanism.

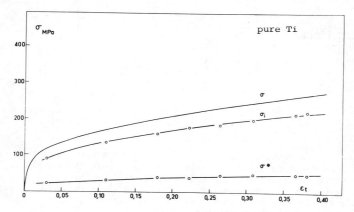

Fig. 3 - Thermal and internal components of the
flow stress σ, at 295 K, versus strain, for
high purity titanium.

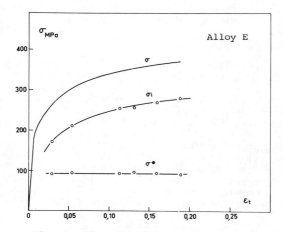

Fig. 4 - Thermal and internal components of the flow
stress σ, at 295 K, versus strain, for Ti-0,26
at % O alloy (alloy E).

Comparing the behavior of the different titanium—oxygen alloys shows that increasing the oxygen content affects both the thermal $\tau^*$ and internal $\tau_i$ shear stress components.

The influence of the oxygen concentration on internal stress at room temperature is depicted in figure 5, where $\tau_i$ is plotted versus the square root of plastic strain. For a given plastic strain $\tau_i$ increases linearly with oxygen concentration. This linear increase is in accord with the model of Mott-Nabarro, [21] [22] based on unlocalized interactions, and the experimental results of Conrad and Okazaki. [8] [23]

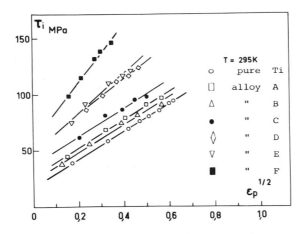

Fig.5 - Internal shear stress component, at 295 K, versus the square root of plastic strain, for high purity titanium and Ti-O alloys.

The internal stress $\tau_i$, then, increases linearly with respect to the square root of the plastic strain $\varepsilon p$, that means that it will also increase linearly with respect to the square root of the dislocation density, for this later is proportional to the plastic strain $\varepsilon p$. [4] [24]

For a given strain the thermal component $\tau^*$ increases proportionately to the square root of the oxygen concentration, as depicted in figure 6 for plastic strain $\varepsilon p = 0.02$. This type of variation is in agreement with the usual models of solid solution hardening involving localized interactions [25] as well as with previous studies concerning the variation of the thermal stress of titanium with the interstitial contents [8] [10] [26 to 29].

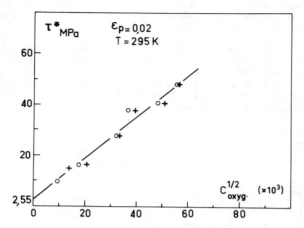

Fig.6 – Thermal shear stress component, at 295 K, versus the square
root of the oxygen concentration, for plastic strain $\varepsilon_p$ = 0.02.

From figure 6 it seems that the value of $\tau^*$ extrapolated to zero oxygen
concentration ($\tau^*_{c=0}$)   is not null. Moreover it has been observed that this
extrapolated value $\tau^*_{c=0}$ increases linearly with the square root of plastic
strain (figure 7).     This result indicates that, for the smaller interstitial solute contents, a non-negligible part of the thermal stress is used to
overcome other obstacles than the solute oxygen atoms. The density of these
obstacles would increase with plastic strain.

This observation supports our hypothesis according to which the cutting
process of "forest trees" by moving dislocations brings a contribution to the
control of the plastic deformation dynamics.

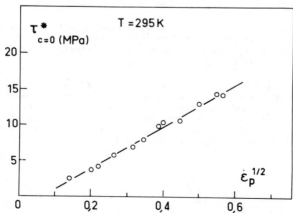

Fig.7 – Relation between plastic strain and the extrapolated   value
of the thermal shear stress to zero oxygen concentration.

These results are consistent with the hypothesis of Kocks [30] concerning the addability of stresses in the case of two superposed rate controlling mechanisms. Accordingly the thermal stress can be written as follows :

$$\tau^* = \tau_1^* + \tau_2^* \tag{8}$$

The $\tau_1^*$ term, which is proportional to the square root of the oxygen concentration, refers to the interaction mechanism between mobile dislocations and oxygen solute atoms, and $\tau_2^*$ represents the extrapolated value of $\tau^*$ to zero oxygen concentration ($\tau_2^* = \tau^*_{c=o}$ ; figure 7) and concerns the "forest" dislocations cutting mechanism. This term becomes negligible, as compared to $\tau_1^*$, for oxygen contents higher than approximately 0.1 at %. The thermal stress would be then independent of strain $\varepsilon p$ (fig. 3 and 4).

These results can be used to calculate the force F exerted by each type of obstacles on the dislocations segments. The theoretical models suggest that the thermal stress $\tau^*$ is related to the obstacle density N and to the force F by the equation [31] :

$$\tau^* = (\frac{F^3}{\mu b^6})^{1/2} \ N^{1/2} \tag{9}$$

If interactions occur between moving dislocations and oxygen atoms, then N is equal to the oxygen concentration ($N=C_{oxygen}$). F can be therefore calculated from the slope of $\tau^*$ versus $C^{1/2}$ straight lines (fig. 6). We find for this type of interaction, at room temperature :

$$F = 24.5 \times 10^{-11} \ \text{Newtons} = 0.08 \ \mu \ b^2$$

In the case of the forest cutting mechanism the obstacles density N would be replaced by $N=b^2\rho$ [25] where $\rho$ is the dislocations density of the forest. The dislocations density varies proportionately to plastic strain : $\rho = K\varepsilon p$, with $K = 5.2 \times 10^{10}$ cm$^{-2}$ according to Brehm [4]. Equation (9) can be therefore written in the following manner :

$$\tau^* = ( \frac{F^3}{\mu b^6})^{1/2} \ b \ K^{1/2} \ \varepsilon p^{1/2} \tag{10}$$

The calculation based on the slope of $\tau^*_{c=o}$ versus $\varepsilon p^{1/2}$ straight lines (fig.7), for this intersection mechanism at room temperature, leads to the following acting force value :

$$F = 76.7 \times 10^{-11} \ \text{Newtons} = 0.25 \ \mu b^2$$

Accordingly, the force exerted by a dislocation of the forest would be, at room temperature, three times higher than that of an oxygen atom.

The variation of the thermal stress with temperature is given in figure 8 for high purity titanium and two titanium-oxygen alloys. The equation

$$\tau^* = \tau_0^* - \alpha T^{1/2} \tag{11}$$

reported by Conrad and al.[6] [27] [28] [32] is found adequate for the representation of this variation. The thermal stress $\tau^*$ increases with oxygen content in the whole temperature range investigated. The influence of oxygen increases when the temperature is lowered.

The critical temperature $T_c$ at which the thermal component $\tau^*$ cancels varies from 450 to 514 K, depending on alloy composition. These values are in agreement with previous ones [6] [32] relative to different titanium grades.

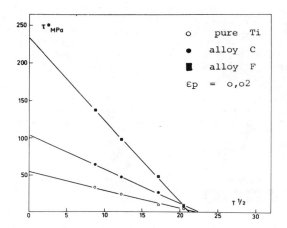

Fig.8 - Temperature dependence of thermal shear stress component for high purity titanium, alloy C (0.11 at % 0) and alloy F (0.32 at % 0).

Variation of activation volume and enthalpy with temperature.

The dependence of activation volume on temperature is depicted in figures 9 and 10. Figure 9 shows the results concerning high purity titanium and figure 10 those of a Ti–0.32 at % 0 alloy. Such data were deduced from stress relaxation experiments made at different temperatures and different plastic strain levels.

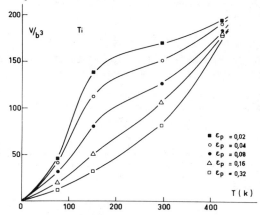

Fig.9 - Temperature dependence of activation volume of high purity titanium, for different plastic strains.

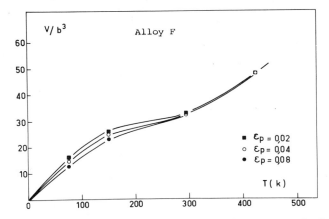

Fig.10 - Temperature dependence of activation volume of Ti-0.32
at % O alloy (alloy F)  for different plastic strains.

The activation volume decreases with temperature and tends toward
zero at absolute zero.

An examination of the curves obtained for the different Ti-O alloys
indicates that the activation volume, at given temperature and strain,
diminishes when the oxygen content increases. This fact is in agreement
with the above mentioned hypothesis of an interaction mechanism between mobile
dislocations and oxygen solute atoms.

These curves furthermore show that  at a given oxygen concentration  the
activation volume is dependent on plastic strain $\varepsilon p$. Such dependence, rather
weak at temperatures equal or higher than room temperature  becomes more important at low temperatures (77 K and 150 K). This behavior indicates an increasing contribution with decreasing temperature of the forest dislocations
cutting mechanism as a plastic deformation rate controlling process. It must
also be kept in mind that low temperature plastic deformation introduces a
notably higher density of dislocations, as it has been suggested by Tang and
Sommer [33].

However, the influence of plastic strain $\varepsilon p$ on the activation volume
diminishes when the oxygen content increases. As an example in the case of
the Ti-0.32 at % O alloy (fig. 10) the role of the plastic strain is particularly weak. Consequently, for oxygen contents of this order of magnitude
or higher, the plastic deformation dynamics is essentially controlled by the
interactions between moving dislocations and oxygen atoms.

The activation volume does not vary in a simple manner with temperature
especially for the lower strains and/or the lower oxygen concentrations
(fig.9). Curves of complex shapes but different of the present one have been
also obtained by several authors [10] [11] [34] and were tentatively interpreted with the assumption of a low temperatures transition (200 - 2504)
between two succesive interaction mechanisms. This question has been discussed in some details by Baur [16] who showed that the shape of the activation
volume variation with temperature depends essentially on the force distance

profile, which is characteristic of the nature of the involved obstacles. The shape can be complex, even in the case of a single interaction mechanism.

In the present case, the shape of the curves established for the lowest plastic strain ($\varepsilon p$ = 0.02) can be interpreted on the basis of a quasi hyperbolic force-distance profile which would be representative of the interactions between dislocations and oxygen atoms [16]. Figure 11 shows the variation of the activation enthalpy with respect to temperature for the lowest studied plastic strain ($\varepsilon p$ = 0.02). The activation enthalpy has been calculated using the relation furnished by Conrad and Widersich [35] :

$$\Delta H = -VT \left(\frac{\partial \tau}{\partial T}\right)^{*}_{\gamma_\rho}$$

This simple linear variation of $\Delta H$ with temperature is also consistant with their interpretation of a single interaction mechanism between dislocation and oxygen solute atom for this low plastic strain level ($\varepsilon p$ = 0.02). This mechanism is characterized, for high purity titanium, by a critical temperature $T_C$ = 500 K to which corresponds an activation enthalpy $\Delta H$ = 0.9 eV.

This value is in agreement with the values obtained by Conrad and al [2] [8] [28] on several titanium grades of different purities ($\Delta H$ = 0.95 to 1.24 eV).

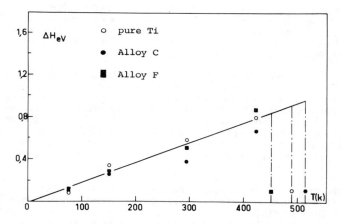

Fig. 11 - Temperature dependence of activation energy for high purity titanium, Ti −0.11 at % O (alloy C) and Ti −0.32 at % O (alloy F).

## Conclusions

The following conclusions can be drawn concerning the effect of solute oxygen on the plastic deformation dynamics of polycristalline high purity titanium and dilute titanium-oxygen alloys.

1) Increasing the oxygen content affects both the thermal and the internal components of the flow stress. The activation volume increases when the oxygen concentration decreases.

2) At room temperature  the thermal stress component and the activation volume appear fairly independent of the plastic strain  if the oxygen concentration of the Ti-O alloys is higher than 0.1 at %. It can be concluded that, in this case, the interactions between the oxygen solute atoms and the moving dislocations constitute the single rate controlling mechanism of the plastic deformation.

3) On the other hand, for oxygen concentrations lower than 0.1 at %, a slight increase of the thermal stress and an important decrease of the activation volume with increasing plastic strain are observed at room temperature. It must be concluded, therefore, that in these conditions  the intersection mechanism of the "forest trees" by the mobile dislocations brings also a contribution to the control of the plastic deformation dynamics.

4) This last contribution becomes increasingly important as temperature decreases. At 150 K, as an example, the activation volume remains plastic strain dependent  for oxygen concentrations up to approximately 0.32 at %. Consequently, at low temperatures and (or) low oxygen concentrations, the plastic deformation dynamics appears to be controlled by two superposed mechanism related to the overcoming of two types of obstacles : the oxygen solute atoms and the dislocations of the forest.

5) The activation volume increases in no simple manner with temperature. Its temperature dependence is affected both by oxygen concentration and plastic strain.

### References

1. R.N. Orava, G. Stone, H. Conrad, Trans. Quaterly, 59 n° 2, 171 (1966)
2. G. Sargent and H. Conrad, Scripta Met., 3, 43 (1969)
3. J. Kratochvil and H. Conrad, Scripta Met., 4, 815 (1970)
4. C. Brehm and P. Lehr, Métaux Corrosion Industrie, 551, 253 (1971) ; 553, 325 (1971) ; 359 (1971)
5. G. Baur and P. Lehr, C.R. Acad. Sci. Paris, Série C, 273, 332 (1971)
6. T. Tanaka and H. Conrad, Acta Met. 20, 1019 (1972)
7. G. Baur and P. Lehr, Comm. aux 3èmes Journées d'Etudes du Titane et de ses alliages (E.N.S.M. Nantes, France, 17-18 Mai 1973)
8. K. Okazaki and H. Conrad, Acta Met, 21, 1117 (1973)
9. S.P. Agrawal, G.A. Sargent and H. Conrad, Metall. Trans. 4, 2613 (1973)
10. V.A. Moskalenko and V.N. Puptosova, Mater. Sci. Eng. 16, 269 (1974)
11. A. Akhtar and E. Teghtsoonian, Metall. Trans., 6A, 2201 (1975)
12. J. Bigot C.R. Acad. Sci. Paris, série C. , 279, 67 (1974)
13. J. Bigot C.R. Colloque Européen de Métallurgie sous vide , Lille (France) (7-9 Oct. 1975). Suppl. à la Revue "Le Vide" N° 177.
14. M. Déchamps, A. Quivy, G. Baur et P. Lehr, Scripta Met, 11, 941 (1977)
15. C. Brehm and P. Lehr : Met. Cor. Ind. N° 551 to 554 (1971)
16. G. Baur : *"Etude de la dynamique de la déformation plastique du titane et des solutions solides diluées Titane-oxygène"*. Thèse de Doctorat es Sciences (Université Paris VI) 8 Janvier 1980
17. G. Baur et P. Lehr, Scripta Met, 11, 587 (1977)
18. G. Baur et P. Lehr, Mém. Scient. Rev. Met. 7-8, 551 (1975)
19. G.A. Sargent, Acta Met. 13, 663 (1965)
20. S.R. Mc Ewen, O.A. KUPCIS and B. RAMASWAMI. Scripta met, 3, 441, (1969)
21. N.F. MOTT and F.R.N. Nabarro. Report on strength of solids — The Physical Society (London) p. 1 (1948)
22. N.F. MOTT : *"Imperfections in nearly perfect crystals"*, Wiley, New York, p. 173 (1952)

23. H. Conrad and K. Okazaki, Scripta Met. 4, 111 (1970)
24. R.L. Jones and H. Conrad, Trans AIME 245, 779 (1969)
25. T. Suzuki. 2nd Int. Conf. on the strength of metals and alloys
    237 (30 Août–4 Sept. 1970)
26. H. Conrad, M. Doner and B. de Meester : *"Titanium Science and Technology"*
    Ed. R.J. Jaffee, H.M. Burte, Plenum Press, 2, 969 (1973)
27. H. Conrad – Acta Met. 14, 1631 (1966)
28. K. Okasaki, K. Morinaka and H. Conrad – Trans. Jap. Int. Metals 14,
    470 (1973)
29. K. Okasaki, M. Kanokogi and H. Conrad – J. Jap. Inst. Met. 37,1250 (1973)
30. U.F. Kocks, A.S. Argon and M. F. Ashby *"Thermodynamics and Kinetics of
    slip"* Progress in materials Science, Pergamon Press 19, 160 (1975)
31. J.W. Christian *"The interactions between dislocations and point defects"*
    edited  by B. L. EYRE, AERE Report 5944 (Harwell G.B.) vol. III,
    604 (1968)
32. K. Okazaki and H. Conrad – Trans. Jap. Inst. Metals 13 (1972) 205
33. P.P. Tung and A.W. Sommer. Metal. Trans. 947 (1970)
34. E.D. Levine – Trans. Met. Soc. AIME 236, 1558, (1966)
35. H. Conrad and H. Widersich – Acta Met. 8, 128 (1960).

# DEFORMATION AND FRACTURE OF Ti-5Al-2.5Sn ELI ALLOY AT 4.2K~291K

Takeshi Kawabata, Shigetaka Morita and Osamu Izumi

The Research Institute for Iron, Steel and Other Metals,
Tohoku University, Sendai, 980, Japan

## Introduction

Ti-5Al-2.5Sn ELI alloy which has high strength and ductility even at liquid herium temperature, is one of the most reliable low temperature structural materials [1]. Usually, it is believed that high strength and high ductility result in high fracture toughness. However titanium alloys show conflicting behaviours, e.g., Shannon and Brown showed that the tensile properties of unnotched specimens furnace- and air-cooled were essentially identical, while fracture toughness of notched specimen air-cooled was superior about 25% to furnace-cooled one [2]. The deformation and fracture behaviours of the alloys are complicated because of the occurrence of deformation twins at cryogenic temperatures [3]. The purpose of present studies is to make clear the effect of material parameters on mechanical properties. In this report, the effect of grain shape on the mechanical properties of Ti-5Al-2.5Sn ELI alloy at cryogenic temperatures will be given.

## Experimental Procedures

The chemical composition of material supplied from Nippon Stainless Steel Co. is Al: 5.13, Sn: 2.42, C: 0.003, H: 0.0002, O: 0.07, N: 0.003 and Fe: 0.03 wt%, respectively, and Ti: remainder. The tensile specimens which were machined from the material hot-forged below $\beta$-transus temperature have a gage part of 3 mm in diameter and 30 mm in length. The specimens were annealed at temperatures above and below $\beta$-transus, i.e., at 1323 and 1088K for 1.8 and 3.6 ks, respectively. The structures obtained were martensitic ($\alpha'$) in the former and equi-axed ($\alpha$) with grain size of about 5 $\mu$m in diameter in the latter. Tensile tests were done at temperatures from 4.2 to 291K at the initial strain rate of $1.3 \times 10^{-3} s^{-1}$ using an Instron type testing machine. Testing was carried out by dipping the specimen in liquid herium and liquid nitrogen and by controlling the temperature using a heater. Fracture surfaces were observed using JSM-T20 by the stereo-graphic method.

## Results

1. Stress-strain curves at cryogenic temperatures.

Fig. 1 shows the examples of stress-strain curves of $\alpha$ and $\alpha'$ materials. The curves at temperatures from 291 to 60K showed smooth ones with small work-hardening. The changes of flow stress with strain of $\alpha$ and $\alpha'$ materials at 50K were wavy-like. At 40K, the wavy flow curves changed into serrated ones form the middle of the curves. At the temperatures lower than 20K, the serration occurred from the initial stage of deformation. The stress amplitude in serrated flow of $\alpha$ material was nearly constant from the initial stage of deformation to fracture, but that of $\alpha'$ material was small at the initial stage then increased with strain.

2.    Strain rate dependence of flow curves.

2.1.    Flow curves of α material at 4.2 and 40K.

Fig. 2 shows the change of stress-strain curves of α material with strain rate at 4.2K. The shapes of serration at $\dot{\varepsilon} = 1.3 \times 10^{-4}$ and $1.3 \times 10^{-3} s^{-1}$ are similar while at $\dot{\varepsilon} = 1.3 \times 10^{-2} s^{-1}$, fracture occurred immediately after a first large serration. The fracture strain is largest at the middle strain rate of

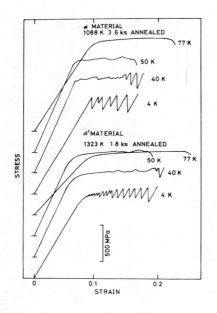

Fig. 1.    Flow curves of α and α' materials at cryogenic temperatures.

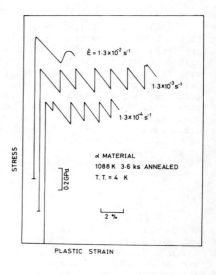

Fig. 2.    Change of flow curve by strain rate in α material at 4.2K.

$1.3 \times 10^{-3} s^{-1}$ and smallest at the highest one of $1.3 \times 10^{-2} s^{-1}$. The effect of strain rate on the shape of flow curves at 40K is shown in Fig. 3. At the strain rate of $1.3 \times 10^{-3} s^{-1}$, serration appeared irregularly after about 5% strain. However, no serration was observable at higher or lower strain rate range. Fracture strain decreased with increasing strain rate.

2.2.    Flow curves of α' material at 4.2K.

Fig. 4 shows the strain rate dependence of flow curve in α' material at 4.2K. Differently from α material, the flow curves of α' material changed into the serrated ones with acute, saw-tooth shape after an initial smooth flow of about 0.5%, and the stress amplitude of serration increased with strain. When the strain rate increased to $1.3 \times 10^{-2} s^{-1}$, the serrated flow became disappeared even at 4.2K, though some irregularities were observed on the curve. The fracture strain tends to decrease with strain rate.

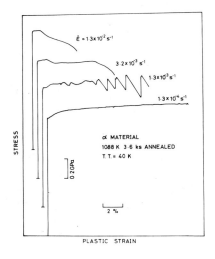

Fig. 3.   Change of flow curve by
          strain rate in α material
          at 40K.

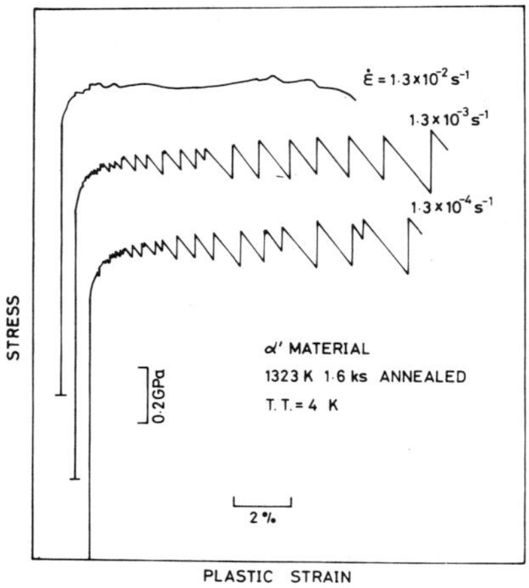

Fig. 4.   Change of flow curve by
          strain rate in α' material
          at 4.2K.

3.   Temperature dependence of mechanical properties.

     The temperature dependence of the proof stress at 0.2% strain ($\sigma_{0.2}$),
the ultimate tensile stress ($\sigma_{uts}$), the plastic strain at $\sigma_{uts}$ and the nominal fracture strain ($\varepsilon_f$) in α and α' materials are shown in Fig. 5.

3.1.   $\sigma_{0.2}$ and $\sigma_{uts}$.

     $\sigma_{0.2}$ and $\sigma_{uts}$ of α material were always larger than those of α' one.
$\sigma_{0.2}$ of α and α' increased with decreasing temperature down to 130K.   Below
77K, the difference of stress levels between α and α' became large rapidly.

3.2.   $\varepsilon_{uts}$ and $\varepsilon_f$.

     $\varepsilon_{uts}$ and $\varepsilon_f$ of α and α' materials were nearly constant from 291 to 130K
but at 77K they increased.   $\varepsilon_{uts}$ and $\varepsilon_f$ of α material at temperatures between
291 and 130K were larger than those of α' material, however, the situation
was reversed below 77K.   $\varepsilon_f$ at temperatures lower than 50K in α material decreased with decreasing temperature.   Values of $\varepsilon_f$ in α' material below 77K
were larger than above 130K.

4.   Effects of temperature and strain rate on necking profile.

     The plastic instability phenomenon (serration) occurred at temperatures
below 40K.   The apparent multiple neckings were observed in the specimen
showing serration.   The diameter of specimens after tensile testing was measured along the tensile axis using a microscope.

## 4.1.   Effect of temperature.

Fig. 6 shows the necking profiles of α material when deformed at each temperature.   The initial profile of specimen is shown by broken lines. Fracture occurred at the end of gage part at 4.2K.   The multiple neckings were pronounced with decreasing temperature, especially at 4.2K, and the number of neckings corresponded to the number of serrations.

The necking profile of α' material is shown in Fig. 7.   Similarly to α material, the multiple neckings are observed in the specimens showing serrations.

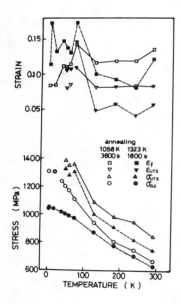

Fig. 5.
Temperature dependence of
mechanical properties in
α and α' materials.

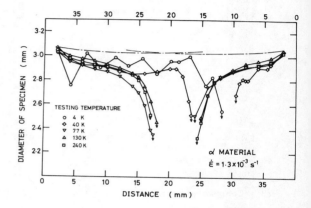

Fig. 6.   Necking profiles of α material
deformed at cryogenic temperatures.

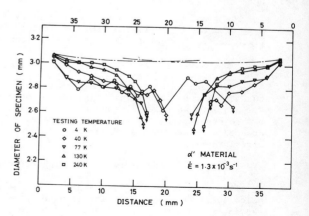

Fig. 7.   Necking profiles of α' material
deformed at cryogenic temperatures.

## 4.2.  Effect of strain rate.

Figs. 8, 9 and 10 show the effect of strain rate on necking profile of α material at 4.2 and 40K and of α' material at 4.2K, respectively.  It is shown in Fig. 9 that at strain rates of $1.3 \times 10^{-4}$ and $1.3 \times 10^{-3} s^{-1}$ the number

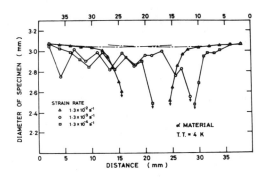

Fig. 8.
Effect of strain rate on necking profile of α material at 4.2K.

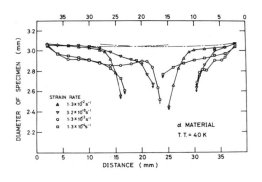

Fig. 9.
Effect of strain rate on necking profile of α material at 40K.

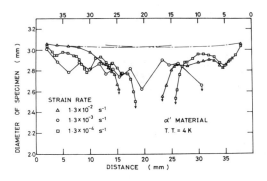

Fig. 10.
Effect of strain rate on necking profile of α' material at 4.2K.

of multiple neckings also corresponds to the number of serrations. Only one necking is seen at $1.3 \times 10^{-2} s^{-1}$, corresponding to the deformation of this specimen (see Fig. 3). That is, the increase of strain rate results in the localization of deformation.

At 40K, the trend of the localization is pronounced with increasing strain rate, as shown in Fig. 9.

In $\alpha'$ material at 4.2K, the multiple neckings are always observable at any strain rate tested (Fig. 10). However, the number of multiple necking decreases with increasing strain rate. The number of serration in Fig. 3 is not coincident with the number of multiple necking at any strain rate tested. From above, it can be said also in $\alpha'$ material that the trend of localization of deformation increases with increasing strain rate, but the trend is not so remarkable as in $\alpha$ material. (See the profile at $1.3 \times 10^{-2} s^{-1}$, near the end of left side, there exists a region being little deformed).

5.   Evidence of deformation twin and initiation site of void and cracking.

It was observed that grains were deformed uniformly at the deformed portion, even near fractured edge, as shown in Fig. 11 a. In $\alpha$ material, deformation twins were frequently observed with decreasing temperature (Fig. 11 b). But the volume fraction of grains containing deformation twin was not so large even in the temperature range where serration occurred. In $\alpha'$ material, the occurrence of deformation twin could not be confirmed by microscopic observation because of the complicated microstructure.

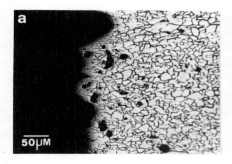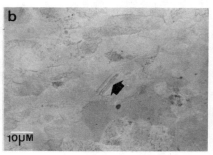

Fig. 11.   Optical micrographs showing the feature of deformation near fractured edge and the initiation site of voids (a) and deformation twins (b) in $\alpha$ material at 4.2K.

The void formation and subsequent cracking in $\alpha$ material occurred at grain boundaries as shown in Fig. 11 a. In $\alpha'$ material, voids were observed at an interface of martensitic leaves through which cracks were grown transversely by linking voids and microcracks (Fig. 12).

6.   Fractography

Fig. 13 a and b are scanning electron micrographs of fracture surface of $\alpha$ and $\alpha'$ materials tested at 4.2K, showing the outskirt and the center, respectively. $\alpha$ and $\alpha'$ material showed a typical cup and cone type fracture at whole temperatures tested even at 4.2K, except $\alpha'$ material at 20K which was shear type fracture with inclined 45° to the tensile axis. At the center

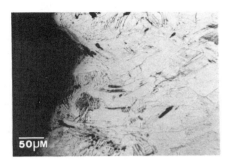

Fig. 12.   Optical micrographs showing initia-
           tion sites of void and microcrack
           and the feature linking the micro-
           cracks near the fractured edge in
           α' material at 4.2K.

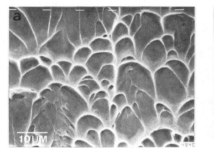 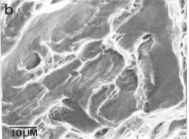

Fig. 13.   Scanning electron micrographs of fracture surfaces in α (a)
           and α' (b) materials tested at 4.2K, showing the outskirt
           and the center, respectively.

portion of fracture surface, highly ductile dimples containing ripple pattern
partly and a large number of deep holes were observed.  Dimples at the portion
of shear lip were shear type.  Fracture surface of α' material was irregularly
rougher than that of α material.  Relatively flat portions seemed to be due
to the separation of interface of martensite were rarely observed.


## Discussion

1.  Cause of plastic instability phenomenon.

1.1.  Possibility due to deformation twin.

    In α material tested at 4.2K, the number of serration coincides with the
number of necking and deformation twins were observed.  However, the deforma-

tion twins were found only in a small number of isolated grains which were not necessarily restricted in a region near fractured edge. Therefore, it is not adequate to expect that the strain due to twinning is enough to cause serrations on a flow curve even at 4.2K.

## 1.2. Possibility due to adiabatic deformation.

According to Basinski [4], the adiabatic local pre-deformation accelerates the subsequent deformation due to thermal feed back. Because the specific heat decreases and the strength (and also the temperature dependence of strength) increases with decreasing temperature, the effect of adiabatic deformation is pronounced at low temperature, thus resulting in much rise of temperature and much decrease of flow stress. Therefore, the plastic insta-bility can occur.

The discontinuous flow in $\alpha$ material at 4.2K in which the number of neck-ing corresponded to that of serrations, can be explained by the adiabatic deformation.

Under the condition of adiabatic deformation the specimen is deformed rapidly and locally. But, when the rapid decrease of flow stress is large enough, the condition of adiabatic deformation becomes to be unsatisfied, resulting in the deformation to stop. Before beginning the next discontinuous flow, the pre-deformed portion is cooled enough and strengthening by the strain hardening is larger than weakening by the reduction of sectional area, the next adiabatic deformation will initiate at different portion, then multiple neckings occur.

## 2. Possibility of lowering of fracture toughness due to adiabatic deformation.

High fracture toughness means that a large energy is necessary for frac-turing and therefore the material should have a large plastic zone. Fracture under the condition of the adiabatic deformation is equal to the deformation and fracture of material with a negative work hardening rate. Now we consider the fracture of notched specimen. Under the adiabatic deformation, the size of plastic zone formed at notch root is smaller than in usual deformation, and the load at which fracture initiates will also decrease. That is, at very low temperature, the adiabatic deformation causes the lowering of fracture toughness.

## Conclusion

Tensile tests of $\alpha$ material with equi-axed grains and $\alpha'$ material with martensitic structure were done at cryogenic temperatures, and the following results were obtained.

$\sigma_{0.2}$ and $\sigma_{uts}$ were higher in $\alpha$ material than in $\alpha'$ material at whole testing temperatures.

$\varepsilon_{uts}$ and $\varepsilon_f$ were larger in $\alpha$ material than in $\alpha'$ material above 130K, but they became reversed below 77K.

The initiation site of voids was grain boundaries for $\alpha$ material and the interface of martensite leaves for $\alpha'$ material.

Fractography showed highly ductile dimples containing ripple pattern partly and deep holes at the center of fracture surface and highly ductile shear type dimples at the shear lip in both materials at whole testing tem-peratures.

From the point of view of toughness as for the structural usage, the mechanical properties of $\alpha'$ material especially at temperatures below 77K are superior to those of $\alpha$ material.

## Acknowledgements

We wish to express our thanks to Nippon Stainless Steel Co. Ltd. for supply of materials.  This work was supported partly by Grants-in Aid for Fundamental Scientific Research from Ministry of Education.

## Reference

1.  W. F. Brown, Jr. et al edts., Aerospace Structural Metals Handbook 4 (1965), Code 3706, P6, Mechanical Properties Data Center, Belfour Stulen, Inc., Michigan.
2.  J. L. Shannon, Jr., and W. F. Brown, Jr. Proc. ASTM 63 (1963), 809.
3.  R. H. Van Stone, J. R. Low, Jr., and J. L. Shannon, Jr., Met. Trans. 9A (1978) 539.
4.  Z. S. Basinski, Proc. Roy. Soc. A240 (1957), 229.

# DEFORMATION AND FRACTURE OF Ti-15Mo-5Zr ALLOY AT 4.2-291K

Takeshi Kawabata, Shigetaka Morita and Osamu Izumi

The Research Institute for Iron, Steel and Other Metals,
Tohoku University, Sendai, 980, Japan.

## Introduction

β-type titanium alloys have high strength and low thermal conductivity which are favorable for the usage as cryogenic structural materials.

Ti-15Mo-5Zr alloy belongs to β-type one having excellent characteristics in cold-workability, corrosion resistance, weldability and strength [1,2]. However, ductility of this alloy is somewhat lower than other α and α+β type alloys and is brittle at cryogenic temperatures.

In order to develop new cryogenic structural titanium alloys with high strength and high toughness, it is valuable to study the brittleness of β-type alloys at cryogenic temperatues.

The purpose of this study is to make clear the deformation behavior and the cause of brittleness of the Ti-15Mo-5Zr alloy at cryogenic temperatures.

## Experimental procedures

The chemical composition of the alloy which was supplied from Kobe Steel, Ltd., is Mo: 14.0, Zr: 5.05, Fe: 0.036, N: 0.0043, O: 0.165, H: 0.0024 wt%, and Ti: remainder. The material was hot-forged from 30 to 10 mm in diameter at temperature below $\alpha + \beta \sim \beta$ transus (1048±15K[3]) after holding at 1120K for 1.8 ks. Tensile specimens which have a gage part of 3 mm in diameter and 30 mm in length were sealed in quartz tubes and heat-treated in three ways: i.e., the first (H1) and the second (H2) were annealed at 1053K for 1.8 and 180 ks, respectively, then quenched into water, and the third (H3) was additionally aged at 773K for 1.8 ks after the same annealing treatment as H1. Tensile tests were done at temperatures from 4.2 to 291K, at initial strain rate of $1.3 \times 10^{-3} s^{-1}$ using an Instron type testing machine (TENSILON UTM-I-10000CW). At 77, 4.2K and above and below 77K, the specimens were tested in liquid nitrogen and helium, and in gas atmosphere of nitrogen and helium controlled with heater. The strain distribution along tensile axis was measured using a microscope. Deformed structures of thin foils and voids and crackings were observed using TEM (JEM-200B and JEM-1000) and an optical microscope. Fractography was observed using a SEM (JSM-T20) and stereo-pairs of micrographs.

## Results

1. Microstructures.

Microstructures of H1, H2 and H3 were equiaxed grains with small second phase particles of α. Grain boundaries in H1 and H3 were curved because of the insufficient annealing and α particles were elongated to the tensile direction by forging. H2 had straight grain boundaries and α particles were remained at grain boundaries. The grain sizes of H1, H2 and H3 were 14, 17 and 14 μm. respectively.

2.    Temperature dependence of stress—strain curves.

    Fig. 1 shows the typical examples of stress—strain (flow) curves in specimen
H1.   The flow curves at 291 ~ 77K are smooth with a small work-hardening rate.
At 50K, although the flow curve is not shown, it becomes irregularly wavy-like,
and at 35K it changes into serrated flow with large and rapid stress drops.
Below 20K, the specimens are failured suddenly even at the elastic region, i.e.,
the behavior seems to be completely brittle at least on the flow curves.

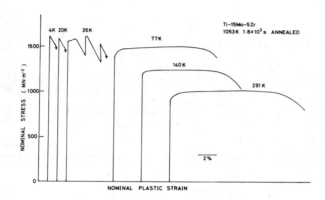

Fig. 1.    Examples of stress—strain curve at cryogenic
           temperatures.

3.    Temperature dependence of mechanical properties.

    (1)   Specimen H1.   Mechanical properties of specimen H1 are summarized in
Fig. 2.   The 0.2% proof stress ($\sigma_{0.2}$) and the ultimate tensile stress ($\sigma_{uts}$)
increase monotonously with decreasing temperature.   In the temperature range
where the discontinuous flow or the apparent brittleness occurred the curve of
$\sigma_f$ closely approaches to that of $\sigma_{0.2}$.   It suggests that fracture occurres immediately after yielding.   Down to 77K from room temperature, $\varepsilon_{uts}$ and $\varepsilon_f$ are
maintained at the levels of 6 ~ 9 and 10 ~ 15%, respectively.   Below 50K they
decrease and especially at 20 and 4.2K they show little ductility ( ~ 1%).   From
291 to 50K the reduction of area (R.A.) shows higher level more than 30% and
even at 4.2K it is about 10%.   The large value of R.A. is comparable to that
of other titanium alloys.   The fracture mode was a cup and cone type at whole
testing temperatures except at 20K, at which it was shear type and the separation occurred along the plane inclined about 45° to the tensile axis.   The change
of fracture mode would result in small R.A..
    (2)   Specimen H2.   Fig. 3 shows the results on specimen H2.   $\sigma_{0.2}$, $\sigma_{uts}$,
$\varepsilon_{uts}$ and $\varepsilon_f$ show a similar tendency to H1 above 77K.   At 50 and 20K, R.A. decreases to about 3%.   But at 4.2K it increases again to a large value of 23%
and becomes larger than that of specimen H1 with smaller grain size.   The increase of R.A. will be caused by the decrease of the amount and the size of
remained $\alpha$ particles.
    (3)   Specimen H3.   Fig. 4 shows the results on specimen H3.   Though the
plots are not so many, the trend seems to be similar to H1.   Effect of aging on
$\sigma_{0.2}$ is not observed.   On the contrary, decreasing of $\sigma_{uts}$ was observed because

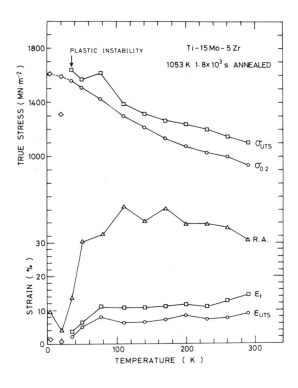

Fig. 2.
Temperature dependence of
mechanical properties in
specimen H1.

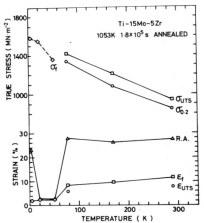

Fig. 3.    Temperature dependence of
mechanical properties in
specimen H2.

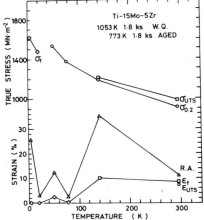

Fig. 4.    Tempearture dependence of
mechanical properties in
specimen H3.

aging results in brittleness.   $\varepsilon_{uts}$ is not shown since fracture occurred suddenly on the way of increasing of stress.   The ductility of aged material (H3) is the same level as that of as-quenched material (H1).   At 77K $\varepsilon_f$ decreases to nearly zero, though the decrease of $\varepsilon_f$ in specimen H1 occurs at 35K.   That is, the transition temperature at which ductility becomes small, is changed to higher temperature by aging treatment.   It is interesting that R.A. shows a high value of 25% even at 4.2K.

4.   Strain distribution.

Fig. 5 shows strain distribution in specimen H1 tested at every temperature. The dotted line represents the profile of undeformed specimen.   At high temperature range (291K and 140K in the figure), specimens are deformed at whole gage part with a large local necking.   Deformation is more uniform at 77K than at 140 and 291K.   Localization of deformation becomes remarkable with decreasing temperature.   Especially at 4.2K, severe deformation is concentrated only in the necked portion.

Profiles of deformed specimens of H2 are shown in Fig. 6.   The whole tendency is similar to H1.

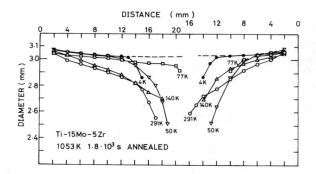

Fig. 5.   Strain distribution in specimen H1 at various cryogenic temperatures.

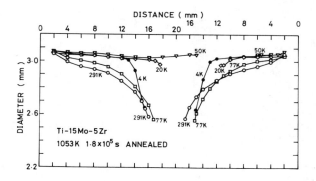

Fig. 6.   Strain distribution in specimen H2.

Fig. 7 shows strain distribution of specimen H3. Double neckings seem to occur at room temperature (291K) because fracture initiated near the end of gage part. Trend of necking profile is similar to H1.

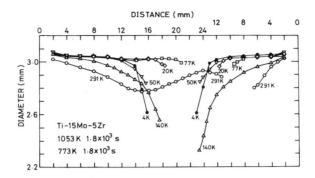

Fig. 7. Strain distribution in specimen H3.

The effect of strain rate and temperature on necking profile is shown in Fig. 8. At 50K, higher strain rate results in narrower necking. Multiple neckings are observed on the specimen tested at 35K at strain rate of $1.3 \times 10^{-3}$ $s^{-1}$, corresponding to the discontinuous flow as shown in Fig. 1. The necking region fractured at 35K is narrower than at 50K. That is, the trend localizing deformation increases with decreasing temperature and with increasing strain rate.

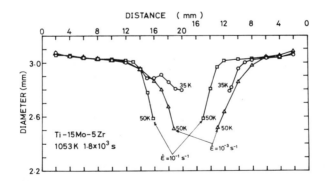

Fig. 8. Strain rate and temperature dependence of strain distribution at 50 and 35K in specimen H1. The strain rate at 35K was $1.3 \times 10^{-3} s^{-1}$.

5. Effect of strain rate on flow curve.

During deformation at 77K, the pulling rate was successively changed as, 2, 5, 2, 0.5, 5, 2, 20, 0.5, and 2 mm/min as shown in Fig. 9. At the initial stage of deformation the increase of pulling rate from 2 to 5 mm/min makes the stress level increase only 1%. Furthermore, at the later stage of deformation or after $\varepsilon_{uts}$ was attained, the increase of stress corresponding to the increase of pulling rate becomes smaller. That is, it is said that the flow stress of this alloy is not sensitive to the strain rate especially at large strain.

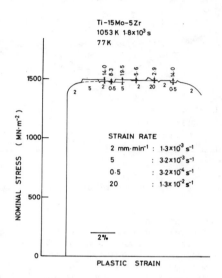

Fig. 9.   Strain rate dependence of flow stress
at 77K in specimen H1

## 6.  Fractography

Fracture mode was a cup and cone type at whole testing temperatures except at 20K where exhibiting the 45° shear type and minimum R.A..

Figs. 10 a-d are scanning electron micrographs showing fracture surface of specimen H1 at 4.2 (a and b) and 291K (c and d). At outskirts (Figs. 10 b and d) ductile but relatively shallow dimples, and at center portions (Figs. 10 a and c) highly ductile dimples and deep holes are observable. Especially low temperatures (e.g. at 4.2K) plate-like patterns are observed at bottom of dimples, and also the shape of dimples changes to the rectangular-like along the plate-like patterns.

Figs. 11 a and b are scanning electron micrographs of the center portion of specimen H3 fractured at 4.2 and 291K, respectively. The shape of dimples in H3 does not vary by temperature, i.e., dimples are highly ductile type and the plate-like pattern is not observable. The feature of dimples at outskirt in H3 was similar to H1, and also fracture was a cup and cone type. Deep holes were observed at whole fracture surface and at any testing temperature.

From the present results on the necking profiles (Fig. 5 ~ 8) and the fractographies (Figs. 10, 11), this alloy is not brittle but essentially ductile.

## 7.  Microstructures in deformed specimens.

Optical micrographs showed that voids were formed at the α particles near

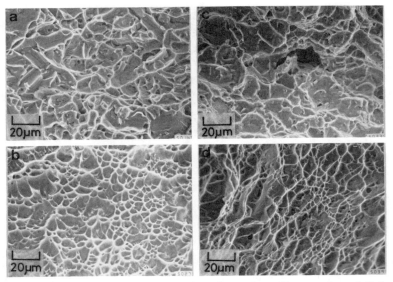

Fig. 10.   Scanning electron micrographs of specimen H1 tested at 4.2 (a
and b) and 291K (c and d) showing the centers (a and c) and
the outskirts (b and d).

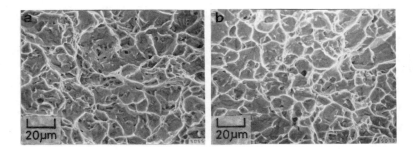

Fig. 11.   Scanning electron micrographs of specimen H3 tested at 4.2 (a)
and 291K (b) observing at center portion of fracture surface.

the fracture part, but the number of voids was relatively small comparing to
other titanium alloys (e.g. Ti-5Al-2.5Sn EL1 alloy).  In specimen H1, deformation twins were observed frequently at low testing temperatures, although their
volume fraction was small as shown in Fig. 12.  In H3 no evidence of deformation
twin was confirmed.

Transmission electron micrographs showed piled-up dislocations and the
heavily deformed region around α particles, suggesting one of the initiation
sites of voids to be the interface of the particles.

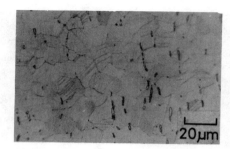

Fig. 12.   Optical micrograph of specimen
H1 tested at 4.2K showing
deformation twins.

## Discussions

1.   Deformation mechanism.

In specimen H1 deformation twins were observed more frequently with decreasing temperature, although their volume fraction was small (see Fig. 12). But in specimen H3 which was aged, any deformation twin could not be observed.

At the bottom of rectangular dimples in specimen H1, plate-like patterns were observed (see Fig. 10 a). The patterns would not be formed by α particles, because α particles have a rod-like shape.

The size of plate-like patterns was nearly equal to that of the deformation twins. Therefore, the plate-like patterns would be the deformation twins.

The volume fraction of twins observed in specimen H1 was only a few per cent or less. Therefore, the contribution of deformation twins to the strain at necked portion would be unexpectable.

On the other hand, R.A. at necked portion in specimen H1 was about 10% at 4.2K (see Fig. 2). Therefore, it is concluded that most of the deformation in specimen H1 would be done mainly by dislocation glide.

In specimen H3, evidence of deformation twin could not be obtained so that deformation would occur only by dislocation glide.

2.   Fracture mechanism.

2.1.   Effect of deformation twins.

Van Stone et al [4] presented evidences that voids nucleated along twins and grain boundaries at offsets made by intersection of slips. In their cases, plate-like patterns which were similar to specimen H1, could be observed at the bottom of dimples.

From the present fractography, the void nucleation in specimen H1 is mainly attributed to the intersection between twins and slip bands especially at lower temperatures.

2.2.   Effect of α particles.

Nucleation of voids at interface of α particles was rarely observed in optical micrographs. But traces of α particles could not be observed at the bottom of dimples. While deep holes were seen frequently on the whole fracture surface. The voids nucleated at their interfaces might develop to the deep holes, because the interface of α particles will separate easily.

3.  Cause of serrations.

In the former section, we concluded that dislocation glide is main mechanism of deformation.  Therefore, the serrations observed at 35K should not correspond to twinning but to dislocation gliding.  It is said that at cryogenic temperature the effect of adiabatic deformation will be pronounced [5].  As shown in Figs. 1 and 5, serrations at 35K can be correlated to multiple neckings which corresponding to dislocation glide.

As 20 and 4.2K, the deformation behavior was completely brittle, but, fractography showed ductile fracture with necking.  Therefore, it can be said that the brittleness at 20 and 4.2K is only an apparent phenomenon and fracture may be caused during a large drop of stress by the adiabatic deformation as proposed by Basinski [5].

## Conclusions

Tensile tests were done at 291-4.2K in Ti-15Mo-5Zr $\beta$ type alloy heat-treated in three ways.  The following results were obtained.
1.  Below 35K flow curves showed that whole specimens fractured in completely brittle manner.
2.  From the fractographic observation, whole specimens showed ductile fracture with necking at whole testing temperatures even at 4.2K.
3.  In specimen H1 deformation twins were more observable with decreasing temperature.  On the contrary no deformation twin could be observed in specimen H3.
4.  Contribution of deformation twins to the deformation of specimen H1 was estimated to be negligible even at 4.2K.
5.  The brittle behavior at low temperatures is considered to be caused by the adiabatic deformation as proposed by Basinski.

## Acknowledgement

The authors are most grateful to Kobe Steel, Ltd., for supply of materials. This work was supported partly by Grant-in Aid for Fundamental Scientific Research from the Ministry of Education.

## Reference

1.  S. Ohtani and M. Nishigaki, J. Japan Inst. Metals 36 (1972), 90.
2.  T. Nishimura, M. Nishigaki and S. Ohtani, J. Japan Inst. Metals 40 (1976), 219.
3.  S. Ohtani, T. Nishimura and M. Nishigaki, J. Japan Inst. Metals 36 (1972), 1105.
4.  R. H. Van Stone, J. R. Low, Jr., and J. L. Shannon, Jr., Met. Trans. 9A (1978), 539.
5.  Z. S. Basinski, Proc. Roy. Soc. A240 (1957), 229.

# LOW TEMPERATURE PECULIARITIES
# OF PLASTIC DEFORMATION IN TITANIUM
# AND ITS ALLOYS

V. A. Moskalenko, V. I. Startsev, V. N. Kovaleva

Physico-Technical Institute of Low Temperatures
UkrSSR Academy of Sciences, Kharkov, USSR

## Introduction

Peculiar low temperature physical conditions of plastic deformation result in a number of unexpected phenomena in the mechanical behaviour of metals which may be attributed to the specific dislocation multiplication, motion and interaction in this temperature range. These phenomena include non-monotonic temperature dependence of deforming stresses, a serrated character of plastic flow, softening under the superconducting transformation, etc. [1]. Theoretical interpretations of each of the phenomena presently available are not satisfactory. Their understanding requires further theoretical and experimental research.

This paper based on the data obtained from studies of strength and plasticity characteristics of different purity titanium and its binary alloys of aluminium, zirconium, vanadium, niobium and oxygen from 4.2 to 293 K, presents the analysis of some of the above peculiarities and factors causing them.

## Serrated Deformation of Titanium
## and its Alloys at Low Temperatures

Regularities of low temperature serrated flow. One of the specific features of plastic deformation of ma-

terials at low temperatures ($\leqslant$ 20 K) is the macroscopic instability of plastic flow under quasi-static loading. The deformation
curve shows this as sharp drops of load amounting sometimes to
15-20% of the applied (Fig. 1). In this section we consider the
effect produced by a number of factors (alloying, preceding deformation, sample sizes and mechanical twinning) on the principal
parameters of the serrated plastic deformation (onset temperature, amplitude and number of drops of loads, etc.) of titanium and
its alloys.

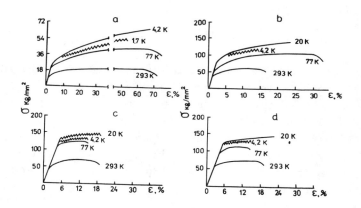

Fig. 1 Diagrams of «stress-strain» tension of titanium
and its alloys at low temperatures: a) titanium
iodide (0.05 at% $O_{eq}$); b) commercial titanium
(0.70 at% $O_{eq}$); c) Ti — 2.4 w/o Zr — 1.2 w/o Mo
alloy (prestrained); d) Ti — 2.4 w/o Zr — 1.2
w/o Mo alloy (annealed).

Alloying effects. In tensile tests ($\dot{\varepsilon} = 1.2 \cdot 10^{-4} sec^{-1}$) on
titanium iodide with 0.05 at% of interstitial impurity, the ser-
rated plastic flow was observed only at 1.7 K, being absent at
4.2 K. Meanwhile, titanium with a higher interstitial atom concentration ($>$0.7 at.% $O_{eqv.}$)[1] displayed several drops of loading before rupture at 20 K (Fig. 1). The titanium iodide alloy-

ing with substitution elements such as Al, Zr, V or Nb leads to
a more intensive serrated behaviour, though their effect is much
lower than that of interstitial elements. Thus for high-purity
titanium ( $\approx$ 0.08 at.% $O_{eqv.}$ ) first drops of load at 4.2 K appear
only after strain $\mathcal{E}_{serr} > 2\%$, while the addition of 1–2 wt.% of any
of the above substitution shifts $\mathcal{E}_{serr}$ (at the same temperature) practically down to zero. An increasing concentration of alloying elements causes an appriciable change in the serrated
process кinetics with the degree of deformation: the number of
jumps per unit deformation dN/d$\mathcal{E}$ becomes lower and the relative
value of drop of load dP/P higher (P is the applied load, dP –
the drop of load), (Fig. 2). It is seen, however, that at deformations close to rupture these parameters become approximately the
same as it is the case for non-alloyed titanium and its binary
alloys.

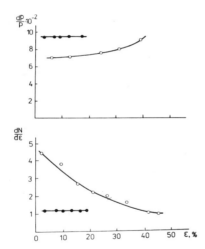

Fig. 2 Parameters of serrated deformation (dP/P is the
relative drop of load, dN/d$\mathcal{E}$ is the number of
jumps per unit deformation) vs. strain at 4.2 K
for titanium (o) and Ti – 3 w/o Al alloy (•).

---

[1] The interstitial impurities (nitrogen, oxygen, carbon, hydrogen) are reduced to the oxygen equivalent $O_{eqv.}$ taкing into account N 2 0, C 2/3 0 and H 0. These equivalents are based on
the relative atom effects of each element on the yield stress of
$\alpha$-titanium.

Degree of plastic deformation. A peculiar feature of the
serrated plastic flow in the materials studied is a gradual decrease in the number of jumps per unit deformation and an increase in their amplitude with a growing degree of deformation. In
this respect the effect of preliminary deformation on the serrated flow parameters is similar to that of alloying, in more
concentrated alloys the dependence of the above parameters on
the degree of deformation being much weaker than that in non-
alloyed titanium (Fig. 2). The priliminary plastic deformation
results in an appreciable higher onset temperature of the serrated flow. It has been found that for samples of commercial $\alpha$ -
alloys of titanium which were not annealed after technological
processing, the drops of load in the stress-strain curve were
observed during tension even at 20 K, while after annealing the
load drop temperature falls down to $\approx 5$ K (Fig. 1c, d) [2] .

Sample size. Another factor greatly influencing the serrated plastic flow is the size (thickness or diameter) of the
strained sample. For example, the experiments on Ti-1.7 Al-1.5
Mn alloy show that as the thickness was reduced from 3.6 mm to
0.65 mm the relative drop of load dP/P fall approximately from
10 to 4%, the average number of drops per unit deformation dN/
d$\varepsilon$ decreases from 3.0 to 1.5. Similar results were obtained on
non-alloyed titanium and its alloys.

Mechanical twinning. Among the possible mechanisms responsible for instable plastic deformation, deformation twinning is
also mentioned, which means a formal extension of the single
crystal results to polycrystalline samples. A qualitative estimate of the share of twinned material $V_{tw}$ show that in high purity Ti samples (0.05 at.% $O_{eav.}$ ) and in Ti-0.3 at.% Nb alloys
strained by 12% at 20 K, $V_{tw} \approx 40\%$, while in Ti of 0.7 at.% $O_{eav.}$
$V_{tw} \approx 18\%$. At the same time serrated deformation in the latter
material is observed even at 20 K, while in the former one only
at 2 K. Thus, the mechanical twinning intensity was not found
to be related to the plastic flow instability.

## Possible Reasons for Plastic Deformation
## Instability at low temperature

According to the dislocation model [3] , at low temperatures where thermal fluctuations become negligible and the plastic deformation induced by dislocations overcoming local obstacles is hampered, arrays of dislocations are formed, which then overcome the barrier assisted by high stresses at the front of the arrays. This concept is supported by the above results for the interstitial impurity effect on the onset temperature of serrated deformation in titanium. So long as the motion of single dislocations overcoming obstacles is sufficient to maintain the sample strain rate equal to that of the testing machine, arrays (if there are) produce no essential influence upon the plastic flow character. Higher impurity concentrations ( e. g., interstitial atoms in Ti), however, increase the number of barriers and their capacity [4] . Therefore, in impure materials the possibility of the dislocation array formation (and hence the flow instability) may arise at higher temperatures.

The experimentally observed intensification of the serrated flow in Ti with growing strain or with alloying is consistent with Seeger's hypothesis. Both alloying and preliminary strain split dislocations and thus hamper any mechanism of obstacle surmounting requiring a split dislocations contractions (e. g., cross-slip, etc.). This will lead to a more rapid growth of stresses at the front of arrays approaching the shear strength level and facilitating the flow instability.

After overcoming an obstacle the velocity of the dislocation motion is quite great. Owing to the transformation of the plastic deformation work into heat under the conditions of low heat capacity and thermal conductivity. the local adiabatic temperature rise is possible [5] , which initiates further avalanche-like dislocation motion through a barrier. The experimental results described are also consistent with this concept. The observed lower relative drop of load with smaller sample thick-

ness seems to arise from the temperature and range of the heating domain, which is dictated by the scale effect and means that
with small sample thickness the rate of heat dissipation approaches the heating rate. As the sample thickness increases, the heating temperature rises sinceless heat is lost through the surface.

Thus, the low temperature plastic flow instability in Ti
and its alloys may be accounted for qualitatively by a combination of the hypotheses of the dislocation array overcoming of a
barrier and the following local heat leading to a stress fall in
the stress-strain curve.

## Non-monotonic Temperature Dependence of Deforming Stresses in Ti and its Alloys

The experimental results for the temperature dependence of
mechanical properties of titanium with different contents of interstitial elements and its binary alloys based on solid $\alpha$-
substitution solutions with Al, Zr, V, Nb under uniaxial static
tension at 4.2–293 K are shown in Figs. 3 and 4 [6]. Common to
both titanium and its alloys is the increase in the ultimate
strength and yield stress by a factor of two or three as temperature falls from ambient to 20 K. Below 20 K the temperature dependence of deforming stresses is non-monotonic.

Both for different purity titanium and its alloys the sign
of the temperature dependence derivative for the ultimate strength d $\sigma_b$/dT reverses in this temperature region. The 16° temperature fall (from 20 to 4.2 K) decreases the ultimate strength by
7–10%, no correlation between the type of alloying element, its
concentration and the anomaly observed being found.

The temperature dependence of the yield stress $\sigma_{0.2}$ is more
complicated (Figs. 3, 4). For high-purity titanium ($\lesssim 0.07$ at.%
interstitial impurity) $\sigma_{0.2}$ grows continuously as temperature
falls down to 1.7 K. Alloying of this material with sufficiently
pure (with respect to interstitial impurities) substitution ele-

ments up to concentrations corresponding to a homogeneous solid
solution does not deteriorate the above dependence and sometimes
makes it even stronger (Ti–Al alloys). Meanwhile, an increase in
the interstitial atom concentration in titanium lowers (down to
vanishing) the temperature sensitivity of the yield stress in
region of $\leqslant 20$ K (Fig. 3). And finally, a decrease in the yield
stress is observed below $\approx 20$ K for alloys in which the concentration of substitution elements corresponds to the solubility limit (or somewhat exceeds it). In this case the change of the
$d\sigma_{0.2}/dT$ sign observed for Ti–3 wt.% V and Ti–5.5 wt.% Al alloys
(Fig. 4) may be attributed to the probable presence of centres
of a new phase (or some heterogeneity of the solid solution).

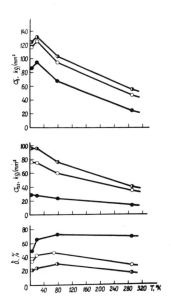

Fig. 3 Temperature dependence of mechanical properties
($\sigma_b$ is the ultimate strength, $\sigma_{0.2}$ the yield
stress, $\delta$ the relative elongations) of titanium
with different contents of interstitial impurities (in at.% $O_{eq}$): ● – 0.07, o – 0.40, o – 0.70.

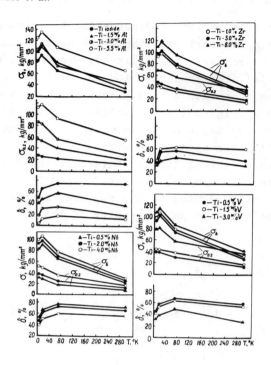

Fig. 4 Temperature dependence of mechanical properties of
       binary alpha-titanium alloys with aluminium, nio-
       bium, zirconium, and vanadium.

This anomalous change in $\sigma_{0.2}(T)$ is not understood even
qualitatively in terms of the conventional thermal fluctuation
concepts of dislocation motion, qualitatively they do not seem
to be inconsistent with the concepts suggested in refs. [7]and
[8], according to which the effects in question may occur at
the pinning point concentration of $\nless 0.1$ at.% (i. e. $\angle < 10^{-6}$ –
$10^{-5}$ cm). The intensity of energy dissipation is to a great extent dictated by the length of the dislocation segment. Therefore, this can essentially be reduced through increasing the
concentration of interstitial impurity atoms in Ti which become
pinning points controlling the plastic deformation rate [4].
Besides, at low temperatures a noticeable reduction of the ef-

fective drag constant B is observed. Under these conditions the
dislocation segment damped by point obstacles will vibrate with
respect to the static equilibrium position realizing its inertia
and weak damping. This is the possibility that was used in Granato's theoretical models [7] . The weak damping condition reali-
zable at low temperature may be written as

$$BL < 2\pi\sqrt{AC} \tag{1}$$

Here $A = \rho b^2/\pi$ is the effective mass of the dislocation unit
length, where $\rho$ is the density of the material, B – the drag
constant, $C = \mu b^2/2$ the linear tension of a bow dislocation.

Thus at low temperatures a decrease in any of the factors
in the lefthand side of the inequality will enchance inertia effects which may appear significant in estimating the yield stress.
Then with decreasing length L the inertia mechanism will be more
effective and appear at higher temperature in an impure metal than
that in a pure one.

We shall estimate the effect of yield stress temperature
sensitivity lowering at a higher concentration of interstitial
impurity atoms for titanium ($T \lesssim 20$ K). According to eq. (1), the
inertia forces, all other parameters being the same, become essential in the material where the distance between obstacles is

$$L < 2\pi\sqrt{AC}/B = 2\pi b^2\sqrt{\rho\mu / 2\pi B^2}$$

For titanium this means $L < 5 \cdot 10^{-6}$ cm ($\rho = 4.5$ g/cm$^3$, $B = 3 \cdot 10^{-4}$
poise and $\mu_{0°K} = 4.8 \cdot 10^{11}$ dyne/cm$^2$).

The experiment suggests that the anomalous change in $\sigma_{0.2}$
with temperature below 20 K starts to occur for titanium when its
concentration of interstitial atoms is $c \gtrsim 0.07$ at.%, i. e. the
mean distance between pinning points of the dislocation segment
is $L < 1.5 \cdot 10^{-6}$ cm. This is in good agreement with the above
$L \lesssim 5 \cdot 10^{-6}$ cm.

Thus, the non-monotonic temperature dependence of deforming stresses below $\approx$20 K for different purity titanium, which is discussed in this paper, results from the specific character of dislocation motion and interaction in this temperature region due to increasing importance of inertia effects and a decreasing role of pure thermal fluctuation effects.

## References

1. V. I. Startsev, V. Y. Ilichev, V. V. Pustovalov. Plastichnost i prochnost metallov i splavov pri nizkikh temperaturakh, M., Metallurgiya, 1975.
2. R. A. Ulianov, V. A. Moskalenko. MiTOM, No. 10, 48, 1966.
3. A. Seeger. Dislocations and Mechanical Properties of Crystals, New York, 1957, p. 243.
4. V. A. Moskalenko, V. N. Puptsova. Mater. Sci. and Engng., 1974, 16, p. 269.
5. Basinski Z. Proc. Roy. Soc., London, 1957, A240, p. 229.
6. V. A. Moskalenko, V. N. Puptsova, R. A. Ulianov. MiTOM, No. 6, 17, 1970.
7. N. V. Granato. Phys. Rev. Lett., 1971, 21, p. 660.
8. K. Kamada, I. Yoshizawa. J. Phys. Soc. Japan, 1971, 31, p. 1056.

THE INFLUENCE OF STRAIN-RATE HISTORY AND TEMPERATURE
ON THE SHEAR STRENGTH OF COMMERCIALLY-PURE TITANIUM

Abdel-Salam M. Eleiche

Department of Mechanical Design and Production,
Faculty of Engineering, Cairo University,
Cairo, Egypt

## Introduction

Deformation mechanisms of titanium under uniaxial stress conditions have
been studied by a great number of investigators. Details are available in such
areas as anisotropy [1,2], temperature effects [3-16], strain-rate effects [16-
19], stress relaxation [5,20], grain size effects [20-22], impurity content [9,
21-24], twinning [2,25,26], recrystallization [27], creep [6,7], fatigue [8]
and fracture [22]. Very much less information exists on the effects of thermal
and strain rate histories.

No doubt, a 'mechanical equation of state' for metals [28], relating stress
with strain rate and temperature, of the form $\tau = f (\gamma^P, \dot{\gamma}^P, T)$, where $\tau$ is
the applied shear stress, $\gamma^P$ the plastic shear strain, $\dot{\gamma}^P$ the plastic shear
strain rate and T the temperature, will be a valuable tool for predicting material behaviour. However, increased experimental evidence summarized recently by
Duffy [29] indicate that such an equation is not valid in general for all fcc
metals. For hcp and bcc metals, the picture is not yet clear.

Few investigations were involved in studying the effects of thermal history
on the general flow behaviour of titanium [4,9,11,30]. On the other hand,
previous studies of the influence of strain-rate history have been performed at
low rates, using rate changes of one or two orders of magnitude [9-11,21,23,24,
31]; only one involved large changes to high rates [32], and apparently none was
undertaken at large strains and over a wide range of temperatures.

In the present work, rate changes of up to 6 orders of magnitude, from $10^{-3}$
to $10^3$ s$^{-1}$, could be imposed during a strain increment of order $10^{-2}$. Very
large changes in deformation history was therefore possible, so that deviations
from a mechanical equation of state could be detected. The inclusion of temperature as a variable ranging from -150 to 400 °C permitted an approximate correction to be made to take account of the effects of adiabatic heating, as well
as the determination of various thermal activation parameters. The use of pure
shear rather than tensile or compressive straining eliminated the need for conversion to true stress and true strain, and also enabled deformation to be continued to large strains. The material chosen was titanium of commercial purity,
intermediate in interstitial content.

## Experimental Procedure

All tests were conducted on a horizontal torsional split Hopkinson-bar
apparatus, particulars of which together with the associated instrumentation
were described in detail elsewhere [33-35]. As shown in Fig. 1, the apparatus
incorporated two collinear elastic input and output bars, made of 6Al-4V titanium (type IMI 318), with the thin-walled tubular specimen sandwiched between
them, thus being based essentially on the principle of the Kolsky bar [36].

The specimen was strained at a dynamic rate of order $10^3$ s$^{-1}$ by a large amplitude torsional wave introduced in the system by the release of a pre-stored torque in a clamped portion of the input bar. By suitably instrumenting the input and output bars with strain gauges, the stress history of the wave before and after its passage through the specimen was recorded on an oscilloscope. Introducing these stress histories into simple reduction formulae allows the average stress, strain and strain rate histories in the specimen to be calculated, and hence a dynamic stress-strain curve to be obtained [37].

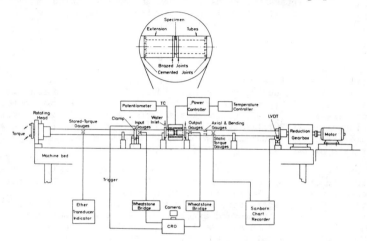

Fig. 1. Schematic layout of torsional Hophinson-bar apparatus

Quasi-static loading was achieved on the same apparatus by means of a slow rotational-drive unit attached to the far end of the output bar. This allowed the specimen to be pre-strained at a rate of order $10^{-3}$ s$^{-1}$ up to the start of the dynamic test, thus obtaining a very rapid strain-rate increase of about six orders of magnitude. Torque and angle of twist in the quasi-static straining were measured separately by means of strain gauges and a differential transformer (LVDT), and traced on a chart recorder.

A furnace was used for testing at 200 and 400 °C, and was continuously kept filled with dry argon to prevent oxidation. For testing at -50, -100 and -150 °C, cooling was achieved by controlling the flow of liquid nitrogen in a tufnol container surrounding the specimen [35].

Specimens were machined from 1.03 inch diameter, hot-rolled annealed rod of commercially-pure titanium (type IMI 130) purchased from Imperial Metal Industries (Kynoch) Ltd., having the following typical maximum metallic and interstitial impurity levels shown in Table 1. The interstitial content expressed in terms of the equivalent atomic percent of oxygen [38] was approximately 0.7 .

Table 1   Chemical composition (Ppm by weight)

| Al | 500 | Cr | 10 | Mn | 500 | Sn | 500 |
|----|-----|----|-----|----|-----|----|-----|
| As | 5 | Cu | 200 | N | 90 | Ta | 15 |
| C | 200 | Fe | 300 | Ni | 15 | U | 10 |
| Cd | 5 | H | 30 | O | 2000 | V | 500 |
| Co | 5 | Mg | 20 | Si | 200 | | |

After machining to the nominal configurations shown in Fig. 2 and measurement of the critical dimensions, the specimens were stress relieved at 700 °C for 1½ hour and furnace cooled under a vacuum of $10^{-4}$-$10^{-5}$ torr. The mean grain density was about 75 mm$^{-2}$. This corresponds to about 3 grains across the wall thickness of the specimen; even with this low value, however, there were about 1400 grains on the cross section.

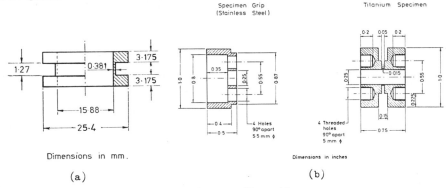

Fig. 2. Specimen configurations

Before testing, each annealed specimen was firmly attached to the inner ends of the two torsion bars, thus reducing the system at any testing temperature to a continuous wave guide of constant mechanical impedance, except for the short tubular gauge length of the specimen. In tests conducted at room and low temperatures, Araldite epoxy cement was used satisfactorily to bond the specimen of Fig. 2(a) directly to the torsion bars. At elevated temperatures, each specimen of Fig. 2(b) was first mechanically attached to two short stainless steel grips of suitable dimensions which were then brazed to a pair of stainless steel extension tubes properly designed to eliminate the wave distortion effects of the thermal gradient along their length [35,39]. This assembly was then attached with Araldite to the torsion bars, as shown in the insert of Fig. 1.

A representative collection of test records for specimens loaded at a variety of strain-rate histories and temperatures, as well as calibration and data reduction procedures have been presented and discussed elsewhere [35].

### Results and Discussion

Constant-strain-rate response

Quasi-static and dynamic stress-strain curves for each of the six testing temperatures are plotted in Figs. 3 and 4 respectively. Each curve is the average of two or more tests, the results of which varied by no more than 5 %. It should be noted that the dynamic test results at the two highest temperatures were obtained at slightly lower strain rates than those at other temperatures, due to the difference in the specimen configuration adopted in each case as shown in Fig. 2. The material exhibits a well-defined yield point at all temperatures and at both strain rates. Also, the flow stress continuously increases with strain up to fracture, except in some of the dynamic tests where the strain hardening rate becomes zero at large strains.

Both yield and strain hardening can be seen to depend on temperature. Also, at the lowest temperatures, the work hardening apparently tends to remain

constant instead of decreasing steadily with strain; the stress-strain curve
thus approaches linearity over extended strain intervals.

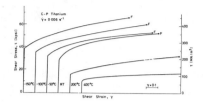

Fig. 3. Quasi-static stress-strain
curves for titanium at
various temperatures

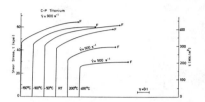

Fig. 4. Dynamic stress-strain
curves for titanium at
various temperatures

From the extensive research on titanium by Reed-Hill and his associates
[13-15] and by Doner and Conrad [10], it is evident that dynamic strain ageing
is a very significant factor in the plastic deformation of impure titanium.
This seems to occur in the temperature range 700 to 800 $^\circ$K; the exact temperature apparently tending to rise with increasing purity [12], thus for a titanium
alloy of commercial purity this blue brittle temperature was located at 750 $^\circ$K
[14]. This dynamic strain-ageing phenomenon was also found to manifest itself
in the occurence of yield points, serrations in the stress-strain curves, a rise
in the flow stress with increasing temperature, minima in the total elongation
versus temperature curves and maxima in the strain-hardening rate versus temperature curves [10]. In all these investigations, tensile specimens were used
and low strain rates ranging from $10^{-5}$ to $10^{-2}$ s$^{-1}$ were applied. The same phenomenon is also indicated by the enhancement of creep strength [6,7] and high
cycle, long-life fatigue properties [8] in commercial-purity titanium in the
temperature range 600 to 850 $^\circ$K. The present tests were limited to a maximum of
673 $^\circ$K, and as far as could be discerned from the test records at the high temperatures none of the above manifestations of dynamic strain ageing occured.

Stress-strain curves for each temperature are also shown in Fig. 5 (a)-(f).
In each figure, the variation of strain rate and adiabatic temperature rise are
shown for the dynamic test. This adiabatic temperature rise $\Delta$T was computed on
the assumption that the work of plastic deformation is totally converted to
heat. Overall, titanium exhibits significant rate sensitivity at all testing
temperatures.

Figure 6(a) shows the variation of yield and flow stresses with temperature
at the low strain rate. The behaviour is similar to that observed with many
other materials, viz. temperature sensitivity at low temperatures and relatively
temperature-insensitive behaviour at high temperatures. The yield stress and
work-hardening rate increase steadily as the temperature decreases from 673 to
297 $^\circ$K. The yield stress continues to increase as the temperature is reduced to
to 123 $^\circ$K; the work-hardening rate, however, decreases with fall of temperature
below 297 $^\circ$K, increasing again at the lowest temperatures. This behaviour may
be associated with the onset of twinning as an important mode of deformation.
It is also seen from this figure that, at this strain rate, titanium exhibits an
increase in the temperature dependence of the flow stress with strain.

Figure 6(b) shows the yield and flow stresses as functions of temperature
at the high strain rate. At such dynamic rate, flow stress values are expected

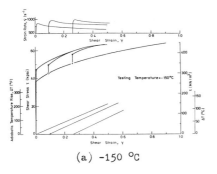

(a) -150 °C

Pre-strains 0.090, 0.258.

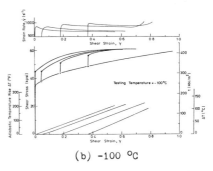

(b) -100 °C

Pre-strains 0.045, 0.178, 0.368.

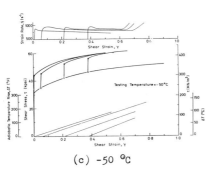

(c) -50 °C

Pre-strains 0.050, 0.213, 0.378.

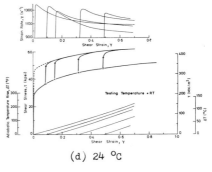

(d) 24 °C

Pre-strains 0.088, 0.150, 0.314, 0.482.

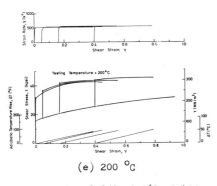

(e) 200 °C

Pre-strains 0.050, 0.163, 0.400.

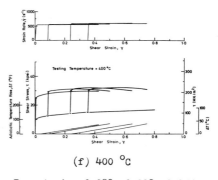

(f) 400 °C

Pre-strains 0.088, 0.238, 0.355.

Fig. 5. Results of incremental strain-rate experiments with titanium

to be reduced because of temperature rise which accompanies rapid plastic deformation.  The measured flow stress value at any strain thus represents the specimen strength at a temperature slightly higher than that existing at the start of
the deformation.  Assuming that all the mechanical work is converted into heat,
and that the effect of temperature history can be neglected, a correction can be
made as shown in the figure.  It is seen that the resulting change in flow
stress is negligible at low strains or high temperatures, and amounts to a maximum of about 10 % at a strain of 0.5 and at the lowest temperatures covered;
the actual corrections will be somewhat lower because of heat loss during deformation and the storage of some energy in the deformed lattice.  Both yield and
flow stresses steadily increase with decreasing temperature from 673 to 123 °K;
the temperature dependence being slightly affected by strain.

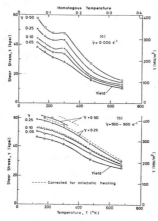

Fig. 6. Temperature dependence of
flow stress of titanium
(a) low strain rate
(b) high strain rate

From the curves of Fig. 6(a) and the corrected curves of Fig. 6(b), the rate
dependence of the flow stress may be determined.  The mean apparent strain-rate
sensitivity may be defined as

$$\mu_{12} = ( \tau_2 - \tau_1 ) / \ln ( \dot\gamma_2 / \dot\gamma_1 ) \qquad (1)$$

where $\tau_1$ and $\tau_2$ are the flow stresses measured at an arbitrary strain $\gamma$ and temperature T in the quasi-static and dynamic tests at rates $\dot\gamma_1$ and $\dot\gamma_2$, respectively.  Calculated values of $\mu_{12}$ are plotted against strain for different temperatures in Fig. 7 (full lines).  $\mu_{12}$ shows a moderate increase with increasing
strain; it also increases with temperature up to 200 °C, falling somewhat at
400 °C.

Strain-rate-change response

The response to dynamic loading after various amounts of quasi-static pre-
straining is shown in Fig. 5 (a)-(f).  Each plotted curve is the average of at
least two curves differing by about 3 % in flow stress and strain rate and by a
maximum of 0.015 in pre-strain.  Also included are the average strain rate and
temperature rise as functions of strain, for the high-rate part of each test.

At no testing temperature does the rate increase produce transient stress
maxima similar to the yield points observed by Santhanam et al. [31] after a
one-decade change in strain rate during quasi-static tension.  The flow stress
increases rather smoothly and continuously.  This increase is characterized by
a well-defined elastic increment followed by gradual transition to a value

similar to that obtained in a test entirely at the dynamic rate. Both the elastic stress increment and the rate of the subsequent transition depend on the testing temperature. This behaviour is qualitatively similar to temperature-history effects observed on commercially-pure titanium in tension [4,30]. To a first approximation, the behaviour exhibited in Fig. 5 may be described by a mechanical equation of state at room and elevated temperatures; at lower temperatures, deviations still exist. Such behaviour is completely different than that exhibited by other materials such as copper (fcc) and mild steel (bcc) [40].

At any given pre-strain, the increment in flow stress following the change in strain rate may be taken as representative of the intrinsic rate sensitivity of the material $\bar{\mu}_{12}$. This is defined as before for $\mu_{12}$ in Eq. (1), but here $\tau_2$ is the yield stress at the end of the elastic increment prior to work hardening. The variation of $\bar{\mu}_{12}$ with strain is shown in Fig. 7 (broken curves), from which values for a strain of 0.1 at different temperatures have been estimated and plotted in Fig. 8.

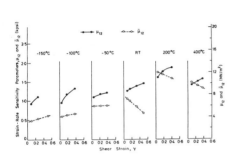

Fig. 7. Strain dependence of apparent and intrinsic strain-rate sensitivities, at constant temperatures

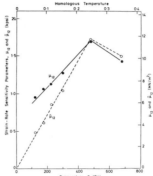

Fig. 8. Temperature dependence of apparent and intrinsic strain-rate sensitivities, at $\gamma = 0.1$

According to the theory of thermally-activated plastic flow in metals, summarized by Li [41], the intrinsic rate sensitivity is related to the activation volume V by the equation

$$\bar{\mu}_{12} = k\,T\,/\,V \qquad (2)$$

where k is Boltzmann's constant and T is the temperature. In deriving this equation, it is assumed that the density of thermal activation sites remains constant during the change in strain rate, and that the activation volume is independent of temperature. From Fig. 8, it is seen that at temperatures up to 473 °K, $\bar{\mu}_{12}$ is proportional to T, as required for a constant activation volume. The slope of the plotted line corresponds to a value $V = 0.54$ (nm)$^3$ or 22.4 b$^3$, where b is the magnitude of the Burgers vector. The value of $\bar{\mu}_{12}/T$ drops considerably at $T = 673$ °K, and corresponds to $V = 37.3$ b$^3$, which suggests that a single activated event takes place below that temperature, similar to previous observations [3,9]. Assuming that V is governed by the dislocation density, a significant degree of dislocation rearrangement and mutual annihilation seems to take place during the pre-straining period at temperatures above 473 °K. At strains larger than 0.1, the behaviour is more complex since $\bar{\mu}_{12}$ decreases with increasing strain even at room temperature.

## Micrographic Examination

Garde et al. [26] showed that in commercially-pure titanium deformed at 77 °K, the volume fraction twinned increases linearly with true strain, reaching 50 % at a tensile strain of about 0.35, i.e. a shear strain of about 0.6. On the other hand, observations by Harding [16] showed that at a given temperature the density of twinning in titanium increases with rate of strain.

Figure 9 shows a number of micrographs of specimens tested at low temperatures. Fig. 9(a) gives a general view of the unstrained flange and the heavily deformed gauge length, including the fracture region. Fig. 9(b) shows the detail of the microstructure in the gauge length adjacent to the flange; twinning is evident in nearly all the grains, and in some grains two families of twins have been developed. Figs. 9(c) and (d) show part of the gauge length of specimens subjected to two different amounts of pre-strain. Comparing Figs. 9(b), (c) and (d), it appears that the density of twinning decreases with increasing pre-strain while the width of the individual twin lamellae increases. This effect is also seen in Figs. 9(e) and (f), which relate to tests at -150 °C without and with pre-straining. Comparison of Figs. 9(b) and (e) shows the increasing intensity of twinning as the temperature is reduced.

It is clear that twinning is an important mechanism of plastic straining in the present tests on titanium as expected, and that some changes take place in the amount and nature of twinning within the range of test conditions used. A quantitative analysis of these changes has not been attempted, but they may well account for some of the secondary features of the observed flow behaviour, viz. the reduction in the 'jump' stress level with increasing pre-strain.

## Summary and Conclusions

Commercially-pure titanium shows a well-defined yield point and no yield drops at all testing temperatures (-150 °C to 400 °C) and strain rates ($10^{-3}$ and $10^{3}$ s$^{-1}$). The strain-hardening rate is small, decreasing with increasing temperature, but varying little with strain rate. Adiabatic heating causes some reduction of flow stress at large strains, but no flow instabilities.

In rate-jump tests there is a large elastic increment of stress at all temperatures, the value of which varies little with pre-strain but increases with temperature. At high temperatures, the flow stress slightly exceeds that obtained at a constant high rate. Strain-rate history is much less important than the instantaneous strain rate; thus, the behaviour approximates closely to that corresponding to a mechanical equation of state. The activation volume for a strain of 0.1, calculated from the response to the strain-rate change, is essentially constant for temperatures up to 200 °C; at 400 °C a larger value is found which may be attributed to recovery processes occuring during pre-straining and apparently causing a reduction in dislocation density.

Micrographic examination confirms the importance of twinning as a deformation mechanism at low temperatures. The distribution and size of twins varies with the amount of pre-strain in rate-jump tests.

## Acknowledgments

This paper is based on research carried out at the University of Oxford in cooperation with Dr. J.D. Campbell, late Reader in Engineering Science. It was sponsored in part by the Air Force Materials Laboratory (LLN), United States Air Force, under Grant No. AFOSR 71-2056.

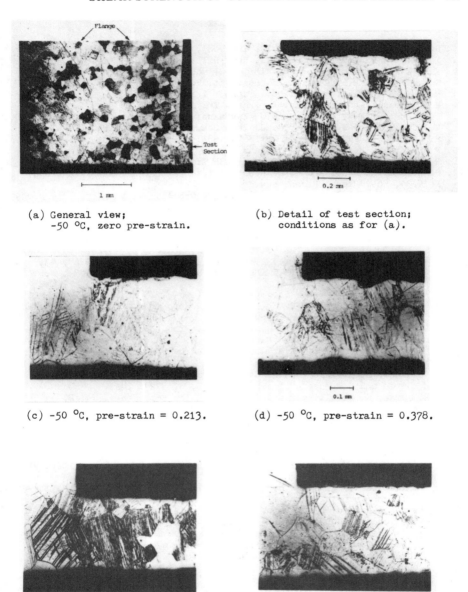

(a) General view;
    -50 °C, zero pre-strain.

(b) Detail of test section;
    conditions as for (a).

(c) -50 °C, pre-strain = 0.213.

(d) -50 °C, pre-strain = 0.378.

(e) -150 °C, zero pre-strain.

(f) -150 °C, pre-strain = 0.258.

Fig. 9. Micrographs of dynamically strained specimens (longitudinal section).

840    *A.M. Eleiche*

## References

1.  D.H. Rogers: Int. J. Mech. Sci., 10(1968), 221.
2.  Y. Lii, V. Ramachandran and R.E. Reed-Hill: Met. Trans., 1(1970), 447.
3.  K.R. Evans: Trans. Met. Soc. AIME, 242(1968), 648.
4.  G.W. Geil and N.L. Carwile: J. Res. Nat. Bur. Stand., 54(1955), 91.
5.  P.P. Tung and A.W. Sommer: Met. Trans., 1(1970), 947.
6.  W.R. Kiessel and M.J. Sinnott: Trans. TMS-AIME, 197(1953), 331.
7.  D.R. Luster, W.W. Wentz and D.W. Kaufmann: Mat. and Methods, 37(1953), 100.
8.  N.G. Turner and W.T. Roberts: J. Less-Common Metals, 16(1968), 37.
9.  H. Conrad: Canad. J. Phys., 45(1967), 581.
10. M. Doner and H. Conrad: Met. Trans., 4(1973), 2809.
11. R.N. Orava, G. Stone and H. Conrad: ASM Trans., 59(1966), 171.
12. H. Conrad, M. Doner and B. de Meester: *Titanium Science and Technology*, 2(1973), 969. Plenum Press, New York.
13. S.N. Monteiro, A.T. Santhanam and R.E. Reed-Hill: *The Science, Technology and Application of Titanium*, (1970), 503. Pergamon Press, New York.
14. A.M. Garde, A.T. Santhanam and R.E. Reed-Hill: Acta Met., 20(1972), 215.
15. A.T. Santhanam and R.E. Reed-Hill: Met. Trans., 2(1971), 2619.
16. J. Harding: Dep. Eng. Sci., Oxford U., (1974), Rep. 1108/74.
17. J.E. Lawson and T. Nicholas: J. Mech. Phys. Solids, 20(1972), 65.
18. K. Tanaka and K. Ogawa: Proc. 19th Japan Cong. Mat. Res., (1976), 43.
19. M.C.C. Tsao and J.D. Campbell: Dep. Eng. Sci., Oxford U., (1973), Rep. 1055/73.
20. G. Sargent and H. Conrad: Scripta Met., 3(1969), 43.
21. H. Conrad and R. Jones: *The Science, Technology and Application of Titanium*, (1970), 489. Pergamon Press, New York.
22. H. Conrad, M.K. Keshavan and G.A. Sargent: Proc. *Boston ICM-II*, Special Volume(1978), 538. ASM, Metals Park.
23. K. Okazaki, M. Momochi and H. Conrad: *Titanium Science and Technology*, 2(1973), 1131. Plenum Press, New York.
24. K. Okazaki, K. Morinaka and H. Conrad: Trans. JIM, 14(1973), 470.
25. A.M. Garde and R.E. Reed-Hill: Met. Trans., 3(1972), 2411.
26. A.M. Garde, E. Aigeltinger and R.E. Reed-Hill: Met. Trans., 4(1973), 2461.
27. K. Okazaki and H. Conrad: Met. Trans., 3(1972), 2411.
28. J.H. Hollomon: Metals Tech. AIMME, (1946), Tech. Paper 2034.
29. J. Duffy: To be publ. 2nd Int. Conf. *Mechanical Properties of Materials at High Rates of Strain*, Oxford, March 1979.
30. T.S. DeSisto and F.L. Carr: Proc. ASTM, 64(1964), 636.
31. A.T. Santhanam, V. Ramachandran and R.E. Reed-Hill: Met. Trans., 2(1971), 2619.
32. K. Tanaka and K. Ogawa: Proc. *Boston ICM-II*, (1976), 1598. ASM, Metals Park.
33. A.M. Eleiche and J.D. Campbell: Air Force Materials Laboratory, Wright-Patterson Air Force Base, Ohio, (1976), Rep. AFML-TR-76-90.
34. A.M. Eleiche and J.D. Campbell: Exp. Mech., 16(1976), 281.
35. A.M. Eleiche: Sci. Eng. Bull. Fac. Eng. Cairo U., 1979/4(1979), 109.
36. H. Kolsky: Proc. Phys. Soc. Lond., 62 B(1949), 676.
37. U.S. Lindholm: J. Mech. Phys. Solids, 12(1964), 317.
38. H. Conrad: Acta Met., 14(1966), 1631.
39. A.M. Eleiche: J. Phys. D: Appl. Phys., 8(1975), 505.
40. A.M. Eleiche: Proc. *1st Cairo U. MDP Conf.*, (1979) Papers MECH-14, MECH-15.
41. J.C.M. Li: *Dislocation Dynamics*, (1967), 87. McGraw-Hill, New York.

# STUDY OF TITANIUM ALLOYS DEFORMATION PROCESS
## BY ACOUSTIC EMISSION

V. P. Vodolazsky, V. K. Kataja, V. K. Aleksandrov,
A. Z. Kaganovich, V. A. Volkov, E. I. Nefed'ev

All-Union Institute of Light Alloys, USSR

Plastic deformation of metals is known to be accompanied by continuous microcrack formation, which appear already at the initial stage of plastic deformation.' As the deformation proceeds, microdefects grow in number and size, forming a microcrack [1] . The crack remains stable for some time, but on further deformation it reaches critical size, loses its stability and the material fractures [2] . To establish reasonable regimes for high-temperature deformation of titanium alloys it is necessary to have unbiassed information about their ductility: the ability to be deformed without fracture. The extent of deformation, accumulated by the material to the moment of fracture, is taken as the measure of ductility. One of the most acceptible in the engineering plan ductility characteristics of ductility is the extent of shear deformation, accumulated by the metal particle before fracture, determined from the well-known formula:

$$\Lambda = \int_{0}^{t} H d\tau$$

where H — intensity of shear deformation rates,
$t$ — deformation time to fracture.

The above-mentioned technique [3] is based on the fact of obligatory crack formation on the free surface of tested specimens, which can be visually observed, while for more accurate ductility measurement it is necessary to monitor them at early stage, with the crack size lower than critical, still preventing the integrity and the material functional capacity. For example,

on warm rolling of difficult to deform titanium alloys sheets
fracture inside the rolled metal takes place in spite of the fact,
that in calculation of reduction regimes data of the above-mentioned technique were taken into consideration. Character of crack
propagation makes it possible to assume, that they formed in slip
planes earlier, than the cracks on the rolled metal surfaces [4] .

The present work deals with the study of the possibility of
using acoustic emission (AE) to detect the initiation of deformed
metal fracture in the process of its ductility tests. The main
task in using AE for earlier detection of metal fracture in deformation is the separation of signals from initiating crack in
the base metal matrix from those of microfractures of various
inclusions, generated on the tool – specimen and other contacts.
AE intensity was studied at tension and compression. According
to the paper [5] , the number of AE pulses under similar deformation conditions amounts to $N = 10^6$ at tension and $N = 10^4$ at compression. The difference for two orders of magnitude is due to
the fact, that fracture of various brittle inclusions occurs at
tensile stresses, which in case of upsetting are transverse and
are much lower than at compression, providing that stresses acting from the tool are equal. That is why upsetting was chosen as
the main test, which facilitated the separation of the required
signal against the background of secondary emissions.

Tensile tests were conducted with the aim of mastering the
technique of monitoring acoustic emission from the deformation
area through intermediate media. Universal machine «Instron 1255»
was used in the experiments. To record the AE signals Dunegan-
Endevco «3000» series system was used.

The sequence of blocks to transform AE signals into electric
ones was the following:
    piezotransducer – preamplifier – amplitude detector- analyzer
    – two-pen recorder.

For synchronious recording signal from testing machine was
conveyed to the second pen of the recorder. During tensile tests
transducers were fixed on the end faces of cylindrical specimens

heads. Compression tests were conducted in a special container. During preliminary experiments the possibility was established to monitor AE signals from deformation area through intermediate media, by using special configuration specimens, eliminating slip on contact surfaces and additional area of placing transducers. Such recording scheme allowed to detect, decode and separate pure acoustic emission in the deformation area from the one, generated on the tool-specimens interface.

The use of AE method in high-temperature tests was complicated by the problem of monitoring acoustic emission from high-temperature area [6] . This problem was solved by using special compound punch, which two parts are fitted in to each other. On heating the punch upper part with constantly attached transducers was removed and returned to its place before upsetting. The level of AE signals discrimination was chosen so that weak signals from secondary emissions don't be picked up by the recorder. Deformation (compression or tension) ceased after a certain level of AE was reached, and specimens were studied to detect inner and outer fracture.

Comprising loading-time diagram with AE signals intensity on tension commercial titanium specimens (Fig. 1) it may be noted, that the character of AE intensity change in the process of loading is similar to one, described in the paper [5] and consists of two picks. On elastic portion of the loading diagram the main contribution is made by «noises», generated on the tool-specimen contact, on plastic portion the pick corresponds to the sources inside the deformation area. It was found, that recording AE signals as a total number of pulses per unit time, facilitates the detection of fracture. Fig. 2 shows the character of loading and AE pulses total number change on Ti–6Al–3Mo–1V specimen tension.

Tension process was stopped in the moment «a» on receiving AE signal at the loading plastic portion.

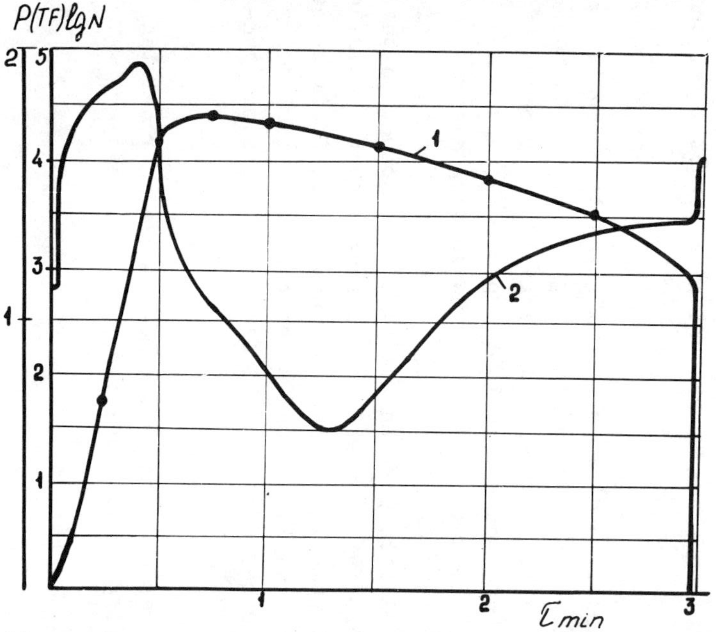

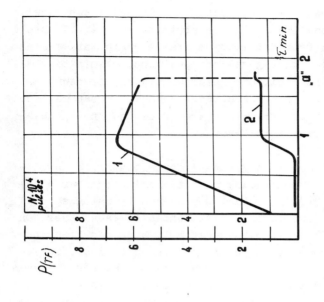

Fig. 2 Change of AE pulses total number (curve 2) during loading
(curve 1) on tension of Ti-
–6Al–3Mo–1V alloy.

Fig. 1 Character of AE intensity change (curve 2)
during loading (curve 1) on tension of commercial titanium.

Despite the absence of fracture exterior signs, micropores
formed inside the specimen along the shear plane lines, resulting
in crack propagation (Fig. 3).

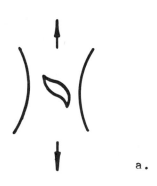 

a.                                                        b.
                                                          x500

Fig. 3 Micropores development (b) along the shear plane,
resulting in crack propagation (a).

Fig. 4 shows loading-time diagram and AE signals diagram
of Ti-6Al-2Zr-1.5V-1.5Mo alloy, which ductility was determined
by compression at 850°C. At the initial stages of deformation
(deformation extent-10%) both full deformation loads and AE intensity, which is due to «noises» on contact surfaces.

Characteristic of this stage is the absence of local deformation zones. Further deformation is accompanied by the appearance of hampered deformation cones with their continuous
approach, which localizes deformation area and limits the development of uniform plastic deformation.

Macrocracks were formed in the intensive flow zones on
reaching 40% deformation extent on some specimens, which was
accompanied by the appearance of additional signals on AE diagram (Fig. 4, 5).

Exterior signs of cracking were observed at deformation extent of 60%, but the change of AE intensity wasn't noted. At the
final stage of upsetting (deformation extent 80% and higher) specimens fractured.

Thus, based on the conducted experiments technique for detecting acoustic emission in the deformation area during high-temperature ductility tests of titanium alloys was developed, the dependence between AE parameters and crack initiation on plastic deformation was traced. The use of AE method for ductility tests of metals will make it possible to obtain additional information about the beginning and the character of fracture depending on deformation temperature, rate, and extent.

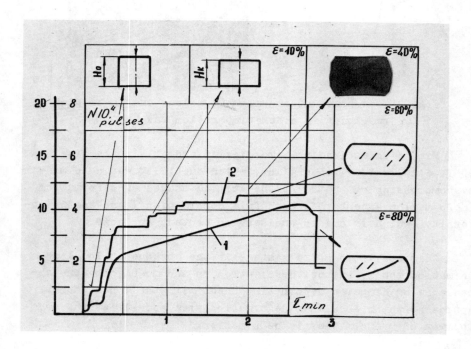

Fig. 4 Character of AE pulses total number change (curve 2) during loading (curve 1) on compression of Ti–6Al––2Zr–1.5V–1.5Mo alloy specimens.

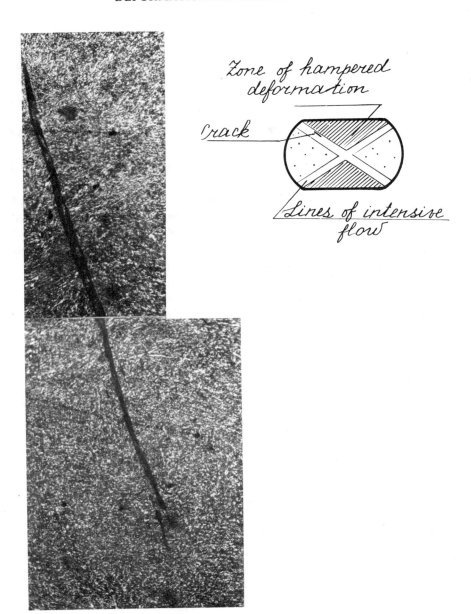

Fig. 5 Crack view (a) along the line of intensive flow.
Upsetting. Deformation temperature — 850°C.

## References

1. V. L. Kolmogorov, A. A. Bogatov, B. A. Migachev et al. In book: Ductility and Fracture, M., Metallurgia, 1977.
2. B. M. Finkel. Fracture Physics. Crack Growth in Solids,M., Metallurgia, 1970.
3. V. L. Kolmogorov. Stresses, Deformation, Fracture,M., Metallurgia, 1970.
4. G. Gelman. Mechanics (periodic collection of foreign articles translations), 1962, N 5, p. 72.
5. S. H. Carpenter, F. P. Higgins. Sources of Acoustic Emissions Generated During the Plastic Deformation of 7075 Aluminium Alloy, Metallurgical Trans., 1978, N 10, pp. 1629–1632.
6. V. E. Vainberg, Kantor A. Sh. Temperature Studies of Acoustic Emissions, Defectoscopy, 1975, N 6.

# RESISTANCE TO DEFORMATION AND DYNAMIC CHANGES IN THE STRUCTURE OF TWO-PHASE TITANIUM ALLOYS UNDER HOT DEFORMATION

G. V. Shakhanova, I. B. Rodina, F. V. Tulyankin,
A. L. Pilipenko, N. A. Sharshagin, N. V. Bukharina

All Union Institute of Light Alloys

USSR

## Introduction

The works dealing with the effect of initial structure of two-phase titanium alloys on their resistance to deformation as well as final structure, resulting from deformation, have shown the expediency of using for forging the billets with globular structure [1, 2]. However, the investigations of that kind, as a rule, do not take into account a variety of the structures of two-phase titanium alloys, caused not only by the morphological features of the $\alpha$-phase, but by the difference in the quantitative parameters of the structural constituence as well.

This was the main task of the present work to study the effect of initial structure, differing in both morphology of the $\alpha$-phase and quantitative characteristics of the plate-like structure, upon the resistance to deformation and final alloy microstructure, determined by its dynamic changes during hot deformation. A wide range of the temperature-rate conditions of deformation was investigated, covering both the regimes of superplasticity and those of conventional hydraulic forging.

## Material and Research Procedure

The experiments were carried out with the $(\alpha+\beta)$-alloys of the BT6ч type (Ti-6Al-4V) and those of BT3-1 type (Ti-6Al-2Mo--2Cr). The temperature of polymorphous transformation $(t_{\alpha+\beta/\beta})$ of the BT6ч alloy was equal to 990°C, that of BT3-1 - 980°C. The specimens Ø 20x30 mm were prepared from the rods 25 mm in dia., rolled at the temperature of $(\alpha+\beta)$-region and subjected to various kinds of heat-treatment to obtain the desirable type of initial structure. The quantitative characteristics of initial structures are given in table.

Table

Characteristics of initial structure at room temperature

| Alloy | Type of structure | Conventional designation | Dimensions, m | | | |
|-------|------------------|--------------------------|-----------------|------------------|-----------|----------|
| | | | β-grains | α-colonies (thickness) | α-plates | α-grains |
| BT6ч | globular | 1 | – | – | – | 5 |
| | plate-like | 2 | 500 | 120 | 6 | – |
| | « | 3 | 500 | 10 | 0.5 | – |
| | « | 4 | 2000 | 250 | 3 | – |
| BT3-1 | globular | 1 | – | – | – | – |
| | plate-like | 2 | 700 | 160 | 3 | – |

The specimens were compressed using the MTS installation, equipped with the sectional furnace to provide for isothermal deformation conditions and special device, enabling to quench the specimens in one second after deformation. The temperature of deformation $(t_d)$ was equal to 850-1100°C, initial rate of deformation $(\mathcal{E})$ $-10^{-4}$ $-10^0 s^{-1}$, reduction $(\mathcal{E})$ - 10-80%. The reduction was determined by a relative change of the specimens' height. The flow stress was calculated using the ratio: $\sigma = \frac{P}{S}$, where R is the load of deformation, S - maximum cross-section of specimens.

The structure of alloys was studied, which forms on heating before deformation and after deformation, using the methods of metallography, electron microscopy and X-Ray analysis.

## Experimental Results
### Strain-stress diagrams

Fig. 1 a-c shows the examples of the curves «true flow stress vs. strain» when compressing the specimens from BT6u and BT3-1 alloys.

The character of the diagrams for both alloys is quite resembling each other, and discussion, in main, refers to the BT6u alloy. The level of the flow stress is determined first of all by the known effect of temperature and rate of deformation.

The case, when the deformation takes place at the rate of $10^{-4}s^{-1}$ of the alloy with the globular structure presents the exception — with increasing deformation temperature from 950°C to 970°C the $\sigma$ level increases, rather than decreases. This is due to the deviation from the conditions of superplastic deformation, which correspond to the temperature interval over 925-950°C for the globular structure of BT6u alloy. These are the features of the superplastic deformation: low values of flow stress ($\sigma < 1$ kgf/mm$^2$), the coefficient of the rate sensitivity of the flow stress $m > 0.3$ and characteristic shape of the «flow stress-strain» curves (Fig. 1d). The «m» coefficient values were determined for the beginning of the plastic flow as the tangents of the slope angle for the $\lg\sigma$ /$\lg\varepsilon$ curves.

The effect of the reduction upon the flow stress, as seen from Fig. 1, can not be unambiguously characterized for all the deformation temperature-rate conditions investigated. When $t_d > t_{\alpha+\beta/\beta}$ according to the views accepted for the one-phase alloys [3] the horizontal region is observed on the strain diagrams, determining the area of stationary stage of deformation, when $\sigma$ does not change with increasing $\varepsilon$ . There appears another shape of the strain diagram for $t_d < t_{\alpha+\beta/\beta}$: almost in all cases,

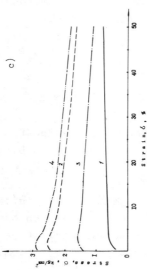

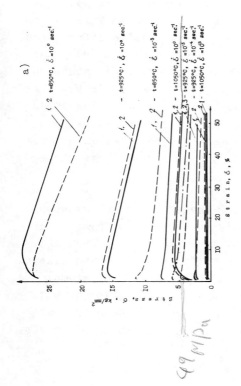

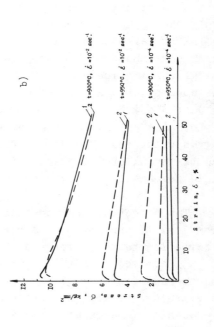

Fig. 1

Examples of the strain diagrams:
a) - BT6ч -alloy; b) BT3-1-alloy;
c) BT6ч -alloy at deformation temperature
of 925°C and deformation rate of $10^{-4}$ S$^{-1}$.
Designations: initial structure of globular
type (1)
initial structure of plate-like
type (2)
initial structure of plate-like
type (3)
initial structure of plate-like
type (4)

irrespective of initial structure, the decrease of $\sigma$ is observed with increasing $\mathcal{E}$ after strengthening, reaching its peak at
$\mathcal{E} \approx 1 \div 6\%$. The intensity of softening increases with decreasing deformation rate. Some slight strengthening is observed
with increasing reduction for initial globular structure under
conditions of superplasticity.

Effect of initial structure on the flow stress of alloys is
characterized by a variety of features, which were revealed by
means of changing the deformation temperature-rate conditions
and kinds of initial structure over a wide range. It was established (Fig.1) that at relatively low temperatures and high deformation rates over the ranges studied the alloys with globular
structure exhibit more resistance to deformation, than those
with plate-like structure. This ares is below the boundary AA
shown in Fig. 2. Above this boundary more resistance to deformation is exerted by the alloy, possessing initial plate-like
structure. The difference in resistance to deformation of alloys
having different types of structure depends upon the size of ß-
grains, and especially on that of the $\alpha$ -colonies in the plate-
like structure. The lower their dimensions, the less is the difference between the flow stress of the alloy with plate-like
structure and that of alloy with globular structure (Fig. 1d).
This regularity is apparently caused by that the $\alpha$ -colonies,
being the areas of the same orientation of the $\alpha-$ phase, behave
on plastic deformation as the uniform $\alpha$ -grains. The difference
in the flow stress of the alloys, possessing differing initial
structures is maintained up to $\mathcal{E} \geqslant 50\%$, decreasing in the run
of the deformation (Fig. 1).

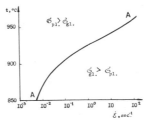

Fig. 2 Diagram of the temperature-rate regions of deformation,
determining the character of influence of the structural type of the BT6y alloy on its resistance to deformation.

## Structural Changes on Hot Deformation

When heating the specimens up to $t_d < t_{d + \beta/\beta}$ the quantitative ratio of the phases is changed, thereby quantitative parameters of initial structure being affected, while its type is not changed. So the size of $\beta$-grains in the BT$\alpha$ alloy with initial globular structure increases from 2 to 40 $\mu$ m with increasing heating temperature from 800 up to 970°C, while the diameter of $\alpha$ - grains thereby decreases from 5 down 1 $\mu$ m. This change is more sharply pronounced at temperatures above 925°C. The thickness of the $\alpha$-plates and $\beta$-layers between them changes in the same manner when heating the specimens with initial plate-like structure.

The investigations, carried out in the present study, have shown the structural changes, occuring during deformation, to be caused by several processes, which are characteristic of $\alpha$ and $\beta$ phases.

The polygonization takes place in the $\beta$-phase under the action of deformation over the whole temperature range studied, which is accompanied by dynamic recrystallization at $t_d > 920$°C and $\varepsilon < 10^{-2} s^{-1}$. The intensity of recrystallization thereby increases with increasing $t_d$ within the two-phase region and decreasing $\varepsilon$. New $\beta$-grains in the plate-like structure are placed as the layers between the deformed $\alpha$ -plates (Fig. 3a), their size being comparable to the width of the $\beta$-layers. This determines its dependence upon initial structure of the alloy and sharp increase with increasing $t_d$.

The dynamic polygonization and recrystallization are observed in $\alpha$ -phase, this being the same as in $\beta$-phase (Fig. 3c, d). The character of the effect of the temperature-rate deformation conditions upon the subgrain structure of the $\alpha$-phase, irrespective of the shape of its grains, is the same as that for $\beta$-phase: the subgrain size increases with decreasing $\varepsilon$ and increasing $t_d$. The $\alpha$ -subgrains are thereby $\sim$ 10 times less as compared to those of $\beta$.

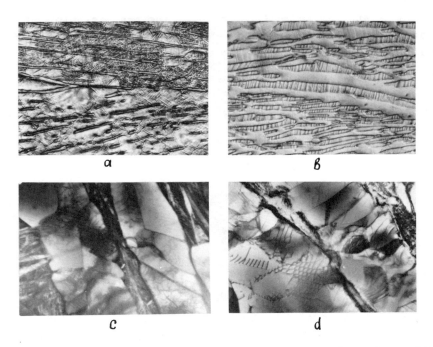

Fig. 3 Stucture of deformed alloys BT3-1 (a, b) and BT6u
(c, d) ($\mathcal{E} = 50\%$). Water cooling after deformation:
a - $t_d = 950^\circ C$ ,     $\mathcal{E} = 10^0 s^{-1}$ , x750;
b - $t_d = 925^\circ C$ ,     $\mathcal{E} = 10^0 s^{-1}$ , x1000;
c - $t_d = 925^\circ C$ ,     $\mathcal{E} = 10^{-1} s^{-1}$ , TEM, x17000;
d - $t_d = 925^\circ C$ ,     $\mathcal{E} = 10^{-4} s^{-1}$ , TEM, x18000.

Together with the coarsening of subgrains with increasing
$t_d$ and decreasing $\mathcal{E}$ the tendency is observed for the polygonal
walls to arrange normally to the boundaries of the $\measuredangle$ -plates,
i. e. formation of the so-called «bamboo structure» (Fig. 3d).

The recrystallization of the $\measuredangle$ -phase, detected by the
X-Ray method is apparently dynamic one, since no $\measuredangle$ -grains
free of dislocations were revealed in the structure of deformed
alloys. Low density of dislocations in $\measuredangle$ -grains is observed
only in the specimens having initial globular structure after
deformation at 925-950°C at the rate of $10^{-4} s^{-1}$, i. e. under the
superplasticity conditions. Fig. 4 shows the lower boundaries of

the temperature-rate regions, determining the beginning of dynamic recrystallization of BT6ч alloy with globular (1-1 line) and plate-like (2-2 line) structures.

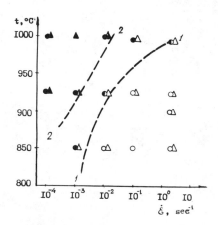

Fig. 4 Diagram of dynamic recristallization of the $\alpha$ -phase
in the BT6ч   alloy specimens with initial globular
(O) and plate-like ($\triangle$) structure (2).
Designations:  o △  - no recrystallization;
               ø▲  - beginning of recrystallization;
               ●▲  - more than 50% of the $\alpha$ -phase
volume is recrystallized.

As seen from the diagram, the recrystallization in the globular $\alpha$ -phase proceeds more readily than it does in the plate-like phase, this being consistent with the data obtained earlier [2] .

In contrast to the ß-phase the deformation in the $\alpha$ -phase in both alloys under study is effected not only by slip, but also by the twinning process. The deformation twins are those of the $(101)_\alpha$ type, most widely met in the h.c.p. metals. The tendency towards twinning becomes more pronounced with decreasing temperature and increasing rate of deformation (especially when going away along this direction from the dotted lines in Fig. 4) Resulting from deformation the twin density in the $\alpha$ -phase increases by the order or two. The width of the deformation twins, the largest of which are easily seen in the microstructure (Fig. 3b) varies within one and the same specimen from 0.01 up to $3 \mu$m. Being intensively developed in the $\alpha$-phase, the twinning contributes little to the total deformation. So, calculations show, that after 50% deformation at $t_d = 925^\circ$C and $\mathcal{E} = 10^\circ S^{-1}$, when the volume percent of the twins in the BT6u alloy specimens reaches 10%, the twins' contribution to the total deformation does not exceed 2%.

The deformation twins in the $\alpha$-phase are distributed rather unevently and may differ in the adjucent microregions as much as $\sim$100 times. The periodicity of this nonuniformity is determined by the size of the $\alpha$-colonies, in those specimens, having plate-like structure and $\alpha$-grain size in the specimens with globular structure. When the local strain is mainly caused by the twinning process, then polygonization and recrystallization are either rather weak or do not develop at all. The nonuniformity in the twins' distribution is, apparently, one of the reasons for considerable nonuniformity of the $\alpha$-phase fine structure.

The tendency to the twinning is also affected by initial structure of alloys, increasing with coarsening the $\alpha$-phase. The tendency of the $\alpha$-phase to recrystallization is respectively less in the case of the coarse-plated structure as compared to that of the fine-plate structure or globular one in particular (Fig. 4).

On deformation the process of the transformation takes place of the plate-like structure into globular one. The plate-like structure is known to be thermodynamically unstable: the equilibrium of surface tension between the phases is easily violated with such a shape of the phase constituents, thus leading to the spheroidization of the plates by means of the dissolution-precipitation mechanism of diffusion. The equilibrium of the surface tension between the phases is violated in those sites, where the twins come out to the interfacial boundary of internal boundaries of $\alpha$-plates: those of subgrains, recrystallized grains and twins.

So, all the internal boundaries ennumerated, promote the spheroidization. The investigation carried out has shown the twin boundaries in two-phase titanium alloys to be most effective in promoting the spheroidization. Since the twinning process, as mentioned earlier, proceeds rather unevenly through the alloy volume, the concomittant process of spheroidization of the $\alpha$-plates inherites this nonuniformity (Fig. 5).

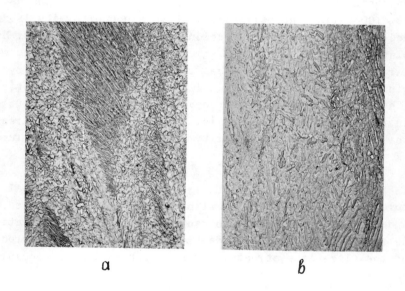

a        b

Fig. 5 Microstructure of the BT6ч alloy with the plate-
like initial structure 2(a) and 3(b) after deformation according to the regime: $t_d = 925^\circ C$;
$\mathcal{E} = 10^{-4}s^{-1}$, $\mathcal{E} = 50\%$, (x200).

The character of the structure, formed on spheroidization, depends upon the type of the boundaries, promoting the spheroidization. When spheroidization takes place, caused by the twinning, the chain is formed of the $\alpha$-particles which are oriented along one and the same direction and repeat the orientation of the matrix $\alpha$-plate. The $\alpha$-particles, considerably disoriented form only when the spheroidization takes place, connected with the recrystallization of the $\alpha$-phase. Thus, the number of disoriented $\alpha$-grains is proportional to the extent to which recrystallization develops in the $\alpha$-phase.

The spheroidization process is characteristic of the globular initial structure to the same extent, to which it is for the plate-like structure, since resulting from the deformation each $\alpha$-grain in the globular structure looses its equiaxity and aquires the shape of the rod or plate, the internal structure of which has no principal difference with respect to that of deformed plates.

## Conclusions

Let us now compare the typical structural changes accompanying hot deformation of two-phase titanium alloys, with respective changes of the flow stress. The comparison of that kind is of primary interest for the $(\alpha + \beta)$ deformation. At $960^\circ C \le t_d \le$ $\le t_{\alpha + \beta/\beta}$ the $\beta$-phase dominates, and the structural changes which are characteristic of this phase, notably – dynamic recrystallization and polygonization – determine the run of the plastic flow. The dynamic recrystallization may thereby promote the decrease of the alloy strength, this being not due to removing of «work hardening», which continuously recovers on deformation, but because of refining the $\beta$-grain size. The process of dynamic recrystallization as that one decreasing the strength, is counteracted by the formation of subgrain structure, which should assist in strengthening the alloy as compared to its annealed state. At high rates of deformation $(\mathcal{E} > 10^{-2} s^{-1})$ the polygonization is apparently dominating of these two processes, which determines a weak strengthening of the alloy, while dynamic recrystallization dominates at low rates, causing slight decrease of the strength.

At lower temperatures of two-phase regimes $(925-950^\circ C)$ where $\alpha$ and $\beta$ phase are comparable in amount, the structural changes of $\alpha$-phase begin to contribute significantly. Spheroi-

dization of the  $\alpha$ -phase and dynamic recrystallization of the
$\beta$-phase over the temperature range above AA line in Fig. 2 promoting the development of the grain and interface boundaries
should also cause the decrease of the alloy's strength, which can
be dominating over the subgrain strengthening.

Spheroidization of the plate-like structure and slow grain
growth in the globular structure when deforming at the rate of
$\mathcal{E} \leqslant 10^{-3} s^{-1}$ and $t_d = 925-950^{\circ}C$ may account for the shape of respective diagrams using the well-known ideas, concerning the me-
chanism of superplastic deformation [4] .

If the twinning is not accompanied by spheroidization of the
$\alpha$ -phase $(\mathcal{E} \geqslant 10^{-1} s^{-1})$ it may possibly exert its effect upon the
flow stress according to the mechanism of geometrical softening
[5] , simultaneously preventing the formation of the subgrain
structure.

The polygonization and recrystallization of the $\beta$-phase, as
well as twinning, polygonization and recrystallization of the $\alpha$ -
phase take place on hot deformation of two-phase alloys irrespective of their initial structure. The latter only exerts some influence upon relative intensity of these processes: the coarser
the structure of the initial  $\alpha$ -phase, the more is its tendency
to twinning on deformation, the slower is the recrystallization
process in it and less uniform these processes occur over the
alloy volume.

At high deformation rates the structure of alloys does not
change its initial type. The dynamic transformation of the plate-
like structure into globular one proceeds only at low rates of
deformation over the temperature range, providing for considerable
diffusion mobility of atoms. This range corresponds to the upper
part of the $(\alpha + \beta)$ region for the alloys studied, i. e. above
$920^{\circ}C$. The globular structure thereby is obtained being more disoriented when the deformation temperature is higher; this may be
due to more intense recrystallization of $\alpha$ and $\beta$ phases, preceding spheroidization of the $\alpha$ -plates. The process of transformation of the plate-like structure into globular one proceeds un-

evenly through the alloy volume. This nonuniformity is caused by nonuniform and non-simultaneous formation in various microregions of stable internal boundaries inside the $\alpha$-plates, along which the spheroidization of the $\alpha$-plates occurs and among those boundaries the twin boundaries are of primary importance. This is also the reason for dependence of the nonuniformity in spheroidization upon the initial structure of alloys and upon the size of $\alpha$-colonies, in particular. Resulting from this nonuniformity there is no complete transformation of the plate-like structure into globular one for any of the deformation conditions studied. The coincidence of the flow stress for any deformation stage within the two phase region for different types of initial structures does not indicate the identity of these structures, which maintain their inherited features up to $\varepsilon \geqslant 50\%$.

## References

1. W. G. Spiegelberg and F. N. Lake. Quarterly Engineering Reports, Army Weapon Command Contract DAAF-03-73-C-0093, January, 1974.
2. C. C. Chen. Proceedings of 3rd International Conference on Titanium. M., VILS, 1973, 239.
3. H. P. Stüme. Acta Metallurgica, 1963, V13, No. 12, 3.
4. M. V. Grabsky. The Structural Superplasticity of Metals. M., Metallurgia, 1975.
5. Structure and Mechanical Properties of Metals. Proceedings of the Conference in Teddingtown, January 1963 (in Russ.), M., Metallurgia, 1967.

# MATERIAL PARAMETERS OF IMPORTANCE IN THE STRETCH FORMING OF Ti-6Al-4V SHEET AT 300-950 K

K. Okazaki*, M. Kagawa** and H. Conrad

Metallurgical Engineering and
Materials Science Department
University of Kentucky
Lexington, Kentucky 40506, U. S. A.

## Introduction

Forming Limit Diagrams are of considerable value in the practical assessment of sheet stretch-forming operations. Experience has established that material parameters such as the strain hardening exponent n (=$d\ln\sigma/d\ln\varepsilon$), the anisotropy strain ratio R (=$\varepsilon_2/\varepsilon_3$) and the strain rate sensitivity exponent m (=$d\ln\sigma/d\ln\dot{\varepsilon}$) play important roles in stretch forming. For instance, the material with a higher n value gives a more uniform strain distribution in stretch-forming (1). That the strain rate sensitivity exponent has an effect on stretch-forming was recognized by several investigators (2-4) who found that higher m values yield more even strain distribution and slower rate of drop of load beyond the maximum. Furthermore, Hecker (5) found that the cup height for a number of metals tested in the modified Olsen cup test could not be predicted from n values alone. He pointed out one reason for this was that the strain rate sensitivity was not included.

The combined effect of strain hardening and strain rate sensitivity on sheet metal forming has recently been taken into account by Ghosh (6), who used a simplified constitutive equation of $\sigma = K\varepsilon^n\dot{\varepsilon}^m$ to analyze the experimental data and identified that n is the most important factor in the distribution of strain prior to the onset of diffuse necking and the presence of a small but positive strain rate sensitivity m exercises an important stabilizing influence on the deformation beyond that point. However, no quantitative assesment of the combined effect on stretch formability was made. Conrad (7) introduced the following equation for the effect of materials parameters on the limiting strain $\bar{\varepsilon}*/Z$ in stretch-forming:

$$\bar{\varepsilon}*/Z = n + \alpha m \tag{1}$$

where Z is the critical subtangent (8) which is a function of the stress (or strain) ratio and $\alpha$ is a constant given by $d\ln\dot{\varepsilon}/d\ln\bar{\varepsilon}$ and refers to the element which eventually failed. He then showed for existing data that the cup height correlates better with $\bar{\varepsilon}*/Z$ than with n, since the parameter $\alpha m$ is added to n to give $\bar{\varepsilon}*/Z$. In a subsequent paper, Conrad, Okazaki and Yin (9) confirmed this correlation for cup tests conducted by them on steel and titanium, and also showed that the experimental values of $\bar{\varepsilon}*/Z$ for both uniaxial tension and cup tests conformed with eq. (1).

More recently the present authors (10) experimentally identified three plastic instability strains in A-75 Ti sheet tested in uniaxial tension at 300 to 700 K (m=0.002 - 0.055) and strain rate of $10^{-4}$ to $10^{-2}$ s$^{-1}$:

*Presently with Allied Chemical Corporation, Morristown, New Jersey 07960, U.S.A.
**Former Graduate student, now with Hitachi Cable, Ltd., Tsuchiura, Ibaraki, 317, Japan

(a) $\varepsilon_I^*$, the initiation of strain concentration leading to a diffuse neck,
(b) $\varepsilon_{II}^*$, the restriction of strain to the diffuse neck region and
(c) $\varepsilon_{III}^*$, the initiation of local necking.
The effects of materials parameters on the critical strains for each of these
instability conditions were given reasonably well by the following expressions.

$$\varepsilon_I^* = n_I + \partial \ln A_0 / \partial \ln \varepsilon \tag{2}$$

$$\varepsilon_{II}^* = \varepsilon_I^* + 3.3 \, m_u^{2/3} \tag{3}$$

$$\varepsilon_{III}^* = 2(n_u + m_u(\alpha_{III} + \varepsilon_u)) \approx 2 \, (n_u + \alpha_{III} m_u) \tag{1a}$$

where $A_0$ is the original specimen cross-sectional area and the subscripts indicate the strain at which the material parameters were determined.

The objectives of the present paper were: (a) first, to generate forming
limit diagrams (FLD) for the Ti-6Al-4V alloy at temperatures and punch speeds
similar to those employed in practice (i.e. 300 to 950 K and $10^2$ to $10^{-2}$ mm/s),
(b) secondly, to carry out uniaxial tension tests to determine the necessary
material parameters in the same temperature and strain rate range and (c)
finally, to correlate the combined effects of the material parameters on the
stretch formability of Ti-6Al-4V sheet.

## Experimental Procedure

Tensile specimens of Ti-6Al-4V sheet with a gauge length of 38.1 mm, a
width of 6.35 mm and a thickness of 1.28 mm were provided by Battelle.  All
tensile specimens had the tensile axis parallel to the rolling direction of
the sheet.   The specimens used for the punch tests to determine the FLD were
127x127 mm blanks with a thickness of 1.28 mm from the same lot as that for
the tensile tests.   For the negative side of the FLD, 12.7 mm wide and 127 mm
long strip specimens were machined from the 127x127 mm blanks with the longitudinal direction parallel to the rolling direction.

· The uniaxial tension tests were performed with a Gilmore servoloop hydraulic testing machine at room temperature   (300 K), 1000 F (811 K) and 1250 F
(950 K) at various ram speeds to give the nominal strain rates of $2.8 \times 10^{-4}$ to
$2.8 \times 10^{-1}$ s$^{-1}$.   At elevated temperatures the specimens were protected from
oxidation by coating the surface with a graphite suspension type lubricant
(Formkite) prior to the test and employing an argon atmosphere of 7.5 cm$^3$/s
flow rate during the test. To obtain the strain rate sensitivity exponent m,
strain rate cycling tests were carried out by making 10:1 or 6:1 incremental
changes in strain rate from each base strain rate (except at the highest rate).
m was calculated from the strain rate cycling data using

$$m = \partial \ln \sigma / \partial \ln \dot{\varepsilon} = \ln(\sigma_2 / \sigma_1) / \ln(\dot{\varepsilon}_2 / \dot{\varepsilon}_1) = \ln(P_2/P_1)/\ln(V_2/V_1) \tag{4}$$

where $P_1$ and $P_2$ are the loads prior to, and following, the ram speed change
from $V_1$ to $V_2$.[2] The strain hardening exponent n was calculated as a function
of strain from the true stress-strain curves through

$$n = (d \ln \sigma / d \ln \varepsilon) = (\varepsilon/\sigma)(d\sigma/d\varepsilon). \tag{5}$$

The FLD tests were conducted using the Gilmore testing machine at constant
punch speeds of 0.025-25 in./min ($1.06 \times 10^{-3}$ to 1.06 cm/s) and over the same
temperature range as the uniaxial tension tests.  For the positive side of the
FLD, modified Olsen cup tests were performed on the 127x127 mm blanks (onto

which circular grids of 2.54 mm dia. had been electro-chemically etched) without
and with lubricants of the MoS$_2$ type (Silver Goop) and of the graphite-suspension
type (Kalgard T-50 and Formkite).  For the negative side of the FLD, 12.7 mm wide
strip specimens were tested also with and without lubricants.  For the tests at
elevated temperatures, the punch and die were first placed in a furnace and
preheated to the desired temperature.  The specimens were then firmly bolted
down onto the preheated die and placed back into the furnace, through which
argon gas was passed to reduce oxidation during heating and testing.  The time
required to heat the combined specimen and die to the desired temperature was
of the order of 20 min.  It was also found that a thick coating of Kalgard applied to the gridded surface (on the side opposite that to be contacted by the
punch) before setting the blank onto the die was very effective in preventing
the grids from becoming obscure as a result of oxidation.

For constructing the FLD, the grid elements above and below the fractured
element were measured at positions parallel and transverse to the rolling direction.  When the measured grid included any neck groove or was non-symetrically
distorted, the measurement was classified "neck-affected"; otherwise it was
designated "neck-free".  The major strain $\varepsilon_1$ and minor strain $\varepsilon_2$ were determined using a measuring tape which was graduated to 1% strain.  In addition
to the measurements at the position parallel and transverse to the rolling
direction, measurements were also made on grid elements adjacent to the fracture in the circumferential direction to determine any anisotropy effects
which might exist.  To determine the anisotropy strain ratio, $R(=\varepsilon_2/\varepsilon_3)$, tensile
specimens were cleaned with a $HNO_3$-HF mixture, washed in running water and dried
in air.  The cleaned specimens were then electrochemically etched to induce
square grids (1.27 mm on a side).  The specimens were then strained in an
Instron machine at room temperature to predetermined strains and unloaded.
One group of specimens was strained at the strain rate of $0.87 \times 10^{-3} s^{-1}$ and
another at $0.87 \times 10^{-2} s^{-1}$.  The deformed specimens were unloaded and transferred
onto an optical microscope to measure the grids.  Within the uniform deformation range, three measurements were made: one at the center of the gauge and
two at points 12.7 mm from the center.  In the longitudinal direction, 10 grid
blocks were measured as a unit, and 4 grids in the width direction.  During
the uniform deformation, the R values were averaged for the three locations,
but when visible necking was observed, the measurement from the neck region
was excluded from the average.

## Experimental Results

The effects of temperature and strain rate on the engineering stress-
strain curves are presented in Fig. 1.  The yield stress decreases with
increasing temperature and decreasing strain rate.  The total elongation
increases with increase in temperature at each strain rate and with decrease
in strain rate at each temperature, superplastic behavior occurring at 950K
for a strain rate of $2.8 \times 10^{-4} s^{-1}$.  On the other hand, the effect of strain
rate on the flow stress does not necessarily follow the same trends as for
the yield stress.  At 300K it is expected that this may be due to adiabatic
heating effects at the high strain rates, whereas at 950 K superplasticity
occurs for the lowest strain rate.  Worthy of mention regarding the results in
Fig. 1 is that the maximum load appears at rather small strains and is quite
strain-rate sensitive within the ranges of temperature and strain rate
presently studied.

Fig. 2 shows the strain hardening exponent $n_u$ near the maximum load
plotted against the ram speed as a function of temperature.  At each temperature $n_u$ decreases with increase in strain rate, and at a fixed strain rate it

decreases with decrease in temperature.  As seen in Fig. 2, the $n_u$ values for
this alloy in the ranges of temperature and strain rate studied are comparatively
small except for the slowest strain rate at elevated temperatures.  Thus, the
maximum load appears at rather small strains, implying that the critical
strains for diffuse necking for this alloy are not as large as those for A-75
Ti (10).

Depicted in Fig. 3 is the strain rate sensitivity exponent $m_u$ versus the
nominal strain rate as a function of deformation temperature.  At elevated

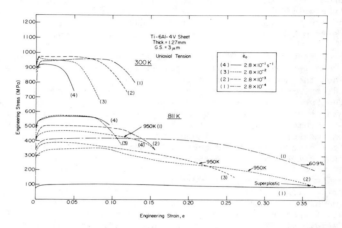

Fig. 1 Engineering stress–strain curves for Ti-6Al-4V sheet
as a function of temperature and strain rate.

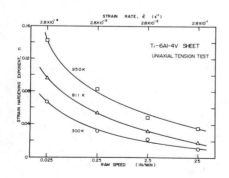

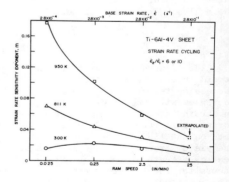

Fig. 2 Strain hardening exponent $n_u$ at
the maximum load versus strain
rate as a function of temperature.

Fig. 3 Strain rate sensitivity exponent
$m_u$ at the maximum load versus
strain rate as a function of
temperature.

temperatures $m_u$ decreases monotonically with increasing base strain rate. However, at room temperature $m_u$ is not influenced as much by the base strain rate, and $m_u$ appears to go through a maximum at $\dot{\varepsilon}=2.8\times10^{-3}$ s$^{-1}$. The $m_u$ values at the highest base strain rate of $2.8\times10^{-1}$ s$^{-1}$ were estimated simply by extra-polating the curves through the data points at lower strain rates at each temperature to $2.8\times10^{-1}$ s$^{-1}$. The $m_u$ values thus estimated were cross-checked with those obtained from the slope of log-log plots of the yield stress versus strain rate and the two were found to be in reasonable accord with each other.

Fig. 4 shows the results of the R value measurements at room temperature, where it is seen that the R value increases from approximately 0.3 at yielding to 1.0 at the maximum load. Within the experimental errors it is difficult to discriminate the data for the low strain rate tests from those at the higher rate. It should be noted that the R value varies with strain so that a unique R value can only be assigned for a particular strain in the construction of the FLD. Based on the stress-strain curves in Fig. 3, it was decided to con-sider the R value at 3% strain (the maximum uniform strain at the highest strain rate and lowest temperature in the present study) as a function of temperature and strain rate. The results obtained are presented in Fig. 5, where it is seen that the R values at 811 K tend to be higher (having a value of approximately 1.0) than those at 300 and 950 K (which range between 0.6 and 0.8), relatively independent of the strain rate. Based on the results shown in Figs. 4 and 5 and the scatter in the R values, it was decided that R values of 0.5 to 1.0 would be tentatively assigned in evaluating the FLD's for this material in the temperature and strain rates ranges studied.

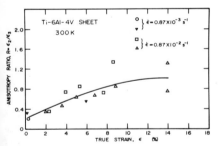

Fig. 4 Anisotropy strain ratio for Ti-6Al-4V measured as a func-tion of strain at 300 K.

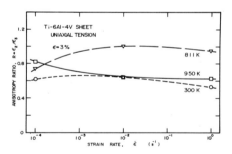

Fig. 5 Anisotropy strain ratio versus strain rate measured at 3% strain as a function of tem-perature for Ti-6Al-4V.

Forming limit diagrams in terms of the true major and minor strains determined from the present punch tests are given in Fig. 6 as a function of the punch speed at three temperatures. It is clearly seen that an increase in temperature for a given punch speed shifts the FLD upward; also, a decrease in punch speed shifts the FLD upward at each temperature. The only exception to this trend are the results for the 25 and 2.5 in/min punch speeds at 300 K, which lie on the same FLD. Although no identification is made here, the data for both the "neck-affected" elements and the "neck-free" elements lie along a single curve; also the data for the position 90° from the rolling direction (transverse position) fell along the curve for results parallel to the rolling direction. No data are available at 300 and 950 K for the punch speed of 0.025 in/min, because at 300 K the load exceeded the capacity of the machine and superplastic behavior occurred at 950 K.

## Discussion

Let us consider the effective strain from the data sets of $(\varepsilon_1, \varepsilon_2)$ shown in Fig. 6 using the equation by Ghosh and Backofen (11) for biaxial tension as

$$\bar{\varepsilon}^* = \sqrt{\frac{2}{3} \frac{(R+2)(R+1)}{(2R+1)}} \left(\varepsilon_1^2 + \varepsilon_2^2 + \frac{2R}{1+R} \varepsilon_1 \varepsilon_2\right)^{\frac{1}{2}} \tag{5}$$

where $\varepsilon_1$ and $\varepsilon_2$ are the major and minor strains respectively. Fig. 7 shows examples of the strain ratio dependence of the effective strain at three different temperatures. The curves through the data points exhibit parabolic behavior with minima nearly at the strain ratio Y $(=\varepsilon_2/\varepsilon_1) = 0$ in all cases. The choice of R value between 0.5 and 1.0 does not change the value of $\bar{\varepsilon}^*$ very much in the range of $-0.3 < Y < 0.3$, but it does affect $\bar{\varepsilon}^*$ $|Y| > 0.4$, the effect being opposite in the positive and negative regions.

To check the strain ratio dependence of the limiting strain $\bar{\varepsilon}^*/Z$ for plastic instability, $\bar{\varepsilon}^*/Z$ was calculated using the subtangent Z introduced by Keeler and Backofen (8), which is given as

$$Z_N = 2(1 - X + X^2)^{\frac{1}{2}}/(1 + X) \tag{6}$$

for the negative region and

$$Z_p = 4(1 - X + X^2)^{3/2}/(1 + X)(4 - 7X + 4X^2) \tag{7}$$

for the positive region, respectively where X is the stress (or strain) ratio

$$X = (\sigma_2/\sigma_1) = (\varepsilon_1 + 2\varepsilon_2)/(2\varepsilon_1 + \varepsilon_2). \tag{8}$$

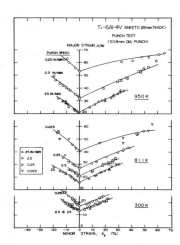

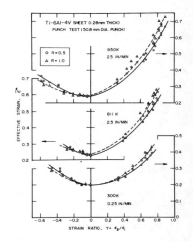

Fig. 6 Forming limit diagram for Ti-6Al-4V constructed for tests at 300 to 950 K as a function of punch speed.

Fig. 7 Effective strain versus strain ratio at selective temperatures and strain rates for Ti-6Al-4V.

Examples of the strain ratio dependence of the limiting strain for the punch tests of Ti-6Al-4V sheet at three temperatures with two R values of 0.5 and 1.0 are depicted in Fig. 8.  As seen in the figure, the limiting strain $\bar{\varepsilon}*/Z$ is relatively constant below the strain ratio $Y\leq0.4$ and starts to increase at higher values of Y.  The choice of R values does not appreciably affect the $\bar{\varepsilon}*/Z$ value below $Y\leq0.4$ (the scatter is within the experimental error).  The deviation from the constant $\bar{\varepsilon}*/Z$ line at Y higher than 0.4 reflects either a true increase in the effective failure strain or that the equations used for $\bar{\varepsilon}*$ and Z are not valid at large values of Y.

Collected in Table 1 are the mechanical properties and average limiting strain $\bar{\varepsilon}*/Z$ for Ti-6Al-4V sheet tested at 300, 811 and 950 K.  It is here seen

Table 1 Mechanical properties and limiting strains

| Temp (K) | Punch speed (in/min) | Yield stress (Mpa) | Tensile strength (MPa) | Strain hardening exponent | Strain rate sensitivity | Limiting strain (Y<0.4) [+++] α | $\bar{\varepsilon}*/Z$ | $(\bar{\varepsilon}*Z)-n$ |
|---|---|---|---|---|---|---|---|---|
|      | 25    | 907 | 972  | 0.011 | 0.011[+] | 8.09 | 0.10 | 0.09 |
| 300  | 2.5   | 887 | 983  | 0.022 | 0.016 | 6.25 | 0.12 | 0.10 |
|      | 0.25  | 852 | 1028 | 0.032 | 0.023 | 6.09 | 0.17 | 0.14 |
|      | 0.025 | 809 | 1044 | 0.067 | 0.016 | – | – | – |
|      | 25    | 484 | 615  | 0.018 | 0.020[+] | 7.00 | 0.16 | 0.14 |
| 811  | 2.5   | 432 | 609  | 0.031 | 0.031 | 5.48 | 0.20 | 0.19 |
|      | 0.25  | 398 | 537  | 0.054 | 0.044 | 5.00 | 0.27 | 0.22 |
|      | 0.025 | 331 | 491  | 0.096 | 0.070 | 4.29 | 0.40 | 0.30 |
|      | 25    | 414 | 492  | 0.035 | 0.031[+] | 5.16 | 0.20 | 0.16 |
| 950  | 2.5   | 353 | 441  | 0.048 | 0.059 | 4.75 | 0.33 | 0.28 |
|      | 0.025 | 268 | 320  | 0.083 | 0.102 | 4.31 | 0.52 | 0.44 |
|      | 0.025[++] | 77 | 112  | 0.133 | 0.177 | – | – | – |

[+]Estimated from plots of m versus log $\dot{\varepsilon}$ (Fig. 3).
[++]In superplastic range: n=0.28, m=0.25
[+++] $\alpha = \{ (\bar{\varepsilon}*/Z)-n\}/m$

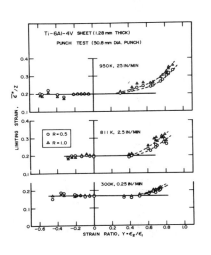

Fig. 8 Limiting strain versus strain ratio for the punch tests of Ti-6Al-4V at three temperatures with    R values of 0.5 and 1.0.

that the limiting strain increases with increasing strain hardening exponent
n or strain rate sensitivity exponent m.  The effect of the strain hardening
exponent n on the limiting strain $\bar{\epsilon}*/Z$ for Y<0.4 is illustrated in Fig. 9,
where a curve from the origin through the data points is parabolic but the
scatter is rather large and beyond experimental error.  This implies that the
limiting strain cannot be described by n alone, as previously pointed out by
Hecker (5), although it clearly depends on n.  Depicted in Fig. 10 is the
strain rate dependence of the limiting strain $\bar{\epsilon}*/Z$, where a linear
relationship exists with less scatter than between $\bar{\epsilon}*/Z$ and n.

Considering the sequence of deformation to fracture, diffuse necking
occurs first, which is closely related to the strain hardening exponent n.
Local necking follows, which is closely related to the strain rate sensitivity
exponent m.  Thus, one might expect the limiting strain to depend on both n
and m.  One  theory which relates the limiting strain in stretch forming to the
material parameters has been introduced by Conrad (7).  Briefly it derives the
final form given by eq.(1) as follows.  Assuming that the consititutive equation
is given by

$$\bar{\sigma} = \bar{\sigma}(\bar{\epsilon},\ \dot{\bar{\epsilon}},\ Y,\ T) \tag{9}$$

where $\bar{\sigma}$ is the effective flow stress, which depends on the prior deformation
history and the instantaneous values of the equivalent strain $\bar{\epsilon}$, strain rate
$\dot{\bar{\epsilon}}$, $Y(=\epsilon_2/\epsilon_1)$ and temperature T, the condition for the instability criterion
is given by (3, 8).

$$(\frac{\partial\bar{\sigma}}{\partial\bar{\epsilon}}) + (\frac{\partial\bar{\sigma}}{\partial\dot{\bar{\epsilon}}})(\frac{d\dot{\bar{\epsilon}}}{d\bar{\epsilon}}) + (\frac{\partial\bar{\sigma}}{\partial Y})\ (\frac{dY}{d\bar{\epsilon}}) + (\frac{\partial\bar{\sigma}}{\partial T})(\frac{dT}{d\bar{\epsilon}}) \leq \frac{\bar{\sigma}}{Z} \tag{10}$$

For isothermal deformation we obtain upon first dividing both sides by $\bar{\sigma}$ and
then multiplying both sides by $\bar{\epsilon}$

$$(\frac{\partial\ln\bar{\sigma}}{\partial\ln\bar{\epsilon}}) + \bar{\epsilon}\ (\frac{\partial\ln\bar{\sigma}}{\partial\ln\dot{\bar{\epsilon}}})(\frac{d\ln\dot{\bar{\epsilon}}}{d\bar{\epsilon}}) + \frac{\bar{\epsilon}}{\bar{\sigma}}\ (\frac{\partial\bar{\sigma}}{\partial Y})(\frac{dY}{d\bar{\epsilon}}) \leq \bar{\epsilon}/Z \tag{11}$$

Taking $n'=\partial\ln\bar{\sigma}/\partial\ln\bar{\epsilon}$ and $m=\partial\ln\bar{\sigma}/\partial\ln\dot{\bar{\epsilon}}$ and substituting into eq. (11) gives

$$n' + m\frac{d\ln\dot{\bar{\epsilon}}}{d\ln\bar{\epsilon}} + \frac{\bar{\epsilon}}{\bar{\sigma}}\ (\frac{\partial\bar{\sigma}}{\partial Y})(\frac{dY}{d\bar{\epsilon}}) \leq \bar{\epsilon}/Z. \tag{12}$$

For uniaxial tension at a constant extention rate, we have up to a maximum load

$$n' = n + \bar{\epsilon}\ m. \tag{13}$$

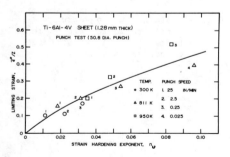

Fig. 9 Strain hardening dependence
of the limiting strain in
Ti-6Al-4V.

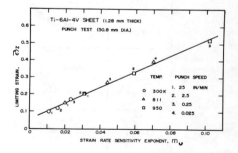

Fig. 10 Strain rate sensitivity
dependence of the limiting
strain in Ti-6Al-4V.

For the metals and conditions considered here, n' ≈ n, and hence the instability initiates at the strain $\bar{\varepsilon}*$ given by

$$\bar{\varepsilon}*/Z = n + m \frac{d\ln\bar{\varepsilon}}{d\ln\bar{\varepsilon}} + \frac{\bar{\varepsilon}}{\bar{\sigma}} \left(\frac{\partial\bar{\sigma}}{\partial Y}\right) \left(\frac{dY}{d\bar{\varepsilon}}\right).$$   (14)

Since the quantity $(\partial\sigma/\partial Y)(dY/d\bar{\varepsilon})/\bar{\sigma}$ is expected to be small (3), the strain $\bar{\varepsilon}*$ for the initiation of necking is finally given by

$$\bar{\varepsilon}*/Z \approx n + \alpha m$$   (14a)

where $\alpha = (d\ln\dot{\bar{\varepsilon}}/d\ln\bar{\varepsilon})$.

Assuming that Eq. 14a applies, one can derive the value of α for the present tests from the data given in Table 1. The values of α so derived range between 4.3 and 7.0 (see Table 1), which is within the range of values determined by direct experimental measurements on a number of metals (7, 9, 10, 12).

For the present tests the product $m_u\alpha$ is significantly larger than $n_u$. It is therefore concluded that the limiting strain in a punch test of Ti-6Al-4V at 300 to 950 K and strain rates of $10^{-1}$ to $10^{-3}$ s$^{-1}$ is determined by both the strain hardening exponent n and strain rate sensitivity exponent m, the effect of m on $\bar{\varepsilon}*/Z$ being larger than that of n.

## Acknowledgement

Support for this research was provided by the Battelle Columbus Laboratories through Air Force contract No. F-33615-77-C-5059.

## References

1. S. P. Keeler, Sheet Metal Industries, July, (1971), 511
2. W. Hosford, Proc. III International Congress on Strength of Metals and Alloys, Cambridge, England, (1973), 18
3. A. K. Ghosh, Met. Trans., 5 (1974), 1607
4. Z. Marciniak, K. Kuczynski and T. Pokora, Intl. J. Mech. Sci., 15 (1973), 789
5. S. S. Hecker, Met. Eng. Quart., 14 (1974), 30
6. A. K. Ghosh, J. Eng. Matl. and Tech.; Trans ASME, 99 (1977), 264
7. H. Conrad, J. Mech. Working Technology, 2 (1978), 67
8. S. P. Keeler and W. A. Backofen, Trans ASM., 56 (1963), 25
9. H. Conrad, K. Okazaki and C. Yin, Proc. 6th North American Metalworking Res. Conf. (NAMRC VI), (1978), 264
10. K. Okazaki, M. Kagawa and H. Conrad, Acta Met., 27 (1979), 301
11. A. K. Ghosh and W. A. Backofen, Met. Trans., 4 (1973), 1113
12. H. Conrad, M. Y. Demeri and D. Bhatt, Formability: Analysis, Modeling and Experimentation, TMS-AIME (1978), 208.

# HIGH TEMPERATURE DEFORMATION OF Ti-6Al-4V*

S. M. L. Sastry, P. S. Pao, and K. K. Sankaran

McDonnell Douglas Reserach Laboratories,
St. Louis, MO 63166, U.S.A.

## 1. Introduction

The high-temperature flow and fracture of alpha-beta titanium alloys are strongly influenced by the alloy composition, grain size, volume fractions of constituent phases, and crystallographic texture of the alloys[1-5]. Thus, to obtain a suitable constitutive equation for high-temperature flow of titanium alloys, the effects of the above metallurgical variables on the temperature, strain, and strain rate dependences of flow stress must be known. Furthermore, the continuous changes in alloy microstructure with temperature and time should be given proper consideration in obtaining the constitutive equation.

In the present investigation, a systematic investigation of the flow behavior of Ti-6Al-4V under creep, high-temperature high-strain-rate, and superplastic conditions was conducted with the objectives of modeling the flow behavior of the two-phase system and obtaining a data base for developing constitutive equations for high-temperature processing and forming of titanium alloys. Since Ti-6Al-4V is a two-phase system in which the volume fractions and flow behavior of the constituent phases are temperature dependent, the effects of compositions and volume fractions of alpha and beta phases on the overall flow behavior of mill-annealed, recrystallization-annealed, and beta-annealed Ti-6Al-4V were correlated with the deformation characteristics of single-phase alpha Ti-Al-based alloys and beta Ti-V-based alloys.

Typical two-phase microstructures of Ti-6Al-4V consist of 5-20 $\mu$m size alpha grains and transformed-beta grains containing either retained meta-stable beta or Widmanstätten alpha+beta (Fig. 1). The room-temperature yield stress of such alloys is governed by some form of superposition of the flow stress of equiaxed alpha and the composite flow stress of transformed-beta (Fig. 2). At the processing temperatures of 800-950°C, however, the alloy microstructure is a two-phase mixture of nearly-equiaxed alpha and beta grains with the volume fraction of beta increasing from $\approx$ 0.18 at 800°C to $\approx$ 0.55 at 950°C (Fig. 3) and the compositions of alpha and beta phases changing with temperature as shown in Table 1. To model the flow behavior of Ti-6Al-4V at 800-950°C, the flow characteristics of alpha and beta alloys of the compositions shown in Table 1 are required. In the present study, binary Ti-Al, Ti-5Al-2.5Sn, and Ti-15V-3Al-3Sn-3Cr alloys were used to approximate the required compositions.

---

*This research was conducted under the McDonnell Douglas Independent Research and Development program.

## 2.   Experimental Procedure

Alloys of the compositions shown in Table 2 were prepared by arc melting in a purified dry argon atmosphere and were processed into 12.5-mm, 3-mm, and 1.5-mm plates.   The high-temperature high-strain-rate deformation characteristics of the alloys were determined by compression testing of 8.9-mm diam and 12-mm long cylindrical specimens at 700–950°C and at strain rates of 0.005–0.5s$^{-1}$.   Creep parameters at 400–650°C were determined from constant-stress creep tests at 100–200 MPa.   The superplasticity parameters at 850–950°C were determined by incremental-strain-rate and constant-stress tests on specimens having 50.0 x 12.5 x 1.5-mm gauge sections.   The deformation substructures developed during high-temperature deformation were studied by transmission electron microscopy.

## 3.   Results and Discussion

### 3.1   High-Temperature, High-Strain Rate, Deformation Characteristics of Ti-6Al-4V

The true-stress/true-strain curves at 800°C, 850°C, and 900°C and for a strain rate of 0.05 s$^{-1}$ of the alpha, beta and alpha+beta titanium alloys are shown in Figs. 4a–4c.   For each of the alloys, the true stress increases with strain to a maximum and then decreases to a plateau, indicating the occurrence of flow-softening with the rate of softening being lowest in the alpha alloy and highest in the alpha+beta alloy.   The strain-rate dependences of flow stress of the alloys are shown in Fig. 5.   Also shown in this figure is the flow stress of Ti-6Al-4V calculated from the rule of mixtures for the superposition of high-temperature flow stresses.   The flow stress at 800°C of Ti-6Al-4V is slightly higher than that of alpha and beta alloys, but at higher temperatures, the alpha+beta alloy has considerably lower flow stress. Because of large differences in the deformability of alpha and beta phases at high temperatures, and consequently differences in the amounts of strains at any given instant in the two phases, the agreement between experimentally determined flow stresses and the values calculated from the rule of mixtures is not satisfactory.

At strain rates of 0.005 – 0.5 s$^{-1}$, the strain rate sensitivity of flow stress decreases from ≈ 0.2 at 850–950°C to ≈ 0.1 at 700°C, implying a change from dislocation-climb dominated deformation at high temperatures to a dislocation-glide controlled deformation at lower temperatures.

The temperature dependence of strain rate of the alloys at stresses of 100 and 200 MPa is shown in Fig. 6.   The apparent activation energies of 204 kJ/mol for the alpha alloy and 166 kJ/mol for the beta alloy agree with activation energies for self diffusion in alpha and beta titanium[7], and hence the high-temperature flow of the alpha and beta alloys in this region is controlled by dislocation climb.   The alpha+beta alloy, however, has a significantly higher activation energy, indicating the possible occurrence of dynamic recrystallization in the two-phase alloy.

The true-stress/true-strain curves at a strain rate of 0.05 s$^{-1}$ and at various temperatures for as-cast, beta-annealed, and mill-annealed Ti-6Al-4V are shown in Figs. 7a–7c.   The highest yield stress is observed in beta-

annealed samples containing a fine Widmanstätten structure, and the lowest
values are observed in mill-annealed samples that have a fine-grain, equiaxed,
two-phase microstructure.  The beta-annealed and as-cast specimens derive
their high strength from the small slip length in the transformation
structures consisting of either martensite or Widmanstätten alpha-beta
plates.  Greater initial softening is observed in as-cast and beta-annealed
specimens than in mill-annealed specimens because of the coarsening of the
transformation substructure.  While the flow stress of mill-annealed specimens
saturates at small strains, the flow stresses of the as-cast and beta-annealed
specimens decrease continuously with increasing strain for strains up to
approximately 50%, which suggests that the replacement of the Widmanstätten
microstucture by the equilibrium microstructure requires large deformations.

The deformation substructure of beta-annealed and mill-annealed Ti-6Al-4V
specimens deformed at a strain rate of 0.05 s$^{-1}$ at 700 and 900°C is shown in
Figs. 8 and 9.  At 700°C, extensive dislocation activity in the alpha phase,
continuity of slip across beta phase, and the absence of polygonization result
in profuse shearing of the beta phase in the beta-annealed Ti-6Al-4V and a
highly tangled dislocation structure in the alpha phase of the mill-annealed
Ti-6Al-4V (Fig. 9a).  Above 850°C, both dynamic recovery and recrystallization
occur, as evidenced by hexagonal networks of dislocations in the alpha phase,
the formation of small equiaxed alpha, and the absence of shearing of beta
phase (Figs. 8b and 9b).

3.2  Creep Deformation of Ti-6Al-4V

Figure 10a shows the comparison of the stress dependence of steady-state
creep rates of two-phase Ti-6Al-4V with single-phase Ti-8Al and Ti-10Al
alloys.  The two-phase alloy has a significantly higher creep rate than the
single-phase alpha alloys.  In single-phase Ti-Al alloys, increasing the
aluminum concentration results in a reduction of creep rate.  The stress
exponents of the alloys decrease from 9 at high stresses to 3 at lower
stresses, with the transition from the high stress-exponent region to the low
stress-exponent region occurring at higher stresses in Ti-6Al-4V than in
single-phase alloys.

The effects of heat treatment on the stress dependence of steady-state
creep rates of Ti-6Al-4V shown in Fig. 10 indicate the significant role of the
elongated Widmanstätten alpha+beta plates on the creep resistance of the
alloys.  Thus, although the mill-annealed and recrystallization-annealed
alloys have greater volume fractions of creep resistant alpha phase, the creep
rates in these alloys are higher because of the increased grain boundary
sliding in the fine grained alloys.

The temperature dependences of the steady-state creep rate of Ti-6Al-4V and
Ti-8Al alloys are shown in Figs. 11a and 11b.  The activation energy of 188
kJ/mol observed for Ti-6Al-4V at low stresses at 450-600°C agrees with the
activation energy for self diffusion in titanium, and hence creep in this
region is controlled by dislocation climb.  The rate-controlling creep
mechanism associated with the high activation energy at higher stresses,
however, is not clear at present.  The higher activation energy for creep of
single-phase Ti-8Al alloy compared with Ti-6Al-4V indicates that creep in Ti-
8Al alloys is perhaps controlled by glide processes in which the short-range
ordered regions and solute atmospheres generally present in Ti-8Al alloys

exert drag on moving dislocations. In agreement with this, the dislocation substructure in Ti-6Al-4V specimens deformed in creep at 550°C consists of a/3 ⟨1120⟩ dislocation networks formed by dislocation cross-slip and climb (Fig. 12), and the dislocation substructure in the single phase Ti-8Al alloy consists of a high density of tangled and relatively homogeneously distributed dislocations characteristic of dislocation-glide controlled processes.

3.3   Superplastic deformation characteristics of Ti-6Al-4V

Figures 13a-13c show the comparisons of strain rate dependence of flow stress at strain rates less than $10^{-3}s^{-1}$ of the alpha, beta and alpha+beta alloys. The alpha and beta alloys have significantly higher flow stresses and lower strain rate sensitivity values than the two-phase alloy Ti-6Al-4V. This is due to the extensive grain growth occurring in the single-phase alloys at higher temperatures and slow strain rates, where the flow stress increases with increasing grain size. Under these conditions, the rule of mixtures fails to predict the flow stress of Ti-6Al-4V because of larger differences in grain growth kinetics of the alloys at the superplastic temperatures.

The strain-rate dependence of flow stress of Ti-6Al-4V at strain-rates of $10^{-5} - 10^{-1}$ $s^{-1}$, encompassing the superplastic range and conventional high-temperature-forming strain rates, can be accurately represented by straight lines in each of three regions, A, B, and C, corresponding to strain-rate sensitivities of approximately 1, 0.5, and 0.2, respectively (Fig. 14). The three different straight-line regions indicate that the deformation mechanism changes from Nabarro-Herring creep [7] or diffusion-accommodated grain-boundary-sliding[8], both of which give m = 1, at low strain rates to an interface-reaction-controlled grain-boundary-sliding[9] or dislocation-motion-accommodated grain-boundary-sliding[10], for which m = 0.5, at intermediate strain-rates to dislocation-climb-controlled deformation with m = 0.2 (or stress exponent of 5 in creep tests) at large strain-rates. Near the transitions from A to B and B to C, the deformation is governed by the superposition of the deformation mechanisms of two adjoining regions, and no discontinuity of slope of the flow-stress/strain-rate curve actually occurs.

The deformation substructures observed for the three strain-rate regions are consistent with the different deformation characteristics. Whereas extensive dislocation tangles and cell structures are produced at strain-rates of 0.001 $s^{-1}$, alloys deformed in region A show no dislocation activity. Confirmatory evidence for the three-region, high-temperature deformation behavior of Ti-6Al-4V was obtained by texture measurements. The initial texture is retained in specimens deformed in region C because of the dominance of dislocation processes, but a reduction in texture sharpness and randomization of the texture occur in specimens deformed in region A because of extensive grain rotations arising from grain boundary sliding.

## 4.   Summary and Conclusions

The high-temperature creep and high and slow strain rate deformation characteristics of Ti-6Al-4V were correlated with the high-temperature deformation parameters of single-phase alpha and beta alloys. The rule of mixtures was found to be inadequate for describing the high-temperature flow of the two-phase Ti-6Al-4V alloy. The strain-rate sensitivities of flow stress were correlated with operative deformation mechanisms in the different temperature and strain-rate regions.

## References

1. D. Lee and W. A. Backofen, Trans. Met. Soc. AIME, 239, (1967), p. 1034.
2. N. J. Grant, W. Ioup, and R. H. Kane, in The Science, Technology, and Application of Titanium, ed. by R. I. Jaffee and N. E. Promisel, (Pergamon Press, Oxford, 1970), p. 607.
3. A. Arieli and A. Rosen, Met. Trans., 8A, (1977), p. 1591.
4. C. R. Whitsett, S. M. L. Sastry, J. E. O'Neal, and R. J. Lederich, McDonnell Douglas Report MDC Q0654 (31 May 1978), Technical Report, ONR Contract No. N00014-76-0626.
5. S. M. L. Sastry, R. J. Lederich, P. S. Pao, and J. E. O'Neal, McDonnell Douglas Report MDC Q0684 (31 May 1979), Technical Report, ONR Contract No. N00014-76-0626.
6. C. M. Libanti and S. F. Diment, J. Mater. Sci., 3, (1968), p. 349.
7. A. K. Mukherjee, in Treatise on Materials Science and Technology, Vol. 6, ed. by R. J. Arsenault, (Academic Press, New York, 1975), p. 163.
8. M. F. Ashby and R. A. Verrall, Acta Met., 21, (1973), p. 149.
9. G. Rai and N. J. Grant, Met. Trans. 6A, (1975), p. 385
10. A. K. Mukherjee, Annual Review of Materials Science, Vol. 9, (1979), p. 191.

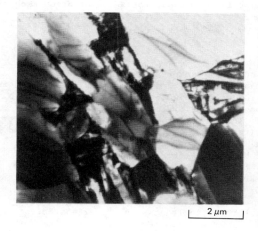

**Fig. 1 Two-phase microstructure of Ti-6Al-4V.**

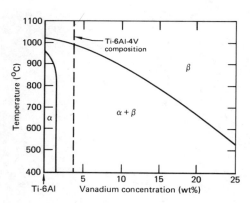

**Fig. 2 Schematic phase diagram of Ti-6Al-4V system.**

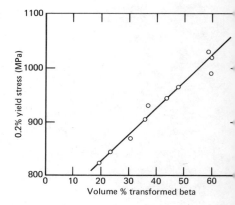

**Fig. 3 Dependence of room-temperature yield stress of Ti-6Al-4V on volume % transformed beta.**

Table 1   Volume fractions and compositions of alpha and beta phases
of Ti-6Al-4V at 800-900°C

| Temperature (°C) | Amount of Alpha (%) | Amount of Beta (%) | Composition of Alpha | | Composition of Beta | |
|---|---|---|---|---|---|---|
| | | | Al (wt%) | V (wt%) | Al (wt%) | V (wt%) |
| 800 | 80 | 20 | 6.1 | 2.8 | 4.7 | 10.5 |
| 850 | 76 | 24 | 6.2 | 2.6 | 5.2 | 8.2 |
| 900 | 65 | 35 | 6.6 | 2.2 | 5.8 | 6.0 |

Table 2   Nominal compositions of the alpha, beta and
alpha+beta titanium alloys

| Alloy type | Nominal composition |
|---|---|
| Alpha | Ti-8Al |
| Alpha | Ti-10Al |
| Alpha | Ti-5Al-2.5Sn |
| Beta | Ti-15V-3Al-3Sn-3Cr |
| Alpha+beta | Ti-6Al-4V |

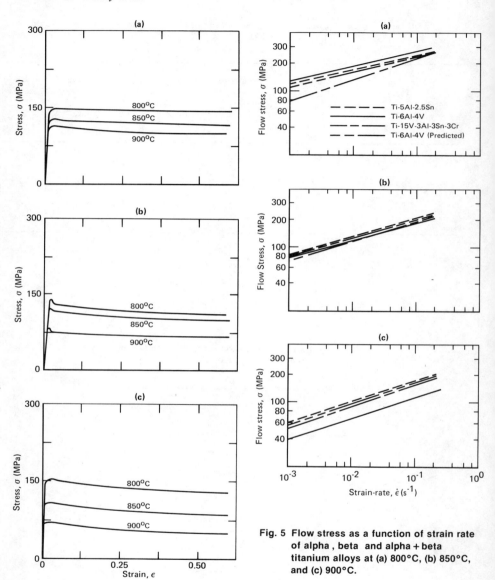

Fig. 4 True-stress as a function of
true-strain for (a) Ti-5Al-2.5Sn,
(b) Ti-15V-3Cr-3Sn-3Al, and (c) Ti-6Al-4V.

Fig. 5 Flow stress as a function of strain rate
of alpha , beta and alpha + beta
titanium alloys at (a) 800°C, (b) 850°C,
and (c) 900°C.

$\dot{\varepsilon} = 0.05/s$

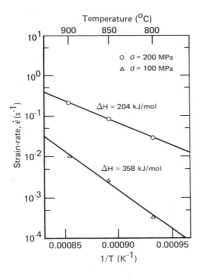

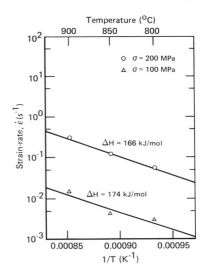

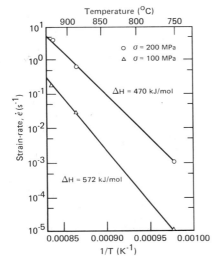

**Fig. 6 Temperature dependence of strain rates of (1) Ti-5Al-2.5Sn, (b) Ti-15V-3Cr-3Sn-3Al, and (b) Ti-6Al-4V.**

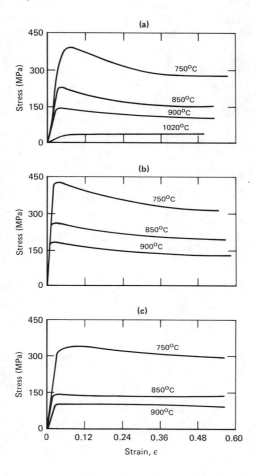

Fig. 7 True-stress as a function of true-strain
of (a) as-cast Ti-6Al-4V, (b) beta-annealed
Ti-6Al-4V, and (c) mill-annealed Ti-6Al-4V.

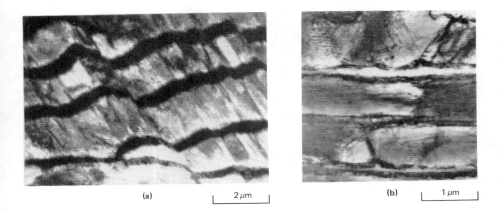

(a)          $2\,\mu m$                     (b)          $1\,\mu m$

Fig. 8 Deformation substructures in beta-annealed Ti-6Al-4V deformed in
compression at an initial strain rate of 0.05 s$^{-1}$ at (a) 700°C and
(b) 900°C.

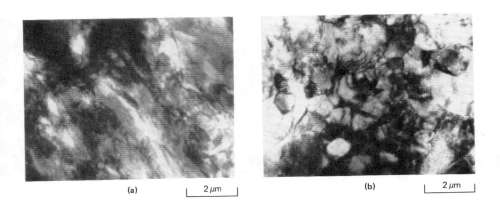

(a)          $2\,\mu m$                     (b)          $2\,\mu m$

Fig. 9 Deformation substructures in mill-annealed Ti-6Al-4V deformed in
compression at an initial strain rate of 0.05 s$^{-1}$ at (a) 700°C and
(b) 850°C.

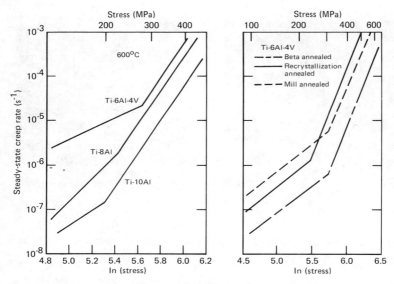

**Figure 10. Stress dependence of steady-state
creep rates of Ti-6Al-4V and Ti-Al
alloys at (a) 600°C and (b) 550°C.**

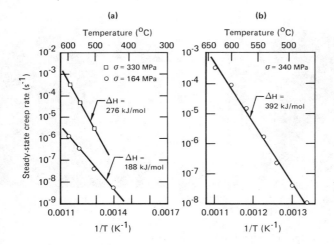

**Fig. 11 Temperature dependence of steady-
state creep rates of (a) Ti-6Al-4V,
and (b) Ti-8Al.**

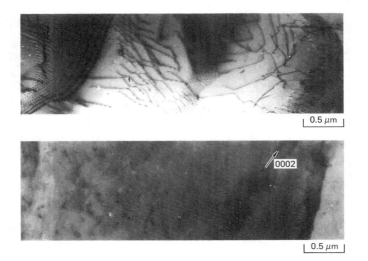

Fig. 12. Deformation substructures in mill-annealed Ti-6Al-4V deformed in creep to 2% strain at 550°C and at stress of 100 MPa.

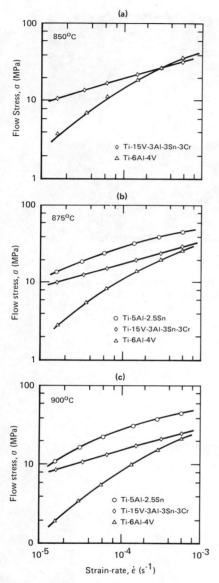

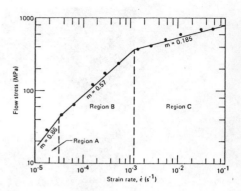

Figure 14.    Flow stress at 850°C as a function of strain rate in recrystallization-annealed Ti-6Al-4V.

Fig. 13. Strain-rate dependence of flow stress of Ti-5Al-2.5Sn, Ti-15V-3Sn-3Cr-3Al, and Ti-6Al-4V at (a) 850°C, (b) 875°C, and (c) 900°C.

# THE INFLUENCE OF DEFORMATION MECHANISM ON YIELDING CHARACTERISTICS IN A SINGLE CRYSTAL OF Ti-Al-Sn $\alpha$ -ALLOY

N.V.Ageev, E.B.Rubina, A.A.Babareko,
V.A.Moskalenko, V.N.Kovaleva

Baikov Institute of Metallurgy, Moscow;
Physico-Technical Institute,
Ukr.SSR Academy of Science, Kharkov

Today, a rather well-defined formulation is available for the theoretical principles of texture formation in the $\alpha$ -phase of titanium alloys /I,2/. Textured products exhibit different anisotropic behavior of strength, ductility and endurance depending on the type of texture and alloy composition /3-7/. It is possible to predict the level of texture strengthening in materials at various service conditions using the data on the crystallographic mechanism of plastic yielding in single crystals and the orientational dependence of their yield stresses. The use of titanium alloys in cryogenic apparatus enhances the interest to the study of deformation mechanism at low temperatures.

The crystallographic deformation mechanisms in Ti of high and commercial purity and Ti-Al alloys have been reffered to /8-I0/. The present study dealt with the deformation mechanism in Ti-Al-Sn alloys in compression and tensile test of " c " and " a " type single crystals in a wide temperature range.

## Material and Experimental

The single crystals of Ti with I.4%-3% Al and I.7% Sn (weight) alloys with the loading axis being within 0-5° from $\langle II\bar{2}0 \rangle$ , $\langle I0\bar{I}0 \rangle$ and $\langle 000I \rangle$ were mechanically tested using the

Instron machine at temperatures 293-4.2 K. The specimens were cut
from one single crystal grown by the zone electron-beam melting
technique. The impurity contents in the single crystal were
0.009% C, 0.14% O, 0.0001% H. Aluminum content variations from
1.4% to 3% Al were observed along the crystal length. Aluminum
and tin concentrations were monitored by the X-ray electron probe
analysis and by chemical analysis. The authors acknowledge their
gratitude to V.V.Shishkov for the test material and to R.Milevs-
kii and V.Khlomov for their assistance in making the study.

The compression cubic specimens with edges 4 or 7 mm were
tested at a rate of $2-4 \times 10^{-3}$ $s^{-1}$. The tensile specimens with
gage length of 8 mm and section of 1.5x1.5 mm were tested at a
rate of $7 \times 10^{-4}$ $s^{-1}$. The crystallographic orientation of the slip,
primary and secondary twinning planes as well as the twinning
shear value were determined using GM-2 gonic-microscope. The compression specimens were additionally studied for the diffraction
X-ray spectrum variations related to the twinning reorientations,
using the special design with specimen oscillation and rotation
during exposure.


## Experimental Results

Tensile test along the $\langle 0001 \rangle$ axis. The main deformation
mechanism in specimens tested at 4.2-293 K is the $\{10\bar{1}2\}$ $\langle \bar{1}011 \rangle$
twinning. There is a weak temperature effect and a very signifi-
cant strengthening aluminum effect on the resolved shear stresses
(Fig.1). The yield strength in $\sim 2\%$ Al crystals is 99-109 kg/mm$^2$
and decreases to 65-80 kg/mm$^2$ at 1.4-1.5% Al. At lower testing
temperatures the twins become thinner. The internal stress accomodation occurs by twinning in the matrix at room temperature and
by slip and twinning in the twins at cryogenic temperatures
(Fig.2). Twins up to 50 $\mu$m in width with curved edges without indications of internal twinning are formed at the ambient temperature, the deformation occurs in two $\{10\bar{1}2\}$ systems with necking.
The twins in one system do not interfere with the twins in another
system (Fig.2a). At 77 K the width of twins does not exceed 15-20
$\mu$m, while thin twins (5 $\mu$m) with flat parallel edges can also be

found. The accomodation of stresses by prismatic slip in the $\{10\bar{1}2\}$ twins appears as coarse slip bands. While the secondary twinning in the matrix makes it fir-looking (Fig.2b). At 20 K the prismatic slip in the twins is encountered less often and has an appearance of thin traces in the crystal with 2.1% Al, while in the crystal with 1.5% Al the relaxation of internal stresses occurs through the secondary $\{11\bar{2}4\}$ twinning in the primary $\{10\bar{1}2\}$ twin (Fig.2c). The coarser of the primary twins display the internal structure of the secondary $\{11\bar{2}4\}$ twins, which in turn are twinned on the $\{10\bar{1}2\}$ planes. The effective secondary twinning systems in the primary twins have a near-zero Schmidt factor relative to the applied load. Besides the main twinning mechanism, additional twinning types in the $\{10\bar{1}1\}$ system at the ambient temperature, in the $\{11\bar{2}1\}$ , $\{11\bar{2}2\}$ and $\{11\bar{2}4\}$ systems at 77 and 4.2 K were observed in the specimen clamping area due to the plane strain condition.

Tensile test along the $\langle 10\bar{1}0 \rangle$ and $\langle 11\bar{2}0 \rangle$ axes. In the temperature range 293-4.2 K the deformation by prismatic slip leads to strong negative dependence of the yield strength on temperature similar to pure titanium. The critical resolved shear stresses in the specified temperature range vary from 21 to 43 kg/mm$^2$. At 4.2 K twinning traces were observed in addition to prismatic slip, however, this twinning can be induced by an included grain in crystal due to a possible constrain by the grain boundary.

Compression test along the $\langle 0001 \rangle$ axis. In the temperature range 293-20 K in the specimens with 2-3% Al twinning occurs in the system $\{10\bar{1}1\}$ $\langle \bar{1}012 \rangle$ with high resolved shear stresses. A negative temperature dependence of shear stresses is the same as in pure titanium. At a temperature of 4.2 K the specimen with increased aluminum content displays the $\{11\bar{2}2\}$ $\langle \bar{1}\bar{1}23 \rangle$ twinning and a yield strength as high as 180 kg/mm$^2$, while a decrease in aluminum content to 1.6% leads to $\{11\bar{2}4\}$ twinning with low resolved shear stresses 60 kg/mm$^2$ at a given temperature. Aluminum content 2-2.3% in crystals of a given orientation is critical in terms of aluminum effect on strength, since there is a transfer from the high to low resolved shear stress twinning for an alloy

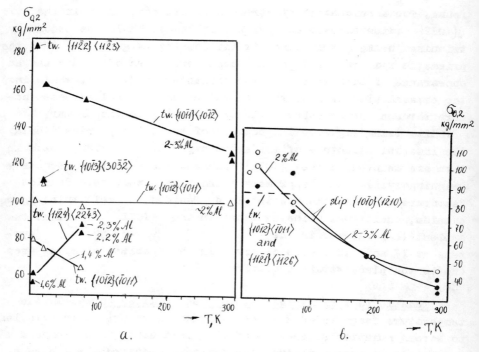

Fig.I. The temperature dependence of yield stresses of
Ti-Al-Sn single crystals: a - deformation along $\langle 000I \rangle$
- ▲-▲- ▲ - compression; △- △ - △ - tension; b -
deformation along $\langle II\bar{2}0 \rangle$ or $\langle I\bar{I}00 \rangle$ : ●-●-● - compression, o-o-o - tension

of this composition in the temperature range < 77 K. At 77 K  the
2.3% Al crystals displayed twinning of the $\{II\bar{2}4\}$  type with a
yield strength $\sim$ 90 kg/mm$^2$ at 20 K - the $\{IO\bar{I}3\} \langle \bar{3}032 \rangle$ twinning
with a yield strength $\sim$ IIO kg/mm$^2$. At these temperatures in the
crystals with increased aluminum content the  $\{IO\bar{I}I\}$ twinning
acts and a yield strength of I60 kg/mm$^2$ is reached. These results
indicate that at least 3% Al is required in the alloys used at
cryogenic temperatures.

The internal stress relaxation mechanism depends on a test
temperature and a primary twinning mode. With the  $\{IO\bar{I}I\}$  type
twinning at the ambient temperature the relaxation occurs by slip
in the matrix (Fig.3a). At lower testing temperatures in specimens

with increased aluminum content additional deformation twins appear in the $\{11\bar{2}1\}$ , $\{11\bar{2}2\}$ and $\{10\bar{1}2\}$ systems, with some of those being of the straight accomodation twin type. The low-strength specimens having $\{11\bar{2}4\}$ (Fig.3b,c) or $\{10\bar{1}3\}$ as their main twinning types at low temperatures, relax the internal stresses through twinning in the primary twins and in the matrix. In the primary twins of the $\{11\bar{2}4\}$ type it is shown secondary $\{10\bar{1}2\}$ twinning having a low Schmidt factor relative to the applied stresses (Fig.3c). The development of primary twinning in several equal-stress systems is typical. With the same twinning type the $\{11\bar{2}4\}$ twin width is greater after the 4.2 K test than after the 77 K test; in case of twinning in the $\{10\bar{1}1\}$ systems the twin width is related to the yield stress levels.

Compression test along $\langle 11\bar{2}0\rangle$ and $\langle 10\bar{1}0\rangle$ axes. At the temperatures 293-77 K in tested pieces it is observed prismatic slip similar to the one found during the tensile deformation, with the yield stress levels and the temperature dependence of resolved shear stress during compression and tensile test being similar. A considerable difference in the deformation mechanism and yield stress levels appears at testing temperatures below 77 K. In case of compression test at < 77 K it is observed $\{10\bar{1}2\}\langle 10\bar{1}1\rangle$ twinning with resolved shear stress independent of temperature. The crystallograpgy of the low-temperature twinning in the "a" axis compression test agrees with that in the "c" axis tensile test, and the crystals have the same yield stress level, 80-100 $kg/mm^2$.

The shape change and failure of crystals during the uniaxial tests. At the prismatic slip the plane deformation occurs with the lattice rotation about the "c" axis. In the crystals compression tested along the "a" axis at the ambient temperature after a 10% deformation the "c" side face looks like a rhomb covered with prismatic shear traces. It is observed the crystallographic shear failure after large strains.

The development of deformation twinning in the "a" - oriented crystals, at low temperatures assists in an axisymmetric deformation. It is nearly always present in the "c" -oriented crystals, however, sometimes these display deformation only in

Fig.2. Twins in Ti-Al-Sn single crystals after tension along
⟨0001⟩ axis: a - twins {10$\bar{1}$2} at 293 K, mag. x120;
b - twins {10$\bar{1}$2} at 77 K, mag. x240; c - twins {10$\bar{1}$2}
with secondary twins {11$\bar{2}$4} at 20 K, mag. x480

two twinning systems symmetrical related test axis. This deformation approaches the planar type (the {10$\bar{1}$1} twinning at the ambient temperature). The lack of a common regularity in the shape
change of the twinning "c" (or "a") crystals suggests a statistical character of the origination of deformation twinning in the
crystallographically identical systems with a same resolved shear
stresses.

One of the tested specimens failed at a tensile strain less
than 0.2% and one at a 2.5% compression strain; 29 crystals were
deformed without apparent failure up to 10-20% during the compression test and in the range of 0.9-2.2% during the tensile test.
The failed specimens contained included grains which had a strain
incompatible with the matrix strain. At the included grain boundary it was observed the accomodation twinning, while the whole specimen was deformed by prismatic slip (compression and tensile loading along the "a" axis). When the included grain {12$\bar{3}$3} is
oriented on the basal surface of the specimen and the {11$\bar{2}$4} and

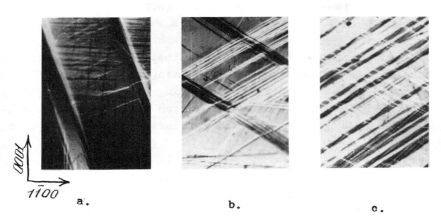

Fig.3. Twins in Ti-Al-Sn single crystals after compression
along ⟨000Ī⟩ axis: a- twins {10Ī1} at 293 K,
mag. x240; b - twins {11Ī4} at 77 K, mag. x120; c -
twins {11Ī4} at 4.2 K, mag. x360

{12Ī1} on the prismatic surfaces during the compression test
along the ⟨10Ī0⟩ axis one observed the plane strain by prismatic
slip in the matrix crystal and the bulging of the included grain
of the basal face of specimen. The accomodation at the boundary
of included grain was by {10Ī2} and {10Ī1} twinning which
contributed deformation along the "c" axis. The lack of accomodation in the included grain (low Schmidt factor for systems producing contribution in the plane strain similar to that found for
the crystal) leads to cracking. The specimen failed after a 2.5%
reduction, while the specimens with the included grain {10Ī1}
oriented on the basal surfaces had higher ductility in the compression test along the ⟨10Ī0⟩ and ⟨000Ī⟩ axis of the matrix, and
the presence of such included grain did not have a significant effect on the yield strength and the main matrix deformation mechanism.

The determination of the twinning shear magnitude. The magnitude of the shear $s = b/d$ where b is the shift vector in the twinning shear direction related to a single atomic layer and d, the
interplanar distance for the twinning plane, were determined for
the most typical "c" axis tensile twinning mechanism {10Ī2} . The
experimental value, $s_{\{10\bar{1}2\}}$ = 0.14—0.18, shows a good agreement

with twinning elements for this system, $K_2 - \{10\bar{1}2\}$ ; $\eta_2 - \langle 10\bar{1}\bar{1}\rangle$, and the theretical value of s /2/. The s value for the low-tempe-rature twinning type during the " c " axis compression test, $\{11\bar{2}4\}$ was determined for the first time. The determination was complicated due to heavy secondary twinning in the twins. The ob-tained value s ᵥ 0.2 supports the twinning elements $K_2 - \{11\bar{2}\bar{2}\}$ and $\eta_2 \langle 11\bar{2}3\rangle$ , found theoretically from the atomic model des-cribed by the homogeneous shear with simultaneous rotation of ato-mic pairs.

### Comparison with the deformation mechanism in polycrystal.

Transversal tensile specimens were cut from the textured sheets of Ti-Al-Sn and Ti-Al-V alloys. Given the basal texture of sheet this corresponded to the $\langle hki0\rangle$ axis type test. The monotonous increasing of yield stresses for the Ti-Al-Sn alloy at lower test-ing temperatures corresponds to the deformation primarily by pris-matic slip, which is confirmed by the lack of twins in the micro-structure, the shape change and the unaltered basal texture in the specimen plane. Also, for the Ti-Al-V alloy at the ambient temperature the prismatic slip is the main deformation mechanism, nevertheless, at lower testing temperatures the role of deforma-tion twinning greately increases. Its contribution decreases the necking tendency, increases the workhardening rate (when the rela-xation by the secondary twinning in twins is lacking and changes the specimen texture. The comparison of experimental pole figures of the alloy with the calculated ones for various twinning types enabled to establish /II/ that the deformation at 77 and 20 K occurs with the assistance of twinning in the $\{11\bar{2}3\}$ $\langle 11\bar{2}\bar{2}\rangle$ system, and at 4.2 K - the primary twinning $\{11\bar{2}4\}$ $\langle \bar{2}\bar{2}43\rangle$ with the secondary one (in the twins) $\{10\bar{1}2\}$ . The optical microsco-pic examination supports the presence of the primary and seconda-ry twinning, the fraction of which reaches 50% before the specimen failure. The increasing yield stress in the alloy with decreasing deformation temperature from 77 K to 20 K suggests a negative tempe-rature dependence of shear stresses for the twinning of the $\{11\bar{2}3\}$ type. The decrease in yield stresses, when the twinning mechanism is changed for $\{11\bar{2}4\}$ at 4.2 K, is similar to the ef-fect found in the single crystals of Ti-Al-Sn alloy with low alu-minum content in the "c" axis compression test. The data obtained on the crystals and polycrystals suggest a strong orientation dependence of the alloy deformation mechanism. The yield stresses during the deformation twinning of crystals with the "c" loading axis orientation determine the texture strengthening tendency of the alloy. The transfer effect at temperatures 77-4.2 K to the deformation twinning of the $\{11\bar{2}4\}$ and $\{10\bar{1}3\}$ types with low shear stress is of great interest. This determines a temperature limitation of the texture strengthening for the titanium alloys.

## Conclusions

I. The orientational dependence of yield stresses in the alpha phase of titanium alloys, i.e. their potential for the texture strengthening, varies with the testing temperature and aluminum content in the alloy in connection with the sensibility of shear stresses during slip and twinning to ones.

2. The anisotropy of the yield strength depends on the sign of applied stresses, and the ratio of yield strengths at the ambient temperature in the specimens for "c" and "a" axis tensile tests in the 2-3% Al alloy is $\sim 2$, while for the compression along "c" and "a" their ratio increases to $\sim 3.5$ due to the increased yield strength of the crystal with the "c" loading axis.

3. The shear stresses for prismatic slip have a strong negative temperature dependence, while the shear stresses during twinning have a weaker dependence on temperature and can have either negative or positive temperature dependence for different individual twinning systems. This produces a decrease in the anisotropy of the yield strength at lower testing temperatures down to $\sim 2$ and even $\sim I$ depending on the effective deformation mechanism at 77 K - 4.2 K.

4. The weakening found in crystals with $< 3\%$ Al in the "c" axis compression test, when temperature is reduced down to $77 \div 4.2$ K, is caused by the deformation twinning of the $\{11\bar{2}4\}$ or $\{10\bar{1}3\}$ type.

5. Secondary twinning in the twins is effective for the relaxation of internal stresses during the deformation in the cryogenic temperature range.

6. The crystallography of twinning acting as the main mechanism during the "c" axis tensile test agrees with that during the "a" axis compreßiön test. The accomodation twinning in the matrix does not follow this behavior.

## References

I. N.V.Ageev, A.A.Babareko, A.I.Khorev.  Structural and Process Metallurgy. The 3rd International Conference on Titanium. M., 1978, v.3, pp. 97-102 (in Russ.).

2. Ya.D.Vishnyakov, A.A.Babareko, S.A.Vladimirov, I.V.Egiz. The Theory of Texture Forming in Metals and Alloys. M., Nauka Publishers, 1976, p. 344 (in Russ.).

3. A.Sommer, M.Kriger.  The 3rd International Conference on Titanium. Structural and Process Metallurgy. M., 1978, v.3,pp.87-96 (in Russ.).

4. A.W.Bower.  Acta Met., 1978, 21, pp. 1423-1433.

5. N.V.Ageev, A.A.Babareko, E.B.Rubina et al. Izv.AN SSSR, Metally, 1973, No 5, pp. 150-159.

6. A.Hashegava, T.Nishimura, M.Fukuda.  The 3rd International Conference on Titanium. M., 1978, v.3, pp. 161-164 (in Russ.).

7. A.I.Khorev, A.I.Krasnozhen, A.A.Babareko et al.  In: Alloying and Heat Treatment of Titanium Alloys. M.,1977, ONTI,pp.242-252 (in Russ.).

8. H.Conrad, M.Doner, B. de Neester.  Titanium Science and Technology. New York - London, Plenum Press, 1973, v.2, pp. 969-1005.

9. N.E.Paton, J.C.Williams, I.P.Rauscher.  Titanium Science and Technology, New York - London, Plenum Press, 1973, pp. 1049-1069.

10. T.Sakai, M.E.Fine.  Ser.Metal., 1974, v.8, pp. 541-545.

II. N.V.Ageev, E.B.Rubina, A.A.Babareko, S.Ya.Betsofen, L.A.Bunin. FMM, 1979, 48, No 3, p. 594.

# B. Texture

# EFFECT OF TRANSFORMATION ON THE TEXTURE OF TITANIUM

S. Fujishiro* and S. Nadiv**

*Metals and Ceramics Division
Air Force Materials Laboratory
W-PAFB, OH 45433

**Department of Materials Engineering
Technion-Israel Institute of Technology
Haifa, Israel

Texture plays an important role in the mechanical properties of Ti and its alloys because of the anisotropic crystalline struc- ture of the alpha grain. Texturing of titanium originates from hot working processes, which produce a combination of deformation, re- crystallization and phase transformation textures. It is difficult to eliminate textures in commercial fabrication processes, since the deformation reductions and the working temperatures fall in limited ranges for practical reasons. In the present work, studies were limited to an understanding of the effect of cooling rates on the transformation texture of commercial pure titanium and the influence of this texture on the mechanical properties.

## Introduction

A literature survey on the thermal expansion of pure titanium gives the conflicting observations that either expansion [1,2] or contraction [3] occurs upon allotropic transformation at $882^\circ C$. The reported linear change was on the order of 0.1%. English and Powell reported that, upon the phase transformation of Ti-7Al-3Cb alloy, three types of dimensional changes, i.e., $\Delta l$ = -, + and zero were observed, attributing this observa- tion to texture and the resulting change in the ratio of alpha and beta phases. The dimensional changes were less than $\pm$ 0.2%. Rudman's discus- sion [4], suggested that the volumetric contraction due to the alpha to beta phase transformation based on X-ray data and high pressure studies, would be as much as 0.55% and 2.6%, respectively, which is equivalent to 0.2% and 0.9% in linear contraction.

In the present study, it was found that a linear expansion took place on cooling in many cases, but the reverse case was not uncommon. This variance is attributed mainly to the transformation texture, which was in- herited from the texture prior to the transformation. This explanation can be further supported by the fact that a slower heating or cooling rate resulted in a greater linear dimensional change, since the formation of a fewer number of transformation grains causes an anisotropic effect.

## Experimental

1. Dilatometric Measurement

A cylindrical specimen, 12 mm long and 8 mm in diameter with a 2 mm longitudinal hole, was prepared from an iodide titanium rod, and was heated by an induction unit in a vacuum of $5 \times 10^{-6}$ Torr. Thermal cycling the

specimens was controlled by a combination of a "Thermac" temperature
controller and a "Data'Track" programmer manufactured by Research Inc.
With full power input of 1 kW, the specimen can be heated by 15°C/sec.
A cooling rate of 10°C/sec. was obtained when the power was shut off.  A
faster cooling rate could be achieved when the cylindrical specimen,
with a thinner wall, was quenched with chilled He gas after the power was
shut off.  However, He gas quenching was not used in the present study,
since it caused nonuniform temperature distribution.

The specimen was held horizontally by a pair of quartz tubes and one
of the tubes was attached to the core rod of a LVDT.  The output from the
LVDT was amplified and recorded on a strip chart recorder and a X-Y recorder
simultaneously.  The electronic noise of this system was very low.  A
sensitivity of $10^{-4}$ cm and an accuracy of $10^{-3}$% in the linear dimensional
change were easily obtained.  In order to reduce the textural effect of the
starting material, the specimens were first rapidly cycled through the
transition temperature several times, and held at 1080°C or 850°C for a few
hours, so that large beta or alpha grains were grown.

## 2.  Metallographic Analysis and X-ray Analysis

In order to observe the surface topographic features caused by the
transformation stress relief, the flat surface of the specimens was smoothed
with 600 grit emery papers.  They were then progressively polished with 9
micron, 3 micron and 0.05 micron metallographic pastes using nylon cloths,
until no polishing scratches were observed at 600X with a metallographic
microscope.  The specimens were annealed at 850°C for two hours and then
heated very slowly above the transition temperature, so that the initial
flatness of the surface was preserved during the alpha to beta transformation.  After annealing these specimens at 1080°C for a minimum of two hours,
they were either slowly cooled or quenched to produce the stress relief
pattern.  The dimensional changes were measured.  The specimen size was
not large enough to allow a pole figure study.  However, a number of X-ray
Laue patterns were taken along the specimen longitudinal axis, to obtain
sufficient information on the average grain orientations.  These specimens
were also used to determine the Young's modulus by compression tests.

## Results

## 1.  Dilation Behavior

Figure 1 represents typical thermal dilation curves obtained when
titanium was rapidly heated and cooled through the transformation temperature.  $H_i$ and $C_i$ in all the figures of this report indicate ith and jth
heating and cooling cycles.  Rapid heating at a minimum rate of 10°C/sec.
has caused a sharp singularity minimum point in the linear dimension change
at the transformation temperature, as shown in Fig. 1, followed by a normal
expansion rate.

The output signal from a LVDT was suppressed at the lowest and highest
isothermal temperatures, so that these curves can be separated from each
other.  The ordinate provides the relative dimensional change.  Since repeated thermal cycling would eliminate possible preferred orientations,
prior to the final heating, the difference in the heating curves in the
alpha and beta regions is considered to be due to the change in the atomic
volumes of these two polymorphs.  This linear contraction on heating,
determined from the extrapolation of these curves was approximately 0.07%;

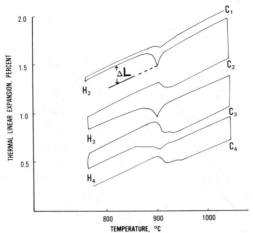

Fig. 1.   Change in the dimension of titanium as a result of consecutive
rapid heating and cooling.  ΔL is a contraction due to atomic
"shuffle".

this value is smaller than anticipated from X-ray data [4].  The singular
contraction of 0.15% at the transition temperature cannot be easily interpreted, although it is conceivable that atomic "shuffle" [5] during the
accelerated phase change creates, momentarily, random positioning of atoms.
The very rapid heating may not allow an orderly coordinated movement of the
lattice during the phase transition.  During cooling, this phenomenon was
not observed, but the expansion took place, probably because natural cooling was not fast enough to cause such "shuffle".  In addition, the anisotropic effect of the grain growth in slow cooling may override the "shuffle"
contraction, if this is what actually happens.

X-ray Laue patterns and transmission electron microscopic studies of
these specimens, which had been slowly heated through the transition temperature and quenched from 1080°C, revealed that the overgrown prior beta
grains transformed to the superimposed microtwins of fine alpha lamellae
with a width of 500 to 2000 Å.  The surface stress relief topograph due to
the multiple microtwin formation is shown in Fig. 2.

Intermediate cooling rates provide a greater textural effect on the
linear dimensional change and repeated thermal cycles implement textural
effects progressively as shown by the cooling curves in Fig. 3.  Figure 4
exhibits linear contractions which occurred on consecutive cooling at a
very slow rate.  This behavior is in contradiction to the previous observation shown in Fig. 3 and is attributable to preferentially oriented alpha
grain growth with the {10$\bar{1}$1} pole, parallel to the specimen axis,
as discussed later.

It is noteworthy that the transformation frequently occurred in accordance with texture memory, although such memory was readily disturbed or
lost by changing the holding time at the beta temperatures, and the heating
and cooling rates, or an accidental minimal displacement of the specimen
with respect to the induction coil.

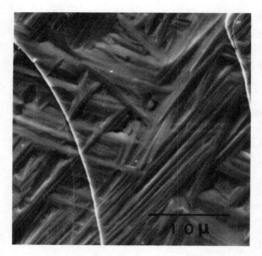

Fig. 2.   Surface stress relief topography due to transformation microtwin formation.

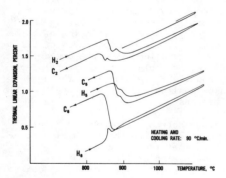

Fig. 3.   Thermal expansion on cooling.

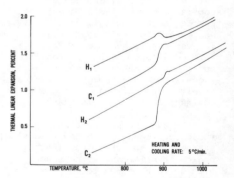

Fig. 4.   Thermal contraction on cooling.

When the specimen was annealed at 1080°C for a few hours to permit the formation of large beta grains and then slowly cooled, enormous linear expansion frequently occurred upon transformation, as shown in Fig. 5. As much as 2% linear expansion was not uncommon. X-ray Laue pattern analysis revealed that such specimens consisted of macrograins with the basal pole preferentially oriented parallel to the axis.

As shown in Fig. 6, curved parallel striations were frequently observed on the surface of specimens which had been annealed at 850°C for a few hours after slow cooling from 1080°C. During annealing, a contraction of approximately 0.1% was always observed, indicating annihilation of the transformation dislocations. If these striations represent terrace steps at the surface, coplanar dislocations, generated in the process of the transformation due to the misfit of two structures, should have moved to the surface and created these steps.

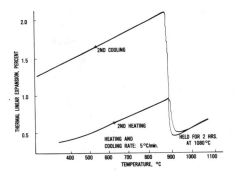

Fig. 5.   Enormous thermal expansion on slow cooling.

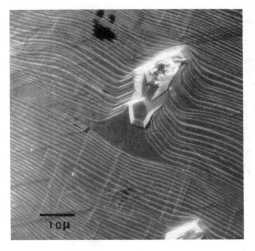

Fig. 6.   Surface terrace steps generated by movement of transformation
          dislocations.

## Discussion

Since Burgers first reported the "Burgers Relationship", {110} //
{0001} and <111> // <11$\bar{2}$0>, on the beta to alpha phase transformation for
zirconium [5], the same relationship has been confirmed a number of times
on the titanium transformation by many researchers. There are six {110}
planes in a b.c.c. structure and only one at a time of these six planes can
transform to the basal plane of a h.c.p. structure. The remaining five are
restricted to grow into one of the prismatic planes, {10$\bar{1}$0} and four of the
first order pyramidal planes, {10$\bar{1}$1} under Burgers' constraints.   Figure 7

represents a superimposed standard stereogram for the b.c.c. and h.c.p. structures with Burgers relationship.

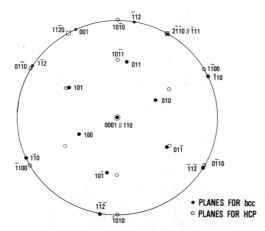

Fig. 7   Superimposition of b.c.c. and h.c.p. standard stereograms under Burgers relationship   constraints.

The plane relationships and the difference in the corresponding lattice parameters are tabulated in Table 1.

Table 1   Lattice Parameter of Titanium

| b.c.c. | | h.c.p. | $d_{hkil}$ (Å) |
|---|---|---|---|
| $d_{110} = 2.31$ Å | | | |
| (110) | → | (0001) | 2.342 |
| ($\bar{1}$10) | → | (1$\bar{1}$00) | 2.557 |
| (101) | → | (01$\bar{1}$1) | |
| (10$\bar{1}$) | → | ($\bar{1}$011) | 2.244 |
| (01$\bar{1}$) | → | (0$\bar{1}$11) | |
| (011) | → | (10$\bar{1}$1) | |

The difference in the lattice parameters suggests that if one of the {110} poles for a beta single crystal is parallel to the directions along which measurements were made, and transforms strictly either to the basal plane, the prismatic plane, or the first order pyramidal plane, the linear dimensional changes, $\Delta l$ will be 1.38%, 10.70% or -2.86%, respectively.  In a real case, no alpha single crystals will grow on cooling, yet slower cooling produces macroalpha grains with preferred orientations.  This is why in some cases, contraction takes place and, in other cases, expansion occurs on cooling.  Depending on the type of textural structure resulting from the phase transformation, zero linear dimensional change could also occur, as reported by English and Powell [3].

In the present study, it was found that the slower the cooling rate, the greater the resulting dimensional change. On heating, such a trend was less pronounced, since the transformation product has a more isotropic structure. Table 2 lists compressive properties determined for specimens which were cooled at two different rates. The faster cooling rate caused a minimal dimensional change, while the slower cooling rate caused approximately 2%. The Young's modulus resulting from the faster and the slower cooling rate were 114 GPa vs. 138 GPa. When a contraction occurred on slow cooling, the Young's modulus was 125 GPa. X-ray Laue pattern analysis revealed that the rapidly cooled specimens consisted of a small number of macrograins whose basal poles were within 15° of the compression axis, while the fast cooled specimens were nearly randomly oriented. Consequently, it appears that the dilatometric measurements are not suitable to study the intrinsic volumetric difference, in the beta and alpha structures, since linear dimensional change due to anisotropic effect prevails.

Table 2 Young's Moduli of Titanium Cooled at Different Rates

| Specimen No. | Cooling Rate | E (GPa) | $\sigma_{YS}$ (MPa) |
|---|---|---|---|
| 5 | 10°C/sec. | 114.4 | 248 |
| 6 | 10°C/sec. | 113.8 | 250 |
| 7 | 10°C/sec. | 115.8 | 245 |
| 8 | 5°C/min. | 133.7 | 272 |
| 9 | 5°C/min. | 139.3 | 277 |
| 10 | 5°C/min. | 134.4 | 278 |

With specimen ends in contact with the quartz tube, the temperature of the end surface is always cooler than the specimen body. Therefore, a temperature gradient from the ends develops along the axial direction of the specimen after the power is shut off. This temperature gradient causes the first nucleation of alpha grains with the C-axis growing preferentially in the longitudinal direction of the specimen. This situation explains a good part of the observed 2% expansion on a very slow cooling.

## Conclusion

The present study demonstrated that either expansion or contraction can occur in pure titanium on cooling; this phenomenon was caused by transformation texturing. Under controlled heating and cooling rates, it was found that a slower cooling rate enhanced textural effects and the basal pole texture frequently developed along the direction of the temperature gradient upon cooling. The linear dimensional change due to texturing is so great that atomic volume change during the beta to alpha phase transformation cannot be detected. Therefore, dilatometric measurement is not suitable to determine such change.

## Acknowledgement

Part of this work was performed by S. Nadiv during his leave from the Technion at the AFML under the National Research Associateship. The opportunity given to him by the NRC and the AFML is acknowledged by the authors.

## References

1.  Thermophysical Properties of High Temperature Solid Materials, ed. by Y.S. Touloukian, 1(1967), 1005.  MacMillan Co., NY.
2.  H.E. McCoy, Jr., Trans. ASM, 57(1964), 743.
3.  J.J. English and G.W. Powell, Trans. AIME, 236(1966), 1467.
4.  P.S. Rudman, Trans. AIME, 233(1965), 864.
5.  J.W. Christian, The Theory of Transformations in Metals and Alloys, (1965), 746.  Pergamon Press, New York.

# PREDICTION OF EARING BEHAVIOUR OF COMMERCIAL PURITY TITANIUM THIN SHEETS WITH DIFFERENT CRYSTALLITE ORIENTATION DISTRIBUTION FUNCTIONS

P. DERVIN[*], M. PERNOT and R. PENELLE
Laboratoire de Métallurgie Physique, Bât. 413, Université de Paris-Sud,
Centre d'Orsay,91405 Orsay Cedex (France)
[*]   Now at Laboratoire Léon Brillouin, Saclay , 91190 Gif sur Yvette (France)

## Introduction

The earing behaviour in thin sheets has been related to the R =f($\alpha$) curves, where R is the ratio $\varepsilon_{22}$ /$\varepsilon_{33}$ for a tensile test along a direction in the sheet plane, rotated of an angle $\alpha$ from the rolling direction  - RD - [1].

Generalisation of TAYLOR model [2]  to anisotropic behaviour has been introduced by HOSFORD and BACKOFEN [3] , it can be used with the Crystallite Orientation Distribution Function (CODF) [4]    , which describes quantitatively crystallographic textures ,  to theoretical predictions of the  curves R=f($\alpha$ ).

For cubic metals, results [5 - 8]  have shown that a quantitative agreement is not observed between theoretical and experimental values of R,because of the simplicity of hypothesis of TAYLOR model. Nevertheless theoretical and experimental curves R=f($\alpha$) , exhibit  the same outline,  so results can be used for prediction of ears position and qualitatively evolution of ears height as a function of mechanical and thermal treatments [9] .

Application of this model to hexagonal metals, particularly to titanium, is of great interest, since with this material textures giving deep drawings, can be obtained [10] .

## Modeling

### 1. TAYLOR hypothesis

TAYLOR assumed that the plastic behaviour of a metal could be described by using following hypothesis:

- Plastic deformation is homogeneous $[E]$ = $[\varepsilon]$ if $[E]$ and $[\varepsilon]$ are macroscopic and microscopic strain tensors.
- Plastic deformation occurs by slip. Five independent slip systems are necessary to accommodate any deformation in a grain.
- Slip occurs when the shear stress on the slip plane in the slip direction reaches a critical value $\tau$ . Hardening is isotropic.
- Among the sets of five slip systems accommodating the deformation, the ones which take  place are those for which strain energy is minimum.

### 2. Scheme of calculation

For a strain increment, the strain tensor in the principal axes can be written :

$$[E] = \delta E_{11} \begin{bmatrix} 1 & 0 & 0 \\ 0 & \dfrac{-R}{1+R} & 0 \\ 0 & 0 & \dfrac{-1}{1+R} \end{bmatrix} \qquad (1)$$

As the deformation is homogeneous :

$$E_{ij} = \varepsilon_{ij} = \sum_k \alpha_{ij}^k \gamma_k \qquad (2)$$

where $\gamma_k$ is the shear strain on the slip system k, and $\alpha_{ij}^k$ orientation factors.

For a given strain $[E]$, the quantities $\gamma_k$ can be calculated for each set of five independent slip systems. The strain energy is :

$$\delta W = \tau \sum_{n=1}^{5} |\gamma_n| \qquad (3)$$

Active slip systems are chosen by minimisation of $\delta W$. For a crystal of given orientation and strain, the TAYLOR factor can be defined :

$$M = \frac{\delta W_{minimum}}{\tau \cdot \delta E_{11}} \qquad (4)$$

BISHOP and HILL [11] have pointed out that the strain energy can be calculated by using the yield surface of the single crystal and the Principle of Maximum Work. We have used this analysis which permits a simplification of numerical computations and gives the same results that Principle of Minimum Work.

## 3. Anisotropic polycrystal

For a polycrystal with a crystallographic texture, it is assumed, that for a given strain in a direction $\alpha$ ; the averaged TAYLOR factor is the sum of TAYLOR factors of crystallites, weighted by their volume fraction :

$$\overline{M} (R) = \int_g M(R,g) \cdot F(g) dg \qquad (5)$$

where $M(R,g)$ is the TAYLOR factor of a crystallite of orientation g and with deformation defined by R; $F(g)$ is the value of the CODF.

HOSFORD and BACKOFEN [3] have proposed a generalisation of the Principle of Minimum Work, that is the deformation R of the polycrystal that we have to select is this one which minimises the averaged TAYLOR factor.

The same calculations in different directions $\alpha$ give the curve $R=f(\alpha)$.

### Application to titanium

In the case of titanium most often observed slip systems are prismatic $\{10\bar{1}0\}$, basal $\{0001\}$ and pyramidal $\{10\bar{1}1\}$ with a direction $<11\bar{2}0>$ as slip direction. Then it is not possible to accommodate the term $\varepsilon_{33}$ of strain tensor.

The slip $\{11\bar{2}2\}<11\bar{2}3>$ seems to have very little contribution to deformation below 300°C [12 - 14] , but several twinning systems have been observed, the most common are $\{11\bar{2}2\} <11\bar{2}3>$ giving a contraction along c axis and $\{10\bar{1}2\} <10\bar{1}1>$ giving an extension [12, 15 - 20] .

It has been admitted that twinning is governed by critical shear stress law in order to use the model [21] .

BISHOP and HILL analysis has been used,by building the yield surface [22] in a similar way that the one used by THORNBURG and PIEHLER [23]. It is assumed that deformation is accommodated by four slip systems,excepted the term $\varepsilon_{33}$ which is accommodated according to its sign, by one or the other of the twinning systems.

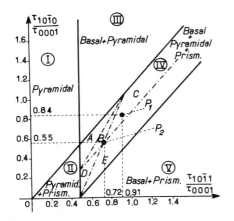

Fig. 1 - Diagram of competition
between slip systems
as function of CRSS
ratios.

In figure 1 it is shown a diagram due to CHIN and MAMMEL [24] presenting
the competition between the different slip systems as function of ratios of
critical resolved shear stresses (CRSS), and five different areas.

Results founded in the literature point out a great influence of the amount
of interstitial elements on the values of CRSS, and most the works have been
done on high purity titanium [12, 15 - 17, 25, 26]. For a titanium with about
1000 ppm of interstitials, that is the same purity as the titanium used in this
work, CHURCHMAN has founded :

$$\tau_{10\bar{1}0} = 9.19 \quad kg/mm^2$$

$$\tau_{0001} = 10.9 \qquad "$$

$$\tau_{10\bar{1}1} = 9.9 \qquad "$$

Those values, corresponding to point $P_1$ on figure 1, have been retained in a
first step.

Differences between the three CRSS determined by CHURCHMAN are probably not
significant and in fact, they are nearly equal, this is not in agreement with
most of the observations showing prismatic slip as preponderant.

We used an other point in the diagram, favouring prismatic slip. In figure 1
region IV, is divided in five subcases in which the yield surfaces are different
[22], so point $P_2$ has been chosen in the same subcase.

Because of the lack of data of twinning CRSS, different values were used.

## Results

### 1. Texture {11$\bar{2}$5} <10$\bar{1}$0>

Computation of curves R=f($\alpha$) has been done for a sheet of commercial
purity titanium (UT 40), of thickness 1 mm. Maximum value of CODF of this sheet
is 5 for the orientation {11$\bar{2}$5}<10$\bar{1}$0> , that represents a crystallite which c
axis is rotated of about 35° from the normal direction (ND) towards transverse
direction (TD), but this texture exhibits a large spread [4] .

Values of R have been measured for different angles between the tensile
direction and the rolling direction.

Experimental curve, and calculated curves with different values of twinning CRSS are presented in figure 2 for point P₁    and on figure 3 for point P₂.

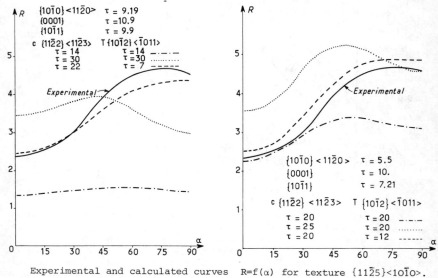

Experimental and calculated curves  R=f(α) for texture {11$\bar{2}$5}<10$\bar{1}$0>.

Fig.2  – for point P₁                    Fig.3 – for point P₂

A good agreement can be observed between experimental and theoretical curves but with different twinning CRSS for point P₁  and point P₂ .

The experimental curve does not present a strongly localised maximum, but a plateau from 60° to 90°, so this sheet gives drawing ears in this range, this result has been experimentally confirmed by cylindrical drawing cups.

## 2. Basal texture

We have simulated a texture  {0001} <10$\bar{1}$0> , that is with c axis parallel to DN, but with a spread shown by the pole figures of figure 4 and 5. Maximum of pole density of the figure  {0001} is 13 and 6.5  for the pole figure {10$\bar{1}$0} ,this correspond to a maximum of the CODF of about 33.

Values of R,  computed with twinning CRSS giving a good fit in figures 2 and 3,  are about 13 ,  little variations are observed with angle α .

This result is in good agreement with experimental values founded in literature, for example in the work of FISHBURN, ROBERTS and WILSON.

Basal texture corresponds to a good drawing behaviour, with a weak percentage of ears.

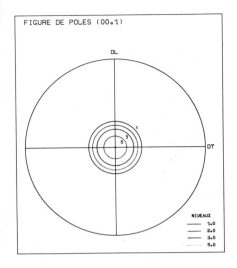

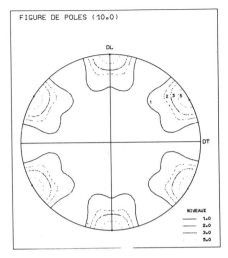

Fig. 4                                Fig. 5

Pole figures of simulated basal texture.

### Discussion

In figures 2 and 3, the theoretical curves which are in good agreement with experimental one, show that CRSS values of point $P_2$ seem to be more realistic than CRSS values of point $P_1$ .

For point $P_1$ the CRSS of the three slip systems are too close, to obtain an anisotropic behaviour near experimental one; that requires twinning CRSS in a ratio about 3, with twinning CRSS for extension lower than slip CRSS. When at point $P_2$ anisotropy is introduced partly by the low CRSS of prismatic slip, so twinning CRSS are in a ratio of about 1.5 and both are higher than slip  CRSS.

LEE and BACKOFEN have pointed out from tensile and compressive tests on a sheet with a basal texture, that extension twinning $\{10\bar{1}2\}$ $<10\bar{1}1>$ is easier than contraction twinning  $\{11\bar{2}2\}$ $<11\bar{2}3>$ , this is very coherent with a tension twinning CRSS lower than the contraction twinning CRSS.

High values of R, for basal texture, is also in agreement with a high contraction   twinning CRSS.

Figure 6 presents in a stereographic   projection the regions in contraction and the regions in extension as a function of R,  X being the tensile direction and Z parallel to ND, so the localisation of c  axis in this referencial indicates if the deformation happens in tension or in contraction. In regions with hachures $\varepsilon_{33}$ changes of sign when R  varies.

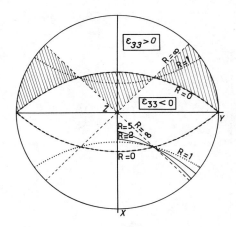

Fig. 6 - Stereographic    projection giving the sign of the term $\varepsilon_{33}$  as a
function of the value of R and of the localisation of the c axis
of the crystallite in the reference system, tensile direction X,
sheet plane XY.

For the texture with c axis tilted towards TD, when the sheet is strained
at $\alpha = 0°$ (along RD), the c axis of crystallites corresponding to the maximum of
CODF, have to strain in contraction. But the system is not very far from a
symmetrical one by respect to $\varepsilon_{22}$  and $\varepsilon_{33}$ , that can explain the low values of
R at $\alpha = 0°$.

When the sheet is strained at $\alpha = 90°$ (along TD), because of the spread of
texture some crystallites have to be deformed in contraction, but others have
to be deformed in extension.  It seems that can explain the fact that value of
R at $\alpha = 90°$, increases when the tension twinning CRSS decreases, as it can be
seen in figures 2 and 3. The metal behaves in such a way that the largest
number possible of crystallites are strained in tension, so the deformation work
decreases and in order to increase the range in tension, R has to increase too.

The basal texture which has been simulated, is only a first approximation
of a real texture, because in figure 6 it can be seen that the anisotropy of the
spread around ND changes the outline of the curve $R=f(\alpha)$  and also the formation
of drawing ears, since it has been shown previously that R value is more
influenced by the spread than by the strength of the texture [9 ].

References

1.  W.T. Lankford, S.C. Snyder and J.A. Bauscher : Trans. A.S.M., 42(1950),1197.
2.  G.I. Taylor : J. Inst. Metals, 62 (1938), 307.
3.  W.F. Hosford and W.A. Backofen  : " Fundamentals of Deformation processing"
    Syracuse Univ. Press (1964), 259.
4.  S. Naka, R. Penelle, R. Valle and P. Lacombe : in this Conference.
5.  H.J. Bunge and W.T. Roberts : J. Appl. Cryst.,2 (1969), 116.
6.  G.J. Davies, J.S. Kallend and T. Ruberg : Met. Science, 9 (1975), 421.
7.  M. Pernot : Thesis, Université Paris XI, Orsay (1977).
8.  M. Grumbach, P. Parnière, L. Roesh and C. Sauzay : Mém. Sc. Rev. Mét., 72
    (1975), 241.

9. O. Ferreira, P. Dervin, M. Pernot and R. Penelle :   In G. Gottstein and
   K. Lucke (ed), ICOTOM 5, Aachen, Springer Verlag, vol 1.(1978),337.
10. D.V. Wilson : J. Inst. Metals, 94 (1966), 84.
11. J.F.W. Bishop and R. Hill : Phil. Mag., 42 (1951), 414.
12. N.E. Paton and W.A. Backofen  : Met. Trans., 1 (1970), 2839.
13. D. Shechtman and D.G. Brandon : J. Mat. Science,8 (1973),1233.
14. T.R. Cass : " The Science, technology and Application of Titanium ", Pergamon Press, New-York (1970), 459.
15. P.G. Partridge : Met. Reviews, 118 (1968),169.
16. H. Conrad, M. Doner and B. de Meester : " Titanium Science and Technology"
   (R.I. Jaffee and H.M. Burte ed.),Plenum Press, New-York (1973),V.2,p.969.
17. A. Akhtar : Met. Trans., 6A (1975),1105.
18. R.A. Fishburn, W.T. Roberts and D.V. Wilson : Metals Technology, 7 (1976),310 .
19. R.P. Arthey and W.T. Roberts : Metals Technology, 7 (1976), 317.
20. D. Lee and W.A. Backofen  : Trans. AIME, 236 (1966),1969.
21. G.Y. Chin, W.F. Hosford and D.R. Mendorf : Proc. Roy. Soc., A 309 (1969), 433.
22. P. Dervin : Docteur Ingénieur Thesis, Université Paris XI, Orsay (1978).
23. D.R. Thornburg  and R. Piehler : Met. Trans. , 6 (1975), 1511.
24. G.Y. Chin and W.L. Mammel : Met. Trans. 1 (1970), 357.
25. A.T. Churchman : Proc. Roy. Soc., A 226 (1954), 216.
26. A. Akhtar and E. Teghtsoonian : Met. Trans. 6A (1975),2201.

# THE FORMATION OF RECRYSTALLIZATION TEXTURE
## IN $\alpha$-ALLOYS OF TITANIUM

R. A. Adamescu, P. V. Geld, D. A. Scryabin

The Urals Polytechnical Institute, USSR

The investigation of texture formation processes during the recrystallization of metallic materials is significant both from the scientific and practical point of view. It is due to the fact that the analysis of preferred orientations enables to better understand the mechanism of the processes being developed. On the other hand, the texture is responsible, as a rule, for anisotropy of physical and mechanical properties. However, the information about regularities of texture formation during recrystallization of titanium is limited, and these regularities are not fully covered in scientific literature.

This paper includes the main results of studying the formation of recrystallization texture in cold-rolled sheets of binary $\alpha$-alloys of titanium by Al, Sn and Mn, and a number of commercial alloys (BT1-0, BT5-1, OT4-1) as well.

## Material and Technique

The chemical composition of binary alloys investigated is given in Table 1.

Cold rolling of 4 mm billets was carried out in one direction with total reductions of 20 to 95 pct. The specimens were annealed with various holdings in a vacuum furnace having $10^{-4}$ torr vacuum at temperatures ranging from $t_r^o$ +50°C to $t_{cr}$ -50°C where $t_r^o$ is the temperature at the onset of recrystallization, and $t_{cr}$ is the temperature of polymorphic transformation.

The investigation of the texture was conducted by analyzing the pole figures {0002} and {1010} which were plotted using the data of difractometric X-raying for «reflection». In order to increase the number of grains participating in reflection the X-raying was carried out with a reciprocal movement of a specimen.

The temperature at the onset of recrystallization was estimated according to the disintegration of Debye lines, X-raying being carried out with defocusing.

Table 1.

Chemical Composition of Investigated Alloys

| Alloy No. | Inclusions and alloying elements, weight pct | | | | | | | | |
|---|---|---|---|---|---|---|---|---|---|
| | Al | Sn | Mn | Fe | Si | V | Cr | H | O |
| 1 | 0.36 | — | — | 0.03 | 0.060 | 0.01 | 0.012 | 0.004 | 0.08 |
| 2 | 0.59 | — | — | 0.05 | 0.065 | 0.01 | 0.014 | 0.004 | 0.09 |
| 3 | 1.04 | — | — | 0.04 | 0.055 | 0.01 | 0.013 | 0.005 | 0.08 |
| 4 | 2.01 | — | — | 0.04 | 0.060 | 0.01 | 0.012 | 0.004 | 0.07 |
| 5 | 3.58 | — | — | 0.05 | 0.055 | 0.01 | 0.010 | 0.004 | 0.08 |
| 6 | — | 0.53 | — | 0.03 | 0.055 | 0.01 | 0.012 | 0.004 | 0.08 |
| 7 | — | 1.10 | — | 0.04 | 0.055 | 0.01 | 0.015 | 0.004 | 0.12 |
| 8 | — | 2.02 | — | 0.11 | 0.045 | 0.01 | 0.013 | 0.004 | 0.05 |
| 9 | — | 3.68 | — | 0.03 | 0.060 | 0.01 | 0.012 | 0.003 | 0.06 |
| 10 | — | 5.04 | — | 0.03 | 0.060 | 0.01 | 0.013 | 0.003 | 0.06 |
| 11 | — | 6.91 | — | 0.03 | 0.100 | 0.01 | 0.015 | 0.003 | 0.06 |
| 12 | — | — | 0.72 | 0.12 | 0.055 | 0.01 | 0.022 | 0.004 | 0.06 |
| 13 | — | — | 1.30 | 0.14 | 0.070 | 0.01 | 0.023 | 0.004 | 0.08 |

Experimental Results and Discussion

The results of studying the temperatures at the onset of recrystallization are given in Table 2. The experimental data have shown that these temperature increased while alloying tita-

nium by aluminium when Al-content is over 1%. The $t_r^o$ was not ma-
terially affected when Al-concentration was less than 1%. On the
contrary, the sharp increase of $t_r^o$ was found to be due to alloying
titanium by Sn, though Sn-content was rather low (1-2%).

The results obtained show that the influence of Mn on $t_r^o$ is
greater than that of Al and Sn when the concentration of alloying
elements is less than 1%. It is due to the inhibition of recrystallization process in Ti-1.30% Mn-alloy by disperse particles of the
second phase.

Table 2

Temperature at the Onset of Recrystallization ($0^oC$)
for Ti-Alloys

| Alloy | The amount of deformation, % | | | | | |
| No. | 20 | 40 | 60 | 80 | 90 | 95 |
|---|---|---|---|---|---|---|
| 1 | 570 | 530 | 520 | 480 | 480 | 470 |
| 2 | 570 | 540 | 510 | 480 | 475 | 470 |
| 3 | 540 | 530 | 530 | 500 | 490 | 480 |
| 4 | 600 | 570 | 560 | 530 | 520 | 510 |
| 5 | 670 | 530 | 580 | 560 | 550 | 550 |
| 6 | 540 | 520 | 500 | 480 | 470 | 470 |
| 7 | 530 | 510 | 490 | 480 | 470 | 470 |
| 8 | 580 | 560 | 530 | 500 | 500 | 500 |
| 9 | 590 | 560 | 560 | 520 | 510 | 500 |
| 10 | 610 | 570 | 570 | 530 | 520 | 520 |
| 11 | 640 | 580 | 550 | 540 | 540 | 530 |
| 12 | 550 | 520 | 510 | 500 | 470 | 470 |
| 13 | 610 | 570 | 520 | 520 | 420 | 520 |
| Commercial titanium | 530 | 510 | 490 | 480 | 470 | 460 |

The significant rising of $t_r^o$ when increasing the concentration
of Al and Sn is associated with a considerable decrease of stacking fault energy [1] . In fact, the polygonization processes in
metals having low stacking fault energy are known to be inhibited

which complicates the formation of recrystallization nuclei and results in an increase of $t_r^o$. In addition, the increase of $t_r^o$ during alloying is due to the energy growth of interatomic relations and the decrease of diffusion mobility of atoms.

The investigation of dilute Ti-alloys has shown that after annealing for 1 hr at $t_r^o$ +50°C their texture is similar to that of unalloyed titanium. The texture is generally characterized by (0001) $\pm \varphi$ ND - TD [10$\bar{1}$0] -component for all amount of reduction where $\varphi$ is inclination of basal planes with respect to the plane of the sheet; ND is the direction of the normal to the surface of the sheet; TD is the transverse direction in the sheet with respect to the direction of rolling. Increasing the deformation during cold rolling only results in the intensity growth of recrystallization (Fig. 1). The texture is not practically affected by an increase of annealing time (till 5 hr) at these temperatures, as well.

Similar regularities of forming the recrystallization texture are typical for binary alloys having up to 1 pct Al, less than 3.7 pct Sn and up to 1.3 pct Mn.

Greater Sn-content results in forming the two-component texture, such as (0001) $\pm \varphi$ ND - TD [10$\bar{1}$0] and (0001) [10$\bar{1}$0], an increase of Sn-content in the alloy resulting in higher intensity of (0001) [10$\bar{1}$0] -component (Fig. 2). On the contrary, greater amounts of reduction results in the intensity growth of (0001) $\pm \varphi$ND - TD [10$\bar{1}$0] -component.

In Ti-Al-alloys an increase of Al-content up to 2 pct results in the formation of (0001) [10$\bar{1}$0] -texture after annealing at $t_r^o$ +50°C (Fig. 3), the exception being the specimens deformed 90-95% which have both (0001) [10$\bar{1}$0] - and (0001) $\pm \varphi$ND - TD [10$\bar{1}$0] -components. But the texture of Ti-3.58% Al-alloy is characterized (0001) [10$\bar{1}$0] -component after annealing for all investigated amounts of reduction (Fig. 3). General regularities established for binary alloys are typical for commercial alloys BT1-0, BT5-1, OT4-1 as well. The recrystallization texture is formed in BT1-0 alloy like in dilute binary alloys, but the main texture components in BT5-1- and OT4-1-alloys are the same as in alloys having greater content of alloying elements (Fig. 4).

Calculated pole figures plotted according to the technique described in [2] were used for establishing the orientation relations between the deformation and recrystallization textures. While plotting these figures, it was taken into account that the mobility of grain boundaries having disorientation angles of $15-45^{\circ}$ is the greatest. The comparison of calculated pole figures with the experimented ones leads to the conclusion that there is a good agreement between them.

It means that the texture intensity peaks corresponding to the poles of basal planes of crystallites are located in the region of «allowed» orientations. Hence, the grains of similar orientations are in favourable condition for their growth with respect to the deformed matrix, and they «survive» during the process of competitive growth. It indicates that general peculiarities of recrystallization texture of $\alpha$-alloys of titanium may be accounted for by the theory of orientated growth.

However, this theory cannot explain the appearance, in some cases of weaker components of recrystallization texture, and the fact that certain orientations being present in the deformed matrix do not develop during recrystallization, e. g. $(0001) \pm \varphi ND - RD [10\bar{1}0]$. The cause may be that the mobility of boundaries depends not only on disorientation of neighbouring grains but on the type and quantity of alloying elements and inclusions [3]. In addition, orientation peculiarities of recrystallization nucleation may contribute to that, as well.

## References

1. A. S. Shishmakov, R. A. Adamescu, P. V. Geld. Proc. of the
   3rd International Conference on Titanium, M., VILS, 1977,
   v. 1, pp. 423-428.

2. D. B. Titorov. Prognozirovaniye tekstury rekristallizatsii, FMM, 1973, v. 36, p. 91.

3. G. Glater, B. Chalmers. Large-Angle Grain Boundaries. M., «Mir», 1975.

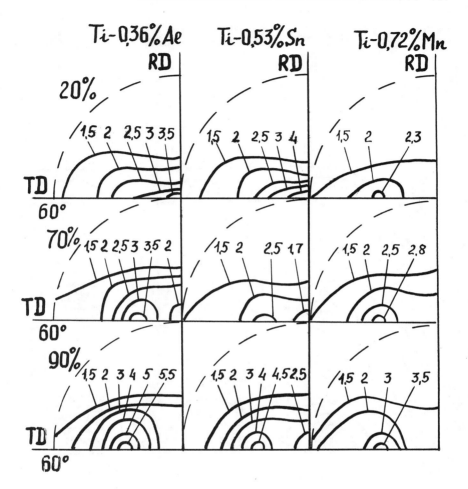

Fig. 1  {0002} pole figures of dilute alloys of titanium after annealing at $t_r^o$ +50°C.

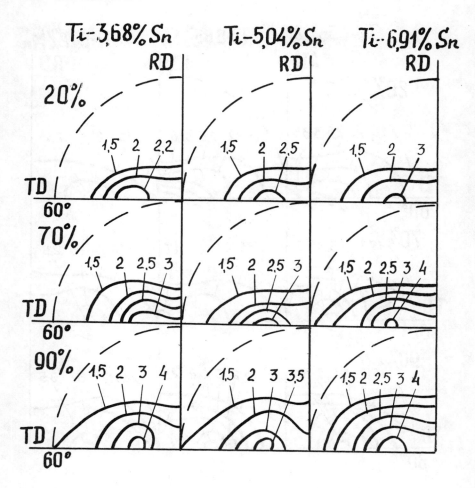

Fig. 2  {0002} pole figures for Ti–Sn alloys after annealing at $t_r^o$ +50°C.

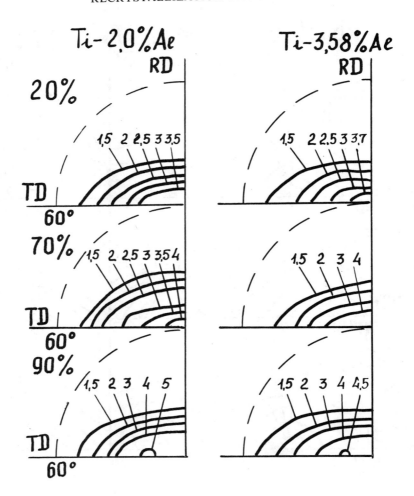

Fig. 3   {0002} pole figures for Ti-Al-alloys after annealing at $t_r^o$ +50°C.

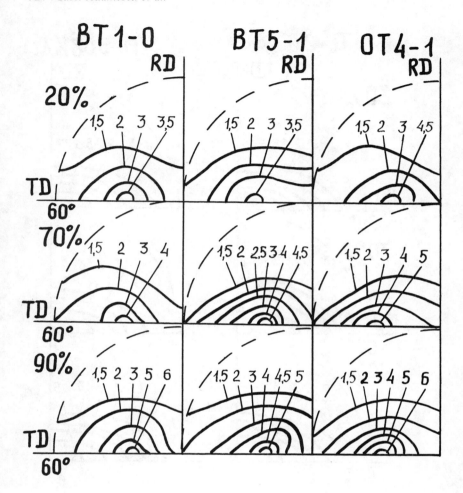

Fig. 4 {0002} pole figures for BT1-0-, BT5-1-, and OT4-1-alloys after annealing at $t_r^o$ +50°C.

# CONTROL OF MICROSTRUCTURE AND TEXTURE IN   Ti-6Al-4V

M. Peters and G. Luetjering

Ruhr-University Bochum, West-Germany

## Introduction

It is well recognized that both microstructure and texture have a pronounced influence on the mechanical properties of the Ti-6Al-4V alloy[1].
Unfortunately, it is often difficult to separate their contributions[2],
which limits their use as a tool to optimize the alloy for a given practical
application.  This is mainly due to the fact, that it is not always possible
to control microstructure and texture successfully during the commercial production of the alloy.

Depending on the detailed thermo-mechanical treatment a large variety
of microstructures can be obtained[3,4].  It is convenient to classify these
microstructures roughly into three categories depending on the geometrical
arrangement of the two phases $\alpha$ and $\beta$: lamellar, equiaxed, or a mixture of
both (bi-modal).

Lamellar structures can be readily controlled by heat treatment.  Slow
cooling into the two-phase region leads to nucleation and growth of the $\alpha$-
phase in plate form starting from $\beta$-grain boundaries.  The resulting lamellar
structure is fairly coarse and is often referred to as "Widmanstatten" or
"$\beta$-annealed" structure.  Water-quenching from the $\beta$-phase field followed by
annealing in the ($\alpha+\beta$)-phase region leads to a much finer lamellar structure
over a martensitic transformation and subsequent formation of $\beta$ at the martensitic plate boundaries.  This structure is usually called "$\beta$-quenched".

Equiaxed microstructures are obtained by mechanical working the material in the ($\alpha+\beta$)-phase field.  Subsequent annealing at about 700°C produces
the so-called "mill-annealed" microstructure which can vary to a large extent depending on the exact deformation procedure[1].  A better heat treatment (4 hours at 925°C followed by slow-cooling) was recently developed to
obtain a more reproducible equiaxed microstructure[1,4].  This so-called
"recrystallization-annealed" structure is fairly coarse with an $\alpha$-grain size
of about 15-20µm.

Bi-modal type microstructures consisting of isolated primary $\alpha$-grains
in a lamellar matrix are best approximated by the so-called "solution treated
and aged (or overaged)" microstructures obtained by an one hour anneal at
955°C followed by water-quenching and aging at 600°C.  The resulting primary
$\alpha$-grain size is again usually about 15-20µm[1,4].

Intense textures are found in the Ti-6Al-4V alloy especially if the
material was heavily deformed in the ($\alpha+\beta$)-phase field.  Quite different
textures are observed depending on deformation mode and on deformation temperature[5-9].  The most prominent textures are the BASAL texture, where the
basal planes are parallel to the rolling plane and which is achieved by
cross-rolling at low temperatures[5].  Rolling at high temperatures produces
a texture, where the basal planes are perpendicular to the rolling plane and

parallel to the rolling direction (TRANSVERSE texture)[6-8]. Further, mixed textures having a BASAL as well as a TRANSVERSE part are often observed[8,9].

The purpose of the present work was to control the α-grain size more readily for equiaxed and bi-modal microstructures with the goal to obtain grain sizes well below 10μm. Since heavy deformations would be involved in achieving these fine microstructures it was tried at the same time to control the resulting textures.

## Experimental Procedure

The composition of the material mainly used in this study was 6.4% Al, 4.0% V, 0.19% oxygen (wt.-%). The alloy was made by TIMET and forged in the β-phase field to a plate thickness of 30mm by Ladish Company. Experiments dealing with microstructural control were also performed on two additional alloys made by Fa. Krupp with the following compositions: 6.1% Al, 3.9% V, 0.20% oxygen (plate thickness 25mm), and 6.2% Al, 4.0% V, 0.17% oxygen (plate thickness 30mm). No significant differences were found between these alloys with respect to the resulting microstructures.

The standard starting heat treatment for all experimental tests was 15 min. at 1050°C followed by water-quenching. Rolling was used as deformation method for controlling microstructure and texture. Prior to rolling the specimens were heated for 5 min. at the deformation temperature. The rolling procedure was done usually in 4 steps at temperatures above 800°C and in 4-10 steps at 800°C and below always with intermediate heating for 2 min. After the last rolling step the material was water-quenched. The final annealing temperature for all specimens dealing with the control of microstructure and texture was 800°C followed by water-quenching. Specimens for tensile tests were subsequently aged for 24 hr at 500°C and air-cooled. All heat treatments were performed in an argon atmosphere.

For light microscopy the specimens were etched in a 2ml HF, 3ml $HNO_3$, 95ml $H_2O$ solution. For transmission electron microscopy thin foils were prepared using a double-jet TENUPOL apparatus, Fa. Struers, and an electrolyte consisting of 90ml perchloric acid, 525ml butanol and 900ml methanol, the temperature was -30°C and the voltage 30V. The texture investigations were performed on a Philips goniometer using the reflection method. The results are given in (0002) pole figures. Tensile tests were done at room temperature on round specimens with a diameter of 4mm and a gage length of 20mm using a strain rate of 8.3 x $10^{-4}s^{-1}$.

## Results and Discussion

### 1. Microstructure

To investigate the formation of the equiaxed microstructure from a lamellar structure unidirectional rolling was performed as a function of deformation degree mainly at two temperatures (800°C and 950°C). After rolling and subsequent water-quenching the material always exhibited still the lamellar starting microstructure now in a plastically deformed state. An example for a deformation degree of $\phi = -0.9$ at 800°C is shown in the light micrograph of Figure 1a and the transmission electron micrograph of Figure 2a.

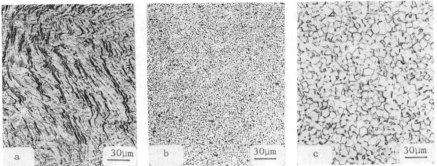

Figure 1:    Development of equiaxed microstructures from fine lamellar struc-
             tures,(LM).
             15 min 1050°C/WQ, φ = -0.9 at 800°C/WQ,
             a) As deformed, b) 1h 800°C, c) 96h 800°C.

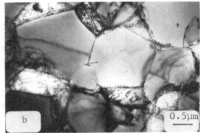

Figure 2:    Development of equiaxed microstructures from fine lamellar struc-
             tures, (TEM).
             15 min 1050°C/WQ, φ = -0.9 at 800°C/WQ
             a) As deformed, b) 1h 800°C.

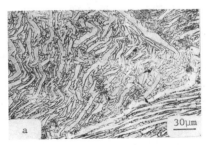
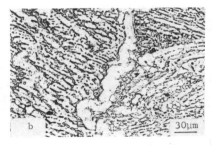

Figure 3:    Effect of coarse lamellar starting structure on equiaxed
             microstructure, (LM).
             15 min 1050°C  1°C/min 950°C/WQ, φ = -0.9 at 950°C/WQ,
             a) As deformed, b) 15h 800°C.

If the deformation degree was high enough the subsequent annealing treatment at 800°C caused the formation of the equiaxed microstructure by a recrystallization process. The resulting structure is illustrated in Figure 1b and 2b. In this example an annealing time of 1 hour at 800°C was long enough to result in complete recrystallization. The equiaxed microstructure was quite uniform (Figure 1b) and the α-grain size was about 2μm. The transmission electron micrograph (Figure 2b) shows further the low dislocation density within the α-grains after recrystallization. A coarser equiaxed microstructure can be readily obtained by longer annealing times at 800°C causing grain growth. An example is shown in Figure 1c for an annealing time of 96 hours leading to an α-grain size of about 12μm.

The qualitative interrelation between minimum annealing time at 800°C to achieve complete recrystallization and with it complete equiaxed microstructures, deformation degree, deformation temperature, and resulting α-grain size was in accordance with a normal recrystallization process. The tendencies were as follows (the detailed values are given elsewhere[10]): With decreasing deformation degree the necessary minimum annealing time increased and the resulting α-grain size increased. With increasing deformation temperature the deformation degree had to be increased to achieve the same α-grain size. A critical deformation degree, increasing with increasing deformation temperature, was necessary to result in a complete equiaxed microstructure. The approximate values for the critical deformation degree were as follows: $\phi = -0.15$ at room temperature, $\phi = -0.4$ to 0.5 at 800°C, $\phi = -0.5$ to 0.6 at 950°C, $\phi = -1.2$ at 1050°C. The last value means that deformation in the β-phase field at 1050°C and water-quenching also gave an equiaxed microstructure upon annealing at 800°C. The same general results were found qualitatively on the Ti-6%Al-6%V-2%Sn alloy[10].

Up to now the starting microstructure before deforming the material in the (α+β)-phase field was kept constant and as fine lamellar as possible through the heat treating schedule: 15 min at 1050°C/WQ, 5 min at temperature before starting deformation. To investigate the effect of a coarser lamellar starting structure specimens were slowly cooled into the (α+β)-phase field and then deformed. An example is shown in Figure 3 for cooling the material at a rate of 1°C/min to 950°C. The light micrograph in Figure 3a shows the deformed state of the coarse lamellar microstructure. Upon annealing at 800°C both phases α and β recrystallized and α-grains could grow into the original β-plates because the volume fraction of α had to increase in going from 950°C to 800°C. The recrystallized β-grains therefore became quite equiaxed whereas the appearance of the α-phase still remained more lamellar (Figure 3b). The results of this series with coarse lamellar starting microstructures showed that the width of the original α-plates had a pronounced influence on the α-grain size of the equiaxed structure and on the appearance of the structure (lamellar vs. equiaxed) even if the material recrystallized completely[10].

If after deformation the recrystallization treatment was carried out in the temperature region between 945°C and 965°C the volume fraction of both phases α and β became comparable. Although the temperature was quite high the grain sizes remained small because in such micro-duplex microstructures grain growth is generally slow[11]. This offers the possibility of generating a microstructure consisting of small equiaxed α-grains surrounded by a fine lamellar structure (bi-modal microstructure). An example of such a bi-modal structure is shown in Figure 4. The size of the primary α-grains

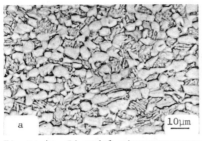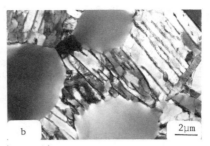

Figure 4:   Bi-modal microstructure, a) LM, b) TEM.
15 min 1050°C/WQ, φ = -1.4 at 800°C/WQ,
1h 955°C/WQ, 1h 800°C.

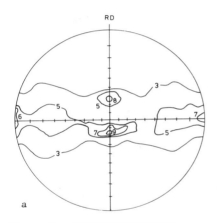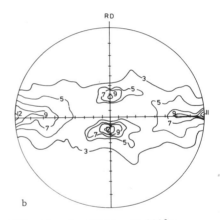

Figure 5:   BASAL/TRANSVERSE texture, unidirectional rolling at 800°C,
(0002) pole figures.
15 min 1050°C/WQ, φ = -1.4 at 800°C/WQ,
a) As deformed, b) 1h 800°C.

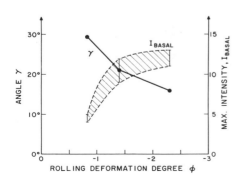

Figure 6:   BASAL/TRANSVERSE texture,
unidirectional rolling.
Maximum intensity of the
BASAL part and angle γ
between rolling plane and
basal planes as a function
of deformation degree.
15 min 1050°C/WQ,
φ at 800°C/WQ, 1h 800°C.

was about 6μm.  The β-grains at 955°C had about the same size.  Upon quenching from 955°C the β-phase transformed martensitically and upon subsequent
annealing at 800°C a fine lamellar structure was formed.  This can be readily
seen in the transmission electron micrograph of Figure 4b.  In such a bi-
modal structure the length of the individual plates in the lamellar portion
is limited by the former β-grain size of the recrystallized micro-duplex
structure at 955°C and approximately equal to the distance between primary
α-grains.

2.  Texture

The textures which were present in the as-received plates were eliminated by the initial heat treatment of 15 min. at 1050°C followed by water-
quenching.  After this heat treatment the material exhibited essentially no
crystallographic texture.

The type of texture which developed upon unidirectional rolling at 800°C
is shown in Figure 5a.  Two different maxima in intensity were observed in the
(0002) pole figure.  One belonged to basal planes which were aligned perpendicular to the rolling plane and parallel to the rolling direction (TRANSVERSE
part of texture).  The second belonged to basal planes which were aligned more
parallel to the rolling plane (BASAL part of texture) but actually formed an
angle of about 20° with the rolling plane.  Because of this mixed character
the texture was termed BASAL/TRANSVERSE texture (B/T - texture).  This texture develops as a α-phase deformation texture in the presence of β[12].  The
microstructure in this as deformed state was still lamellar as was shown in
Figures 1a and 2a.  Upon annealing at 800°C for 1 hour which caused the
formation of the equiaxed microstructure the type of texture was not changed
as can be seen from Figure 5b.  Only the maximum intensities increased slightly by the recrystallization process.  Further annealing at 800°C for up to 96
hours did not change the texture significantly.

With increasing deformation degree at 800°C the maximum observed intensities increased for both the BASAL and the TRANSVERSE part of the texture.
This is illustrated in Figure 6 for the BASAL part.  Also shown in the figure
is the angle γ between the maximum intensity of basal planes and the rolling
plane.  It can be seen that with increasing deformation degree the angle γ
decreased but had still a value of about 15° even after the high deformation
degree of $\phi = -2.3$.

The effect of increasing the deformation temperature to 900°C and above
on the texture is shown in Figure 7.  It can be seen that deformation at 900°C
produced a material with only a slight texture (Figure 7a).  Increasing the
temperature for unidirectional rolling still further, for example to 960°C
(Figure 7b), resulted in the development of a pure TRANSVERSE texture
(T - texture).  The maximum observed intensities as a function of deformation
temperature for a constant unidirectional rolling degree of $\phi = -1.4$ are
summarized in Figure 8.  The tendencies can be explained as follows:  The
deformation texture of the α-phase (B/T - texture) is declining with increasing the deformation temperature to 900°C and above.  The then reappear-
ing higher intensities of the transverse part (T - texture) are actually due
to a new α-phase texture development mechanism from a bcc β-phase deformation
texture.  Fredericks[13] proposed that if the β to α transformation takes
place under constraints from the rolling deformation only one of the six
highly textured {110}-planes is favored in transforming to the (0002)-plane

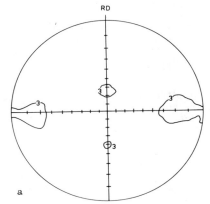

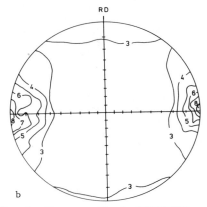

a
b

Figure 7:   Unidirectional rolling, (0002) pole figures, 15 min 1050°C/WQ,
            $\phi$ = -1.4 at T°C/WQ, 1h 800°C,
            a) T = 900°C, b) T = 960°C (TRANSVERSE texture).

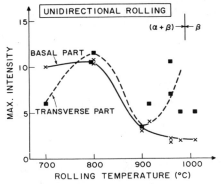

Figure 8:   Maximum intensities for
            the BASAL and the TRANS-
            VERSE parts as a function
            of deformation temperature,
            $\phi$ = -1.4.

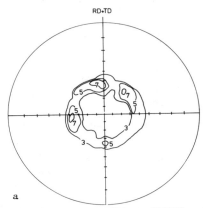

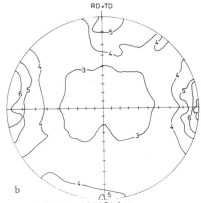

a
b

Figure 9:   Cross-rolling, (0002) pole figures, 15 min 1050°C/WQ,
            $\phi$ = -1.4 at T°C/WQ, 1h 800°C,
            a) T = 800°C (BASAL texture), b) T = 980°C.

of the hexagonal lattice leading to the observed T-texture. This explanation is in agreement with the result that upon deformation in the β-phase field at 1010°C a pure T-texture was developed (Figure 8). If the textured β-phase was allowed to recrystallize first and was then transformed to the α-phase all six {110}-planes are equally favored in transforming to the basal plane and a normal β to α transformation texture was observed accompanied by a lamellar microstructure[12].

If the rolling method was changed from unidirectional rolling to cross-rolling, that means the rolling direction was changed after each pass by 90°, all textures showed more radial symmetries. For deformation temperatures below 900°C textures of a pure BASAL type (B – texture) were developed, an example is shown in Figure 9a for a deformation temperature of 800°C. The TRANSVERSE part of the mixed B/T – texture for unidirectional rolling was completely suppressed by cross-rolling because upon changing the rolling direction by 90° the basal planes belonging to this texture part are in an unfavorable position perpendicular to the rolling direction and should disappear[12].

For rolling temperatures above 900°C the principal type of texture did not change in going from unidirectional to cross-rolling. Still pure TRANSVERSE textures were developed, only in case of cross-rolling two maxima appeared 90° apart. An example is shown in Figure 9b for a deformation temperature of 980°C. The results for cross-rolling as a function of deformation temperature are summarized in Figure 10 showing the maximum observed intensities for the BASAL and the TRANSVERSE part for a constant deformation degree of $\phi = -1.4$.

3.   Combination of microstructure and texture

It is readily possible to combine equiaxed as well as bi-modal microstructures with different texture types because the textures are developed during the deformation process whereas the desired microstructures can be achieved by the subsequent heat treatment.

The type of texture is primarily determined by deformation mode and deformation temperature, for example: unidirectional rolling at temperatures below 900°C gives a B/T – texture (Figure 5), cross-rolling below 900°C a B – texture (Figure 9a), unidirectional rolling above 900°C a T – texture (Figure 7b). Both parameters have no large influence on the possibility of obtaining different microstructures as long as the dislocation density is high enough after finishing the deformation process.

The different microstructures can then be obtained by varying annealing time and temperature, for example: short annealing times at 800°C produce a fine equiaxed structure (Figure 1b), long annealing times at 800°C a coarse equiaxed structure (Figure 1c), short annealing times at 955°C followed by water-quenching and 1 hour at 800°C give a bi-modal structure (Figure 4). Again these parameters do not change significantly the already present texture of the α-phase.

This opens up the possibility to evaluate the influence of texture (at constant microstructure) or of microstructure (at constant texture) on mechanical properties[14].

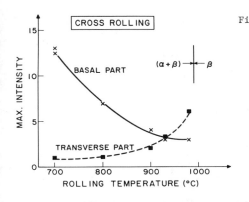

Figure 10: Maximum intensities for the BASAL and the TRANSVERSE parts as a function of deformation temperature, $\phi = -1.4$.

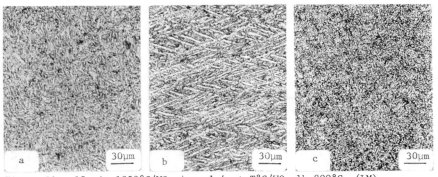

Figure 11: 15 min 1050°C/WQ, $\phi = -1.4$ at T°C/WQ, 1h 800°C, (LM)
a) T = 960°C, b) T = 975°C, c) T = 990°C

Table 1:  Tensile properties

|  | E (GPa) | $\sigma_{0.2}$ (MPa) | $\sigma_F$ (MPa) | $\varepsilon_F$ |
|---|---|---|---|---|
| B/T - RD | 107 | 1120 | 1650 | 0.62 |
| B/T - 45° | 113 | 1055 | 1560 | 0.76 |
| B/T - TD | 123 | 1170 | 1515 | 0.55 |
| T - RD | 113 | 1105 | 1540 | 0.57 |
| T - 45° | 120 | 1085 | 1610 | 0.76 |
| T - TD | 126 | 1170 | 1665 | 0.70 |
| B | 109 | 1120 | 1505 | 0.70 |

For once combination of microstructure and texture, namely for combining a pure T – texture with a fine equiaxed microstructure having an α-grain size smaller than about 3μm, a possible difficulty might arise if the deformation temperature is not closely monitored. This is illustrated in Figure 11. The material was deformed by unidirectional rolling at 960°C (Figure 11a), 975°C (Figure 11b), and 990°C (Figure 11c). For all three conditions a pure TRANS-VERSE texture was obtained. The microstructures after 1 hour at 800°C show that for the rolling temperature of 975°C the structure still appeared lamellar although both α and β exhibited equiaxed recrystallized grains. The reason for this appearance is, that the α-plates of the starting lamellar structure were growing too much during the 5 min. holding time at 975°C before starting the deformation process. This difficulty vanished upon lowering the deformation temperature to 960°C or increasing the deformation temperature into the β-phase field to 990°C (Figure 11c). It is still possible to obtain also an equiaxed looking structure for the material deformed at 975°C by increasing the annealing time at 800°C but the smallest α-grain size would then be larger than 3μm, the thickness of the α-plates in the starting lamellar structure. This example points out once again the importance of starting structure for the resulting equiaxed structure.

## 4. Tensile properties

Tensile tests were performed on specimens having all the same fine equiaxed microstructure (α-grain size about 2μm) but different texture types. For the texture types B/T and T derived by unidirectional rolling at 800°C and 960°C tests were done with the stress axis parallel to the rolling direction (RD), parallel to the transverse direction (TD), and in between under 45° (Table 1). The BASAL texture which was obtained by cross-rolling at 800°C was tested only in one direction (RD). The rolling degree was $\phi = -1.4$ and all specimens were annealed for one hour at 800°C followed by water-quenching and aging at 500°C for 24 hours. The results in Table 1 show that the modulus of elasticity E had the lowest value when the stress axis was parallel to the basal planes (B/T-RD, T-RD, B). The highest modulus was obtained for specimens with the stress axis perpendicular to the basal planes (B/T-TD, T-TD). The values for specimens inclined under 45° were lying in between. The same tendency was true for the yield stress except that the specimens lying under an angle of 45° showed the lowest values, because the basal planes are in the most favorable position for slip. These general tendencies for the dependence of modulus and yield stress on texture are in agreement with the literature and are discussed there already in detail[6,15]. Further, it seems that texture has no large effect on the tensile fracture process[16]. Fracture strain and fracture stress were approximately constant for the various textures tested (Table 1). It should be noted, that the fracture strains were very high although the specimens were aged at 500°C for 24 hours leading to precipitation of $Ti_3Al$-particles in the α-phase and therefore to a high yield stress. This high ductility is due to the small α-grain size[14,17]. More extensive results on the influence of microstructure and texture on mechanical properties especially on fatigue strength and fatigue crack propagation are given elsewhere[12,14] or will be published soon.

## Conclusions

1. Equiaxed microstructures are obtained from lamellar structures by a recrystallization process.
2. Fine equiaxed and bi-modal structures with α-grain sizes from 2 to 12μm

were achieved in this work by controlling the amount of "cold deformation", recrystallization time and temperature, and the starting lamellar
structure.
3. The retained amount of "cold deformation" is determined by deformation
degree, deformation temperature, and cooling rate from the deformation
temperature.  The deformation rate will be another important parameter.
4. For fine equiaxed microstructures the starting lamellar structure should
be as fine as possible (β-quenched condition).
5. Deformation in the β-phase field also can lead to equiaxed microstructures.
6. The type of texture is mainly determined by deformation mode and
deformation temperature.
7. Since texture and microstructure can be adjusted separately various combinations are possible.
8. Fine equiaxed microstructures can be age-hardened to a high yield stress
without losing the high ductility.

## Acknowledgement

The assistance of I. Moritz and K. Rittner in the experimental work is
gratefully acknowledged.  One alloy was kindly provided by Fa. Fuchs,
Meinerzhagen.  The part on microstructural control was supported by the
Deutsche Forschungsgemeinschaft.  The part on texture control was supported
by the Electric Power Research Institute, Palo Alto, under contract RP 1266-1.

## References

1. N. E. Paton, J. C. Williams, J. C. Chesnutt, and A. W. Thompson: AGARD-
   CP-185, (1975), 4-1.
2. A. W. Bowen:  Scripta Met., 11 (1977), 17.
3. R. A. Wood and R. J. Favor (Eds.):  Titanium Alloys Handbook, MCIC-HB-02,
   (1972), Battelle, Columbus.
4. J. C. Chesnutt, A. W. Thompson and J. C. Williams: AFML-TR-78-68, (1978).
5. S. F. Frederick and G. A. Lenning:  Met. Trans., 6B (1975), 601.
6. F. Larson and A. Zarkades:  MCIC-74-20, (1974), Battelle, Columbus.
7. A. W. Sommer and M. Creager:  AFML-TR-76-222, (1977).
8. M. J. Blackburn, J. A. Feeney, and T. R. Beck:  Advances in Corrosion
   Science and Technology, 3 (1973), 67, Plenum Press, New York.
9. M. F. Amateau, D. L. Dull, and L. Raymond: Met. Trans., 5A (1974), 561.
10. M. Peters, G. Luetjering, and G. Ziegler:  to be published in Z. Metall-
    kunde.
11. E. Hornbogen:  Fundamental Aspects of Structural Alloy Design, (1977),
    389, Plenum Press, New York.
12. M. Peters:  Ph.D. - Thesis, (1980), Ruhr-University Bochum.
13. S. F. Frederick:  AFML-TR-73-265, (1973).
14. M. Peters, A. Gysler, and G. Luetjering:  Proc. Fourth Intern. Conf. on
    Titanium, (1980), Kyoto, Japan.
15. A. W. Bowen:  Mat. Sci. Eng., 29 (1977), 19.
16. A. W. Bowen:  Mat. Sci. Eng., 40 (1979), 31.
17. G. Terlinde, A. Gysler, and G. Luetjering:  Proc. Third Intern. Conf. on
    Titanium, (1976), Moscow, USSR, in press.

# THE FORMATION OF HOT ROLLED TEXTURE IN COMMERCIALLY PURE TITANIUM AND Ti-6Al-4V ALLOY SHEETS

Akiyoshi Tanabe, Takashi Nishimura and Masahito Fukuda
Kobe Steel Ltd., Kobe, Japan

and

Kenichi Yoshida* and Junji Kihara
Faculty of Engineering, Tokyo University, Tokyo, Japan

## Introduction

In HCP titanium, most common slip modes are $\{0001\}\langle11\bar{2}0\rangle$, $\{10\bar{1}0\}\langle11\bar{2}0\rangle$ and $\{10\bar{1}1\}\langle11\bar{2}0\rangle$. Since slip direction is perpendicular to c-axis there is no operative slip system for yielding in c-axis. As the result, mechanical anisotropy is caused by the crystallographic orientation. The preferred orientation, the texture, in hot or cold rolled c.p.Ti sheet is described as $(0002)\pm\alpha°[10\bar{1}0]$, i.e., (0002) poles concentrated at $\pm\alpha°$ from the sheet normal in the transverse direction and $[10\bar{1}0]$ parallel to the rolling direction. The texture is affected by rolling temperature [1], rolling direction [2,3], reduction in thickness [4], heat treatment [5] and impurities. It's well known that great biaxial strength is observed in the case of Ti-Al alloy having the ideal basal texture, 'texture hardening' [3,6]. This phenomenon suggests the possibility of texture control and it's important to obtain the knowledge for the effects of various deformation and heat treating conditions on the formation of texture. This manuscript is an experimental study for the effects of initial crystallographic orientation, rolling temperature and reduction on the formation of hot rolled texture in commercially pure titanium sheet and for the effect of rolling temperature on the formation of hot rolled texture in Ti-6Al-4V alloy sheet.

## Experimental Procedure

Commercially pure Ti sheets with three kinds of initial textures and Ti-6Al-4V alloy sheets cut from slab forged at 1050°C were used for the starting materials. Fig.1 shows (0002) pole figures of these starting materials. Material A (c.p.Ti, O:1000ppm, Fe:400ppm) was cut from forged slab and its irregularly distributed (0002) poles were due to the large grain size (about 130μ). Material B (c.p.Ti, O:900ppm, Fe:500ppm) and material C (c.p.Ti, O:700ppm, Fe:600ppm) were hot rolled and annealed sheets. Initial textures of material B and C showed main peaks of (0002) poles at about 25° and 40° from the sheet normal towards the transverse direction respectively.

All starting sheets with 5mm thick were hot rolled at temperatures above 400°C and below 1100°C approximately 50% by one pass with a high speed rolling mill (roll dia. 100mm, roll speed 100m/min.). To obtain the fixed rolling temperature, the cooling curve during material handling from the furnace to the rolling mill was measured and then the furnace temperature was determined.

The (0002) pole figure was measured for all as-rolled sheets using full automatic pole figure diffractometer. The $(10\bar{1}0)$ pole figure was measured for some annealed sheets.

---

*Present address: Government Industrial Research Institute, Chugoku, Kure, Japan

## Results and Discussion

### 1.   Hot Rolled Texture of c.p.Ti

a. Texture Developed in α –temperature Region (below 850°C):   In material A with initial random orientation, the texture was affected by rolling temperature (Fig.2).   At temperatures below 700°C, the texture with (0002) pole concentrated at 20-30° from the sheet normal in the rolling direction was developed. This split RD texture had a spread of (0002) poles to the transverse direction. At temperatures above 750°C, additional peak of (0002) poles concentrated at 50-90° from the sheet normal in the transverse direction was developed (split TD texture).   In this split TD texture, basal poles tended to tilt from the sheet normal to the transverse direction with increasing rolling temperature.

While, in material B and C initially having stable rolling texture, the texture developed in α-temperature region was essentially similar to the initial texture (Fig.3 shows only the texture developed at 700°C).   In these case, the hot rolled texture was composed of only split TD type and no split RD texture was developed.

It was considered that the split RD texture was caused by random initial orientation because this component was developed only in material A.   As shown in Fig.4, this split RD texture in material A was found only at low reductions (below 73.2%) and was changed to split TD texture at reductions above approximately 80%.   Another interesting feature of the split RD texture was that this texture was changed to the split TD type by annealing at 700°C as shown in Fig.5. These results indicate that the split RD texture is an intermediate component developed only at low reductions and is unstable for annealing.

The split RD texture was also developed at low reductions in cold rolling. Both split RD and TD textures were developed at 34.4% cold reductions in material A (Fig.6).   Theoretical investigation based on twinning explains that new basal poles appear near the sheet normal and in the transverse direction at early stage in cold rolling [7].   It's possible to account for the discontinuous reorientation in cold rolling on the basis of the operation of twinning.   As critical shear stress for twinning is higher than that for slip at elevated temperatures [8], it seems difficult to account for the split RD component of the hot rolled texture in Fig.2 by the operation of twinning.

b. Texture Developed in β –temperature Region (above 900°C): At temperatures above 900°C, which HCP structure varies to BCC structure, the hot rolled texture was entirely different from that developed in α-temperature region and no significant difference caused by the initial crystallographic orientation was observed (Fig.7).   These basal pole figures in Fig.7 are considered $(110)_\beta$ pole figures transformed on the basis of Burgers' relationship, $(110)_\beta // (0001)_\alpha$ and $[111]_\beta // [11\bar{2}0]_\alpha$.   In general, the rolling texture of BCC metal is described as $(001) \pm \alpha° [110]$, i.e., [110] parallel to the rolling direction and (001) rotated $\pm\alpha°$ about [110] from the sheet normal in the transverse direction.   In Fig.7 (b) and (d), $(110)_\beta$ poles corresponding to (001)[110], $(\bar{1}14)[110]$ and $(\bar{1}12)[110]$ are also indicated.   According to the positions of $(110)_\beta$ poles belonging to (001) $\pm\alpha°$ [110] texture in Fig.7 (b) and (d), the β –rolling texture of c.p.Ti is regarded as to be mainly composed of $(001)\pm0-30°[110]$ texture.

The β-rolling texture of c.p.Ti in Fig.7 is formed through following three steps: 1) development of transformation texture (α→β) based on Burgers' relationship during heating, 2) development of rolling texture of β phase and 3) development of transformation texture (β→α) during cooling.   Fig.8 shows the first transformation texture observed for material B and the predicted positions of basal poles based on Burgers' relationship. The observed texture coincides with the predicted one and many peaks of basal poles appear on heating in β-temperature region. This is the reason why there was no significant effect of the initial crystallographic orientation on the formation of β-rolling texture.

The predicted texture in Fig.8 was determined for the initial orientation of (0002)±30°[11$\bar{2}$0]. As [10$\bar{1}$0] was parallel to the rolling direction in the initial orientation of material B, it was expected that the rotation about c-axis occurred from [10$\bar{1}$0]//RD to [11$\bar{2}$0]//RD during heating before transformation. From the (10$\bar{1}$0) pole figures for material B heated at temperatures above 700°C (Fig.9), it was observed that the rotation about c-axis took place on heating at 800° and 850°C. At 900°C, however, different texture was developed by the transformation. Hu and Cline [9] obtained the cold rolling and annealing textures for iodide titanium sheet and reported that rotations about c-axis and about sheet normal were caused by annealing at temperatures in recrystallization region. In c.p.Ti, however, our results indicated that the rotation about c-axis occurred only at temperatures above 800°C in grain growth region.

## 2. Hot Rolled Texture of Ti-6Al-4V Alloy

Since the relative volume ratio of α and β phases varies continuously with increasing temperature, the hot rolled texture of this alloy is affected by rolling temperature in different manner from the case of c.p.Ti. Fig.10 shows basal pole figures of Ti-6Al-4V alloy sheet rolled at temperatures above 500°C.

At temperatures below 850°C, the texture composed of both split RD and TD components was developed, which was the most popular texture for this alloy [2, 3]. This is almost the α-rolling texture, because the amount of β-phase is about 15-30% in this temperature region [10] and therefore there is a little amount of the transformed β-rolling texture. The split RD component in this texture is not changed to split TD type by annealing or by rolling at reductions over 90%, different from the case of c.p.Ti, and is considered a component caused by alloying elements.

At 900°C, the amount of β phase was approximately 50% and both α-rolling and transformed β-rolling textures were developed. Comparing this texture with that developed at temperatures below 850°C and above 1000°C, (0002) poles in Fig. 10 (d) lying in the transverse direction are regarded as α-rolling texture. (0002) poles lying in the rolling direction and at about 50° from the sheet normal in the transverse direction are regarded as transformed β-rolling texture. On the other hand, (0002) poles concentrated at about 30° from the sheet normal in the rolling direction are considered that α-rolling and transformed β-rolling textures coexist.

At temperatures above 1000°C in β region, only the transformed β-rolling texture was developed similarly to the case of c.p.Ti. According to the positions of (110)$_\beta$ poles belonging to (001)±α°[110] texture in Fig.10 (e) and (f), the β-rolling texture of Ti-6Al-4V is considered to be mainly composed of (001) ±40-60°[110] and the larger rotation of (001) about [110] axis is characteristic in comparison with the case of c.p.Ti.

## Summary

1. α-rolling texture of c.p.Ti was affected by initial orientation and rolling temperature. Especially, in the case of random initial orientation, split RD texture was developed at low reductions and was changed to split TD type (stable rolling texture in c.p.Ti) at reductions over 80% and by annealing at 700°C.

2.  For c.p.Ti rolled in β-temperature region, β-rolling texture regarded as (001)±0-30°[110] was developed. In this case, rotation about c-axis occurred at temperatures above 800°C and [11$\bar{2}$0] reoriented parallel to the rolling direction

3.  Hot rolled texture of Ti-6Al-4V developed below 850°C was almost α-rolling texture composed of both split RD and split TD components.

4.  β-rolling texture of Ti-6Al-4V was regarded as (001)±40-60°[110] and the larger rotation of (001) about [110] was characteristic in comparison with c.p. Ti.

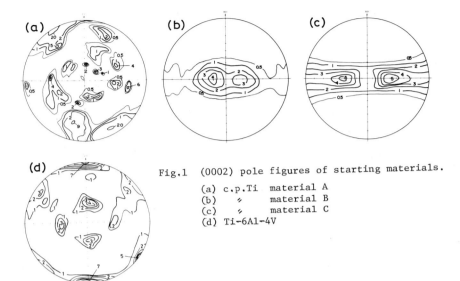

Fig.1 (0002) pole figures of starting materials.

(a) c.p.Ti  material A
(b)   "      material B
(c)   "      material C
(d) Ti-6Al-4V

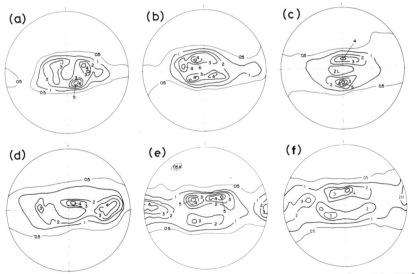

Fig.2 (0002) pole figures of material A rolled at (a) 400°C, (b) 500°C, (c) 700°C, (d) 750°C, (e) 800°C and (f) 850°C.

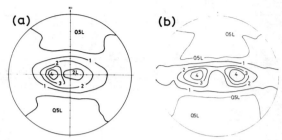

Fig.3   (0002) pole figures of material B and C rolled at 700°C.
(a) material B, (b) material C

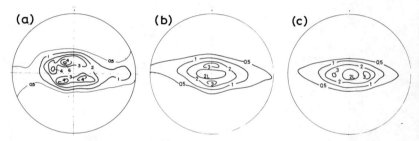

Fig.4   The effect of reduction on the hot rolled texture of
c.p.Ti.   (0002) pole figures of material A rolled (a) 39%,
(b) 73.2% and (c) 86.4% at 500°C.

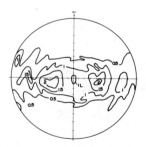

Fig.5   The effect of annealing on the hot rolled texture of
c.p.Ti.   (0002) pole figure of material A rolled at 700°C
followed by annealing at 700°C.

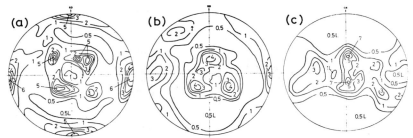

Fig.6  (0002) pole figures of material A cold rolled (a) 11.2%, (b) 20.5% and (c) 34.4%.

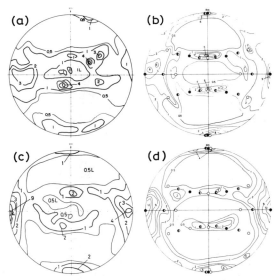

Fig.7  (0002) pole figures of material A and B rolled at temperatures in β-region.  (a) material A rolled at 900°C, (b) material A rolled at 950°C, (c) material B rolled at 900°C and (d) material B rolled at 950°C. (110)β poles with following orientations belonging to (001)±α° [110] texture are also indicated in (b) and (d).

●:(001)[110]   α=0°,   ◑:(1̄14)[110]   α=19.5°,   ○:(1̄12)[110]   α=35.3°

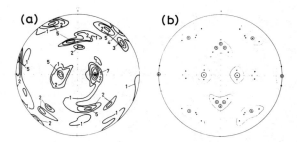

Fig.8  Observed and predicted transformation textures of material B.
(a) Observed (0002) pole figure of material B heated at 1050°C  and
(b) predicted (0002) pole figure based on Burgers' relationship for
the initial orientation of (0002)±30°[11$\bar{2}$0].  The encircled points
indicate superposition of 2 or 8 poles with the number increasing with
size of circle.

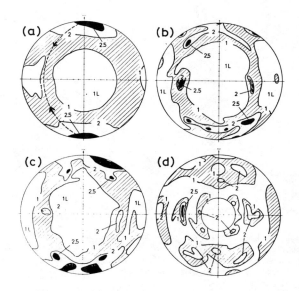

Fig.9  Rotation about c-axis during heating.  (10$\bar{1}$0) pole figures
of material B heated at (a) 700°C, (b) 800°C, (c) 850°C and   (d)
900°C.  The dotted line in (a) represents the rotation about c-axis.

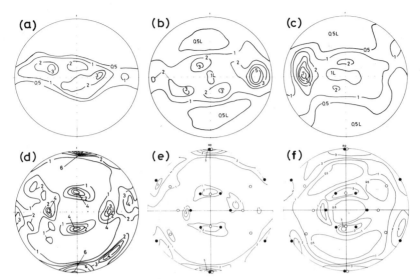

Fig.10   (0002) pole figures of Ti-6Al-4V rolled at (a) 500°C, (b) 800°C,
(c) 850°C, (d) 900°C, (e) 1000°C and (f) 1100°C.
(110)$_\beta$ poles with following orientations belonging to (001)±α°[110] texture are also indicated in (e) and (f).
O:($\bar{1}$12)[110] α=35.3°,  ●:($\bar{1}$11)[110] α=54.7°

## References

1.  C. J. McHargue and J. P. Hammond : J. Met., 57 (Jan. 1953)
2.  D. L. Dull and M. F. Amateau : Tech. Doc. News, No. 1925-13 (Feb. 1970)
3.  F. A. Crossley : ASM Tech. Paper, 1970 Western Metal and Tool Conference
    and Exposition, 10-13 (Mar. 1969)
4.  V. V. Mukhayev, R. A. Adamesku and P. V. Gel'd : AN SSSR. Izvestiya. Metally,
    No. 6, 98 (1968)    Translated Report No. FDT-HT-23-41-70
5.  J. H. Keeler and A. H. Geisler : J. Met., 80 (Feb. 1956)
6.  A. D. Hatch : Trans. AIME, 233, 44 (Jan. 1965)
7.  F. R. Larson, A. Zarkades and D. H. Avery : Titanium Science and Technology,
    Plenum Press, New York-London, 2, 1169 (1973)
8.  H. Conrad, M. Doner and B. de Meester : Titanium Science and Technology,
    Plenum Press, New York-London, 2, 969 (1973)
9.  Hsun Hu and R. S. Cline : Trans. Met. Soc. AIME, 242, 1013 (June 1968)
10. S. Sasano, S. Komori and H. Kimura : J. Japan Inst. Met., 38(3), 199 (1974)

# CONTROLLED TEXTURED TITANIUM ALLOYS - SOME COMMENTS ON THEIR PRODUCTION AND INDIRECT ASSESSMENT

A. W. Bowen

Materials Department, Royal Aircraft Establishment,
Farnborough, Hants., GU14 6TD, UK

## Introduction

In an earlier paper [1] the importance of crystallographic texture in the characterization of α-based titanium alloys was emphasized and the discussion indicated that titanium metallurgy could be advanced significantly if texture was controlled in a consistent and predictable manner. It must be admitted at the outset that this is no easy task [2]; nevertheless, the problems posed at present by uncontrolled textures (for example, scatter in all mechanical properties [3], one vivid example of which can be found in Fig 4 of [4]), should act as a considerable incentive to control this parameter.

To be in a position to specify a particular texture it is first of all necessary to know the types of textures likely to be produced during thermo-mechanical processing and heat treatment. Zarkades and Larson [2,5] have shown that there are five major types of (0002) textures for α-based titanium alloy sheet and plate. A relatively wide range of mechanical properties has been measured for most of these textures [2], and hence it is not too difficult, knowing the properties that are required, to prescribe an appropriate texture. Closer study of these textures and their properties, however, indicates that the situation is simplified somewhat since three of the five textures would seem to cover most of the property requirements demanded of α-based titanium alloys. These are: basal; edge; and random (Fig 1).

At present the only texture capable of production on a commercial scale is that of the basal type [6]*. But recent work by Sommer and Craeger [8] has indicated the conditions necessary to produce the edge texture which has been observed on occasions in some UK [9] and US [10] commercially processed plate. Furthermore, Sommer and Craeger [8] and others [11,12] have indicated some heat treatments likely to produce random textures; similar isotropic textures can apparently be produced by β working [2,13].

If we consider this work in an optimistic light we may already be in a position to produce all three of the above textures consistently and reproducibly. One may then ask: How can they be assessed? A pole figure, although not unambiguous since it is two-dimensional, would show the three textures quantitatively. However, there is a reluctance to use this technique of texture determination in a routine way, possibly on the grounds of cost, time and expertise [14,15]. This is unfortunate since the judicious use of the time and effort spent in crystallographic analysis of α-based titanium alloys will, in many instances, be well worthwhile; it may also resolve anomalies [1]. But for the three textures shown in Fig 1 pole figure analysis is not imperative if only an indirect indication of texture is required – and this situation might arise, for example, in the quality assurance of sheet or

---

*The only publicized commercial applications of such textures are for pressure vessels [2] and a helicopter application [7].

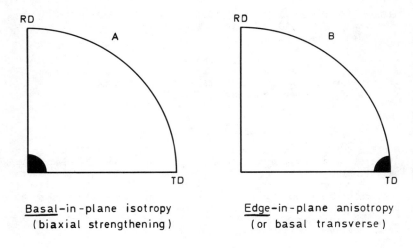

Basal-in-plane isotropy
(biaxial strengthening)

Edge-in-plane anisotropy
(or basal transverse)

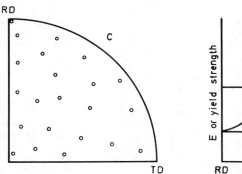

Random - isotropy

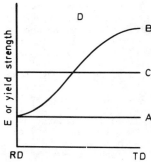

Fig 1    (a)-(c)  The 3 (0002) pole figures likely to satisfy the
         property requirements of α-based titanium alloys.
         (d) Variations in Young's modulus E and yield strength for
         textures (a) to (c). (RD and TD indicate rolling and
         transverse directions respectively.)

plate produced to a specified texture. For the relatively simple textures considered here the relationship between Young's modulus E and the basal pole orientation γ can be utilized to show the texture qualitatively, as will be described below. However, it cannot be stressed too strongly that this relationship should not be used as a substitute for (0002) pole figures, for reasons given later, nor should it detract attention from the determination of pole figures themselves, and their analysis in terms of crystallite orientation distribution functions, charts for which are already available for h.c.p. metals [16].

## Background

The relationship between E and γ for h.c.p. metals [17] shows that E is symmetrical about the c-axis of the hexagonal cell and can be related to γ by:

$$\frac{1}{E} = S_{11}(1-\cos^2\gamma)^2 + S_{33}\cos^4\gamma + (2S_{13} + S_{44})\cos^2\gamma\,(1-\cos^2\gamma) \qquad (1)$$

where $S_{11}$ etc are the standard compliance functions. If values for these compliances [18,19] are substituted into equation (1) it reduces to:

$$\frac{1}{E} = 9.97\; -2.4\cos^2\gamma - 0.67\cos^4\gamma \quad \text{(Flowers et al [18])} \qquad (2)$$

and

$$\frac{1}{E} = 9.59 - 1.56\cos^2\gamma - 1.01\cos^4\gamma \quad \text{(Fisher and Renkin [19])} \qquad (3)$$

in units of $m^2/TN$.

The variation in E values given by equations (2) and (3) is shown in Fig 2, and although large is, in fact, one of the smallest exhibited by h.c.p. metals because of the near ideal c/a ratio of α-titanium [20]. Rigorous application of equations (2) and (3) is limited to single crystals of unalloyed titanium, but in spite of this, these equations can still be used to predict the approximate behaviour of α-based titanium alloys because: (a) polycrystalline behaviour can be accommodated by assuming constant stress (which is the more convenient) or constant strain between adjacent grains [21,22]. (b) the elastic constants, at least for the Ti-6Al-4V alloy, are more or less identical with those of α-titanium [23]. (c) the background of randomly oriented grains and the presence of second phase particles can, to a first degree, be neglected.

## Method and Discussion

The present proposal is somewhat different from those given earlier [11,24-27]; it can, for instance, complement non-destructive evaluation (NDE) techniques [14,15,26,27]. It involves simply reversing equations (2) and (3) so that γ can be determined from the E values. This is most conveniently carried out by first registering the E values on a standard stereographic projection, as follows. For a longitudinal test E will be a maximum at L, Fig 3 (when stressing is normal to the basal plane) and gradually decrease according to equations (2) and (3) until it is a minimum along ST, Fig 3

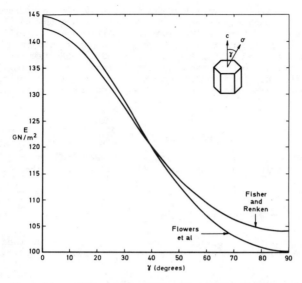

Fig 2    Variation of Young's modulus E with basal pole orientation γ
using the single crystal data of Flowers et al [18] and
Fisher and Renkin [19].

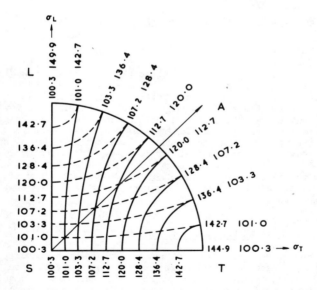

Fig 3    Stereographic projection showing great circles of equal modulus
(in GN/m²) for longitudinal (dotted lines) and transverse
(solid lines) tests. (L, S and T denote longitudinal, short
transverse and transverse directions respectively.)

(when stressing is parallel to the basal plane)(see dotted lines, Fig 3).
Likewise, for a transverse test E will show a similar trend with a maximum at
T, Fig 3 and a minimum along SL, Fig 3 (see solid lines, Fig 3). Along SA E
will be equal for both tests. Over most of the range of $\gamma$ values equations (2)
and (3) will yield very similar results, and in a recent study equation (3),
although the lesser used equation in the literature, was found to represent
more closely the texture changes due to heat treatment of the Ti-6Al-4V alloy
[12].

Thus the orientation of the basal pole peak in an $\alpha$-based titanium alloy
can be located if longitudinal and transverse tests are carried out (ie at 90°
to one another) and E values determined: it will lie at the intersection of the
great circle of each of the respective E values. Examples illustrating the
method are shown schematically in Fig 4 for two of the three conditions
considered here. The only texture requiring additional comment is that of the
third type – random – since this illustrates a limitation of the proposed
method. Random textures are likely to be characterized by E values of
~115 GN/m² [12,20]. Clearly this is not a single basal pole peak situation,
and yet it is possible to interpret this as a strong basal pole orientation
near Fig 4(e-f) (one could differentiate between these two types of textures
either by using a NDE technique to determine whether there are any
concentrations of basal poles (as there would be in Fig 4(e-f), or by further
measurement of E at other angles. Note, however, that the overall mechanical
properties for both textures would be fairly similar).

Such ambiguity can arise when there is more than one basal pole peak
because this technique cannot distinguish between a true single basal pole peak
and a summation of peaks. As an example of this, consider the results obtained
by Olsen and Moreen [28]. These workers summed a weighted mean of all
orientations obtained from a pole figure and arrived at a single 'summation'
peak which could be assigned a single E value, rather than a number of
individual peaks all with different E values. Thus a complex texture of known
orientation can always be related to a single E value, but the process can be
reversed only for simple single basal pole textures. And in these situations
comparison with the limited data in the literature, particularly the review by
Zarkades and Larson [29], shows that the proposed method does predict (0002)
pole figures (but not any others) relatively accurately and consistently.

This technique, then, is seen as a means of monitoring, not only single
basal pole peak textures, but also random textures of $\alpha$-based titanium alloys
by specifying, for example, upper and lower limits to E values determined during
quality assurance. It should also prove useful as a quick guide to indicate
whether edge or basal textures have changed during heat treatment [12]. For
ease of interpretation it should be limited to sheet or plate, but this is not
seen as a serious disadvantage in view of the extensive use made of these types
of products. In the case of complex shapes, probably the best that can be
achieved at present is to attempt to arrive at the required texture only in
critical areas; unless one can utilize, for instance, the diffusion bonding of
textured sheet.

Unfortunately, textures more complex than those considered here are only
too possible in current practice and unless one is deliberately producing one
of the textures dealt with in this paper, caution must be exercised in the
extension of this technique to general use. Hopefully, if some urgency is
attached to the translation and extension of the work of Sommer and Craeger [8]
on to a commercial scale it should be possible to demonstrate clearly that the
mechanical properties of a controlled textured $\alpha$-based titanium alloy are much
more closely defined and reproducible than those of the corresponding uncontrolled textured titanium alloy – merely a reduction in scatter of properties would

Stereographic plot of E
values for longitudinal
and transverse tests

(0002) pole figures

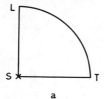

a

$E_L = 100.3 \text{ GN}/\text{m}^2$;
pole peak must lie on ST
$E_T = 100.3 \text{ GN}/\text{m}^2$;
pole peak must lie on LS
hence peak is at S

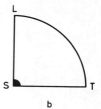

b

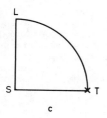

c

$E_L = 100.3 \text{ GN}/\text{m}^2$;
pole peak must lie on ST
$E_T = 144.9 \text{ GN}/\text{m}^2$;
pole peak must lie at T
hence peak is at T

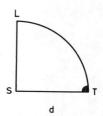

d

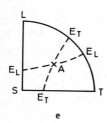

e

$E_L = E_T = 112.7 \text{ GN}/\text{m}^2$;
hence pole peak lies at
intersection of $E_L$ and
$E_T$ great circles as
shown at A

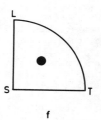

f

Fig 4   Schematic representation of method for determining orientation of (0002) pole peak from E values measured in longitudinal ($E_L$) and transverse ($E_T$) tests.

be a significant achievement; compare, for example, the fatigue results for alloys showing consistent [30] and variable [31] textures. A much more reliable data base will then become available, thus enabling further design exercises to be carried out, such as those reported recently by Sommer et al [32].

## References

1.   A. W. Bowen:  Scripta Met, 11(1977), 17.
2.   F. R. Larson and A. Zarkades:  Metals and Ceramics Information Centre Report 74-20(1974).
3.   Boeing SST programme, quoted in [2].
4.   M. J. Harrigan:  Met Eng Q, 14(1974) (4), 16.
5.   A. Zarkades and F. R. Larson:  Titanium Science and Technology, 2(1973), 1321.  Plenum Press, New York.
6.   S. F. Frederick and G. F. Lenning:  Met Trans, 6B(1975), 601.
7.   Technology **Forecasts:** Met Prog, 111(1977)(1), 36; 113(1978)(1), 38.
8.   A. W. Sommer and M. Craeger:  AFML-TR-76-222(1977).
9.   A. W. Bowen:  Microstructure and Design of Alloys, 1(1973), 446. Inst. of Metals, London.
10.  R. G. Baggerly:  Boeing Document D2-36002-1, Aug(1964).
11.  A. Zarkades and F. R. Larson:  AMMRC-TR-69-32(1969).
12.  A. W. Bowen:  Met Sci and Eng, 29(1977), 19.
13.  M. J. Blackburn, J. A. Feeney and T. R. Beck:  Corrosion Science and Technology, 3(1973), 67.  Plenum Press, New York.
14.  B. R. Tittman and G. A. Alers:  Met Trans, 3(1972), 1307.
15.  B. R. Tittman, G. A. **Alers** and L. J. Graham:  Met Trans, 7A(1976), 229.
16.  G. J. Davies, D. J. Goodwill and J. S. Kallend:  J App. Cryst., 4(1971), 193.
17.  E. Schmid and W. Boas:  Plasticity of Crystals, (1950), 21. F. A. Hughes and Co Ltd, London.
18.  J. W. Flowers Jr, K. C. O'Brien and P. C. McEleney:  J. Less Common Metals, 7(1964), 393.
19.  E. S. Fisher and C. J. Renkin:  Phys Rev, 135A(1964), 482.
20.  W. T. Roberts:  J. Less Common Metals, 4(1962), 345.
21.  G. A. Alers and Y. C. Liu:  Trans TMS-AIME, 236(1966), 482.
22.  Y. C. Liu and G. A. Alers:  ibid, 489.
23.  F. R. Larson:  AMRA TR-65-24, Oct(1965).
24.  M. J. Harrigan, A. W. Sommer and G. A. Alers:  North American Rockwell Corpn Rpt NA-69-909(1969).
25.  M. J. Harrigan, A. W. Sommer, P. G. Reimers and G. A. Alers:  ref [5], 2, 1297.
26.  C. Feng:  Lockheed Georgia Company Report RM-308 June(1968).
27.  S. F. Frederick:  AFML-TR-73-265(1973; see also Mater. Eval,(1975) Sept, 213.
28.  R. H. Olsen and H. A. Moreen:  Met Trans, 4(1973), 701.
29.  A. Zarkades and F. R. Larson:  AMMRC-TR-71-60(1971).
30.  F. R. Larson and A Zarkades:  Texture and the Properties of Materials (1976), 210.  Metals Soc, London.
31.  A. W. Bowen:  ref [5], 2, 1271.
32.  A. W. Sommer, M. Craeger and S. Fujishiro:  J. of Metals, 28(1976)(12), A36.

## Acknowledgements

Crown Copyright  © Controller HMSO LONDON 1980.

# RECRYSTALLIZATION TEXTURE DEVELOPMENT IN
# A COMMERCIAL PURITY TITANIUM THIN SHEET

*S.NAKA, **R.PENELLE, *R.VALLE and **P.LACOMBE

*Office National d'Etudes et de Recherches Aérospatiales
29, Av. de la Division Leclerc, 92320 Chatillon France
**Laboratoire de Métallurgie Physique, Bat 413, Université
de Paris-Sud, Centre d'Orsay, 91405 Orsay France

## 1. Introduction

The purpose of this study was to identify the recrystallization mechanisms of the commercial purity titanium (T 40) which lead to different
recrystallization textures. This investigation took into consideration
different values of the amount of reduction by cold-rolling, the temperature and the time of annealing.

The problem was approached in two different ways. On a statistical
view-point, the three-dimensional analysis of textures was used to
describe quantitalively the texture evolution during the primary recrystallization. The orientation distribution fonction (ODF) was calculated
for each metallurgical state : in the as-rolled state and after annealing
for different values of the temperature and the time.

On a microscopic scale, the structural evolution was observed
directly during in-situ annealing performed in the high voltage
electron microscope (1MV). These observations were made on specimens
cold-rolled to various degree of deformation. The observed behaviour was
correlated to the results obtained by the three-dimensional analysis of
textures, confirming the recrystallization mechanisms suggested by this
analysis [1].

## 2. Experimental

### 2.1. Material

A titanium thin sheet of commercial purity (Table 1) was unidirectionally cold-rolled without reversing nor lubrication to 30 %, 40 %,
60 %, 80 % and 90 % reduction (for a specimen thickness t, the amount
of reduction by cold rolling is given by the relation $(1-t/t_0) \times 100$
where $t_0 = 3.3$ mm is the initial thickness).

Table 1

| Type | Thickness (mm) | Chemical composition (% wt) | | | | | | Grain size (μm) |
|------|------|------|------|------|------|------|------|------|
| | | C | N | H | O | Fe | Ti | |
| UT 40 | 3.3 | 0.08 | 0.06 | 0.01 | 0.25 | 0.25 | 99.35 | 30 |

## 2.2. Textures analysis

Annealing treatments were performed in high vacuum ($10^{-6}$Torr) in the temperature range : 450°C - 850°C. For each temperature, the annealing time after the heating-up period varied from 1 to $10^4$ minutes.

The pole figures which are necessary for the calculation of ODF were obtained by X-ray diffraction using the reflection-transmission method. The specimens were chemically polished to a thickness approaching 0.1 mm. In the particular case of titanium, the CuKα radiation was used for the reflection part of the {10$\overline{1}$0} pole figure in order to minimize the defocussing effect. The MoKα radiation was used in reflection-transmission for {0002} , {10$\overline{1}$1} and {10$\overline{1}$2} pole figures and only in transmission for the {10$\overline{1}$0} pole figure. It is from the complete normalized pole figures possessing the orthotropical symmetry that the ODF was calculated for a 16 th order truncation /2/.

## 2.3. In-situ annealing tests

In-situ annealing tests were performed in the 1MV high voltage electron microscope. The influence of the oxygen content on the recrystallization temperature of titanium is well known. In the case of in-situ annealing, the observation at high temperature (600°C - 800°C) may last from several minutes to an hour, thereby necessitating a high vacuum to minimize the formation of a metal oxygen solid solution. The highest admissible pressure of oxidizing agents (oxygen and water vapour) was determined through the investigation of the oxidation process at low pressure and high temperature. The required pressure (5 x$10^{-8}$Torr) has been attained in the high vacuum specimen chamber designed at ONERA /3/. Static recordings from the same area at different times were made on photographic films. A low light level and high resolution video system /4/ was used for a continuous recording of the recrystallization process.

Thin foils for the in-situ annealing test were prepared not only from the conventional rolling plane section, but also from the transverse section /5/.

## 3. Results

### 3.1. Statistical approach

From the direct pole figures determined by X-ray diffraction, the volume fraction F(g) of crystallites possesing a given orientation (g) is calculated by comparison with a random orientation distribution sample of the same material :

$$\frac{dV}{V_o} = K \ F(g) \ dg \qquad (1)$$

where F(g) is the orientation distribution function (ODF)
  g = $\psi$, $\theta$ , $\phi$ = Euler's angles, rotation which relates the sample axes to the crystal ones
  Vo the total volume of the sample
  dg = d$\psi$ sin $\theta$ d$\theta$ d $\phi$
 and K = $1/8\pi^2$ a normalization constant

The principle of ODF calculation which was used is briefly recalled below ; The experimental pole densities q $(\eta, \chi)$ are related to the ODF F(g) through the basic relation.

$$q\ (\eta, \chi) = \frac{1}{2\pi}\ \int\ F(g)\ d\gamma \qquad (2)$$

where $\eta$ and $\chi$ are the spherical coordinates of the normal to the diffracting plane in the crystal axes and $\gamma$ the rotation about this normal.

Then the pole densities are expanded as a series of surface spherical harmonics $Y_1^m(\eta, \chi)$,

$$q\ (\eta, \chi) = \sum_{l=o}^{\infty}\ \sum_{m=-1}^{+1}\ Q_{lm}\ Y_1^m(\eta, \chi) \qquad (3)$$

and F(g) is expanded as a series of generalized spherical harmonics,

$$F\ (\psi, \theta, \phi) = \sum_{l=o}^{\infty}\ \sum_{m=-1}^{+1}\ \sum_{n=-1}^{+1}\ f_{lmn}\ T_{mn}^l(\psi, \theta, \phi) \qquad (4)$$

where $T_{mn}^l = e^{-im\psi}\ Z_{mn}^l(\cos \theta)\ e^{-in\phi}$.

By substituting (3) and (4) in (2) and by integrating, the following linear relation is obtained :

$$Q_{lm} = 2\pi\ (\frac{2}{2l+1})^{\frac{1}{2}}\ \sum_{n=-1}^{+1}\ f_{lmn}\ Y_1^{*n}(\alpha, \beta) \qquad (5)$$

where $\alpha$ and $\beta$ are the spherical coordinates of the normal to the diffracting plane in the crystal axes.

The coefficients $Q_{lm}$ are obtained from the pole densities, so by using the relation (5), it is possible to calculate the coefficients $f_{lmn}$ by resolution of the system of linear equations, then the ODF can be computed.

Evolution of rolling texture
------------------------------

For the various amounts of reduction by cold-colling al ready mentioned, the main component (2115) $\underline{/0110/}$ of the rolling texture is stable in orientation and becomes more pronounced for high amounts of reduction (Fig. 1). So, for specimens cold-rolled to 90 % of reduction, the ODF may attain a maximum value of 12.

Evolution of recrystallization texture
----------------------------------------

The variation of F(g) as a function of the annealing temperature is represented on Fig. 2 in the case of 80 % of reduction and annealing time of 10 minutes. For specimens annealed at low temperature (up to 550°C), the recrystallization texture is similar to the rolling texture. On the contrary, for specimens annealed at high temperature, the main component of the recrystallization texture is found to be near the ($\bar{1}$103) $\underline{/11\bar{2}0/}$ orientation.

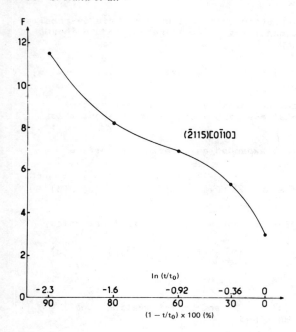

Fig. 1 Evolution of the value of ODF for the component of the rolling texture as a function of the amount of reduction.

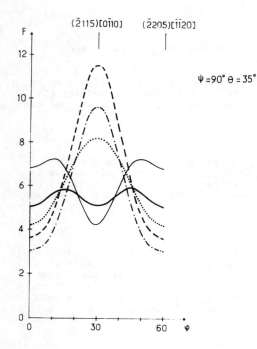

Fig. 2 Evolution of the components $(\bar{2}115)$ $\underline{/0\bar{1}10/}$ and $(\bar{2}205)$ $\underline{/\bar{1}\bar{1}20/}$ for the samples rolled to 90 % as a function of annealing temperature : --- rolled to 90 %; -·-· 450°C,10 mn; ... 550°C,10 mn; —— 700°C, 10 mn; — 800°C,10 mn.

The influence of the amounts of reduction by cold-rolling on the recrystallization texture was also investigated in the range from 30 % to 90 %. The annealing conditions were as follow s  : 10 minutes at a temperature of 800°C (Fig. 3).

For samples cold-rolled to 30 % of reduction, the recrystallization texture is similar to the rolling texture, but for specimens deformed to 90 %, the major component of the recrystallization texture is near the (1̄103) /112̄0/ orientation.

Thus, a low amount of reduction by cold-rolling associated to a high annealing temperature has the same effect on the recrystallization texture as a low annealing temperature associated to a high amount of reduction. In both cases, the rolling texture is maintained. This may be attributed either to the so-called in-situ recrystallization phenomenon /6/ i.e. the growth of subgrains inside "hard" grains /7/ having a high Taylor factor M, or to the strain induced boundary migration (S.I.B.M.) of a hard grain /8/.

The formation of the (1̄103) /112̄0/ component may be attributed to the growth of pre-existing nuclei. This assumption is confirmed by the fact that this component is already important in the rolling texture i.e. F(g) _ 4 (Fig. 2). This component may be supposed to develop at the expense of the (2̄115) /011̄0/ component, these two components being related by a rotation of 30° about the /0001/ direction.

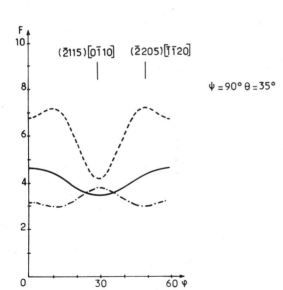

$\psi = 90° \theta = 35°$

(2̄115)[011̄0]    (2̄205)[1̄1̄20]

Fig. 3 Evolution of the components (2̄115) /011̄0/ and (2̄205) /1̄1̄20/ for the samples rolled to

30%  - · - · -
60%  ———
90%  - - - -

and annealed at 800°C for 10 mn.

## 3.2. Microscopic approach

Three dimensional analysis of textures and electron microscope observations have shown that the recrystallization mechanisms were similar in thin sheets deformed to 60 % or 80 % of reduction. In specimens cold-rolled to 80 % reduction and prepared from the rolling plane section, the microstructure is indistinct and the grain boundaries are hardly visible (Fig. 4-a). This may be attributed to the high dislocation density but also to the nature of the sectioning which leaves most of the cell boundaries parallel to the foil. However, a well defined cell structure was observed in specimens prepared from sections orthogonal to the sheet plane after 60 % of deformation. These small size cells were elongated in the rolling direction with a cell thickness of approximately 0.2 um (Fig. 5-a).

In specimens prepared from the rolling plane section, the formation of a new grain was observed in the region (M) after in-situ annealing at 600°C. In the present case, it was not possible to elucidate the mechanism of formation of this new grain (G) (Fig. 4-b). However, in specimens prepared from sections orthogonal to the sheet plane, the recrystallization process observed in the microscope may be attributed to a subgrain growth mechanism (FiG. 5-b), but the subgrain coalescence mechanism was not observed.

The recrystallized grains observed during the in-situ annealing tests were comparable in size with the grains observed in bulk annealed samples. The specimens of the in-situ annealing were approximately 5000 - 10000 Å in thickness. Such a thickness is superior to the limit proposed by ROBERTS and LEHTINEN under which the recrystallization phenomenon may be inhibited by surface effects such as anchoring of grain boundaries on the grooves generated by surface diffusion during heating (thermal grooving) /9/.

In specimens cold-rolled to 40 % of reduction, the grain boundaries are still visible (Fig. 6). The strain induced boundary migration was observed during the in-situ annealing tests performed in the microscope. The bulging out boundary observed after an annealing time of 10 minutes at 750°C is quite typical of the S.I.B.M. mechanism /10/. A polygonized sub-structure may also be observed on Fig. 6. The formation of a new grain at the triple point junction was observed in the same area. According to DOHERTY and CAHN /11/, the formation of this grain may also be attributed to the S.I.B.M. and may be explained as follows . After deformation, the grains are elongated in the rolling direction. The equilibrium angle of 120° at the triple point junction will be re-established with a reduction in total grain boundary energy. The subgrains involved in this process may become potential nuclei for the S.I.B.M.

Fig. 7 shows the microstructure of another specimen, after 40 % deformation. During in-situ annealing at a temperature of 800°C (Fig. 8), the gradual rearrangement of the sub-structure was observed inside each deformed grain. This observation may be attributed to the so-called in-situ recrystallization process.

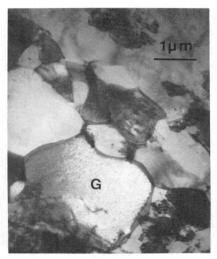

(a)

Typical microstructure of the
80% as-rolled specimen,
prepared from the rolling
plane section.

Fig. 4

(b)

The same area as (a) after in-
situ annealing at 600°C.

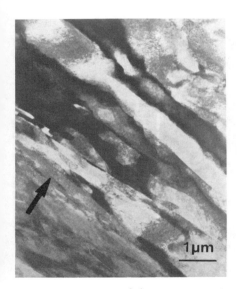
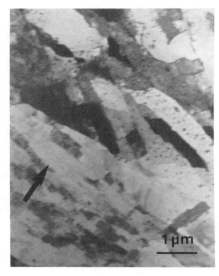

(a)

Typical microstructure of the
60% as-rolled specimen,
prepared from the transverse
section.

Fig. 5

(b)

The same area as (a) after in-
situ annealing at 700°C.

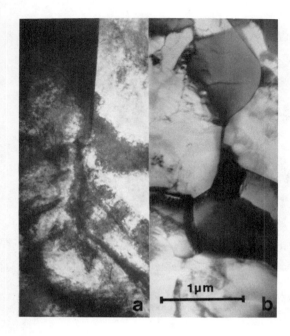

Fig. 6
(a) Typical microstructure
    of the 40% as-rolled
    specimen.
(b) The same area as (a)
    after in-situ annealing
    at 750°C.

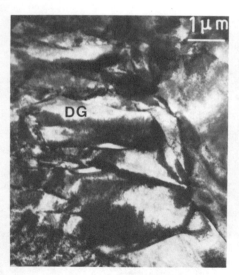

Fig. 7   Another microstructure of
the 40% as-rolled specimen.

Fig. 8   The same area as Fig. 7
after in-situ annealing at 800°C.

## 4. Conclusion

The following conclusions may already be derived from the complementary results obtained through the three-dimensional analysis of texture and the in-situ annealing tests performed in HVEM on specimens prepared from both the rolling plane section and the transverse section.

The primary recrystallization of alpha titanium is controlled by the growth of pre-oriented nuclei. This process is in good agreement with the fact that all the major orientations after annealing were already represented in the as-rolled state. Depending on the amount of reduction by cold-rolling and the annealing temperature, the following cases may be considered :

— For deformations superior to 60 % of reduction and annealing temperatures over 550°C, the recrystallization phenomenon is essentially controlled by cell growth. In this case, the major component of the recrystallization texture is near the $(\bar{1}103)$ $/11\bar{2}0/$ orientation.

— For deformations inferior to 50 % of reduction and annealing temperature over 550°C, the recrystallization process may be attributed either to the strain induced boundary migration or to the in-situ recrystallization. In the present case, the rolling texture and the recrystallization texture are similar i.e. $(\bar{2}115)$ $/0\bar{1}10/$. It shoud be noted that after 80 % of deformation and annealing at lower temperatures than 550°C, the recrystallization texture was also similar to the rolling texture.

### References

1.    S. NAKA : Thesis of Docteur-Ingénieur, Orsay (1978).
2.    P. DERVIN, J.P. MARDON, M. PERNOT, R. PENELLE and P. LACOMBE : J. Less Common Metals, 55 (1977), 25.
3.    R. VALLE, B. GENTY, A. MARRAUD et P. REGNIER : Proc. 5th Int. Conf. HVEM, Kyoto (1977), 137.
4.    R. VALLE, B. GENTY, A. MARRAUD et B. PADRO : Proc. 5th Int. Conf. HVEM, Kyoto (1977), 163.
5. a W.B. HUTCHINSON and R.K. RAY : Phil. Mag., 28 (1973), 831.
   b R.K. RAY, W.B. HUTCHINSON, F.M. BESAG and R.E. SMALLMAN : J. Microscopy, 97 (1973), 217.
6.    C. CRUSSARD : Mém. Sci. Rev. Métal., 41 (1944), 118.
7.    R. PENELLE, J. MILLON et P. LACOMBE : 7ème Colloque de Métallurgie, Ecrouissage, Restauration, Recristallisation, Paris (1963), 153.
8.    I.L. DILLAMORE and H. KATOH : Met. Sci., 8 (1974),73.
9.    W. ROBERTS and B. LEHTINEN : Phil. Mag., 26 (1972), 1153.
10.   P.A. BECK and P.R. SPERRY : J. Appl. Phys., 21 (1950), 150.
11.   R.D. DOHERTY and R.W. CAHN : J. Less Common Metals, 28 (1972), 279.

# Texture Hardening of Ti-6Al-4V Alloy --Formation of Basal Plane Texture by Dentate Rolling

Y. Murayama, Y. Suzuki and M. Shimura

The Research Institute for Iron, Steel
and Other Metals, Tohoku University
Sendai, Japan

## Introduction

For biaxially stressed applications such as pressure vessels, a sheet texture with the basal plane poles normal to the plane of the sheet would be very desirable. Such a basal plane texture shows "texture hardening" and very high biaxial strengths are able to be realized. For Ti-6Al-4V alloy, there is possibility of obtaining much higher biaxial strengths by texture control. However, the sheet texture of Ti-6Al-4V from commercial heats with standard processing shows a split basal pole with considerable scatter in both rolling and transverse directions. Frederick and Lenning [1] produced the basal plane texture by cross rolling and round rolling in the $\alpha+\beta$ field. However, in both procedures, the work piece must be rotated every pass, 45deg for round rolling and 90deg for cross rolling. Moreover, both procedures have the demerits not to be able to take the large width and length of sheets and to be difficult to produce the flat sheets.

Kumazawa and et al. [2] indicated that unidirectional rolling with both dentate roll and conventional smooth roll has the same effect as multidirectional rolling for the (100)[011] texture formation of silicon steel sheets. This procedure has not above demerits revealed in cross rolling and round rolling, because of unidirectional rolling. In the present investigation, this procedure was applied to the formation of the basal plane texture in Ti-6Al-4V alloy sheets.

In general, the influential factors on the texture formation in Ti-alloy sheets are mainly (1) rolling procedure, (2) prior texture, (3) rolling temperature, (4) total reduction and (5) alloying elements. In the present investigation, the texture development in Ti-6Al-4V alloy sheets produced by unidirectional rolling with both dentate roll and smooth roll was compared with one by conventional rolling and cross rolling. The texture development was evaluated by the (0002) plane pole figure and the mechanical properties under biaxial stresses were deduced by uniaxial tensile tests, Knoop hardness tests and through-thickness compression tests.

## Procedure

In order to remove the effect of prior texture, the specimens, which were cut from the 6mm thickness sheets produced by standard processing, were vacuum annealed for 20min at 1030°C, bofore rolling. The chemical compositions of the specimens are shown in Table 1.

Table 1   Chemical compositions

| Elements | Al | V | N | O | C | H | Fe |
|----------|------|------|--------|-------|-------|-------|-------|
| wt pct | 6.02 | 3.98 | <0.001 | 0.079 | 0.003 | 0.011 | 0.099 |

The schematic presentation of the unidirectional rolling procedure with both dentate roll and smooth roll (dentate rolling), is shown in Fig. 1-a.   The 2-high rolling mill with 130mm diam. rolls, whose top roll is a dentate roll and bottom roll is a conventional smooth roll, was used. In Fig. 1-b, the surface profile of the dentate roll is also represented. The numerous V-shape grooves were notched in the circumferential direction. The work piece was turned upside-down every pass.   At the same time that the top surface was indentated by the top roll, the bottom surface was smoothed by the bottom roll, so that the sheet thickness decreased gradually.   After such a process of dentate rolling, the sheets were rolled to

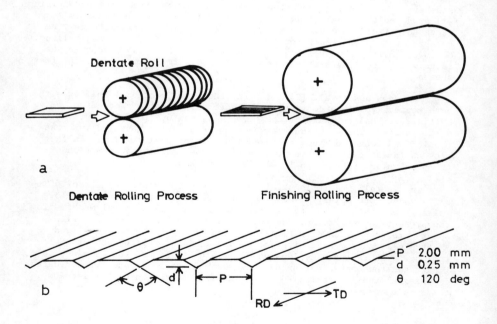

Fig. 1.   Schematic presentation of a: dentate rolling process and b: surface profile of dentate roll.

the required thickness by the 2-high rolling mill with 250mm diam. smooth rolls.  This finishing rolling process is simultaneously the smoothing process to flatten the sheets which have the numerous straight projection on one side.  In the dentate rolling process, such a smoothing process is important for obtaining the same effect as multidirectional rolling, as well as the process using the dentate roll [2].  In the present investigation, the reduction for finishing rolling process was limited within the range from 30pct to 40pct.  Conventional rolling and cross rolling were performed on the 2-high rolling mill with 250mm diam. smooth rolls.

The specimens, vacuum annealed for 20min at 1030°C, were rolled by above three rolling procedures respectively at the temperatures of 550°C, 650°C, 750°C and 900°C.  The specimens rolled 82pct were vacuum annealed for 2hr at 700°C respectively, and then they were offered to evaluate the texture development and to carry out the various mechanical tests.

The (0002) plane pole figures for the midplane of the sheets were determined by the reflection technique.  Chromium $K\alpha$ radiation was used with vanadium filter.  The specimens cut from the sheets were polished mechanically and chemically repeatedly to the final thickness of 0.085mm. The exposure by the reflection technique covered the $\alpha$ range ($\alpha$ is the angle between the rolling plane and the (0002) plane pole) from 65deg to 90deg.  Normalization of intensities was achieved by comparison with the reflection intensity from a random powder sintered sample.

The uniaxial tensile tests were carried out in order to determine the uniaxial yield strengths and the r values in the direction of rolling and transverse respectively.  The specimen size follows the JIS No. 7 specimen.  The r value was determined from $\varepsilon_1$ and $\varepsilon_w$, which were measured from the specimen size after extension up to $\varepsilon_1 = 0.08$.  The yield locus was derived from these uniaxial yield strengths and the r values based on Hill's anisotropic yield criterion [3].

The results of the Knoop hardness tests [4] made on three orthogonal surfaces were represented on the yield locus diagram based on Hill's anisotropic theory, so that the biaxial yield strength was deduced in such a manner.

Moreover, the through-thickness compression tests were carried out in order to estimate directly the effect of "texture hardening".  The specimens were cut from the textured sheets, and then they were machined into the shape of 5mm squares and 1mm height (as-rolled thickness).  The compression direction was normal to the rolling plane and two sides of the specimen were perpendicular to the rolling and transverse directions respectively.  Johnson wax No. 111 was used for lubrication.

### Result and Discussion

Fig. 2 shows the (0002) plane pole figures of the sheets, which were rolled 82pct at 550°C by three different rolling procedures respectively. The specimen, which was vacuum annealed for 20min at 1030°C before rolling, displayed less intensity of basal pole than 1/2X (1X represents the intensity for powder sintered sample) in the $\alpha$ range from 65deg to 90deg.  The (0002) plane pole figure for conventional rolling (Fig. 2-a) shows the longitudinal split texture, which has the maximum density of basal pole at about 10deg from the sheet normal direction, SN, to the rolling direction, RD. This texture development looks like the magnesium or zinc type. Since there were considerable β phase more than 15pct in the specimens before rolling, such β phase seems to affect the texture development in Ti-6Al-4V alloy sheets. F. R. Larson and A. Zarkades [5] have indicated that sufficient amount of β phase at room temperature will cause a texture transition, so that the (0002) plane pole figure of the α+β Ti-alloy looks like the magnesium or zinc type. Since the specimen displayed considerably weak intensity of basal pole before rolling and no mechanical twin was observed after rolling, the texture development seems to be achieved by slip deformation of α phase and the texture transition. For cross rolling (Fig. 2-c), the high density of basal pole tilts not only in the rolling direction but also in the transverse direction. In both directions, the angle of the tilt is about 10deg. The (0002) plane pole figure for cross rolling seems to pile the pole figure for conventional rolling (Fig. 2-a) up one rotated 90deg. Therefore, the effect of cross rolling on the texture formation consists not in wholly different texture formation, but in two directional rolling. Fig. 2-b shows the (0002) plane pole figure for dentate rolling. In this case, two peaks of the high density of basal pole are shown. The peak of the maximum density exists at the very near position to SN. This peak is absent from other pole figures for cross rolling (Fig. 2-a) and conventional rolling (Fig. 2-c).

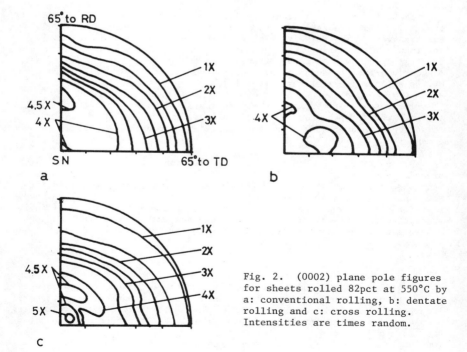

Fig. 2. (0002) plane pole figures for sheets rolled 82pct at 550°C by a: conventional rolling, b: dentate rolling and c: cross rolling. Intensities are times random.

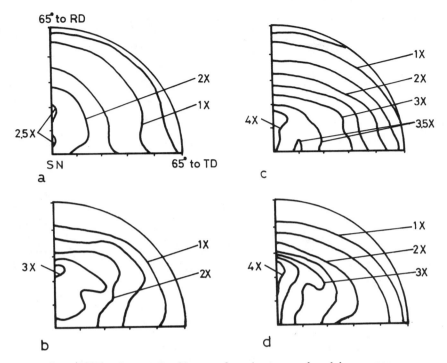

Fig. 3.  (0002) plane pole figures for sheets produced by a: conventional rolling 81pct at 650°C, b: conventional rolling 81pct at 750°C, c: dentate rolling 82pct at 650°C and d: dentate rolling 81pct at 750°C.
Intensities are times random.

Fig. 3 shows the effect of the rolling temperature on the texture formation.  In the case of conventional rolling, it is regarded that the sheet texture would not alter essentially in the temperature range up to 750°C, and the longitudinal split texture is also shown.  In the case of dentate rolling, the high density of basal pole seems to extend toward SN.  For rolling at 650°C, this tendency becomes clear.  However, the difference between dentate rolling and conventional rolling is not so distinct as shown in the rolling temperature of 550°C (Fig. 2).  Therefore, it is regarded that the procedure of dentate rolling would not have the remarkable effect above a certain rolling temperature.

According to Hill's anisotropic theory, the yield criterion for two dimension may be expressed as follows:

$$r_y(1+r_x)\sigma_x^2 + r_x(1+r_y)\sigma_y^2 - 2r_x r_y \sigma_x \sigma_y = r_y(1+r_x)X^2 = r_x(1+r_y)Y^2 \qquad (1)$$

where  x, y    :principal stress axis in rolling and transverse direction
       X, Y    :uniaxial yield strength in x and y direction
       $r_x$, $r_y$   :r value in x and y direction
       $\sigma_x$, $\sigma_y$   :principal stresses in the plane of the sheet

For $r_x=r_y=1$, above yield criterion is equivalent to isotropic yield criterion of von Mises, as follows:

$$\sigma_x^2 + \sigma_y^2 - \sigma_x\sigma_y = X^2 = Y^2 \tag{2}$$

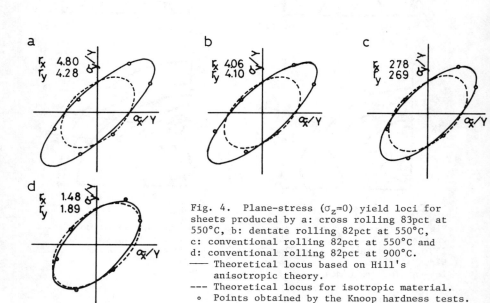

Fig. 4.  Plane-stress ($\sigma_z=0$) yield loci for sheets produced by a: cross rolling 83pct at 550°C, b: dentate rolling 82pct at 550°C, c: conventional rolling 82pct at 550°C and d: conventional rolling 82pct at 900°C.
—— Theoretical locus based on Hill's anisotropic theory.
--- Theoretical locus for isotropic material.
∘ Points obtained by the Knoop hardness tests.

Fig. 4 shows the theoretical yield loci and circles derived from the Knoop hardness tests.  The solid line represents the theoretical locus based on Hill's anisotropic theory (equation (1)), and then the dashed line represents the theoretical locus for isotropic material (equation (2)). Results of the Knoop hardness tests agree reasonably well with prediction of yield locus based on Hill's anisotropic theory, particularly in tension-tension quadrant.  It is indicated that the sheets produced by dentate rolling and cross rolling at 550°C show considerable "texture hardening". The biaxial yield strengths of the sheets can be deduced from Hill's anisotropic yield locus and the points derived from the Knoop hardness tests.  By Hill's anisotropic theory, the ratio of the balanced biaxial yield strength to the uniaxial yield strength, $\sigma_b/Y$, is represented as follows:

$$\frac{\sigma_b}{Y} = \sqrt{\frac{r_x(1+r_y)}{r_x+r_y}} \qquad (3)$$

For sheets produced by dentate rolling and cross rolling, $\sigma_b/Y=1.59$ and 1.67, are predicted respectively.

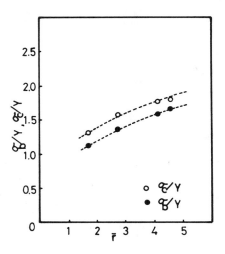

Fig. 5. Correlation between the degree of strengthening under biaxial stresses and average $\bar{r}$ values. $\sigma_c/Y$ is derived from the through-thickness compression test. $\sigma_b/Y$ is derived from Hill's anisotropic theory.

   In Fig. 5, the ratio of the compression yield strength obtained from the through-thickness compression tests, to the uniaxial yield strength, $\sigma_c/Y$, is plotted as against the average $\bar{r}$ value ($\bar{r}=(r_x+r_y)/2$). Also plotted in the figure is the ratio, $\sigma_b/Y$, based on equation (3). The values of $\sigma_c/Y$, are generally higher about 13pct than the values of $\sigma_b/Y$. Firstly, such deviation seems to be responsible for the effect of friction. Secondly, the sheet thickness is much smaller than the width (thickness/width<1), so that the condition of uniaxial compression would not be satisfied. However, the trend line for the relation between the value of $\sigma_c/Y$ and the $\bar{r}$ value can be drawn, and such tendency corresponds to Hill's anisotropic theory. Since the through-thickness compression test is directly related to the sheet strength under biaxial stress condition, it is regarded that considerable "texture hardening" is achieved in the sheet produced by dentate rolling as well as the sheet produced by cross rolling.

   In the case that the width of the sheet is sufficiently large, the deformation process in conventional rolling is considered to be the plane strain deformation process, because the friction on the contact surface between the roll and the sheet constrains the deformation in the transverse direction. It is considered that such macroscopic constraint would affect the texture development. It has been established that the basal plane texture can be produced by multidirectional rolling such as cross rolling or round rolling. The aim of such procedures seems to consist in getting the near isotropic deformation in the sheet plane.  Such deformation would provide the high degree of freedom for the microscopic deformation to produce a crystallographic texture.  It is regarded that such high degree of freedom would be very important for the formation of the basal plane texture.

In dentate rolling process, the numerous projection on the rolling plane are repeatedly to flatten and to rise in every pass, so that, in microscopic point of view, the high degree of freedom for the microscopic deformation seems to exist.  Therefore, it is regarded that this dentate rolling procedure would be effective to produce the basal plane texture as well as the multidirectional rolling procedures.

## Acknowledgements

The authors wish to thank Kobe Steel, Ltd. for providing the Ti-6Al-4V alloy sheets.

## References

1.  S. F. Frederick and G. A. Lenning:  Metall. Trans., Vol. 6B, no. 12 (1975), pp. 601-05.
2.  M. Kumazawa, Y. Nakagawa and T. Sekine:  Tetsu to Hagane, Vol. 63, no. 12 (1977), pp. 1828-37.
3.  R. Hill:  Mathematical Theory of Plasticity, London, Oxford University Press (1950), pp. 317-40.
4.  H. W. Babel and S. F. Frederick:  J. Metals, Vol. 20, no. 10 (1968), pp. 32-38.
5.  F. R. Larson and A. Zarkades:  Advances in Deformation Processing, J. J. Burke, V. Weiss, eds., Available from Plenum Press (1978), pp. 321-49.

# EFFECTS OF MICROSTRUCTURE AND CRYSTAL ORIENTATION ON THE MECHANICAL PROPERTIES IN AN α-β TITANIUM ALLOY

Takashi Nishimura, Yasushi Sugimura,
Yoshimasa Ito and Hidetake  Kusamichi

Kobe Steel, Ltd., Kobe, Japan

## Introduction

As the α-β titanium alloys are most widely  used  in aircraft
industry and  other areas,  the good secondary mechanical properties such as
notched stress-rupture strength, fracture toughness and fatigue strength as
well as the primary mechanical properties such as tensile properties are required.  In order to obtain the good secondary mechanical properties, it is
necessary to control the microstructure and crystal orientation of the alloy
by means of processing as well as controlling the composition.  The α-β titanium alloy Ti-6Al-4V is a two-phase alloy with about twenty percent beta phase
in the annealed condition, and shows many kinds of microstructures depending
upon deformation and heat treatment.  For example, in the case of the thermomechanical treatment in the α-β field, the equiaxed alpha structure is usually
obtained, while the acicular alpha structure in the case of deformation or heat
treatment in the beta filed.  There are many kinds of structures in the
equiaxed and the acicular alpha structures.  In this study the new method is
developed for the expression of microstructure by utilizing the shape of the
alpha grain,  the prior-beta grain and the beta grain size.  Using these micro-
structual parameters this investigation was carried out to make quantitatively
clear the correlation of microstructure to mechanical properties, especially
the notched stress-rupture strength.

In addition, the simple method for determination of preferred crystal orientation was developed in order to study the relationship between mechanical
properties and crystal orientation in the α-β titanium alloy.

## Experimental Procedure

Materials used for this study were hot rolled bars with 22 mm diameter,
forged bars with 16 mm diameter and forgings with various kinds of shapes such
as a large die forging (105x245x1100mm), a forged block (80x80x500mm) and a closed
forged disk (30t x 400mm diameter).  Fifteen materials were used for this
study.  All materials were tested in this investigation after annealing at
700°C for 2 hours and air cooling.  Hot rolled bars were heated at 900°C in
the higher α-β field and 1030°C in the beta field, and cooled in the three
kinds of cooling rates that are water quenching (2000°C/min), air cooling
(250°C/min) and furnace cooling (2.5°C/min).  After these heat treatment these
materials were annealed.  These heat treatments were performed to obtain the
various kinds of microstructures and to make clear the effect of heat treatment
on mechanical properties and microstructures.  Chemical compositions of used
materials were as follows:  Al : 6.0-6.8%, V : 3.8-3.9%, Fe : 0.07-0.08%,
Si : 0.02-0.03%, C : 0.008-0.010%, O : 0.08-0.12%, N : 0.014-0.04%, H : 0.004-
0.008%.

For the determination of the notched stress-rupture strength (NRS), AMS
4928 No.4 specimens were used.  The stress was increased with an increment of 2.5

kg/mm$^2$ in every two hours until the specimen was broken, in order to determine quantitatively the notched stress-ruptures strength.   In the case of being broken after keeping for A hours at a stress of $\sigma_1$, the notched stress-rupture strength (NRS) was defined as follows;

$$NRS = \sigma_1 - 2.5 \times (1 - A/2)$$

The smooth stress-rupture strength was also determined in the same manner.  The smooth and notched tensile tests were carried out using the same specimen size as the stress-rupture tests.  The compression tests were carried out on the specimen of 14x14x14mm or 10x10x10mm.  Microstructural observation was carried out with optical microscope method.  In order to represent the various kinds of microstructural features in the Ti-6Al-4V alloy, the mean width of the primary alpha grain ($x_1$), the mean length of the primary alpha grain ($x_2$), the mean distance between primary alpha grains ($x_3$) and the mean prior-beta grain diameter were measured.  The prior-beta grain diameter in a equaxed alpha structure was measured by using Margolin's method [1].  By utilizing these data the relationships between mechanical properties and microstructural components were calculated by means of the multiple regression analysis as follows:

$$Y = \sum a_i \, F \, (x_i) + b$$

where          Y : Mechanical properties
                (NRS, TS, YS etc.)

            $x_i$: Microstructural components

        a, b: Constant

    $F(x)$: $x$, $\sqrt{x}$, $1/\sqrt{x}$, $1/x$

    The simple observation method for crystal orientation was developed in this study.  By the X-ray diffractometer method the diffraction intensities of (0002), (10$\bar{1}$0), (10$\bar{1}$1) were measured.  Instead of measuring the random specimen, the intensity standard $\bar{I}_\alpha$ was determined as follows:

$$\bar{I}_\alpha = \left\{ I_\alpha(10\bar{1}0) + I_\alpha(0002) + \frac{3}{10} I_\alpha(10\bar{1}1) \right\}/3$$

The relative intensities $\bar{I}_\alpha$ (hkil) of the three alpha planes (hkil) in the measured sections were also determined as follows:

$$\bar{I}_\alpha \, (hkil) = d \cdot I_\alpha \, (hkil)/\bar{I}_\alpha$$

where, d is constant which is 1 for (0002) and (10$\bar{1}$0)  , and 3/10 for (10$\bar{1}$1).

From this equation

$$0 \leqq \bar{I}_\alpha \, (hkil) \leqq 3$$

is obtained.
For random material $\bar{I}_\alpha$ (10$\bar{1}$0) = $\bar{I}_\alpha$(10$\bar{1}$1) = $\bar{I}_\alpha$(0002) = 1.  For hot rolled bars and forged bars the crystal orientation was determined by measuring the relative intensities of the alpha planes for three perpendicular sections which are parallel  or parpendicular to the deforming axis in the hot rolled and the forged bars.

Results and Discussion

## 1. Effect of Heat Treatment on the Mechanical Properties

It was observed that the notched stress-rupture strength decreased with increasing heat treating temperature and with decreasing cooling rate, as shown in Fig.1.  The other strength properties such as the smooth stress-rupture strength, the notched and the smooth tensile strength, the tensile and the compressive yield strength (0.2%) showed the same tendency as the notched stress-rupture strength.  On the other hand, the tensile ductility such as elongation and reduction of area showed the higher values on the materials heat treated in the higher temperature of the $\alpha$-$\beta$ field because of the softening, and the lower values in the conditions heat treated in the beta field.  It is well-known that the equiaxed structure shows the higher ductility than the acicular structure in the $\alpha$-$\beta$ titanium alloys.

## 2. Correlation between Mechanical Properties

It was found that there was the significant positive correlation between the notched stress-rupture strength and the other mechanical properties such as the smooth stress-rupture strength, the notched and the smooth tensile strength, the tensile and the compressive yield strength (0.2%), while not significant relationship between the notched stress-rupture strength and the tensile ductility and charpy impact values, as shown in Fig.2.

## 3. Relationship between Mechanical Properties and Microstructure

Equations representing the correlation mechanical properties and micro-structure in the $\alpha$-$\beta$ alloy Ti-6Al-4V were obtained by the multiple regression analysis, as shown in Table 1.  From this result it was found the notched

Table 1   Relationship between mechanical properties and
microstructure in Ti-6Al-4V alloy

| property * | | equation | standard deviation |
|---|---|---|---|
| NRS | (kg/mm$^2$) | $y = -1.70x_1 - 0.090x_2 + 0.666x_3 - 0.007x_4 + 144.7$ | 5.33 |
| SRS | (kg/mm$^2$) | $y = 25.6/\sqrt{x_2} + 82.6$ | 4.45 |
| NTS | (kg/mm$^2$) | $y = 67.5/\sqrt{x_2} + 5.14/\sqrt{x_3} + 120.2$ | 4.33 |
| TS | (kg/mm$^2$) | $y = -3.35\sqrt{x_1} - 1.22\sqrt{x_2} + 3.18\sqrt{x_3} + 108.0$ | 4.76 |
| YS  0.2% | (kg/mm$^2$) | $y = -1.14\sqrt{x_2} + 2.61\sqrt{x_3} + 94.3$ | 5.49 |
| El | (%) | $y = -0.376\sqrt{x_4} + 22.5$ | 3.83 |
| RA | (%) | $y = 3.56\sqrt{x_3} - 0.928\sqrt{x_4} + 44.3$ | 6.37 |
| CYS 0.2% | (kg/mm$^2$) | $y = 49.8/\sqrt{x_2} - 58.2/\sqrt{x_4} + 84.5$ | 5.39 |

$1.0 \leqq x_1 \leqq 8.60$ , $2.3 \leqq x_2 \leqq 225$ , $0.1 \leqq x_3 \leqq 8.4$ , $10 \leqq x_4 \leqq 1500$

*NRS: notched stress-rupture strength   YS : yield strength

SRS: smooth stress-rupture strength   El : elongation

NTS: notch tensile strength   RA : reduction of area

TS : tensile strength   CYS: compressive yield strength

stress-rupture strength increased with decreasing both primary alpha grain size and prior-beta grain size and with increasing distance between the primary alpha grains.  It is clear that the smaller grain size result in the increases

of strength.  The area between the primary alpha grains usually showed the
secondary acicular alpha structure which was consisted of two phases, alpha
and beta, while it showed only the beta phase when cooled very slowly, FC.
It was considered that the secondary acicular alpha area reduced the stress
concentration of the primary alpha grains and improved the strength and the
toughness.  The other strength properties were mainly dependent upon the primary alpha grain size and were improved with the smaller primary alpha grains.
 On the other hand, the tensile ductility depended mainly upon the prior-beta
grains.  It was considered that in the structure with the larger prior-beta
grain the acicular alpha formed preferrentially in the specified directions
under Burgers' relationship, and that this structure had the low ductility.
The standard deviations of these experimental equations will become smaller
by introducing the texture components into the equation and reducing the deviation of the chemical compositions.

4. Preferred crystal orientation of bars

      The result of measuring the preferred crystal orientation of hot rolled
bars was shown in Fig.3.  In the material annealed at 700°C the intensity of
the basal plane (0002) was very low for the transverse section perpendicular
to the rolling axis, and was very high for the longitudinal section parallel
to the rolling axis.  The material heat treated at 900°C  in the α-β field
showed the same tendency.  On the contrary, in the material heat treated at
1030°C in the beta field it was found that the randomization occurred to reduce the intensity of (0002) for the longitudinal section and increase that
for the transverse section.  The intensity of (10$\bar{1}$0) plane perpendicular to
the basal plane showed the opposite tendency to that of the basal plane.  From
this result it was found that the basal plane of the alpha phase was preferrentially oriented in parallel to rolling axis for rolled bars, and that this
preferred orientation became random by beta heat treatment.

      The result on forged bars was shown in Fig.4.  The material forged at the
lower temperatures in the α-β field showed the more preferred orientation of
the alpha phase, while the material forged at the higher temperatures in the
α-β field had the more random orientation and the more secondary acicular alpha
in the structure.  This became from that the more beta phase transformed to
the acicular alpha after forged in the higher temperatures.

5. Relationship between mechanical properties and preferred crystal orientation.

      The correlation between the intensity of (0002) and the compressive yield
strength was shown in Fig.5 (a).  The material which was heated at 700°C and
had the more preferred orientation of the basal plane perpendicular to the
loading axis showed the higher compressive yield strength in the transverse
direction.  It could be explained from the slip mode in the alpha phase, since
there is no operative slip system for yielding in the slip direction perpendicular to the basal plane.  It was also found that the anisotropy of the compressive yield strength had the close   relation with the difference between
the relative intensities of the basal plane, as shown in Fig.5 (b).

                              Summary

1. The equations were obtained by the multiple regression analysis to represent
the relationship between mechanical properties and microstructural components.
2. The simple determination method for preferred crystal orientation was developed.  It was found mechanical properties such as the compressive yield
strength were closely related to the preferred orientation of the basal plane.

Reference

1. M.A. Geenfield, P.A. Farrar, and H. Margolin: Trans. AIME, 242 (1968), 755.

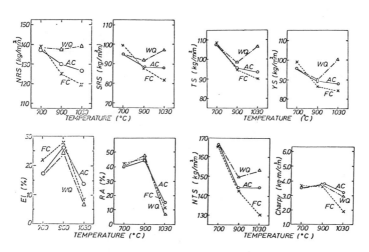

Fig.1   Effect of heat treatment on the mechanical properties of Ti-6Al-4V hot rolled bars, annealed at 700°C, and heat treated at 900°C and 1030°C, and subsequently annealed at 700°C.

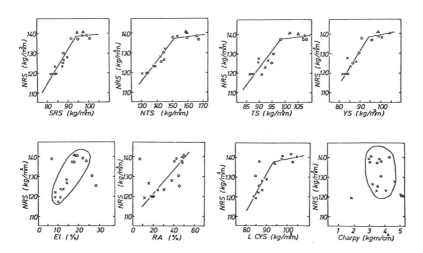

Fig.2   Correlations between the notched stress-rupture strength (NRS) and the other mechanical properties in hot rolled bars, forged bars and forgings of Ti-6Al-4V.
x acicular, Δ acicular + equiaxed, o equiaxed

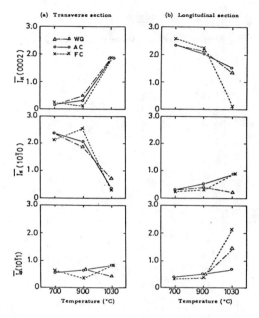

Fig.3   Relative intensities of the alpha planes in the transverse
        and longitudinal sections of the Ti-6Al-4V hot rolled bars,
        annealed at 700°C, and heat-treated at 900°C and 1030°C, and
        subsequently annealed at 700°C.

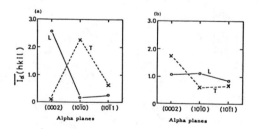

Fig.4   Relative intensities of the alpha planes in the transverse
        sections (T) and longitudinal (L) sections of Ti-6Al-4V alloy
        bars, (a) forged at the lower temperatures 900 - 650°C, and
        (b) forged at the higher temperatures 950 -850°C.

(a)                                    (b)

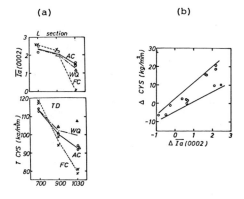

Fig.5   Correlation between compressive yield strength and the
        relative intensity of the basal plane.

(a)     Transverse compressive yield strength (T CYS) and $\bar{I}\alpha$ (0002)
        in the longitudinal section.

(b)     $\Delta \bar{I}\alpha$ (0002)  = $\bar{I}\alpha$ (0002) (LS) − $\bar{I}\alpha$ (0002) (TS)
        $\Delta$ CYS        = CYS (TD) − CYS (LD)

# C. Superplasticity and Creep

DEFORMATION BEHAVIOR OF Ti-6Al-4V ALLOY
UNDER SUPERPLASTIC CONDITION

Yoshimasa Ito,  Atsushi Hasegawa

Kobe Steel, Ltd., Kobe, Japan

## Introduction

In the aerospace industry, titanium alloys play a very important role due to good characteristics of the strength-density ratio, fracture toughness and high temperature property.  However, because of its high strength and lower e-lastic modulus, it is difficult to fabricate its components and therefore, much fabricating cost is required.

In recent years, in order to solve these difficulties, considerable effort has been made in making clear superplastic behavior of titanium alloys and in developing new industrial technologies using its properties. ( 1)~6) )

In this report, following two tests were made for Ti-6Al-4V alloy in order to apply superplastic characteristics with excellent ductility and lower flow stress to practical working process.

### Test I

The purpose of the test I is to study an influence of metallurgical factors ( microstructure and alpha grain diameter ) and deformation conditions ( strain rate and deformation temperature ) on the superplastic behavior of Ti-6Al-4V alloy.

### Test II

This test was made for studying the characteristics of tensile property and microstructure of Ti-6Al-4V alloy deformed under isothermal and superplastic conditions.  The effects of initial microstructure and deformation condition on the properties of isothermally deformed materials were studied.

## Experimental procedure

### Test I

In this test, four kinds of specimens with an equiaxed structure with aver-age alpha grain diameter of $3.3\mu$, $10.7\mu$ and $20.0\mu$, and an acicular one were pre-pared.  The four specimens with different microstructures were designated to be A,B,C and D as shown in Photo. 1.  They were made from two different lots of ingot.  The chemical compositions of these two lots are given in Table 1.

Table 1  Chemical compositions (wt%)

| Lot | Specimen | Al | V | Fe | O | H | N | C |
|---|---|---|---|---|---|---|---|---|
| 1 | A,B,D | 6.12 | 4.06 | 0.214 | 0.154 | 0.0010 | 0.0061 | 0.0150 |
| 2 | C | 6.16 | 4.12 | 0.208 | 0.151 | 0.0025 | 0.0027 | 0.0190 |

The tensile specimen size were 4.0 mm diameter and 25 mm length of reduced section.  All the specimens were tested on an Instron testing machine with triple zone electric furnace.  The temperature during test was monitered by the Pt-Rh thermocouple directly attached to the specimens.   Tensile tests were carried out at a temperature range from 750°C to 1050°C and at a strain rate of $7.0 \times 10^{-4}$ to $1.28 \times 10^{-2}$ sec$^{-1}$.

In order to estimate the superplastic properties under various metallurgical and deforming conditions, the flow stress and tensile elongation after fracture were measured, and the microstructure near the fractured part of a tensile specimen was observed.

### Test II

The materials having an equiaxed structure with average alpha grain diameter of 3.7µ and 9.0µ and an acicular one were used.  They were fabricated from a same ingot and designated as specimen A, B and C respectively.  The chemical compositions of as-received billet are shown in Table 2.

Table 2   Chemical compositions (wt%)

| Al | V | Fe | O | H | N | C |
|------|------|-------|-------|--------|--------|--------|
| 6.37 | 4.15 | 0.248 | 0.170 | 0.0042 | 0.0069 | 0.0130 |

The test specimen with 50 mm diameter and 50 mm height were used.  The compressive test was carried out on an universal testing machine capable of 200 tons with a electric furnace.  The test pieces were compressed ( forged ) to 40, 60 and 80% reduction in height by one stage operation at fixed temperature.  The testing temperature was in the range above 800° below 980°C and three initial strain rates were between $7.0 \times 10^{-4}$ to $2.56 \times 10^{-2}$ sec$^{-1}$.

After isothermally deformed and annealed at 700°C x 1 Hr, 4.0 mm diameter tensile pieces were taken from the center parts of the specimen.

### Results and discussion

### Test I

#### Effect of microstructure

The influences of initial microstructure and deformation temperature on the tensile elongation are summarized in Fig. 1.  These results were obtained by testing at a fixed strain rate of $1.28 \times 10^{-3}$ sec$^{-1}$.

The specimen A with a grain diameter of 3.3µ showed remarkably higher ductility than those of the other specimens.  The difference of elongation between the specimen A and the other equiaxed specimen B and C with grains larger than 10.7µ was very large especially when deformed at a temperature below 900°C.  It is apparent that the tensile elongation is considerably dependent on the initial microstructure and alpha grain diameter.  With decreasing of grain diameter, particularly under 10µ, tensile elongation increases remarkably.

For the specimen A, the maximum elongation over 800% was obtained at a temperature of 850° to 900°C and strain rate of $1.28 \times 10^{-3}$ sec$^{-1}$.  Even if the deformation temperature decreases below 800 °C, it shows elongation slightly over 500%.  On the contrary, at a temperature above 950°C, the elongation was rapidly reduced to about 200% approximately equal to those of the other specimens.  Thus it is recognized that the deformation behavior at a higher temperature is different from that at a lower temperature in view point of void formation and localized necking.  Namely, at a lower temperature, void formation appeared to be obvious, but relatively uniform deformation occurred.  On the other hand, at a higher temperature, the voids were not observed, but sharply localized necking occurred.

For the specimen B, with mean grain diameter of 10.7μ it showed a low peak of about 250% elongation when deformed between 900° to 925°C. Above that temerature, it was observed to have almost constant elongation of 200%, but below 900° C, the elongation decreased rapidly.

The specimen C with mean grain diameter of 20.0μ showed to be lower ductility at all over the temperature range comparing with the specimen B.

The temperature dependence of elongation of the specimen D with an acicular structure, is similar to that of the specimen B. But its maximum elongation was not over 300%.

## Effect of strain rate

As mentioned above, the influence of initial microstructure and deformation temperature on tensile elongation of Ti-6Al-4V is extremely large. At the same time, it was observed to depend on the strain rate. The effects of the strain rate on it are shown in Fig. 2 and 3. This effect in relation to the specimen B is shown in Fig. 2, and Fig. 3 exhibits the difference of the strain rate dependence of tensile elongation among four initial microstructure at a temperature of 900°C. It was observed clearly in Fig. 2 that as the strain rate is reduced from $1.28 \times 10^{-3}$ sec$^{-1}$ as shown in Fig. 1 to $7.0 \times 10^{-4}$ sec$^{-1}$, much increase of elongation occurred evidently below temperature between 850° to 925°C. However, it does not come up to the level for the specimen A, but shows the maximum elongation of about 400%.

On the other hand, at a higher strain rate of $5.12 \times 10^{-3}$ sec$^{-1}$, the elongation decreased under all the temperature range and elongation maximum was not observed.

The change in elongation within the strain rate range from $1.28 \times 10^{-3}$ sec$^{-1}$ to $1.28 \times 10^{-2}$ sec$^{-1}$ is most pronounced in the specimen A in different from the others. The difference of the change in elongation among them was very small, but the specimen B is seemed to have improvement in elongation at the lower strain rate of $7.0 \times 10^{-4}$ sec$^{-1}$ as shown in Fig. 2. This rise of elongation at a lower strain rate was observed to be slight in the specimen C with larger grain diameter of 20.0μ, but no increase occurred in the specimen D with an acicular structure. On the other hand, the reduction of elongation occurred at the strain rate of $7.0 \times 10^{-4}$ sec$^{-1}$ in the specimen A is considered to be grain growth during the test.

## Effect of microstructure on flow stress

The initial microstructure not only affect greatly on the tensile elongation as above mentioned, but also have influence over the flow stress. Fig. 4 shows the strain rate dependence of flow stress at 900°C for four specimens. The flow stress of the specimen A was about one fourth as high as that of the specimen B at a strain rate of $1.28 \times 10^{-3}$ sec$^{-1}$. On the other hand, the specimen D with an acicular alpha structure showed about six times as high as that of the specimen A. This difference should be considered to be very large. In view point of the strain rate sensitivity, m value calculated at a strain rate of $1.28 \times 10^{-3}$ sec$^{-1}$ to $1.28 \times 10^{-2}$ sec$^{-1}$ was 0.66, 0.39, 0.263 and 0.282 for the specimen A, B, C and D respectively. It is obvious from these data that the material with a fine alpha grain structure has lower flow stress and higher strain rate sensitivity of flow stress.

## Void formation

After tensile fracture, the microstructure and void formation was examined. Photo. 2 and 3 shows the microstructure just near the fractured part for the specimen A, B and D tested at a strain rate of $1.28 \times 10^{-3}$ sec$^{-1}$ and various temperatures.   The top and bottom direction in Photo. 2 and 3 is parallel to the tensile axis.   The figures noted in the parenthesis means the tensile elongation values.

As for the specimen A in Photo. 2, the grain growth was observed to occur during testing especially at a higher temperature.   It is recognized that there is apparent difference in the void formation behaviors in the two temperature regions above 900°C and below 850°C.   This is, when deformed at a lower temperature of 750° to 800°C, relatively fine voids were observed to form and grow up along the tensile axis.   However, no or only a few void formation was found above 850°C.   Therefore, it is assumed that the different deformation mode operates in those two different temperature regions, considering the localized necking behavior and void formation and that at this peak temperature, the void formation is restricted and the resistance against the localized necking increase.

Photo. 3 shows the microstructure for the specimen B and D.   Comparing with that for the specimen A, the specimen B showed the different appearance in the following three view points.   Firstly, the size of one void is very large.   Secondly, the void formation is observed when deformed not only at 800°C, but also at 850° to 900°C and lastly, the voids tend to extend toward the direction perpendicular to the tensile axis at 800° to 850°C.   It is considered that these behaviors such as the void formation and growth, cause to reduce the elongation drastically in different from the specimen A.   At a higher temperature, no void formation was observed and the appearance resembled to the specimen A.

Further, it becomes more apparent for the specimen B that this void formation occurs at and near alpha/beta phase boundaries.

The specimen C showed similar behaviors to the specimen B.

As for the specimen D, at 800°C where only 40% elongation occurred, grain boundary cracking was observed at prior beta grain boundary.   At 900°C, the acicular structure was broken and fine equiaxed alpha recrystallization occurred.

On the base of these results, we studied the characteristics and its utility of Ti-6Al-4V alloy deformed under superplastic and isothermal conditions, and also examined the influence of initial microstructure and deformation condition on the properties.   These results are as follows.

Test II

Effect of microstructure

Fig. 5 shows room temperature tensile properties of three Ti-6Al-4V alloy specimens which were deformed up to 80% reduction in height at a strain rate of $1.28 \times 10^{-3}$ sec$^{-1}$ and various temperature.   The solid lines show the present results and the dotted lines the properties of conventionally forged Ti-6Al-4V alloys.   The open and solid triangle marks show the properties of materials having initial equiaxed structure and acicular one respectively.

With a rise of deformation temperature, tensile and yield strength of all three specimens tend to decrease.   This tendence is remarkable especially for the specimen C with the initial acicular alpha structure.   On the other hand, in spite of the difference in initial microstructure between the specimen A and B, both materials exhibits   a similar behavior of the strength.   This corresponds to the structural similarity of both deformed specimens.

As to tensile elongation, both specimen A and B showed almost no change above 850°C.   But for the specimen C, it tends to decrease slightly with a rise of temperature.

It is recognized that the present data almost agree with those obtained by conventional forging process in view points of the strength levels and of temperature dependence of strength.

Effect of deformation reduction and strain rate

The influence of the deformation reduction and strain rate on the tensile properties are shown in Fig. 6 and 7, respectively. Fig. 6 shows one of the results exhibiting the effect of deformation reduction up to 80% at 900°C and $1.28 \times 10^{-3}$ sec$^{-1}$.

With an increase of reduction, the tensile and yield strength of both specimen A and B were observed to increase slightly, but the specimen C showed a considerable increase of strength. The amount of this increase was ~8 kg/mm$^2$ of yield strength and ~6kg/mm$^2$ of tensile strength. As for elongation, both specimen A and B are appeared to be almost independent of the reduction, but the specimen C was improved up to the level of the specimen A and B with an increase of reduction.

On the other hand, in the case of the deformation at 800°C, even the specimen A showed a considerable increase of yield strength. The influence of deformation reduction of the strength is considered to be related with structural change.

Fig. 7 shows the strain rate dependence of tensile properties under fixed conditions as described in it. It can be seen that the strength of all three specimens tend to increase with the rise of strain rate. These behaviors are seemed to be dependent upon the difference of alpha grain diameter for the specimen A and B, and upon the extent of recrystallization from an acicular to a fine equaxed alpha structure for the specimen C.

Structural characteristics

The microstructure of isothermally deformed specimens are given in Photo. 4 and 5. These photographs show the microstructural change for the specimen A and C, respectively. It can be seen that the microstructural change is considerably affected by the deformation temperature. For the specimen A in Photo. 4, very fine recrystallized alpha grains were observed to occur near alpha/beta phase boundaries. This behavior depends on the deformation temperature and reductions. The formation of this fine grain was most pronounced at a temperature of 800° to 900°C, but was not observed above 950°C. In this view point, similar behavior was observed for the specimen B. It is considered that this is the reason why both specimen A and B showed similar temperature dependence of tensile properties in spite of their different initial alpha grain diameter. For the specimen C in Photo. 5, it was observed that the acicular structure was almost broken and fine equiaxed alpha structure was produced at 800° to 900°C. The microstructure at 950°C was characterized by an acicular structure with irregular recrystallization.

Conclusion

1)   The tensile elongation and flow stress of Ti-6Al-4V alloy at elevated temperatures are considerably dependent upon the initial microstructure and the alpha grain diameter. For the material with average alpha grain diameter of 3.3μ, very large elongation over 800% at 850° to 900°C and lower flow stress occurred. These behaviors are considered to be corresponds to the limitation of void formation and to resistance against the localized necking.
2)   The properties of isothermally deformed Ti-6Al-4V are as good as that of conventional forgings.

After deformation, very fine recrystallized alpha grains were observed to occur near alpha/beta phase boundaries. The formation of this fine grain was most pronounced at a temperature of 800° to 900°C, but was not observed above 950°C.

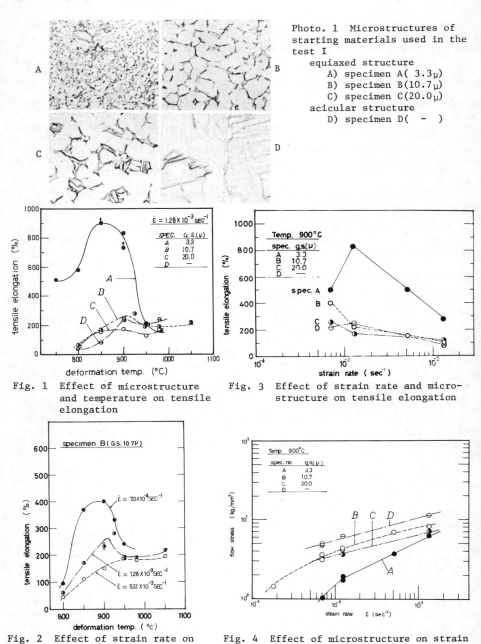

Photo. 1   Microstructures of
starting materials used in the
test I
    equiaxed structure
        A) specimen A( 3.3μ)
        B) specimen B(10.7μ)
        C) specimen C(20.0μ)
    acicular structure
        D) specimen D( - )

Fig. 1   Effect of microstructure
and temperature on tensile
elongation

Fig. 3   Effect of strain rate and microstructure on tensile elongation

Fig. 2   Effect of strain rate on
tensile elongation of the
specimen B

Fig. 4   Effect of microstructure on strain
rate dependence of flow stress

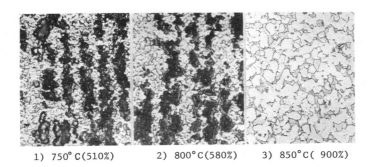

1) 750°C(510%)    2) 800°C(580%)    3) 850°C( 900%)

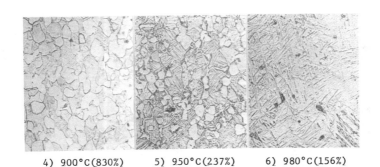

4) 900°C(830%)    5) 950°C(237%)    6) 980°C(156%)

Photo. 2  Microstructures of longitudinal section for the
specimen A after tensile fracture ( X400 )
( strain rate = $1.28 \times 10^{-3}$ sec$^{-1}$ )

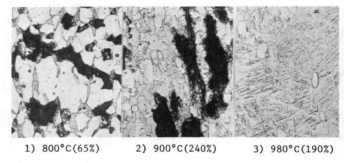

1) 800°C(65%)      2) 900°C(240%)      3) 980°C(190%)

1) Specimen B

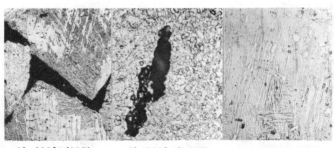

1) 800°C(40%)      2) 900°C(255%)      3) 980°C(240%)

2) Specimen D

Photo. 3   Microstructures of longitudinal section for the
          specimen B and D after tensile fracture (x400)
          ( strain rate = $1.28 \times 10^{-3}$ sec$^{-1}$ )

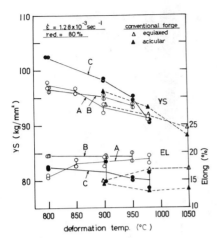

Fig. 5  Effect of microstructures
        on tensile properties of
        isothermally deformed
        specimens

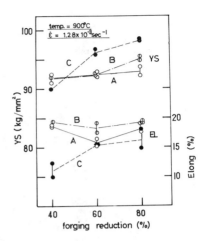

Fig. 6  Effect of deforming reduction
        on tensile properties of iso-
        thermally deformed specimens

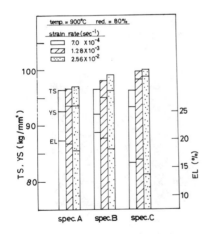

Fig. 7  Effect of strain rate on
        tensile properties of iso-
        thermally deformed specimens

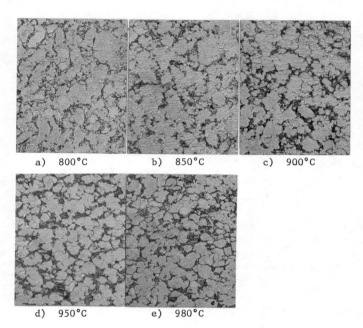

a)   800°C          b)   850°C          c)   900°C

d)   950°C          e)   980°C

Photo. 4   Microstructures of the specimen A after isothermally
           deformed at indicated temperatures ( X400 )
           ( strain rate = 1.28x10⁻³ sec⁻¹, reduction = 80% )

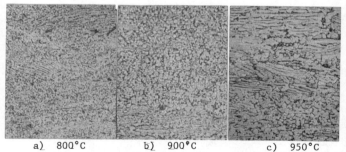

a)   800°C          b)   900°C          c)   950°C

Photo. 5   Microstructures of the specimen C after isothermally
           deformed at indicated temperatures ( X400 )
           ( strain rate = 1.28x10⁻³ sec⁻¹, reduction = 80% )

## Reference

1.   C.C.Chen and J.E.Coyne   : Met. Trans., 7A(1976)1931
2    A.Arieli and A.Rosen     : Met. Trans., 8A(1977)1951
3.   N.E.Paton and C.H.Hamilton : Met. Trans., 10A(1979)241
4.   A.K.Ghosh and C.H.Hamilton : Met. Trans., 10A(1979)669
5.   M.V.Hyatt et al.   : AFML-TR 77-81
6.   H.M.Burte et al.   : J. Metals, 30, 9(1978)7

# FACTORS INFLUENCING THE DUCTILITY OF SUPERPLASTIC Ti-6Al-4V ALLOY

N. Furushiro[*], H. Ishibashi[**], S. Shimoyama[**] and S. Hori[*]

* Department of Materials Science and Engineering, Faculty
of Engineering, Osaka University, Suita 565, Japan

** Graduate School, Osaka University, Suita 565, Japan

## Introduction

It is well known that the ductility in superplastic deformation is remark-ably influenced by the temperature of deformation, the strain rate and the metallographic conditions of the specimen used [1]. In many investigations of the effects of these conditions on the ductility, metallographic factors were determined on the specimen before the test, under the assumption that the structure remained unchanged during the deformation [1~3]. However, changes in the structure were often observed in the actual superplastic deformation [4]. So, the strain dependence of the structual factor should be taken into consideration, especially when the elongation is chosen as a measure of the ductility.

Ti-6Al-4V alloy is characterized by its high strength, so that high tem-peratures are required in the forming operation. However the great elongation achieved in the superplastic state makes this alloy valuable. In the super-plastic deformation state the structure of this alloy consists of dual phases of $\alpha$ and $\beta$, the configurations of which will vary according to the heat treat-ment and to the conditions of the deformation [2,3,5]. There are few reports on the effects of metallographical factors on the ductility of this alloy which take changes in structure during the deformation process into consideration. In the present study, therefore, the strain dependence of the structural fac-tors and the strain rate sensitivity index of flow stress $m$ ($\sigma = K\dot{\varepsilon}^m$, $\sigma$: flow stress, $\dot{\varepsilon}$: strain rate, K: constant) of a Ti-6Al-4V alloy during superplastic deformation were investigated together with the examination of the effects of deformation conditions on the superplastic behaviour. Further, the correla-tion between the structure and $m$, and the influence of these factors on the ductility were also studied.

## Experimental

The mill plate of a Ti-6Al-4V alloy containing 6.15% Al, 4.31% V, 0.13% Fe, 0.14% O, 0.0044% H and 0.0088% N supplied by Kobe Steel Ltd. was used. Tensile specimens of gauge length 20 mm and width 4 mm were machined from the 1 mm plate and then were vacuum-annealed under various conditions. This was fol-lowed by water-quenching. The annealing conditions employed were a tempera-ture of 700°C for a period of 60 h (Specimen A) and temperatures of 800, 900 and 1050°C, respectively (Specimen B, C and D), each for 4 h.

Tensile tests were carried out in a purified argon atmosphere at elevated temperatures using an Instron type machine. After being set on the machine the specimens were heated at a rate of 20°C/s in an infared ray furnace and were held for a suitable period before measurements were taken. The deforma-tion temperature and the initial strain rate ranged from 600 to 900°C and from $8.3 \times 10^{-5}$ to $4.2 \times 10^{-2}$ $s^{-1}$, respectively. Values of $m$ were calculated from the stresses at a strain rate and that when the strain rate was doubled during the deformation.   For optical and electron microscope examinations, the deformation process was interrupted when predetermined strain levels were reached and the specimens were then air-cooled.   The specimens were polished mechanically and then electrically using a solution of acetic anhydride and perchloric acid (19:1).   Specimens for optical microscopy were etched in a solution of hydrofluoric acid, nitric acid and ethyl alcohol (1:2:6).

## Results

Grain size and directionality of α phase were measured on optical micrographs as shown in Fig. 1.   The directionality was defined as the mean value of the ratios of the distance at the widest area to the distance normal to the widest area.   Table 1 lists these results.   Four kinds of specimens were tensiled at 600, 700, 800 and 900°C and at strain rates of $4.2 \times 10^{\bar{3}}$ and $4.2 \times 10^{4}$ $\bar{s}^{1}$.   Typical stress-strain curves obtained at 800°C are shown in Fig. 2.   It can be seen in Fig. 2 that the curve for Specimen D passes a maximum during the initial stage of the deformation then falls rapidly during the subsequent deformation resulting in an elongation of only 50%.   The other specimens showed lower maximum stresses followed by smaller decreases during the subsequent deformation leading to larger elongations than that of Specimen D.   All elongations are plotted as a function of the test temperature in Fig. 3.   The temperature dependence of *m* at a given strain rate for a specimen was similar to that of the elongation as shown in Fig. 4. However, no clear correspondence between the elongation in Fig.3 and the *m* value in Fig. 4 under the same conditions was found, though a good correspondence of the above two factors was common for superplasticity [1].

Then, in order to examine the strain dependence of *m*, Specimens A and B were tensiled at temperatures from 700 to 900°C with an interval

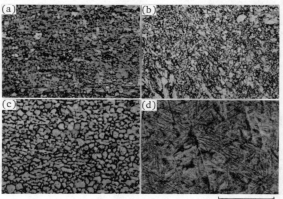

Fig. 1. Optical microstructures of specimens used; (a) Specimen A, (b) Specimen B, (c) Specimen C and (d) Specimen D. Conditions of the heat treatment are shown in Table 1.

Table 1. Volume fraction, grain size and directionality of α phase measured using the optical micrographs of specimens which were water-quenched from each temperature of heat treatments.

| Specimen | Heat treatment | Volume fraction | Grain size | Directionality |
|----------|----------------|-----------------|------------|----------------|
| A | 700°C - 60 h | 0.84 | 3.2 μm | 1.50 |
| B | 800°C - 4 h | 0.67 | 3.6 μm | 1.20 |
| C | 900°C - 4 h | 0.60 | 7.4 μm | 1.35 |
| D | 1050°C - 4 h | — | — | — |

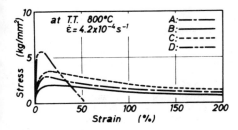

Fig. 2. Stress-strain curves at 800°C and a strain rate of $4.2 \times 10^{-4} \bar{s}^1$ for various specimens.

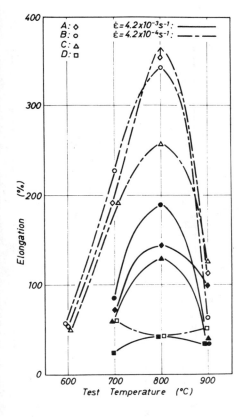

Fig. 3. Influence of test temperature on elongation for various specimens deformed at strain rates of $4.2 \times 10^{-3}$ and $4.2 \times 10^{-4} \bar{s}^1$.

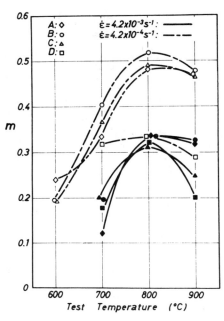

Fig. 4. Influence of test temperature on $m$ of various specimens deformed at strain rates of $4.2 \times 10^{-3}$ and $4.2 \times 10^{-4} \bar{s}^1$.

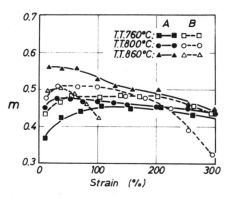

Fig. 5. Change in $m$ during the deformation of Specimens A and B at temperatures of 760, 800 and 860°C and at a strain rate of $4.2 \times 10^{-4} \bar{s}^1$.

of 20°C and at $4.2 \times 10^{-4}$ s$^{-1}$.    Typical results are shown in Fig. 5, in which curves can be classified into three types; (1) the value of $m$ decreases with an increase in strain, *e.g.* the curve for Specimen A at 860°C, (2) a almost constant value of $m$ follows during the deformation, after a slight initial increase, *e.g.* the curve for Specimen B at 760°C, though the $m$ value at later stages, and (3) the $m$ value increases with the strain up to a fairly large strain and then decreases, *e.g.* the curve for Specimen A at 760°C.    The variation of $m$ values during the deformation, as shown in Fig. 5, may result in the lack of simple relation between the magnitude of $m$ and the elongation. Values of $m$ at strains of 50 and 200% in Fig. 5 are shown as a function of the test temperature in Fig. 6.    Figure 7 shows the elongation at various temperatures.    These revealed that the temperature at which the maximum elongation was obtained did not agree with that of the maximum value of $m$ at 50% strain, but agreed with the corresponding temperature at 200% strain for both of Specimens A and B.

## Discussion

### 1. Effect of temperature

Figures 3 and 7 show that the elongation has a maximum against the test temperature.    Such maxima were

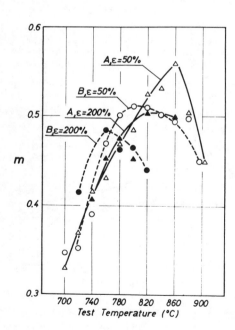

Fig. 6. Relation between test temperature and $m$ at strains of 50 and 200% for Specimens A and B, as shown in Fig. 5.

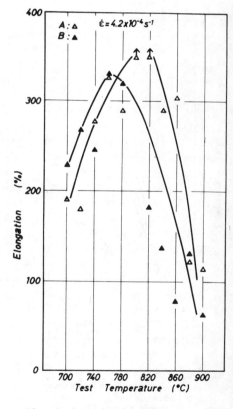

Fig. 7. Relation between test temperature and elongation in the deformation as shown in Fig. 6.

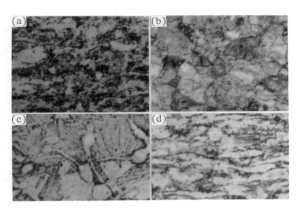

Fig. 8. Optical micro-
structures of Specimen B
fractured at various tem-
peratures  and  strain
rates;
(a) 600°C and $4.2 \times 10^{-4} s^{-1}$,
(b) 800°C and $4.2 \times 10^{-4} s^{-1}$,
(c) 900°C and $4.2 \times 10^{-4} s^{-1}$
and
(d) 800°C and $4.2 \times 10^{-2} s^{-1}$,
respectively. A tensile
axis is horizontal.

20μm

often reported for the superplastic behaviour in the temperature of a dual
phase region [6,7]    In those papers, it was considered that the maximum
should be related to the amount of the phase boundary, since the temperature
for the maximum agreed with that where the volume fractions of    both phases
were equal.   The present results, however, does not support the above consideration because such agreement was not observed, the temperature for the
same volume fraction of the both phases being higher than 900°C as shown in
Table 1.
    Optical micrographs of Specimen B fractured under various conditions are
shown in Fig. 8.    These observations were carried out on the central cross
section parallel to a tensile axis of the specimen and not so close to the
fractured end.    It is clearly shown in (a), (b) and (c) of Fig. 8 that the
grain growth tended to become remarkable with the rise of the temperature.
The grain growth was considered as a factor which inhibited the superplastic
behaviour, while the temperature was one of the important factors which made
easy to operate the superplastic deformation [1].    It was considered, therefore, that the relation between the elongation and the test temperature was
determined by the competition    between the above two factors.

2. Effect of strain rate

    Figure 9 shows the relationship between the elongation and the strain
rate for Specimen B at 800°C, indicating a maximum elongation at a strain rate
of $8.3 \times 10^{-4} s^{-1}$.    Observation under an optical microscope revealed that grains
grew a little at a low strain rate of $8.3 \times 10^{-5} s^{-1}$.    It was considered that the
growth may be resulted from the long holding period because of the slow strain
rate.    But, it was not clear if this growth was the reason for lowering of
the elongation at lower strain rates.    While the fractured specimen at a
higher strain rate of $4.2 \times 10^{-2} s^{-1}$ than that as maximum elongation consisted of
elongated grains toward a tensile axis, as shown in Fig. 8 (d).    It is well
known that superplasticity appeared under an optimum set of conditions, especially, of the strain rate [1,8].    It was also known that when the strain
rate increases, dislocation creep deformation becomes a dominant process, and
thus the $m$ value and the elongation decrease [1,9 10].    The dislocation
structure during the deformation was not able to be observed in detail,because
β phase of this alloy was unstable at room temperature due to the martensitic
tranformation.    However, it seems that the elongated grain may be resulted
from a process governed by the dislocation motion.    It is, therefore,suggest-

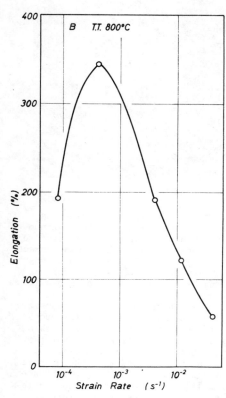

Fig. 9. Strain rate dependence of elongation of Specimen B deformed at 800°C.

2μm

Fig. 10. An example of electron micrograph on Specimen B cooled in an atmosphere of argon from 800°C without loading.

ed that the decrease in the elongation at high strain rates was given by the increase of the strain due to the dislocation motion and by the decrease in the $m$ value.

## 3. Effect of grain structure

Extremely small values of elongation and $m$ value were obtained for Specimen D as shown in Figs. 3 and 4. The structure of Specimen D during the deformation was unknown, because of the transformation of β phase to other phases at room temperatures as shown in Fig. 1 (d), however, it was possible to assume from Fig. 1 (d) that Specimen D consisted of larger grains at higher

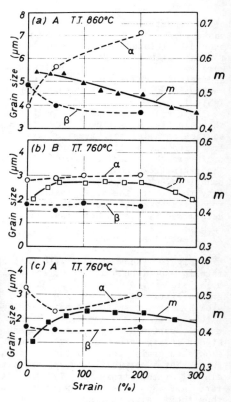

Fig. 11. Changes in grain sizes of α and β phases and $m$ during the the deformation under typical conditions as shown in Fig. 5.

temperature.   Specimen C was supplied to a test also in a state of larger grains than that of Specimens A and B, as can be seen in Table 1.   These showed Specimens C and D which had relatively large initial grain sizes did not give any typical superplasticity, though, as discussed later, the structure  given  by  the  each preliminary heat treatment may be alterable during the high temperature deformation.   It may be also pointed out that the directionality has an important effect on the elongation since the elongation appeared in Fig. 3 shows a correspondence to the directionality rather than grain size which are shown in Table 1.

The curves in Fig. 5  imply  the difficulty of representing $m$ as a single value under a given set of conditions of superplasticity.   It seems that this difficulty may be partly resulted from the difference between the temperature of the test and that of the preliminary heat treatment.   Therefore, electron microscope observations on the specimen after the interruption of the deformation were carried out in order to examine the correspondence between the $m$ value and the structure under the typical condition of the three types in Fig. 5.   Grain sizes of $\alpha$ and $\beta$ phases were measured using the electron micrograph obtained.   It was thought that $\beta$ phase was transformed into martensite or $\alpha$ phases during cooling after the deformation, so that the area of $\beta$ phase at the high temperature should consist of the more complex structure than that of $\alpha$ phase, as shown in Fig. 10.   The grain size of $\alpha$ phase, as shown in Fig. 11 (a), increased with lowering the $m$ value, while $\beta$ became fine.   Figure 11 (c) showed that the grain size of $\alpha$ phase decreased first and then increased during the subsequent deformation.   Nearly constant values of $m$ and the grain sizes were observed in Fig. 11 (b).   These suggested that the $m$ value changed closely with the variation of the grain size of $\alpha$ phase.

According to the metallographic examination, specimens under steady state deformation were free of void in the present experiment, while in the close neighborhood of the fractured ends the voids were observed almost always.   The possibility of void formation in the area was considered to be related to the amount of the elongation of the specimen.   The above observation implies that there may be a relationship between the fracture under the present conditions of superplastic deformation and the void formed, because of the locally high strain and strain rate in the fracturing region.

## Conclusions

The superplastic behaviour of the Ti-6Al-4V alloy is influenced by many factors which are the temperature and the strain rate of deformation and the metallographic conditions of the specimen.   The temperature and the strain rate dependences of the ductility are explained by introducing the effect of grain growth and that of dislocation motion, respectively.   It is impossible that a single value of $m$ represents the superplastic behaviour under a given set of the deformation, because the $m$ value varies much with the strain. This strain dependence is considered to be related to the change in the grain size of $\alpha$ phase during the deformation.   It is also pointed out that the strain dependence of the metallographic factor or $m$ is important for the investigation of the effect of each factor on the superplastic behaviour.

## Acknowledgement

The authors are grateful to Dr. Y. Moriguchi of Kobe Steel Ltd. for his effort in providing the Ti-6Al-4V alloy.

## References

1.  J.W. Edington, K.N. Melton and C.P. Culter: Progress in Materials Science, 21(1976), 61.
2.  D. Lee and W.A. Backofen: Trans. AIME, 239(1967), 1034.
3.  A. Arieli and A. Rosen: Met. Trans. A, 8A(1977), 1951.
4.  S. Hori and N. Furushiro: *pro. 19th Jap. Congr. Mater. Research*, Soc. Mater. Sci. Jap., (1976), 1.
5.  N.E. Paton and C.H. Hamilton: Met. Trans. A, 10A(1979), 241.
6.  D.M.R. Taplin and S. Sagat: Mater. Sci. Eng., 9(1972), 53.
7.  T. Hirano, M. Yamaguchi and T. Yamane: Met. Trans., 5(1974), 1249.
8.  S. Hori and N. Furushiro: Bull. J. Jap. Inst. Metals, 14(1975), 673.
9.  N. Furushiro and S. Hori: Scripta Met., 12(1978), 35.
10. N. Furushiro and S. Hori: Scripta Met., 13(1979), 653.

# CHARACTERIZATION OF SUPERPLASTIC DEFORMATION PROPERTIES OF Ti-6Al-4V

C.H. Hamilton and A.K. Ghosh

Rockwell International Science Center
Thousand Oaks, California 91360

## Abstract

The deformation behavior of a superplastic Ti-6Al-4V alloy at 927°C has been characterized by means of constant strain-rate tensile tests up to large plastic strain. Significant hardening has been recorded in the course of deformation, and microstructural studies on deformed samples indicate the occurrence of simultaneous strain-rate induced grain growth, which explains nearly all of the hardening. As a result of concurrent grain growth, the strain-rate sensitivity is found to decrease with strain, thus indicating that stress-strain rate behavior determined initially may not be applicable after large amounts of plastic strain. A series of tests were conducted in which the strain-rate sensitivity, m, was measured for a fixed strain (0.15) and grain sizes (7.7 μm) and for a range of strain rates encompassing that of maximum superplasticity. The results showed general agreement with those obtained by the step/strain-rate method.

## Introduction

Subsequent to the first demonstration of superplasticity in titanium alloys more than a decade ago [9], increasing interest has been developing in pursuit of the technology whereby superplastic forming of these alloys may be achieved [2-6]. Numerous benefits results from this process have been indicated, including reduced cost, improved structural performance and attendant weight reduction, reduced lead time, etc. In fact, few metal-working developments have offered the prospect of such wide-ranging implications in structural design as superplastic forming (SPF) and superplastic forming with combined diffusion bonding (SPF/DB) to the aerospace industry. Because of the ability to form large complex structures, it is possible to design with considerably greater latitude and imagination, and a repeating result is fewer component pieces and greatly reduced fasters.

While the benefits of these processes are impressive, it is nonetheless an important recognition that these forming processes are "high technology" processes, requiring a great deal of understanding of both material and the process to achieve repeated success in their implementation. Variations in materials of forming parameters can readily lead to both success and failure for any given component [5]. It is at least in part an acknowledgement of this which has lead to the increased research in this area recently [7-11]. This research continues to address a range of titanium alloys, but the Ti-6Al-4V alloy has received the greatest attention, at least in the U.S. This alloy has been widely used in the aerospace industry, and has been found to exhibit excellent superplastic properties in the conventionally produced form, a combination of factors which have stimulated the interest in its superplastic formability.

However, several microstructural features which can influence the superplasticity of the alloy have been found to vary from heat to heat [8]. Also, the measurement of superplasticity as indicated by strain rate sensitivity, m, has not been clearly defined [10,11]. The primary intent of the work, therefore, was to establish a fairly accurate characterization of the flow properties of the alloy for large plastic strains at 927°C. The "strain hardening" characteristics are evaluated and associated microstructural features correlated with the hardening. An experimental test was conducted in an effort to provide an accurate measurement of m for a fixed grain size and strain level, both of which have been cited as changing during strain rate change measurements and therefore raising questions about the validity of such tests [12-14].

## Experimental

### Materials

Conventionally produced Ti-6Al-6V alloy sheet (1.63 mm thick) was used for this study. Three initial grain sizes were developed for the experimental program. These grain sizes were 6.4 μm 9.0 μm, and 11.5 μm, respectively. While the first one represents the as-received material, the others were obtained by holding the as-received material at 955°C in vacuum for 2 and 7 hours, respectively. Each of these reported grain sizes were actually measured from tensile specimens that were raised to the test temperatures of 927°C in the test fixtures, held for 10 min, and quenched, and therefore, represent the grain sizes immediately prior to the tensile test.

### Tensile Tests

In testing superplastic materials, constant crosshead speed generally leads to a decreasing strain-rate within the specimen gage length, which additionally complicates the measurement of their mechanical behavior. An effort to maintain constant strain-rate included the following: (1) the specimen filet region was reduced to a minimum to reduce its contribution to the overall extension, and (2) the crosshead speed was programmed to increase with specimen elongation so as to maintain a constant strain rate (assuming the elongation to arise entirely out of the gate length). This latter function was performed by attaching a variable speed stepping motor to the drive gear of the Instron machine and using a programmable profiler (Versatrak) to alter the input voltage to the motor.

Temperature during the test was held at 927 ±2°C with less than a degree variation along the specimen length. A total of four strain rates were investigated. For two of these, $2 \times 10^{-4}$/sec and $10^{-3}$/sec, three specimens were tested at each rate, and subsequently unloaded and quenched at true strain levels of 0.3, 0.66, and 1.0, respectively. For the other two strain rates, $5 \times 10^{-5}$/sec and $5 \times 10^{-3}$/sec, one specimen was tested at each rate and was quenched after unloading. The quenching was performed by injecting argon gas directly on to the specimen at high velocity and was intended to "freeze" the high temperature microstructure. A temperature drop of 600°C within 2 min was achieved by the argon quench.

Although the material was found to be fairly isotropic with respect to superplastic properties, all tests were conducted in the transverse orientation in order to exclude the directionality variable.

A 2.54 mm diameter circle grid pattern etched on the specimen prior to testing was utilized for strain measurements upon unloading. Strain variation was observed to increase with increasing strain-rate, but the specimen center exhibited strain rates, close to the desired rate. Areas selected for subsequent metallographic examination were therefore taken from specimen centers.

The rolling planes (i.e., planes parallel to the sheet surface) were examined for grain size in each specimen after quenching and strain measurement. The static grain growth characteristics were available for comparison from previous work (15) and micrographs are shown in Fig. 1. These samples were held for various times (without deformation) at 927°C and water quenched. All grain size measurements in this work were carried out by three methods: (1) linear intercepts along the rolling direction, (2) linear intercepts along the trnasverse direction, and (3) Hilliard circle (20 cm). Both $\alpha$ and $\beta$ grains are considered for grain size evaluation. Grain sizes reported are average grain diameters, as determined by the following relation [9]: 1.68 L/MN, where N = number of intercepts, M = magnification and L = length of line. Bars on data points are used in the next section (Fig. 2) to incorporate this directional variation, the grain size from Hilliard circle lying within this range. Since prior studies have indicated that the volume fraction of two phases may influence the superplastic characteristics [8], the volume fraction of beta was measured as a function of time at 927°C using a linear intercept method.

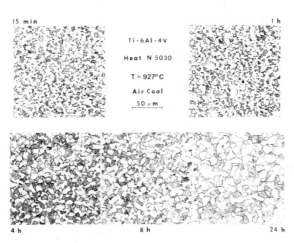

Fig. 1.   Static grain growth shown after different furnace holding times and air cooling.

An evaluation of the Ti-6Al-4V alloy was also conducted with the objective of determining the strain-rate sensitivity exponent, m, in which grain size and strain variations were not a factor. In this evaluation, it was intended to measure m for constant strain and grain size, and over the strain rate range encompassing that for which superplastic deformation occurs. Constant strain-rate tests were conducted with a technique which was previously described. With the definition of strain-influenced grain growth kinetics (Fig. 2), it was therefore possible to conduct such a test in which the grain size at a given strain is fixed.

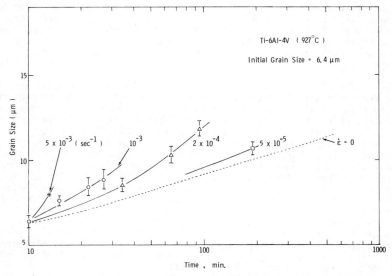

Fig. 2   Grain growth kinetics at four different tensile strain rates
compared with static kinetics for an initial grain size
of 6.4 µm.

In the evaluation, five tensile specimens were tested at 927°C (1700°F)
under constant strain rate conditions at strain rates of $2.7 \times 10^{-5}$,
$1.8 \times 10^{-4}$, $3.5 \times 10^{-4}$, $6.5 \times 10^{-4}$, and $2 \times 10^{-3}$ s$^{-1}$. At a strain of about
0.15, the strain rate was instantaneously changed by 40%; this strain rate
sustained for a relatively small strain (~.02) increment, after which the
strain rate was instantaneously returned to its original value.

The grain sizes of each of the five samples were made approximately
equivalent at the moment of the strain rate change by holding the samples
for various times at the test temperature before commencing the test.   In
order to achieve equivalent grain sizes at the time of the strain rate
change test, the influence of strain rate on grain growth kinetics had to be
considered.   Therefore, the extent of grain coarsening expected during deformation at a given strain rate was considered independently of that expected during static exposure conditions.   Verification of grain size was
made by conducting metallographic examination of the test specimens which
were quenched in room temperature argon flow immediately after the strain
rate change test.   The grain sizes for the test specimens were measured to
be in the range of 7.5 to 8.3 µm.

## Results and Discussion

The sheet alloy was found to exhibit an equiaxed microstructure, as can
be seen in the grain growth sequence of Fig. 1.   Grain coarsening does occur
for this alloy at elevated temperautres, and the rate of coarsening
increases with temperature as shown in Fig. 2 [15].   The beta volume
fraction at 927°C appears to stabilize fairly rapidly at about 45 to 50% as
shown in Fig. 3.

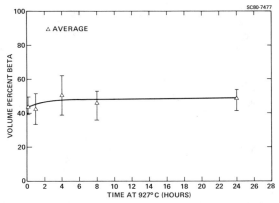

Fig. 3   Volume fraction of the beta phase as a function of time at 927°C.

During tensile tests at constant strain-rate, a continuous load rise was observed at the slower strain-rates and load drop at the higher strain rates. Because of the high degree of strain uniformity here, it is reasonable to calculate instantaneous cross section from axial strain assuming volume constancy and thereby compute the instantaneous flow stress even though the load drops. True stress-strain curves for all three initial grain sizes are shown in Fig. 4 for strain-ratesof $2 \times 10^{-4}$/sec and $10^{-3}$/sec while those for 6.4 μm material are also shown for $5 \times 10^{-5}$ and $5 \times 10^{-3}$/sec. The solid parts of the curves indicate good strain uniformity

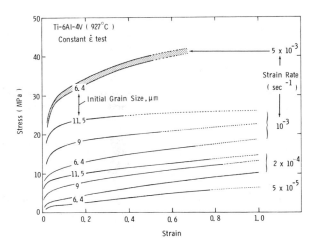

Fig. 4   True stress-strain curves for three initial grain sizes of 6.4, 9, and 11.5 μm, respectively. The bottom-most curve shown is for a strain rate of $5 \times 10^{-3}$/sec, and the top-most one for $5 \times 10^{-3}$/sec, both for the 6.4 μm grain size material. The dotted part indicates significant strain gradient developing in the specimen.

in the specimen, while the finely dotted parts indicate the uncertainty due to increasing nonuniformity. A serrated stress-strain curve was shown for a strain-rate of $5 \times 10^{-3}$/sec, perhaps indicating the occurrence of dynamic recrystallization. The shaded band envelopes the maxima and minima of such a curve.

All of these curves exhibit hardening of the material with increasing deformation. It appears that the extent of hardening with strain decreases with increasing strain rate up to near $10^{-3}$/sec and then increases again for $5 \times 10^{-3}$/sec. The stress level is also found to increase with increasing grain size in agreement with previous findings.

The observed hardening appears to be related to grain growth occurring during tensile test, as illustrated in the micrographs in Fig. 5. Figure 5 shows that the initial grain size of 6.4 μm grows on the average to 7.6 μm, 8.4 μm, and 8.9 μm following straining to 0.3, 0.6 and 1.0, respectively, at $10^{-3}$/sec. Measurable grain elongation has also been observed generally increasing with increasing strain. As much as 20% grain elongation has been recorded at a strain of 1.0, even though these measurements may not be highly reliable.* Grain elongation at $2 \times 10^{-4}$/sec appeared to be no more than 12% at $\varepsilon = 1.0$. Similar grain growth characteristics were observed for the other initial grain sizes.

After testing at $\dot{\varepsilon} = 10^{-3} \, \mathrm{S}^{-1}$ (927°C)

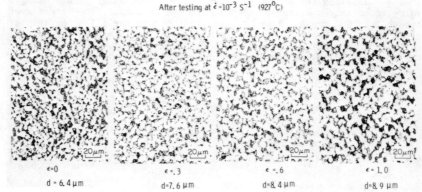

Fig. 5   Grain growth observed after different amounts of strain at $10^{-3}$/sec (927°C) and quenched in argon. Light phase is α and the dark phase is transformed β.

Grain growth kinetics plots of Fig. 2 illustrate the average grain size vs. time of exposure with and without concurrent deformation. Figure 2 is for an initial grain size of 6.4 μm at all the strain rates under study. It is clear from these plots that grain growth is enhanced by the imposed strain rate and despite some scatter in the data, the trend persists for all starting grain sizes. Stress-assisted or strain-rate assisted grain boundary mobility have been observed before in Pb [16] and Cu-Al-Fe alloys

---

*It is possible that some growth of the alpha phase occurred on cooling even though the samples were quenched in argon. Such growth of the alpha phase would probably result in "rounding" of the grains because of energy considerations and thereby reduce the aspect ratio.

[17] and relates to the enhanced grain growth observed in the present investigation.

With evidences of hardening as well as grain growth during deformation available, an attempt is now made to relate the two in Fig. 6. This is done by cross-ploting flow stress from Fig. 4 against grain size during deformation, obtainable from the grain growth kinetics plots such as that of Fig. 2 (for each applied strain-rate). The solid curves in Fig. 6 are thus grain growth hardening curves for initial grain sizes of 6.4, 9, and 11.5 μm for various applied strain-rates. The flow stress corresponding to each initial grain size is indicated by a data point taken from the "knee" between the elastic and grossly plastic parts of the load-extension plots and represents as much as 4% total strain.

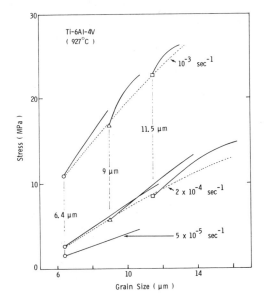

Fig. 6  The change in flow stress as a function of grain size changing during tensile tests indicated by solid curves. The data points indicate the three initial grain sizes from which tests were started. The appropriate strain rates are shown on the plot.

The locus of initial flow stresses for each material, shown dotted, defines the basic, or initial, grain size dependence of flow stress for the indicated strain-rate. The solid curves rising above this indicates that deformation produces a small amount of hardening in excess of what might be expected from grain growth alone. This hardening may arise from the fact that grains do elongate in the course of deformation. On the basis of Ashby and Verral's grain switching model of superplastic flow [18], the departure from equiaxed grain shape is expected to increase the resistance to boundary sliding. Even though their model was proposed for a single-phase material, this would intuitively appear to be true since the ease of grain boundary shear would be maximum when boundaries are at 45° to the tensile axis, which is more likely for an equiaxed grain structure.

Another possible rationale for the excess hardening might lie in the phenomenon of "grain clustering," [19] whereby some of the neighboring like grains (e.g., α-α or β-β) might behave as a unit rather than individual grains while participating in the grain switching mechanism. This may increase the "effective" grain size and, therefore, flow stress. Irrespective of the source, this hardening is a significantly smaller effect in comparison to the primary effect of grain growth hardening.

The grain size dependence of flow stress in Fig. 7 is found to change from ~ $d^2$ dependence for the lower strain-rates to less than ~ d for the higher strain-rates. Intermediate strain-rates actually show a gradual changeover from the higher power to the lower power dependence. While these dependencies are larger than those observed by Lee and Backofen [1], it is interesting to note that diffusional mechanisms of superplasticity to predict grain size dependencies of a similar order. For example, Nabarro Herring model [20,21] of through the grain diffusion predicts $\sigma \propto d^3$; Ashby-Verral's [18] diffusional accommodation of grain switching predicts a ~ $(d^2 + d^3)$ dependence. Dislocation creep mechanisms [23] on the other hand, predict $\sigma \sim d^{1/2}$ and the changeover in slope observed in Fig. 7 is believed to relate to a change from a diffusional mechanism for lower strain rates and/or smaller grain sizes to a dislocation creep mechanism for higher strain rates and/or larger grain sizes.

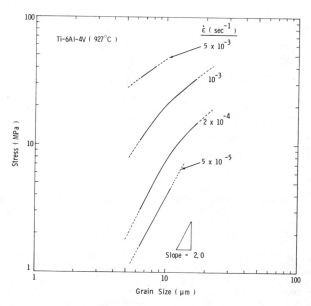

Fig. 7   Grain growth hardening curves at various strain rates obtained by averaging envelope of curves in Fig. 6.

The influence of grain growth on strain rate sensitivity is illustrated in the next series of figures. The stress vs. strain-rate plots in Fig. 8(a) are obtained from step strain-rate test in contrast to the constant strain-rate test discussed so far. The test technique has been illustrated in

Fig. 2, which does cause accumulation for strain as strain-rate is progressively stepped up. Hence the higher strain-rate portions of these plots contain some grain growth hardening. However, since the accumulated strain does not exceed 20% and its hardening contribution at the higher strain-rate is very small, the $\sigma$ - $\dot{\varepsilon}$ curves in Fig. 8(a) are considered to represent the three initial grain sizes fairly well. Results from a previous study on a 20 $\mu$m grain size material is also included [15]. While significant grain growth hardening is exhibited at the lower strain-rate, the curves approach each other at the high strain-rate end, thereby suggesting a drop in m (= d log $\sigma$/d log $\dot{\varepsilon}$) with increasing grain size. Figure 8(b) shows m (i.e., slope from Fig. 8(a) as a function of log $\dot{\varepsilon}$ for the different grain sizes. These plots show a maximum in m at an intermediate strain-rate. The value of maximum m drops with increasing grain size as well as the strain-rate at which $m_{max}$ is reached for each grain size. Data of this nature have been previously reported by Lee and Backofen [1].

These results on different grain sizes compare well with those obained from the same material after different amounts of strain. Figure 8(c) shows this in a material with initial grain size of 6.4 $\mu$m after deforming to a true strain of 0.45. While the initial test accumulated a strain of 0.17, subsequent deformation at $2 \times 10^{-4}$/sec followed by step strain-rate test within the strain range of 0.45-0.52 clearly shows the hardening effect with associated drop in m.

The change in m with deformation is best studied in a constant strain-rate test, with periodic m determinations by strain-rate departures of small magnitude. This is schematically illustrated in Fig. 9 where the strain-rate was incremented by 25% only, maintained for 2-3% plastic strain and brought back to the original rate. It is felt that the small strain rate departures did not alter the pattern of microstructural change that occurs at the original rate. The value of m here is determined from m = log $(P_2/P_1)$/ log $(\varepsilon_2/\varepsilon_1)$, where $P_2$ is the load corresponding to a strain-rate of $\varepsilon_2$ and $P_1$ is the load at $\varepsilon_1$.

These test results, shown in Fig. 10 for all three grain sizes, indicate than m decreases with strain. The values of m are found to be largest at the lowest strain-rate and for the smallest grain size. The drop is more gradual for this case also. Coarser grains give rise to a lower m which also drops more rapidly with strain. During an actual forming operation, strain-rate in a deforming element changes continuously. However, it is this instantaneous value of m that determines its necking resistance at any instant. Thus m determined this way though time-consuming, is perhaps the most meaningful parameter that can be related to ductility. Since m changes with strain as well as strain rates, the question of the variation of m with strain rate for constant structure remains. This question is addressed in the following test sequence.

With such a thorough characterization of grain growth kinetics, as influenced by deformation, it was then possible to design a test sequence whereby the measurement of m could be established for fixed grain size and strain level as described under the Experimental section. An example of a stress-strain curve by which the strain-rate sensitivity was measured as shown in Fig. 11.

This m determined this way provides a measure of the total rate sensitivity, including both stress and strain hardening rates. It is expected that the relatively small change in strain rate, and for its short duration, does not cause any appreciable change in the structure of the material. That this

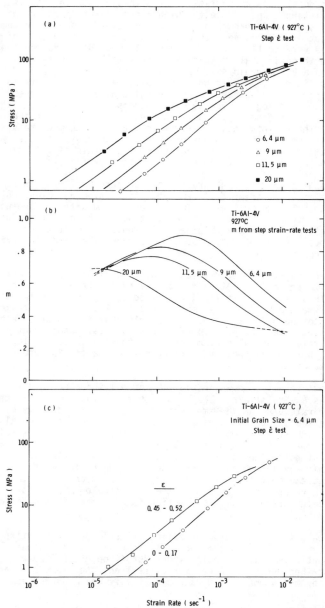

Fig. 8   (a) Stress vs strain-rate plots for four different initial grain
sizes obtained by step strain-rate test. The data for 20 μm grain
size material is from previous unpublished work. (b) Strain-rate
sensitivity, m, given by the slopes of the curves in (a).
(c) Stress vs strain-rate plots for 6-4 μm grain size material
initially (i.e., up to $\dot{\varepsilon}$ = 0.17), and after a strain of 0.45 at a
rate of $2 \times 10^{-4}$/sec, showing hardening contribution due to the
deformation exposure.

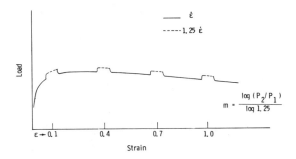

Fig. 9   A schematic representation showing how instantaneous measurements of m was made at periodic intervals during the tensile test, by strain-rate increments of 25%.

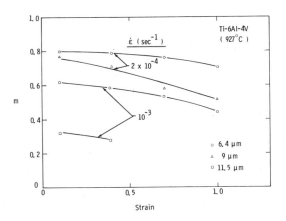

Fig. 10   Instantaneous values of strain-rate sensitivity during tensile test measured by the technique illustrated in Fig. 6, shows decreasing m with increasing strain.

is the case is indicated by the observation of no discontinuity in the stress-strain curve after the strain rate change was made.

Results of these tests are shown in Fig. 12 in which m is presented as a function of log strain rate. Also shown for comparison are the curves for this alloy as determined by the step/strain rate method, which corresponds to a grain size of 6.4, 9, 11.5, and 20 µm, and encompasses a strain range from 0 to about 0.2. These results indicate that the maximum values of m for the Ti-6Al-4V alloy exceed 0.5, and vary continuously with strain rate in a manner similar to that observed for the step/strain rate test.

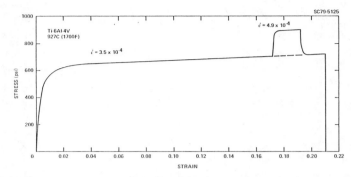

Fig. 11   True stress vs true strain for a constant strain rate of
3.5 × 10⁻⁴/sec.  Strain rate was temporarily increased to
4.9 × 10⁻⁴/sec to permit a determination of m.

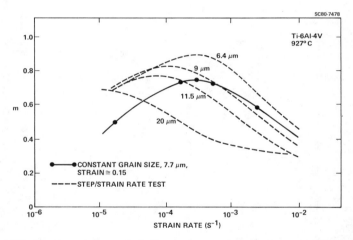

Fig. 12   Strain rate sensitivity exponent, m, vs log strain rate corresponding to approximately constant strain and grain size.
For comparison, results of m determination using the step/
strain rate method are also presented.

The maximum in m vs log $\dot{\varepsilon}$ indicates the presence of an inflection in the
log σ vs log $\dot{\varepsilon}$ curve observed for superplastic materials tested by the step/
strain rate method (Fig. 8).  The characteristic of the m vs log $\dot{\varepsilon}$ curve over
the superplastic strain-rate range is consistent, in principle at least, with
the concepts proposed by Ashby and Verrall [18], who suggest, as a mechanism
of superplasticity, a combination of creep by dislocation climb and diffusion
accommodated grain boundary sliding involving a threshold stress.

It is important to note that the results of these tests are for a relatively small strain, and cannot be assumed to necessarily reflect the charac-

teristics of the alloy at much higher strains, since considerable grain growth may occur during straining.  It was previously shown that m decreases with strain, and therefore the m vs log $\dot{\varepsilon}$ curve for constant grain size and strain would also be expected to decrease with increasing strain in the superplastic strain-rate range as observed for the step/strain-rate test (Fig. 8).  It is realized that a complete description of the history-dependent behavior is not provided by either test techniques discussed here.  However, since the structure is not altered by these small strain-rate increments, the present technique offers a truer measure of current rate sensitivity, important in forming studies.

## Conclusions

1. The Ti-6Al-4V alloy undergoes strain hardening at 927°C and during superplastic deformation.  The hardening observed is predominantly, though not entirely, due to grain coarsening which occurs concurrent with the deformation.

2. The strain-rate sensitivity, m, decreases with strain during constant strain-rate deformation at 927°C.  The decrease is consistent with hardening which is observed, and is therefore primarily related to grain coarsening during deformation.

3. Grain growth kinetics at 927°C are accelerated by superplastic deformation, and increase with increasing strain rate.

4. The strain rate sensitivity, m, as measured for constant grain size and strain level varies with strain rate, and exhibits a maximum of 0.75 at about $4 \times 10^{-4}$ s$^{-1}$ for a grain size of about 7.7 $\mu$m.  The m vs log $\dot{\varepsilon}$ curve determined for these conditions is similar to that measured by the step/strain rate method.  The maximum in m provides firm evidence of an inflection in the log $\sigma$ vs. log $\dot{\varepsilon}$, supporting the sigmoidal shape normally observed for this alloy as tested by the step/strain rate method.

## Acknowledgements

The authors are indebted to Mr. L. F. Nevarez who conducted the tensile test, and to Mr. R. A. Spurling who conducted the supporting metallographic evaluation.  The research was conducted under the Independent Research and Development Program of Rockwell Internations.

## References

1.  D. Lee and W. A. Backofen, Trans. AIME, 1967, Vol. 239, pp. 1034-1040.
2.  C. H. Hamilton and G. E. Stacher, Metal Progress, March 1976, pp. 34-37.
3.  N. G. Tupper, J. K. Elbaum, and H. M. Burte, Journal of Metals, September 1978, p. 7.
4.  F. H. Froes, C. F. Yolton, J. C. Chesnutt, and C. H. Hamilton, Proceedings of Forging and Properties of Aerospace Materials, The Chameleon Press, London, (1978), p. 371.
5.  C. H. Hamilton, Proc. Formability: Analysis, Modeling and Experimentation, Chicago, TMS-AIME publication, 1978.
6.  E. D. Weisert and G. W. Stacher, Metal Progress, March 1977, p. 33.
7.  A. K. Ghosh and C. H. Hamilton, Met. Trans. 10A, 1979, p. 699-706.
8.  N. E. Paton and C. H. Hamilton, Met. Trans. 10A, 1979, p. 241-250.

9.   S. P. Agrawal and E. D. Weisert, Proc. North American Metalworking Conference VII, Ann Arbor, Michigan, 1979, SME publication, p. 197.
10.  A. Arieli and A. Rosen, Scripta Met. 10, 1976, p. 471-475.
11.  A. Arieli and A. Rosen, Met. Trans. 8A, 1977, p. 1591-1596.
12.  A. K. Ghosh and R. A. Ayers, Met. Trans. A, 1976, Vol. 7A, pp. 1589-1591.
13.  J. Hedworth and M. J. Stowell, J. Mat. Sci. 6, 1971, p. 1061-1069.
14.  G. Rai and N. J. Grant, Met. Trans. A 6A, 1975, p. 385-390.
15.  C. H. Hamilton and S. P. Agrawal, unpublished research, Rockwell International Corporation, Los Angeles, 1977.
16.  R. C. Gifkins, Trans. AIME, 1959, Vol. 215, pp. 1015-1022.
17.  P. Ducheyne and P. DeMeester, J. Mat. Sci., 1974, Vol. 9, pp. 109-116.
18.  M. F. Ashby and R. A. Verral, Acta. Met., 1973, Vol. 21, pp. 149.
19.  D. J. Dingley, "Trends in Physics," European Physics Soc., Geneva, 1973, p. 319.
20.  F. R. N. Nabarro, Report on Conf. on Strength of Solids, Phys. Soc., London, 1948, p. 75.
21.  C. Herring, J. Appl. Phys., 1950, Vol. 21, p. 437.
22.  R. L. Coble, J. Appl. Phys., 1963, Vol. 34, p. 1679.
23.  N. F. Mott, Proc. Phys. Soc. B, 1951, Vol. 64, p. 729.

# MICROSTRUCTURAL ASPECTS OF SUPERPLASTIC FORMING OF TITANIUM ALLOYS

M.E. Rosenblum*, P.R. Smith[+] and F.H. Froes[+]

*Metcut-Materials Research Group
P.O. Box 33511
W-PAFB, OH 45433

[+]Metals and Ceramics Division
Materials Laboratory, AFWAL
W-PAFB, OH 45433

Superplastic forming of titanium alloys is now a viable production method for fabrication of complex sheet metal parts. The present paper gives a detailed indepth discussion, with practical examples, of the microstructural features important in this process, and how they should be treated. In particular dynamic effects are considered, where microstructure changes occur throughout the forming operation. Microstructural features considered are volume fraction, grain size (mean linear intercept), grain size distribution, grain shape and contiguity. Standardization of the measurements reported should arrest the confusion in this technology area.

## Introduction

The mechanical properties of titanium and titanium alloys make them an attractive choice for use in aerospace systems. However titanium is an inherently expensive material, particularly because of high fabrication and machining costs. In recent years the problem of this high cost has received considerable attention [1] and has resulted in advances in a number of technologies including isothermal forging, superplastic forming (SPF), diffusion bonding (DB), casting and powder metallurgy [2]. The present paper will discuss the SPF aspects of the combined SPF/DB technology as applied to sheet structures. For the combined process savings as high as 50% over conventional fabrication techniques have been indicated [3]. In particular the microstructural aspects will be discussed and suggestions made for improvements in the data developed so that it relates better to SPF/DB behavior.

It is well established that superplastic behavior is a function of the initial state of a material [4,5]. However, changes occurring during the process, which continuously affect material behavior, have not been sufficiently recognized [6,7]. Until recently, it was assumed that once the starting conditions were met, particularly grain size, superplastic forming was essentially controlled by temperature and strain rate. Of metallurgical changes which influence superplasticity during deformation grain growth has been the most extensively documented. When the material is deformed, grain growth is enhanced and microstructural morphology may be changed, for example, some elongation and clustering of alpha may occur [6]. The simple sigmoidal stress-strain rate plot displaying superplastic behavior inadequately describes the actual process because of these dynamic changes. A time dependent parameter, which can be shown as a third dimension on the plot, is required. Thus the superplastic behavior is represented by a surface. A simple consideration of time is insufficient, parameters such as strain and/or grain size are not dependent on time alone, but vary with other factors such as strain rate [6]. In Fig. 1, is shown with

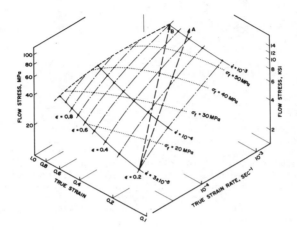

Figure 1.   Three dimensional representation of superplastic behavior.

the third axis as true strain [7], although this may not be the optimum parameter to use. With the identification of a suitable third axis, a constitutive equation can be derived for this surface, as in the original two dimensional work [8], and applied to manufacturing processes.

The microstructural features which are of use in explaining (and therefore predicting) superplastic behavior are those that can be directly measured with relative ease. Once the relationships between these features and behavior are established improvements can be made in application of superplastic forming of complex shapes. Based on observations by various researchers the measurable microstructure features of importance are:

> volume fraction
> grain sizes
> grain size distribution
> grain shape
> contiguity.

Anomalies such as banding and so-called "blocky alpha" may be important [4] but will not be considered here.

Measurement procedures and derivations of microstructure parameters from data developed is critical. Statistically significant measurements of microstructure must be made. Additionally, the material examined must spend time at temperature and must be rapidly cooled so as to retain the high temperature structure. In contrast slow cooling (including air cooling) results in changes in microstructure from that actually present during superplastic forming and must be considered unacceptable (Fig. 2). Further, measurements should be made on all three major planes as a structure deformed differently in 3 directions is being considered. Once sufficient data is obtained it is likely that measurements on one plane will be adequate. Since titanium alloys used for superplastic forming are

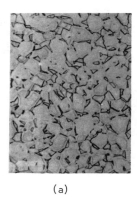
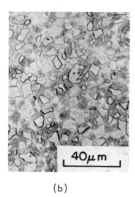

(a)                                          (b)

Figure 2.   Effect of cooling rate on microstructure of Ti-6Al-4V, 925°C(1700°F).
a) Furnace cool (alpha phase grain size, 9μ), b) water quench (alpha
phase grain size, 6μ).

generally two phase, consideration of both phases is important at superplastic
forming temperatures where volume fraction of the phases is significant.

## Experimental Procedures

All materials in the grain growth studies were taken from the same
Ti-6Al-4V mill annealed sheet.  Heat treatments were performed in evacuated
silica capsules with a water quench from temperature.  Quantitative measurements were made on the ground glass screen of a metallograph at 980X actual
magnification.  Area fractions were measured with ten throws on each of five
random areas and a sixteen point grid.  Size and shape were derived from
measurements with a 5.1 cm circle or 10.2 cm crossed lines, three throws
on each of five areas.  Whenever possible, the point and line counting were
done on the same area.  Mean linear intercept is used to represent grain
size.

## Results and Discussions

1.   Volume Fraction

Equilibrium volume fraction of the two phases is a function of alloy
chemistry and temperature.  Experimental evidence indicates that the time
to reach equilibrium, at least starting with a mill annealed, nonequiaxed
structure is generally longer than had been thought (Fig. 3).  It is important to note that these changes in volume fraction with time at temperature may be partly caused by shape changes from the initial structure.

Determining the effects of volume fraction on superplasticity is
difficult.  A change in temperature will also vary other parameters in
the material.  Evidence provided by work in another alloy, Corona-5, where
superplastic forming at 870°C has been successful [9] suggests that a
specific volume fraction may not be critical.  A 50:50 mixture of the two
phases is generally considered optimum for superplastic forming behavior [5]
though this must be modified because the behavior generally improves with
temperature.  Volume fraction of the alpha phase in Ti-6Al-4V at 925°C is
60% to 70% while the volume fraction in Corona-5 formed at 870°C was
measured to be 32%.

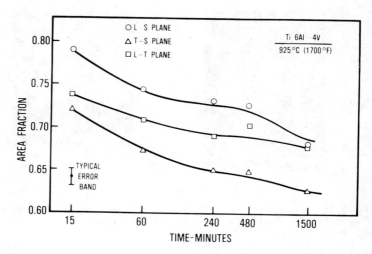

Figure 3.   Variation of area fraction with time, measured on 3 major axes of Ti-6Al-4V mill annealed sheet.

## 2.   Grain Size

Grain size has the greatest microstructural influence on superplastic behavior; as measured grain size increases resistance to necking (or superplastic behavior) decreases. Some effects of grain size distribution have also been noted but the work has been semiquantitative [10]. Grain growth is produced by time at temperature and is enhanced by deformation, with deformation rate also having an effect [6]. In most cases, quoted grain sizes have been the result of measuring all boundaries without discrimination, using the linear intercept method. In some cases data has then been corrected to a so-called "true grain size" following the method proposed by Rostoker [11]. In the modification of this method (for example, [4]) grain size is equated to 1.68 L/MN where L is the length of the line of intercept, M is the magnification and N is the number of intercepts; essentially 1.68 times the mean linear intercept ($\bar{L}_3$). While use of this parameter is not incorrect it is unfortunate that it has been employed since it adds nothing to the definition of the size parameter. Further it causes confusion since it prohibits direct comparison of "grain size" from other work. It is recommended that the mean linear intercept, $\bar{L}_3$, be used to allow direct comparisons of "grain size".

A further confusion which has been added is the practice adopted by some of reporting an as-received "grain size" in mill annealed material. The term "mill annealed condition" is at best ill-defined, at worst totally inadequate (see for example, [12]). The microstructure exhibited in mill annealed material varies from a heavily worked structure to equiaxed grains depending on the conversion practice used. No meaningful grain size can be attached to the worked microstructure. In all cases a grain size, useful in explaining superplasticity, can only be obtained by taking the material to the SPF temperature (allowing recrystallization to occur) water quenching, and measuring the required parameter. Again it is recommended that this practice be adopted to avoid confusion.

In studies of superplasticity in alpha-beta titanium alloys little work has been done to separate effects of the two phases. A common assumption is to regard the microstructure as consisting of alpha particles in a beta matrix. This has been caused partly by using the microstructural information available at room temperature without attempting to preserve the high temperature structure and morphology (the slow cooling combined with manipulation of the microstructure is used for better metallographic definition). At superplastic forming temperatures, Ti-6Al-4V with 60-70% alpha phase must be regarded as a mixture of two phases. In such a configuration the boundaries of both phases can contribute to sliding and accommodation [13], possibly to a differing extent. Consequently, until the controlling mechanisms are defined each type of boundary should be considered separately.

In the derivations that follow it is assumed that the material is isotropic. Certain microstructural properties can be measured without additional special assumption and statistically exact relationships can be derived. These characteristics, including volume fraction, surface area per unit volume and mean linear intercept [14], will be used to arrive at a system for defining measurements relating to superplasticity. Two cases will be considered, measurement of all boundaries(defined as space filling and contiguous) and separation of phases as derived from consideration of dispersed particle interactions. All information can be obtained from volume fraction measurements (grid counting) and linear intercept counting. For the case of "all boundaries" the equations are simple, [15]

$$\bar{L}_3^t = 1/N_L^t = 1/P_L^t$$

$$\bar{L}_3^t = 1/(P_L^{\alpha\alpha} + P_L^{\beta\beta} + P_L^{\alpha\beta})$$

$$S_v^t = 2N_L^t = 2P_L^t$$

$$S_v^t = 2(P_L^{\alpha\alpha} + P_L^{\beta\beta} + P_L^{\alpha\beta}).$$

For the two phase mixture derivations, the complicating factors are volume fraction of the phases and double counting of alpha-beta boundaries,

$$V_v^i = P_p^i/P_t$$

$$\bar{L}_3^i = V_v^i/N_L^i = V_v^i/(P_L^{ii} + \tfrac{1}{2}P_L^{\alpha\beta})$$

$$S_v^i = 4N_L^i$$

$$\phantom{S_v^i} = 4P_L^{ii} + 2P_L^{\alpha\beta}$$

and additionally,

$$\sigma = \bar{L}_3 + \lambda$$

$$\sigma^i = 1/N_L^i$$

$$\phantom{\sigma^i} = 1/(P_L^{ii} + \tfrac{1}{2}P_L^{\alpha\beta})$$

$$\lambda^i = (1-V_v^i)/N_L^i$$

$$\phantom{\lambda^i} = (1-V_v^i)/(P_L^{ii} + \tfrac{1}{2}P_L^{\alpha\beta})$$

Also because of the shared boundaries,

$$S_v^t \neq S_v^\alpha + S_v^\beta$$

With

$S_v^t$ = Total internal surface area per unit test volume

$S_v^i$ = ith phase internal surface area per unit test volume

$\lambda^i$ = Mean free path

$\bar{L}_3$ = Mean intercept length of 3D bodies

$\sigma^i$ = Particle spacing

$P_L$ = Number of intersections with phase surfaces per unit length of test line

$N_L$ = Number of intersections of particles or cells per unit length of test line

$V_v$ = Volume of particles or phase per unit test volume

$P_p$ = Number of test points inside particles or phase

$P^t$ = Total number of test points on counting grid

Again, these equations are valid, in the strictest sense, only for equiaxed structures. If there are suspicions that the initial structure (even after time at temperature) is elongated and elongation occurs during deformation, measurements should be made on all three planes.

Grain size distribution is complex in real structures. The two dimensional distribution observed on a plane of polish is the product of the distribution caused by slicing a three dimensional structure and the actual size distribution in three dimensions. Using the linear intercept method of measurement superimposes a third distribution. It is impossible, at this time, to accurately deconvolute such a combination. Some attempts could be made to provide a semiquantitative comparison of linear intercept distribution shapes. The first attempts in this direction have been made in looking at the size distribution of as-received structures [10].

Several methods of measuring and presenting grain size can be used. The most common of these has been counting all intercepts with boundaries (alpha-alpha, beta-beta and alpha-beta) and using a total grain size concept. As pointed out above alpha and beta grain sizes can be considered separately. When this separation is applied, Fig. 4, to heat treated material no differences in growth behavior is evident.

Drawing a parallel with other systems [16] the grain boundary sliding rates in superplastic forming of Ti-6Al-4V are likely to be highest for the alpha-beta boundaries. Consequently, another measure of grain size, defined using only alpha-beta boundaries, could be considered. This results in a total alpha grain size which ignores alpha-alpha interfaces (alpha grains which are close but with a definite line of demarcation are considered as separate). In Figs. 5 and 6 the results of total alpha counting are presented for Ti-6Al-4V with and without yttrium addition. All three directions of the thermally treated sheet are indicated and seen to behave somewhat similarly with sigmoidal shaped growth curves. With this new grain size the difference

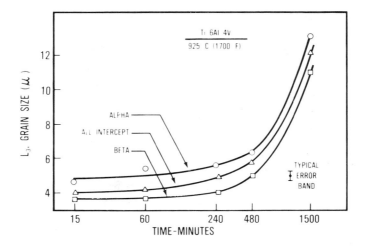

Figure 4.   Grain growth in heat treated mill annealed Ti-6Al-4V, mean linear
            intercept measured on L-S plane.

in growth behavior caused by yttria additions is evident.   Initial grain
size is smaller; growth is retarded for a time, displacing the sigmoidal
curves to longer times.   It has been suggested that ultrafine $Y_2O_3$ precipitates on grain boundaries and thus prevents boundary migration [17].   With
time at temperature, the precipitates could start to dissolve and growth
accelerates [18].   This reaction would explain the increase in growth rate
at long hold times.

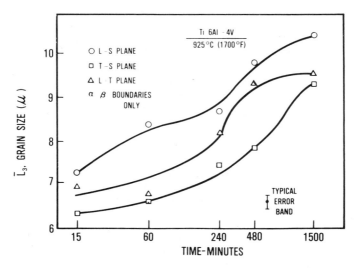

Figure 5.   Grain growth in mill annealed Ti-6Al-4V, mean linear intercept
            of total alpha ($\alpha$-$\beta$ boundaries only).

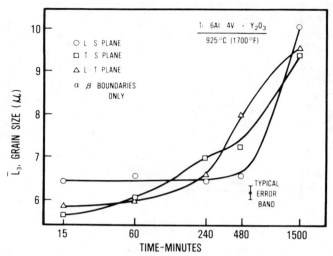

Figure 6.   Grain growth in mill annealed Ti-6A1-4V+Y$_2$O$_3$, mean linear intercept of total alpha ($\alpha$-$\beta$ boundaries only).

Mean linear intercept has been used in this paper to represent grain size, since it is an easily visualized concept. To keep the measure more closely related to specific mechanisms, grain boundary sliding for example, a slightly different parameter could be used: surface to volume ratio. This parameter allows consideration in terms of the phase surface area available for sliding or rotation and to more clearly visualize the process. For the present, however, mean linear intercept should be used.

## 3. Grain Shape

Elongation of the grains occurs during superplastic forming, an effect of deformation which may contribute to the lowering of resistance to neck-ing though no quantitative data has been developed.  Such elongations of up to 20% have been observed.[6]  It appears that elongations in the initial microstructure are more deleterious [10] and those occurring during the process will not have a large effect on performance.  However, the effect should continue to be examined, a simple measurement of linear intercept parallel and perpendicular to the major sheet material direction will pro-vide the necessary information.

## 4. Contiguity

With the volume fraction normally found at superplastic forming temper-atures appreciable contact between each phase is found.  The sharing of boundaries, alpha-alpha and beta-beta can be expressed by utilizing a contiguity (C)$_x$ parameter [19].

$$C_{ii} = \frac{2S_v^{ii}}{2S_v^{ii} + S_v^{\alpha\beta}}$$

or

$$C_{ii} = \frac{P_L^{ii}}{P_L^{ii} + \frac{1}{2}P_L^{\alpha\beta}}$$

these expressions indicate sharing behavior.

Clustering, a phenomena found in heavily deformed structures can also be included here. It has been shown that alpha particles will cluster together with a surrounding denuded region. The extent of effects of this type of grain movement on forming are not yet known. The microstructural study of these effects must be performed with care as the above equations cannot be used because some measure of contextural information must be established. This should not only be a mean or average quantity, but also some localized measurement which can describe this nonuniform structure.

## Summary

A review of microstructural features important in superplastic forming has been made, with particular emphasis on the Ti-6Al-4V alloy. It has been pointed out that starting parameters alone are insufficient to adequately describe the process, and importantly to relate microstructural features to the real world of actual part fabrication. A preliminary "dynamic" model has been described which begins to allow the real, varying, situation to be addressed. Parameters discussed are volume fraction of phases, grain size(s), grain size distribution, grain shape and contiguity. A discussion of a suggested procedure to define grain size, by mean linear intercept, is presented which should arrest the confusion presently occurring because of various methods being followed. Specifically mean linear intercept should be defined, after exposure at temperature, followed by a rapid cool (generally a water quench is sufficient).

It has been shown that

1)  volume fraction takes a significant time to equilibrate,

2)  the separate $\alpha$ and $\beta$ grain sizes do not differ in grain growth behavior from the all intercept grain size

3)  a new grain size measurement, the total alpha concept, can be useful in demonstrating grain growth

4)  the influence of additions of yttria on grain growth, can be demonstrated clearly using the total alpha concept.

## Acknowledgements

The authors would like to express their appreciation to W.R. Kerr, Y. Mahajan, D. Eylon, C.H. Hamilton and C.F. Yolton for helpful discussions. The experimental assistance provided by D. Allen and A. Houston is acknowledged.

## References

1.  Summary of Air Force Industry Manufacturing Cost Reduction Study, Air Force Technical Report, AFML-TM-LT-73-1, January (1973).

2.  N.G. Tupper, J.K. Elbaum and H.M. Burte, Jnl of Metals, Sept.(1978), 7.
3.  C.H. Hamilton and G.E. Stacher, Metal Progress, March (1976), 34.
4.  C.H. Hamilton, G.W. Stacher, J.A. Mill and H.W. Li, AFML-TR-75-62, April (1975).
5.  J.W. Edington, K.N. Melton and C.P. Cutler, Progress in Materials Science, 21(1976), No. 2.
6.  A.K. Ghosh and C.H. Hamilton, Met. Trans. 10A(1979), 699.
7.  W.R. Kerr, Jnl of Materials Science, to be published.
8.  N. Paton, Jnl of Eng. Mat. and Tech., Oct. (1975), 313.
9.  F.H. Froes, J.C. Chesnutt, C.F. Yolton, C.H. Hamilton and M.E. Rosenblum, this conference.
10. N.E. Paton and C.H. Hamilton, Met. Trans. A, 10A(1979), 241.
11. W. Rostoker and J.R. Dvorak: Interpretation of Metallographic Structures, 2nd Edition, (1977), 224.  Academic Press, N.Y.
12. R.R. Boyer and R. Bajoraitis, AFML-TR-78-131, September, (1978).
13. R.C. Gifkins, Met. Trans. A, 7(1976), 1225.
14. E.E. Underwood: Quantitative Stereology (1970), 80.  Addison-Wesley Publishing Co., Reading, MA.
15. E.E. Underwood: Ref 14, p. 34.
16. T. Chandra, J.J. Jonas and D.M.R. Taplin, J. of Mat. Science, 13(1978), 2380.
17. S.P. Agrawal, R.R. Boyer and E.D. Weisert, this conference.
18. M. Hoch, private communication, Dec. (1978).
19. E.E. Underwood: Ref 14, p. 100.

# SUPERPLASTIC FORMING BEHAVIOR OF CORONA 5 (Ti-4.5Al-5Mo-1.5Cr)

F.H. Froes*, J.C. Chesnutt[††], C.F. Yolton[†], C.H. Hamilton[††] and M.E. Rosenblum**

*Metals and Ceramics Division
Materials Laboratory, AFWAL
W-PAFB, OH 45433
††Rockwell International Science Center
Thousand Oaks, CA 91360
†Crucible Materials Center
Pittsburgh, PA 15230
**Metcut-Materials Research Group
P.O. Box 33511
W-PAFB, OH 45433

The application of CORONA 5 (Ti-4.5Al-5Mo-1.5Cr) to superplastic forming (SPF) and diffusion bonding (DB) fabrication techniques has been studied, and compared with Ti-6Al-4V. The CORONA 5 appears to be superior from both mechanical behavior, and perhaps more importantly in the present economic environment, cost considerations. These advantages result from the potential to process the newer alloy by a continuous strip process and to fabricate CORONA 5 at lower temperatures and age it to higher strength levels following the SPF/DB cycle.

## Introduction

During the past few years the strong drive for increased performance in advanced aircraft, military aircraft in particular, has led to a necessity for material with improved mechanical property combinations. The high strength to density ratio of titanium, in conjunction with excellent corrosion resistance and fracture behavior, has resulted in extensive use of this material in advanced systems. The Ti-6Al-4V alloy, developed twenty-five years ago, exhibits attractive mechanical property combinations in varied applications and is the standard against which other titanium alloys are judged.

A recently introduced titanium alloy, CORONA 5 (Ti4.5Al-5Mo-1.5Cr), was originally developed for application in fracture critical components [1-6]. The attractive property combinations exhibited by the alloy have resulted in attempts to fabricate the alloy by advanced techniques such as superplastic forming and diffusion bonding. Optimum mechanical properties are obtained by manipulating the microstructure to meet the requirements of the application. Basically for high fracture toughness a lenticular primary alpha is required while for optimum superplastic forming behavior a fine equiaxed alpha is needed. The design of the CORONA 5 alloy is discussed in detail elsewhere [4-6].

## Superplastic Forming/Diffusion Bonding (SPF/DB)

Because of costly extraction techniques and difficulties with both processing/fabrication and machining, titanium is an inherently expensive material as conventionally used. Thus titanium is selected for use only

when the superior mechanical property characteristics significantly outweigh
the cost disadvantages which exist in comparison with some competitive materials [7]. Recently widespread results have shown that SPF with or without
concurrent DB can result in substantial cost-savings (>50% of conventionally
fabricated components) creating great interest in this fabrication method [8].

To date, most SPF/DB work has concentrated on Ti-6Al-4V in the temperature range 900-925°C [9]. While the behavior of this alloy has been more
than adequate to meet current design concepts, imaginable innovations possible
with SPF/DB have led to the desire for improved performance. Two major directions for advancement are (i) to lower temperature/lower pressure/shorter time
fabrication and (ii) to age to higher strength levels subsequent to the SPF/DB
operation. Consideration of the microstructural requirements for optimum (SPF
and DB) behavior [10] indicate that the first advancement can be achieved by
using a fine equiaxed microstructure with approximately equal properties of
the two phases at the forming temperature. Because of the phase proportions
present at various temperatures in the Ti-6Al-4V alloy, and due to the inherent difficulty in imparting substantial amounts of strain of this material
(which would lead to a very fine equiaxed microstructure on subsequent heat-
treatment), the above mentioned developments are difficult to achieve in this
alloy (though possible with innovative processing [11]). The second advance
is also difficult with Ti-6Al-4V because the alloy is inherently not deep-
hardenable [12], thus only very limited hardening is possible after the SPF/
DB cycle. In contrast, CORONA 5 can exhibit the required phase proportions
at lower temperature (~840°C) [13], can be heavily worked to give a fine equiaxed microstructure [14], and does have the potential to achieve high strength
levels after forming [13,14].

The excellent chance for advancement in material performance using the
CORONA 5 alloy led to a program to investigate the SPF/DB characteristics of
the material. Since the requirement for this application is a fine equiaxed
microstructure, substantial working in the alpha-beta region is necessary
followed by annealing approximately 85°C below the beta transus temperature
so that a fine globular alpha results [4,14].

Initial efforts involved producing sheet by a unidirectional warm rolling process. However, the material produced by this method was highly directional (Fig. 1) with superplastic behavior acceptable in the longitudinal
direction but inadequate in the transverse direction. Two processing routes
were considered to eliminate this directionality: (a) the cross-rolling
sequence, low in the alpha-beta field, used throughout the titanium industry
to produce reasonably isotropic Ti-6Al-4V and (b) impart sufficient plastic
strain to the material so that a subsequent anneal would result in complete
recrystallization and a uniform equiaxed fine microstructure. The cross-
rolling method is obviously not amenable to a continuous strip rolling process--
which can yield lower cost, more uniform (gage and microstructure) product.
Thus to tie into a lower cost processing sequence the second processing route
was explored further. Processing by this route was achieved by modifying the
processing sequence to include a final cold rolling campaign. First attempts
were unsuccessful due to the formation of strain induced orthorhombic martensite during the rolling operation which severely limited the amount of cold
rolling possible (<10%). However, by annealing at 845°C for 4 hrs. prior to
cold rolling, the beta matrix was sufficiently enriched in beta stabilizer
(Mo and Cr) to prevent subsequent formation of orthorhombic martensite during
cold rolling. More recently it has been shown [15] that a simulated strand
anneal (760°C, 5 minutes) sufficiently enriches the beta matrix to allow cold
rolling. This again ties in with a continuous strip process.

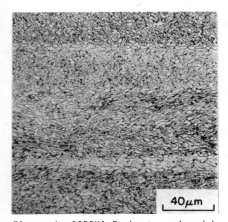

Figure 1. CORONA 5 sheet produced by a warm rolling sequence. Note directionality of microstructure.

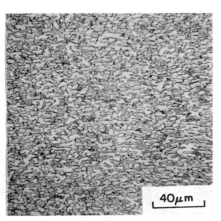

Figure 2. CORONA 5, produced by final cold rolling sequence. Note fine equiaxed microstructure

After cold rolling approximately 40% and annealing at 845°C for 4 hrs., a fine (~2 μm) equiaxed microstructure was produced (Fig. 2). A total of eleven tensile specimens machined from the cold-rolled fine-grained sheet, were tested to determine the superplastic properties [8] of the CORONA 5 sheet. The tests included the step/strain-rate test as well as total elongation test under constant strain-rate conditions. During the total elongation tests, the strain-rate was periodically and temporarily changed by 40% to permit a determination of the strain-rate sensitivity exponent, m, as a function of strain. The m vs. $\dot{\epsilon}$ curves for these data are shown in Fig. 3. At 870°C, the m values appear to be superior to Ti-6Al-4V especially in the intermediate strain rate range ($10^{-4}s^{-1}$-$10^{-5}s^{-1}$), exhibiting somewhat higher maxima in m than the Ti-6Al-4V alloy. In comparison with the Ti-6Al-4V alloy at 925°C, the CORONA 5 m values are somewhat lower. It is, however, significant that the m for CORONA 5 at 870°C is in excess of 0.5 for strain rates from about $6 \times 10^{-4}$ to $10^{-3}s^{-1}$, rates for which corresponding forming times are considered to be practical.

The results of the total elongation tests showed that elongations of 280% to 500% were possible at 870°C for strain rates of $10^{-3}$ and $2 \times 10^{-4}$ respectively. These values confirm that significant superplastic forming of this alloy is possible. This was demonstrated by fabricating a small pan from the CORONA 5 alloy at 870°C using the differential gas pressure superplastic forming process (Fig. 4). The effect of the SPF cycle on the grain structure can be seen by comparing Fig. 5 with Fig. 2.

Preliminary determinations of the diffusion bonding characteristics suggest that similar comparisons exist with Ti-6Al-4V as were observed in the case of superplastic forming behavior. That is, flow stress for CORONA 5 at 870°C will be somewhat higher than for Ti-6Al-4V at 925°C. This would dictate an increase in diffusion bonding pressure and/or time compared to Ti-6Al-4V (the latter alloy at the higher temperature).

To date, the Ti-6Al-4V alloy has only been used in the as fabricated condition after SPF or SPF/DB. This is, in part, a result of the rapid decomposition characteristics of the alloy and hence low potential for aging following cooling from the forming temperature [12]. In contrast, CORONA 5

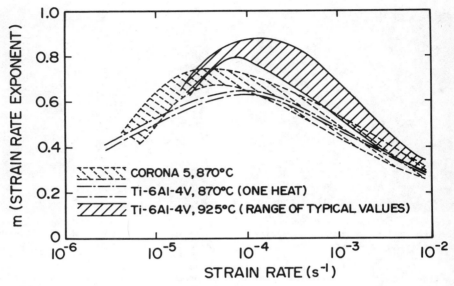

Figure 3. Strain rate exponent (m) as a function of strain rate for CORONA 5 and Ti-6Al-4V.

Figure 4. Small pan fabricated from CORONA 5 at 870°C by the differential gas pressure technique.

Figure 5. Microstructure taken from pan shown in Figure 4.

exhibits somewhat slower beta to alpha decomposition and a greater inherent hardening capability. A comparison of the time-temperature-transformation for CORONA 5 after water quenching from 870°C and for Ti-6Al-4V from 925°C is shown in Fig. 6.[15] These annealing temperatures give attractive SPF/DB behavior for the two alloys (and also are equally below the respective beta

transus temperatures of the alloys). The aging potential for the two alloys
are shown in Fig. 7 after both air cooling and water quenching from their
respective (suggested) superplastic forming temperatures. The substantially
higher strength attainable in the CORONA 5, especially after the much more
practical air cool, is readily apparent.

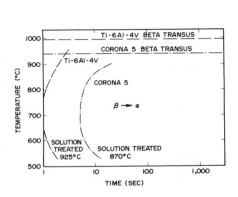

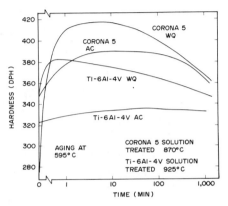

Figure 6. Time-temperature-transformation curves for CORONA 5 and Ti-6Al-4V.

Figure 7. Aging potential of CORONA 5 and Ti-6Al-4V.

In order to directly compare the mechanical properties of the two
alloys, alpha-beta processed [4,5,6] CORONA 5 was given a thermal exposure
to produce an equiaxed microstructure followed by a duplex anneal, slow
cool, and age. In a production situation, the heat-treatments could be
part of one thermal exposure. Tensile results are shown in Table I and
smooth and notched fatigue results in Fig. 8 [5]. The greater than 10%
increase in strength over the Ti-6Al-4V alloy [16], the latter in the as
superplastically formed condition, is reflected in the fatigue results where
the CORONA 5 data is at the top end of the scatter band exhibited by the Ti-
6Al-4V alloy [17]. This strength, and hence fatigue performance, differential could be increased further by decreasing the final aging treatment for
the CORONA 5 alloy.

## Status and Future Work

CORONA 5 can be readily processed to sheet, and fabricated using superplastic forming and diffusion bonding techniques. Indications are that the
alloy is more amenable to fabrication by this technique than Ti-6Al-4V because it can be: (a) processed to a higher quality, lower cost sheet product, (b) formed at a lower temperature and (c) aged to a higher strength
condition after the SPF/DB cycle. By reducing the temperature of superplastic forming for CORONA 5 even further to ~840°C the proportion of the
alpha phase would be ~50% and the behavior may be equally attractive. Further
an adjustment in the aging treatment should allow CORONA 5 to be aged to even
higher strength following the superplastic forming operation. These advances
combined with a continuous, high quality-lower cost strip process are the
obvious avenues for further work.

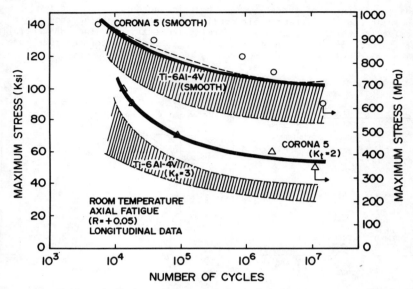

Figure 8.   Fatigue behavior of material intended for post superplastic forming Application.

Table I   Tensile Properties of Material Intended for Superplastic Forming
Applications

| Alloy | YS (MPa) | UTS (MPa) | Elongation (%) |
|-------|----------|-----------|----------------|
| CORONA 5[†] | 990 | 1035 | 13.0 |
| Ti-6Al-4V* | 865 | 950 | 13.0 |
| Ti-6Al-4V[@] | 870 | 925 | 9.0 |

[†]   Present work, material alpha-beta worked, annealed at 870°C - 4 hrs.
to produce a fine equiaxed microstructure and to simulate a superplastic forming thermal exposure.   Materials subsequently annealed
920°C - ½ hr., 845°C - 4 hrs. and aged 650°C - 8 hrs.

*   Superplasticity formed material.

[@]   Minimum values for Ti-6Al-4V per MIL-T-9046.

## Acknowledgements

The authors would like to express their appreciation to Mr. W.T. Highberger for his continued interest and support of the program. In addition helpful discussions with Dr. J.P. Hirth are acknowledged.

## References

1.  R.G. Berryman, J.C. Chesnutt and F.H. Froes: Metal Progress, Dec. (1977), 40.
2.  R.G. Berryman, J.C. Chesnutt, F.H. Froes and J.C. Williams: J. Aircraft, 14(1977), 1182, No. 12.
3.  J.C. Williams, F.H. Froes, J.C. Chesnutt, C.G. Rhodes and R.G. Berryman: ASTM STP 651 (1978), 64.
4.  F.H. Froes, J.C. Chesnutt, R.G. Berryman, G.R. Keller and W.T. Highberger: Proceedings of 10th National SAMPE Technical Conference, SAMPE National Business Office, Azusa, CA, (1978), 522.
5.  F.H. Froes and W.T. Highberger: Journal of Metals, March (1980).
6.  G.R. Keller, J.C. Chesnutt, W.T. Highberger, C.G. Rhodes and F.H. Froes: Proceedings of the 4th International Titanium Conference, May (1980).
7.  N.G. Tupper, J.K. Elbaum and H.M. Burte: Journal of Metals, Sept. (1978), 7.
8.  C.H. Hamilton and G.E. Stacher: Metal Progress, March (1976), 34.
9.  J.C. Chesnutt, C.H. Hamilton, C.F. Yolton and F.H. Froes: to be submitted for publication in Metallurgical Transactions, (1980).
10. J.W. Edington, K.N. Melton and C.P. Cutler: Progress in Material Science, 21(1976), 63.
11. F.H. Froes, C.F. Yolton, J.C. Chesnutt and C.H. Hamilton: Proceedings of Forging and Properties of Aerospace Materials, (1978), 371. The Chameleon Press, London.
12. F.H. Froes, R.F. Malone, J.C. Williams, M.A. Greenfield and J.P. Hirth: Proceedings of Forging and Properties of Aerospace Materials, (1978), 143. The Chameleon Press, London.
13. M.E. Rosenblum, P.R. Smith and F.H. Froes: Proceedings of the 4th International Titanium Conference, Kyoto, Japan, May (1980).
14. F.H. Froes, C.F. Yolton, J.C. Chesnutt, C.H. Hamilton and W.T. Highberger: Proceedings of the 3rd International Conference on Mechanical Behavior of Materials, (1979).
15. C.F. Yolton, F.H. Froes and R.F. Malone: Metallurgical Transactions, 10A(1979), 132.
16. C.F. Yolton and F.H. Froes: work in progress, (1979-80).
17. Preliminary Design Data and Materials Allowables for SPF/DB Structures, AFFDL, Dayton, OH (1978), Report No. NA-78-539-1.

# THE USE OF SUPERPLASTIC FORMING FOR THE MANUFACTURE OF TITANIUM T-A6V ALLOY TANKS

G. Hilaire, E. Budillon, E. Huellec and J. Chanteranne

Aerospatiale, France

For a metal or alloy to show superplastic behaviour, in precise conditions of temperature and deformation speed, it must have an equiaxed fine grain structure. This applies more particularly to industrial sheets of Ti-6 Al-4V titanium alloy where the grain size is generally less than 10 microns.

The state of superplastic strain of this alloy has been determined by a study of the m factor:  the equation between flow stress and strain rate.

Hemispheric and toric parts have been thermoformed by inflation from unwelded sheets or sheets welded by the TIG process.

Taking account of some thinning down caused by the strains occurring during forming, it is possible to envisage thermo-forming of parts, making optimum use of the manufacturing range and the  design of the tool as a function of the desired industrial applications, particularly in the aeronautics and space sector.

INTRODUCTION

In the present context of the titanium alloy industry, techniques which help to reduce metal loss are becoming more and more important.  Among these methods, hot forming in the superplastic state holds a special place when the parts to be manufactured have complex shapes, in which case more tradit- ional methods result in sometimes considerable losses of metal and consequently  very costly machining times.

In this paper we intend to give a summary of the work carried out over several years on T-A6V alloy, leading to the manufacture of parts under highly satisfactory industrial conditions.

CHOICE OF SUBJECT AND OBJECTIVES OF THE STUDY

The subject selected is the manufacture of toric and spherical tanks used for storing nitrogen in the space sector.
These items are quite suitable for:
- calculating strains and monitoring their paths,
- studying the behaviour of a weld seam pulled biaxially during forming,
- on the practical level, easy calculation of theoretical bursting pressures that can be checked by simple tests.
The study contains two main objectives:
- on the theoretical level, determination of the superplastic properties of T-A6V alloy,
- on the industrial level, the manufacture by thermoforming of half-tores with a very small curve radius and half-spheres with different relative thicknesses (thickness:diameter ratios).

Achieving the first objective

Attaining the first objective required the following studies:
- a theoretical analysis of thermoforming in the superplastic state, in order to show the influence of operating conditions on the geometry of the parts obtained,

- a study of the superplastic behaviour of a semi-finished
  T-A6V alloy product forged in the α/β state and processed
  in the α,β or β states (determination of the m factor:
  flow stress to strain rate, and metallographic study),
- traction tests on basic welded test parts, in order to
  study the evolution of the structure of the welded join
  and its mechanical strength.

Most of this theoretical work was carried out in the
Physics and Technology Laboratory at the University of METZ,
by Mr. C. HOMER and Mr. JP LECHTEN, to whom we wish to express
our thanks, as well as to Mr. B. BAUDELET, the Laboratory
Director, for his valuable advice and his competence in this
field.

Achieving the second objective

The second objective was to ensure industrial thermoforming
of elementary  parts, using circular blanks welded or not, in
order to:
- put into concrete terms the possibilities and operating
  conditions of strain, determined in the laboratory, for
  temperatures of ⩾ 900°C,
- monitor the behaviour of a weld seam in biaxial strain in
  the case of mechanically welded preforms,
- optimize the geometry of the weld join made on elementary
  parts intended for subsequent assembly by tungsten inert gas
  (TIG) or electron beam welding,
- demonstrate the possibilities of using an economical tool,
  designed for small scale production, with modular elements
  that can be easily modified during finalization of forming
  in the prototype phase,
- check whether it is possible to ensure the leaktightness
  necessary for good performance of the operation by freeing
  the tool from any external force at the level of the blank
  holder, so it can then be used in an industrial furnace.

DETERMINATION OF THE SUPERPLASTIC PROPERTIES OF THE T-A6V ALLOY

Theoretical analysis of thermoforming

To conduct this analysis, we considered the simple case of inflating a sheet embedded in a circular manner and subjected to uniform pressure. Several writers have already carried out this type of analysis and some have put forward theoretical models of this forming operation: JOUANE (1) and BELK (2) assume that the distribution of thicknesses is uniform during thermoforming, but this assumption can only be admitted in the case of inflation of a complete sphere. HOLT (3) adopts the idea that all points of the chief strains are equal, but this is incompatible with conditions at the boundaries imposed by embedding. The assumptions used here are those adopted by CORNFIELD and JOHNSON (4), i.e. in particular:
- the behaviour law used during uniaxial tests is valid for
  . a multiaxial test-when it is expressed in terms of equivalent stress and strain rate: $\bar{\sigma} = k\,(\bar{\xi}^{\cdot})^{m}$
- the value of the m factor is constant at all points and at all moments of strain
- the deformed or strained area is assumed at all times to be an element of a sphere
- bending stresses and radial stresses are considered to be negligible in comparison with tangential and circumferential stresses.

The main results obtained in this theoretical section are as follows:
- The geometry of the parts is strongly dependent on the rheological factor m, in particular the larger the m factor, the more uniform the final distribution of thicknesses.
- A law related to variation of the applied pressure was determined in order to make the m factor constant in the area undergoing the most strain and deformation, i.e. at the summit of the deformed area (Figure 1).

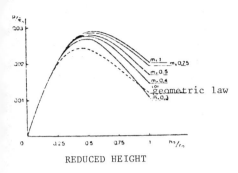

REDUCED HEIGHT

Figure 1 - Theoretical evolution of the forming pressure p as a ratio of the equivalent initial strain $\bar{\sigma}_o$ as a function of the height of the cap $h_o$ as a ratio of the embedding radius $r_o$, such that the equivalent stress at the summit is held constant and for different values of factor m.

- The kinetics of the strain were established as a function of factor m and of the pressure variation law (Figure 2).

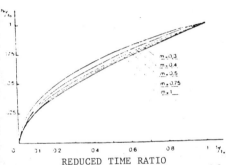

REDUCED TIME RATIO

Figure 2 - Theoretical evolution of the height $h_o$ of the cap as a ratio of the embedding radius $r_o$ as a function of the final characteristic time, for different values of factor m and when the equivalent stress is held constant at the summit.

These results are to be considered when parts are designed and production ranges are established.

## Study of the superplastic behaviour of T-A6V

The state where the m factor is high has been determined on various semi-finished products made of T-A6V alloy (5) and (6). The tests were carried out at three different temperatures: 750, 850 and 950°C. This state is compatible with the conditions under which the temperature must be greater than half the melting temperature and with the technical possibilities of the heating press used in part for the remainder of the study.

For these three temperatures, it is observed that the high values of factor m appear for strain rates of around $2.10^{-2}$ or less. It should be noted that higher values are seen, on the product under study, when the test temperature increases from 750°C to 850°C; on the other hand, lower values are observed when the temperature rises from 850°C to 950°C. (Figure 3).

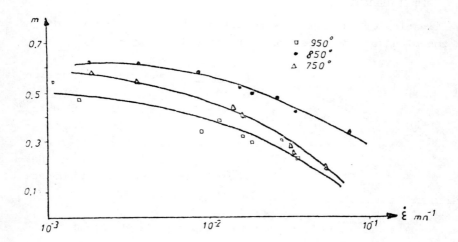

Figure 3

Evolution of the factor as a function of the strain rate for three temperatures.

This is due to the fact that at a temperature of 950°C the structure is fairly unstable and the grains enlarge during traction, possibly bacause of the proximity of the transus point $\alpha + \beta$ , $\beta$.

Depending on the alloys, this transus point can move between 10 and 30 degrees and behaviour may be very different at a given temperature.

Thus with another semi-finished product, we have observed that the highest value of factor m appears at a temperature of 950°C.

In addition, a study was conducted on the alloy thermally treated in state β in order to see whether high values of the m factor, associated with large strains without striction, could appear in materials with a structure that is not equiaxed (7).

The test parts were deformed by traction at temperatures of 850 and 900°C and at constant rates of between $2.7.10^{-3}$ and $2.7.10^{-1}$ min$^{-1}$. Comparative observations on non-deformed and deformed areas revealed a development in the acicular structure characteristic of a heat treatment in state β . Indeed, after deformation the α phase needles show an increase in diameter and a reduction in length for low strain rates (on the order of 30% at 900°C).

Furthermore, measurements taken show that the value of the m factor is close to 0.5 or 0.6. The results show the possibility of observing superplastic behaviour among alloys with structures that are not equiaxed.

## Behaviour of welded test pieces

For particularly complex parts, the formed elements must be welded together or a structure must be formed directly from welded sheets. This latter method raises the problem of the behaviour of the welded area, which we have studied on TIG-welded test pieces pulled by hot traction.

Maximum elongation of the test pieces was limited to 150% by the length of the furnace. After each test, the pieces are air cooled.

All the tests carried out show satisfactory behaviour of the weld, no matter what type of test piece is used and regardless of the temperature and the rate of strain.

In particular, we observed no decoherence at the interface of the weld and the base material, and no break in the welded area.

For a very low elongation rate, the α phase needles are oriented in any direction with respect to that of the traction. For an elongation rate of 90%, the needles are inclined over the direction of traction and a low density of small equiaxed grains appears. After local elongation of 370%, the structure shows small grains with a diameter of around 4 microns. Theoretical studies have explained this gradual passage of the weld to the superplastic state. It may be assumed that during deformation, a slide occurs at the interface between the α phase and the more ductile β phase: this slide leads to reorientation of the needles depending on the direction of traction. The needles then split, at the joins of the α phase grains, into small approximately equiaxed grains, as has been shown elsewhere (8,9).

The results obtained make it possible to envisage forming already welded sheets by superplastic thermoforming.

## EXAMPLES OF INDUSTRIAL ACHIEVEMENTS

### Forming of half-tores

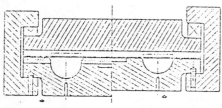

Constitution for test No. 1 with r = 3 tests Nos. 2 and 3 with r = 10

Constitution for tests Nos. 4, 5, 6, 7 and 8

Tool for forming half-tores.

Fig. 4

The half-tores, thermoformed from 2-mm thick circular blanks, intended for the manufacture of toric tanks, have a nominal dia. of 300 mm and a tube dia. of 100 mm. The forming tool (Fig. 4), designed for small scale production, is made of carbon steel. It consists of:
- a machined bottom die with the external shape of a half-tore. It has two air holes with a diameter of 2 mm and is equipped with a thermocouple for monitoring the temperature on the tool and two studs used to center the support pieces, made integral with the T-A6V blank by leaktight welding.

- an outer ring, plus an inner ring, removable intermediate elements used for studying the geometry of the weld join and for optimization of the minimum input radii to be provided on the outer and inner diameters of the die.
- a top plate which acts as a blank holder and is held on the die by 8 corner-clamped stirrup pieces.

The tool thus formed has a mass of 550 kg.

For thermoforming by inflation, supports are machined in a 20-mm thick sheet of T-A6V (Figure 5). After electron beam welding of the blank to the top of the support, these elements make up a leaktight chamber through which pressurized Argon is admitted by means of a metal pipe connected to a bottle of U grade Argon gas.

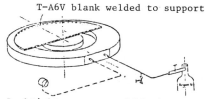

T-A6V blank welded to support

Leaktight support and blank pressurizing circuit.

Fig. 5

A second pipe connects the tool to the manometer controlling the pressure exerted inside the blank support unit.

All tests were carried out in a "Ripoche" furnace with a mobile hearth, where the maximum usable temperature is 1000°C.

The working temperature adopted for industrial tests was selected from curves of theoretical pressure modulation as a function of the height of the deformed area at any time (Fig. 6).

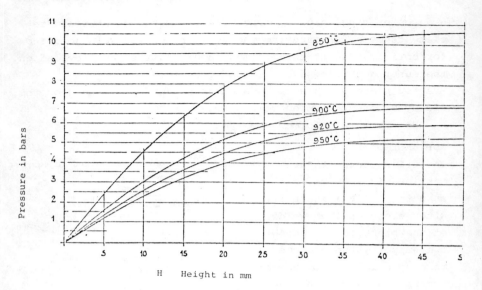

Modulation of pressure as a function of height          Fig. 6
ϴ     850°C to 950°C


     Considering the possibilities of the tools used and the
advantage of using low pressures, we adopted a temperature range
of 900-920°C.

     The chief results are as follows:

     In all cases, the external profile of the half-tores is a
replica of the internal geometry of the die.  In particular, the
quality of the surface condition obtained depends only on the
quality of the internal machining of the die, including the case
of tests on blanks welded according to a given diameter with no
preliminary levelling of the weld join.  On the other hand, the
heterogeneous distribution of thicknesses after forming requires
internal machining of the blanks before assembly, and this
operation takes place during the phase allocated to preparation
of the weld join and thus requires no additional tooling.

Finally, the complete forming operation takes less than two hours, at a maximum pressure on the order of 7 bars.

## Forming of half-spheres

The principle of manufacturing half-spheres by forming in the superplastic state is only slightly different from that used for half-tores, in that, for reasons of profitability, two blanks are produced in the same operation. Thus the tool placed in the mobile-hearth furnace consists of two half-spheres between which are placed two circular blanks, welded at the periphery (Figure 7).

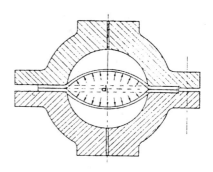

Figure 7

Different geometries have been envisaged for the pieces; here we give the extreme cases tested to date, i.e. the smallest and the largest relative thickness, since this parameter conditions the maximum pressure to be reached during forming.

Sphere with low relative thickness:

Original blank   diameter: 740 mm
                 $e_o$ :    3 mm

Finished part    diameter: D = 580 mm
                 thickness after machining:  1 mm
                 bursting pressure:  180 bars

Relative thickness:   $\dfrac{3}{580} \simeq$ 0.5%

## Forming conditions

temperature: 850°C, time:  1 hr 25 min, pressure:  2 bars

temperature: )20°C, time:  1 hr 15 min, pressure: 0.8 bar

Sphere with high relative thickness:

Original blank     diameter:   350 mm
                   thickness:   15 mm

Finished part      diameter:   200 mm
                   thickness after machining:  5 mm
                   theoretical bursting pressure:  900 bars
                   real (tested) bursting pressure: 940 bars

Relative thickness:   $\dfrac{15}{200}$ = 7.5%

Forming condition

   temperature:  950°C, time:  1 hr 10 min, pressure:  15 bars

The chief data to be drawn from these tests are as follows:
- As in the case of half-tores, the external surface condition of the half-spheres is an exact replica of the quality of the machining of the die and the heterogeneous nature of the thicknesses between base and summit requires internal machining prior to final assembly.
- As a first approximation, it may be considered that the maximum pressure to be attained during forming is a linear function of the relative thickness:

$$Pmax \;=\; k \; ; \; \frac{e_o}{D}$$

- The thermoforming operation has only a slight effect on the static characteristics of the material since the breaking test on the 200-mm dia. sphere gave a bursting pressure of 940 bars for 900 bars calculated with a breaking strength of 900 MPa for the material.

CONCLUSION

The theoretical studies and industrial applications of the superplastic nature of the T-A6V alloy show up the following:
- The metallurgical study shows that during deformation (strain), the acicular structure resulting from welding or from processing in the β phase evolves towards a fine grain structure characteristic of superplastic behaviour.
- For the use of economical tools, industrial tests have allowed verification of  the possibilities of manufacturing parts by thermoforming, using welded or non-welded T-A6V blanks.  These tests were carried out in conditions fairly similar to those envisaged by the preliminary theoretical study.
- Examination of the parts thus formed, and in particular of the thickness distribution, means that certain conditions must be imposed on the use of this manufacturing process.

For aeronautical requirements, always subject to strict tolerances, these parts can only be calibrated blanks which must be reworked by internal machining.

A comparison of manufacturing costs for three processes: forming (die stamping), hot drawing and thermoforming in the superplastic state, clearly shows the economic value of the last named.

## BIBLIOGRAPHY

(1)   F. JOUANE, Int. J. Mech. Sci. 1964, 6, p.303
(2)   J.A. BELK, Int. J. Mech. Sci. 1975, 17, p.505
(3)   D.L. HOLT, Int. J. Mech. Sci. 1970, 12, p.491
(4)   G.C. CORNFIELD, R.H. JOHSON, Int. J. Mech. Sci. 1970, 12, p.479
(5)   B. BAUDELET, Mém. Sci. Rev. Mét. Février 1975, p.101
(6)   B. BAUDELET, A. BOURGEOIS, J.P. BRUSSON, G. SERTOUR, Rapport DGRS, n° 7371757, Juillet 1975
(7)   R.F. MEHL, Metal Hand Book American Society For Metals, 1972, p.327
(8)   G. HERRIOT, M. SUERY and R. BAUDELET, Scripta Met., 6, 1972, p.657
(9)   M. SUERY, B. BAUDELET, B. LABULLE et C. PETITAS, Scripta Met., 8, 1974, p.703

# AN EVALUATION OF THE EFFECTS OF CONCURRENT GRAIN GROWTH DURING SUPERPLASTIC FLOW OF THE Ti-6Al-4V ALLOY

A. Arieli, B. J. MacLean and A. K. Mukherjee

Materials Science Section, Department of Mechanical
Engineering, University of California, Davis, CA 95616

## Abstract

The effect of concurrent grain growth on the appropriate parameters in the constitutive equation for superplastic flow was investigated using a relation put forward by Suery and Baudelet [4]. It was found that although the proposed relation can predict the overall effect, i.e. hardening, it fails to correctly yield acceptable numerical values for the variation of the parameters m, p, Q and A with strain, for this alloy, especially at low strain rates. The probable causes for this deviation are discussed.

## Introduction

The superplastic forming of Ti-6Al-4V alloy is now a success story [1]. However, the modeling of the flow characteristics of this alloy at large strains, as those encountered during superplastic forming, requires adequate knowledge about concurrent grain growth kinetics during superplastic flow [2,3]. The few attempts made until now [2,3] try to correct the experimentally obtained stress-strain rate curves for the instantaneous grain size on a one-to-one basis, i.e. each point on the $\sigma$-$\dot{\varepsilon}$ curve is corrected for the corresponding instantaneous grain size value. The above approach can be improved if a single constitutive relation for large strains and various strain histories can be constructed. Such a constitutive relation has been proposed recently by Suery and Baudelet [4] and is the purpose of this work to compare the proposed relation with experimental data for Ti-6Al-4V alloy.

## The Constitutive Relation

Suery and Baudelet [4] characterized the kinetics of the grain growth during superplastic flow by three parameters, i.e.

$$f = \frac{\partial (\log d^*)}{\partial \varepsilon} \bigg|_{\dot{\varepsilon},T} \tag{1}$$

$$g = \frac{\partial (\log d^*)}{\partial (\log \dot{\varepsilon})} \bigg|_{\varepsilon,T} \tag{2}$$

$$h = \frac{R}{2.3} \frac{\partial (\log d^*)}{\partial (1/T)} \bigg|_{\dot{\varepsilon},\varepsilon} \tag{3}$$

where $d^*$ = instantaneous grain size value, $\varepsilon$ = true strain, $\dot{\varepsilon}$ = true strain rate, T = temperature ($^\circ$K) and R = gas constant.

For the high temperature, diffusion-controlled, deformation mechanisms the stress ($\sigma$) is related to the strain rate ($\dot{\varepsilon}$), temperature (T) and initial grain size ($d_0$) by the following expression:

$$\frac{\dot{\varepsilon}\, k\, T}{D_0\, G\, b} = A \left(\frac{b}{d}\right)^P \left(\frac{\sigma}{G}\right)^n \exp\left(-\frac{Q}{RT}\right) \tag{4}$$

where: $D_0$ = diffusion constant, G = shear modulus, b = Burger's vector, A = dimensionless parameter, Q = activation energy, d = grain size, p = strain rate dependence on grain size coefficient, n = stress sensitivity and is equal to 1/m, where m is the strain rate sensitivity, k = Boltzmann's constant.

Eqn. (4) can be rewritten in the logarithmic form:

$$\log \sigma = m \log\left[\frac{k\, T\, G^{\frac{1-m}{m}}}{A\, D_0\, b^{p+1}}\right] + m \log \dot{\varepsilon} + pm \log d + m\, 2.3\, \frac{Q}{RT} \tag{5}$$

or in the differential form:

$$\partial \log\sigma = m\, \partial\left\{\log\left[\frac{k\, T\, G^{\frac{1-m}{m}}}{A\, D_0\, b^{p+1}}\right]\right\} + m\, \partial\left(\log \dot{\varepsilon}\right) + pm\partial\left(\log d\right) + m\, 2.3\, \frac{Q}{R}\, \partial\left(\frac{1}{T}\right) \tag{6}$$

If the values of m, p, A and Q are determined for the initial structure, i.e. d = $d_0$ = initial grain size value, they are assigned the subscript "o" in Eqn. (6), i.e.

$$\partial \log\sigma = m\, \partial\, \log\left[\frac{k\, T\, G^{\frac{1-m_0}{m_0}}}{A_0\, D_0\, b^{p_0+1}}\right] + m_0\, \partial\,(\log \dot{\varepsilon})$$
$$+ p_0 m_0\, \partial\,(\log d_0) + m_0\, 2.3\, \frac{Q_0}{R}\, \partial\left(\frac{1}{T}\right) \tag{6a}$$

On the other hand, if the parameters m, p, A and Q are determined at an arbitrary strain, i.e. changed microstructure where d = d* = instantaneous grain size value, they are assigned superscript * in Eqn. 6 , i.e.

$$\partial\,(\log \sigma) = m*\, \partial\left\{\log\left[\frac{k\, T\, G^{\frac{1-m*}{m*}}}{A*\, D_0\, b^{p*+1}}\right]\right\} + m*\, \partial\,(\log \dot{\varepsilon})$$
$$+ p*m*\, \partial\,(\log d*) + m*\, 2.3\, \frac{Q*}{R}\, \partial\left(\frac{1}{T}\right) \tag{6b}$$

It can also be written,

$$\partial\,(\log d*) = f\, \partial\, \varepsilon + g\, \partial\,(\log \dot{\varepsilon}) + \frac{2.3\, h}{R}\, \partial\left(\frac{1}{T}\right) + \partial\,(\log d_0) \tag{7}$$

Combining Eqns. (6b) and (7) we obtain:

$$\partial (\log \sigma) = m^* \partial \log \frac{k T G^{\frac{1-m^*}{m^*}}}{A^* D_o b^{p^*+1-1}} + m^* p^* f \partial \varepsilon$$

$$+ m^* \left(1 + p^*\right) g \partial (\log \dot{\varepsilon}) + \frac{2.3}{R} m^* \left(Q^* + p^* h\right) \partial \left(\frac{1}{T}\right) \tag{8}$$

$$+ m^* p^* \partial (\log d_o)$$

Since both Eqn. (6a) and Eqn. (8) describe the same deformation mechanism, the following relations hold:

$$m^* = \frac{m_o}{3} (3 + P_o g) \tag{9}$$

$$p^* = \frac{3 P_o}{3 + P_o g} \tag{10}$$

$$Q^* = \frac{P_o h + 6.9 Q_o}{2.3 (3 + P_o g)} \tag{11}$$

$$A^* = A_o^{\left(\frac{-3}{3 + P_o g}\right)} \times 10^{\left(\frac{P_o f \varepsilon}{3 + P_o g}\right)} \times \left(\frac{D_o b}{k T G^{1/m_o}}\right)^{\left(\frac{6 + P_o g}{3 + P_o g}\right)}$$

$$\times \left(\frac{G}{P_o}\right)_b^{\left(\frac{6}{3 + P_o g}\right)} \tag{12}$$

Eqns. (9) through (12) can now be used to calculate the values of the parameters m, p, Q and A at any arbitrary strain and the calculated values will be compared with the experimental ones in the next section.

### Experimental

Constant strain rate tensile tests were carried out on Ti-6Al-4V to determine the values of parameters $m = 1/n$, p, Q and A (Eqn. (4)) at various strains, temperatures and strain rates. The initial grain size value was 14.0 $\mu$m for all the conditions investigated. Complete details about the material and test procedure used is given in Ref. [5].

Figure 1 shows the stress-strain rate behavior at $T = 1173^\circ K$ and $d_o = 14.0$ $\mu$m. When the data were collected at the peak stress[+] all the data

---

[+]  It was assumed that at the peak stress the deformation becomes fully plastic and only then the strain-enhanced grain growth started. Therefore, the values of the parameters m, p, Q and A obtained with data collected at peak stress are those corresponding to the initial microstructure and are assigned the subscript o.

fall along a single line with the slope 0.39. As the deformation continues ($\varepsilon$ = 0.2 and 0.35) the datum points fall onto two lines of different slopes. At $\varepsilon \geq 5 \times 10^{-4}s^{-1}$ the slope of the line changes very little, but at $\dot{\varepsilon} < 5 \times 10^{-4}s^{-1}$ there is a drastic decrease in the slope.

In addition to generating mechanical data, grain growth during deformation was monitored at various strains, strain rates and test temperature, and its kinetics are given by the relation [5]:

$$d* = d_o + \left( \frac{3.74 \times 10^9}{d_o^{0.75}} \right) (t)^{0.33} \exp \frac{-43,800}{RT} \tag{13}$$

where:  t = time (min.) and d* and $d_o$ in $\mu$m. Both the mechanical and grain growth data, thus generated, were analyzed using the constitutive relation outlined in the previous section.

### Analysis

Figure 2 is a semilogarithmic plot of the instantaneous grain size versus the true strain for three strain rates. The slopes of the curves in Fig. 2 yield the value of parameter f in Eqn. (1). It is evident from Fig. 2 that the f-values decrease as the strain rate increases. The variation of f-values with strain rate is shown in Fig. 3. As the temperature increases the f-value also increases, Fig. 4.

The variation of the instantaneous grain size with strain rate at constant temperature and strain is shown in Fig. 5. The plot in Fig. 5 reveals that at every temperature-strain combination there are two g-values, a higher one at strain rates below $2 \times 10^{-4}s^{-1}$ and a lower one at strain rates above $2 \times 10^{-4}s^{-1}$. The variation of g with strain and temperature is shown in Figs. 6 and 7. g decreases with increasing strain and decreases with increasing temperature.

Figure 8 shows the variation of the instantaneous grain size with the reciprocal of the absolute temperature for two strain rates and two strains. It is evident from Fig. 8 that h is independent of strain regardless of strain rate. h, also, increases as the strain rate decreases.

The above results show that the largest grain size will be found after deformation at the lowest strain rate, highest test temperature and largest strain. This conclusion is supported by experiment [2,5].

The values of the parameters m*, p*, Q*, and A* were calculated using the values for f, g and h determined above. The calculated values are listed in Table 1 together with the experimentally determined ones.

Table 1   Calculated and Experimental m*, p*, Q* and A* values

| Parameter | $\dot{\varepsilon}(s^{-1})$ | $T(^{\circ}K)$ | $\varepsilon$ @ peak stress | $\varepsilon = 0.2$ Calc. | $\varepsilon = 0.2$ Exp. | $\varepsilon = 0.35$ Calc. | $\varepsilon = 0.35$ Exp. |
|---|---|---|---|---|---|---|---|
| m* | $<2\times10^{-4}s^{-1}$ | 1173 | 0.38 | 0.33 | 0.26 | 0.31 | 0.17 |
| | $\geq2\times10^{-4}s^{-1}$ | 1173 | 0.38 | 0.35 | 0.38 | 0.34 | 0.38 |
| p* | $<2\times10^{-4}s^{-1}$ | 1173 | 2.2 | 2.5 | 6.6 | 2.7 | 10.2 |
| | $\geq2\times10^{-4}s^{-1}$ | 1173 | 2.2 | 2.4 | 2.2 | 2.4 | 2.2 |
| $Q*\left(\dfrac{Kcal}{mole}\right)$ | $<2\times10^{-4}s^{-1}$ | | 73.2 | 82.0 | 106.0 | 84.4 | 156.0 |
| | $\geq2\times10^{-4}s^{-1}$ | | 71.2 | 76.7 | 62.1 | 78.0 | 53.3 |
| A* | $<2\times10^{-4}s^{-1}$ | 1173 | $1.5\times10^{4}$ | $3.7\times10^{44}$ | $8.8\times10^{16}$ | $1.4\times10^{45}$ | $7.2\times10^{36}$ |
| | $\geq2\times10^{-4}s^{-1}$ | 1173 | $3\times10^{4}$ | $1.3\times10^{42}$ | $5.15\times10^{2}$ | $5\times10^{42}$ | $2.8\times10^{1}$ |

Comparison between the calculated and the experimental values shows that although the constitutive relation used in this work correctly predicts the trend for the variation of the parameters m, p and Q with strain, the actual calculated values are much lower.

The inability of the constitutive relation to yield correct numerical values for the parameters m, p, Q and A at various strain histories might stem from several reasons, i.e.

(a) the failure of Eqn (7) to accurately describe the grain growth kinetics during superplastic flow of this alloy;
(b) excess hardening results from other changes in the microstructure in addition to grain growth, i.e. grain clustering, expecially at low strain rates, grain elongation (3), nonuniform microstructure (6,7), etc;
(c) probably as the grains grow other deformation mechanisms become operative in addition to the superplastic one. In this case, Eqn. (4) ceases to describe the $\sigma$-$\dot{\varepsilon}$ behavior for the test under consideration and, consequently, the constitutive relation based on it will yield inaccurate results.

## Summary

The overall effect of the concurrent grain growth during superplastic flow of the Ti-6Al-4V alloy is to increase the flow stress, i.e. hardening effect.

A constitutive relation, originally proposed by Suery and Baudelet [4], was used to evaluate this effect. The relation correctly predicts the hardening effect but the calculated values are lower than the experimental ones, especially at low strain rates (below $2\times10^{-4}s^{-1}$).

A better understanding of the deformation mechanisms operative during superplastic flow as well as improved characterization of the concurrent grain growth kinetics might lead to better constitutive relations for superplastic flow where microstructure changes continuously.

## Acknowledgements

This work was supported by Air Force Office for Scientific Research under the Grant AFOSR 79-0069.

## References

1. C. H. Hamilton and G. W. Stacher, Metal Progress, March (1976).
2. A. Arieli and A. Rosen, Met. Trans., 8A(1977), 1591.
3. A. K. Ghosh and C. H. Hamilton, Met. Trans., 10A(1979), 699.
4. M. Suery and B. Baudelet, Rev. Phys. Appl. 19(1978), 53.
5. A. Arieli, B. J. Maclean and A. K. Mukherjee, to be published.
6. N. E. Patton and C. H. Hamilton, Met. Trans., 10A(1979), 241.
7. R. R. Boyer and J. E. Magnuson, Met. Trans., 10A(1979), 1191.

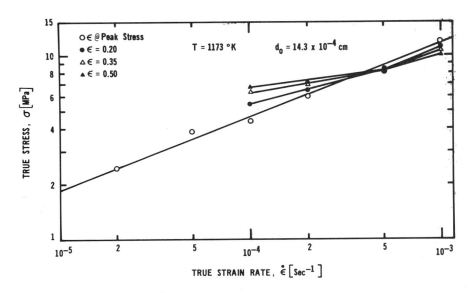

<u>Fig. 1</u>    Log-log plot of the stress versus strain rate at constant temperature and initial grain size and various strains.

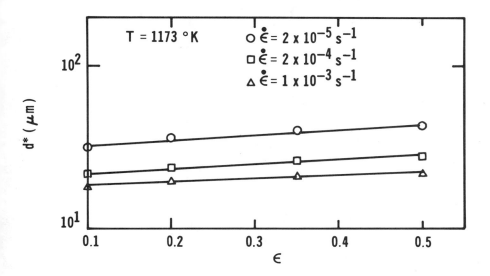

<u>Fig. 2</u>    The increase in the instantaneous grain size with strain as a function of strain rate.

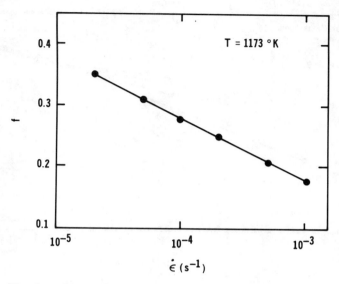

<u>Fig. 3</u>    The dependence of parameter f on strain rate.

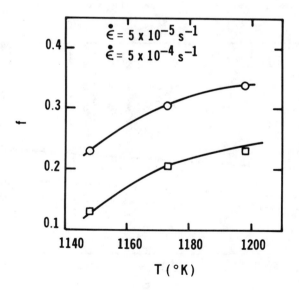

<u>Fig. 4</u>    The dependence of parameter f on temperature at two strain rates.

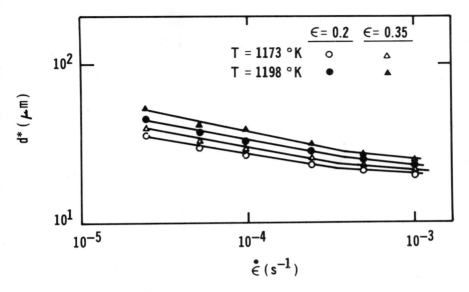

<u>Fig. 5</u>   The variation of d* with strain rate for 4 temperature-strain combinations.

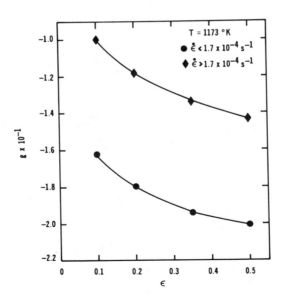

<u>Fig. 6</u>   The dependence of parameter g on strain.

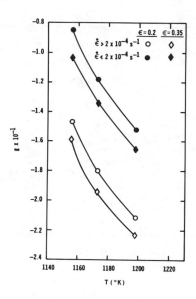

<u>Fig. 7</u>   The dependence of parameter g on temperature for 4 strain-strain rate combinations.

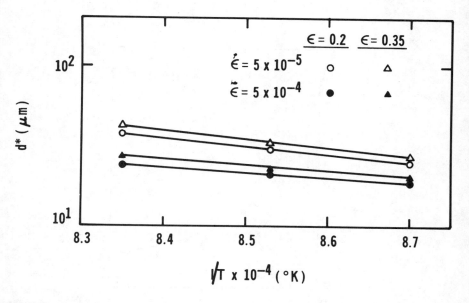

<u>Fig. 8</u>   Plot of the instantaneous grain size value vs. the reciprocal of the absolute temperature.

# EFFECT OF SMALL AMOUNTS OF YTTRIA
## ON THE SUPERPLASTIC BEHAVIOR OF Ti-6Al-4V

Suphal P. Agrawal, Rodney R. Boyer* and Edward D. Weisert

Rockwell International Corporation
North American Aircraft Division
P. O. Box 92098
Los Angeles, California   90009

*Boeing Commercial Airplane Company
Boeing Technology Services
P. O. Box 3707
Seattle, Washington   98124

## Abstract

Successful superplastic forming of Ti-6Al-4V alloy sheet into large complex shapes requiring long times and large elongations may ultimately be limited by the grain growth which occurs during processing and by the attendant decrease in superplasticity of the material. This paper summarizes the results of a preliminary investigation of the role of yttria, a conventional grain refiner, in controlling such in-process grain growth and improving the superplastic formability of the baseline Ti-6Al-4V. Additions of 0.018 and 0.025 wt pct yttria to Ti-6Al-4V alloy were observed to retard the rate of grain growth both under the static conditions of thermal exposure alone and the combined dynamic conditions of strain and heat. Yttria was found to be particularly effective in inhibiting the deformation-assisted grain growth. Improved superplastic formability was demonstrated in the yttriated alloys by cup- and hat-shaped cavity-forming tests. Two sizes of yttria particles, ~250-500 nm and ~10-20 nm, were observed, although the larger particles appeared to have little overall effect on the superplastic behavior of the parent metal. Further effort is warranted to precisely define the role of these particles in retarding grain growth and improving superplastic formability of Ti-6Al-4V and to optimize their concentration, size and distribution to maximize the benefit.

## Introduction

Following the initial work over a decade ago[1-4], the superplastic behavior of Ti-6Al-4V alloy has become the subject of extensive investigation within the past few years[5-13]. The impetus for such interest has been the development of a new process for fabricating titanium alloy sheet structures combining superplastic forming and diffusion bonding (SPF/DB). Using this combined process, particularly suited to Ti-6Al-4V sheet, complex hardware can be produced as monolithic structures at a lower cost and reduced weight compared to conventional fabrication techniques[14]. Sheets of Ti-6Al-4V alloy exhibit pronounced superplastic behavior in their off-the-shelf condition corresponding to the normal aerospace specifications, i.e., with standard mill-processing treatments. Large ($\geq$500%) elongations can be achieved over a range of strain rates and temperatures. Superplastic forming of Ti-6Al-4V, both with and without the concurrent diffusion bonding, has developed to the point that the process is beginning to be transitioned into production for several commercial applications[14-17]. Initial development of the SPF and SPF/DB processes was aimed at airframe components, but they are finding widespread usage in fabricating engine, space and missile components[18] due to the cost and weight savings possible.

In transitioning these processes into production, current development efforts are aimed at achieving close process control as well as at maximizing the process tolerance for the variations in day-to-day operations. Consequently, fine-tuning of the process parameters on one hand and optimization of the starting material on the other have become necessary. While a remarkable variety of structures has been demonstrated using standard Ti-6Al-4V stock, limitations have occasionally been encountered. In particular, grain growth as a result of certain SPF/DB processing conditions has been responsible for some fabrication failures. This is because the in-process grain coarsening is accompanied by a diminished superplastic behavior and results in severe local thinning and, in extreme cases, rupture prior to completion of forming. The occurrence of grain growth under conditions of superplastic deformation processing and its influence on the subsequent formability have been discussed in detail in a recent review[13].

Two practical means of alleviating in-process grain growth and its attendant effects on the superplastic behavior of Ti-6Al-4V alloy may be considered. One involves adjusting of the process parameters, such as temperature, strain rate, forming time, etc., to minimize grain growth. Some measures of this nature, such as reduced temperatures, have already been adopted. Extensive parametric adjustments are to be avoided, however, since they tend to reduce the tolerance of the process to demands of and variations in daily shop practice. An alternate solution is to optimize the basic superplastic performance of the Ti-6Al-4V alloy via alloy modifications, an approach which has not yet been seriously explored. The primary goal of such optimization would be to significantly increase the stability of the grain size under process conditions. A second goal would be to significantly decrease the grain size of the as-received sheet. Small additions of rare earth oxides, such as yttria, have been previously shown[19] to act as grain refiners and to thereby improve the hot workability of the Ti-6Al-4V alloy through the ingot breakdown and secondary mill-processing stages. The present study was conducted to determine if the yttria particles were also beneficial in retarding grain growth during SPF processing and to determine any effects these particles might have on the superplastic behavior of this alloy.

## Material

Two sheets of yttriated Ti-6Al-4V and one of the standard Ti-6Al-4V alloy were used. The chemical compositions and the as-received grain size measurements of the three alloys are shown in Table 1. Each alloy had the typical Ti-6Al-4V composition. The yttriated sheets also contained 0.014 and 0.020 weight percent yttrium, not present in the reference alloy. Assuming this yttrium to be present as yttria, the amount of yttria present will be 0.018 and 0.025 weight percent, respectively. (That yttrium was indeed present primarily as yttria was confirmed by selected-area diffraction and will be discussed later.)  The reference Ti-6A1-4V alloy (sheet 1) showed an average grain size of 6.8 μm, measured by the Hilliard circle method and defined as:  Grain Size = 1.68L/MN [20], where L = length of the line of intercept, M = magnification and N = number of intercepts.  Measured similarly, the two yttriated sheets had a grain size of 5.5 μm (sheet 2) and 5.7 μm (sheet 3).  All grain size measurements were made in the plane of the sheet (i.e., in the short-transverse view).  Some of the grain boundaries appeared jagged in structures of sheets 2 and 3, a feature which was not apparent in the microstructure of sheet 1.

Table 1

Chemical Composition and As-Received Grain Size of
Various Alloy Sheets Investigated (Weight %)

| Heat No. | Alloy No. | O | N | H | C | Al | V | Fe | Y | Ti | (Y$_2$O$_3$) | As-received G.S. (μm) |
|---|---|---|---|---|---|---|---|---|---|---|---|---|
| N5030 | 1 | 0.130 | 0.012 | <80ppm | 0.018 | 6.0 | 4.2 | 0.070 | — | Bal | — | 6.8 |
| 891926 | 2 | 0.138 | 0.011 | 77ppm | 0.03 | 5.7 | 3.8 | 0.15 | 0.014 | Bal | (0.018) | 5.5 |
| 801746 | 3 | 0.116 | 0.016 | 51ppm | 0.02 | 5.9 | 3.8 | 0.18 | 0.020 | Bal | (0.025) | 5.7 |

A promising trend is immediately noted.  While no special attempt was made during mill processing to obtain fine grain in the finished sheet, the yttriated heats show finer grains than the control heat.  Their grain sizes are also smaller than 8.2 μm which was the average value for 6 heats of Ti-6A1-4V previously evaluated for superplastic forming[11].

## Experimental Results and Discussion

The elevated-temperature flow properties were determined for each sheet at 927 C using a step-strain-rate method in uniaxial tension.  The results are shown in Fig. 1 as log-log plots of flow stress, $\sigma$ , and strain rate, $\dot{\epsilon}$ .  Sheets 1 and 2 exhibit comparable flow stresses in the range of strain rates from 5 x 10$^{-5}$ sec$^{-1}$ to approximately 1 x 10$^{-2}$ sec$^{-1}$. Somewhat lower flow stresses were observed for sheet 3 over the entire range of strain rates investigated.  The corresponding m (= $\partial \log \sigma / \partial \log \dot{\epsilon}$ ) values as a function of strain rate are presented in Fig. 2.  Also shown here for comparison is a similar plot representing an average of six Ti-6A1-4V heats evaluated elsewhere[11] for superplastic forming.  These curves show that while the magnitude of peak m values does not vary appreciably among the three sheets studied in the present work, the strain rates at which peak m occurs vary by nearly an order of magnitude.  The strain rate values corresponding to maximum m were approximately 5 x 10$^{-5}$ sec$^{-1}$ for sheet 1 (the reference alloy), 1.5 x 10$^{-4}$ sec$^{-1}$ for sheet 3 and 5 x 10$^{-4}$ sec$^{-1}$ for sheet 2 .  A narrower range of strain rates corresponding to the maximum m values is obtained when the behavior of the two yttriated sheets of the present work is compared to the average of six Ti-6A1-4V sheets of reference 11.  Relatively higher peak m values of the yttriated sheets also become apparent.  The results of Fig. 2 are particularly significant in view of the implications that (i) higher forming rates could be employed using the yttriated alloys than would be possible with the reference Ti-6Al-4V alloy without sacrificing the inherent material capability for neck-resistance, and (ii) larger superplastic elongations could be developed in the yttriated alloys.

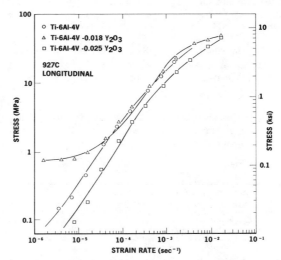

Fig. 1.  Flow stress as a function of strain rate for the three sheets described in Table 1.

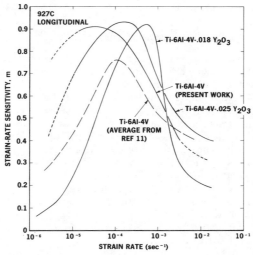

Fig. 2.  Strain-rate sensitivity index m = ($\partial \log \sigma / \partial \log \dot{\epsilon}$) as a function of strain rate for the three sheets evaluated in the present study.  Also shown here for comparison is a similar plot representing an average of six Ti-6Al-4V heats evaluated in Ref. 11.

The grain growth kinetics of these sheets were studied as a function of time and temperature over temperatures ranging from 870C to 954C. Each sample was held at a given temperature for a given time, air cooled, and placed in metallurgical mounts, again with a short-transverse view. The grain size was measured as discussed previously. The baseline grain growth was established in a previous study [9]. Sheet 2 was analyzed over the full range of times and temperatures and sheet 3 was only spot-checked. The results for sheets 1 and 2 are presented in Fig. 3. The solid curves are from the present study on sheet 2 and the dashed curves are from the previous study [9] on the base metal (sheet 1). For each temperature the initial grain growth rate of the yttriated material was less than that of the baseline, to a minimum of about 4 hours. After this initial period, the yttriated material grew at a faster rate than its base-metal counterpart, and, at the three lower temperatures investigated, actually crossed over the base-metal curves. The retardation is significant in that the SPF times are of the order of 2-4 hours, so it should be effective through the duration of the forming cycle. Except for the 954 C data the "starting" or the short-time-exposure grain size was less for the yttriated material; this may be attributable to its smaller starting grain size.

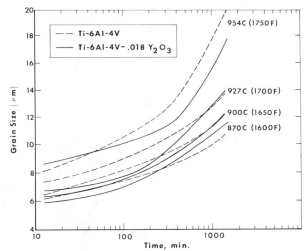

Fig. 3.    Grain size of sheets 1 and 2 plotted as a function of time at various temperatures. Solid curves are from the present study on sheet 2. The dashed curves are on the baseline alloy from Ref. 9 and are superimposed here for comparison.

The rapid rise of sheet 2 curves after approximately 4 to 5 hours of exposure to a given temperature suggests that whatever phenomena were responsible for retardation in grain growth prior to this time are no longer active. Indeed, the crossing over of some of these curves indicates that the growth rate is now faster in sheet 2 than in sheet 1.

In an effort to determine the nature and distribution of the particles in the yttriated alloy sheets, and thereby to understand their influence on the grain growth behavior, several thin foils were examined by transmission electron microscopy (TEM). A bi-modal distribution of yttria

or assumed-to-be-yttria particles was observed.  Figure 4 shows typical
observations in the as-received condition of the two yttriated alloy
sheets.  The large particles, seen near the grain boundaries, were indexed
as BCC yttria by selected-area diffraction technique.  Their size (~250-
500 nm) is comparable to the powder particles added to the melt in the
state-of-the-art mill practice[21].  It is, therefore, believed that these
are the originally added particles which dispersed near the grain boundaries as the melt solidified, and managed to retain their location through
the various mill processes.  While their preferential location near grain
boundaries may suggest an important role in retarding the grain growth
kinetics of these alloys (Fig. 3), two observations negate such possibility.  One is that in order for these particles to play an important role in
moderating the grain growth kinetics, a larger number of these particles
would be required than was observed.  Secondly, while in several instances
grain boundaries were observed to be kinked, no such particles were to be
seen in the immediate vicinity.  This suggests the possibility of a second
kind of particles which are much smaller in size, albeit large in numbers,
and perhaps quite difficult to detect.  A high resolution TEM technique,
called the weak-beam dark field imaging technique [22], was employed to
determine the presence of these particles.  Although that study is not yet
complete, there are sufficient indications of small second-phase (~10 - 20
nm in size) particles being present in the as-received structure of the
yttriated alloy sheets.  The particles associated with the dislocations,
indicated by arrows in Fig. 4, may indeed represent this second family of
particles.  The source of evolution of such particles may lie in partial
dissolution of the larger yttria particles (those added to the melt) and
subsequent precipitation during cool down[21].  Such partial dissolution
may occur in the melt column.  Once precipitated, the fine particles
apparently grow only very slowly.

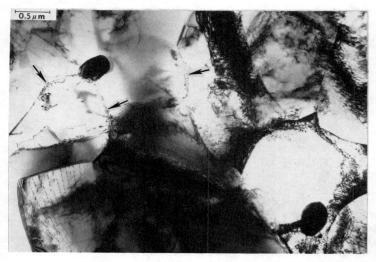

Fig. 4.   Transmission electron micrograph of typical as-received micro-
          structure of sheets 2 and 3.  In addition to the large yttria
          particles, presence of a second family of particles is suggested
          at locations indicated by the arrows.

After a four-hour exposure at 927C, most of the larger particles were no longer at grain boundaries, but were contained within the alpha grains. The finer particles were also resolved only within the alpha grains, although their absence near the grain boundaries could not be clearly established. These observations suggest that the slower initial grain growth in the yttriated material (Fig. 3) could be related to the interaction between yttria particles and grain boundaries. With continued thermal exposure, however, grain boundaries eventually break away and the rate of grain growth increases.

The effect of yttria on the actual superplastic formability was evaluated using cup forming tests. Cups were formed under argon gas pressure at 927C using the base metal and the yttriated alloy sheets. Identical test conditions were used for both materials. The formed cups were then sectioned and evaluated for grain size in the flange and the cup-radius areas. Examination at these two locations permits evaluation of grain size as a function of strain — the flange area being unstrained and the radius representing the maximum strain condition. The results are contained in Table 2. There are three key observations to be made from these data:

1. The grain size of the yttriated material after SPF processing is smaller than that of the base metal, particularly in the formed region. In the underformed area, the sheets exhibited comparable growth rates which are in agreement with the curves of Fig. 3.

2. Strain and strain-rate assisted growth was observed in the baseline material, as also reported previously[9,12]. This effect was virtually nonexistent in the yttriated material. If this represents typical microstructural response of these materials to SPF deformation, it could have significant implications in the SPF processing.

3. The yttriated material formed a tighter radius, indicating that more forming had occurred in this material than in the baseline material.

Table 2

Grain Size and Cup Radius Comparison in Sheets 1 and 3

| Sheet No. | Initial G.S. ($\mu$m) | Flange G.S. ($\mu$m) | Radius G.S. ($\mu$m) | Cup Radius |
|---|---|---|---|---|
| 1 | 6.8 | 9.7 | 12.1 | 1.7 $t_o$ |
| 3 | 5.7 | 9.0 | 9.2 | 1.2 $t_o$ |

In another test to compare formability of the base metal with its yttriated counterparts, several cavities (hat-shaped in cross-section) were superplastically formed at constant strain rate under argon gas pressure.

This test represented two improvements over the cup test just described. First, any possible effects due to the difference in the initial grain sizes of the sheets were eliminated by using a sheet from another Ti-6Al-4V heat with the initial grain size ($5.7 \mu$m) virtually equal to those in the two yttriated sheets. Secondly, a constant strain rate (rather than a constant rate of gas pressurization) was maintained in each forming sheet segment. In each case, a sheet coupon was loaded into a hot die, and each cavity was formed by introducing argon gas pressure in accordance with an analytical pressure-time profile designed to maintain a constant strain rate ($\sim$5 x $10^{-4}$ sec$^{-1}$). Since these cavities represented different levels of resultant strain, a hold-time at temperature after forming was used in order to maintain the total time exposure to temperature constant for each sheet segment, thus ensuring identical time-and-temperature re-lated grain growth in each case.

The cross-sections of various cavities thus formed are shown in Fig. 5, with the final radius formed in each case indicated as a function of the initial sheet thickness, $t_o$. It is evident that, similar to the cup tests, a smaller radius was obtained with the yttriated alloy sheet in all cases, again indicating that this sheet was superplastically more formable than the Ti-6Al-4V sheet under equivalent conditions.

Grain size and true thickness strain were measured at various loca-tions along the cross-section of one of these cavities (that shown at the far right in Fig. 5) for both sheets. Figure 6 shows their relationship. The circles and triangles represent the data points obtained from the hat-shaped cavity tests for the yttriated Ti-6Al-4V and the base-metal sheets, respectively. The squares represent the data points obtained from the cup tests. While both the yttriated and the base-metal sheets show appreciable grain growth with large strains, the tendency for the deformation-assisted growth is clearly much less in the yttriated sheets.

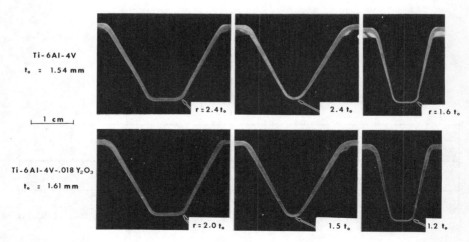

Fig. 5.   A comparison showing improved superplastic formability in a yttriated alloy sheet (sheet 2) over the standard Ti-6Al-4V for several geometries at 927 C.

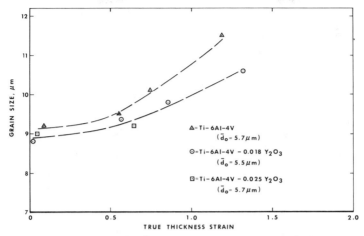

Fig. 6. A comparison of the grain growth kinetics of the yttriated
alloy sheets (sheets 2 and 3) with the standard Ti-6Al-4V
after exposure to the dynamic conditions.

## Summary and Conclusions

Our study shows that small additions of yttria to the Ti-6Al-4V alloy
retard its tendency for grain growth under SPF processing conditions and
result in an overall improvement in the superplastic performance of the
baseline alloy. It is important to note that no attempt, whatsoever, was
made to optimize the concentration, size or distribution of the yttria particles in the two yttriated sheets used in this investigation. Therefore,
while absolute magnitudes of some of the individual effects of yttria additions to the Ti-6Al-4V alloy may appear to be small, they are quite significant in that they consistently demonstrate an overall improvement in
the superplastic response of the reference alloy. They become even more
significant in view of the implication that a systematic optimization of
the concentration, size and distribution of the yttria particles may further enhance the superplastic characteristics of this alloy. An additional
benefit is expected in the diffusion bondability, since a fine stable
grain structure is also an inherent requirement of that process.

The following conclusions are drawn from the results of the present
observations:

1. Small yttria additions (up to 0.025 wt pct) to Ti-6Al-4V alloy
   retard grain growth due to thermal exposure alone.

2. The retardation in grain growth is significantly more pronounced
   under dynamic conditions where influence of strain and heat is
   combined.

3. Two sizes of yttria particles ($\sim$250-500 nm and $\sim$10-20 nm) appear to exist in the microstructure of the yttriated Ti-6Al-4V
   alloys. Sparse concentration of the larger particles suggests
   their little overall effect on the superplastic behavior of
   these alloys.

4. The superplastic behavior of the yttriated alloys is superior to that of the baseline metal as evidenced by the more tightly formed radii in the cup- and hat-shaped cavity tests.

## Acknowledgments

Two of the authors (SPA and EDW) acknowledge the financial support of Rockwell Independent Research and Development Program and the third author acknowledges support of Boeing Independent Research and Development Program.

## References

1. J.E. Lyttle, G. Fischer and A.R. Marder:  J. Metals, vol. 17 (1965), p. 1055.
2. D. Lee and W.A. Backofen:  Trans. TMS-AIME, vol. 239 (1967), p. 1034.
3. D.A. Woodford:  Trans. Quart. ASM, vol. 62 (1969), p. 291.
4. R.H. Johnson:  Metallurgical Reviews, vol. 15, no. 146 (1970), p. 115.
5. N. E. Paton:  J. Engr. Mater. Tech., vol. 97 (1975), p. 313.
6. A. Arieli and A. Rosen:  Scripta Met., vol. 10 (1976), p. 471.
7. A. Arieli and A. Rosen:  Met. Trans., vol. 8A (1977), p. 1591.
8. C. H. Hamilton:  Proc. Formability:  Analysis, Modeling and Experimentation, Chicago, 1977, TMS-AIME publication, 1978.
9. S. P. Agrawal and C. H. Hamilton:  Unpublished research, Rockwell International Corporation, Los Angeles, 1977.
10. R. R. Boyer and J. E. Magnuson:  Met. Trans., vol. 10A (1979), p. 1191.
11. N. E. Paton and C. H. Hamilton:  Met. Trans., vol. 10A (1979), p, 241.
12. A. K. Ghosh and C. H. Hamilton:  Met. Trans., vol. 10A (1979), p. 699.
13. S. P. Agrawal and E. D. Weisert:  Proc. North American Metalworking Conference VII, Ann Arbor, Michigan, 1979, SME publication, p. 197.
14. E. D. Weisert and G. W. Stacher:  Metal Progress, (March 1977), p. 33.
15. C. H. Hamilton and G. W. Stacher: Metal Progress, (March 1976), p. 34.
16. C. H. Hamilton, G. W. Stacher, J. A. Mills and H. W. Li:  Superplastic Forming of Titanium Structures, Technical Report AFML-TR-75-62, April 1975, Air Force Materials Laboratory, Wright-Patterson Air Force Base, Ohio.
17. E. D. Weisert, G. W. Stacher  and B. W. Kim:  Manufacturing Methods for Superplastic Forming/Diffusion Bonding, Technical Report AFML-TR-79-4053, May 1979, Air Force Materials Laboratory, Wright-Patterson Air Force Base, Ohio.
18. E. D. Weisert:  Proc. Tri-Service Metals Manufacturing Technology, Program Status Review, 25-27 September 1979, Daytona Beach, Florida.
19. RMI Company:  Yttrium Enhancement of Titanium Alloys, Technical Report, August 1976.
20. W. Rostoker and J. R. Drorak:  Interpretation of Metallographic Structures, p. 224, Academic Press, 1976.
21. Private Communication with H. Bomberger, RMI Company, 1978.
22. R. M. Allen and J. B. Vander Sande:  Met. Trans., vol. 9A (1978), p. 1251.

# STRAIN RATE SENSITIVITY OF FLOW STRESS AND EFFECTIVE STRESS IN SUPERPLASTIC DEFORMATION OF Ti-8Mn ALLOY

N. Furushiro and S. Hori

Department of Materials Science and Engineering,
Faculty of Engineering, Osaka University,
Suita 565, Japan

## Introduction

The most fundamental characteristic of fine grain superplasticity is a remarkably high strain rate sensitivity of flow stress, compared with that in usual high temperature deformation [1,2]. The strain rate sensitivity is expressed as $m$ in the following equation [3],

$$\sigma = K \dot\varepsilon^{m} \quad \cdots\cdots\cdots\cdots (1)$$

where $\sigma$ is flow stress, $\dot\varepsilon$ is strain rate and K is a constant. In order to make clear the deformation mechanism of superplasticity, it is first necessary to understand the reason for the high strain rate sensitivity during the deformation [4].

The flow stress $\sigma$ for a range of metals can be expressed in the following form [5],

$$\sigma = \sigma_i + \sigma_e, \quad \cdots\cdots\cdots\cdots (2)$$

where, $\sigma_i$ and $\sigma_e$ are the internal stress and the effective stress, respectively. Ti-8Mn alloy is of a ($\alpha + \beta$) type, in which $\beta$ phase is stable at room temperature. Superplasticity of a Ti-8Mn alloy was reported by P. Griffiths and C. Hammond [6], who found that high values of $m$ were obtained at temperatures in the ($\alpha + \beta$) phase region and also even in the single phase region of $\beta$. In the latter region the grain size was larger by about two orders than that of ordinary superplastic materials and subgrain networks were formed in the grains during the deformation. It was also pointed out that even though $m$ was close to unity, elongations obtained were less than 140% for sheet specimens. These results were much different from those in the literature on the usual superplasticity [1,2,7]. So, Ti-8Mn alloy has become of interest in connection with a systematic investigation on superplasticity.

Therefore, the strain rate sensitivity of $\sigma_i$ and $\sigma_e$ at temperatures in the ($\alpha + \beta$) and $\beta$ phase regions were investigated.

## Experimental

The mill plate of a Ti-8Mn alloy containing 7.8% Mn, 0.27% Fe, 0.104% O, 0.0119% H and 0.0118% N supplied by Kobe Steel Ltd. was used in this experiment. Tensile specimens of gauge length 20 mm and width 4 mm were machined from the plate of thickness 1 mm, then vacuum-annealed in a quartz tube for 2 h at various temperatures in the ($\alpha + \beta$) and $\beta$ phase regions of the equilibrium diagram, followed by air-cooling. Tensile tests were carried out in a purified argon atmosphere at elevated temperatures using an Instron type tensile machine. After being set to the machine the specimen was heated at a rate of 20°C/s in an infared ray furnace and held for a suitable period, and then loaded. The deformation temperature and the initial strain rate ranged from 580 to 900°C and from $8.3×10^{5}$ to $1.7×10^{3}$ $\bar{s}^{1}$, respectively.

The stress $\sigma_s$ and the period of the deformation stagnation following the stress dipping after steady state deformation were measured and then $\sigma_i/\sigma$ was determined by an extrapolation method on the graph [4,8,9]. Values of $m$ were calculated from the stress at a strain rate and that when the strain rate was doubled during the deformation.

## Results and Discussion

Figure 1 shows the relation between the test temperature and the $m$ value measured at ~50% strain of the deformation at the same temperature as the previous heat treatment. It was shown that $m$ value increased with the temperature up to 650°C, then decreased, and above 680°C $m$ value appeared to be constant, about 0.43. All $m$ values obtained in Fig. 1 satisfy the common criterion of superplasticity [7], *i.e.* $m > 0.3$. The temperature of β transus is ~780°C. So, it is realized that high values of $m$ were also obtained at the temperature of β single phase region, where elongations were not so large.

The value of $\sigma_i/\sigma$ was almost independent of the strain as shown typically in Fig. 2. Measurement of $\sigma_i/\sigma$ was, therefore, carried out at strains less than ~50%. Various specimens heat-treated at different temperatures were supplied for the measurement of $\sigma_i/\sigma$ at 600, 700 and 800°C. Results obtained as a function of $m$ under the deformation are shown in Fig. 3. The value of $\sigma_i/\sigma$ is much dependent on temperature and strain rate, as shown in the previous paper [4]. Under certain conditions of temperature and strain rate, it can be seen from Fig. 3 that $\sigma_i/\sigma$ at 600°C was negatively proportional to $m$, while values of $\sigma_i/\sigma$ at 800°C did not show a specific relation to $m$. It is still unclear if the difference of the $m$ value dependence of $\sigma_i/\sigma$ can be directly correlated to the constituent phases and the deformation mechanism operated. In this connection, values of $\sigma_i/\sigma$ at 700°C, which is in the ( α + β ) region, appear on the curve at 600°C in Fig. 3.

The negative proportionality between $\sigma_i/\sigma$ and $m$ was also found for Al-Cu

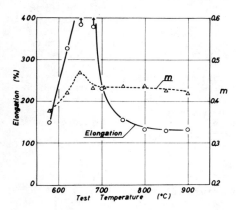

Fig. 1. Changes in the elongation and the strain rate sensitivity index of flow stress, $m$, of a Ti-8Mn alloy vs test temperature, which is the same as that of the previous heat treatment.

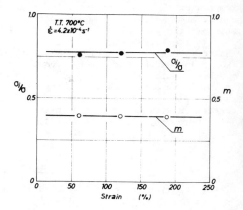

Fig. 2. The effect of strain on $\sigma_i/\sigma$ of a Ti-8Mn alloy deformed at 700 °C and a strain rate of $4.2×10^{4}$ $s^{-1}$.

[4] and Pb-Sn alloys [10]. This implies that the internal stress takes place under super-plastic deformation and that the high strain rate sensitivity of flow stress may be based on either the internal stress or the effective stress.

Then, in order to examine the strain rate sensitivity of $\sigma_i$ and $\sigma_e$, specimens heat-treated at 700°C were tensiled at 600°C and at various strain rates. Values of $\sigma$, $\sigma_i$ and $\sigma_e$ obtained are plotted as a function of the strain rate in Fig. 4. According to equation (1), the slope of the curves in Fig. 4 represents $m$. It is clear that the strain rate sensitivity of $\sigma_e$ is higher than that of $\sigma_i$. This tendency was also observed in the results at 800°C. These suggest that the strain rate sensitivity of flow stress owes much to that of $\sigma_e$. The strain rate sensitivity index of $\sigma_e$, $m(\sigma_e)$ was close to unity, while $m(\sigma_i)$ was about 0.3.

The relation between the strain rate and the stress in dislocation creep deformation may be expressed in the following form,

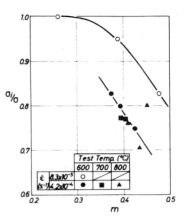

Fig. 3. The relationship between $\sigma_i/\sigma$ and $m$ of specimens heat-treated and tested under various conditions.

$$\dot{\varepsilon} = K'\sigma^n, \quad \cdots\cdots\cdots\cdots(3)$$

where $K'$ is a constant and $n$ is $3 \sim 5$ for the steady state region. As mentioned previously [4], the superplastic deformation with a nearly constant flow stress is regarded as a kind of steady state creep deformation and the correlation between $n$ and $m$ is expressed by

$$1/n = m, \quad \cdots\cdots\cdots\cdots(4)$$

where $1/n$ was calculated to be $0.25 \sim 0.33$ and those values were nearly equal to $m(\sigma_i)$.

It is considered that a diffusion controlled process is included in the mechanism, from which $\sigma_e$ results since $m(\sigma_e)$ is close to unity. However, the experimental difficulty due to the slight oxidation of the specimen surface during the deformation prevented a clear and exact mechanism, leading to the high $m(\sigma_e)$ values, from being shown. A detailed explanation on these feature will be given in near future, being partly based on the metallographic observation of the interior of grains.

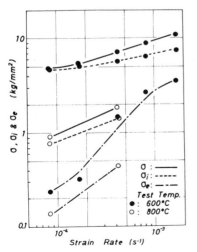

Fig. 4. The strain rate dependence of $\sigma$, $\sigma_i$ and $\sigma_e$ for deformations at 600 and 800°C, for specimens heat-treated at 700 and 800°C, respectively.

### Conclusion

Superplastic deformation of a Ti-8Mn alloy was observed at temperatures in dual and single phase regions of the phase diagram. The high value of $m$ is due to $\sigma_e$, which is very strain rate sensitive, while $m(\sigma_i)$ is as small as the strain rate sensitivity of the stress under dislocation creep deformation.    Therefore, the $m$ value may depend on the ratio of $\sigma_e$ to $\sigma$ under certain conditions of the deformation.

## Acknowledgement

·    The authors are grateful to Dr. Y. Moriguchi of Kobe Steel Ltd. for his effort in providing the Ti-8Mn alloy. They are also indebted  to Mr. S.Shimoyama for his experimental collaboration.

## References

1.  R.H. Johnson : Met. Rev.   15(1970), 115.
2.  S. Hori and N. Furushiro: Bull. J. Jap. Inst. Metals  14(1975), 673.
3.  R.L. Bell and T.G. Langdon: " *Interfases Conference* ", Edited by R.C. Gifkins, Butterworths, 115 (1969).
4.  N. Furushiro and S. Hori: Scripta Met., 12(1978), 35.
5.  S. Takeuchi and A.S. Argon : J. Mater. Sci., 11(1976), 1542.
6.  P. Griffiths and C. Hammond : Acta Met., 20(1972), 953.
7.  J.W. Edington, K.N. Melton and C.P. Culter : Progress in Materials Science, 21(1976), 61.
8.  K. Toma, H. Yoshinaga and S. Morozumi : J. Jap. Inst. Metals, 38(1974), 170.
9.  K. Toma, H. Yoshinaga and S. Morozumi : J. Jap. Inst. Metals, 39(1975), 621.
10. N. Furushiro and S. Hori : Scripta Met., 13(1979), 653.

# ON THE NATURE OF "MATERIAL CONSTANTS"

## IN THE SHERBY-DORN AND LARSON-MILLER PARAMETERS
## FOR CREEP IN TITANIUM

H. W. Rosenberg
TIMET, Pittsburgh, PA 15230

## Introduction

Creep parameters have found two important uses: 1) to effectively guide test conditions for exploring the creep-rupture characteristics of a material so as to shorten testing time and 2) to derive design stress allowables. The generalized equation for creep in metals is given by

$$\dot{\varepsilon} = \dot{\varepsilon}_o(T, \sigma,)\exp(\frac{-G(T,\sigma)}{kT}) \tag{1}$$

where $\dot{\varepsilon}$ tensile strain rate $\dot{\varepsilon}_o$ is a pre-exponential factor that includes the mobile dislocation density, the area swept out per activation event and the frequency of attempt. G is the Gibbs free energy of activation. T is absolute temperature, $\sigma$ is applied tensile stress while k is Boltzmann's constant. In the most general case $\dot{\varepsilon}_o$ and G each depend on temperature and stress.[1-8]

Larson and Miller[9] expressed the rate equation in the form

$$\dot{\varepsilon} = \dot{\varepsilon}_o \exp(\frac{-Q(S)}{kT}) \tag{2}$$

In their words, "Q(s) is the activation energy for the process under the conditions considered.....".

Equations(1)and(2)can each be written in the equivalent form

$$\dot{\varepsilon} = \dot{\varepsilon}_o \exp(\frac{-H}{kT}) \tag{3}$$

where H is the activation enthalpy and $\dot{\varepsilon}_o$ now contains an entropy term. Integrating equation(3), between appropriate limits, taking logs and rearranging leads to

$$\frac{H}{k} = T(\ln \frac{\dot{\varepsilon}_o}{\varepsilon} + \ln t) \tag{4}$$

for any given stress.

Larson and Miller[9] showed empirically that equation(4)in the form

$$\log \sigma = A + B (T(20 + \log t)) \tag{5}$$

where T is in degrees Rankin, holds for a wide variety of materials. Here A and B are the intercept and slope respectively on a plot of log $\sigma$ against T(20 + log t). In practice, therefore, Larson-Miller parameter methodology makes use of three material constants. Larson and Miller did not explain the meanings of the material constants. It is significant, however, that they found A and B to be "constants" for a given alloy over limited domains of stress in many cases.

Sherby and Dorn(10,11)showed that creep data for pure metals and simple alloys can be correlated by

$$\epsilon = f(\sigma, \theta) \tag{6}$$

where $\theta = t \exp(-H/kT)$ is a temperature compensated time function. In application, $\theta$, or log $\theta$, has been plotted against $\epsilon$, ln $\epsilon$, $\sigma$ or ln $\sigma$.(12-14) The log $\theta$-$\sigma$ plot is equivalent to employing the parameter

$$\sigma = A + B \left(\frac{C}{T} - \ln t\right) \tag{7}$$

where A, B and C are material constants. Orr et al(12)correlated results from a number of materials and also found the correlations hold only over limited domains of stress. The $\theta$-$\epsilon$ plot has the shape of a creep curve.(13)

It is a purpose of this paper to examine the nature of the material constants in the Larson-Miller and Sherby-Dorn parameters as their meanings and values are affected by the assumption of linear or logarithmic stress.

## The Models

Table 1 illustrates the connections between the parameters and the creep models from which they derive. According to Sherby and Burke(13), model 1 typically holds in the low temperature-high stress region of creep where $\dot{\epsilon}/D > 10^9$. Here D is diffusivity. The mechanism acting in this case involves glide but otherwise is not well understood. The quantity n in model 1 is a material constant.

Model 2 assumes that the free energy of activation is modified by stress along the lines pioneered by Basinski(1)and developed by others.(2-6)   In this model, n has the meaning of an activation volume, $V = \partial H/\partial\sigma|_T - T\partial S/\partial\sigma|_T$ where S is the entropy of activation. This is discussed further below.

Model 3 is the classical power law where creep is thought to be controlled by dislocation climb(13). In this model n is a material constant ranging in typical value from 3 to 5.

Model 4 derives from the original work by Larson and Miller(9)who used log $\sigma$ to correlate their parameter. Possibly, it works so well in practice because $\sigma^{n/kT}$ and $(\sigma/E(T))^n$ have qualitatively similar trends with stress and temperature. The latter form where E is modulus holds for a number of metals.(15,16)  The essential difference between Larson-Miller and Sherby-Dorn parameters is that a kT term appears in stress functions of the Larson-Miller but not in the Sherby-Dorn models.

There are a number of limitations in the use of Larson-Miller and Sherby-Dorn parameters for titanium or any other metal. They relate to the assumptions that go into each parameter. Some of these are:

1. The stress function and the model chosen are assumptions unless there is a theoretical basis for the choice of a model and a demonstration that the related parameter fits best.

2. Linear creep or creep rupture is assumed. In fact, this is usually not even a good approximation for short time creep tests to small deformations. (The Sherby-Burke(13)method of correlating $\theta$ with $\epsilon$ gets around this problem.)

Table 1.   Relationship Between Creep Models and Parameters

| Model | Parameter | A | B | C |
|---|---|---|---|---|
| 1. $\dot{\epsilon} = \dot{\epsilon}_0 \exp(n\sigma) \exp\left(-\frac{H}{kT}\right)$ | $\sigma = A + B\left(\frac{C}{T} - \ln t\right)$ | $\frac{1}{n}(\ln \epsilon/\dot{\epsilon}_0)$ | $\frac{1}{n}$ | $\frac{H}{k}$ |
| 2. $\dot{\epsilon} = \dot{\epsilon}_0 \exp\left(\frac{n\sigma}{kT}\right)\exp\left(-\frac{H}{kT}\right)$ | $\sigma = A + BT(C - \ln t)$ [1] | $\frac{H}{n}$ | $\frac{k}{n}$ | $\ln \epsilon/\dot{\epsilon}_0$ |
| 3. $\dot{\epsilon} = \dot{\epsilon}_0\, \sigma^n \exp\left(-\frac{H}{kT}\right)$ | $\ln \sigma = A + B\left(\frac{C}{T} - \ln t\right)$ | $\frac{1}{n}(\ln \epsilon/\dot{\epsilon}_0)$ | $\frac{1}{n}$ | $\frac{H}{k}$ |
| 4. $\dot{\epsilon} = \dot{\epsilon}_0\, \sigma^{\frac{n}{kT}} \exp\left(-\frac{H}{kT}\right)$ | $\ln \sigma = A + BT(C - \ln t)$ [1] | $\frac{H}{n}$ | $\frac{k}{n}$ | $\ln \epsilon/\dot{\epsilon}_0$ |

(1) Since $\ln \frac{\epsilon}{\epsilon_0}$ is negative, these forms are equivalent to $F(\sigma) = A - BT(C + \log_{10} t)$ which is the traditional format.

3. Constant stress is assumed when, in fact, during the early to middle portions of a test the true tensile stress changes as(1 + e)where e is the engineering tensile strain. Later, necking may ensue to complicate the stress state further.

4. In the case of rupture plots, constant strain to rupture is assumed when, in fact, rupture strain is not constant.

5. Temperature and time are assumed to compensate for each other. This is only true within the temperature-stress-strain rate domain where a given creep mechanism is operative.

6. Parameter constants are usually assumed to be constant for a variety of strains. This cannot be strictly so because one of the constants in every parameter contains a strain function.

7. The pre-exponential factor $\dot{\epsilon}_0$ is assumed to be independent of temperature. In fact, under constant applied stress and at low homologous test temperature, $\dot{\epsilon}_0$ necessarily depends on temperature through the modulus which controls the back stress observed by dislocation sources subject to activation under a given applied stress.

Some of these assumptions can lead to serious errors, others may be of no consequence. It is both a little surprising and gratifying that the parameters work as well in practice as they do. One important reason for this, of course, is that in practice a designer is usually only interested in a relatively narrow time-temperature-stress regime. Furthermore, the stress-temperature maxima in cyclic service often occur together for short times so even limited data, properly chosen, can be quite sufficient. Even more important from the system reliability viewpoint is the fact that a poor choice of parameter leads to added scatter in the data and thus to overly conservative design statistics. This results in a design less efficient than it otherwise might be. Of course, reliability is enhanced.

There follows a brief illustration of how the selection of the stress function and creep parameter can affect the standard error in a design stress calculation.

## Results and Discussion

Table 2 presents some typical creep-rupture results for beta annealed Ti-6Al-2Sn-4Zr-2Mo-.1Si which are analyzed by model in Table 3. In the method used here, $\dot{\epsilon}_0$, n and H are best fits in terms of the parameter, not of the creep rupture model. Note that H varies by 50%, depending on choice of model. And the parameter constants of best fit vary significantly, depending on the stress function chosen. The best fit, according to the standard error for stress, is provided by model 4, the Larson-Miller parameter using the ln σ plot. Note also that the standard error for stress attending model 2 is 79% greater than that for model 4.

A word about the Table 3 analysis. Experimentally, rupture time is the dependent variable. However, stress was taken as the dependent variable in a least squares fit to each parameter. For the standard error to have meaning, the design variable must be taken as the dependent variable in a least squares fit of the data regardless of the experimental conditions.

Table 4 presents creep data for beta annealed Ti-6Al-2Sn-4Zr-2Mo-.1Si. The four creep parameters were fit to these data by least squares with the results of best fit shown in Table 5. The Sherby-Dorn parameter fits best at small strains, the Larson-Miller best at large strains. This may mean that the creep mechanism is strain dependent. Possibly, also the results are an artifact of the strain-time laws, or are simply fortuitous owing to the rather small amount of data.

Table 2.   Stress Rupture Data For Beta Annealed
Ti-6Al-2Sn-4Zr-2Mo-.1Si

| Temperature $^\circ$K | Stress MPa | Time to Rupture Hours |
|---|---|---|
| 783 | 524 | 674 |
| 783 | 565 | 99 |
| 839 | 400 | 101 |
| 839 | 427 | 41 |
| 867 | 241 | 343 |
| 867 | 345 | 44 |
| 922 | 138 | 518 |
| 950 | 103 | 500 |
| 978 | 103 | 102 |

Table 3.   Data of Table 2 Analyzed For Parameter Goodness of Fit

| | Model | | | |
|---|---|---|---|---|
| | 1 | 2 | 3 | 4 |
| Stress Function | $\sigma$ | $\sigma$ | $\ln \sigma$ | $\ln \sigma$ |
| Parameter | $B(\frac{C}{T} - \ln t)$ | $BT(C-\ln t)$ | $B(\frac{C}{T} - \ln t)$ | $BT(C-\ln t)$ |
| Standard Error of $\sigma$, MPa | 25.3 | 35.7 | 31.2[1] | 19.9[1] |
| $\dot{\varepsilon}_0$, HR$^{-1}$ | $7.0 \times 10^{22}$ | $2.6 \times 10^{28}$ | $6.1 \times 10^9$ | $3.1 \times 10^{21}$ |
| n | .225 | 411 | 6.73 | 12300 |
| H,K Joules/Mole | 511 | 611 | 398 | 603 |
| Parameter constant,C | 257 | 29.7[2] | 200 | 22.8[2] |

(1) Standard error non linear, average value at average stress.

(2) Given as common log for comparison with literature.

The Larson-Miller parameter constant, C of Table 1, depends on the stress law assumed, the temperature range over which it is derived and the creep or average rupture strain. These dependencies appear in Table 6. The total variation is more than a multiple of two. Only for the $\sigma$ $^{n/kT}$ stress law and the wider temperature range is the Larson-Miller constant reasonably stable. A common feature is that the Larson-Miller constant for the creep data increases with strain in the four cases illustrated. While this is in accord with theory, the range appears excessive in three cases. An ability to display data for a material in the form of isostrain curves over wide ranges of stress and strain using a parameter with a single valued constant

Table 4.      Creep Data For Beta Annealed Ti-6Al-2Sn-4Zr-2Mo-.1Si

| Temperature $^\circ$K | Stress MPa | Hours to % Creep Strain | | |
|---|---|---|---|---|
| | | .1 | .2 | .5 |
| 783 | 448 | 1 | 7 | 103 |
| 811 | 345 | 4 | 20 | 95 |
| 811 | 414 | 1 | 4 | 32 |
| 839 | 173 | 29 | 110 | 1080 |
| 839 | 173 | 25 | 91 | 761 |
| 839 | 241 | 11 | 34 | 84 |
| 839 | 310 | 2 | 7 | 29 |
| 867 | 172 | 12 | 31 | 136 |
| 867 | 207 | 3 | 11 | 38 |
| 867 | 276 | 1 | 2 | 10 |
| 922 | 69 | 10 | 49 | 710 |
| 922 | 103 | 2 | 4 | 42 |
| 922 | 138 | 1 | 3 | 12 |
| 978 | 69 | 2 | 7 | 35 |

Table 5.      Creep-Rupture Models of Best Fit Over The
              783-978$^\circ$K Temperature Range

| Strain | .001 | .002 | .005 | Rupture |
|---|---|---|---|---|
| Stress Law (Model)[1] of Best Fit | $\exp(n\sigma)$ | $\exp(n\sigma)$ | $\sigma^{\frac{n}{kT}}$ | $\sigma^{\frac{n}{kT}}$ |
| Standard Error, MPa | 20.0 | 25.8 | 11.0 | 19.9 |

(1) $\exp(\overset{n}{n\sigma})$ is the Sherby-Dorn parameter-vs-linear stress(Model 1)
    $\sigma^{\overline{kT}}$ is the Larson-Miller parameter-vs-log stress(Model 4)

Table 6.      Effect of Stress Law Assumed, Temperature and
              Strain on the Larson-Miller Constant      (C, Table 1)

| Stress Law | Stress Plot | Temperature Range-$^\circ$K | Larson-Miller Constant$(Log_{10}\frac{\varepsilon}{\dot{\varepsilon}_0})$ | | | |
|---|---|---|---|---|---|---|
| | | | .001 | .002 | .005 | Rupture(1) |
| $\exp(\frac{n\sigma}{kT})$ | $\sigma$ | 783-978 | 15.9 | 18.3 | 26.3 | 29.7 |
| $\exp(\frac{n\sigma}{kT})$ | $\sigma$ | 783-867 | 16.8 | 20.9 | 26.8 | 29.7 |
| $\sigma^{\frac{n}{kT}}$ | $\ln\sigma$ | 783-978 | 21.4 | 21.9 | 22.9 | 22.8 |
| $\sigma^{\frac{n}{kT}}$ | $\ln\sigma$ | 783-867 | 13.7 | 17.6 | 22.4 | 23.2 |

(1) Derived from Table 2.

is often desired. A constant of 22 would yield master curves for the $\sigma^{n/kT}$ law over the 783-978° range. Of course, use of an averaged constant enhances scatter and is not consistent with deriving error bar minima.

In Table 7, the effects of stress law, temperature range and strain on $\dot{\varepsilon}_0$ and H arising from the Larson-Miller parameter are given. All the H values are a few multiples higher than the enthalpy of self diffusion which is about 150 K Joules/Mole.(17,18) Such high values are explainable if dynamic strain aging is involved.(19) If this is so, then the question of which parameter to use is empirical.

The Sherby-Dorn parameter yields somewhat smaller values for H as shown in Table 8. The $\sigma^n$ stress law produces the least strain sensitive constants, particularly for the wider temperature range. The Sherby-Dorn parameter constant inside the parentheses is H/k. For the $\sigma^n$ stress law over the 783-987°K range, the average Sherby-Dorn parameter constant is 206 degrees(not to be confused with application temperature).

### Some Theoretical Limitations

Although nominally a constant, $\dot{\varepsilon}_0$ in practice can vary widely. This is because $\dot{\varepsilon}_0$ contains the mobile dislocation density. At temperatures where plastic flow proceeds by glide, the mobile dislocation density may be only a small fraction of the total dislocation density. The mobile dislocation density is extremely sensitive to stress.

$\dot{\varepsilon}_0$ is also dependent on temperature. After Pharr and Nix(8)one can write

$$L = Nb = \frac{\alpha\mu b}{T-T_b} \tag{8}$$

where $\alpha$ is a constant near unity(20), b is the Burgers vector, L is the distance between two pinning points in a dislocation source that can just be activated by the effective shear stress $T - T_b$, N is a "critical source constant" that defines L in terms of b, u is the shear modulus, $T$ is the applied shear stress and $T_b$ is the opposing back stress that arises from the dislocation structure surrounding the source. The point is that $\dot{\varepsilon}_0$ will be independent of temperature only if N is. Differentiating equation(9)with respect to temperature and equating to zero gives

$$\frac{d\mu}{dT} = \frac{\mu}{T-T_b} \frac{\partial T_b}{\partial T}\Big|_T \tag{9}$$

as the condition for which N is independent of temperature. Since $T$ can be varied at will, N, in general, will depend on temperature. $\dot{\varepsilon}_0$ is, therefore, temperature dependent in the general case, even for a given flow mechanism.

So $\dot{\varepsilon}_0$ can hardly be a true constant in any of the models. However, in fitting the various parameters to creep rupture data, if $\dot{\varepsilon}_0$ is found to change with strain by orders of magnitude, one should suspect either the data or the applicability of that parameter and model. See Tables 7 and 8 for examples.

The quantity n presents similar difficulties and has different meanings in each model. Consider model 2. Here n has the theoretical meaning of an activation volume, V, defined by

$$V = \frac{\partial G}{\partial T}\Big|_{T,T_b} \tag{10}$$

Table 7.   Effects of Stress Law Assumed, Temperature and Strain on the Arrhenius Constants Derived Algrebraically From Least Squares Coefficients for Larson-Miller Parameter

| Strain | Temperature Range °K | $\dot{\varepsilon}_o$, HR⁻¹ exp $\frac{n_\sigma}{ki}$ | $\sigma \frac{n}{kT}$ | H, K Joules/Mole exp $\frac{n_\sigma}{ki}$ | $\sigma \frac{n}{kT}$ |
|---|---|---|---|---|---|
| .001 | 783-978 | $8.3 \times 10^{12}$ | $2.7 \times 10^{18}$ | 308 | 573 |
|  | 783-867 | $6.8 \times 10^{13}$ | $4.6 \times 10^{10}$ | 320 | 439 |
| .002 | 783-978 | $2.2 \times 10^{15}$ | $1.6 \times 10^{19}$ | 360 | 590 |
|  | 783-867 | $1.6 \times 10^{18}$ | $7.5 \times 10^{14}$ | 396 | 518 |
| .005 | 783-978 | $9.7 \times 10^{23}$ | $4.1 \times 10^{20}$ | 523 | 636 |
|  | 783-867 | $3.3 \times 10^{24}$ | $1.2 \times 10^{20}$ | 507 | 622 |
| Rupture[1] | 783-978 | $2.8 \times 10^{28}$ | $3.4 \times 10^{21}$ | 610 | 702 |
|  | 783-867 | $2.7 \times 10^{28}$ | $1.1 \times 10^{22}$ | 580 | 743 |

(1) From Table 2 data, $\bar{\varepsilon} = .06$

Table 8.   Effects of Stress Law Assumptions and Strain on the Arrhenius Constants Derived Algebraically From Least Squares Fits to the Sherby-Dorn Parameter

| Strain | Temperature Range °K | $\dot{\varepsilon}_o$, HR⁻¹ exp$(n\sigma)$ | $\sigma^n$ | H,K Joules/Mole exp$(n\sigma)$ | $\sigma^n$ |
|---|---|---|---|---|---|
| .001 | 783-978 | $1.1 \times 10^{12}$ | $8.4 \times 10^{6}$ | 294 | 404 |
|  | 783-867 | $6.6 \times 10^{10}$ | $7.6 \times 10^{-4}$ | 272 | 221 |
| .002 | 783-978 | $1.6 \times 10^{14}$ | $2.3 \times 10^{7}$ | 336 | 404 |
|  | 783-867 | $1.2 \times 10^{15}$ | $4.5 \times 10^{9}$ | 346 | 292 |
| .005 | 783-978 | $7.8 \times 10^{17}$ | $3.9 \times 10^{7}$ | 413 | 420 |
|  | 783-867 | $1.3 \times 10^{21}$ | $1.0 \times 10^{5}$ | 452 | 381 |
| Rupture[1] | 783-978 | $7.4 \times 10^{22}$ | $1.5 \times 10^{4}$ | 510 | 396 |
|  | 783-867 | $6.7 \times 10^{22}$ | $1.9 \times 10^{9}$ | 489 | 395 |

(1) From Data of Table 2, $\bar{\varepsilon} = .06$

This is analogous to the pressure derivative of the Gibbs free energy in thermodynamics. Expanding equation(1)in terms of its shear stress partial derivative leads to

$$kT \left.\frac{\partial \ln \dot{\gamma}}{\partial T}\right|_{T,T_b} = KT \left.\frac{\partial \ln \dot{\gamma}_o}{\partial T}\right|_{T,T_b} + T \left.\frac{\partial S}{\partial T}\right|_{T,T_b} - \left.\frac{\partial H}{\partial T}\right|_{T,T_b} \qquad (11)$$

where use has been made of

$$G = H - TS \qquad (12)$$

and shear strain rate $\dot{\gamma}$ has been substituted for $\dot{\varepsilon}$. From equation (12)

$$\left.\frac{\partial G}{\partial T}\right|_T = \left.\frac{\partial H}{\partial T}\right|_T - T \left.\frac{\partial S}{\partial T}\right|_T \qquad (13)$$

For a reversible process, the thermodynamic equation relating mechanical work and entropy to enthalpy is

$$dH = TdS + VdT \qquad (14)$$

where stress has been substituted for pressure, P. Fixing temperature, equation(14) can be written

$$V = \left.\frac{\partial H}{\partial T}\right|_T - T \left.\frac{\partial S}{\partial T}\right|_T \qquad (15)$$

which, in view of equation(13), provides confirmation of equation(10). The validity of equations(10)and(15), of course, depends on the assumptions that pressure and stress serve equivalent thermodynamic functions and that metal flow proceeds reversibly.

For an ideal gas, $\partial H/\partial P|T=0$. For real gases, $\partial H/\partial P|T$ is usually finite. Thus, the activation volume for metal flow is partitioned into two parts, the ideal $T\partial S/\partial T|_T$ and the non-ideal $\partial H/\partial T|_T$. It is evident from equation (11)that $\partial H/\partial T|_T$, if it exists, can only be found upon extrapolation to $0°K$ and then only if $\partial \ln \dot{\gamma}_o/\partial T|_T$ behaves nicely. As an experimental quantity derived from model 2 then the apparent activation volume as usually measured has more to do with $\partial \ln \dot{\gamma}_o/\partial T|_T$ than with $\partial G/\partial T|_T$.

A practical, yet common, limitation encountered has to do with the data base. At the least, several stress levels at each of the several temperatures are required with replication to assess test error. Test error should be independent of the dependent variable.(21)

A final caution. Modern creep resistant alloys are highly complex and may employ several metallurgical techniques to achieve high creep and rupture strength. Deformation mechanisms may operate in tandem or in parallel. They may overlap in temperature and in stress. And strain rate may be a factor. Ideally, a designer would have a time-temperature-stress parameter map with error bars in the dependent variable of choice. In this connection also, interpolations will be more accurate than extrapolations, even for well behaved linear functions.(21)

## Conclusions

1. The parameter options should be surveyed and the one that minimizes the standard error should be used in the absence of good reason not to.

2. The parameter exhibiting the most stable constants as stress, temperature and strain are varied is likely to have the best theoretical basis.

3. Caution should be exercised in relating any constants arising from parameter fitting to the theoretical quantities in the underlying model.

4. Care should be taken when using a parameter beyond its domain of validity and wherever possible to interpolate rather than extrapolate.

## References

1. Z. R. Basinski, Acta Met, vol 5, p 684, 1957.

2. H. Conrad and H. Wiedersich, Acta Met, vol 8, p 128, 1960.

3. G. B. Gibbs, Phys. Stat. Sol., vol 5, p 693, 1964.

4. G. Schoeck, Phys. Stat. Sol., vol 8, p 499, 1965.

5. J. C. M. Li, "Dislocation Dynamics", McGraw-Hill, NY, p 87, 1968.

6. J. P. Hirth and W. D. Nix, Phys. Stat. Sol., vol 35, p 177, 1969.

7. R. Gasca-Neri and W. D. Nix, Acta Met, vol 22, p 257, 1974.

8. G. M. Pharr and W. D. Nix, Acta Met, vol 27, p 433, 1979.

9. F. R. Larson and J. Miller, Trans ASME, 765, July, 1952.

10. O. D. Sherby and J. E. Dorn, Trans AIME, vol 194, p 959, 1952.

11. O. D. Sherby and J. E. Dorn, Journal of Metals, vol 5, p 324, 1953.

12. R. L. Orr, O. D. Sherby and J. E. Dorn, Trans ASM, vol 46, p 113, 1954.

13. O. D. Sherby and P. M. Burke, Progress in Materials Science, vol 13, p 325, 1968.

14. Linda C. Hitzel and O. D. Sherby, "Time-Temperature Parameters for Creep Rupture Analysis", ASM Publication, No. D8-100, Oct., 1968.

15. O. D. Sherby, Acta Met, vol 10, p 135, 1962.

16. C. R. Barrett, A. J. Ardell and O. D. Sherby, Trans AIME, vol 230, p 200, 1964.

17. C. M. Libanati and A. F. Dyment, Acta Met, vol 11, p 1263, 1963.

18. A. F. Dyment and C. M. Libanati, Journal of Materials Science, vol 3, p 349, 1968.

19. H. W. Rosenberg, "Titanium Science and Technology", vol 4, Plenum Pub. Company, NY, p 2127, 1974.

20. J. P. Hirth and J. Lothe, "Theory of Dislocations", McGraw-Hill, NY, p 683, 1968.

21. Mary G. Natrella, "Experimental Statistics", National Bureau of Standards Handbook 91, Issued August 1, 1963, U.S. Government Printing Office, Ch. 5.

# CREEP BEHAVIOR OF CAST Ti-6Al-4V ALLOY

Amiya K. Chakrabarti and Edwin S. Nichols

Materials Research and Engineering, Detroit Diesel Allison

Division Of General Motors Corporation
P. O. Box 894, Indianapolis, Indiana 46206, U. S. A.

## Introduction

Creep behavior of cast plus HIP processed (1172°K/103.5MPa/2 Hrs.) Ti-6Al-4V alloy was investigated within a temperature range of 394 to 727°K in air. The microstructures of the test material consisted of discontinuous grain boundary alpha and colonies of transformed beta which contained packets of parallel oriented alpha-platelets separated by a thin layer of aged beta. Low temperature (394-616°K) creep deformation at stress levels close to the elastic limit consists of primary creep in which creep rate decreases until it reaches a value of essentially zero. The initial creep rates within a temperature range of 319-616°K at stress levels above elastic limits are very high; however, the creep rates decrease steadily with time and between 50 to 200 hrs. the creep rates attain a zero value (a creep saturation or exhaustion occurs at this stage). At the higher temperature range (above 616°K) both the primary and the secondary creep rates are significant. The results have been analyzed in terms of stress-temperature-time relationships and an attempt has been made to identify the possible mechanisms of creep deformation and the creep saturation phenomena.

## Experimental Procedure

### 1. Materials

Cast, HIP processed (1172°K/103.5MPa/2 Hrs.), and aged (950°K/1.5Hrs.) Ti-6Al-4V alloy material from the hubs of centrifugal compressor impellers produced by Precision Cast Parts Corporation were used for this study. The general microstructure of this material is shown in Figure 1. The microstructure consists of transformed beta grains with discontinuous grain boundary alpha and colonies of transformed beta which contained packets of parallel oriented alpha-platelets separated by a thin layer of aged beta. Specimens of two different average grain sizes were selected for the creep tests. The fine grained specimens had a grain size (average linear intercept) of 1.0 mm whereas the coarse grained specimens had an average grain size of 6.0 mm.

### 2. Specimen Preparation, Creep and Tensile Testing

Specimen blanks of approximately 5.72cm X 0.95cm X 0.95cm section size with the long axis oriented tangential to the hub section of impellers were machined to 0.150 in. (3.81mm) diameter standard type creep specimens. The specimens were lathe-turned followed by 320 grit size emery paper polish. The creep rupture tests were performed within a temperature range of 394 to 727°K using dead-load type creep frames in air covering a stress range of 40 to 80 ksi (275.8-551.6 MPa). The strain (or the extension) as a func-

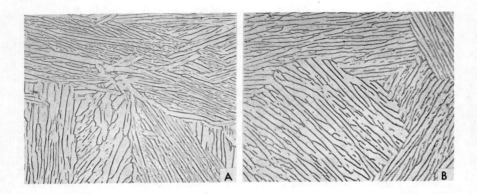

Figure 1.    Microstructure of cast/HIP-Processed and aged
            Ti-64 Impeller material.  (Etched in Keller's
            Reagent)  A -250X showing grain boundary alpha
            and colonies of transformed beta.  B -(400X)
            showing colonies of transformed beta.

tion of time was recorded by LVDT type strain gauges for each specimen using
a multi channel x-y type of recorder.  Time to 0.1, 0.2 and 1.0% creep
strains were recorded for most specimens.  Most of the tests were discon-
tinued at 500 hrs.; however, a few low temperature tests were continued to
1000 hrs.

Room and elevated temperature tensile tests were also conducted per
applicable ASTM specification to determine the proportional limit, 0.2%
off-set yield strength, ultimate tensile strength, percentage elongation and
reduction in area.  Specimens were machined from coarse grain sections of an
impeller casting which had been processed as described previously.  The tensile results are indicated in Table 1.

## Experimental Observations

### 1.  Effect Of Temperature and Stress Level

The creep-test results for all the specimens are summarized in Table 2.
The initial creep rates at 75 to 80 ksi (517-552 MPa) at temperatures above
477°K were very high.  0.1, 0.2 and 0.3% creep strains in all such cases
were attained within one hour.  It may be observed from Table 2 that at a
stress level of 75 ksi (517 MPa) 0.1% creep strain is reached within 10 hrs.
for the 394°K test.

Low temperature creep behavior of wrought Ti-6Al-4V(Ti64) alloy has
been observed by a number of workers at stress levels even below the elastic
limit of the material [1-4] .  Creep in other titanium alloys at low temperatures has also been noted. [5-6] .  Creep rates of the order of $10^{-5}$ (in/in
creep strain per hour) were observed in wrought Ti-64 material at room temperatures and at stress levels much below the proportional limits [2]

Table 1 - Tensile Test Results On Cast Ti 6-4 Alloy

| Test Temp. $^\circ F$ ($^\circ K$) | Proportional Limit, Ksi (MPa) | 0.2% Offset Y.S. Ksi (MPa) | Ultimate Stress Ksi (MPa) | % Elonga- tion | % RA |
|---|---|---|---|---|---|
| Room Temp (298$^\circ$K) | 96.2 (663.3) | 125.1 (862.6) | 133.0 (917.0) | 10.1 | 18.8 |
| Room Temp | 92.1 (635.1) | 116.7 (804.6) | 123.2 (849.5) | 10.4 | 20.4 |
| Room Temp | 88.0 (606.8) | 114.3 (788.1) | 123.8 (853.6) | 7.1 | 28.1 |
| Room Temp | 101.8 (701.9) | 119.1 (821.2) | 124.4 (857.7) | 9.1 | 23.2 |
| Room Temp | 96.9 (668.1) | 124.7 (859.8) | 132.5 (913.6) | 4.2 | 9.5 |
| 250$^\circ$F (394) | 79.1 (545.4) | 98.7 (680.5) | 112.6 (776.4) | 9.5 | 23.2 |
| 300$^\circ$F (422) | 78.2 (539.2) | 94.5 (651.6) | 109.1 (752.2) | 6.3 | 19.1 |
| 400$^\circ$F (477) | 68.4 (471.6) | 85.0 (586.1) | 97.7 (673.6) | 10.3 | 32.2 |
| 500$^\circ$F (533) | 61.8 (426.1) | 76.6 (528.2) | 88.5 (610.2) | 10.8 | 24.6 |
| 600$^\circ$F (589) | 52.1 (359.2) | 65.9 (454.4) | 79.4 (547.5) | 11.2 | 30.7 |
| 700$^\circ$F (644) | 43.30 (298.5) | 56.1 (386.8) | 69.3 (477.8) | 17.5 | 38.9 |
| 800$^\circ$F (700) | 49.65 (342.3) | 59.5 (410.3) | 71.5 (493.0) | 11.0 | 36.6 |

A few typical creep strain versus time relations are shown in Figures 2 through 6. Figure 2 shows the creep strain versus time relation at 75 Ksi (517 MPa) for 394 to 616$^\circ$K tests. At a constant stress level, the creep strain at a fixed time increases as the test temperature increases. How- ever in Figure 2, it is shown that the creep strain at 616$^\circ$K at a fixed time is lower than that at 477$^\circ$K. This apparent anomaly is due to the fact that there is a high plastic strain on loading at higher temperatures and the initial creep rate (within first few minutes) is very high. So the creep strain that is measured versus time is related to the selection of the starting time (or zero-time) for creep deformation. In order to avoid this arbitrary situation, it would be more appropriate to plot the total plastic strain (i. e. plastic strain on loading plus the creep strain) as a function of time. Figure 3 shows this relation for the 394, 422, 450, 477 and 616$^\circ$K creep tests at 517MPa (75 Ksi). It may be observed (Figure 3) that the

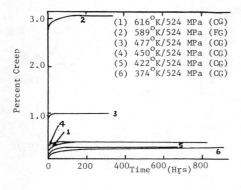

Figure 2    Percent creep strain versus time
            relation for 524 MPa strain creep
            tests

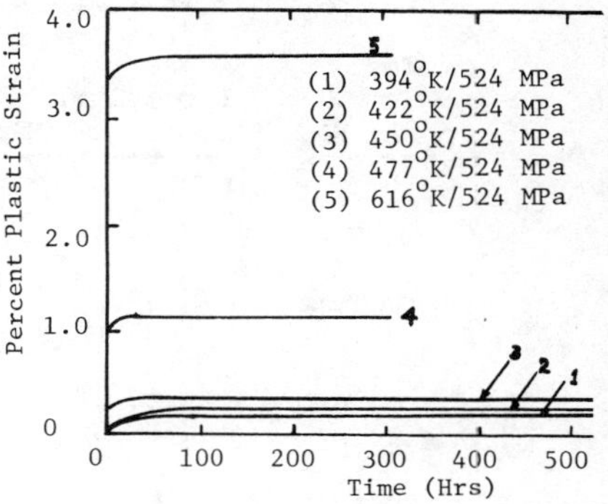

Figure 3    Percent total plastic strain
            versus time relation for the
            524 MPa stress creep tests

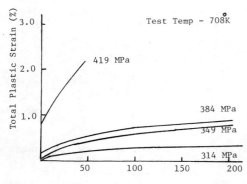

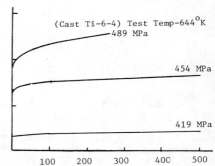

Figure 4    Total plastic strain versus time
            relation for 700°K creep tests

Figure 5    Time (Hrs.)
            Total plastic strain versus time
            relation for the 644°K creep tests

CG = Coarse Grain
FG = Fine Grain

Table 2 – Creep Test Results For The Cast Ti 6-4 Materials

| Temp. °F (°K) | Stress ksi | MPa | Plastic Strain On Loading (%) | Test Duration Hrs. | Time In Hrs.to Reach (% Creep) 0.1 | 0.2 | 1.0 | Final Creep % | Grain Size |
|---|---|---|---|---|---|---|---|---|---|
| 850 (727) | 40.0 | (275.8) | 0 | 611.2 | 2.0 | 9.6 | 610.0 | 1.002 | FG |
| 800 (700) | 40.0 | (275.8) | 0 | 500.0 | 15.0 | 60.0 | – | 0.49 | FG |
| 800 (700) | 50.0 | (344.8) | 0 | 297.5 | 3.5 | 11.0 | 291.5 | 1.00 | FG |
| 750 (672) | 65.0 | (448.2) | 0.7 | 251.4 | 7.5 | 22.0 | – | 0.549 | FG |
| 700 (644) | 60.0 | (413.7) | 0.3 | 500 | 240.0 | – | – | 0.146 | FG |
| 600 (589) | 75.0 | (517.1) | 2.04 | 330.9 | 0.02 | 0.04 | 0.1 | 3.1 | FG |
| 500 (533) | 77.5 | (534.4) | 2.1 | 307.9 | 0.01 | 0.02 | 0.1 | 3.05 | FG |
| 400 (477) | 80.0 | (551.6) | 0.56 | 138.0 | 0.1 | 0.13 | 1.5 | 2.45 | FG |
| 400 (477) | 77.0 | (530.9) | 0.8 | 18.2 | 0.02 | 0.04 | 0.16 | 3.0 | FG |
| 350 (450) | 75.0 | (517.1) | 0.01 | 1006.0 | 0.4 | 2.2 | – | <.4 | CG |
| 300 (422) | 75.0 | (517.1) | | 500 | 0.25 | 1.2 | – | – | CG |
| 300 (422) | 75.0 | (517.1) | | 500 | 1.7 | 12.2 | – | – | FG |
| 250 (394) | 75.0 | (517.1) | 0.0 | 1006.1 | 9.8 | 160.0 | – | <3% | CG |

total plastic strain at a certain fixed time increases as the test temperature increases.  It is observed from Table 2 that the plastic strain on
loading for 517 MPa at 394°K and 422°K is zero.  The plastic strain
on loading to the same stress at 450°K is only 0.01% but it increases rapidly as the temperature increases.  Figure 4 shows the total plastic strain
versus the time relation for the creep tests conducted at 700°K and at various stress levels.  It may be observed from Figure 4 that the creep-rate
as well as the total plastic strain at a fixed time increases as the stress
level increases.  This increase is accelerated at the 414 MPa stress level.
It may be observed from Table I that the stress levels 345 and 379 MPa are
below the yield stress level for this material.  Higher creep rate at 700°K
and 144 MPa can be rationalized in terms of the creep stress level (which is,
in this case, nearly equal to the yield strength level).

It may be observed from the Figures 2, 3 and 6  that for the creep
tests up to 589°K the creep rates decrease with time; and after about 200
hours, the creep rates become equal to zero, i. e., the creep strain does
not increase within the next 100 to 500 hours.  The only exception is the
test that was carried out at 394°K and at a stress level of 517 MPa.  Here
although the creep rate is very low ($1 \times 10^{-4}$/hr.) the creep strain increas-

es slowly with time up to 1000 hours. The initial creep strain versus time relation for most materials exhibits a primary type creep behavior where the initial rate of creep strain decreases gradually until a steady creep-rate is achieved (steady state creep or secondary creep). In the present case, up to 589°K creep tests indicate a primary type of creep with a steady decrease in the creep rate until a zero creep rate is attained. The specimens did not exhibit a secondary or steady state creep. At 644°K and above (Figures 4 and 5) strain versus time plots show the characteristic primary and the steady-state creep; a positive secondary creep was observed. The difference in the creep behavior between the temperature ranges 394-589°K and 644-727°K suggests that the creep mechanism in these two cases are different beyond the primary creep regime.

The stage at which the creep strain (or the total plastic strain) does not increase with time (Figures 2, 3 and 6) may be considered as a creep-saturation condition. It may be observed from Table 3 that the total plastic strain at which creep saturation takes place increases steadily with the increase in test temperature. The time to reach creep saturation decreases as the temperature increases from 394°K to 450°K and then it remains within a bound of 100 to 200 hours. It may also be observed that the saturation plastic strain at the same temperature and stress level is higher for the fine grain material compared to the coarse grain material. The significance of the saturation plastic strain in terms of the mechanism of creep deformation was not well understood. It is apparent that a stable dislocation substructure may be achieved at a fixed temperature and stress level. The mechanism of creep deformation and substructure formation is being discussed in a separate section for specimens tested at temperatures up to 505°K.

Table 3 - Cast and HIPed Ti 6-4 Impeller Material Creep Data

| Test Temp °F (°K) | Creep Stress ksi | (MPa) | Plastic Strain At Which Creep Saturation Takes Place (%) | Time To Reach Saturation Creep (Hrs.) | Grain Size |
|---|---|---|---|---|---|
| 250 (394) | 75.0 | (517.1) | 0.26 | 1000 Hrs. | CG |
| 300 (422) | 75.0 | (517.1) | 0.26 | 114 | CG |
| 350 (450) | 75.0 | (517.1) | 0.37 | 44 | CG |
| 350 (450) | 78.0 | (537.8) | 1.25 | 70 Hrs. | CG |
| 400 (477) | 75.0 | (517.1) | 1.13 | 138 | CG |
| 400 (477) | 80.0 | (551.6) | 1.15 | 191 | CG |
| 650 (616) | 75.0 | (517.1) | 3.65 | 145 | CG |
| 400 (477) | 77.0 | (530.9) | 3.90 | – | FG |
| 400 (477) | 80.0 | (551.6) | 2.92 | 42 | FG |
| 500 (533) | 77.5 | (534.4) | 5.1 | 190 | FG |
| 600 (589) | 75.0 | (517.1) | 5.2 | 166 | FG |

2. Effect Of Grain Size On Creep Strain

Figure 6 shows the creep strain versus time relation for 552 MPa stress at 477°K for a fine grain as well as a coarse grain structure. It is observed that the creep strain within the first 50 hours in the fine grain material is greater by a factor of 2.5 compared to the coarse grain material. It is also observed from Table 2 that the plastic strain on loading is high-

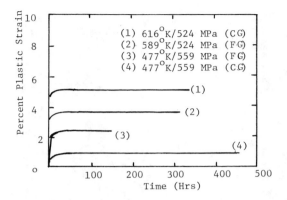

Figure 6    Percent total plastic strain versus time relation
for fine grain and coarse grain materials.

er for the fine grain material compared to that in the coarse grain material.
Figure 6 also shows the total plastic strain versus time relation for 616°K/
517 MPa (coarse grain) and 589°K/517 MPa (fine grain) creep tests. It may
be observed that the total plastic strain in the coarse grain material tested at 616°K/517 MPa is much lower than that for the fine grain material
tested at a slightly lower temperature (589°K) and at the same stress level
(517 MPa). It appears from Figure 6 that the higher plastic strain on loading and creep deformation in the fine grain material is probably due to contribution from the grain boundary or colony boundary deformation. Smaller
grain size material (also having smaller colony size) having more grain
boundary areas may contribute more to the plastic strain. This is supported
by the microstructural observations on the creep-deformed specimen discussed
in the following section.

## Studies On Creep Deformation Mechanisms

### 1. Specimen Preparation and Testing

In order to study the creep deformation mechanism, a flat-gauge section
specimen was specially designed. These specimens are 3.81 mm thick with two
parallel flat sides of 6.35 mm width and 2.86 cm uniform gauge length (Figure 7A). The flat sides of these specimens were machined, ground and metallographically polished carefully using a supporting fixture to avoid any
possible distortion (or straining). The surfaces were etched with Keller's
reagent to reveal the surface microstructures. Light scratches were in-
scribed on the specimen surfaces 1.27 mm apart along and across the loading
axis of the specimens. Photographs of the specimen surfaces were taken at
5X for initial records (Figure 7B). The specimens were deformed in creep
(in identical manner as described earlier) up to a predetermined strain
level at the end of which the specimens were unloaded for metallographic observations. The details of the test conditions are listed in Table 4. It
may be noted that the plastic strain on loading at 450°K would be of the

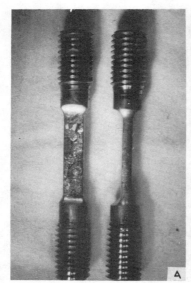

Figure 7    A-photographs of creep specimen for deformation mechanism studies
(1X).   B-photographs of the polished, etched and lightly
scratched surface of the above specimen (5X) showing general
macrostructure

Figure 8    Metallographic picture of 0.85% creep deformed specimen
showing (A) colony boundary deformation (d, 100X); B-relative
displacement (ds) and coarse slip-line formation on surface
(100X).

order of 0.01%; whereas for 505°K the plastic strain on loading would be of the order of 1.0% based on the other tests (Table 1).

Table 4 - Creep Test Condition For Mechanism Studies

| Specimen No. | Test Temp. °F (°K) | Stress, Ksi (MPa) | % Creep Deformation | Remarks |
|---|---|---|---|---|
| 1 | – | – | 0% deformation | Undeformed specimen |
| 2 | 350 (450°K) | 75 (517) | 0.38* | Round specimen |
| 3 | 350 (450°K) | 78 (538) | 0.80 | Flat gauge-section |
| 4 | 350 (450°K) | 78 (538) | 1.25 | Flat gauge-section |
| 5 | 450 (505°K) | 80 (552) | 5.20* | Flat gauge-section |
| 6 | 350 (450°K) | – | 0.80 | Deformed in tensile machine-short time |

## 2. Metallographic Observations

Metallographic observations were done on the flat surfaces of the creep deformed specimens to observe grid distortions and other surface appearances which might be helpful to understand the plastic deformation mechanisms. Figure 8 shows the surface appearance of the specimen deformed to 0.8% creep strain. Grid distortions are observed at the colony boundaries (8A), indicating possible deformation along colony boundaries or rotation of the colonies. Relative displacement ($d_s$) of colony boundaries are observed in Figure 8B. Some evidence of coarse slip across colonies are also observed in photomicrograph (8B). The slip lines are noticed to be shifted at the colony boundaries which is due to the relative displacement at the colony boundaries. Figure 9 shows the surface appearance of the specimen deformed to 1.25% creep strain. Evidence of grain boundary sliding is observed in Figure 9A; in which a shear type grid displacement is apparent across the grain boundary (Figure 9B). Grain boundary and colony boundary deformation along with crystallographic slip lines within grains are observed in Figure 9C. Evidence of extensive grain boundary-colony boundary deformation and crystallographic slip are observed in 5.2% creep deformed specimen (Figure 9D). Evidence of severe grain boundary and colony boundary deformation and relative displacements are observed in Figures 9D and E. Evidence of cross slip and slip-band (or slip localization) formation are also observed on the surfaces of this specimen (Figure 10A and 10B respectively). Cross-slips along directions 60° to one another are observed in many grains; an example of this cross slip is shown in Figure 10B. Slip off-sets produced by slip localization along planes approximately 60° to the long axis of the alpha platelets can be observed in Figure 10B and C.

*Tests were terminated at the saturation creep strain.

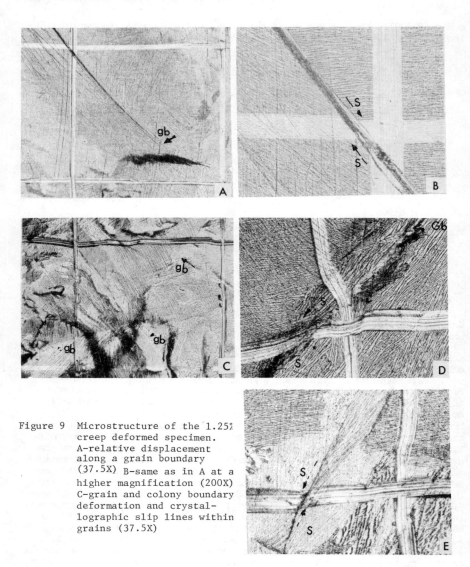

Figure 9    Microstructure of the 1.25%
creep deformed specimen.
A-relative displacement
along a grain boundary
(37.5X) B-same as in A at a
higher magnification (200X)
C-grain and colony boundary
deformation and crystal-
lographic slip lines within
grains (37.5X)

D & E    Metallographic picture of 5.2% creep deformed specimen. D-
severe deformation along grain boundaries (150X). E- Relative
displacement along a colony boundary (150X)

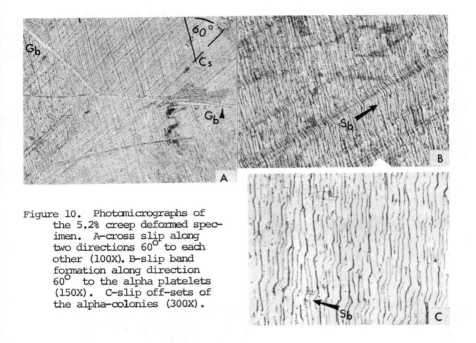

Figure 10.  Photomicrographs of
the 5.2% creep deformed specimen.  A-cross slip along
two directions 60° to each
other (100X). B-slip band
formation along direction
60° to the alpha platelets
(150X).  C-slip off-sets of
the alpha-colonies (300X).

3.  Transmission Electron Microscopic Observations

Thin sections of creep deformed specimens were wet polished to 0.005 in
(0.127 mm) thickness and were electropolished in a methanol-butynol alcohol-
perchloric acid electrolyte at 233 to 213 °K until thin areas, as required
for electron diffractions, were obtained.  The specimens were examined in a
100 KV JEM Electron Microscope.  The microstructure of the undeformed
specimen consists of large colonies of Widmanstätten alpha interleaved with
thin plates of beta (Figure 11A). Figure 11B is a dark field photomicrograph
of a different area.  The alpha-beta interface phase was identified to be
a fcc structure in which the relative orientations of the various phases are
as follows:

$$(0001)_\alpha \;\|\; (1\bar{1}1)_{fcc} \;\|\; (110)_\beta$$
$$\text{and} \quad [11\bar{2}0]_\alpha \;\|\; [1\bar{1}0]_{fcc} \;[111]_\beta \;.$$

Similar type of orientation relations were also proposed by Rhodes and
Paton [7].  All the alpha platelets in a given colony were found to have the
same Burger's orientation with respect to the beta phase.

A bright field photomicrograph of 0.38% creep deformed specimen is presented in Figure 12A.  The diffraction pattern (corresponding to Figure 12A)
is represented in Figure 12B which is rotated by 4.5° with respect to Figure
12A to compensate for the rotation in the electron microscope.  The [0001]

Figure 11    TEM photomicrograph of underformed specimen.
A-bright field photograph (10,000X) and B-dark
field photograph (8,500X) showing the alpha
platelets (white area, 11A) and the beta platelets
(white line, 11B) and the interface region.

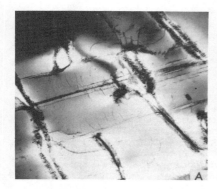
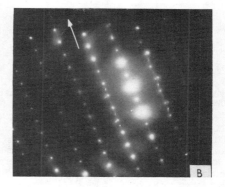

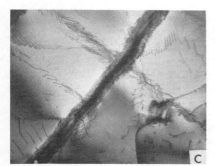

Figure 12    Photomicrographs
(TEM) of 0.38% creep
deformed specimen.
A-bright field photo-
graph (8500X) showing
slip traces and dis-
location structure.
B-diffraction pattern
for the area A showing
(0001)   direction.
C-a typical bright
field photomicrograph
of 0.8% creep deformed
specimen showing
several non-parallel
slip traces (8000X).

direction is indicated in Figure 12B which is observed to be perpendicular to the slip traces in Figure 12A. Since the dislocations along the trace of the $(0001)_\alpha$ plane are observed to be edge-on, it can be concluded that the slip lines (Figure 12A) correspond to traces of the basal plane. General observations on this sample indicate that the slip is confined to a few slip bands.

A typical bright field photomicrograph of the 0.8% creep-deformed specimen is shown in Figure 12C. Several non-parallel slip traces are observed in this photomicrograph indicating the presence of more than one operating slip system. It may be noted that although several systems may operate, one slip system appears to be predominate in a given colony.

Figure 13A and B show the bright field photomicrographs of a 1.25% creep deformed specimen from two different areas. The diffraction pattern corresponding to Figure 13A is shown in Figure 13B. It is not possible from the available information to ascertain whether the slip is basal, $\{10\bar{1}0\}$ or $\{10\bar{1}1\}$, because the slip off-sets shown in Figure 13A could be produced by any of these planes, all of which would have a $\langle 11\bar{2}0\rangle$ slip direction parallel to the traces shown. Two slip systems are observed in Figure 13C; it is quite likely that the dominant system (the one which traverses the $\alpha$-platelets) is either a $\{10\bar{1}0\}$ or a $\{10\bar{1}1\}$ slip system (as both of these planes are a part of the $\langle 11\bar{2}3\rangle$ zone axis). Figure 13D is the bright field photomicrograph of the 0.8% tensile tested specimen, showing a much greater number of slip bands compared to the 0.38% and 0.8% creep deformed specimens. The stress range at 0.8% plastic strain is of the order of 100 ksi (689.5 MPa) which is much higher than the creep test stress level and consequently, a greater number of slip systems is activated at this stress level.

## Summary And Discussion Of The Results

The experimental observations can be summarized as follows:

(1) Creep deformation is observed in cast and HIP processed Ti-64 alloy at temperatures $394^\circ$K and above and at stress levels both above and below the elastic limit of the material.

(2) Within the temperature range of 394 to $589^\circ$K only a primary type of creep phenomena is observed in which the creep rate decreases with time, followed by a zero creep rate (creep-saturation or creep exhaustion). No secondary creep is observed within the stress range examined in these experiments.

(3) Within the temperature range of 644 to $727^\circ$K, both primary and secondary (or steady state) creep are observed in which the creep rate increases with temperature or stress level.

(4) Grain and colony boundary deformation and sliding along with crystallographic slip and cross-slip markings are observed in specimens creep tested at $450^\circ$K.

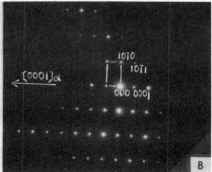

Figure 13.   TEM photomicrograph of 1.25% creep deformed specimen
A-bright field photograph (10,000X) showing parallel
sets of slip traces,   B-diffraction pattern of the
same area,   C-bright field photograph of another area
showing non-parallel sets of slip traces (10,000X).
Figure 13D shows the bright field photograph of 0.8%
tension tested specimens (10,000X) showing a higher
density of slip traces compared to the creep deformed
specimen.

Although quantitative measurements of the grain and colony boundary
sliding were not accomplished in the present experiments, an appreciable portion of the creep deformation was shown to be due to grain and colony boundary sliding.   Evidence of grain boundary sliding during creep in other titanium alloys has been previously reported [8] ; although the creep tests in
that case were at a higher temperature (811°K), it was shown that both prior
beta grain boundaries and martensitic platelet boundaries contributed to the
sliding, with the prior beta grain boundaries being the more prevalent location.   The higher creep strain at the same temperature and stress level for

the fine grain material in our work also indicates that a considerable proportion of creep strain is due to grain boundary and colony boundary sliding and deformation.

Odegard and Thompson [4] have suggested that the low temperature
creep in Ti 64 alloy can be considered as components of a tensile test at an
extremely slow rate which maximizes contributions from thermally activated
processes [9]. They also observed that the low temperature creep in
Ti-5-2.5 alloy exhibited similar activation energies as reported for tensile deformation of a variety of titanium alloys [5]. They further suggested that the low temperature deformation behavior depends on thermally
activated overcoming of interstitial obstacles, according to the model proposed for this process by Okzaki and Conrad [10].

The evidence of creep saturation (or creep exhaustion) at temperatures
394 to 589°K supports the observation and prediction proposed by Odegard
and Thompson [4]. It seems that at a fixed temperature and stress level, a
limited number of dislocation sources are activated to overcome the barrier. As the temperature or the stress level is increased, a relatively
greater number of sources are activated and consequently, a greater saturation creep strain is observed prior to the creep exhaustion (Table 3). The
sources of dislocations which can be activated at stress level below the
elastic limit and at low temperatures have not been identified in the present observations; however, the alpha-beta interfaces within the colony,
the colony boundaries and also the grain boundaries, may act as dislocation
sources. It has been reported in literature that a prior straining followed by creep causes an early creep exhaustion in Ti 64 alloy [4]. Similar
observations were made in this Laboratory for specimens which were deformed
in fatigue prior to creep tests. This supports the idea of the exhaustion
of available dislocation sources to explain a creep saturation phenomena.

Margolin [11] suggested that diffusion of oxygen and subsequent strain
ageing by pinning the mobile dislocations may lead to a creep exhaustion
phenomena. Evidence of such low temperature strain ageing in titanium alloys
have been observed by a number of workers [12-15]. Either mechanism, strain
ageing or exhaustion of dislocation sources, seems to be adequate to ra-
tionalize the phenomena of creep saturation observed in the present tests;
however, more detailed investigation would be necessary to identify the
actual process.

It is observed from these TEM and metallographic studies that as the
temperature and stress level of the creep test were increased, slip due to
more than one slip system and slip localization both became evident, which
is consistent with other observations. This type of slip localization in
titanium alloys has been previously reported by a number of workers [16-18]
under a number of different testing conditions.

## Acknowledgment

The authors wish to thank Professor H. Margolin and M. Young of
Polytechnic Institute of New York, Brooklyn, N. Y. for the Transmission
Electron Microscopic work and for valuable technical discussions.

## References

1.  W. H. Reimann, Journal of Metals, JMLSA, vol. 6, No. 4 (1971), 926.
2.  M. A. Imam and C. M. Gilmore, Met. Trans. vol. 10A (1979), 419.
3.  W. C. Harrigan, Jr., Met. Trans. vol. 5 (1974), 565.
4.  B. C. Odegard and A. W. Thompson, Met. Trans. vol. 5, 1974), 1207.
5.  A. W. Thompson and B. C. Odegard, Met. Trans. vol. 4 (1973), 899.
6.  A. J. Hatch, J. M. Patridge and R. G. Broadwell: J. Matir., vol. 2 (1967), III.
7.  C. G. Rhodes and N. E. Paton, "Mechanical Behavior of Titanium Alloys", Final Report to office of Naval Research on Contract No. N00014-76-C-0598, April 1979.
8.  N. E. Paton and M. W. Mahoney, Met. Trans. vol. 7A, (1976) 1685.
9.  R. Lagneborg. Int. Met. Rev. vol. 17, (1972), 130.
10. K. Okzaki and H. Conrad, Acta Met, vol. 21, (1973), 1117.
11. H. Margolin, Unpublished research, Department of Physical and Engineering Metallurgy, polytechnic Institute of New York, Brooklyn, N. Y. 11201, private communication.
12. W. C. Harrigan, Jr., Met. Trans. vol. 5, (1974), 565.
13. G. H. Narayanan and T. F. Archbold, Met. Trans., vol. 2 (1971), 1264
14. R. J. Wasilewski, Trans. ASM, vol. 56, (1963), 221.
15. F. D. Rost and F. C. Perkins, Trans. ASM, vol. 45, (1953), 792.
16. D. Eylon, J. A. Hall, C. M. Pierce and D. L. Ruckle, Met. Trans. vol. 7A (1976), 1817.
17. D. Eylon and J. A. Hall, ibid, vol. 8A (1977), 981.
18. A. K. Chakrabarti and B. A. Ewing, IR&D Report Materials Research and Engineering, Detroit Diesel Allison, Division of General Motors Corporation, 1979.

EFFECT OF SUPERPLASTICITY ON THE DIFFUSION WELDING OF Ti-6Al-4V ALLOYS

Toshio Enjo, Kenji Ikeuchi, Naofumi Akikawa and Makoto Ito

Welding Research Institute of Osaka University,
Suita, Osaka, Japan

## Introduction

Ti-6Al-4V alloys, to which diffusion welding is extensively applied, deforms superplastically in a temperature range from 850° to 950°C. It has been reported that the superplastic behaviour of this alloy is influenced markedly by its microstructure[1,2]. On the other hand, in the bonding process of diffusion welding the plastic flow of base metal is a very important factor for the attainment of intimate contact between faying surfaces[3,4]. According to some previous papers[5,6,7], the diffusion welding of Ti-6Al-4V alloy is successfully carried out in the welding temperature range where this alloy deforms superplastically. Consequently, it is considered that the diffusion welding of Ti-6Al-4V alloy should be affected markedly by the microstructure of base metal. However, there is no available information on the effect of the microstructure although the effects of welding conditions, faying surface roughness and insert-metal have already been reported[5,6,7].

In the present investigation, the effect of the superplastic behaviour of base metal on the bonding process of Ti-6Al-4V alloy has been investigated by varying the microstructure of base metals. The microstructure of base metal has been varied by annealing treatment prior to welding. The diffusion welding of Ti-6Al-4V alloy to titanium also has been carried out in order to examine the effect of an insert-metal of Ti-6Al-4V alloy on the diffusion welding of titanium.

## Experimental Procedure

Base metals used in this investigation were prepared from Ti-6Al-4V alloy and commercially pure titanium of round bar 20mm in diameter. Their chemical compositions are shown in Table 1. The base metal used for microscopic

Table 1 Chemical compositions of base metals used.

| Materials | composition, wt% | | | | | | | |
|---|---|---|---|---|---|---|---|---|
| | C | Fe | N | O | H | Al | V | Ti |
| Ti-6Al-4V | 0.015 | 0.214 | 0.0061 | 0.154 | 0.0010 | 6.12 | 4.06 | Bal. |
| Titanium | 0.018 | 0.036 | 0.0057 | 0.083 | 0.0026 | - | - | Bal. |

observation of bonding zone had a cubic shape 4mm in edge length and that for tensile test was cylindrical rod 20mm in diameter and 37mm in edge length. The faying surfaces of both base metals were finished by polishing with 1500 grade emery paper unless otherwise stated and degreased by washing in acetone just before the welding.

The welding of joint for microscopic observation were carried out using a high temperature optical microscope( Union Kogaku Ltd., Type HM ) equiped with a compressing device. Joints for tensile test were welded with a vacuum diffusion-welding chamber reported in a previous paper[8]. The bonding zone was heated with radiant resistance heater of molybdenum foil and the welding pressure onto the bonding interface was applied with a hydraulic press. The welding was carried out in a vacuum environment of $10^{-4} \sim 10^{-5}$ mmHg. The welding temperature was monitored with thermocouple percussion-welded at a point 2mm from bonding interface.

The etchant for the observation of microstructure was the Kroll's reagent. In order to fix the etching condition, etching temperature and time were kept to be 0°C and 30s. A tensile test on the joint was carried out using Instron type machine by the deformation rate of $1.7 \sim 10^{-2}$ mm/s at room temperature. Specimens for the tensile test were prepared by machining the joint after welding. The gauge length and diameter of the specimen were 36mm and 8mm, respectively.

A compression test at a high temperature was performed using the high temperature optical microscope equiped with a compressing device in a vacuum environment of $10^{-4} \sim 10^{-5}$ mmHg. Specimens for the compression test were rectangular parallelepiped $3 \times 3 \times 6$ mm.

## Results and Discussion

### 1. Microstructure and Superplasticity of the Base Metal

In order to vary the microstructure of base metal, the base metal of Ti-6Al-4V alloy was annealed in vacuo at 900°C ( alpha-beta field ) for 4hrs or at 1010°C( beta field ) for 1hr prior to welding. Fig. 1 shows the microstructures of the base metals subjected to the annealing treatment. As shown in this figure, appreciable grain growth occured by the annealing at 900°C compared with the as-received, and coarse plate-like alpha grains were observed in the base metal annealed at 1010°C. The mean grain size of alpha phase was estimated to be $2 \sim 3 \mu$m for the as-received base metal and $8 \mu$m for the 900°C-annealed base metal.

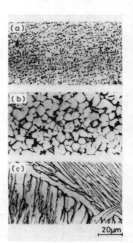

A compression test at 850°C was carried out in vacuo in order to examine the deformation behaviour of the base metals in welding temperature range. As shown in Fig. 2, the strain rate of the as-received base metal was highest and that of the 1010°C-annealed base metal lowest. The value of strain rate sensitivity factor m, which is defined by $m = d \ln \sigma / d \ln(\dot{\varepsilon}/\dot{\varepsilon})$ ( $\sigma$ = compressive stress, $\dot{\varepsilon}$ = strain rate ), was estimated to be 0.6 for the as-received and 900°C-annealed base metal, and 0.3 for the 1010°C-annealed base metal. The values of m suggest that the as-received and 900°C-annealed base metal deform superplastically, but the 1010°C-annealed base metal does not have the superplasticity.

Fig. 1 Microstructures of the as-received (a), 900°C-annealed (b) and 1010°C-annealed base metal (c).

In this investigation, the as-received, 900°C-annealed and 1010°C-annealed base metal were diffusion welded.

2. Diffusion Welding of Ti-6Al-4V Alloy
   with Different Microstructures

Fig. 3 shows microstructures for the
bonding zones of the as-received, 900°C-
annealed and 1010°C-annealed base metal.
As shown in the figure, microvoids were
observed at the bonding interface and the
amount of these microvoids decreased as
the grain size of the base metal became
finer. Fig. 4 shows the variation of void
ratio q in the bonding interface with
welding temperature. The void ratio q was
defined by q = $l$/L, where $l$ was the
summation of void length along the
bonding interface and L was the length of
bonding interface in a micrograph as shown
in Fig. 3. As shown in Fig. 4, the void
ratio decreased with the rise of welding
temperature and was influenced markedly by
the microstructure of base metal; the void
ratio in the joint of 1010°C-annealed base
metal was indeed more than 70% at the
welding temperature of 900°C, whereas that
in the joint of the as-received base metal
was nearly zero at the same welding
temperature. Fig. 5 shows the tensile
strength of the joint of the as-received
and 900°C-annealed base metal. As shown in
the figure, the joint strength of the as-
received base metal increased more rapidly
than that of the 900°C-annealed. These
effects of the microstructure of base
metal on the void ratio and joint strength
indicate that the bonding process of
Ti-6Al-4V alloy developes more rapidly in
the base metal having finer microstructure.

It is said that the plastic flow in
the vicinity of bonding interface is an
important factor for the attainment of
intimate contact between the faying
surfaces having microasperities[3,4].
It is considered that the effect of faying
surface roughness on the bonding process
should vary with the microstructure of
base metal. Fig. 6 shows the effect of
the faying surface roughness on the void
ratio for each base metal. The faying
surface roughness was varied by grinding
with 220~1500 grade emery paper. As shown
in Fig. 6, the void ratio had a tendency
to increase as the faying surface became
rougher. A remarkable increase in the
void ratio was observed between the 600
and 400 grade in the joint of 900°C-
annealed base metal and between the 400

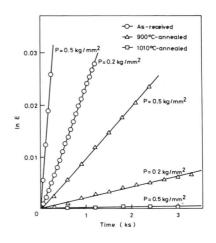

Fig. 2 Semilogarithmic plots of
strain $\varepsilon$ against time on a
compression test at 850°C for each
base metal. P denotes the
compressive stress.

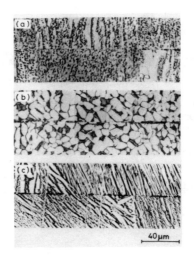

Fig. 3 Microstructures of the
bonding zones of the as-received (a),
900°C-annealed (b) and 1010°C-
annealed base metal. The welding
temprature $T_W$, pressure $P_W$ and time
$t_W$ are 850°C, 0.2kg/mm$^2$ and 10min,
respectively.

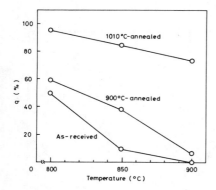

Fig. 4 Variation of the void ratio q with welding temperature for each base metal. $P_W$ and $t_W$ are 0.2kg/mm² and 10 min, respectively.

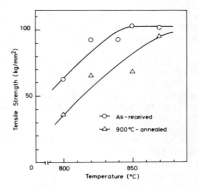

Fig. 5 Variation of tensile strength with welding temperature for the joint of the as-received and 900°C-annealed base metal. $P_W$ and $t_W$ are 0.2kg/mm² and 10min, respectively.

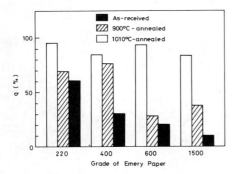

Fig. 6 Effect of faying surface roughness on the void ratio q for each base metal. $T_W$, $P_W$ and $t_W$ are 850°C, 0.2kg/mm² and 10min, respectively.

and 220 grade in that of the as-received base metal. As shown in Fig. 7, the joint strength of the as-received and 900°C-annealed base metal also decreased as the faying surface became rougher.

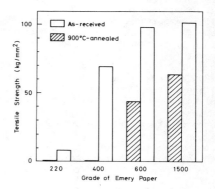

Fig. 7 Effect of the faying surface roughness on the tensile strength of the joint of the as-received and 900°C-annealed base metal. $T_W$, $P_W$ and $t_W$ are 850°C, 0.2kg/mm² and 10 min, respectively.

The joint strength decreased markedly between 600 and 400 grade for the 900°C-annealed base metal and between 400 and 220 grade for the as-received base metal in accordance with the variation of void ratio as shown in Fig. 6. It is considered that these results are caused by the increase in deformation rate of the base metal which promotes the intimate contact between faying surfaces.

The following three factors which are considered to be most important for the diffusion welding process are discussed to account for the effects of the base metal microstructure on the bonding process; (1) diffusion rate, (2) oxide film on the faying surface and (3) deformation behaviour of the base metal. That is, the diffusion-welding process proceeds more rapidly as the diffusion rate becomes higher, the oxide film is disrupted more easily and the base metal deforms at lower flow stress.

(1) Diffusion rate: Diffusion in metals can be divided into two processes depending on the diffusion path. They are volume diffusion and grain boundary diffusion. However, the contribution of grain boundary diffusion to the diffusion rate is regarded as negligibly small compared with that of volume diffusion at the welding temperature range because the volume diffusion occurs very rapidly. Therefore, only the contribution of volume diffusion is discussed here. In Ti-6Al-4V alloy, the diffusion rate increases with the rise of volume fraction of beta phase since the diffusion coefficient for beta phase is higher than that for alpha phase at a same temperature[9]. However, if the volume fraction of alpha and beta phase approaches very quickly the equilibrium value at a welding temperature, the diffusion rate should be independent of the microstructure of base metal before welding.

On the other hand, if the volume fraction does not get so quickly to the equilibrium value at a welding temperature, it can be concluded that the weldability of this alloy can not be determined by the diffusion rate. Namely, the volume fractions of alpha and beta phase for 900°C-annealed and 1010°C-annealed base metal are considered to be nearly equal because they were cooled in a furnace after the annealing. The as-received base metal was annealed at 700°C for a long time so as to adjust the volume fraction to the equilibrium value at 700°C. Consequently, the volume fraction of beta phase for the as-received base metal is considered to be smaller than those of 900°C-annealed and 1010°C-annealed base metal. However the as-received base metal has the best weldability as shown in Figs. 3~7. Therefore, the effect of the base metal microstructure on the bonding process is not due to the difference in the diffusion rate between the base metals.

Some investigators[2,10] have pointed out that the grain boundary diffusion is important for the rate-controlling process of superplasticity. However, this contribution of grain boundary diffusion is involved in the factor (3).

(2) Oxide film on the faying surface: It has been pointed out by several investigators[3,4,8,11] that the oxide film on faying surfaces is a very important factor for the bonding process of diffusion welding. However, the properties of the oxide film is not considered to be influenced markedly by the microstructure of base metal.

(3) Deformation behaviour of the base metal: Fig. 8 shows the void ratio q as a function of welding deformation for each base metal. The welding deformation is defined as a fractional increase in cross sectional area of base metal in the vicinity of bonding interface. As shown in Fig. 8 the void ratio can be described as a function of welding deformation regardless of the microstructure of base metal and the void ratio is nearly equal to zero for a welding deformation more than 1%. As shown in Fig. 9, elongation of joint on tensile test can also be described by a function of welding deformation regardless of the microstructure of base metal. When the welding deformation was more than 1%, the joint fractured at the base metal. This welding deformation is in good accord with that for which the void ratio was nearly zero. From these results, it can be concluded that the effects of the microstructure of base metal on the void ratio and joint strength are due to the difference in the deformation behaviour between the base metals. That is, the diffusion welding of Ti-6Al-4V alloy is accelerated by taking advantage of the superplasticity as the microstructure of base metal becomes finer.

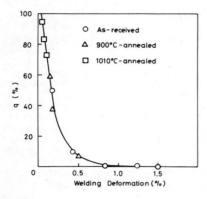

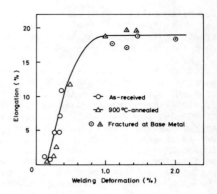

Fig. 8 Void ratio q versus welding
deformation for the joint of each
base metal.

Fig. 9 Elongation versus welding
deformation for the joint of as-
received and 900°C-annealed base
metal.

## 3. Diffusion Welding of Ti-6Al-4V Alloy to Titanium

Ti-6Al-4V alloy( as-received ) was diffusion-welded to titanium in order
to examine the possibility to improve the joint efficiency of titanium by
applying Ti-6Al-4V alloy having the superplasticity as an insert-metal. As
described in a previous paper[12], the diffusion-welded joint of titanium did
not have enough strength to fracture at base metal unless the welding
temperature was higher than $\alpha \rightleftarrows \beta$ transformation( 880∼900°C ). However, at a
welding temperature higher than the transformation, the mechanical properties
of the base metal are considered to be reduced because the base metal became a
coarse Widmanstätten structure. Consequently, it is to be desirable that
satisfactory joint efficiency can be obtained at a welding temperature lower
than $\alpha \rightleftarrows \beta$ transformation temperature by taking advantage of an insert-metal.

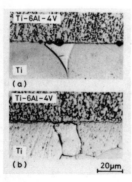

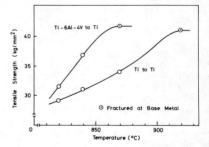

Fig. 10 Microstructures of the bonding
zones between Ti-6Al-4V alloy and
titanium at the welding temperature
of 840°C (a) and 870°C (b). $P_W$ and $t_W$
are 0.5kg/mm$^2$ and 10min, respectively.

Fig. 11 Tensile strength versus
welding temperature for the joint of
Ti-6Al-4V alloy to titanium and
titanium to titanium. $P_W$ and $t_W$ are
0.5kg/mm$^2$ and 10min, respectively.

Fig. 10 shows microstructures for the bondong zone between Ti-6Al-4V alloy and titanium. As shown in this figure, some voids remained in the bonding interface at the welding temperature of 840°C, but no void was observed at 870 °C. As shown in Fig. 11, the joint strength of Ti-6Al-4V alloy to titanium increased more rapidly than that of titanium to titanium and reached an enough value to fracture in titanium base metal at the welding temperature of 870°C. Thus, satisfactory joint efficiency, which was evaluated from the void ratio and joint strength, was obtained at a welding temperature lower than $\alpha \rightleftarrows \beta$ transformation for the diffusion welding of Ti-6Al-4V alloy to titanium. Consequently it can be concluded that Ti-6Al-4V alloy with fine microstructure is successfully applied to the diffusion welding of titanium as an insert-metal.

## Summary

The effect of superplasticity on the diffusion welding of Ti-6Al-4V alloy was investigated using the as-received base metal( grain size of alpha phase $d_a$ = 2~3$\mu$m ), base metal annealed at 900°C for 4hrs( $d_a$ = 8$\mu$m ) and base metal annealed at 1010°C for 1hr( coarse plate-like alpha grain ). The development of the bonding process was evaluated from void ratio at bonding interface and tensile strength of joint. Results obtained are summarized as follows:

(1) The strain rate sensitivity factor m obtained from a compression test at 850°C indicates that the as-received and 900°C-annealed base metal deformed superplastically( m = 0.6 ) but the 1010°C-annealed base metal did not have the superplasticity( m = 0.3 ). The strain rate increased as the microstructure of base metal became finer.

(2) The bonding process of Ti-6Al-4V alloy developed more rapidly in the base metal having finer microstructure. This fact is attributed to the difference in the superplastic behaviour between the base metals described in (1), because the void ratio at bonding interface and joint strength could be described as a function of welding deformation regardless of the microstructure of base metal.

(3) The diffusion welding of Ti-6Al-4V alloy( as-received ) to titanium resulted in an improvement in the void ratio and joint strength compared with that between titanium. This result suggests that the Ti-6Al-4V alloy having fine microstructure( superplasticity ) is successfully applied to the diffusion welding of titanium as an insert-metal.

## References

1. D. Lee and W. A. Backofen: Trans. TMS-AIME, 239(1967), 1034.
2. A. Arieli and A. Rosen: Met. Trans., 8A(1977), 1591.
3. R. F. Tylecote: *The Solid Phase Welding of Metals,* (1968), 301. Edward Arnold, London.
4. M. M. Schwartz: *Modern Metal Joining Techniques,* (1969), 370. John Wiley & Sons, New York.
5. K. R. Perun: Weld. J., 46(1967), 385-s.
6. R. J. Rehder and D. T. Lovell: Weld. J., 49(1970), 213-s.
7. H. G. Kellerer and L. H. Milacek: Weld. J., 49(1970), 219-s.
8. T. Enjo, K. Ikeuchi, M. Kanai and T. Maruyama: J. Japan Weld. Soc., 46 (1977), 82.
9. F. S. Buffington, K. Hirano and M. Cohen: Acta Met., 9(1961), 434.
10. T. H. Aden: Trans. ASM, 61(1968), 559.
11. P. M. Bartle: Weld. J., 54(1975), 799.
12. T. Enjo, K. Ikeuchi, N. Akikawa and T. Maruyama: J. High Temp. Soc., to be published.

# MECHANICAL PROPERTIES

# A. Mechanical Properties I

ON THE STRENGTH OF THE COMMERCIALLY
PURE TITANIUM SHEET MADE IN JAPAN

Motohiko Nagai[1], Koichi Kashida[2], Keizo Murase[3],
Yasuo Moriguchi[4], Naosuke Tsuruoka[5], Toshio Kimura[6]
Seiitiro Kashu[7], Kimiyoshi Ono[8], Masaki Koizumi[9],
and Kenji Mori[10]

The Strength subcommittee, Technical Committee,
The Japan Titanium Society, Tokyo, Japan

## Introduction

It is well known that the chemical composition is the important factor to
the mechanical properties, such as yield strength, tensile strength, elongation
and hardness of commercial pure titanium sheet and plate.

Other effective factors on the mechanical properties of titanium are the
the grain size, texture, internal stress and others, which are induced by
variations of the manufacturing processes, shown in table 1.

We have investigated how these factors have effects on the mechanical
property of titanium sheets and plates produced by four Japanese makers, and
will report the results on effects of the chemical composition, especially in
care of designing the mechanical property.

*1  Nippon Stainless Steel Co., Ltd.
*2  Nippon Mining Co., Ltd.
*3  Nippon Stainless Steel Co., Ltd.
*4  Kobe Steel Ltd.
*5  Furukawa Metals Co., Ltd.
*6  Mitsubishi Metal Co., Ltd.
*7  Vacuum Metallurgical Co., Ltd.
*8  Osaka Titanium Co., Ltd.
*9  Toho Titanium Co., Ltd.
*10  The Furukawa Electric Co., Ltd.

Table 1   Causes of variations of factor

| Factor | Cause of variation | Factor | Cause of variation |
|---|---|---|---|
| Chemical composition | Raw material of metallic titanium<br><br>Aiming of chemical composition<br><br>Melting method and condition<br><br>Segregation of chemical composition<br><br>Method and accuracy of chemical analysis | Grain size and Texture | Hot rolling condition (heating temperature rolling finishing temperature, rolling ratio, number of heating times)<br><br>Cold rolling condition (rolling ratio, intermediate annealing condition, rolling direction) |
| Internal stress | Annealing condition<br><br>Descaling condition (shot blasted and others)<br><br>Flattening condition (stretching, leveling and others)<br><br>Preparation of test specimen | | Annealing condition (temperature, time) |

## 1.   Results of Experiments

The test specimens were prepared from the 2D finished titanium sheets of 1 mm thick and No.1 finished plates of 5 mm thick.  These sheets and plates consist of 3 grades, which are Grade 1, Grade 2 and Grade 3 of J.I.S. H 4600. These titanium sheets and plates were all produced in four Japanese titanium makers.  On these test specimens, the tests of mechanical property and hardness, chemical analysis and grain size measurement were performed.  The test results are shown in Table 2.

The type of test specimen employed conforms to the requirement of No. 5 tension test piece of JIS Z 2201.  The types of tension testing machines employed are three Amsler tensile test machines (maximum load 10 ton or 30 ton, oil pressure, automatic or manual speed control, response time of X-Y recorder 4 sec., electronic or swing weighing system) and an Olsen (maximum load 10 ton, screw loading, manual speed control, response time of X-Y recorder 1 sec., air-micro weighing system).

Table 2   Results of experiments

| Grade Thick-ness | Maker (Test-er) | Speci-men No. | Sampl-ing | Yield strength Kg/mm2 | | | | Tensile strength Kg/mm2 | | | | Grain size No. |
|---|---|---|---|---|---|---|---|---|---|---|---|---|
| | | | | 3) 0.15 | 3) 0.30 | 3) 0.70 | 3) 1.5 | 3) 0.15 | 3) 0.30 | 3) 0.70 | 3) 1.5 | |
| | A(C) | 3 | L1 | 26.7 | 26.8 | 26.9 | 27.5 | 38.2 | 38.0 | 38.0 | 38.0 | 8 |
| | A(D) | 4 | T | 22.5 | 22.7 | 24.8 | 25.2 | 39.7 | 39.5 | 39.6 | 39.5 | 8 |
| JIS | B(B) | 11 | L1 | 27.9 | 27.8 | 28.2 | 28.4 | 38.1 | 37.5 | 37.9 | 38.2 | 8 |
| G1 | B(A) | 12 | T | 24.8 | 25.4 | 25.8 | 28.2 | 38.6 | 39.4 | 39.7 | 39.1 | 8.5 |
| | C(D) | 19 | Le | 18.0 | 18.3 | 19.0 | 19.3 | 33.8 | 33.7 | 33.4 | 33.8 | 4 |
| 1mm | C(C) | 20 | T | 19.6 | 20.0 | 20.2 | 21.2 | 31.9 | 32.2 | 32.1 | 32.1 | 4 |
| | D(A) | 27 | Le | 23.9 | 23.9 | 23.6 | 24.4 | 34.5 | 35.1 | 34.3 | 33.9 | 5 |
| | D(B) | 28 | T | 22.2 | 22.2 | 22.8 | 23.1 | 34.2 | 34.5 | 34.9 | 34.2 | 5 |
| | A(A) | 1 | Le | 22.8 | 22.8 | 22.7 | 23.3 | 35.7 | 37.4 | 37.7 | 36.8 | 6 |
| | A(B) | 2 | T | 25.1 | 25.9 | 26.4 | 26.8 | 36.7 | 36.8 | 37.1 | 35.3 | 6 |
| JIS | B(D) | 9 | Le | 25.5 | 25.6 | 25.7 | 26.5 | 39.5 | 39.8 | 39.3 | 39.5 | 9.5 |
| G1 | B(C) | 10 | T | 25.5 | 25.6 | 25.7 | 26.0 | 39.5 | 39.8 | 40.1 | 39.4 | 9 |
| | C(B) | 17 | L1 | 20.6 | 20.8 | 20.2 | 24.7 | 30.8 | 31.1 | 30.9 | 31.3 | 7 |
| 5mm | C(A) | 18 | T | 17.5 | 17.1 | 17.4 | 18.1 | 31.1 | 31.5 | 31.4 | 31.4 | 6.5 |
| | D(C) | 25 | L1 | 20.5 | 20.7 | 21.1 | 21.4 | 35.3 | 35.4 | 35.2 | 35.3 | 8 |
| | D(D) | 26 | T | 22.4 | 23.2 | 23.9 | 24.4 | 35.0 | 35.5 | 36.0 | 35.4 | 8 |
| | A(B) | 7 | Le | 28.6 | 29.2 | 32.1 | 33.8 | 43.6 | 43.9 | 44.2 | 43.7 | 7 |
| | A(A) | 8 | T | 26.9 | 28.2 | 29.1 | 30.0 | 44.4 | 42.8 | 43.3 | 43.3 | 8.5 |
| JIS | B(C) | 15 | Le | 29.6 | 30.3 | 31.2 | 32.1 | 45.7 | 46.3 | 46.3 | 46.9 | 10 |
| G2 | B(D) | 16 | T | 30.3 | 29.3 | 29.5 | 29.8 | 46.8 | 47.2 | 46.6 | 45.9 | 10 |
| | C(A) | 23 | L1 | 29.9 | 30.1 | 31.3 | 33.0 | 40.9 | 41.3 | 41.3 | 40.8 | 6.5 |
| 1mm | C(B) | 24 | T | 26.4 | 26.7 | 27.1 | 29.6 | 40.4 | 39.6 | 40.5 | 40.0 | 6.5 |
| | D(D) | 31 | L1 | 24.6 | 25.1 | 26.0 | 26.2 | 42.3 | 42.4 | 42.2 | 42.4 | 7.5 |
| | D(C) | 32 | T | 28.0 | 28.5 | 29.9 | 30.6 | 40.7 | 40.6 | 40.6 | 40.7 | 7.5 |
| | A(D) | 5 | L1 | 27.1 | 27.3 | 28.4 | 29.0 | 44.3 | 44.3 | 44.3 | 44.0 | 6.5 |
| | A(C) | 6 | T | 29.6 | 30.0 | 31.1 | 32.1 | 43.1 | 42.7 | 42.8 | 42.8 | 6.5 |
| JIS | B(A) | 13 | L1 | 32.6 | 33.0 | 33.5 | 34.5 | 47.6 | 48.2 | 48.6 | 47.6 | 9.5 |
| G2 | B(B) | 14 | T | 28.6 | 29.9 | 31.4 | 34.0 | 44.8 | 45.3 | 46.0 | 47.6 | 8.5 |
| | C(C) | 21 | Le | 31.0 | 31.8 | 33.0 | 33.6 | 43.4 | 43.5 | 44.4 | 44.6 | 9 |
| 5mm | C(D) | 22 | T | 23.5 | 24.1 | 24.8 | 25.4 | 42.0 | 41.7 | 42.1 | 42.6 | 9 |
| | D(B) | 29 | Le | 29.3 | 29.5 | 30.5 | 31.9 | 42.8 | 42.0 | 42.5 | 41.8 | 6 |
| | D(A) | 30 | T | 30.1 | 30.5 | 31.2 | 31.5 | 42.1 | 44.3 | 44.2 | 42.1 | 6 |
| | A(B) | 77 | Le | 39.2 | 39.4 | 39.8 | 40.6 | 50.2 | 49.6 | 49.6 | 49.6 | 8 |
| | A(A) | 88 | T | 30.6 | 30.2 | 31.6 | 34.5 | 49.7 | 50.2 | 50.8 | 48.7 | 9 |
| JIS | B(C) | 155 | Le | 34.2 | 34.7 | 35.8 | 39.5 | 50.9 | 50.6 | 50.9 | 54.2 | 8 |
| G3 | B(D) | 166 | T | 35.6 | 35.9 | 37.1 | 37.3 | 52.7 | 52.3 | 52.0 | 52.7 | 8 |
| | C(A) | 233 | L1 | 44.4 | 44.7 | 45.5 | 46.9 | 57.4 | 57.9 | 57.8 | 57.0 | 8 |
| 1mm | C(B) | 244 | T | 39.3 | 40.0 | 40.4 | 42.7 | 55.7 | 56.2 | 56.3 | 55.2 | 7 |
| | D(D) | 311 | L1 | 34.8 | 35.5 | 36.2 | 36.3 | 50.0 | 49.9 | 49.3 | 49.3 | 8 |
| | D(C) | 322 | T | 31.1 | 32.1 | 32.5 | 33.7 | 49.3 | 48.8 | 49.6 | 49.6 | 8.5 |
| | A(D) | 55 | L1 | 41.0 | 40.8 | 42.2 | 43.8 | 56.0 | 56.1 | 56.5 | 56.4 | 9 |
| | A(C) | 66 | T | 42.7 | 43.6 | 44.7 | 45.5 | 54.7 | 54.8 | 54.8 | 54.9 | 8.5 |
| JIS | B(A) | 133 | L1 | 41.6 | 40.6 | 42.0 | 42.8 | 51.7 | 52.1 | 52.6 | 53.2 | 8.5 |
| G3 | B(B) | 144 | T | 30.3 | 32.3 | 33.1 | 34.9 | 51.0 | 50.8 | 51.7 | 50.6 | 8.5 |
| | C(C) | 211 | Le | 42.8 | 43.3 | 44.4 | 45.5 | 58.9 | 58.0 | 58.6 | 58.5 | 8.5 |
| 5mm | C(D) | 222 | T | 42.8 | 44.2 | 45.4 | 46.1 | 57.9 | 58.3 | 58.6 | 57.5 | 9 |
| | D(B) | 299 | Le | 41.7 | 42.1 | 42.2 | 42.5 | 51.5 | 50.8 | 51.0 | 50.5 | 8.5 |
| | D(A) | 300 | T | 45.1 | 47.7 | 48.4 | 50.0 | 53.9 | 54.9 | 55.2 | 53.6 | 9 |

| Speci-men No. | Elongation % | | | | Hardness Hv[2] | | | | Chemical composition % | | | | |
|---|---|---|---|---|---|---|---|---|---|---|---|---|---|
| | 3) 0.15 | 3) 0.30 | 3) 0.70 | 3) 1.5 | Surface | | Section | | Fe | N | C | H | O |
| | | | | | 1* | 5* | 1* | 5* | | | | | |
| 3 | 42.4 | 40.6 | 42.0 | 43.2 | 135 | 131 | 133 | | 0.060 | 0.006 | 0.009 | 0.0016 | 0.060 |
| 4 | 38.0 | 37.6 | 38.8 | 38.4 | 140 | 132 | 126 | | | | | | |
| 11 | 38.2 | 40.0 | 39.0 | 38.6 | 128 | 124 | 131 | | 0.034 | 0.007 | 0.006 | 0.0012 | 0.057 |
| 12 | 44.0 | 40.0 | 40.0 | 42.0 | 131 | 128 | 128 | | | | | | |
| 19 | 50.6 | 50.8 | 50.8 | 50.4 | 119 | 110 | 105 | | 0.039 | 0.002 | 0.004 | <0.0010 | 0.050 |
| 20 | 49.6 | 50.0 | 49.6 | 49.4 | 122 | 112 | 112 | | | | | | |
| 27 | 50.0 | 50.0 | 50.0 | 52.0 | 132 | 124 | 125 | | 0.032 | 0.002 | 0.009 | 0.0018 | 0.067 |
| 28 | 43.0 | 44.6 | 43.0 | 44.8 | 136 | 123 | 130 | | | | | | |
| 1 | 46.0 | 42.0 | 42.0 | 44.0 | | 194 | | 145 | 0.041 | 0.008 | 0.015 | 0.0018 | 0.084 |
| 2 | 43.4 | 42.6 | 42.6 | 40.6 | | 205 | | 129 | | | | | |
| 9 | 39.8 | 40.2 | 40.8 | 41.0 | | 146 | | 125 | 0.076 | 0.010 | 0.005 | <0.0010 | 0.058 |
| 10 | 41.4 | 41.6 | 40.0 | 41.0 | | 148 | | 131 | | | | | |
| 17 | 54.8 | 51.0 | 53.2 | 51.0 | | 143 | | 122 | 0.042 | 0.003 | 0.007 | 0.0010 | 0.069 |
| 18 | 45.0 | 44.0 | 48.0 | 44.0 | | 147 | | 132 | | | | | |
| 25 | 44.0 | 43.4 | 43.8 | 44.0 | | 183 | | 122 | 0.023 | 0.003 | 0.007 | 0.0020 | 0.068 |
| 26 | 46.2 | 44.2 | 41.6 | 42.0 | | 164 | | 117 | | | | | |
| 7 | 35.6 | 34.4 | 36.0 | 36.5 | 170 | 164 | 169 | | 0.060 | 0.007 | 0.018 | 0.0024 | 0.10 |
| 8 | 37.0 | 38.0 | 38.0 | 39.0 | 171 | 166 | 141 | | | | | | |
| 15 | 36.8 | 35.6 | 34.6 | 33.2 | 166 | 160 | 158 | | 0.095 | 0.006 | 0.008 | 0.0017 | 0.10 |
| 16 | 35.6 | 35.6 | 36.8 | 34.8 | 158 | 162 | 153 | | | | | | |
| 23 | 40.0' | 40.0 | 38.0 | 38.0 | 155 | 149 | 144 | | 0.082 | 0.004 | 0.008 | <0.0010 | 0.096 |
| 24 | 41.3 | 43.4 | 40.8 | 42.6 | 155 | 145 | 149 | | | | | | |
| 31 | 35.2 | 34.6 | 35.0 | 34.6 | 156 | 143 | 138 | | 0.050 | 0.008 | 0.013 | 0.0019 | 0.089 |
| 32 | 37.4 | 37.0 | 38.0 | 38.4 | 157 | 144 | 150 | | | | | | |
| 5 | 33.4 | 33.2 | 33.2 | 34.6 | | 217 | | 148 | 0.058 | 0.007 | 0.023 | 0.0020 | 0.11 |
| 6 | 36.2 | 36.2 | 35.6 | 36.4 | | 214 | | 149 | | | | | |
| 13 | 34.0 | 34.0 | 34.0 | 34.0 | | 174 | | 152 | 0.095 | 0.014 | 0.010 | <0.0010 | 0.097 |
| 14 | 41.2 | 40.0 | 39.6 | 37.4 | | 173 | | 146 | | | | | |
| 21 | 36.2 | 35.0 | 37.2 | 38.0 | | 183 | | 154 | 0.070 | 0.005 | 0.007 | <0.0010 | 0.12 |
| 22 | 37.8 | 38.8 | 36.4 | 38.0 | | 170 | | 150 | | | | | |
| 29 | 36.0 | 36.8 | 35.0 | 34.2 | | 240 | | 149 | 0.027 | 0.004 | 0.009 | 0.0021 | 0.14 |
| 30 | 38.0 | 40.0 | 38.0 | 38.0 | | 242 | | 151 | | | | | |
| 77 | 33.8 | 34.0 | 34.0 | 34.8 | 223 | 212 | 204 | | 0.067 | 0.005 | 0.043 | 0.0013 | 0.15 |
| 88 | 38.0 | 38.0 | 36.0 | 36.0 | 205 | 205 | 199 | | | | | | |
| 155 | 30.6 | 28.6 | 30.0 | 28.6 | 174 | 170 | 174 | | 0.15 | 0.013 | 0.015 | 0.0016 | 0.14 |
| 166 | 30.8 | 30.6 | 30.8 | 30.6 | 188 | 175 | 169 | | | | | | |
| 233 | 32.0 | 30.0 | 30.0 | 32.0 | 217 | 210 | 202 | | 0.17 | 0.004 | 0.012 | 0.0012 | 0.22 |
| 244 | 33.4 | 32.0 | 32.6 | 33.2 | 203 | 207 | 197 | | | | | | |
| 311 | 25.6 | 26.0 | 26.8 | 26.6 | 179 | 181 | 161 | | 0.032 | 0.003 | 0.012 | 0.0044 | 0.15 |
| 322 | 34.0 | 34.0 | 35.4 | 35.2 | 187 | 163 | 170 | | | | | | |
| 55 | 31.8 | 31.8 | 31.6 | 33.0 | | 225 | | 175 | 0.11 | 0.007 | 0.039 | 0.0015 | 0.17 |
| 66 | 35.2 | 34.6 | 35.0 | 34.8 | | 228 | | 193 | | | | | |
| 133 | 26.0 | 30.0 | 30.0 | 30.0 | | 203 | | 159 | 0.087 | 0.007 | 0.008 | <0.0010 | 0.14 |
| 144 | 34.4 | 36.6 | 30.8 | 31.9 | | 188 | | 150 | | | | | |
| 211 | 34.6 | 34.4 | 34.6 | 35.0 | | 261 | | 197 | 0.099 | 0.006 | 0.012 | 0.0011 | 0.26 |
| 222 | 33.0 | 34.0 | 32.8 | 34.0 | | 253 | | 200 | | | | | |
| 299 | 33.7 | 36.0 | 35.0 | 33.3 | | 234 | | 181 | 0.11 | 0.007 | 0.039 | 0.0021 | 0.18 |
| 300 | 36.0 | 34.0 | 34.0 | 36.0 | | 233 | | 198 | | | | | |

Note   (1)   $L_1$ : Specimen taken from the middle part of sheet and
                  plate on longitudinal direction.

            Le : Specimen taken from the outside part of sheet and
                  plate on longitudinal direction.

       (2)   Using Vickers hardness tester.

       (3)   Strain rate:  %/min, the mean strain rate through
                           0.2% yield strength measuing.

        *    Thickness of sheet and plate.

## 2.  Analysis of Test Results

2-1   Results of the Analysis of Variance

    Each chemical composition, Fe, O, N and H, as the characteristic value,
factors and levels of factors in table 3 are arranged to a orthogonal array
table L32, and then the analysis of variance is calculated from the test
results.

    From the analysis of variance of testing results, each contribution ratio*
of factor, that is significant with 5 percent and more significance in the
factors for each chemical composition, is shown in Table 4.

Table 3   Factors and levels of factors

| Mark | Factor | Level of factor |
|------|--------|-----------------|
| A | Maker | 4 companies |
| B | Grade | 3 Grades (JIS G1, G2 and G3) |
| C | Manufacturing process | 2 kinds (No.1[*1] and 2D[*2] finish) |
| D | Longitudinal and transverse direction | 1 direction (longitudinal) |
| F | Test performance company | 4 companies |
| G | Longitudinal, center and side parts | 2 parts (center and side part) |

Note:  *1   No.1 finish ----- hot rolled and annealed

       *2   2D finish   ----- cold rolled and annealed

Table 4  Contribution ratio* of each chemical
composition on results of analysis of variance, %

| Factor | Fe | N | C | H | O |
|--------|----|----|----|----|----|
| A | 15.1 | | 30.7 | 51.5 | 7.8 |
| B | 45.7 | | 36.5 | 4.9 | 80.4 |
| C | | | | 1.8 | 2.5 |
| D | | | | | |
| F | | | | | 0.1 |
| G | | | | | 0.6 |
| A × B | | | 13.2 | 16.7 | 10.8 |
| A × C | | | | | 0.9 |
| A × D | | | | | |
| B × C | | | | 6.8 | 0.2 |
| B × D | | | | | |
| C × D | | | | | |
| D × G | | | | | |
| composition of each grade | % | | % | | % |
| $B_1$ | 0.043±0.013 | | 0.008±0.003 | | 0.064±0.002 |
| $B_2$ | 0.067±0.013 | | 0.012±0.003 | | 0.106±0.002 |
| $B_3$ | 0.103±0.013 | | 0.023±0.003 | | 0.176±0.002 |

Note * contribution ratio = (correlation coefficient)$^2$, %

2-2  Relationship between Chemical Composition and Mechanical Properties

From table 2, the relations between chemical compositions and respective mean values of yield strength, tensile strength and hardness are induced and shown in figure 1.  It is recognized, from figure 1, that the relations between each chemical composition, Fe, C, H and O, and respective mean values of each mechanical property are roughly linear functions.

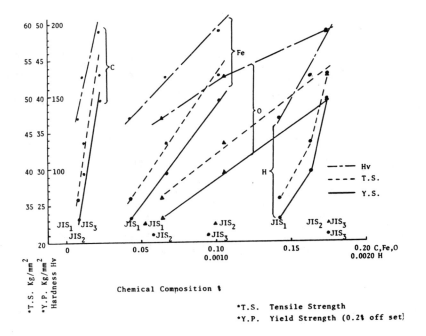

Fig. 1.   Relation Between chemical composition and respective mean value of yield strength, tensile strength and hardness.

2-3  Presumption of Material Strength by Chemical Composition

From the results of clauses 2-1 and 2-2, it is assumed that, there are the strong correlations between chemical compositions and each mechanical property The regression formulas therefore, which are to calculate each mechanical property from the chemical compositions, are induced by two methods, the method of least squares and simulation techniques, and then shown in table 5.

Nitrogen, on this examination, had no effect on each mechanical property, but the formulae shall considerably include the effect of nitrogen, because the chemical performances of nitrogen are as same as oxygen.

Each formula, shown in table 5, is checked out on the accuracy of himself by calculation on the contents of chemical compositions in table 2, and then the calculated values by the formulas are approximately equal to the experimental values of each mechanical property in table 2, as shown in table 6.

Therefore, added a calibration for the fluctuation of grain size to each formula, the accuracy of each formula in table 5 could be better.

Table 5  Regression formula for mechanical property

| Character | Simulation technique | Least squares method |
|---|---|---|
| Yield strength $Kg/mm^2$ | $\sigma s = 8.6843 + 0.625Fe + 4.25N$ $+ 1.50C + 2.18175H +$ $0.8438O$ | $\sigma s = 11.52 + 0.4204Fe$ $+ 3.5114N + 1.2347C$ $+ 0.88753H +$ $0.9693O$ |
| Tensile strength $Kg/mm^2$ | $\sigma a = 22.65 + 0.625Fe +$ $5.0N + 0.1875C +$ $4.375H + 0.25O$ | |
| Elongation % | $E\ell = 52.38 - 0.50Fe - 2.5N$ $-0.3125C - 3.75H - 0.425O$ | |
| Hardness Hv | $Hv = 82.654 + 0.9375Fe +$ $17.50N + 7.734C + 2.875H$ $+ 3.125O$ | |

Note:  Unit of content of chemical element;

Fe, N, C and O    0.01%

H                0.001%

Table 6   Comparison between experimental
values and calculated values of
mechanical properties

Table 6-1   Comparison between experimental values and
calculated values of yield and tensile
strength from chemical compositions

| Grade | Maker (Thickness) | Sample | | Y.S.* Kg/mm$^2$ | | | T.S.* Kg/mm2 | | Grain size |
| | | No. | Direction | Ex* | Eq.1* | Eq.2* | Ex* | Cal.* | ASTM No. |
|---|---|---|---|---|---|---|---|---|---|
| | D | 25 | L | 20.7 | | | 35.4 | | 8 |
| | | 26 | T | 23.2 | 22.5 | 22.7 | 35.5 | 35.3 | |
| | (5) | | | | | | | | |
| JIS | B | 9 | L | 25.6 | | | 39.8 | | 9〜9.5 |
| | | 10 | T | 25.6 | 25.5 | 25.4 | 39.8 | 38.2 | |
| G1 | (5) | | | | | | | | |
| | A | 1 | L | 22.8 | | | 37.4 | | 6 |
| | | 2 | T | 25.9 | 27.9 | 27.6 | 36.8 | 39.5 | |
| | (5) | | | | | | | | |
| | B | 13 | L | 33.0 | | | 48.2 | | 9 |
| | | 14 | T | 29.9 | 32.4 | 32.0 | 45.3 | 42.6 | |
| JIS | (5) | | | | | | | | |
| G2 | D | 31 | L | 25.1 | | | 42.4 | | 7.5 |
| | | 32 | T | 28.1 | 28.8 | 28.6 | 40.6 | 40.6 | |
| | (1) | | | | | | | | |
| | C | 233 | L | 44.7 | | | 57.9 | | 7〜8 |
| | | 244 | T | 40.0 | 44.0 | 43.9 | 56.2 | 46.3 | |
| JIS | (1) | | | | | | | | |
| G3 | D | 311 | L | 35.5 | | | 49.9 | | 8〜8.5 |
| | | 322 | T | 32.1 | 36.0 | 33.9 | 48.8 | 49.6 | |
| | (1) | | | | | | | | |

Note:   *   Y.S.;   Yield strength      T.S.;   Tensile strength

         Ex;   Experiment          Cal.;   Calculation

         Eq.1;   Formula from simulation techniques

         Eq.2;   Formula from method of least squares

Table 6-2　Comparison between experimental values
and calculated values of elongation and
hardness from chemical compositions

| Grade | Maker (thick-ness) | Sample | | Elongation % | | Hardness Hv | | Grain size ASTM No. |
|-------|------|-----|-------|-------|-------|-------|-------|------|
| | | No. | Direction | Ex.* | Cal.* | Ex.* | Cal.* | |
| JIS G1 | D | 25 | L | 43.4 | 42.7 | 122 | 122 | 8 |
| | (5) | 26 | T | 44.2 | | 117 | | |
| | B | 9 | L | 40.2 | 41.5 | 125 | 132 | 9～9.5 |
| | (5) | 10 | T | 41.6 | | 131 | | |
| | A | 1 | L | 42.0 | 40.1 | 145 | 144 | 6 |
| | (5) | 2 | T | 42.6 | | 129 | | |
| JIS G2 | B | 13 | L | 34.0 | 38.9 | 152 | 157 | 9 |
| | (5) | 14 | T | 40.0 | | 146 | | |
| | D | 31 | L | 34.6 | 39.2 | 138 | 145 | 7.5 |
| | (1) | 32 | T | 37.0 | | 150 | | |
| JIS G3 | C | 233 | L | 30.0 | 35.3 | 202 | 187 | 7～8 |
| | (1) | 244 | T | 32.0 | | 197 | | |
| | D | 311 | L | 26.0 | 31.3 | 161 | 160 | 8～8.5 |
| | (1) | 312 | T | 34.0 | | 170 | | |

Note 1: Ex. *; Experiment　　　　　　Cal. * Calculation

2: Compare the experimental values of tensile strength and elongation with the calculated values of them on table 6-1 and 6-2, both values are approximately equal, except JIS G3, if the grain size could be considered.

On table 6-2, the experimental values of hardness are approximately equal with the calculated values.

## 3   Conclusion

Through the test results of mechanical properties of titanium sheets and plates produced by 4 Japanese titanium makers, we may declare as follows;

It is clear that the chemical compositions, Fe, N, C and O, contribute mainly to the mechanical properties of titanium sheets and platets.

Therefore, when chemical compositions are decided, in accordance with the formulae induced by the simulation techniques in table 5, the strength of titanium sheet and plate can be pressumed with a considerable accuracy.

## Reference

Japan Titanium Society; Titanium Zironium, Vol.20, No.5 (1972) p.258

# MECHANICAL PROPERTIES OF COMMERCIALLY PURE TITANIUM SHEET

Poul Kvist

Department of Mechanical Technology
Technical University of Denmark
Lyngby, Denmark

## Introduction

In this work the mechanical properties of commercially pure titanium sheet have been investigated according to the procedure normally used when investigating sheet metals.

This procedure contains the following steps:

- determination of the stress-strain curve by tensile testing and by bulge testing
- investigation of the sensitivity of the stress-strain curve to temperature and deformation speed
- analytical representation of the stress-strain curve
- investigation of the state of anisotropy
- correction of the stress-strain curves due to the anisotropy
- determination of the forming limit curve

## Equipment and Procedure

The tensile testing was carried out on an 0.1 MN Amsler tensile test machine. Before testing, the surface of the specimens were photochemically marked with a circular grid of 2.00 mm diameter. During testing the tensile force was registrated on a digital display. With appropriate time intervals the display and the specimen were photographed. By this procedure all necessary informations concerning the analysis could be read directly on the photographs, fig. 1.

On the basis of the photographs mentioned above the true stress-natural strain curve and the state of anisotropy were calculated for all tensile tests.

The bulge testing was carried out on a bulge test machine developed at the department, fig. 2. This machine calculates automatically the stress-strain curve of the sheet material during testing.

The stress-strain curve and the forming limit curve were determined on this equipment by using one circular and three elliptical dies.

Fig. 1   Tensile testing

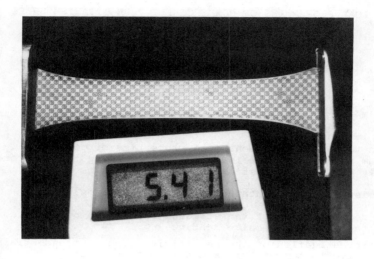

Fig. 2   Bulge test machine

## Material

The material used in this investigation was commercially pure titanium sheet, thickness 0.81 mm. The cast analysis is tabulated in Table 1.

Table 1   Cast analysis

| C | $O_2$ | $H_2$ | Fe | N |
|---|---|---|---|---|
| 0,009 | 0,055/58 | 0,002 | 0.04 | 0.007 |

## Experimental Results

1.  Tensile test, stress-strain curves

Five series of experiments were carried out to determine the sensitivity of the titanium sheet due to temperature raise caused by deformation and due to strain rate. Three cross head speeds were examined. In two series the testing was carried out in distilled water to eliminate some of the influence of the deformation heat on the test results. The influence of the test conditions on the proof stress, tensile strength and elongation to fracture are listed in Table 2.

Table 2   Influence of test conditions
on test results

| Series | Test conditions | Orientation to rolling direction | Proof stress $N/mm^2$ | Tensile strength $N/mm^2$ | Elongation % |
|---|---|---|---|---|---|
| 1 | $v$ = 6 mm/min $\bar{\varepsilon}$ = 2.1·$10^{-3}$/s Air 20 $^{o}$C $\pm$ 1$^{o}$ | 0$^{o}$ | 245 | 333 | 42 |
| | | 45$^{o}$ | 260 | 324 | 44 |
| | | 90$^{o}$ | 290 | 334 | 43 |
| 2 | $v$ = 0.6 mm/min $\bar{\varepsilon}$ = 1.3·$10^{-4}$/s Air 20 $^{o}$C $\pm$ 1$^{o}$ | 0$^{o}$ | 205 | 296 | 51 |
| | | 45$^{o}$ | 220 | 284 | 51 |
| | | 90$^{o}$ | 240 | 311 | 50 |
| 3 | $v$ = 60 mm/min $\bar{\varepsilon}$ = 1.7·$10^{-2}$/s Air 20 $^{o}$C $\pm$ 1$^{o}$ | 0$^{o}$ | 275 | 325 | 38 |
| | | 45$^{o}$ | 280 | 321 | 33 |
| | | 90$^{o}$ | 315 | 352 | 37 |
| 4 | $v$ = 6 mm/min $\bar{\varepsilon}$ = 1.4·$10^{-3}$/s Water 20 $^{o}$C $\pm$ 0,5 | 0$^{o}$ | 280 | 358 | 50 |
| | | 45$^{o}$ | 275 | 340 | 48 |
| | | 90$^{o}$ | 275 | 340 | 48 |
| 5 | $v$ = 0.6 mm/min $\bar{\varepsilon}$ = 1.7·$10^{-4}$/s Water 20 $^{o}$C $\pm$ 0,5 | 0$^{o}$ | 290 | 356 | 50 |
| | | | | | |
| | | | | | |

The experiments in series 1 and 4 were repeated  twice. The proof stress was reproduced within 5%, the tensile strength within 2% and the elongation within 1%.

The stress-strain curves determined by tensile testing are shown in fig. 3.

The stress-strain curves for series 1 and 4 were reproduced within 2%.

Fig. 3   The influence of the test conditions
on the stress-strain curves

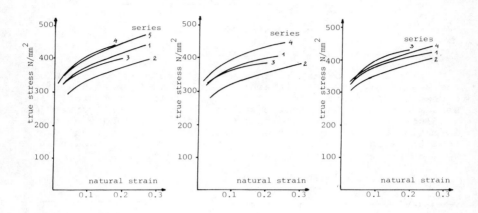

The accuracy of representing the stress-strain curves by
an equation of the form $\sigma = C \cdot \varepsilon^n$ where n is the strain hardenin
exponent was investigated. Fig. 4 shows the stress-strain curve
and the corresponding log-log plot of stress and strain obtaine
from the experiments in series 4.

Fig. 4   Stress-strain curves and
log-log plots of stress and strain

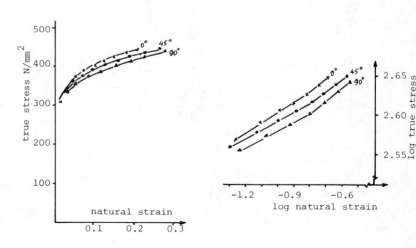

On the basis of fig. 4, representative for all the tensile test experiments, it is clear that the strain hardening characteristic cannot be represented by one single strain hardening exponent, independent of the strain. It is obvious that there is a change in the strain hardening characteristic at a strain of approximately 0.15. The strain hardening exponent n is apparently constant at lower as well as higher strains.

The n-values at low and high strain for all tensile test experiments are listed in Table 3.

Table 3   n-values at low and
high strain

| Series | Orientation to rolling direction | n at strain lower than $0.15 \pm 0.02$ | n at strain higher than $0.15 \pm 0.02$ |
|--------|----------------------------------|------------------------|-------------------------|
| 1 | $0^{\circ}$ | 0.12 | 0,17 |
|   | $45^{\circ}$ | 0.11 | 0,12 |
|   | $90^{\circ}$ | 0.10 | 0,15 |
| 2 | $0^{\circ}$ | 0.15 | 0,20 |
|   | $45^{\circ}$ | 0.13 | 0,20 |
|   | $90^{\circ}$ | 0.11 | 0.18 |
| 3 | $0^{\circ}$ | 0.11 | 0.12 |
|   | $45^{\circ}$ | 0.08 | 0.09 |
|   | $90^{\circ}$ | 0.18 | 0.12 |
| 4 | $0^{\circ}$ | 0.14 | 0,17 |
|   | $45^{\circ}$ | 0.11 | 0.15 |
|   | $90^{\circ}$ | 0.11 | 0.18 |
| 5 | $0^{\circ}$ | 0.11 | 0.23 |

## 2.   Tensile test, anisotropy

The ratio of width strain to thickness strain R was calculated for all the tensile test pieces. In fig. 5 the calculated R-values are plotted against strain and it is obvious that the R-values depend on the strain. In similar cases for other metals IDDRG /1/ recommends that the material is given the R-value at a strain of 20% $\pm$ 1%. This recommendation has been used in fig. 5 where the R-values are plotted against the orientation to rolling direction.

## 3.   Bulge test

The condition for bulge testing with hydraulic oil on one side is expected to correspond best with the tensile testing condition in series 4, the specimen here being surrounded by distilled water. Because of this correspondance the testing speed of the bulge testing is chosen so that the average strain rate at the pole corresponds with that of the tensile testing in series 4 i.e. $\dot{\varepsilon} \simeq 10^{-3}$/s. The stress-strain curve found by bulge testing, fig. 7, is calculated under the assumption that the

Fig. 5  R-value vs strain

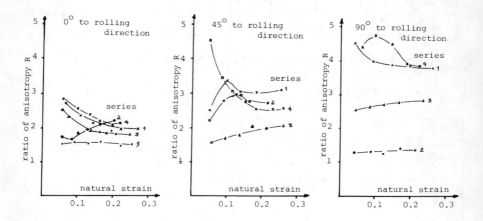

Fig. 6  R-value vs orientation
to rolling direction

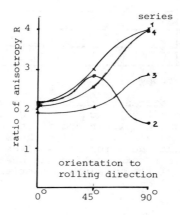

material is isotropic. The log-log plot shows that the strain hardening exponent is a constant (n = 0.32) for strains higher than 0.18.

The dotted curves in fig. 7 are the stress-strain curves determined by tensile testing in series 4.

Fig. 7   Stress-strain curve and log-log
plot of stress and strain

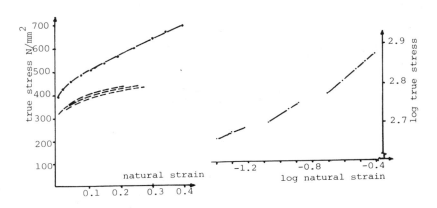

Due to Hill's general theory of anisotropy /2/ there should
be a correlation between the stress-strain curves determined by
bulge testing and by tensile testing. This was   examined for
titanium sheet by Bramley and Mellor in 1966 /3/. They found good
agreement between the bulge test curves found experimentally and
predicted from the tensile tests. Bramley and Mellor used the
relations

$$\sigma_z = (\frac{1 + \bar{R}}{2})^{\frac{1}{2}} \cdot \sigma, \quad \varepsilon_z = (\frac{2}{1 + \bar{R}})^{\frac{1}{2}} \cdot \varepsilon \tag{1}$$

where $\bar{R} = \dfrac{R_{0°} + R_{90°} + 2 \cdot R_{45°}}{4}$   is the normal anisotropy.

The validity of these relations (1) were checked in this
work by trying to predict the stress level and the slope of the
tensile test curves from the experimental bulge test curve fig.
8.

4.   Forming limit curve (FLC)

Four points on the forming limit curve were determined on
the bulge test machine using one circular and three elliptical
dies fig. 9.

The experiments with the circular die indicated that the
formability of the material was lowest perpendicular to the
rolling direction (fracture in rolling direction). On the basis
of this the rolling direction of the specimens for the experiments with the elliptical dies were alligned with the major
axes of the dies before testing. In the tensile testing region
one point was calculated from the experiments perpendicular to
the rolling direction series 4.

The method proposed by C. C. Vermann /4/ for calculating
major and minor strains were used. The determined FLC is plotted
in fig. 10.

Fig. 8   Experimental and predicted
tensile test curves $\bar{R} = \dfrac{2.1 + 3.9 + 2 \cdot 2.6}{4} = 2.8$

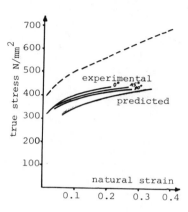

Fig. 9   Dies used to determine the FLC

The dotted FLC in fig. 10 is found for a 0.61 mm commercially pure titanium sheet to determine the influence of thickness on the level of the FLC. The cast analysis for the 0.61 mm titanium sheet is listed in Table 4.

Table 4   Cast analysis

| C | $O_2$ | $H_2$ | Fe | N |
|---|---|---|---|---|
| 0.008 | 0.07 | < 100 ppm | 0.04 | 0.007 |

Fig. 10  Forming limit curve

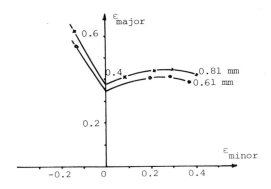

## Conclusion

The tensile test experiments show that the mechanical properties i.e. proof stress, tensile strength, elongation, stress-strain curve and state of anisotropy are so close related to the testing conditions that even small changes in these have a significant influence on the test results.

The log-log plots of stress and strain shows that representation of the stress-strain curves from yield point to necking point by an equation of the form $\sigma = C \cdot \varepsilon^n$ is only possible if the strain is devided in intervals. This is independent of the state of stress.

The correlation of the experimental stress-strain curves found by tensile testing and the one predicted from bulge testing due to anisotropy is poor. The main reason for this is possibly that the state of anisotropy in the titanium sheet does not fit the assumption made for the calculations.

The forming limit curves show that the necking strains of the 0.81 mm sheet are about 0.05 higher than those of the 0.61 mm sheet. This as well as the shape of the curves are in good agreement with forming limit curves for other sheet metals.

## References

1.  W. Schmidt: Blech Rohre Profile 25, 6, p. 271 (1978).
2.  R. Hill: Proc. R. Soc. A 193, p. 281 (1948).
3.  A. N. Bramley, P. B. Mellor: Int. J. Mech. Sci.10, p.211 (1968)
4.  C. C. Vermann: 7th Biennial Congress of IDDRG, 19.1, Amsterdam 1972.

RELATION BETWEEN BRINELL HARDNESS OF TITANIUM
AND IMPURITIES ($O_2$, Fe, N AND C)

Kameji Shimasaki, Kimiyoshi Ono and Takeo Tsuruno

Osaka Titanium Co., Ltd.
Amagasaki, Japan

## Introduction

Hardness of titanium which is closely related to its mechanical properties varies depending on the kind and the quantities of the impurities contained. According to the results of the researches conducted by T.D. McKinley,
W.L. Finlay, Kobe Steel, Ltd., and by a group in U.S.S.R., etc., considerable
differences are noted in the correlation between the impurities and the corresponding Brinell hardness ($H_B$) depending on the differences in the quali-
ties of base materials and the testing conditions employed.

As a manufacturer of titanium sponge, we have conducted a set of experiments with the objectives of clarifying the relation between the impurities
and the hardness and utilizing the relation for the manufacturing process control of titanium sponge and quality assurance of shipping lots. Through the
experiments, the following equation has been obtained on 4 elements: $O_2$, Fe, N
and C, which have prominent influences.

$$H_B = 411\ O_2\ \% + 128\ Fe\ \% + 675\ N\ \% + 380\ C\ \% + 65.6$$

## Experimental Procedure

1. Test Samples and Preparation of Samples

1.1 Ti sponge as base material

From the Ti sponge containing less impurities as shown in Table 1, 112
test samples each weighing 80 g were prepared.

Table 1 Chemical Composition of Test Sample

| Impurity (Wt %) | | | | |
|---|---|---|---|---|
| Fe | $O_2$ | N | C | $H_B$ |
| 0.01 | 0.048 ∿ 0.055 | 0.004 ∿ 0.006 | 0.006 | 89 ∿ 92 |

1.2 Additives

Oxygen ($O_2$ % = 7)

14 g of high purity $TiO_2$ (99% or higher) was mixed with 66 g base Ti
sponge. After melting twice in the test furnace, a button was crushed in a
mortar into granules of 5 ∿ 20 mesh.

Iron  (Fe % = 2.1)

24 g of electrolytic iron (purity 99% or higher) was mixed with 56 g of the same Ti sponge material and melted in the same manner.  After crushing, 1.6 g of the granules were mixed with 78.4 g Ti sponge and melted again.  Then, the product was turned on a lathe and used for the test in a form of fine turned chips.

Nitrogen  (N % = 0.35)

125 g of the same Ti sponge material was heated at 950°C for 2 hours in a resistance furnace to produce nitride while feeding nitrogen gas.  With 10 times of the base Ti sponge, it was melted in the furnace and turned on the lathe and used in a form of fine chips.

Carbon  (C % = 2.0)

6.4 g TiC (C % = 19.6) was mixed with 73.6 g of the Ti sponge, melted in the furnace and used in a form of fine turned chips.

## 1.3   Procedure for preparing test pieces

The above described impurity additives were added to the base Ti sponge in stages, and melted twice in the laboratory furnace to produce button pieces each weighing about 80 g.  After the pieces were made flat, dents were formed at the 3 points as specified on each of the test pieces.  Brinnel hardness was measured and then the contents of the impurities were analyzed.  Two test pieces each were prepared and measured for hardness for each stage.

## 2.   Test Results and Comments

## 2.1   When a single impurity element is added in stages:

Results of the measurements of $H_B$ and the determinations of $O_2$, Fe, N and C contents are shown in Figs. 1, 2, 3 and 4 respectively.  The test pieces were prepared for 8 stages in the range of $0.05 \sim 0.15\%$ for $O_2$, for 10 stages in the range of $0.01 \sim 0.3\%$ for Fe, for 8 stages in the range of $0.004 \sim 0.07\%$ for N and for 10 stages in the range of $0.006 \sim 0.06\%$ for C, respectively.

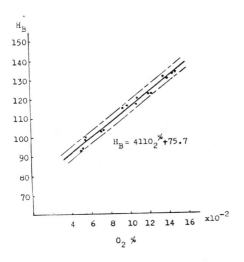

Fig. 1   Relation between $H_B$ and $O_2$

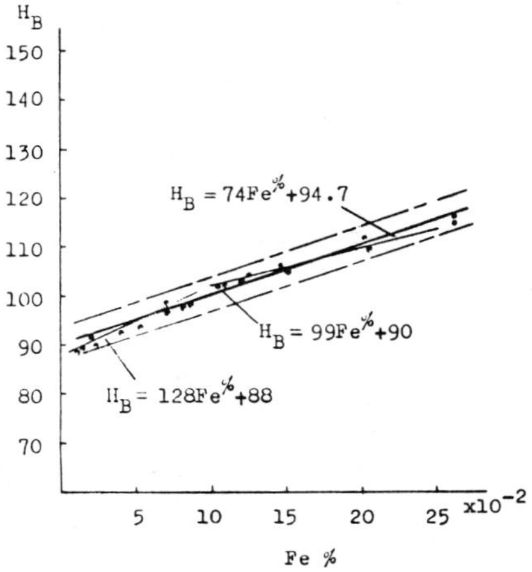

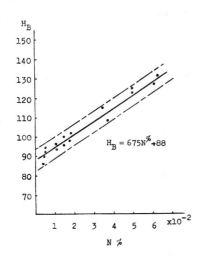

Fig. 2   Relation between $H_B$ and Fe

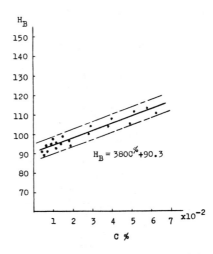

Fig. 3   Relation between $H_B$ and N

Fig. 4   Relation between $H_B$ and C

A high degree of correlation was noted upon the regression analysis. The coefficients of correlation and the values of the standard deviations obtained are shown in Table 2.

Table 2

|  | $H_B$ vs $O_2$ | $H_B$ vs Fe | $H_B$ vs N | $H_B$ vs C |
|---|---|---|---|---|
| Coefficient of correlation | 0.99 | 0.97 | 0.98 | 0.98 |
| Standard deviation | 1.6 | 1.95 | 2.8 | 2.28 |

The following equations have been established through the measurements and the determinations:

$$H_B = 411\, O_2\,\% + 75.7 \quad \dots\dots\dots\dots\dots (1)$$
$$H_B = \quad 99\, Fe\,\% + 90 \quad \dots\dots\dots\dots\dots (2)$$
$$H_B = 675\, N\,\% + 88 \quad \dots\dots\dots\dots\dots (3)$$
$$H_B = 380\, C\,\% + 90.3 \quad \dots\dots\dots\dots\dots (4)$$

In the case of Fe, the gradient tends to become smaller as the Fe content increases. Following are the relations between hardness and impurity contents obtained for the ranges below and over 0.08%, which is considered to be the critical point for the change in the size of crystalline granules.

$$\text{Below } 0.08\% \quad H_B = 128\, Fe\,\% + 88 \quad \dots\dots\dots\dots\dots (2')$$
$$\text{Over } 0.08\% \quad H_B = \quad 74\, Fe\,\% + 94.7 \quad \dots\dots\dots\dots\dots (2'')$$

In practical applications, Fe contents are usually 0.08% or lower. Consequently, the equation (2') was employed and combined with other equations obtained through the determinations on the individual impurity elements.

Each equation is corrected as follows in order to exclude the influence given to the constants by other impurities.

From equation (1)   $H_B = 411\, O_2\,\% + 66.06$
From equation (2)   $H_B = 128\, Fe\,\% + 60.37$
From equation (3)   $H_B = 675\, N\,\% + 61.71$
From equation (4)   $H_B = 380\, C\,\% + 63.91$

Averaging these constants, equation (5) is formulated as follows:

$$H_B = 411\, O_2\,\% + 128\, Fe\,\% + 675\, N\,\% + 380\, C\,\% + 63 \quad \dots\dots\dots\dots\dots (5)$$

2.2  When the impurity elements were added simultaneously:

Four kinds of the impurity additives were added to the base Ti sponge simultaneously so that the hardness of approximately 95, 100, 105, 115, 120 and 135 would be obtained. Values for $H_B$ were calculated through the equation (5) on $O_2$, Fe, N and C respectively. Table 3 shows the comparison between these values and the $H_B$s obtained by the actual measurements. The table indicates that the mixing ratios were very close to the values intended while the $H_B$ values actually measured were higher than the calculated values on all of the impurities.    This is considered to be attributable to the influence of the interactions between the impurity elements.

Table 3

| Analyzed Value (Wt %) | | | | $H_B$ | | | Constant Obtained through Measured Value |
|---|---|---|---|---|---|---|---|
| $O_2$ | Fe | N | C | Measured Value | Calculated Value | Difference | |
| 0.048 | 0.021 | 0.009 | 0.006 | 96 | 94 | 2 | 64.8 |
| 0.062 | 0.029 | 0.008 | 0.006 | 103 | 100 | 3 | 65.6 |
| 0.072 | 0.033 | 0.013 | 0.006 | 110 | 108 | 2 | 65.2 |
| 0.083 | 0.052 | 0.010 | 0.008 | 117 | 114 | 3 | 66.1 |
| 0.105 | 0.062 | 0.006 | 0.011 | 125 | 122 | 3 | 65.4 |
| 0.118 | 0.085 | 0.013 | 0.012 | 139 | 136 | 3 | 66.5 |
| | | | | | $\bar{x}$ | 2.7 | 65.6 |

The average of constants obtained with the equation (5) based on the $H_B$ values actually measured for each of the test pieces is 65.6, which is greater than the constant of the equation (5) by 2.6.    On the other hand, as for the constants obtained with the equation (5) on 230 shipping lots made in past, the average was 65.7, which quite agrees with the constant obtained in the case of simultaneous addition of the impurity additives.

Conclusion

The equations indicating correlation between the 4 impurities; $O_2$ , Fe, N and C and the hardness of titanium were obtained.

A high degree of correlation was noticed on each of the equation obtained independently.    Although the constant obtained upon the integration of these equations and the constant obtained when four impurities were added simultaneously differed by 2.6, 65.6 was finally employed as constant in consideration of the value obtained from the actual shipping lots.

As a Ti sponge manufacturer, we have been succesfully making use of the data obtained from the experiments for calculation of the mixing ratio and the quality assurance of the shipping lots, to meet various specifications demanded by a number of customers.

# MECHANICAL PROPERTIES OF TITANIUM-BASED
## AMORPHOUS ALLOYS WITH METALLOIDS

Akihisa Inoue, Hisamichi Kimura, Shuji Sakai*
and Tsuyoshi Masumoto

The Research Institute for Iron, Steel and Other Metals,
Tohoku University, Sendai 980, Japan

## Introduction

In recent years, the splat-quenching technique has attracted special interest for the direct production of a continuous amorphous[1] or crystalline [2] tape, the final shape for many applications. It has also been reported [1,2] that these rapidly quenched tapes possess high strength as well as good bend ductility. However, almost all the investigations[3] have so far been focused on iron-, cobalt- and nickel-based alloys, despite the expectation that the application of the splat-quenching technique to titanium-based alloys may produce ductile tapes possessing very high specific strength without any intermediate process. Polk et al.[4] and Tanner and Ray[5,6] have recently reported that they were able to produce an amorphous phase in binary Ti-Si and ternary Ti-Ni-Si, Ti-Be-Si and Ti-Be-Zr systems. Up to date, however, systematic information about the amorphous phase formation of titanium-base alloys containing transition metals such as V, Cr, Mn, Fe, Co, Nb and Ta etc. and their mechanical properties has not been reported in spite of the high engineering potential due to superconductivity, good corrosion resistance and hydrogen storage, etc. of these alloys. Good mechanical properties would be of value for applications in such engineering fields. From this point of view, we have applied the splat-quenching technique to Ti-M-Si and Ti-M-B(M=V, Cr, Mn, Fe, Co, Ni, Cu, Zr, Nb, Mo and Ta) ternary alloys and succeeded in producing their amorphous alloys in the form of long continuous ribbons. This paper deals with the aspects involving the composition range for the formation of the amorphous phase in these systems and the mechanical properties and thermal stability of these amorphous alloys. Another paper in this Conference [7] deals with the crystallization behavior of titanium-based amorphous alloys.

## Experimental Procedure

Mixtures of 99.5 wt% pure titanium, 99.999 wt% pure silicon, 99.5 wt% pure boron and 99.5 wt% pure V, Cr, Mn, Fe, Co, Ni, Cu, Zr, Nb, Mo and Ta were melted in an arc furnace on a water-cooled copper hearth with a tungsten electrode. The melting was accomplished in a purified and gettered argon atmosphere at a pressure of about $8 \times 10^4$ Pa. The alloys were melted repeatedly to ensure complete mixing of the three elements. Weight loss was typically less than 15 mg in a 30 g ingot, and the compositions reported are the nominal ones.
Continuous ribbon specimens of about 1.5 mm width and 0.03 mm thickness were prepared from these alloys under a protective argon atmosphere using a modified single roller quenching apparatus designed for high melting alloys by

---

* Permanent address, Hitachi Cable Ltd., Tsuchiura 300, Japan.

the present authors.  A schematic illustration of the apparatus is shown
in Fig. 1.  The alloy was levitation
melted at an argon pressure of about
3 MPa, and by opening the shutter and,
at the same time, by cutting the coil
current off, the molten alloy was e-
jected, by the difference in pressure
between the two chambers, through the
nozzle at the end of a quartz tube
onto the surface of a copper roll
which was rotating at a high speed in
the lower evacuated room.  The ejected
melt was solidified on the roll surface in the form of a continuous ribbon.  Typically, the amount of alloys
melted in one run was about 5 g and
the rotation speed of the roll(20 cm
in diameter) was about 4000 rpm.  This
modified liquid-quenching apparatus is
useful for refractory metals with high
melting temperatures and for active
metals which easily react with a
quartz tube and/or oxygen in air.

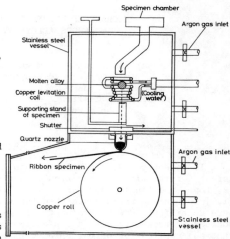

Fig. 1   Schematic illustration of the
rapid quenching apparatus used
in the present work.

Identification of the as-quenched
phases was made by conventional X-ray diffraction method using nickel filtered
Cu Kα radiation.  The ribbons were classified amorphous when the X-ray intensity as a function of the diffraction angle showed a typical liquid like
structure.  The as-quenched alloys were also examined by transmission electron
microscopy.  Hardness and strength of the specimens were measured by a Vickers
microhardness tester with a 100 g load and an Instron-type tensile testing
machine at a strain rate of $1.7 \times 10^{-4}$/s, respectively.  The crystallization
temperatures of the alloys were examined at a heating rate of $8.33 \times 10^{-2}$ K/s
by a differential thermal analyzer(DTA).  The ductile-brittle transition behavior was tested for the specimens annealed for 1 h at various temperatures
in evacuated quartz capsules.  Ductility was evaluated by measuring the radius of curvature at fracture in a simple bend test[8].

Results and Discussion

1.   Composition range for amorphous phase formation

As typical examples, Fig. 2 (a)-(d) shows the composition ranges in
which the amorphous phase formed without any trace of crystallinity for the
Ti-Fe-Si, Ti-Co-Si, Ti-V-Si and Ti-Nb-Si ternary alloy systems.  These
alloy systems are especially important since they possess high strength and/or
superconductivity[9-12].  The amorphous phase forms in the wide range of 0-27
at%Fe, 0-37 at%Co, 0-32 at%V and 0-43 at%Nb and the silicon content is limited
to 13-22 at%.  As an example, a typical electron micrograph of the amorphous
phase in the thinned $Ti_{70}Nb_{15}Si_{15}$ alloy is shown in Photo. 1, together with
the selected area diffraction pattern.  Lack of contrast in the bright field
image(a) and the diffuse haloes in the diffraction pattern(b) indicate clearly that the material is amorphous.  No evidence of crystalline inclusions was
found by means of dark field electron microscopy for the alloys within the
composition ranges described above.  In addition, the formation ranges of
amorphous single phase for the $Ti_{85-x}M_xSi_{15}$(M=V, Cr, Mn, Fe, Co, Ni, Cu, Zr,
Nb and Ta) and $Ti_{90-x}M_xB_{10}$(M=Cr, Mn, Fe, Co and Ni) alloys are represented in
Fig. 3.  The composition ranges of amorphous phase formation for $Ti_{85-x}M_xSi_{15}$

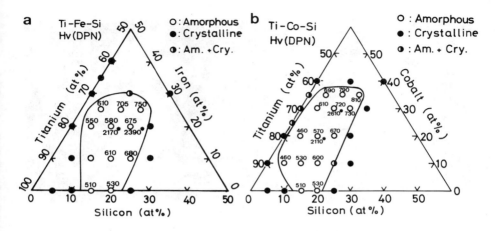

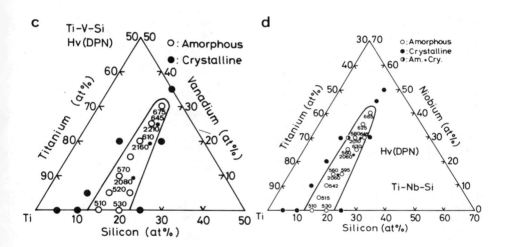

Fig. 2   Composition range for the formation of the amorphous phase and the
changes of Vickers hardness(Hv) and tensile fracture strength($\sigma_f$)
in the (a) Ti-Fe-Si, (b) Ti-Co-Si, (c) Ti-V-Si and (d) Ti-Nb-Si
systems.

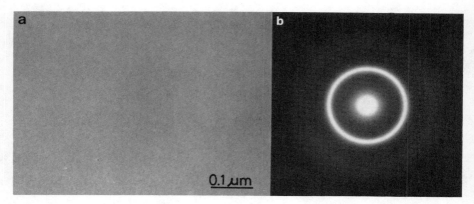

Photo. 1   Transmission electron micrograph and selected area diffraction pattern showing the as-quenched structure of Ti$_{70}$Nb$_{15}$Si$_{15}$ alloy.

alloys become narrower in the order of Zr, Nb, Ta, Ni, Co, Mn, V, Cu, Cr and Fe and no amorphous phase was found for Ti$_{85-x}$Mo$_x$Si$_{15}$ alloys. Further, one can notice that silicon is more effective than boron to form the amorphous phase. This is in contrast to the tendency of the late transition metal-metalloid amorphous alloys such as Fe-B[13], Co-B[14] and Ni-B[15]. Additionally, the Ti-M(M=Fe, Co and Ni) binary alloys showed a crystalline phase and ·no amorphous phase was found in the present investigations. This result differs from the previous one [4] that Ti-Ni binary alloys quenched rapidly by the piston and anvil method showed an amorphous single phase.

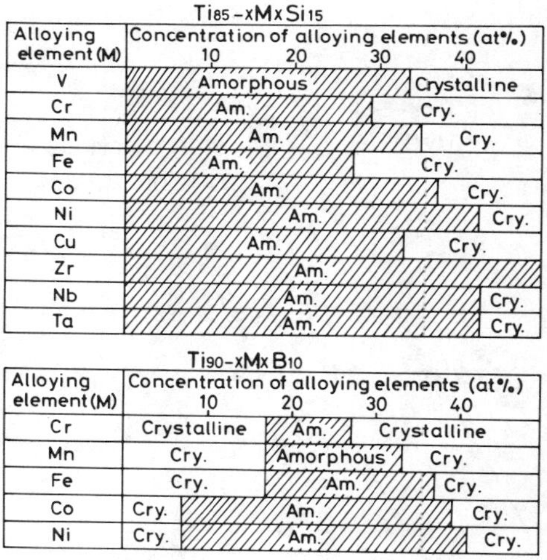

In general, the composition range of amorphous alloys obtained by liquid-quenching technique is located around a deep eutectic in the alloy phase diagram. Turnbull[16] has noted that a deep eutectic can be interpreted in terms of a comparatively large negative heat of formation of the liquid alloy. Hence, it is predicted that alloys having compositions near a deep eutectic are most prone to the formation of amorphous phase. Each phase

Fig. 3   Composition range for the formation of the amorphous phase in Ti$_{85-x}$M$_x$Si$_{15}$(M=V, Cr, Mn, Fe, Co, Ni, Cu, Zr, Nb and Ta) and Ti$_{90-x}$M$_x$B$_{10}$(M=Cr, Mn, Fe, Co and Ni) alloys.

diagram[17,18] of the Ti-Si, Ti-B, Ti-M(M=Cr, Mn, Fe, Co, Ni and Cu) and M-Si (M=V, Zr, Nb and Ta) binary systems features a eutectic reaction. As is evident from Fig. 2 (a)-(d) and these phase diagrams, the present amorphous-forming composition ranges fall near the trough of the eutectics in their ternary systems, as is common in other liquid-quenched amorphous alloys, and thus the ability to form an amorphous phase seems to be closely related to the large negative heat of formation of the liquid alloy.

## 2. Mechanical properties

The Vickers hardness(Hv) and tensile fracture strength($\sigma_f$) of Ti-M-Si(M= Fe, Co, V and Nb) amorphous alloys are shown in Fig. 2 (a)-(d), wherein the values marked by an asterisk are the tensile fracture strengths expressed in units of MPa.  As is clear from the figure, Hv rises gradually with the amount of the alloying element(M) or silicon and attains about 750 DPN for $Ti_{60}Fe_{25}$-$Si_{15}$, 810 DPN for $Ti_{50}Co_{35}Si_{15}$, 675 DPN for $Ti_{55}V_{30}Si_{15}$ and 665 DPN for $Ti_{45}$-$Nb_{40}Si_{15}$.  The fracture strengths are about 2300 MPa for Ti-Fe-Si alloys, 2350 MPa for Ti-Co-Si alloys, 2150 MPa for Ti-V-Si alloys and 2050 MPa for Ti-Nb-Si alloys.  These values are much higher than those(350 DPN and 1150 MPa)[19] for conventional high strength titanium alloys.  The crystallization temperature (Tx), critical fracture temperature($T_f$), specific strength($\sigma_f/\rho$) and Hv/$\sigma_f$ are presented in Table 1 together with the Vickers hardness(Hv) and tensile fracture strength($\sigma_f$).  $T_f$ is the aging temperature at which the alloy fractured by a perfect bending after aging for 6000 s.  Hv and $\sigma_f$ of Ti-Cr-Si, Ti-Mn-Si, Ti-Ni-Si and Ti-Cu-Si amorphous alloys are almost of the same order of Ti-M-Si(M=Fe, Co, V and Nb) alloys.  Further, the specific strength($\sigma_f/\rho$) of all the titanium-base amorphous alloys is in the range of 35-60 × $10^3$ m and is higher than those[20] of conventional high strength materials such as maraging steel.  The mean value of Hv/$\sigma_f$ is about 2.8, similar to those of a number of amorphous alloys[21].  This implies that the titanium-based alloys also possess a plastic-rigid behavior[22].

Tensile fracture occurred on the shear plane at 45-55 deg to the tensile axis in the direction of thickness, and the fracture surface consists of a smooth part produced by shear slip and a vein-like part produced by plastic instability as shown in Photo. 2, similar to the characteristics[23] of the fracture morphology for the late transition metal(Fe, Co, Ni)-metalloid amorphous alloys.

Further, the alloys possess a good bend ductility.  As an example, the deformation structure of $Ti_{55}Nb_{30}Si_{15}$ amorphous alloy bent completely by pressing against the edge of a razor blade is shown in Photo. 3.  Numerous deformation markings can be seen near the bent edge, and no cracks are observed even after such a severe deformation.

Ductile-brittle transition behavior for Ti-M-Si(M=Fe, Co, V and Nb) amor-

Table 1  Mechanical and thermal properties for representative titanium-based amorphous alloys

| Alloy system (at%) | Tx(K) | Hv(DPN) | $\sigma_f$(MPa) | $T_f$(K) | $\sigma_f/\rho \times 10^3$(m) | Hv/$\sigma_f$ (DPN/kg·mm$^{-2}$) |
|---|---|---|---|---|---|---|
| $Ti_{85}Si_{15}$ | 702 | 510 | 1960 | 670 | 47.8 | 2.6 |
| $Ti_{80}Si_{20}$ | 702 | 530 | 1910 | 660 | 47.8 | 2.7 |
| $Ti_{70}Cr_{20}Si_{10}$ | 892 | 590 | – | – | – | – |
| $Ti_{70}Mn_{20}Si_{10}$ | 826 | 580 | 2740 | 580 | 57.2 | 2.1 |
| $Ti_{70}Fe_{20}Si_{10}$ | 822 | 580 | 2170 | 620 | 44.6 | 2.6 |
| $Ti_{70}Co_{20}Si_{10}$ | 768 | 570 | 2110 | 660 | 41.6 | 2.7 |
| $Ti_{60}Co_{30}Si_{10}$ | 841 | 720 | 2610 | 580 | 47.4 | 2.7 |
| $Ti_{70}Ni_{20}Si_{10}$ | 742 | 510 | 2370 | – | 46.8 | 2.1 |
| $Ti_{70}Cu_{20}Si_{10}$ | 727 | 510 | 2080 | 630 | 40.9 | 2.4 |
| $Ti_{75}Co_{20}B_5$ | 674 | 470 | 1970 | 640 | 38.1 | 2.3 |
| $Ti_{65}Co_{30}B_5$ | 716 | 540 | 1980 | 610 | 35.3 | 2.7 |

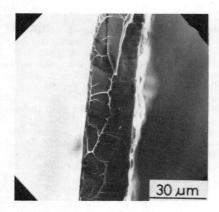

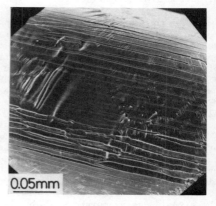

Photo. 2  Scanning electron micrograph showing the tensile fracture appearance of Ti$_{55}$Nb$_{30}$Si$_{15}$ alloy.

Photo. 3  Scanning electron micrograph showing deformation markings at the tip of Ti$_{55}$Nb$_{30}$Si$_{15}$ amorphous alloy bent through 180°.

phous alloys was examined as a function of the aging time and temperature. As an example, their embrittlement behavior during isochronal aging for 1 h is shown in Fig. 4, wherein the results[8] of Fe$_{80}$P$_{13}$C$_7$ and Fe$_{78}$Si$_{10}$B$_{12}$ amorphous alloys are also shown for comparison. The strain on the outer surface required for fracture, $\varepsilon_f$, is estimated from the equation $\varepsilon_f = t/(2r-t)$, where r is the radius of curvature of bent sample at fracture and t is the thickness of the ribbon specimen. The temperature for starting of embrittlement is about 700 K for Ti-Co-Si and Ti-Nb-Si alloys and is much higher than that(570K) for the iron-based amorphous alloys. Judging from the crystallization temperatures of these amorphous alloys and the difference in the thermal history for measuring Tx and T$_f$, it is presumed that these alloys remain ductile until the precipitation of crystalline phases. On the other hand, T$_f$ for Ti-M-Si(M=V, Cr, Mn and Fe) alloys is in the range of 580 to 620 K as shown in Fig. 4 and Table 1, indicating the tendency that the replacement of titanium by the middle transition metals in the periodic table such as Cr, Mn and Fe etc. decreases T$_f$. Thus, the change in T$_f$ by mixing depends largely on the kind of constituent elements.

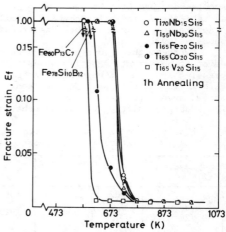

Fig. 4  Change in fracture strain by annealing for 1 h at various temperatures for Ti$_{85-x}$M$_x$Si$_{15}$ (M=Fe, Co, V and Nb) amorphous alloys. Data for Fe$_{80}$P$_{13}$C$_7$ and Fe$_{78}$Si$_{10}$B$_{12}$ amorphous alloys are also included for comparison.

From Fig. 2 (a)-(d) and Table 1, it is noticed that the hardness values increase with the amount of alloying elements. Figure 5 shows the change in Hv for $Ti_{85-x}M_xSi_{15}$(M=V, Cr, Mn, Fe, Co, Ni, Cu, Zr, Nb and Ta) amorphous alloys with the amount of M elements. When compared at 10 at% concentration of the M elements, the effectiveness of alloying elements on the increase of Hv becomes large in the order of Zr, Cu, Ni, Nb, Cr, V, Ta, Co, Fe and Mn. Among the elements(V, Cr, Mn, Fe, Co, Ni and Cu) which belong to the same periodicity, the effectiveness increases with decreasing average outer electron concentration(e/a) of metallic atoms in amorphous alloys except for $Ti_{75}V_{10}Si_{15}$ and $Ti_{75}Cr_{10}Si_{15}$. This tendency agrees well with the previous results[24,25] that the smaller the values of e/a the larger is Hv. This result indicates that a strong chemical bonding between metal and metalloid atoms plays an important role in determining the hardness of the amorphous alloys. Further, it is noted that the addition of refractory metals such as Zr, Nb and V is ineffective for increasing Hv. As a result, the data of $Ti_{85-x}M_xSi_{15}$(M=V, Zr, Nb and Ta) alloys deviate largely from the above-described tendency. The reason for such a deviation may be due to the following two facts: (1) Zr, Nb and Ta belong to different periods and have a much larger atomic size. The increase of Hv for $Ti_{85-x}M_xSi_{15}$(M=Zr, Nb and Ta) amorphous alloys seems to be mainly due to the solid solution hardening because of a large difference in atomic size. Hence, this result suggests that the effect of atomic size can not be neglected for an understanding of the effect of alloying elements which belong to the different periodicity. The importance of atomic size in determining Hv has been pointed out for other amorphous alloys such as (Fe, Ni)-X-$Si_{10}B_{12}$(X=Cr and Nb)[26], Cr-X-C(X=Fe, Co, Ni, Nb, Mo and Ta)[25] and (Fe, Co, Ni)-X-C(X=Cr, Mo and W)[27-29]. (2) The equilibrium phase diagrams of Ti-M(M=V, Zr, Nb and Ta) binary alloys are of the isomorphous type and are different from a eutectic-type for Ti-M(M=Mn, Fe, Co, Ni and Cu) binary alloys[17, 18]. This difference suugests that the compound-forming ability for the former alloy systems is weak compared with that for the latter alloy systems. This may be the reason why V, Zr and Nb atoms showed a low effectiveness on the increase of Hv, in spite that these elements possess a much higher melting temperature.

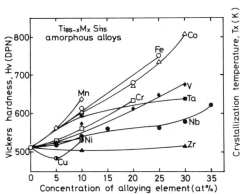

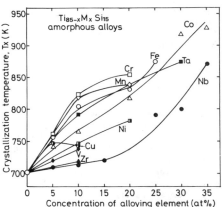

Fig. 5  Change in Vickers hardness(Hv) of $Ti_{85-x}M_xSi_{15}$(M=V, Cr, Mn, Fe, Co, Ni, Cu, Zr, Nb and Ta) amorphous alloys with concentration of the alloying element(M).

Fig. 6  Change in crystallization temperature(Tx) of $Ti_{85-x}M_x-Si_{15}$(M=V, Cr, Mn, Fe, Co, Ni, Cu, Zr, Nb and Ta) amorphous alloys with concentration of the alloying element(M).

3.  Thermal stability

The change of Tx for $Ti_{85-x}M_xSi_{15}$(M=V, Cr, Mn, Fe, Co, Ni, Cu, Zr, Nb and Ta) amorphous alloys by the replacement of titanium with M element is shown in Fig. 6. Also, the data of Tx for $Ti_{90-x}M_xSi_{10}$(M=Cr, Mn, Fe, Co, Ni and Cu) and $Ti_{95-x}Co_xB_5$ amorphous alloys are represented in Table 1. In the figure and table, it is seen that Tx increases with increasing content of the alloying elements(M) and metalloid(Si or B), similar to the composition dependence of Hv, and attains a value as high as 930 K. The effectiveness of alloying elements on the increase of Tx becomes large in the order of Zr, Nb, V, Cu, Ni, Co, Ta, Fe, Mn and Cr. When compared among the elements belonging to the same period, there is a tendency that the effectiveness increases with decreasing e/a of metallic atoms except for $Ti_{75}V_{10}Si_{15}$. This result also indicates that a strong chemical bonding between metal and metalloid atoms plays an important role in determining the thermal stability as well as the mechanical properties of the amorphous alloys.

## Summary

Titanium-based amorphous alloys possessing high strength and good bend ductility have been found in binary Ti-Si system and in ternary systems of Ti-M-Si and Ti-M-B(M=V, Cr, Mn, Fe, Co, Ni, Cu, Zr, Nb and Ta) by using the rapid quenching apparatus designed for high melting alloys. Specimens were produced in the form of a continuous ribbon of 1-2 mm width and 0.02-0.04 mm thickness. Hardness(Hv) increased with increasing content of the transition metal(M) or metalloid(Si or B) and attained a value as high as 810 DPN. Fracture strength ($\sigma_f$) was in the range of 1870-2750 MPa and these alloys possessed high specific strength values($\sigma_f/\rho$) of 35-60 × $10^3$ m. Effectiveness of alloying elements on increasing Hv increased in the order of Cu, Ni, Co, Fe and Mn and the hardness was closely related to the average outer electron concentration(e/a) of metallic atoms in the amorphous alloys. This indicates that a chemical bonding between metal and metalloid atoms plays an important role in determining the hardness of the amorphous alloys.

In conclusion, these amorphous alloys may be of interest as high strength materials because they combine desirable mechanical and thermal properties with a low density. In addition, the easy formation of continuous amorphous tapes with highly ductile nature suggests the prospect of alternative methods for the production of high strength titanium materials.

## References

1.  For example, T. Masumoto and R. Maddin: Acta Met., 19(1971), 725.
2.  For example, T. Minemura, A. Inoue, Y. Kojima and T. Masumoto: Met. Trans., in press; Tetsu to Hagane, 65(1980), No. 2, in press.
3.  T. Masumoto: Sci. Rep. Res. Inst. Tohoku Univ., A26(1977), 246.
4.  D. E. Polk, A. Calka and B. G. Giessen: Acta Met., 26(1978), 1097.
5.  L. E. Tanner and R. Ray: Scripta Met., 11(1977), 783.
6.  L. E. Tanner: Scripta Met., 12(1978), 703.
7.  C. Suryanarayana, A. Inoue and T. Masumoto: Proceedings of this Conference.
8.  A. Inoue, T. Masumoto and H. M. Kimura: J. Japan Inst. Metals, 42(1978), 303; Sci. Rep. Res. Inst. Tohoku Univ., A27(1979), 159.
9.  A. Inoue, H. M. Kimura, T. Masumoto, C. Suryanarayana and A. Hoshi: Phys. Rev. B, to be published.
10. A. Inoue, C. Suryanarayana, T. Masumoto and A. Hoshi: Trans. JIM, to be published.
11. A. Inoue, C. Suryanarayana, T. Masumoto and A. Hoshi: Sci. Rep. Res. Inst. Tohoku Univ., A28(1980), in press.
12. A. Inoue, T. Masumoto, C. Suryanarayana and A. Hoshi: to be presented at

the 4th Intern. Conf. on Liquid and Amorphous Metals, Grenoble, France, July 1980.

13.  L. A. Davis, R. Ray, C. P. Chou and R. C. O'Handley:  Scripta Met., 10 (1976), 541.

14.  A. Inoue, T. Masumoto, M. Kikuchi and T. Minemura:  J. Japan Inst. Metals, 42(1978), 294;  Sci. Rep. Res. Inst. Tohoku Univ., A27(1979), 127.

15.  A. Inoue, A. Kitamura and T. Masumoto:  Trans. JIM, 20(1979), 404.

16.  D. Turnbull:  J. Phys. (Paris) Colloq., 35(1974), 1.

17.  C. J. Smithells:  Metals Reference Book, 5th Edition, Butterworths, (1976).

18.  Metals Handbook, 8th Edition, Structures and Phase Diagrams, ASTM (1973).

19.  Metals Databook, The Japan Inst. Metals, (1974), p. 168, Maruzen, Japan.

20.  Metals Databook, The Japan Inst. Metals, (1974), p. 119, Maruzen, Japan.

21.  A. Inoue, T. Masumoto, S. Arakawa and T. Iwadachi:  Rapidly Quenched Metals III, Ed. B. Cantor, Vol. 1, p. 265, The Metals Society, London (1978).

22.  R. Hill:  The Mathematical Theory of Plasticity, (1967), p. 213, Oxford University Press, London.

23.  T. Masumoto and H. R. Kimura:  J. Japan Inst. Metals, 39(1975), 133.

24.  M. Naka, S. Tomizawa, T. Watanabe and T. Masumoto:  Proc. of 2nd Intern. Conf. on Rapidly Quenched Metals, Section I, p. 273, (1976), MIT press.

25.  A. Inoue, S. Sakai, H. M. Kimura and T. Masumoto:  Trans. JIM, 20(1979), 255.

26.  I. W. Donald and H. A. Davis:  Rapidly Quenched Metals III, Ed. B. Cantor, Vol. 1, p. 273, The Metals Society, London (1978).

27.  A. Inoue, T. Iwadachi and T. Masumoto:  Trans. JIM, 20(1979), 76.

28.  A. Inoue, T. Naohara and T. Masumoto:  Trans. JIM, 20(1979), 330.

29.  A. Inoue, T. Naohara, T. Masumoto and K. Kumada:  Trans. JIM, 20(1979), 577.

# SOLID-SOLUTION STRENGTHENING OF ALPHA TITANIUM ALLOYS

Hisaoki Sasano and Hirozo Kimura

National Research Institute for Metals,
Tokyo 153, Japan

## Introduction

The mechanical properties of substitutional alpha solid-solution titanium alloys have been extensively investigated. The specimens used in these studies contained different amount of interstitial impurities each other. Since the mechanical properties of alpha titanium are significantly affected by the interstitial impurity content, it is difficult to compare the data of an investigator with those of anothers. Collings and co-workers have examined systematically the strength and other physical properties of Ti-Al, Ti-Ga, Ti-Sn and Ti-Mo systems, and they have discussed solid-solution strengthening in titanium alloys from an electronic view point (1,2).

In the present work, the mechanical properties of alpha solid-solution titanium alloys of nine binary systems were systematically examined. To clarify the mechanism of the substitutional solid-solution strengthening of alpha titanium, the interrelations between the solid-solution strengthening rate and atomic misfit, Young's modulus and electrical resistivity were examined.

## Experimental Procedures

Alloys used in this work are given in Table 1. The concentrations of the substitutional alloying elements were within the solubility limit of alpha phase at 900K, and the specimens contained almost the same amount of interstitial impurities. Button ingots weighing 70 g were melted in an arc furnace with an unconsumable electrode in an argon atmosphere. The sheets 0.4 mm thick were fabricated from the buttons by rolling. The sheet specimens were annealed in vacuum at temperatures from 900K to 1100K to have the same grain size of about 16 μm.

Table 1  Chemical analysis of the specimens.

| Alloying element | Nominal composition (at%) | | | | | | Interstitial impurities* (wt ppm) | | |
|---|---|---|---|---|---|---|---|---|---|
| | 0.25 | 0.5 | 1 | 2 | 4 | 8 | O | N | C |
| Commercial purity titanium | — | — | — | — | — | — | 840 | 87 | 110 |
| Zr | 0.19 | 0.45 | 0.92 | 1.78 | 3.96 | 8.32 | 560 | 26 | 140 |
| Hf | 0.17 | 0.45 | 0.89 | 1.89 | 4.07 | 8.34 | 580 | 31 | 140 |
| Al | 0.23 | 0.49 | 0.91 | 2.03 | 4.10 | 8.36 | 550 | 38 | 110 |
| In | 0.24 | 0.47 | 1.13 | 2.11 | 3.81 | 7.83 | — | — | — |
| Sn | 0.22 | 0.47 | 1.07 | 1.87 | 3.73 | 8.05 | — | — | — |
| Nb | — | — | — | — | — | — | — | — | — |
| Ta | — | — | — | — | — | — | — | — | — |
| V | — | — | — | — | — | — | 1100 | 66 | 110 |
| Ag | — | — | — | — | — | — | 680 | 62 | 130 |

\* Values for 2 at% alloys.    — These were not analysed.

The tensile tests were carried out by using an Instron type machine, in the temperature range from 77K to 873K at a strain rate of $2.2 \times 10^{-4} \text{sec}^{-1}$. The gauge length and width of the tensile specimen were 15 mm and 4 mm, respectively. The strain rate change tests were also carried out to measure the activation volume.

Young's modulus data at 300K were obtained by utilizing the measurement of resonance frequency of a simple vibrational mode with an "Elastmat".

Measurements of electrical resistivity were made by measuring the potential drop across the sample in liquid nitrogen. The specimens used had the size of about 5 mm width and 100 mm length cut out from the 0.4 mm thick sheets.

## Results

The mechanical properties of alpha titanium are affected by deformation twinning below room temperature (3) and also affected by dynamical strain aging phenomena from 500K to 700K (4). The effects were different from alloy to alloy. Therefore, the solid-solution strengthening rates adopted here were those obtained from the yield strength at 300K.

Influences of solute species and concentration on the yield strength at 300K are shown in Fig. 1. The yield strength of any systems except for Ti-Al increased almost linearly with the solute concentration. In the Ti-Al system, alloy softening occured at up to about 0.5at%. The alloy softening in the system was observed by Sakai and Fine (5). The solid-solution strengthening rates were derived from the slope between 0.5at% alloy and the highest concentration alloy in each system. The solid-solution strengthening abilities of Sn and Ag were largest, and the abilities decreased in the sequence Al, In, V, Zr, Hf, Nb, Ta. The sequence was also held at 873K, where the yield strength was almost entirely occupied by its athermal component in titanium alloys, as shown in Fig. 2. The solid-solution strengthening rates discussed hereafter are those of 300K.

Typical dislocation structures of Ti-4at%Zr and Ti-4at%Al extended about 3 pct at 300K are shown in Fig. 3. There was only a little difference in configurations and density of dislocations in all the alloys. The majority of the dislocations had screw character. Ti-Al and Ti-Zr shown as examples exhibited the largest difference in the dislocation structure each other.

Activation volume measurements were performed by measuring the stress change $\sigma_1 - \sigma_2$ upon changing the strain rate between $\dot{\varepsilon}_1 = 2.2 \times 10^{-4} \text{ sec}^{-1}$ and $\dot{\varepsilon}_2 = 2.2 \times 10^{-3} \text{ sec}^{-1}$, using the following equation

$$V^* = kT \ln(\dot{\varepsilon}_1/\dot{\varepsilon}_2)/(\sigma_1 - \sigma_2) \tag{1}$$

where k and T are Boltzmann's constant and absolute temperature respectively. The values at 300K of the various alloys are tabulated in Table 2, as multiplies of $b^3$. There was little difference from alloy to alloy.

Table 2   Activation volume $V^*$ in $b^3$ at 300K of various alloys.

| Pure titanium | 4at%Zr | 4at%Hf | 4at%Al | 4at%In | 4at%Sn | 1at%Nb | 1at%Ta | 1at%V | 1at%Ag |
|---|---|---|---|---|---|---|---|---|---|
| 32 | 34 | 32 | 36 | 31 | 33 | 33 | 34 | 34 | 35 |

The lattice constants of 2at%, 4at% and 8at% alloys were measured by X-ray diffraction. The lattice constants a and c linearly varied with increasing solute content. The relative size factors of the solute atoms were estimated from the volume change of the unit cell with alloying. The results obtained are listed in Table 3. Since the lattice constant change was small in Ti-4at%Ta Ti-4at%Ag, Ti-2at%Nb and Ti-2at%V alloys, the listed values for the solute atoms

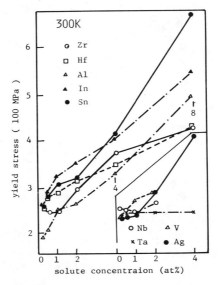

Fig. 1   Influences of solute species and the concentration on the yield strength at 300K.

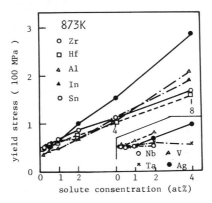

Fig. 2   Influences of solute species and the concentration on the yield strength at 873K.

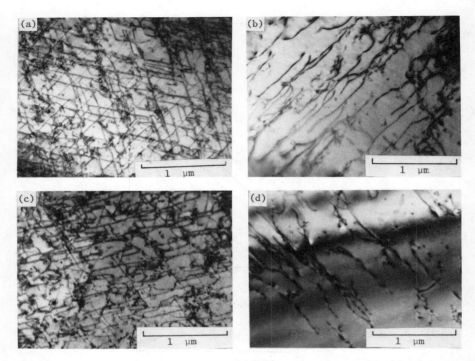

Fig. 3  Dislocation structures of (a) (b) Ti–4at%Al and (c) (d) Ti–4at%Zr,
extended about 3 pct at 300K.
(a) (c)   Incident beam // [0001]
(b) (d)   Incident beam // [11$\bar{2}$0]

are quoted from the table by Laves (6).

Table 3   Relative size factors of the substitutional atoms with titanium.

| Solute atom | Zr | Hf | Al | In | Sn | Nb | Ta | V | Ag |
|---|---|---|---|---|---|---|---|---|---|
| Atomic misfit (%) | 11.9 | 11.3 | -3.8 | 6.9 | 6.3 | 1.4* | 0.7* | -6.2* | -0.7* |

* After F. Laves, " Theory of Alloy Phases ", 1954, ASM.

Young's moduli of various 2at% alloys are tabulated in Table 4.  Young's moduli of Ti-Al alloys increased apparently with increasing Al content, as had been reported by Ogden and co-workers (7).  The addition of Sn, V and In reduced the modulus apparently.  In the other alloys except for Ti-Ta, the modulus reduced a little with alloying.

Table 4   Young's moduli of various 2at% alloys.  ($10^{11}$dyne/cm$^2$)

| Pure titanium | 2at%Zr | 2at%Hf | 2at%Al | 2at%In | 2at%Sn | 2at%Nb | 2at%Ta | 2at%V | 2at%Ag |
|---|---|---|---|---|---|---|---|---|---|
| 10.51 | 10.42 | 10.46 | 10.69 | 10.29 | 10.13 | 10.43 | 10.58 | 10.19 | 10.46 |

The composition dependencies of the electrical resistivity were measured. In all the alloys the resistivity increased linearly with the concentration of the solute.  The resistivity increasing rates are listed in Table 5.  The resistivity increasing rates by non-transition metal elements; Ag, Sn, In and Al, were an order of magnitude higher than those by transition metal elements; Zr, Hf, Nb, Ta and V.  These results agree with the results obtained by Collings and co-workers (1,2).

Table 5   Electrical resistivity increasing rate with alloying.

| Element | Zr | Hf | Al | In | Sn | Nb | Ta | V | Ag |
|---|---|---|---|---|---|---|---|---|---|
| $\Delta\rho/\Delta C$ ($\mu\Omega$-cm/at%) | 2.25 | 1.88 | 14.40 | 15.88 | 17.61 | 3.11 | 2.65 | 3.51 | 19.44 |

Fig. 4 shows the relation between the solid-solution strengthening rate and Young's modulus change with alloying.  There is no systematic correlation between these properties.

The relation between the solid-solution strengthening rate and the electrical resistivity change is shown in Fig. 5.  The addition of non-transition elements ( solid symbols ) apparently strengthens alpha titanium and also largely increases the resistivity, while the addition of transition elements ( open symbols ) affects only a little.

Fig. 6 shows the relation between the strengthening rate and atomic misfit. It is found that roughly linear correlation exists only among the alloys added the transition element, whereas there is no correlation among the alloys added the non-transition element.

## Discussion

There was little difference in the density and configuration of dislocation from alloy to alloy.  The almost same values of about 35 b$^3$ of activation volume were obtained in all the alloys.  These results indicate that the deformation mechanism of the alpha titanium alloys is not much different from commercial purity titanium.  Therefore, the following discussion is restricted to the resistance for dislocation motion due to a solute atom.

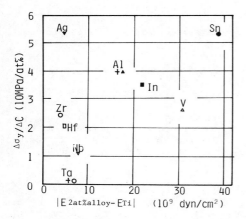

Fig. 4  Relation between strengthening rate and Young's modulus change with alloying.

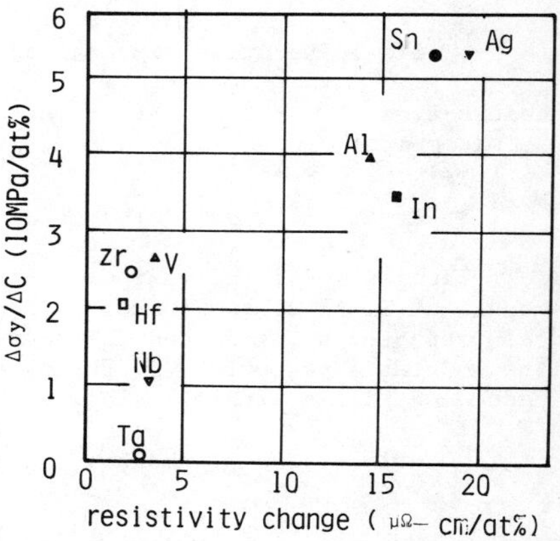

Fig. 5  Relation between strengthening rate and electrical resistivity change with alloying.

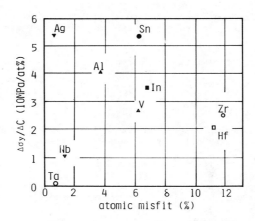

Fig. 6  Relation between strengthening rate with alloying and atomic size misfit.

Several theoretical models have been proposed to explain the effect of substitutional solute on the strength of alloys in terms of dislocation-solute interactions. Mott and Nabarro suggested the model based on the difference in atomic size between solute and solvent atoms (8). Fleisher (9), Takeuchi (10) and Labusch (11) extended the model, adding the effect of shear modulus misfit. Cottrell and co-workers (12) and Sugiyama (13) proposed the strengthening by an electrical interaction of an edge dislocation and a solute atom. In the results of the present work, there is the best correlation between the solid solution strengthening rate and the electrical resistivity change with alloying. And there is no correlation between the strengthening rate and the Young's modulus, and little correlation exists between the strengthening rate and the atomic misfit except for the case of the transition metal element solute. Therefore, the solid-solution strengthening in alpha titanium cannot be regarded to be determined by the shear modulus effect and the atomic size effect. The strengthening must be concerned with the electronic properties of the alloys.

Collings and co-workers first pointed out that there is the good correlation between the solid-solution strengthening rate and the electrical resistivity in alpha titanium alloys (1,2). They suggested from a standpoint of electronic bonding mechanism that the strengthening is related to a pairwise interaction parameter in the regular solution model defined by Fowler and Guggenheim (14). However, to the present authors, it seems that the strength of bonding must reflect to the elastic modulus. The Young's modulus of Ti-Sn alloys, which are strengthened apparently by alloying, decreases with alloying, whereas that of Ti-Al alloys, which are also strengthened apparently with alloying, increases. Therefore, it will be considered that the electrostatic interaction between an edge dislocation and a solute atom is the dominant cause of the solid-solution strengthening of alpha titanium.

## Summary

In order to clarify the mechanism of the substitutional solid-solution strengthening of alpha titanium, the interrelation between the solid-solution strengthening rate and the atomic misfit, the Young's modulus and the electrical resistivity for nine binary alloy systems were examined. Results obtained are summarized as follows;
(1)  There is no systematic correlation between the strengthening rate and the Young's modulus change with alloying. Young's moduli of the alloys except for Ti-Al and Ti-Ta reduce with alloying.
(2)  The addition of the non-transition metal elements; Ag, Sn, Al and In, apparently strengthens alpha titanium and significantly increases the electrical resistivity, while that of the transition metal elements; Zr, Hf, V, Nb and Ta, slightly affects the strength and the resistivity.
(3)  The roughly linear correlation exists between the strengthening rate and the atomic size misfit only among the alloys containing the transition metal elements.
(4)  The above results probably lead to the conclusion that the electrostatic interaction between an edge dislocation and the solute atom is the dominant cause of the solid-solution strengthening of alpha titanium alloys and the little strengthening is also caused by the elastic interaction due to atomic size misfit.

## References

1.  E. W. Collings, J. E. Enderby, H. L. Gegel and J. C. Ho; *Titanium Science and technology*, 2(1973), 801. Plenum Press, New York.

2.   E. W. Collings, H. L. Gegel and J. C. Ho;   US Air Force Technical Report, AFML-TR-72-171(1972).
3.   H. Sasano and H. Kimura;   J. Japan Inst. Metals, 41(1977), 933.
4.   H. Sasano and H. Kimura;   *Proceedings of 3rd International Conference on Titanium,* to be published.
5.   T. Sakai and M. E. Fine; Acta Met., 22(1974), 1359.
6.   F. Laves;   *Theory of Alloy phases,* (1954), ASM.
7.   H. R. Ogden, D. J. Maykuth, W. L. Finlay and R. I. Jaffee;   Trans. AIME, 197(1953), 267.
8.   N. F. Mott and F. R. N. Nabarro;   Proc. Phys. Soc. (GB), 52(1940), 86.
9.   R. L. Fleischer;   Acta Met., 9(1961), 996.
10.   S. Takeuchi;   J. Phys. Soc. Japan, 27(1969), 929.
11.   R. Labusch;   Phys. Status Solidi, 41(1970), 659.
12.   A. H. Cottrell, S. C. Hunter and F. R. N. Nabarro;   Phil. Mag., 44(1953), 1064.
13.   A. Sugiyama;   J. Phys. Soc. Japan, 21(1966), 1873.
14.   R. Fowler and E. A. Guggenheim;   *Statistical Thermodynamics,* (1939), Cambrid Univ. Press, London.

# MICROSTRUCTURE AND MECHANICAL PROPERTIES OF A LARGE Ti-6Al-4V RING FORGED AT THE DUPLEX PHASE TEMPERATURE RANGE

Michio Hanaki and Yoshinori Fujisaki

Technical Research Center, Nippon Mining Co., LTD.
Saitama, Japan

Touru Ishiguro and Masafumi Miyashita

Muroran Plant, The Japan Steel Works LTD.
Muroran, Japan

## Introduction

Ti-6Al-4V, which is the most extensively used among the various titanium alloys, is composed of two phases, $\alpha$ and $\beta$, at the temperature up to about 1000°C, and becomes a single phase $\beta$ above this temperature. The forgeability of this alloy depends strongly upon the consisting phases; i.e. it becomes deteriorated in the duplex phase region and excellent in the single region. The forging at the single phase region, however, is not favorable because of the unsatisfactory ductility resulted from coarse prior $\beta$ grains. This alloy, therefore, must be thermomechanically treated at the duplex phase region in order to improve its mechanical properties. In this work, a large Ti-6Al-4V ring forged from 3-ton ingot at the duplex phase temperature range was studied from respect to its microstructure and mechanical properties.

## Preliminary Study of Hot Working

As a preliminary study in order to obtain proper forging conditions which will result in the equiaxed micro-duplex structure, hot tensile tests over the temperature range of 800-1200°C were conducted. The forged bars, its diameter of 30 mm, were used in the study and its chemical compositions are tabulated in Table 1. The tensile tests were conducted at the strain rate of $3.3 \times 10^{-4}$

Table 1. Chemical composition of Ti-6Al-4V used in the preliminary study. ( % )

| Al | V | Fe | O | H | C | N | Ti |
|------|------|------|-------|--------|-------|-------|------|
| 6.32 | 4.11 | 0.20 | 0.138 | 0.0032 | 0.016 | 0.008 | bal. |

sec$^{-1}$, 3.3 x 10$^{-3}$sec$^{-1}$, 3.3 x 10$^{-2}$sec$^{-1}$ . These specimens were β annealed of 1050 °C for 1 hr in order to uniformalize its microstructures.

Fig. 1 shows the effect of temperature on the elongation and the maximum true flow stress, which are plotted in the relation to the reciprocal of the absolute test temperature for considering the elevated temperature deformation process to be the heat activation process. These data indicate that the flow stress of Ti-6Al-4V increase largely at the temperature below β transus and its deformation ductility becomes deteriorated.

Fig. 2 shows the effect of the strain rate on the elongation and the maximum true flow stress at 950°C in α+β temperature range and 1050°C in β temperature range. Generally. there is a relationship $\sigma = k\dot{\varepsilon}^m$ between stress $\sigma$ and strain rate $\dot{\varepsilon}$. The strain rate sensitivity index m is 0.49 in α+β temperature range and 0.24 in β temperature range. It is suggested, therefore, for the flow stress to be sensitively affected by strain rate in α+β temperature range.

Fig. 3 shows the effect of deformation strain (hot tensile strain) on the fraction of the equiaxed α area. The test results are similar almost over the test temperature, and the deformation tensile strain over 10 % results in partly recrystallization and the deformation tensile strain over 100 % is necessary in order to obtain the equiaxed grain completely.

Fig. 4 shows the effect of the holding time at α+β forging temperature (950°C) on the grain size and the mechanical properties after working (forging ratio: 2.25) at 950°C. From the results, it is not recognized a rapid grain growth and the large variation of mechanical properties. It is, therefore, appeared that the frequent reheating of the work piece is allowed.

## Ring Manufacturing Procedure

On the basis of the preliminary study, the large ring of Ti-6Al-4V was forged. The 3-ton ingot was melted by the double consumable electrode arc melting method and its chemical compositions are tabulated in table 2. Then the ingot was forged and punched in the β phase region with soaking temperature of 1150°C in order to break down coarse solidification structures. The structure refinement due to the hot working was achieved during the enlarging process by mandrel forging. A 8000-ton hydraulic forging press was used for

Table 2.   Chemical composition of Ti-6Al-4V used in the ring manufacturing.                    ( % )

| Al | V | Fe | O | H | C | N | Ti |
|------|------|------|-------|--------|-------|-------|------|
| 6.29 | 4.11 | 0.18 | 0.130 | 0.0040 | 0.010 | 0.008 | bal. |

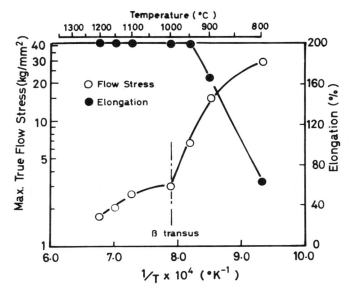

Fig. 1  The effect of temperature on the elongation and the maximum true flow stress.

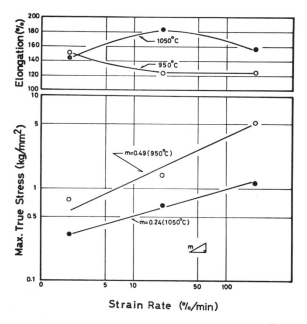

Fig. 2  The effect of strain rate on the elongation and the maximum true flow stress.

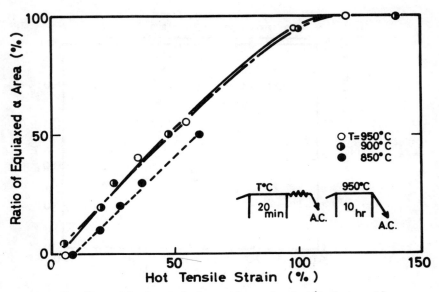

Fig. 3   The effect of deformation strain (hot tensile
strain) on the fraction of the equiaxed α area.

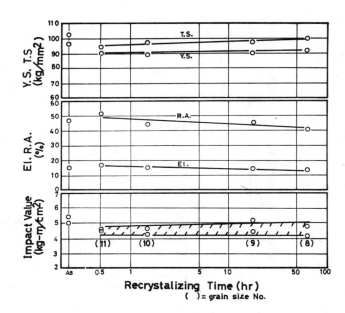

Fig. 4   The effect of the holding time at the α+β
forging temperature on the grain size and
mechanical properties after working at
950°C.

enlargement. The finished shape of this ring is shown in Fig. 5. Metallurgical and mechanical tests were conducted on this ring with regard to tensile properties, impact toughness and fracture toughness.

## Microstructures and mechanical properties

Fig. 6 shows the distribution of mechanical properties of axial direction. The strength tends to decrease slightly from near surface section to internal section. Impact properties are found not to change almost over the cross section. The microstructure distribution of radial direction of the ring is almost uniform and the equiaxed grain exsists in small amount of transformed β matrix as shown in Fig. 7.

The ring machined 40mm in thickness was heat-treated. Table 3 shows the influence of microstructures and aging temperatures on tensile properties and fracture toughness. Fracture toughness of recrystallized equiaxed structures are found to be excellent than that of recrystallized elongated structures in which grain boundary α was generated as shown in Fig. 8. It is proved that grain boundary α has worse fracture toughness, elongation and reduction of area. In summarizing the study it is apparent that the more desirable properties were obtained with specimens heat-treated 640°C for 3 hr after 950°C for 2 hr WQ.

Table 3.   Mechanical properties of the Ti-6Al-4V ring.

| Aging Temperature (°C x 3 hr) | Tensile Properties at R.T. | | | | $K_{IC}$ (kg/mm$^2$.mm$^{\frac{1}{2}}$) |
|---|---|---|---|---|---|
| | Y.S. (kg/mm$^2$) | T.S. (kg/mm$^2$) | El. (%) | R.A. (%) | |
| Equiaxed α* 550 | 90.9 92.2 | 102.1 99.5 | 16.0 12.8 | 37.7 39.4 | 230.4 235.8 |
| 640 | 92.2 97.6 | 101.8 107.8 | 16.0 14.4 | 40.0 44.8 | 238.6 |
| 730 | 87.0 88.3 | 99.0 99.8 | 15.8 15.6 | 35.5 38.3 | 266.7 |
| Elongated α* 550 | 92.9 92.0 | 102.7 101.9 | 10.0 10.3 | 30.9 31.3 | 207.6 |
| MIL Spec. (40mm thickness) | 91 | 98 | 12 | 20 | – |

* Solution heat treatment was conducted as below
    Equiaxed α  : 950°C for 2 hr WQ
    Elongated α : 1050°C for 1hr FC to 950°C for 2 hr WQ

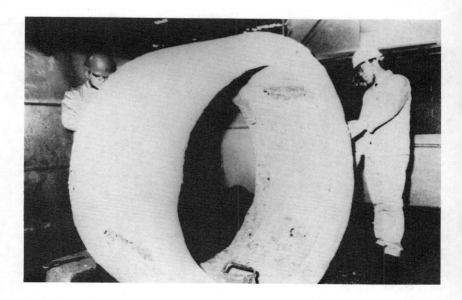

Fig. 5   The finished shape of the Ti-6Al-4V ring.

```
outer diameter : approx. 1500mm
inner diameter : approx. 1200mm
    height      : approx.  500mm
```

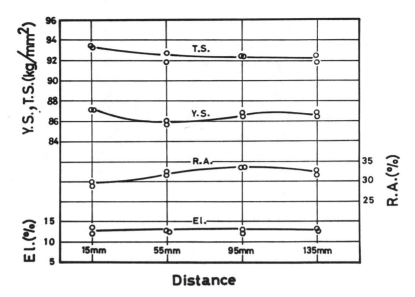

Fig. 6   The distribution of mechanical properties
of axial direction in the Ti-6Al-4V ring.

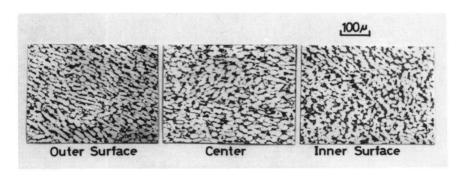

Fig. 7   The microstructures at three locations of
the Ti-6Al-4V ring.

$100\mu$

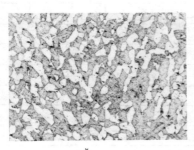
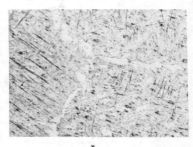

Equiaxed α\*aged at 550°C      Elongated α\*aged at 550°C

Fig. 8   Microstructures of the heat-treated Ti-6Al-4V ring.
( \* see Table 3 )

## Summary

Based on the data generated, it is summarized that in order
to obtain equiaxed α grains and more desirable properties in Ti-
6Al-4V the hot working strain (tensile strain) over 100% at the
α+β temperature range is necessary. The temperature range from 800
°C to 950°C is considered to be favorable for the α+β forging temperature range, because the deformation resistance increases and
hot working ductility decreases at lower temperature below β
transus.
    Then the large ring of Ti-6Al-4V, its outer diameter of
approx. 1500mm and inner diameter of approx. 1200mm and height of
approx. 500mm, was forged successfully over the α+β temperature
range, and excellent mechanical properties and microstructures
were obtained.

# EFFECT OF TRANSFORMED BETA MORPHOLOGY AND OTHER STRUCTURAL FEATURES ON MECHANICAL PROPERTIES IN Ti-6Al-4V ALLOY

Wang Jin-you    Cao Chun-xiao    Shen Gui-qin

The Institute of Aeronautical Materials,
Beijing, China

## Introduction

For the quality control of titanium alloy products, the microstructures are so important that a lot of studies of the relations between microstructures and mechanical properties in titanium alloys have been made. However, some problems need to be further clarified, and some experimental results seemed to be contradictory to each other. For example, there were no systematic studies on the transformed beta morphologies in acicular structures in the past, and in structures containing equiaxed alpha, the effect of transformed beta morphologies was often neglected. Additionally, some papers gave evidence that the fatigue property of acicular alpha structure is lower than equiaxed alpha structure, but other references showed that the fatigue property of the former was similar to or better than that of the latter[1-7]. Hence, an attempt was made to find part of the internal relations, and it will be helpful to control and improve the quality of products.

## Materials and Specimens

The chemical compositions of alloy investigated is given in table I.

Table I. Chemical compositions (wt.%)

| Al | V | Fe | C | Si | $O_2$ | $N_2$ | $H_2$ | Ti |
|----|----|----|----|----|----|----|----|----|
| 6.39 | 4.13 | 0.07 | 0.03 | 0.04 | 0.10 | 0.023 | 0.006 | Bal. |

The Ti-6Al-4V rods, which presented different transformed beta morphologies, primary alpha percentages and effective beta grain sizes, were produced by means of different heating and hot rolling processes from an ingot. All specimens were machined from 20mm diameter rolled rods except fracture toughness specimens. All the rods had been annealed ($800°C/1hr$, AC), and subsequently comparing their microstructures and properties.

## Results and Discussions

Figure 1 illustrates some typical micrographs taken from

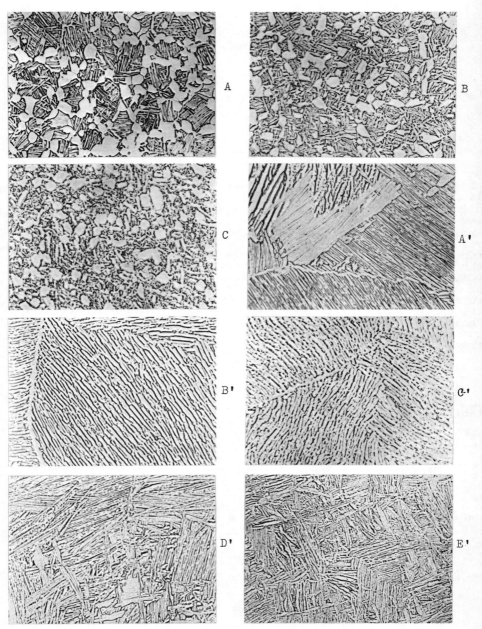

Fig.1 Typical microstructures with various transformed beta
morphologies (A–C in structures containing equiaxed alpha and
A'–E' in acicular structures) in Ti–6Al–4V.  x500.

specimens under different conditions. The changes of transformed
beta morphology (A → B → C in the structures containing alpha,
A'→ B'→ C' in the mainly aligned alpha structures and D'→ E' in
the basket-weave structures) exhibit a gradual reduction in straightness of acicular alpha, continuity of retained beta and integrity of grain boundary alpha, where the last one is only suited
to various acicular structures or the structures containing a few
primary alpha. Comparison of morphology between D' and E' also
shows that the latter possesses more and finer colonies of similar
orientated alpha platelets within a prior beta grain than the
former.

For acicular structures, the "effective beta grain" means "
prior beta grain", and for the structures containing equiaxed alpha, it may be considered as secondary effective recrystallized
beta grain" surrounded by primary alpha.

Fig. 2-7 show the experimental results. In general, the data
of various properties illustrated in these figures are mean
values obtained from 3-5 specimens.

Tensile strength

As shown in Fig.2, there exists a certain relationship
between tensile strength and primary alpha percentage. For example, for the same transformed beta morphologies, the structures
containing high percentage of primary alpha (such as 85%) will
possess the highest tensile strength. Fig.2 also shows that, for
structures containing identical percentage of primary alpha, the
more the reduction in straightness of acicular alpha and continuity of retained beta is (i.e. transformed beta morphology A- B
- C), the greater the improvement of tensile strength.

For acicular structures with the same prior beta grain size,
morphology A' possesses the best tensile strength (see Fig.3).
The tensile strength values of acicular structures are usually
higher than that of the structures containing equiaxed alpha when
the transformed beta structures are all undeformed (i.e. morphology A or A'). But the strength values of the former are usually similar to or lower than that of the latter when the transformed beta structures are all deformed (i.e. morphology B,B' or
C,C'), as shown in Fig.2 and 3.

Tensile ductility

Fig.2 and 3 shows that tensile ductility is very sensitive
to microstructure. For example, when transformed beta morphologies are all A and the primary alpha percentages are lower than
20%, the tensile ductility decreases considerably with the reduction of primary alpha percentage. This result is in agreement .
with that obtained by Ashton and Chambers[6]. In the absence of
primary alpha, all the acicular structures with A morphology
exhibit significant room temperature brittleness which increases
with the prior beta grain size.

This experiment also shows that the tensile ductility not
only depends on equiaxed primary alpha percentage and effective
beta grain size, but also depends on transformed beta morphology.
The broken grain boundary alpha, distorted acicular alpha and
discontinuous retained beta suppress the nucleation and growth of

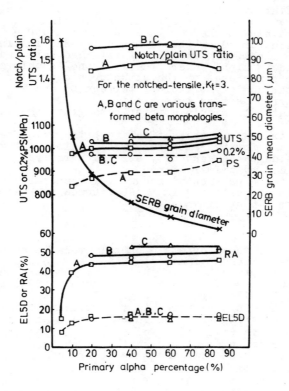

Fig.2 The effect of primary
alpha percentage and transformed beta morphology on tensile
property in Ti-6Al-4V.

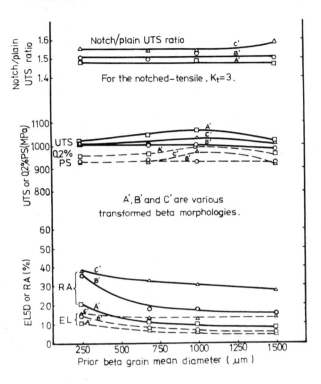

Fig. 3 The effect of prior beta grain
size and transformed beta morphology on
tensile property in Ti-6Al-4V.

voids, and also depress the formation of the weak path. All of these will improve tensile ductility.

Notched-tensile strength

According to reference[8], Ti-6Al-4V forgings produced from different heats show that the tensile strength ratio of the notched to the unnotched will be reduced if the primary alpha percentage is too high. However, our data obtained from the same heat show that this ratio isn't sensitive to primary alpha percentage or prior beta grain size, rather depends on transformed beta morphology consistently. Either the structures containing various percentages of equiaxed alpha or the acicular structures with different prior beta grain sizes, this ratio for morphology B or C is always higher than that for morphology A, and this ratio for morphology B' or C' is also higher than that for morphology A'. It may be suggested that the improvement of tensile ductility benefits the continuous adjustment of stress and strain distribution in the area near the notch.

High cycle fatigue (HCF)

Fig.4 shows the following general tendency: the coarser the effective beta grain is, the more the fatigue property decreases; the higher the primary alpha percentage is, the more the fatigue property increases. At the same time, the transformed beta morphology has a more significant influence upon the HCF. When the primary alpha percentage is 10%, the fatigue strength for morphology A is less than 490 MPa, but that for morphology C is about 570 MPa, which is superior to the structure containing any percentage of primary alpha and with A transformed beta morphology. The transformed beta morphology also presents pronounced effect in acicular structures. When the prior beta grain diameter is 250$\mu$m, the fatigue strength for morphology A' is only 390 MPa, which is much lower than equiaxed structure, but that for morphology C' is about 440 MPa, and that for morphology E' is further increased to 490 MPa, this is equal to or higher than that for the structures containing equiaxed alpha in some conditions.

Fig.4 also shows that when the prior beta grain size is too coarse in the mainly aligned alpha structures (such as an average diameter of 700-1500$\mu$m), though the room temperature tensile ductility may be increased by means of breaking grain boundary alpha, distorting the acicular alpha and separating the retained beta (i.e. changing the transformed beta morphology A' to B' or C'), the fatigue property cann't be improved, the HCF strength for morphology B' or C' is still on a very low level ($\rightarrow$310 MPa). This fact shows that the fatigue property sensibly depends on prior beta grain size in some conditions. Probably because of the different experimental conditions, Lucas' experimental results show that the influence of prior beta grain size upon the fatigue property isn't so distinct as ours(7).

It is known that for HCF, the fatigue crack initiation stage takes the majority of the total fatigue life. Just as Yan Ming-gao pointed out that the emphasis should be given on the study of the effect of metallographic feature on the crack

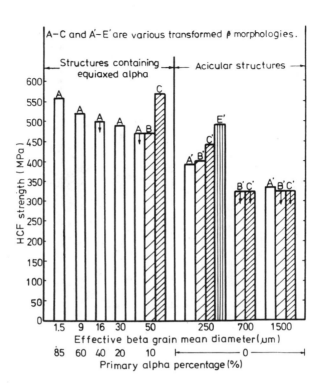

Fig.4 The effect of microstructural
features on high cycle fatigue strength
(10⁷ cycles) in Ti-6Al-4V. R=-1.

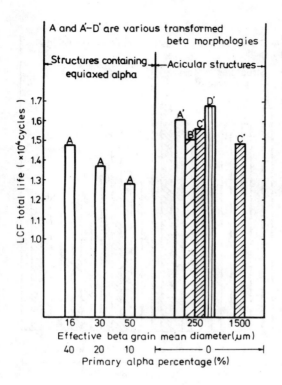

Fig.5 The effect of microstructural
features on low cycle fatigue life
in Ti-6Al-4V. R=0.1, $\sigma_{max}$=600MPa,
$K_t$=2.4, frequency 0.2Hz.

initiation stage(9). The HCF strength (>$10^7$ cycle) mainly depends on crack nucleation resistance of various microstructures, the refinement of grain, reduction of retained beta continuity and increase of equiaxed alpha percentage may be in favor of the mobility of dislocation and the dispersion of strain energy along slip bands., which may be considered as a contribution to the extension of nucleation stage.

Low cycle fatigue (LCF)

Fig.5 shows that LCF isn't so sensitive to the microstructure as HCF. But the tendency has been seen as follows: firstly, in the structures containing equiaxed alpha, the lower the equiaxed primary alpha percentage is, the shorter the LCF life; secondly, the LCF life of acicular structures (particularly the basket-weave structures) is generally longer than that of the structures containing 10-40% equiaxed alpha.

Since LCF is tested under high stress level, the crack propagation stage becomes predominant. The acicular structures which exhibiting tortuous crack propagation path and better fracture toughness may offset or even surpasses its disadvantage of low crack nucleation resistance. Therefore it's possible that the total LCF life of acicular structures is similar to or better than that of structures containing 10-40% equiaxed alpha.

Fracture toughness

The experimental results have shown that K1c is improved with decreasing percentage of equiaxed alpha as all of the transformed beta morphology are A (see Fig.6).

Experimental results have also shown that the transformed beta morphology presents very important effect too. Although the acicular structures with morphology A', D' or E' give better K1c than equiaxed structures, K1c values of the acicular structures with morphology B' or C' aren't better or even worse than that of the equiaxed structures.

Creep

Fig.7 shows that in the structures containing equiaxed alpha and as their transformed beta morphologies are all A, the creep resistance improves with decreasing the percentage of equiaxed alpha. But it's notable that, though the basket-weave structure presents superior creep resistance to the structure containing equiaxed alpha, the mainly aligned structure doesn't like so, whose creep resistance is similar to or even lower than that of the structure containing equiaxed alpha in some conditions.

## Conclusion

The tensile ductility, fatigue strength (HCF and LCF), fracture toughness, creep resistance and other properties of Ti-6Al-4V alloy not only depend on whether it's an acicular or an equiaxed structure, but also depend on primary alpha percentage, prior beta grain size, transformed beta morphology and other conditions. Among these conditions, the transformed beta

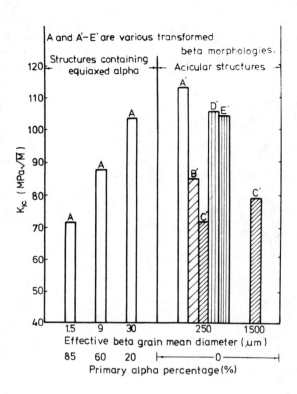

Fig.6 The effect of microstructural
features on fracture toughness in
Ti-6Al-4V.

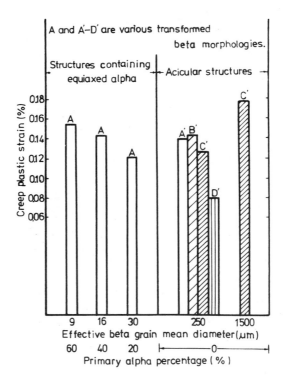

Fig.7 The effect of microstructural features on creep plastic strain( 400°C/29.5MPa/100hrs) in Ti-6Al-4V.

morphology is an especially important factor.

For a certain property, such as fatigue strength, an acicular structure is better than a structure containing equiaxed alpha in one condition, but the former may be similar to or poorer than the latter in another condition. Therefore, it's not correct to compare these two types of structures unconditionally.

The finer basket-weave structure exhibits the best combination of various mechanical properties among all structures mentioned above.

## Acknowledgement

The authors are very grateful to Prof. Yan Ming-gao for his valuable advice and suggestion.

## References

1. R. I. Jaffee: "Titanium Science and Technology", 3(1973), 1665. Plenum Press, New York.
2. J. E. Coyne: "The Science, Technology and Application of Titanium" (1970), 97. Pergamon Press, London.
3. C. M. Gilmore and M. A. Imam: "3rd International Conference on Titanium-Abstracts" (1976), 139.
4. M. A. Greenfield et al. :"Titanium Science and Technology", 3 (1973), 1731. Plenum Press, New York.
5. О. П. Солонина и С. Г. Глазунов : "Жаропрочные Титановые Сплавы" (1976). Москва.

6. S. J. Ashton and L. H. Chambers: "The Science, Technology and Application of Titanium" (1970), 879. Pergamon Press, London.
7. John J. Lucas: "Titanium Science and Technology", 3 (1973), 2081. Plenum Press, New York.
8. "Titanium-Alloy Forgings", OTS PB 151100, DMIC.
9. Yan Ming-gao : The Micromechanisms and Mechanics on the Fatigue Propagation in Metals, HKCL Tech, Reports, No.79-1-1 (1979).

# IMPROVED MECHANICAL PROPERTIES OF α+β Ti ALLOYS BY Pt ION PLATING

S. Fujishiro* and D. Eylon**

*Metals and Ceramics Division
Materials Laboratory, AFWAL
W-PAFB, OH 45433

**Metcut-Materials Research Group
P.O. Box 33511
W-PAFB, OH 45433

Platinum ion plating of high temperature titanium alloys was originally developed to increase their creep and oxidation resistance. Further studies of the effect of this coat on other properties showed that it can improve, also, the high cycle fatigue strength at room and especially at elevated temperatures. The purpose of this paper is to review the work that has been done by the authors in the area of titanium alloy Pt ion plating and the possible use of this type of coat in aerospace systems operating at high temperatures. The mechanisms of creep and fatigue resistance improvement are discussed in terms of surface impregnation, surface hardening and oxygen induced phase transformation. The present work also demonstrates that this coating method is technologically practical and provides the opportunity to extend the temperature range of existing titanium alloys.

## Introduction

Extensive effort to improve high temperature capability of Ti alloys has been undertaken by the aerospace industry. As a result, a number of high temperature Ti alloys including Ti-6246, Ti-6242, Ti-17, Ti-11, Ti-5522, Ti-5524, IMI-685 and IMI-829 have been developed. Some of these are currently used in gas turbine engines but the service temperatures are limited to approximately 450°C for extended service. This limitation is due mainly to the inherent vulnerability of Ti to oxidation at higher temperatures.

The properties required to qualify a Ti alloy for high temperature applications include high modulus to density ratio, high tensile and fatigue strengths, high creep resistance, and metallurgical and chemical stability at the working temperature and environment. Severe oxidation of Ti alloys causes deterioration of the mechanical properties, such as post-creep ductility and fatigue strength. As a result, a number of research programs have been conducted in an effort to improve the oxidation resistance by means of coatings applied by either electrochemical plating or thermal/plasma spray. These coatings included Al, Cr, Ni and their alloys or slurries. The major drawback of such coatings is a severe loss of high cycle fatigue strength [1,2]. Consequently, high temperature Ti alloy components coated by one of these methods cannot be used in fatigue critical applications.

The present work demonstrates that a very thin layer of ion plated noble metal protects the alloy from oxidation and, in addition, improves

both high temperature fatigue and creep strengths.  Advantages of ion plating over conventional coating techniques are two-fold.  First, the implanted
elements become integrated with the substrate alloy surface over which a
coherent plating can subsequently be built up.  Secondly, ion impingement
by the coating materials may produce surface hardening [3] which will improve
the fatigue strength of the substrate.

     The use of this coating was initially investigated by the authors [4]
in an effort to increase the high temperature creep resistance of Ti alloys.
The lower creep rates which were observed in an argon atmosphere [5] and in
vacuum [6] led to investigation of ductile oxidation resistant coatings
which isolated the material from the atmosphere.

     Ion plating of noble metals was selected because it forms a sound and
effective coating without the risk of hydrogen contamination and brittle
intermetallic compound formation.  The negligibly small solubility of
platinum in titanium [7] up to 600°C allows the use of this coating to
that temperature.

## Experimental Procedures and Results

### 1.  Material

     Ti-6Al-2Sn-4Zr-2Mo-0.1Si (Ti-6242S) alloy bars were solutionized at
950°C and annealed at 590°C for 8 hours.  The resulting microstructure
(Fig. 1) consisted of 60 volume pct. equiaxed primary α grains uniformly
distributed in a matrix of transformed β phase.  Some creep experiments
were also performed on CP titanium and the α alloy Ti-5Al-2.5Sn, both cross
rolled to 6 mm thick plate.

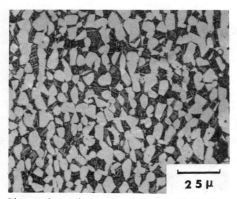

**25 μ**

Figure 1.   Microstructure of Ti-6242S.

### 2.  Coating

     The coating was applied using a 7 kV ion plating apparatus.  The bias
potential between the specimens and the molten metal source was initially
set at 4 kV to obtain a deeper ion implantation and later reduce to 3 kV
for effective plating.  The ion beam current was a maximum of 90 mA for 5
minutes.  Under these conditions the thickness of the coating was found to

be approximately 1 μm by SEM observation of the cross section.  To establish
the effect of coat thickness on oxidation resistance, a thinner coat (0.1 μm)
was also applied by reducing the ion-plating time to less than 1 minute.  The
thin coat thickness was determined by optical interferometry technique.

## 3.  Oxidation Resistance

In applying the oxidation protective coating, the thermal stability
of the coating/substrate interface is of prime importance.  In the preliminary work involving many elements coated on Ti-6242S, Au and Pt were found
to be stable up to 480°C and 590°C, respectively, for at least 500 hours
with no spalling or alloying with the matrix.  When a diffusion barrier such
as W was ion plated as a primary, the external Pt coating was sustained intact to temperatures as high as 700°C for 500 hours.  Table 1 summarizes the
weight gain due to oxidation with various coatings.

Table 1   Weight Gain Rates of Ti-6242S in Air

| Ion-Plating Materials** | Temperature (°C) | Weight Gain Rates $(mg/cm^2 \cdot hr)$ |
|---|---|---|
| No Coating | 590 | $6.9 \times 10^{-2}$ |
| Au | 430 | $2.2 \times 10^{-4}$ |
| Au | 480* | $2.6 \times 10^{-3}$ |
| Pt | 590* | $1.2 \times 10^{-3}$ |
| W/Pt | 650 | $3.3 \times 10^{-4}$ |
| W/Pt | 700* | $1.7 \times 10^{-3}$ |

*Highest temperatures under which no spalling or loss of the coating was de-
  tected after 500 hours.
**Coat thickness is a minimum of 1 μm.

To explore the possibility of reducing the amount of precious coating
materials used in this process, oxidation experiments were also conducted
on the material with only 0.1 μm thick coat.  Table 2 shows that oxidation
rate was increased only by a factor of 1.3 when the thickness was reduced
by a factor of 10.

Table 2   Weight Gain Rates of Ti-6242S With Different Coating Thickness

| Ion Plating Material | Thickness (μm) | Temperature (°C) | Weight Gain Rate $(mg/cm^2 \cdot hr)$ |
|---|---|---|---|
| Pt | 0.1 | 565 | $7.1 \times 10^{-4}$ |
| Pt | 1 | 565 | $5.8 \times 10^{-4}$ |

## 4.  Creep Tests

Creep tests of coated and uncoated α+β alloy (Ti-6242S) and α alloys
(CP titanium and Ti-5Al-2.5Sn), were performed in air on a constant load
testing machine.  The elongations were continuously monitored by an LVDT
transducer.  Vacuum creep tests were performed on a servohydraulic testing
machine equipped with a strain gage extensometer and specimens were heated
by a radiant heater in a vacuum better than $3 \times 10^{-5}$ torr.  All creep tests
reported in Fig. 2 were conducted at 570°C with an applied stress of 241 MN/m$^2$

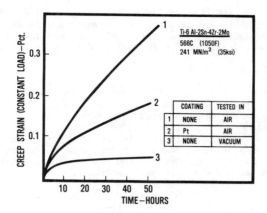

Figure 2.    Creep curves of Ti-6242S in vacuum and in air with and without coating.

on specimens with gage dimensions of 4 mm in diameter by 25 mm long.   The α alloy creep tests shown in Fig. 3 were conducted at 482°C and the creep stresses are indicated in the figure.

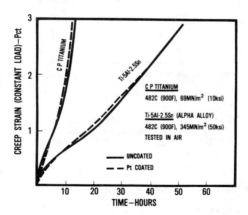

Figure 3.    Creep curves of Ti and Ti-5Al-2.5Sn tested in air with and without coating.

5.   Post-Creep Ductility

Post-creep ductility of Ti alloys is a significant selection criterion for high temperature applications.   Room temperature tensile properties of coated and uncoated Ti-6242S specimens creep exposed at 450°C and 241 MN/m$^2$ for 500 hours are compared with those of uncoated unexposed specimens in Table 3.

6.   Fatigue

Smooth bar fatigue life tests were conducted at 6 mm diameter by 25 mm gage length Ti-6242S specimens at room temperature and 455°C using an axial

Table 3   Pre- and Post-Creep Tensile Properties of Ti-6242S

| Condition | σ YS MN/m² | σ UTS MN/m² | Elongation (4D) pct. | RA pct. |
|---|---|---|---|---|
| Uncoated, prior to creep | 1151 | 1207 | 8.0 | 43.5 |
| Pt coated, post creep | 1145 | 1200 | 8.6 | 41.5 |
| Uncoated, post creep | 1172 | 1220 | 8.2 | 35.8 |

load controlled 6 ton fatigue machine at 2000 cpm and a stress ratio value of R = +0.1.  To reduce the data scatter due to surface conditions, the gage section was progressively hand polished after machining with 5 micron and 3 micron metallographic diamond paste.  The fatigue life data, demonstrating the effect of Pt coating on the fatigue life, is shown in Fig. 4.

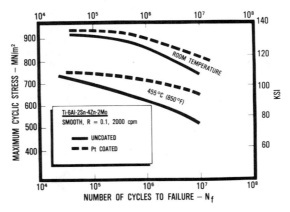

Figure 4.   Fatigue S/N curves of Ti-6242S tested in air with and without Pt coat at room temperature and 455°C.

## Discussion

The oxidation results in Table 1 and 2 indicate excellent oxidation resistance and high thermal stability of the thin Pt ion plating.  The integrity of the coating after 10⁷ runout fatigue cycles at 455°C and 586 MN/m² maximum stress is demonstrated in Fig. 5a.  The SEM image of the creep exposed Ti-6242S specimen with Pt coating (Fig. 5b) shows good coherence with no signs of spalling even after 2 pct. strain.  Table 2 also demonstrates that much thinner coat can effectively protect the material with minimal oxidation rate increase.

The Pt coating provided an effective barrier for oxygen diffusion and resulted in improved air creep strength for the α+β alloys (Fig. 2).  Higher creep strength in vacuum also indicates that oxygen diffusion is one of the controlling factors in the creep behavior of α+β titanium alloys.  In previous work [4] which examined both Ti-6242S and Ti-11 alloys, the accelerated creep behavior of uncoated specimens tested in air was interpreted on the basis of an oxygen induced phase transformation of the metastable β

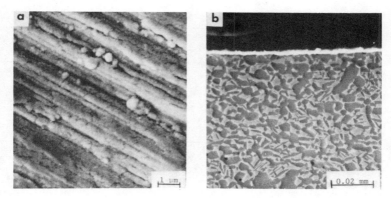

Figure 5.  The surface (a) and the cross section (b) of Pt coated Ti-6242S
           after fatigue and creep exposure, respectively.

phase and the concurrent increase of the mobile dislocation density.  Since
the improvement of creep strength by Pt coating was found only in alloys
containing both the α and β phases, it was then concluded that when unpro-
tected material is exposed to the atmosphere oxygen can deeply penetrate
along the α/β phase boundaries.  The oxygen being a strong α phase stabilizer,
will then transform some of the β into α and generate more mobile dislocations, thus promoting an increase of the secondary creep rate expressed by:

$$\Delta \dot{\epsilon}_s = \bar{b} \cdot v \cdot \rho \, (\sigma, \Delta C_0^{\alpha/\beta})$$

where $\Delta C_0^{\alpha/\beta}$ is the increment of oxygen concentration at the phase boundary.
$\bar{b}$, $v$, $\rho$ and $\sigma$ are the Burgers vector, dislocation velocity, dislocation density and applied stress, respectively.  In this study, no loss in post-creep
ductility was detected in the Pt coated material when compared to uncoated
creep specimens in the exposed and unexposed condition (Table 3).

     The improvement of high cycle fatigue strength by Pt ion plating is
clearly demonstrated in Fig. 4 both for room and an elevated temperature.
The fatigue limits at 455°C and room temperature were improved by 20 and
10 percent respectively.  At the high cycle region of the 455°C S-N curve,
all coated specimens failed from subsurface origin (Fig. 6) while all other

Figure 6.  Fractography of fatigue failed Pt coated Ti-6242S with subsurface
           crack initiation.

specimens (R.T. and 455°C) failed from surface origin. Inferior fatigue
properties of the uncoated specimens at elevated temperatures can be explained by two mechanisms; one is surface degradation by oxidation and the
other is the thermal relief of surface compressive stresses. The ion plating may be causing surface hardening by generating compressive stresses
similar to those generated by shot peening. The increased smooth fatigue
strength at elevated temperatures is therefore attributed to the oxygen
impeding effect of the coating as well as the surface hardening resulting
from Pt ion bombardment.

The improvement of fatigue by Pt coating is consistent with previous
fatigue experiments of noble metals in air and Ti alloy in vacuum and air.
Wadworth and Hutchings [8] found that fatigue life of Cu was strongly dependent on the oxygen partial pressure whereas fatigue life of Au was independent of the environment. In recent work on Ti-6Al-4V Leutjering [9]
reported that room temperature fatigue strength was significantly greater
in vacuum than in air.

The oxidation effect on fatigue crack initiation, and thus the total
fatigue life, will be maximal at higher temperature and lower stress tests.
In those conditions, the long exposure under cyclic stress enhances oxidation
promoting early crack initiation. For this reason, the greatest Pt coat
effect is at the high cycle region of the 455°C fatigue S-N curves (Fig. 4).
In order to distinguish the Pt coating effect from oxidation effect, it is
necessary, however, to investigate the fatigue behavior of Pt coated and
uncoated Ti alloys in vacuum.

1.   Potential Applicaltions and Economics

So far it was demonstrated that the Pt ion plating increased the high
temperature oxidation and creep resistance as well as high cycle fatigue
strengths of $\alpha+\beta$ Ti alloys. The coherent surface protection may also be
valuable in improving wear and fretting fatigue resistance and reducing
corrosion related failures. These suggest potential applications of this
technology to gas turbine engines containing titanium components operating
at elevated temperatures. Many high temperature Ti alloy components in
the hot sections of gas turbine engines are creep and fatigue limited and
subject to erosion, corrosion, stress corrosion cracking, and oxidation.
It appears that Pt ion plating can increase the serviceable temperature of
the Ti components by up to 150°C.

The cost of the coating materials required is nominal. One ounce of Pt
can theoretically cover 1.45 $m^2$ of areas with 1 µm thick coat. In practice,
2 oz. of Pt may be needed because some ions will not impinge the substrate.
After coating, some of it can be recovered from the ion plating chamber. As
shown in Table 3, 0.1 µm coating could be adequate for many applications to
effectively protect the surface from oxidation.

A commercial system employed in this work has the capability to generate
a 7 kV bias potential with maximum current of 100 mA. Such a power source is
adequate to uniformly ion plate an area of 0.3 $m^2$ from a single evaporant
source. In order to produce a 1 µm coating, only a few minutes of evaporation
are needed. This rate is much faster than conventional sputtering of vapor
deposition processes. In the production scale the vacuum chamber size and
the power supply capacity are practically unlimited and can be made to accept
any geometry and size of components by designing a suitable rotating jig.

## Summary

The present study demonstrates that Pt ion plating significantly improves oxidation resistance and creep and fatigue properties of α+β Ti alloys at elevated temperatures.  It appears that application of such a technique to gas turbine engines is not only technically feasible but economically affordable.  The authors are strongly convinced that implementing the present approach to any alloy system, e.g., by exploring a high voltage ion implantation for increased fatigue strength, and proper selection of implanting and coating materials for improving corrosion, erosion, and wear resistance, would be fruitful in meeting the future material needs and requirements.

## Acknowledgement

The authors wish to acknowledge Mr. R. Gordman of Hohman Plating Co. for his assistance in performing the coatings.  Parts of this work were conducted under U.S.A.F. Contracts F33615-76-C-5227 and F33615-79-C-5152.

## References

1.  M. Levy and J.L. Morrosi, Army Materials and Mechanics Research Center, Technical Report 76-4 (1976).
2.  M.T. Groves, NASA Report, CR-134537 (1973).
3.  G. Pearnely and N.W. Hartley, Thin Solid Films, 54(1978), 215.
4.  S. Fujishiro and D. Eylon, Scripta Metallurgica, 11(1977), 1011.
5.  G.S. Hall, S.R. Seagle and H.B. Bomberger, AFML-TR-73-37 (1973).
6.  S. Fujishiro and D. Eylon, 105th AIME Annual Meeting, Abstract Book, p. A14, Las Vegas, Nevada, February 1976.
7.  M. Hansen and K. Anderko, Constitution of Binary Alloys, (1958), 219.  McGraw Hill, N.Y.
8.  N.J. Wadworth and J. Hutchings, Phil. Mag., 3(1958), 1154.
9.  G. Luetjering, EPRI-RP 1266 (1979)

# B. Mechanical Properties II

# INFLUENCE OF ERBIUM AND YTTRIUM ADDITIONS ON THE
# MICROSTRUCTURE AND MECHANICAL PROPERTIES OF TITANIUM ALLOYS

B. B. Rath*, B. A. MacDonald†, S. M. L. Sastry‡,
R. J. Lederich‡, J. E. O'Neal‡, C. R. Whitsett‡

*Naval Research Laboratory, Washington, D. C. 20375 USA
†Office of Naval Research, Arlington, VA 22217 USA
‡McDonnell Douglas Research Laboratories, St. Louis,
Missouri 63166 USA

## Abstract

Additions of small amounts of erbium and yttrium have a significant effect on the microstructure and mechanical properties of titanium alloys. Because alloy additions of yttrium or erbium improve room temperature and high temperature ductility of the alloy, they have the potential of reducing the cost of processing titanium alloys. The effects of small additions of erbium and yttrium on the microstructure, tensile properties, fracture toughness, and high temperature deformation of titanium and the Ti-6Al-4V alloy were investigated.

The erbium and yttrium additions produced 20-100 nm size second phase dispersions in titanium and titanium alloy caused significant grain refinement, suppressed titanium hydride formation, scavenged interstitial oxygen, and enhanced twinning. Accompanying these microstructural and compositional modifications are increased room temperature ductility, improved superplasticity, and reduced susceptibility to hydride embrittlement. The grain refinement in the titanium rare-earth alloys results in increased flow stress in accordance with the Hall-Petch relationship. However, the dispersion strengthening expected from the yttrium and erbium particles is not observed in Ti-Er and Ti-Y alloys because of the scavenging of interstitial oxygen and subsequent softening of the titanium matrix by the additives. The deformation texture of titanium characterized by a forty degree shift of the (0001) on either side of the rolling plane normal along the transverse direction remains unaffected by the erbium and yttrium additions, although the modes of grain reorientation through recrystallization and grain growth are significantly altered.

The addition of 0.1 wt.% erbium or 0.05 wt.% yttrium improves the high temperature formability of Ti-6Al-4V, improves the yield during initial forging of the Ti-6Al-4V ingots, reduces the high temperature flow stress, controls grain size at beta-processing temperatures, and has no significant effect on yield strength and fracture toughness of the alpha-beta processed alloy. The rare earth effects on room temperature mechanical properties are not as pronounced in the strong two-phase microstructure of conventionally processed Ti-6Al-4V as in a single phase titanium alloy.

The microstructural modifications, microscopic deformation mechanisms, and strength and ductility as influenced by yttrium and erbium additions are discussed on the basis of microscopic observations and mechanical property measurements.

## Introduction

Within the last 25 years since rare-earth elements became commercially available, extensive studies have been made of their chemical equilibria with other metals [1] and the improvements in the physical and chemical properties of both ferrous and non-ferrous alloys by the addition of rare-earth elements. For example, Collins, Calkins, and Gurtz [2] demonstrated the extraordinary effects of several rare-earths and yttrium on the workability, grain refinement, resistance to recrystallization at high temperatures, and high-temperature oxidation resistance of Fe-Cr alloys. Furthermore, it has been conclusively shown that small additions of rare-earths or mischmetal to Mg alloys significantly improve the strength and creep-resistance of these alloys [3,4], and such alloys have been produced commercially for aerospace applications.

The limited information presently available on the various effects of rare-earth elements on the properties of Ti and Ti-alloys, because of the lack of a systematic study, leads to perplexing and often contradictory conclusions. In one of the early studies, Liu [5] concluded that the addition of 0.1 to 1.0 at.% of rare-earth elements did not refine the grain structure in $\beta$-Ti. Love [6], in reviewing the results of his investigation of Ti-based "rare-earth alloys," indicated that the addition of either La, Gd, Er, or Y up to 1 at.% did not change the hardness or the tensile properties of Ti. Chapin and Liss [7], on the other hand, showed that Gd and La can act as strengtheners to Ti because of internal oxidation. The results of Samsonov [8], in contrast to those of Liu and Love, show that La and Y are effective grain-refiners in cast Ti.

In a recent study, Buczek, et. al. [9] examined the microstructure and properties of Ti-6Al-4V and Ti-3Al-8V-6Cr-4Mo-4Zr to which had been added Y or $Y_2O_3$. Although they identified the grain-refining effect of the Y or $Y_2O_3$ additives, they did not show a correlation between the microstructural variables and the mechanical properties of the Ti alloys. The enhancement of the transverse ductility and the decrease in toughness parallel to the forging axis, which they observed, can be explained by the influence of crystallographic texture rather than the Y or $Y_2O_3$ additives. Furthermore, they inferred that similar results were obtained by adding Y either as a metal or as the oxide, $Y_2O_3$. This contention is not easily rationalized because $Y_2O_3$ predominantly remains as an insoluble inclusion in the Ti matrix whereas up to 0.04 wt.% Y dissolves in Ti at 875°C [10]. The effects of a solid-solution or second-phase dispersion hardening by Y should be significantly different from the effects of innocuous $Y_2O_3$ particles.

A comprehensive study has been undertaken to evaluate the effects of Er and Y in varying concentrations on the microstructure and properties of low interstitial Ti (ELI) and commercial Ti-6Al-4V (0.16 wt.% $O_2$). The alloys were prepared by consumable-electrode, arc-melting process. The ingots were forged and hot-rolled into strip form for testing.

## Results and Discussion

1.   Effect of Er on the Structure and Properties of Ti

A three-part investigation was conducted to quantitatively establish: 1) the extent of grain refinement, 2) the degree of $TiH_2$ precipitate suppression, and 3) the changes in mechanical properties as a function of Er concentration in Ti. Electrolytic Ti and eight Ti-Er alloys, with Er concentrations varying from 0.02 to 1.0 at.%, were used in this investigation. The alloys were prepared by repeated vacuum arc-melting for homogenization, followed by hot-rolling at 900°C.  The rolled samples were vacuum annealed to reduce the hydrogen concentration to ∿10 ppm by weight.  Each sheet was subsequently reduced by unidirectional cold-rolling to 50% of its original thickness.  To determine the effect of Er on $TiH_2$ precipitation, samples from these cold-rolled sheets were cathodically charged with hydrogen to ∿200 ppm by weight in aqueous 10% $H_2SO_4$ at 0.1 A·cm$^{-2}$.  The hydrogen-charged samples were annealed at 300°C for 3 h to homogenize the dissolved hydrogen, and slow cooled at a rate of $10^{-2}$°C/s to precipitate $TiH_2$.

The grain refinement in Ti as a result of Er addition may be accomplished by one of the following processes:  increasing the frequency of recrystallization nuclei, suppressing the rate of recrystallization, or inhibiting grain growth.  In this investigation, the recrystallization and grain growth behavior in Ti and an alloy containing 0.3 at.% Er were studied.  These processes were examined for the two compositions by quantitative metallography over ranges of temperature and time.  In addition, changes in crystallographic texture, resulting from recrystallization and grain growth, were determined to establish the nature of grain reorientation during these processes.

Recrystallization parameters, such as recrystallized volume fraction, interfacial area of migrating boundaries, average recrystallized grain size, and the number of recrystallized nuclei per unit volume as a function of time, temperature and Er concentration, were determined from linear intercept measurements.  Furthermore, changes in microhardness of the deformed matrix during recrystallization were determined to establish the concurrence of the recovery process with recrystallization [11].

Figure 1 shows the time dependence of the volume fraction recrystallized, $X_v$, for the two compositions, at temperatures from 400 to 650°C.  The sigmoidal curves shown in the figure are typical of the recrystallization behavior in most metals.  The results show that a small Er addition, as little as 0.3 at.%, significantly affects the recrystallization rate.  For example, the recrystallization rate at 550°C and $X_v$ = 0.5 (representing 50% recrystallization) is suppressed about 100 times by an Er addition of 0.3 at.%.

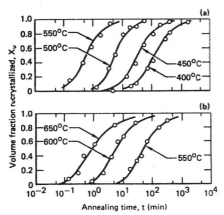

Figure 1.  Recrystallization Behavior in Ti and Ti-Er Alloys, (a) Ti and (b) Ti-0.3 at.% Er

For precise characterization of the recrystallization process, it is essential to determine the time dependence of recovery and recrystallized nuclei frequency, which are frequently assumed to be constant [12,13].  Changes in the microhardness values of the non-recrystallized region of the Ti and Ti-Er alloys, partially recrystallized during annealing, were measured at several tem-

peratures. A sharp drop in the hardness values in the early period of annealing clearly illustrates that a recovery process continues concurrent with recrystallization in these samples. This implies that in the Ti and Ti-Er alloys, the stored-energy of deformation, which provides the driving force for
recrystallization, steadily decreases through a recovery process as recrystallization no longer remains independent of time. Experimentally observed
behavior of the growth of recrystallized grains (determined from the slope of the
change of recrystallized volume fraction with time, $dX_v/dt$, and the interfacial
area of the migrating boundaries, $S_v$) as a function of time at various temperatures in pure Ti is shown in Figure 2. Results show that the growth rate, inversely proportional to time, varies over three orders of magnitude in the temperature range under investigation, pri-marily as a result of concurrent recovery

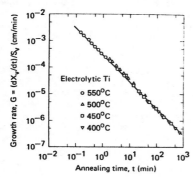

Figure 2. Time Dependence of Growth
Rates of Recrystallized Grains in Ti

The temperature dependence of growth rate of the recrystallizing grains at
$X_v = 0.5$ reveals that the addition of Er to Ti significantly suppresses the
rate of recrystallization as well as increases the activation energy for the
recrystallization process. The measured value of 35 kcal mole$^{-1}$ for electrolytic Ti, comparable with the 30 kcal mole$^{-1}$ reported for self diffusion in Ti
using tracer techniques [14], increases to 65 kcal mole$^{-1}$ by the addition of Er.

It is concluded from these results that the Er second-phase particles present in the Ti matrix in a range of sizes, affect the recrystallization behavior
as follows: 1) the larger particles contribute to an increase in the nucleation
frequency, whereas their small volume fraction does not significantly affect the
stored energy of cold-work and 2) the smaller particles impede recrystallization
presumably by a more stable cell formation and by imposing an effective drag
force on the moving boundary [15,16].

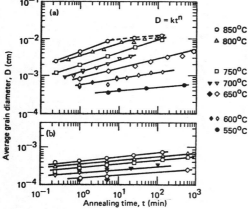

The grain growth kinetics has been
evaluated in Ti and Ti-Er alloys following recrystallization. Results
are shown in Figure 3. The values
of average grain diameter D were
determined from the measurements of
300 to 500 grain diameters in each
sample by optical and electron microscopy. The solid symbols shown in
Figure 3 refer to values obtained
from electron microscopic measurements are in good agreement with
those obtained by optical microscopy.

Grain growth was found to change
abruptly when the grain size reached
a diameter of ~100 μm. This change
is indicated by the dashed line in
Figure 3a. This variation in rate
is attributed to the limiting sample

Figure 3. Grain Growth in Ti and Ti-Er
Alloys (a) Electrolytic Ti and (b) Ti-
0.3 at.% Er

thickness rather than the impurity drag effect, which impedes grain growth in a region where the boundary velocity is below a critical value [17,18].

Recrystallized microstructures in Figure 3 illustrates the profound effects on grain refinement of small quantities of Er in Ti. For example, a grain size of 10 μm, obtained after a 0.25 min anneal at 750°C in pure Ti, is not achieved even after annealing the 0.3 at.% Er for 11 hours. An extrapolation of the grain growth behavior of 0.3 at.% Er suggests that this alloy would reach a grain size of 10 μm after annealing at 750°C for ~70 days. Significant grain refinement is achieved even for an Er concentration of 0.03 at.% in Ti. The effect of Er on the grain refinement in Ti is further illustrated by the micrographs shown in Figure 4. Under identical annealing conditions, the average grain size in Ti is reduced by a factor of 20 as a result of 0.3 at.% Er.

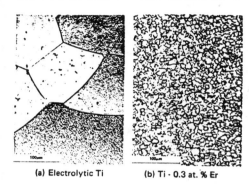

(a) Electrolytic Ti          (b) Ti - 0.3 at. % Er

Figure 4. Effect of Erbium on Grain Refinement in Titanium: 750°C, 600 min

Regarding the time and temperature dependence of grain growth, Okazaki et. al. [19] and Hu and Cline [20] have reported that the time exponent for grain growth, n = 0.33, in pure Ti, is independent of temperature. The results of the present investigation clearly show that n continuously increases with increasing temperature and the rate of increase is dependent on the Er concentration. Figure 5 shows the effects of temperature and Er concentration on the time exponent for grain growth. The presence of Er suppresses the influence of the annealing temperature on grain growth, as indicated by the nearly zero rate of increase of n with temperature for the 0.3 at.% Er alloy. The temperature dependence of n, increasing with temperature towards the limiting values of 0.5, is consistent with the grain growth behavior of most metals and is particularly applicable for Ti and 0.03 at.% Er alloy. The measured values of n, always remaining below 0.5, indicate that the grain growth process in Ti and Ti-Er alloys is predominantly controlled by the diffusion of dissolved impurities in Ti [21].

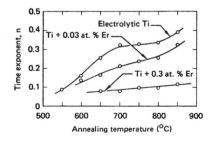

Figure 5. Effects of Er Concentration and Annealing Temperature on the Time Exponent of Grain Growth in Ti

The temperature dependence of grain growth for a constant grain size, analyzed in terms of the Arrhenius relation, suggests that the grain growth in Ti and Ti-Er alloys over the range of temperatures investigated is not a single activated process. The measured energy for grain growth for constant grain size and constant n is about 70 kcal mole$^{-1}$ for Ti and increases to 76 kcal mole$^{-1}$ with 0.03 at.% Er addition. An increase in the Er concentration to 0.3 at.% Er does not alter this value.

The evolution of crystallographic texture in Ti, as a function of annealing temperature, and the effect of Er additions on this texture were examined
by the x-ray diffractometer method using high intensity CuKα radiation and a Ni
filter.  The deformation texture in Ti, following a 50% reduction in thickness
by unidirectional cold-rolling,
can be best described by nearly
symmetrical texture components
represented as 40 deg (0001)
[$\bar{1}$010] and shown in Figure 6.  The
(0001) is shifted by ~40 deg to
either side of the rolling plane
normal along the transverse direction; the [$\bar{1}$010] remains parallel for both components to the
rolling direction.  There is an
appreciable spread of each texture component which is predominantly a result of rotations about
two individual axes.  These rotation axes are the rolling direc-
tion or [$\bar{1}$010] and the (11$\bar{2}$2) plane
normal.  The angular displacement
of the basal pole from the rolling
plane normal has been reported to
be between 27 and 40 deg [22-24].
Hu and Cline [20] have rationalized
this angular variation, which
reaches a value of 40 deg when the
amount of cold reduction approaches
the 90% values.  Results of the present investigation indicate that
the tilt angle is nearly 40 deg
when the samples are deformed only
50% by cold reduction.  The pres-
ence of Er second-phase particles does not    alter this tilt angle, although
the texture spread increases with increasing Er concentration as shown in Figure
6c.  The recrystallization texture, shown by the (10$\bar{1}$0) pole figures in Figure
6b and 6d for Ti and Ti-1.0 at.% Er respectively, indicates that the crystallographic positions of the (10$\bar{1}$0) maxima essentially are unaltered.  However, a
closer comparison of the recrystallization and deformation textures suggests
that a reorientation, characterized by a rotation about the [0001], occurs during
recrystallization.  The extent of this rotation appears to increase with addition
of Er to Ti.  Furthermore, hitherto unrecognized, recrystallization causes a
shift of the basal pole away from the rolling plane normal by about 4 deg.

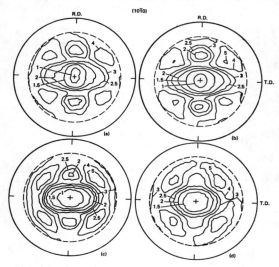

Figure 6.   (a and b) Ti Deformed 50% and
Annealed at 500°C for 100 min (D=5 μm) respectively, (c and d) Ti-1.0 at.% Er Deformed 50% and Annealed at 700°C for 120
min (D=3.5 μm) respectively

Upon grain growth, the texture changes significantly as shown in Figure 7a
which clearly suggests that the grain growth process in Ti is strongly dependent
upon grain orientation.  While the changes in the angular parameters, such as γ
[the angle between the rolling plane and (0001)], and β (the angle between the
rolling direction and the nearest [10$\bar{1}$0]), are consistent with previous results
[20], δ (the angle between the tilt axis and the rolling direction) remains unaffected during grain growth.  The major features of the grain growth texture
are the symmetrical splitting of the {10$\bar{1}$0} intensities  of each recrystallization texture component by an ~23 deg clockwise- and counterclockwise-rotation
about [0001].  Such a texture, however, is not seen in the Ti-Er alloys (Figure
7b) simply because of the inhibition of grain growth in these alloys.

The mechanical properties of Ti and Ti-Er alloys at 25°C under uniaxial strain were determined as a function of grain size and Er concentration.  Standard strip tensile samples, of 0.03 cm² cross-sectional area and 5.0 cm length, were uniaxially strained at a nominal strain rate of 3 x 10⁻⁴ s⁻¹.

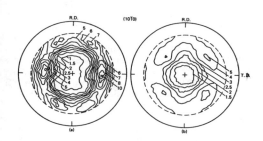

Based on the results of recrystallization and grain growth studies, samples of each composition were annealed at various temperatures to obtain grain sizes ranging from 3 to 130 μm.  An increase of Er concentration drastically decreased the upper grain size range in these alloys [25].

Figure 7.   Effect of Er on the Grain Growth Texture in Ti:   (a) Ti Annealed at 700°C/ 120 min (50 μm) and (b) Ti-1.0 Er Annealed at 860°C/1240 min (9 μm)

It has been well established that the strength of polycrystalline materials is increased as a result of decreasing grain size.  The strength parameters, frequently obtained from uniaxial tension or compression tests performed at fixed strain rates, have been shown to depend on the average grain diameter, D, in accordance with the Hall-Petch relation [26,27]:   $\sigma = \sigma_o + kD^{-1/2}$, where $\sigma$ is the yield stress and $\sigma_o$ and k are experimental constants.  Armstrong et. al. [28] have shown that the Hall-Petch relation is valid over a wide range of plastic strain where $\sigma$ represents the true flow stress at constant strain.  Figure 8 shows the grain size dependence of stress for Ti and Ti-Er alloys at 0.2 and 10.0% offset.  Within experimental scatter, the results for the compositions and strain are in excellent agreement with the Hall-Petch relation, yielding singular values of $\sigma_o$ and k at constant strain.  These values are indicative of the negligible effect of the Er second-phase particles on the hardening of Ti.

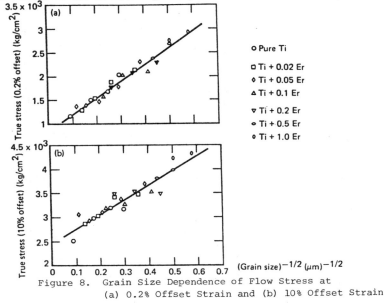

Figure 8.   Grain Size Dependence of Flow Stress at (a) 0.2% Offset Strain and (b) 10% Offset Strain

2.   Effect of Y on the Structure and Properties of Ti

The selection of Y as an alloy additive to Ti was based upon (1) its atomic size being identical to that of Er, (2) its valence state being similar to Er and other rare-earth metals, (3) its limited solubility in Ti, and (4) its abundance in nature and hence lower cost.  Similar to Er grain refinement in Ti was effectively accomplished by the addition of Y.  To systematically characterize the grain refinement in Ti as a function of Y-concentration and thus establish the optimum Y-concentration, the kinetics of recrystallization and grain-growth in Ti-Y alloys were determined over a wide range of annealing temperatures. The recrystallized volume-fraction as a function of time and temperature was determined for each alloy.  Figure 9 shows the typical recrystallization behavior in these alloys when annealed at 550°C (825 K).  Both the incubation period and the recrystallization rate are significantly altered by the addition of Y to Ti.  The sluggishness of recrystallization is most pronounced in the Ti-0.09 Y alloy, and greater addition of Y to Ti have decreasingly less effect.  This behavior is quite analogous to that observed earlier in Ti-Er alloys.

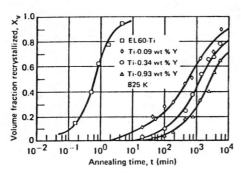

Figure 9.   Recrystallization Behavior of Ti-Y Alloys at 550°C (825 K)

The temperature dependence of the growth rate of the recrystallizing grains at $X_v = 0.5$ shows that the presence of Y in Ti, either as a second phase or in solution, increases the measured activation-energy from 35 kcal/mole (146 kJ/mol) for pure Ti to 72 kcal/mole (300 kJ/mol) for the Ti-Y alloys.

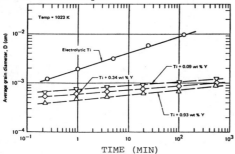

TIME (MIN)

Figure 10.   Effect of Yttrium on the Grain Growth in Titanium at 750°C (1023 K)

For grain-growth measurements, samples were isothermally annealed at five temperatures between 650 and 850°C (925 and 1125 K) and for times from 0.25 to 2500 min.  The average grain-diameter, D, as a function of time and temperature is shown in Figure 10.  Similar to recrystallization the maximum suppression of grain growth was achieved by the addition of 0.09 wt.% Y to pure Ti, and a further increase in the Y concentration contributed only slightly to suppressing the average grain size.

3.   Hydride Density in Ti-Er and Ti-Y Alloys

Various compositions of Ti-Er and Ti-Y alloys were charged to 200 ppm of hydrogen by weight, annealed at 300°C for 3 h, and slowly cooled at $10^{-2}$°C $s^{-1}$ to room temperature.  The quantitative determination of the $TiH_2$ density, expressed in terms of the $Ti-TiH_2$ interfacial area per unit volume, as a function of the Er concentration in shown in Figure 11.  Ti-Y alloys exhibit a similar behavior to $TiH_2$ precipitation the pronounced decrease in $TiH_2$ density with increasing Er or Y concentration, in excellent agreement with previous results [29], shows that 1.0 at.% Er reduces the hydride density to about 3% of that in unalloyed Ti. In Ti, the $TiH_2$ precipitation sites are almost equally divided between the grain and the grain boundaries, assuming an equi-distribution of their size in both locations. Increasing the Er concentration in Ti, however, continues to preferentially suppress the precipitation of intra-granular hydrides, such that in the 1.0 at.% Er alloy, the only hydrides found are located at grain boundary sites [30]. This observation is reasonable since the precipitates having a higher specific volume than the matrix would preferentially select the boundary site for nucleation.

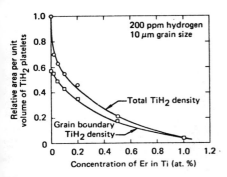

Figure 11. Effect of Er Concentration on $TiH_2$ Density in Hydrogen-Charged Ti-Er Alloys

4.   High Temperature Deformation of Ti-Er and Ti-Y Alloys

Although grain refinement contributed to increased strength, dispersion strengthening expected from the rare earth precipitates was not observed in Ti-Er and Ti-Y alloys as apparent from the Hall-Petch plots. To evaluate the role of rare-earth additive on scavenging of oxygen and consequently contributing to the softening of the Ti-matrix as well as on second-phase dispersion hardening the temperature dependent flow-stress data of Ti-0.5Er and Ti-0.2Y alloys were analyzed on the basis of different assumed combinations of interstitial strengthening and dispersion strengthening. The residual interstitial oxygen level in the various Ti-rare earch alloys was calculated assuming complete oxygen scavenging by the rare-earth elements to form $Er_2O_3$ and $Y_2O_3$. The dependence of flow stress on interstitial-oxygen concentration in Ti was measured, and the results were used to determine the interstitial strengthening attributable to the residual oxygen in the Ti-rare earth alloys. The particle bypass stress can be calculated from the modified Orowan relation [31].

The observed flow-stress differences between Ti and Ti-rare earth alloys as a function of temperature in comparison with the calculated flow stress for different combinations of dispersion- and interstitial-strengthening contributions show that below 325°C (600 K), the observed flow-stress variations agree well with the calculated temperature dependence of the combined interstitial-strengthening and dispersion-strengthening when oxygen scavenging by the rare earths to form $Er_2O_3$ and $Y_2O_3$. However, the flow-stress above 330°C (625 K) are considerably lower than predicted by any combination of strengthening mechanisms. Further evidence for the scavenging of oxygen by Er and Y in Ti is the increased incidence of twinning in the alloys.

5.   Microstructure and Properties of Ti-6Al-4V-Rare Earth Alloys

The as-cast microstructures in Ti-6Al-4V and Ti-6Al-4V-Rare Earth alloys were determined from the center sections of the 20-mm thick slices taken from the cast ingots. Figure 12 shows the typical microstructures observed in the Ti-6Al-4V reference and Ti-6Al-4V-0.1Y. The as-cast microstructure of the reference alloy consists of large grains with extensive coarse-α formed during

cooling at the prior-β grain boundaries.  In contrast, the Y- and Er-containing
alloys exhibit a more homogeneous and grain refined structures.  The preferential α-nucleation at the prior-β grain boundaries observed in the reference
alloy is absent in the Y- and Er-containing alloys, in which the α-phase nucleates uniformly.  The grain size is considerably smaller in the Y- and Er-containing alloys than in the reference alloy, and the prior-β grain size significantly decreases with increasing concentrations of Y and Er.

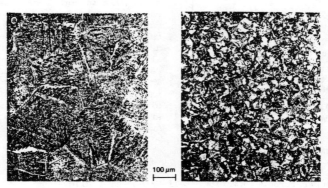

Figure 12.  Microstructures of As-Received, Mill-Annealed [950°C (1225 K)1 h/AC]
Alloys:  (a) Ti-6Al-4V and (b) Ti-6Al-4V-0.1Y

The nature of the dispersoids in selected rare-earth-containing alloys was
studied by electron microscopy of thin foils and extraction replicas.  The rare-
earth phase is spherical, 20-70 nm in diameter, and incoherent with the matrix.
The small size and uniform distribution of the Er and Y dispersoids observed in
Ti-6Al-4V implies that they precipitated during cooling of the alloy.  Both Er
and Y are soluble in all proportions in molten Ti, and the initial-reaction
step in the preparation of the ingots for this study was the dissolution of the
metallic rare-earths in the alloy melts.  Because the negative free-energies of
formation of $Er_2O_3$ and $Y_2O_3$ are higher than those of $TiO_2$ and $Al_2O_3$, the dissolved rare-earths will scavenge oxygen in the melt and to a lesser extent scavenge interstitial oxygen during high-temperature processing of the solid alloy.
Rare-earth oxides formed in the melt will form fine, stable particles before the
alloy is solidified, and it is quite probable that many of the observed dispersoids are oxide precipitates.  The compositions of the dispersoids is not a
factor in their direct influence on the alloy behavior; the important factors
are the small size of the dispersoid particles that are formed from the metallic
rare-earths dissolved in the Ti-alloy melt and their stability at the temperatures for which the dispersoid influence is sought for modifying the alloy behavior.

The influence of β-annealing and α-β annealing treatments on the microstructural changes of alloys hot-rolled under mill conditions was studied.  The
alloys were hot-worked at 1010°C followed by finish rolling at 955°C and heat-
treated both above and below the β-transus temperatures.  Following the high-
temperature heat treatments, the alloy sheets were annealed at 705°C for 2 hours
and air cooled.  The alloys annealed in the β-field, water quenched, and then
annealed for 2 hours at 705°C show the needle-like martensitic α' structure,
produced by diffusionless transformation of the β phase.  Upon annealing the
martensitic structure at 705°C, the α' transforms to equilibrium α+β.  The
time-dependent β-grain growth was studied at two temperatures, $T_β$ + 56°C and

$T_\beta$ + 28°C show nearly an order of magnitude grain-size reduction in alloys containing the concentrations of Er and Y. The results clearly establish that Y and Er not only effectively refine β grains but also retard grain growth at elevated temperatures.

The influence of rare-earth additives on the mechanical properties of β-annealed alloys is shown in Figure 13(a). With increasing rare-earth concentration, the yield stress decreases slightly and the ductility increases. However, the decrease in yield stress is small (<5%) for up to 0.1 wt.% Er or Y.

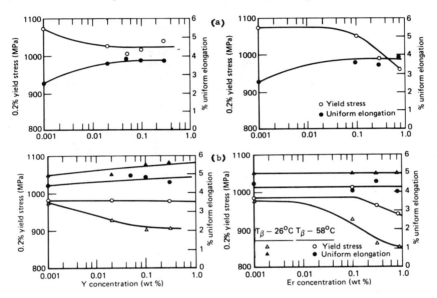

Figure 13.   Yield-stress and Uniform Elongation of (a) Beta and (b) Alpha-Beta Annealed Ti-6Al-4V-rare earth alloys as functions of concentrations of Y and Er

The influence of rare-earth additives on the yield stress and uniform elongation of α-β annealed alloys is shown in Figure 13(b). In the α-β annealed alloys, the rare-earth influence is dependent on the annealing temperature. For Y- and Er-containing alloys annealed at a temperature near $T_\beta$, the yield stress decreases slightly with the addition of rare earths, similar to the β-annealed alloys. However, for the alloys annealed at the lower α-β temperature, the reduction of yield stress with the addition of Er and Y is less.

The results presented in Figures 13(a) and (b) show that when the microstructure is predominantly transformed-β, the yield stress is controlled by the Widmanstatten α+β structure. The rare-earth additives do not contribute any strengthening because of the overriding influence of the transformed-β structure. The slightly reduction in strength observed in rare-earth-containing alloys is likely due to a lower interstitial oxygen content of the Ti-6Al-4V-rare earth alloys. The rare earths because of their greater affinity for oxygen can scavenge part of the dissolved oxygen from the alloy matrix. As the volume fraction of primary-α is increased by annealing at lower α-β temperatures, a strengthening influence of the Y and Er additives is evident that largely compensates the oxygen-scavenging effect.

## Acknowledgment

This research was supported by McDonnell Douglas independent research and development program and by the Office of Naval Research under contract No. N00014-76-C-0626.

## References

1.   C. E. Lundin, The Rare Earths, (1961), 224.
2.   J. F. Collins, V. P. Calkins, and J. A. Gurtz, The Rare Earths, (1961), 499.
3.   T. E. Leontis and D. H. Feisel, Trans. AIME 209(1957), 1245.
4.   J. F. Pashak and T. E. Leontis, Trans. AIME 218(1960), 102.
5.   Y. C. Liu, Watertown Arsenal Laboratory Report, Nov. 1953.
6.   B. Love, WADC Tech. Report No. TR57-666, Part II, Mar. 1959.
7.   E. J. Chapin and R. Liss, Report of NRL Progress, Dec. 1955.
8.   G. V. Samsonov, K. A. Kashuchuk and A. J. Cherkashin, Metal Science and Heat Treatment, (1970), 925.
9.   M. J. Buczek, G. S. Hall, S. R. Seagle, and H. B. Bomberger, AFML-TR-74-255, Nov. 1974.

10.   R. P. Elliott, Constitution of Binary Alloys - First Supplement, (1965), 853
11.   B. B. Rath, R. J. Lederich, and J. E. O'Neal, Grain Boundaries in Eng. Materials, (1975), 39.
12.   R. A. Vandermeer and Paul Gordon, Recovery and Recrystallization of Metals, (1963), 211.
13.   J. E. Burke and D. Turnbull, Progress in Metal Physics 3(1952), 220.
14.   C. M. Libanti and S. F. Dyment, Acta Met. 11(1963), 1263.
15.   R. J. Lederich, J. E. O'Neal, and B. B. Rath, TMS-AIME Abs. Bull. (1974), 119.
16.   B. B. Rath, J. E. O'Neal, and R. J. Lederich, Proc. Electron Microscopy Soc. of America, (1974), 522.
17.   John W. Cahn, Acta Met. 10(1962), 789.
18.   J. P. Frolet and A. Galibois, Met. Trans. 2(1971), 53.
19.   K. Okazaki, M. Momochi, and H. Conrad, Titanium Science and Technology 3 (1972), 1649.
20.   H. Hu and R. S. Cline, Trans. AIME 242(1968), 1013.
21.   H. Hu and B. B. Rath, Met. Trans. 1(1970), 3181.
22.   H. T. Clark, Jr., Trans. TMS-AIME 188(1950), 1154.
23.   C. J. McHargne, S. E. Adair, Jr., and J. P. Hammond, Trans. TMS-AIME 197 (1953), 57.
24.   J. H. Keeler and A. H. Geisler, Trans. AIME 206(1956), 80.
25.   B. B. Rath and R. J. Lederich, Recrystallization in Ti and Ti-Er Alloys, TMS-AIME Abs. Bull. (1974), 64.
26.   E. O. Hall, Proc. Phys. Soc. London B64(1951), 747.
27.   N. J. Petch, J. Iron Steel Inst. 174(1953), 25.
28.   R. W. Armstrong, I. Codd, R. M. Dontwaite, and J. J. Petch, Phil. Mag. 7 (1962), 45.
29.   B. B. Rath, J. E. O'Neal, and T. C. Grimm, ASM-AIME Abs. Bull. (1974), 70.
30.   B. B. Rath, J. E. O'Neal, and R. J. Lederich, Grain Boundaries in Eng. Materials, (1975), 643.
31.   M. F. Ashby, Physics of Strength and Plasticity, (1969), 113.

RELATIONSHIPS BETWEEN MICROSTRUCTURE AND
MECHANICAL PROPERTIES IN Ti-6Al-2Sn-4Zr-2Mo~0.1Si
ALLOY FORGINGS

C. C. Chen and J. E. Coyne
Wyman-Gordon Company, Worcester, Massachusetts   01613

## INTRODUCTION

During recent years, the increasing demands for higher temperature performance and lower weights of jet engine components have brought about the development and commercial use of near-alpha elevated temperature titanium alloys. As a result, the Ti-6Al-2Sn-4Zr-2Mo~0.1Si (designated as Ti-6242Si) alloy has been mostly used for high temperature applications of titanium alloys [1, 2]. The use of Ti-6242Si has extended the useful temperature of titanium alloys to the 900°F(482°C) regime; here both creep and fatigue properties at high temperatures are critical factors in determining the life cycle of the components. Very recently, the optimum processing and heat treating conditions have been further established for producing the alloy forgings that extend the use of forgings to higher operational temperatures (i.e., 1100°F/593°C) [3,4]. However, in spite of the great importance of the Ti-6242Si alloy in jet engine applications and large influence of the microstructure on mechanical properties, the relationships between microstructure and mechanical properties have not yet been well documented. In particular, very limited information has been made available as to how the change in microstructure varies the creep resistance of the alloy forgings.

Metallurgically, the alloy Ti-6242Si is a near alpha, alpha-beta titanium alloy [1]. Density of the Ti-6242Si alloy is 0.164 lb/in$^3$ (4.5 gm/cm$^3$), and the $(\alpha+\beta)/\beta$ transus temperature is about 1820°F(993°C). The 6% Al addition in the alloy is a potent $\alpha$-phase stabilizer, while 2% Mo represents a moderate quantity of $\beta$-phase stabilizer. The Sn and Zr are relatively neutral with respect to stabilization behavior and are solid solution strengthening elements for both $\alpha$- and $\beta$-phases. Mo is added to increase room- and elevated-temperature strength and stability, while the combination of Al, Sn, and Zr is to improve the creep strength of the alloy. Silicon is a strong $\beta$-stabilizer and is more soluble in $\beta$-phase than in $\alpha$-phase. The minor addition of 0.1% Si to this alloy has been shown to significantly enhance the high temperature creep properties of the alloy without any apparent undesirable effects [2].

In forging practices, the $(\alpha+\beta)$-forging combined with $(\alpha+\beta)$-solution treatment has been generally used to manufacture Ti-6242Si alloy forgings. Until recently, however, the $\beta$-forgings was used to acicularize the microstructure which resulted in a reduction in LCF and ductility, but increased the creep strength and fracture toughness. Like other titanium alloys, the control of the nature and distribution of both globular-$\alpha$ and transformed products by forging and heat-treating variables are of great importance in determining the mechanical properties of the forgings.

In the present paper, the relationships between microstructure and mechanical properties for the Ti-6242Si alloy pancake forgings are presented. The experimental effort involved the determination of microstructure-mechanical property correlations and the examination of microstructural features through optical, SEM-, and TEM-microscopies. The variation of creep properties with microstructure was particularly emphasized.

## EXPERIMENTAL VARIABLES

The starting material for this investigation was 9 inch diameter bar of commercial Ti-6242Si grade. Metallographic examination showed that the starting material had a structure characterized by final processing in the ($\alpha$+$\beta$)-phase field. The chemical composition (wt.%) of the as-received bar (TMCA heat number N-4694) as analyzed by Wyman-Gordon is 5.93% Al, 2.00% Sn, 4.20% Zr, 1.80% Mo, 0.11% Si, 0.08% Fe, 0.03% Cu, 0.099% $O_2$, 0.011% C, 0.007% N, and 0.0032% $H_2$. The ($\alpha$+$\beta$)/$\beta$ transus temperature for this heat bar was estimated to be 1820°F(993°C).

In this investigation, various forge and heat-treat combinations were used to produce the forgings with a wide variety of microstructures and a broad range of creep properties. The forging and heat treating variables for the forgings used have been described previously [4]. Briefly, sixteen pancake forgings were produced. The billet multiples were upset from 4 inches(102 mm) for 69% thickness reduction, both isothermally and conventionally, subtransus and supertransus (from 1750 to 2100°F or 954 to 1149°C) to produce pancake forgings of 1.25 inches(32 mm) thick by 16 inches(406 mm) in diameter. Both ($\alpha$+$\beta$)-preforms (a structure of approximately 70% globular-$\alpha$ in a transformed matrix) and $\beta$-preforms (a structure of transformed Widmanstatten colonies and acicular-$\alpha$) were used. All pancakes were sectioned and subjected to selected heat-treat conditions. The effects of solution treatment at temperatures from 1650 to 2000°F (899 to 1093°C) and times ranging from 1 to 8 hours were investigated. Aging treatments were made between 8 and 16 hours at 1100°F(593°C). The effects of cooling rate from forge temperature and solution-treat temperature were also investigated.

The mechanical property evaluations included creep, room temperature tensile, hot tensile, fracture toughness, and post-creep tensile properties. All of the test specimen blanks were cut along circumferential orientation of the pancakes. The fracture toughness measurements were performed only for the forgings having distinctly different microstructures; compact tension fracture toughness specimens were used. The structural features for each condition were examined using optical, scanning, and transmission electron microscopies. Fractographic analyses were made to examine the topographical features in both broken fracture toughness and creep rupture specimens. In some cases, TEM extraction replica was used to identify the nature and distribution of the precipitates.

## EXPERIMENTAL RESULTS AND DISCUSSION

Correlation Between Microstructure
and Creep Property

For some time, the microstructural features at optical magnifications have been used to characterize the mechanical properties of the alloy. Based on the forging microstructural appearance and creep results obtained in this investigation, the high temperature property (creep) of Ti-6242Si alloy can be broadly grouped into eight microstructural categories depending on the nature of primary-α and transformed products (Figure 1). From a creep standpoint, they can be further divided into three different groups, namely high (A and B), intermediate (C, D, E, and F), and low (G and H) creep resistances.

It is seen from Figure 1 that the microstructures characterized by less-defined fine Widmanstatten patches (A) and pseudo-β structures (B) appear to have the highest creep resistance. In practice, such high creep microstructures can be produced by either β-forging or (α+β)-forgings, followed by (α+β)-solution treatment for β-forgings and β-solution treatment for (α+β) forgings. Cooling rate from β-phase field is important in restricting the size of Widmanstatten packets. Previous work has shown that the size and the orientation of Widmanstatten colonies are dominant factors in the fracture behavior of titanium alloy [5].

Both globular-α (G) and transformed martensitic-α (H) appear to significantly degrade the creep properties. Globular-α microstructures are

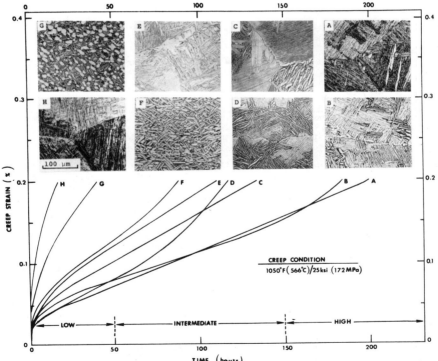

Figure 1 Creep strain versus time curves for eight microstructural conditions of Ti-6242Si alloy pancakes

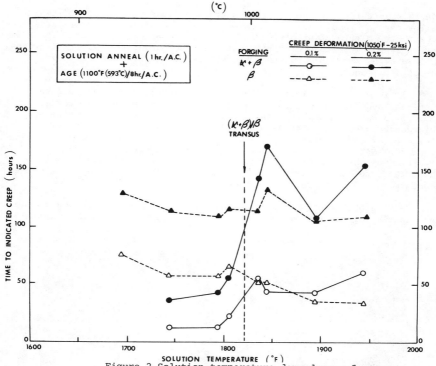

Figure 2 Solution-temperature dependence of creep
resistance for (α+β) and β-forgings

obtained by (α+β)-forging combined with (α+β)-solution treatment, and such a
forging and heat treating sequence is most often used for manufacturing the
Ti-6242Si alloy forgings.  It appears that the conventional practices to
spheroidize the primary-α of the alloy forgings has limited the creep capability of the alloy.  However, the presence of globular-α is known to significantly improve the low cycle fatigue of the alloy.  It also provides a
slight increase in both tensile strength and ductility as compared with
transformed basketweave microstructure.  The martensitic transformed structure
is developed by water-quench from β-solution temperature.  Such a microstructure is known to be beneficial for high cycle fatigue, and often gives significantly higher strength and lower ductility for the forgings.

    The microstructures with intermediate creep resistance (C, D, E, and F)
are generally characterized by coarser Widmanstatten-α amd colonies and/or
pseudo-β microstructures, as compared to the fully transformed microstructures
in high creeps.  Similar to high creep conditions, these microstructures were
developed by either (α+β) or β-forgings followed by either (α+β) or β-solution
treatments, but differences in forge temperature, solution temperature, and/or
cooling rate were applied.  However, a comparison of microstructural appearance in Figure 1 clearly shows that the difference in the microstructural
features between high and intermediate creep resistances cannot be distinguished based on the microstructural control at optical magnifications.

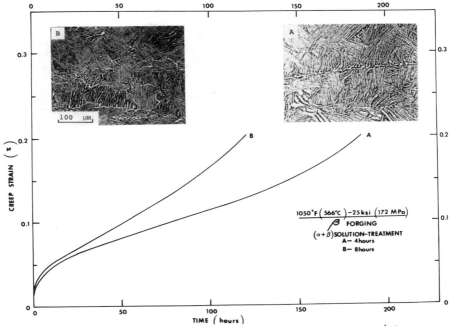

Figure 3 Effect of solution-anneal time on creep resistance
of Ti-6242Si alloy forgings

Thus, the relationships between creep properties and optical microstructures were limited to only several extremely different conditions, i.e.,
transformed martensitic-α, globular-α, and Widmanstatten-α microstructures.

## Processing/Structure/Mechanical
## Property Interactions

The variations of mechanical properties with microstructures of the Ti-
6242Si alloy depend strongly on both forging variables and heat-treating
conditions.  In particular, the creep properties of alloy forgings have a
complicated relation to the heat treat conditions.  Although there is no
obvious difference in optical microstructures, the changes in solution temperature (Figure 2) and time (Figure 3), intermediate anneal (Figure 4), and
cooling method (Figure 5) have significant influence on creep resistance of
the alloy forgings.  The use of (α+β)-solution treatment significantly de-
grades the creep properties of (α+β) forgings, and the (α+β)-solution treatment generally gives a better creep strength than that resulting from β-
solution treatment for β-forgings.

The creep property appeared to be insensitive to aging time.  Within the
forge conditions investigated, the type of forging, forge temperature, and
preform microstructure were found to have an important effect on creep properties [3].  Since solution treatment for this alloy forgings has been generally made at very high temperatures, none of the die temperature, ram rate,
cooling method from the press were detrimental to creep resistance.

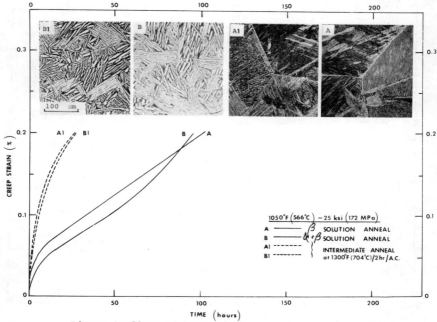

Figure 4 Effect of intermediate anneal on creep resistance
of Ti-6242Si alloy forgings; $\beta$-forged

Earlier investigation [4] had determined the influence of processing
variables on the forging deformation and resultant structures of Ti-6242Si
alloy forgings.   The results demonstrated that the forging deformation and the
resultant structures are strongly dependent on the forge temperature, die
temperature and ram rate, and the effect of preform microstructure was particularly emphasized.   The deformation characteristics under isothermal forging
were quantitatively related to the hot deformation properties of the alloy and
the rate-controlling deformation process in ($\alpha+\beta$)-processing conditions is
attributed to the dynamic softening.   This means that the structural change
occurs during forging deformation.

## Observation of Heterogeneous Precipitates

As discussed earlier, the creep resistance of the alloy forgings varied
strongly with forging schedule and subsequent heat treatment, which cannot be
simply interpreted in terms of microstructural features of the alloy forging.
A wide range of creep properties are often observable within a given microstructure and the variations in creep can be generally related to the processing histories of the forgings.

Figure 6 presents the effect of the nature and distribution of incoherent
precipitates on the creep resistance of the alloy forgings; these submicro-
structures were obtained from electrolytically extracted replica.   X-ray
fluorescent analysis of the particles indicates that these particles are
generally rich in silicon and tin (Figure 7).   Such undesirable particles
are often formed as a result of improper forging and heat treating condi-

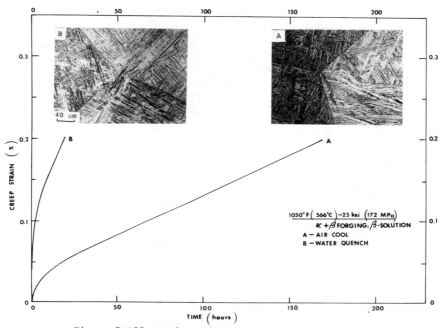

Figure 5 Effect of cooling rate from solution-treat temperature
on creep resistance of Ti-6242Si alloy forgings

tions and do not significantly affect the tensile ductility and strength
properties.  These particles are not observable at optical magnification and
are more confined to the interface region for low creep conditions.  Attempts by thin foil TEM failed to correlate these particles to $\alpha_2$-phases.
It is also seen from Figure 6 that at high creep conditions, an excellent
combination of creep, tensile, and fracture toughness properties could be
achieved for the forgings.

Figure 8 further illustrates examples of fracture surface of broken
creep-rupture samples and fracture toughness samples at high, intermediate,
and low creep conditions.  At both high and low creep resistances, the size
of prior β-grains and Widmanstatten colonies are shown to control the creep
fracture behavior of the alloy (Figure 8a).  Similar to that observed for
Ti-6Al-4V alloy forgings [6], the higher fracture toughness of fully transformed microstructures appears to associate with the extent of crack branching (Figure 8b).

It has been suggested that the immobilization of mobile dislocations by
either fine silicide or silicon atmosphere may dominate the rate controlling
creep mechanism for the high temperature titanium alloys containing silicon
[7,8].  Since Si and probably Sn are effective solid solution strengtheners
for creep resistance, the extraction of these elements from solid solutions
to form incoherent precipitates due to improper thermomechanical history
would obviously affect the state of silicon dissolution, thereby limiting
the availability of silicon in solution for improved creep properties.

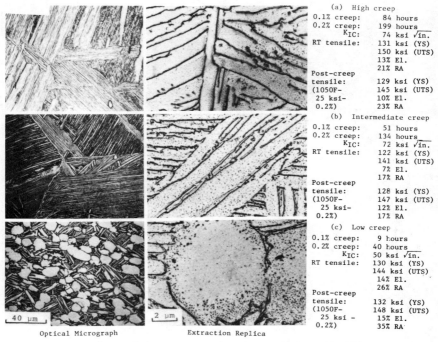

(a)  High creep

| 0.1% creep: | 84 hours |
|---|---|
| 0.2% creep: | 199 hours |
| $K_{IC}$: | 74 ksi √in. |
| RT tensile: | 131 ksi (YS) |
| | 150 ksi (UTS) |
| | 13% El. |
| | 21% RA |
| Post-creep | |
| tensile: | 129 ksi (YS) |
| (1050F- | 145 ksi (UTS) |
| 25 ksi- | 10% El. |
| 0.2%) | 23% RA |

(b)  Intermediate creep

| 0.1% creep: | 51 hours |
|---|---|
| 0.2% creep: | 134 hours |
| $K_{IC}$: | 72 ksi √in. |
| RT tensile: | 122 ksi (YS) |
| | 141 ksi (UTS) |
| | 7% El. |
| | 17% RA |
| Post-creep | |
| tensile: | 128 ksi (YS) |
| (1050F- | 147 ksi (UTS) |
| 25 ksi- | 12% El. |
| 0.2%) | 17% RA |

(c)  Low creep

| 0.1% creep: | 9 hours |
|---|---|
| 0.2% creep: | 40 hours |
| $K_{IC}$: | 50 ksi √in. |
| RT tensile: | 130 ksi (YS) |
| | 144 ksi (UTS) |
| | 14% El. |
| | 26% RA |
| Post-creep | |
| tensile: | 132 ksi (YS) |
| (1050F- | 148 ksi (UTS) |
| 25 ksi - | 15% El. |
| 0.2%) | 35% RA |

| 40 μm | | 2 μm |
|---|---|---|
| Optical Micrograph | | Extraction Replica |

Figure 6 Optical microstructures at three distinctly different creep
resistances and corresponding extraction replicas showing the
nature and distribution of precipitates in Ti-6242Si alloy
forgings
Creep test conditon = 1050°F(566°C)-25ksi(172MPa)
1 ksi = 6.9 MPa, 1 ksi √in. = 1.1 MPa√m

From the above, it is clearly demonstrated that one of the major factors
in limiting the high creep capability of Ti-6242Si alloy is the formation of
incoherent precipitates containing Si and Sn, which, in turn, depend criti-
cally on the forge and heat treat schedules.  The basic problem in the conventional treatments for correlating the microstructure and mechanical properties arises because of the lack in understanding the formation of such
heterogeneous precipitates.

Achievable Properties and
Process Optimization

At the present, the most common forge and heat treat schedule for the Ti-
6242Si alloy comprises:  forge at temperatures high in the (α+β)-field, air
cool, solution treat for one hour at temperatures high in the (α+β)-field, air
cool, stabilization age for 8 hours at 1100°F(593°C), and air cool.  The 0.1%
creep strain at 950°F(510°C)/35 ksi(241 MPa) test condition for these (α+β)-
forgings requires about 35 and 125 hours, respectively, depending on the
specification requirements.  As stated earlier, the β-forgings of this alloy
have been very recently used in some cases to acicularize the microstructure
and have shown to result in increases in creep strength and fracture toughness
with a slight reduction of both strength and ductility.  By using the (α+β)-

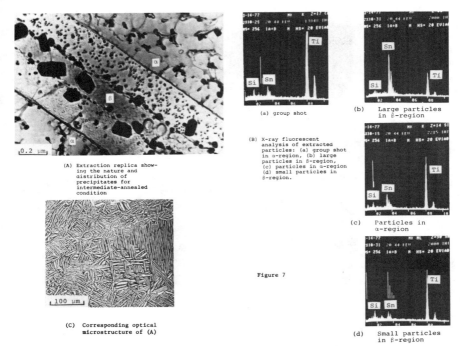

(A) Extraction replica showing the nature and
distribution of
precipitates for
intermediate-annealed
condition

(B) X-ray fluorescent
analysis of extracted
particles: (a) group shot
in α-region, (b) large
particles in β-region,
(c) particles in α-region
(d) small particles in
β-region.

(C) Corresponding optical
microstructure of (A)

(a)    group shot

(b)    Large particles
in β-region

(c)    Particles in
α-region

Figure 7

(d)    Small particles
in β-region

solution treatment and aging, the time required to reach 0.1% creep strain for
β-forging could be increased from (α+β)-forged condition by a factor of 2
quite easily.  The times to reach 0.1% creep at 950°F(510°C)/35 ksi(241 MPa)
test condition for transformed microstructures produced by β-forging are in
the range of 200 to 400 hours; this is estimated equal to about 35 hours at
1050°F(566°C)/25 ksi(172 MPa) test condition.

By optimizing microstructural and submicrostructural features of the
alloy forgings, the present investigation has shown that the creep properties
of Ti-6242Si alloys could be maximized for both (α+β)- and β-forgings through
processing and structural controls.  For (α+β)-forgings, the best creep can be
achieved by a β-solution treatment, followed by a conventional aging at
1100°F(593°C) for about 8 hours, and air cooling.  The solution treatment
should be conducted within the (α+β)-phase range for β-forgings.  These
processing conditions obviously result in mimimizing or eliminating the
formation of incoherent precipitates and in achieving the maximum creep
properties of the alloy.  The range of achievable creep properties through the
above processing conditions obtained in this investigation are 70 to 110 hours
0.1% creep and 160 to 200 hours 0.2% creep at 1050°F(516°C)/25 ksi(172 MPa)
test conditions, and 30 to 50 hours 0.1% creep and 80 to 100 hours 0.2% creep
at 1100F(595°C)/15 ksi(103 MPa).  Other properties associated with these
processing conditions are:   130 to 135 ksi (896 to 931 MPa) yield strength,
145 to 155 ksi (1000 to 1069 MPa) UTS, 8 to 13% elongation, and 15 to 25%
reduction of area.  The fracture toughness values are generally in the range
of 70-75 ksi $\sqrt{\text{in.}}$ (77∿82  MPa $\sqrt{m}$) $K_{Ic}$.  Table I illustrates examples of the
mechanical properties achievable for Ti-6242Si alloy forgings through various
processing controls, and at various property combinations.

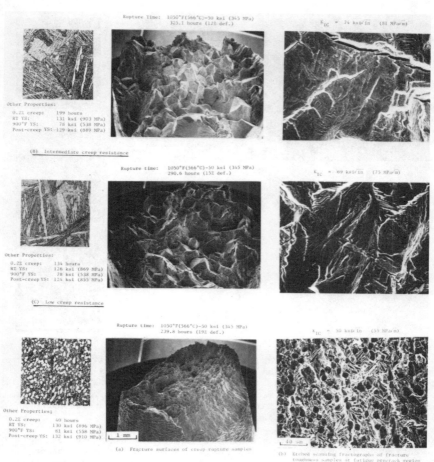

Figure 8 Fractographic features of Ti-6242Si alloy forgings at high, intermediate, and low creep resistances

## SUMMARY

The results of this investigation demonstrated that excellent combination of creep, tensile, and fracture toughness properties of the alloy forgings can be achieved for both (α+β) and β-forgings through optimum control of the microstructures, on both a micro- and submicroscale. The creep resistance of the Ti-6242Si alloy forgings has an obvious variation with optical microstructure only at a broad range of microstructural differences. The microstructure characterized by less-defined fine Widmanstatten colonies appears to have the highest creep resistance if proper heat treatment is used. Both transformed martensitic-α and globular-α appear to significantly degrade the creep and fracture toughness properties. By eliminating and/or reducing the incoherent precipitates through adequate control of processing variables, excellent creep resistance of Ti-6242Si alloy forgings was achievable.

TABLE 1

ACHIEVABLE MECHANICAL PROPERTIES OBTAINED FROM VARIOUS FORGING AND HEAT-TREATING COMBINATIONS

| Forge/ Heat Treat Conditions | Creep* (Hours to 0.2% Strain) | Creep** Rupture Time (Hours) | K_IC ksi√in. (MPa√m.) | RT Tensiles | | | | 900F Tensiles | | | | Post-Creep Tensiles | | | |
|---|---|---|---|---|---|---|---|---|---|---|---|---|---|---|---|
| | | | | Y.S. ksi (MPa) | UTS ksi (MPa) | El. (%) | RA (%) | Y.S. ksi (MPa) | UTS ksi (MPa) | El. (%) | RA (%) | Y.S. ksi (MPa) | UTS ksi (MPa) | El. (%) | RA (%) |
| (α+β)/β | 199 | 325 | 74 (81) | 131 (903) | 150 (1034) | 13 | 21 | 78 (538) | 100 (690) | 13 | 36 | 129 (889) | 145 (1000) | 10 | 23 |
| (α+β)/β | 169 | 296 | 69 (76) | 134 (924) | 156 (1076) | 10 | 15 | 82 (565) | 108 (745) | 11 | 25 | - | - | - | - |
| (α+β)/(α+β) | 56 | 234 | 50 (55) | 138 (952) | 154 (1062) | 14 | 27 | 86 (593) | 112 (772) | 15 | 44 | - | - | - | - |
| (α+β)/β | 19 | - | ~35 (38) | 155 (1069) | 181 1241 | 3 | 5 | 124 (855) | 143 (986) | 5 | 17 | - | - | - | - |
| β/(α+β) | 185 | - | ~70 (77) | 135 (931) | 154 (1062) | 11 | 23 | 84 (579) | 109 (752) | 17 | 38 | - | - | - | - |
| β/(α+β) | 160 | - | ~75 (82) | 136 (938) | 156 (1076) | 10 | 22 | - | - | - | - | - | - | - | - |
| β/(α+β) | 116 | - | 76 (84) | 130 (896) | 150 (1034) | 11 | 25 | 82 (565) | 110 (758) | 15 | 40 | 134 (924) | 152 (1048) | 12 | 23 |
| β/β | 134 | 275 | 72 (79) | 122 (841) | 141 (972) | 7 | 17 | 80 (552) | 104 (717) | 12 | 28 | 128 (882) | 147 (1014) | 12 | 17 |

*  Creep condition:  1050°F (566°C) - 25 ksi (172 MPa)

** Creep rupture:   1050°F (566°C) - 50 ksi (345 MPa)

~  Estimated value

REFERENCES

1. "Metallurgical and Mechanical Properties of Ti-6Al-2Sn-4Zr-2Mo Sheet, Bar, and Forgings", Technical Data Sheet, Titanium Metals Corporation of America, Pittsburgh, Pennsylvania 1966.
2. S. R. Seagle, et al: Metals Engineering Quarterly, p. 48, February 1975.
3. C. C. Chen: Wyman-Gordon Company Report RD-79-116, 1979.
4. C. C. Chen: Wyman-Gordon Company Report RD-77-110, 1977.
5. D. Eylon, et al: Metallurgical Transactions, 7A (1976), 1817.
6. C. C. Chen and J. E. Coyne: "Titanium and Titanium Alloys", P. 383, Plenum Press, New York, N.Y., 1980.
7. H. W. Rosenberg: "Titanium Science and Technology", 4 (1973), p.2127, Plenum Press, New York, N.Y.
8. N. E. Paton and M. W. Mahoney: Metallurgical Transactions, 7A (1976), 1685.

THE RELATIONSHIP OF PROCESSING/MICROSTRUCTURE/MECHANICAL PROPERTIES
FOR THE ALPHA-BETA TITANIUM ALLOY  TI-4.5AL-5MO-1.5CR (CORONA-5)

G. R. Keller,* J. C. Chesnutt,** W. T. Highberger,***
C. G. Rhodes,**** and F. H. Froes*****

*G. R. Keller - North American Aircraft Division, Rockwell
International, Los Angeles, CA.
**J. C. Chesnutt - Science Center, Rockwell International,
Thousand Oaks, CA.
***W. T. Highberger - Naval Air Systems Command, Washington, D.C.
****C. G. Rhodes - Science Center, Rockwell International,
Thousand Oaks, CA.
*****F. H. Froes - Air Force Materials Laboratory, WPAFB, OH

## Abstract

The effect of processing parameters on microstructure and mechanical properties for a high-fracture-toughness titanium alloy, CORONA-5 (Ti-4.5Al-5Mo-1.5Cr), is discussed. Data developed from forgings and plate produced on production equipment are presented and compared with Ti-6Al-4V. Mechanical properties are discussed in terms of microstructure and fractographic features.

## Introduction

Over the past decade, a concentrated effort has been made by producers and aerospace users of titanium alloys to develop improved mechanical properties for advanced-aircraft applications. This effort has been motivated to a large extent by the advent of new design criteria and concepts such as fracture mechanics and design-to-cost. Ti-6Al-4V, the major titanium alloy in aircraft structures, is being pushed to its technological limits to satisfy these requirements and, in an increasing number of instances, it fails to meet desired levels. A new titanium alloy Ti-4.5Al-5Mo-1.5Cr (CORONA-5) has been developed specifically to meet the requirements of aircraft components designed to fracture mechanics criteria while remaining cost effective.

This paper will present the results of a recently completed study designed to define the relationship between processing parameters, microstructure, and mechanical properties for CORONA-5. It will be demonstrated that by careful control of processing conditions, a lenticular primary alpha microstructure can be obtained which yields optimum fracture mechanics characteristics. It will further be shown that this microstructural feature is desirable for high fracture toughness over the strength range 130 to 200 ksi (895 to 1,375 MPa) yield strength.

Data developed from forgings and plate produced on production equipment will be presented. Comparisons of strength, fracture toughness, ductility, fatigue crack growth, fatigue (smooth and notched), and stress corrosion cracking resistance with corresponding data for Ti-6Al-4V will be made. The superior mechanical property combination exhibited by the new alloy will be explained in terms of microstructural and fractographic features.

## Thermomechanical Processing

The CORONA-5 alloy evolved from an early program in which a spectrum of titanium alloys, ranging from alpha-beta to rich metastable beta compositions, was evaluated [1]. In this program, strong consideration was given to chemistry, thermomechanical processing, and final heat treatment. A detailed study was also made of the resultant microstructures and how these affected fracture toughness [2,3].

It was found that the alpha phase within the beta matrix strongly influenced strength and fracture toughness. The morphology and distribution adopted by the alpha relates directly to prior processing and the final heat treatment. Thus, the mechanical properties depend on the thermomechanical history of the material. Fracture toughness, for example, was found to relate to the aspect ratio of the relatively coarse alpha formed at higher temperatures [2,3,4] (within about 300° F [150° C] of the beta transus temperature), while strength level correlated with the finer alpha formed at lower temperatures [4] (800 to 400° F [425 to 205° C] below the beta transus temperature). High-fracture toughness was associated with a high-aspect-ratio (lenticular) alpha, while lower fracture toughness values at the same strength level corresponded to a low-aspect-ratio (globular) alpha. This effect has been related to the tortuosity of the fracture path[4].

Fracture toughness also shows an inverse relationship with strength level for a given microstructure[4]. Thus, by the appropriate selection of thermomechanical processing and final heat-treatment parameters, the alloy may be tailored to optimize fracture toughness, while meeting minimum strength levels dictated by design criteria. Conversely, the alloy may be tailored to provide a lower fracture toughness level, as determined by inspectability imposed limitations on critical crack length, and the strength level may be adjusted accordingly.

Two basic processing options after beta annealing are available to achieve the desired primary alpha morphology: (1) to beta process, where processing is carried out wholly above the beta transus or in which finish processing is completed below the beta transus but at a temperature high enough that very little alpha phase is present, or (2) to alpha-beta process, in which case processing is carried out below the beta transus temperature in the presence of alpha phase. After either processing route, annealing below the beta transus temperature (within 300° F [150° C]) results in a distribution of primary alpha which is related to the processing sequence and annealing temperature. With beta processed material, the desired lenticular alpha morphology is achieved and maintained. With alpha-beta processing the primary alpha can become globular during subsequent heat treatment, with a resulting decrease in fracture toughness. The change in morphology of the alpha phase from lenticular to globular is a direct result of the working (deformation) of the alpha. The strain energy in the alpha is thought to cause the alpha to recrystallize and in the course of recrystallization to relax to the lower surface energy globular configuration[5]. This is demonstrated in Figure 1. The rate at which the lenticular alpha transforms to globular alpha is a function of (1) annealing temperature, (2) annealing time, and (3) the amount of working the alpha has received. Hence, lightly worked alpha remains lenticular longer than heavily worked alpha. Figure 2 illustrates this effect for three levels of deformation. Time-temperature combinations to the bottom right of the appropriate deformation line produce lenticular alpha, while to the top left, globular alpha results. In practice, a partial coarsening of the alpha while maintaining a high aspect ratio appears desirable to fine tune the fracture toughness level[3] in agreement with the findings for other titanium alloys[6,7].

AS ROLLED          5 HOURS          24 HOURS   └─┘
1 μm

Figure 1.   Transformation of Lenticular Alpha to Globular Alpha by Holding at 1,475° F (800° C) After Deforming Material 70 Percent in the Alpha-Beta Field

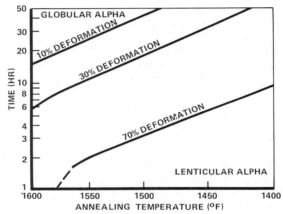

Figure 2.   Relationship of Alpha Morphology to Annealing Time and Temperature After Various Amounts of Alpha-Beta Deformation

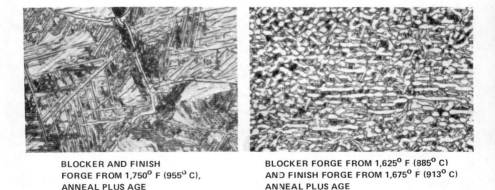

BLOCKER AND FINISH
FORGE FROM 1,750° F (955° C),
ANNEAL PLUS AGE

BLOCKER FORGE FROM 1,625° F (885° C)
AND FINISH FORGE FROM 1,675° F (913° C)
ANNEAL PLUS AGE

Figure 3.   Microstructures of Forged CORONA-5 Illustrating Two Processing Routes

## Microstructure and Mechanical Properties

The understanding of the previously described microstructure development in CORONA-5 has led to two recommended alternate thermomechanical processing routes for the alloy after beta annealing. The first is beta processing which is followed by a high alpha-beta anneal (to produce lenticular primary alpha) and an aging/stabilization treatment (to adjust the strength level). Alternately, the alloy may be alpha-beta processed. In this case, a duplex high-alpha-beta anneal is required; the first anneal is close to the beta transus to globularize the alpha which has been worked and to minimize the volume fraction of this microstructural constituent. This is followed by a second lower temperature anneal to precipitate new lenticular alpha between the globular alpha particles. This duplex treatment is then follwed by an aging/stabilization treatment in a manner similar to the beta-processed material.

The microstructures resulting from these two processing routes are shown in Figure 3 for the same final annealing and aging treatments. Figure 3a illustrates the high-aspect-ratio primary alpha resulting from beta blocking and finish forging in the beta region, followed by alpha-beta annealing and aging. This structure developed a fracture toughness value, $K_Q$, of 141.6 ksi $\sqrt{in}$. (155.8 MPa $\sqrt{m}$) calculated from slow-bend precracked Charpy samples at a strength level of 123.4 ksi (850.2 MPa) TYS. Figure 3b shows the reduced aspect-ratio primary alpha phase surrounded by a structure of fine lenticular alpha formed from alpha-beta blocker forging, followed by finish forging just below the beta transus temperature, $T_B$ = 1,700° F (925° C), annealing and aging to approximately the same strength level as in Figure 3a, 126.4 ksi (870.9 MPa) TYS. A fracture toughness of 129.9 ksi $\sqrt{in}$. (142.9 MPa $\sqrt{m}$) resulted from this microstructure. Similar comparisons obtain for these microstructures when heat treated to higher strength and correspondingly lower fracture toughness levels[6].

Alpha-beta working followed by short-time, single-temperature annealing also gives a lenticular alpha (Figure 2); however, in this case, properties are highly directional and therefore not generally desirable.

The desirability of maintaining high-aspect-ratio lenticular alpha to develop high fracture toughness over a wide strength range in CORONA-5 is illustrated in Figure 4. In this study [8], duplicate longitudinal and transverse tensile and slow-bend, precracked Charpy specimens were tested from 27 CORONA-5 plate samples produced by the three primary processing routes. Beta-processed material was rolled from 1,800° F (982° C), annealed, and aged to develop fracture toughness over a range of strength from 130 to 200 ksi (895 to 1,375 MPa). Similarly, alpha-beta processed and duplex-annealed material was rolled from 1,600° F (871° C), duplex annealed and aged; alpha-beta-processed and alpha-beta-annealed material was rolled from 1,600° F (871° C), single annealed and aged. As shown in Figure 4, higher toughness is associated with processing sequences which result in a more highly lenticular alpha phase morphology; i.e., beta finished plus annealed. It can also be seen that by adjustment of thermomechanical processing sequences and heat treatment, shifts in the trend lines can be produced as desired. This permits an increase in fracture toughness without the attendant loss in strength associated with adjustment in heat-treatment cycles alone.

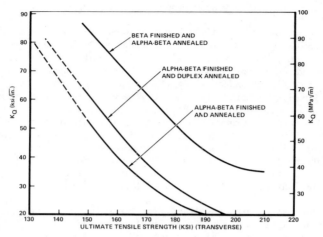

Figure 4.    Effect of Processing Mode on Strength/
Toughness Relationship of CORONA-5

## Mechanical Properties of Production Material

Mechanical property data have been developed for CORONA-5 in the range
135 to 150 ksi (930 to 103 MPa) UTS from production ingots (up to 5,000
pounds [2270 kilograms]) converted on production equipment to forgings [3,8]
and plate [8].    Room-temperature tensile properties, $K_{I_c}$, $K_{I_{SCC}}$, fatigue
crack propagation characteristics (da/dn), and fatigue properties (S-N) have
been obtained.

During the forging program [3], six different forging sequences
were evaluated to determine the effects of varying beta- and alpha-beta processing sequences on resultant microstructures and mechanical properties.
In addition, two final heat-treatment cycles were employed to provide
strength variation.    Under the plate program [8], three rolling sequences
and two heat-treatment cycles were evaluated in the production of 3-inch
plate.    The results of these evaluations indicated that the best combinations
of strength and fracture toughness were produced by processing the material
above its beta transus or at temperatures high in the alpha-beta field as
discussed previously.    Both these processing routes are represented in
strength/toughness data for optimally processed CORONA-5 forgings and plate
presented in Table 1.    Strength and toughness differences between similarly
processed forgings and plate are attributed to differences in oxygen levels,
0.15 and 0.18 weight-percent, respectively, and texturing effects.

The beta-processed materials, in particular, show excellent strength/
fracture toughness combinations and the data also compare favorably with
typical properties for high toughness Ti-6Al-4V, Condition BA[9] and RA[10],
as shown in Table 2.

Table 1   Mechanical properties of CORONA-5

| Product Form | Processing | Dir | Tensile Properties | | | | Fracture Toughness | |
|---|---|---|---|---|---|---|---|---|
| | | | TYS (ksi) | UTS (ksi) | El (%) | RA (%) | KQ [1] (ksi √in.) | KIc (ksi √in.) |
| Forging | Beta forge | L | 117.5 | 132.1 | 15.6 | 23.1 | 137.1 | - |
| | | T | 123.4 | 136.4 | 12.0 | 20.8 | 141.6 | 121.6 [2] |
| | Beta forge + light alpha-beta forge | L | 120.1 | 131.0 | 17.2 | 33.4 | 131.1 | - |
| | | T | 126.4 | 141.7 | 15.3 | 31.3 | 151.2 | 148.0 [2] |
| | Beta forge + light alpha-beta forge | L | 123.9 | 134.9 | 17.4 | 29.3 | 148.9 | - |
| | | T | 129.3 | 145.2 | 13.4 | 20.0 | 162.5 | 145.9 [2] |
| | Alpha-beta forge | L | 124.9 | 133.5 | 17.4 | 27.3 | 96.6 | - |
| | | T | 126.4 | 135.1 | 14.8 | 27.4 | 129.9 | - |
| Plate[3] | Beta roll | L | 129.4 | 136.0 | 17.6 | 43.4 | - | - |
| | | T | 133.5 | 139.2 | 15.4 | 33.3 | - | 124.1 |
| | Beta roll | L | 136.3 | 145.9 | 14.4 | 34.1 | - | - |
| | | T | 140.4 | 150.0 | 13.1 | 26.9 | - | 112.3 [4] |
| | Alpha-beta roll | L | 134.2 | 136.2 | 19.3 | 54.5 | - | - |
| | | T | 139.2 | 139.2 | 18.3 | 50.1 | - | 88.6 |

1. Computed from slow-bend, precracked Charpy specimen
2. Failed thickness requirement of ASTM E399
3. Tensile properties are averaged duplicate specimens
4. Failed precrack shape requirement of ASTM E399

Table 2   Typical mechanical properties of Ti-6Al-4V

| Process | TYS (ksi) | UTS (ksi) | El (%) | RA (%) | KIc (ksi √in.) |
|---|---|---|---|---|---|
| Cond BA | 125 | 135 | 12 | 25 | 90 |
| Cond RA | 125 | 135 | 15 | 30 | 75 |

Limited stress corrosion cracking resistance testing has been conducted for forgings and plate.  Single specimens of 100-hour WOL sample $K_{I_{scc}}$ tests in 3.5-percent NaCl were tested and the results, together with fracture toughness data, are presented in Table 3.

Table 3   Stress corrosion cracking resistance of CORONA-5

| Product | Processing | $KI_{SCC}$ (ksi $\sqrt{in.}$) | $KI_c$ (ksi $\sqrt{in.}$) | Percent of $KI_c$ |
|---------|-----------|-----------------|---------------|----------------|
| Forging | Beta forged | 106.6 | 121.6[1] | 87 |
| Forging | Beta forged + light alpha-beta forged | 127.5 | 148.0[1] | 86 |
| Plate | Beta rolled | 69.0 | 112.3[2] | 61 |
| Plate | Alpha-beta rolled | 68.1 | 88.6 | 77 |

1.  Failed ASTM E399 thickness requirement
2.  Failed ASTM E399 crack shape requirement

The scanning electron microscope (SEM) evaluation of the fractured $KI_{SCC}$ specimens of forged material revealed no stress corrosion cracking; some evidence of plastic deformation was noted in the region of the crack tip, suggesting the specimen geometry for the very high-toughness condition may have been insufficient to prevent extensive crack-tip plasticity. Nevertheless a $KI_{SCC}$ of 60 ksi $\sqrt{in.}$, or larger appears valid for the material and compares favorably with Ti-6Al-4V beta-annealed plate[9] at equivalent oxygen content (0.18 weight-percent).

Fatigue crack propagation (FCP) rate tests were conducted in low-humidity air for four forging conditions and six plate-processing and heat-treatment conditions.  Lowest crack propagation rates resulted from beta-processed material and were found to be relatively insensitive to variations in strength level.  FCP test results for beta-processed plate and forgings are presented in Figure 5 with upper and lower bound lines for comparative Ti-6Al-4V BA, RA, and MA data [9].  CORONA-5 is seen to fall on the lower bound (slow-growth rate) for Ti-6Al-4V MA, but on the upper bound for the BA and RA conditions which represent refinements to the chemistry, processing, and heat treatment of Ti-6Al-4V MA.  Further refinements in chemistry (particularly oxygen) and manipulation of microstructure may similarly enhance the FCP characteristics of CORONA-5.

Notched and unnotched bar axial tension fatigue data are presented for CORONA-5 plate and forgings, compared with 6Al-4V products, in Figure 6. The data are taken from results of tests of plate produced by both processing routes; i.e., beta processed plus alpha-beta annealed and alpha-beta processed and duplex annealed.  Forgings, similarly, were beta forged and alpha-beta annealed.  The data show good uniformity irrespective of testing direction, product form, or processing mode, particularly for the notched (Kt = 3) condition.  The unnotched bar fatigue strength can be seen to be higher than Ti-6Al-4V products compared beyond $3 \times 10^4$ cycles, while the notched bar data fall within the narrow range for Ti-6Al-4V.

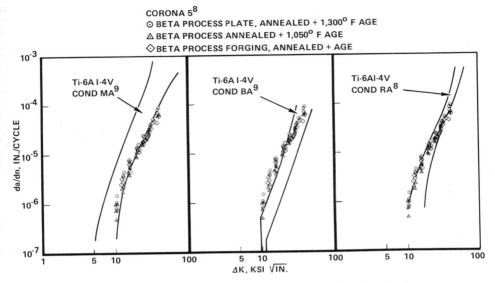

Figure 5.  Fatigue Crack Propagation Data for CORONA-5
Plate Compared to Bound Lines for Ti-6Al-4V

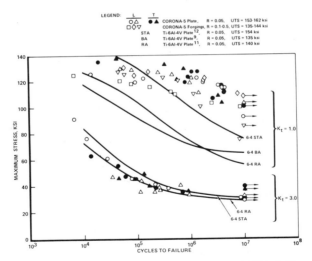

Figure 6.  Axial Tension Fatigue Data for CORONA-5 Plate and
Forgings Compared to Ti-6Al-4V MA, RA, and BA

## Microstructure and Fractography Correlations

The influence of microstructure on mechanical properties of CORONA-5 is most strongly manifested by the fracture characteristics of the alloy, especially fracture toughness and the associated fracture surface topography. Three microstructural conditions of forged material[3,8] are used to illustrate the effect; these include (1) beta blocked and beta forged (Figure 7a), (2) beta blocked and high alpha-beta forged (Figure 7c), and (3) alpha-beta blocked and low alpha-beta forged (Figure 7e). All three were heat treated to tensile yield strengths of 125 ksi (860 MPa); the first two conditions are the first two forged conditions in Table 1. The three microstructures are shown with representative scanning electron micrographs of the fracture surface in Figure 7.

Analysis of the scanning electron micrographs reveals that fracture toughness in this alloy can be directly related to fracture path and to the path-controlling microstructure. The fracture surface of the beta-forged specimen (Figure 7b) is highly irregular; this results from extensive crack branching on the scale of the prior beta grain size. A tortuous fracture path is caused by microstructural features of a different scale in the high alpha-beta forged material (Figure 7d). This path results from the high aspect ratio of the primary alpha (Figure 7c). The elongated alpha particles provide preferential crack paths either along the major axis of the particle or at the interface between the particle and the beta matrix. The increased tortuosity of the crack path for both conditions increases the net area of the crack plane and thereby increases the fracture toughness In contrast, the low alpha-beta forged material, which has considerably lower toughness, exhibits an equiaxed primary alpha microstructure (Figure 7e) and an associated flat, smooth fracture comprised of small equiaxed dimples (Figure 7f).

Similar fracture trends are observed in plate material, but the benefits of the elongated primary alpha can be lost if the particles are not randomly oriented.

## Summary

The mechanical properties, especially fracture toughness, of CORONA-5 depend strongly on the morphology and distribution of the alpha phase. Through variations in thermomechanical processing, the strength and fracture toughness can readily be adjusted. Lenticular alpha formed from one of several processing routes promotes high-fracture toughness. This has been demonstrated for plate and forged material produced on mill production equipment. Tensile strength, fracture toughness, fatigue crack propagation rates, S/N fatigue, and stress corrosion cracking resistance compare favorably with Ti-6Al-4V. Microstructural and fractographic evaluation indicates the high fracture toughness of CORONA-5 results from the tortuous crack path associated with the high-aspect-ratio lenticular alpha developed in the alloy.

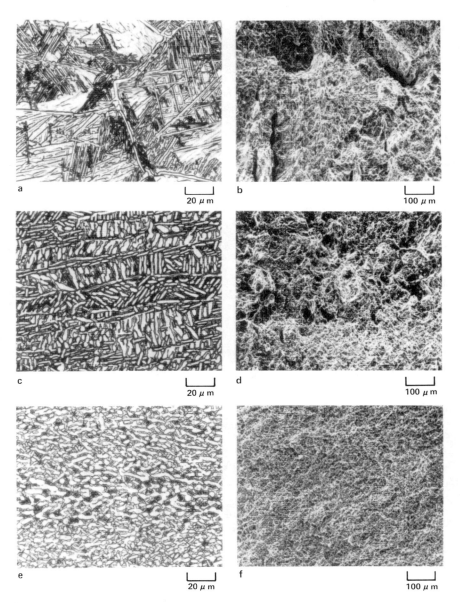

Figure 7.  Microstructure and Fracture Topography in CORONA-5 - (a,b) Beta Forged, (c,d) High Alpha-Beta Forged, and (e,f) Low Alpha-Beta Forged

## References

1. Berryman, R. G., Froes, F. H., Chesnutt, J. C., Rhodes, C. G., Williams, J. C., and Malone, R. F., Final Engineering Report, Naval Air Systems Command Contract N00019-73-C-0355, TFD-74-657, July 1974.

2. Berryman, R. G., Froes, F. H., Chesnutt, J. C., Rhodes, C. G., Williams, J. C., and Malone, R. F., Final Engineering Report, Naval Air Systems Command Contract N00019-74-C-0273, TFD-75-640, July 1975

3. Berryman, R. G., Chesnutt, J. C., Froes, F. H., and Rhodes, C. G., Naval Air Systems Command Contract N00019-75-C-0208, TFD-76-471, June 1976.

4. Hirth, J. P., and Froes, F. H., Met. Trans., Vol 8A, July 1977, p 1165.

5. Froes, F. H., Chesnutt, J. C., and Williams, J. C., Work in Progress, 1979-80.

6. Margolin, H., Greenfield, M. A., and Greenhut, I., Titanium Science and Technology, Plenum Press, New York, 1973, p 1709.

7. Hall, I. W., and Hammond, C., Mat. Sci. and Engr., Vol 32, 1978, p 241.

8. Keller, G. R., Chesnutt, J. C., Froes, F. H., and Rhodes, C. G., Final Engineering Report, Naval Air Systems Command Contract N00019-76-C-0427, NA-78-917, December 1978.

9. Boyer, R. R., and Bajoraitis, R., "Standardization of Ti-6Al-4V Processing Conditions," Technical Report, AFML-TR-78-131, Air Force Materials Laboratory, Wright-Patterson AFB, Ohio, September 1978.

10. Damage Tolerant Design Handbook, MCIC-HDBK-01, Metals and Ceramics Information Center, Battelle Columbus Laboratories.

11. "B-1 Airframe Material Fatigue Design Properties Manual", Technical Report NA72-1088, North American Aircraft Division, Rockwell International, Los Angeles, CA, August 1975.

12. Military Standardization Handbook, MIL-HDBK-5C, "Metallic Materials and Elements for Aerospace Vehicle Structures," Dept of Defense, Washington, D.C., Change Notice 1, 15 December 1978.

# MICROSTRUCTURE AND STRENGTH OF SOLUTION TREATED Ti-8Al-1Mo-1V

S.J. Vijayakar, E.S.K. Menon, S. Banerjee and R. Krishnan

Metallurgy Division, Bhabha Atomic Research Centre,
Trombay, Bombay-400 085
India

## Introduction

The commercial Ti-8Al-1Mo-1V alloy is suitable for high temperature applications due to a combination of several desirable properties like good weldability and formability, high strength and Young's modulus and low density. The
alloy Ti-8Al-1Mo-1V is often described as a "near alpha" alloy which is a relatively accurate description of its microstructure after the heat treatments
used industrially. Although this is a near alpha alloy, an increase in tensile
strength by almost 25% over that of the mill annealed material can be obtained
by an appropriate choice of heat treatment sequence consisting of quenching
followed by an ageing treatment [1] . The main heat treatment variables
involved are the solutionising temperature and the temperature and the duration
of ageing. The solutionising temperature determines the relative proportions
of the constituent phases and also the extent of solute partitioning the enrichment of the alpha phase with alpha stabilising Al and the enrichment of the beta
phase with beta stabilising Mo and V. The formation of the ordered $Ti_2Al$ phase
and the metastable omega phase in the alpha and the beta regions respectively
during the cooling subsequent to solution treatment or during ageing would
further modify the microstructure of the alloy. The influence of different
heat treatments on the structure of this alloy has been studied by many
workers [1-4] . Fopiano and Hickey [1] have shown that at a critical temperature of solutionising (around 800°C), this alloy exhibits a yield strength
minimum and a maximum for ductility and toughness. They have also found that
a plot of elastic modulus versus solutionising temperature shows a minimum in
the same temperature range. Similar observations have been reported in many
other titanium alloys [5-7] and this conspicuous drop in yield strength
has been attributed to the instability of the retained beta phase during plastic
deformation. However, experimental evidence in support of such a stress induced
transformation in this alloy has not been reported so far.

The present work was carried out with a view to correlating the microstructures obtained in the alloy Ti-8Al-1Mo-1V by quenching from different
temperatures in the beta and the alpha + beta phase fields with its mechanical
properties. In order to investigate the stress induced transformation in this
alloy, when quenched from a temperature close to 800°C, during subsequent
plastic deformation, microstructures developed in the deformed samples have
been examined.

## Experimental Procedure

The titanium alloy used in this investigation was obtained from the
Titanium Metals Corporation of America in the form of 2 mm thick sheets in the
hot rolled and annealed condition. The maximum levels of C, O, N and H were
specified to be 0.09, 0.12, 0.06 and 0.015% respectively. Tensile specimens

of 12.5 mm gauge length were obtained from sheets cold rolled to a thickness
of about 0.3 mm. All heat treatments were carried out on samples wrapped
in tantalum foils and then encapsulated in silica capsules under helium
atmosphere. Solution treatments were carried out in the temperature range
1100 to 600°C at intervals of 50°C. After appropriate heat treatments, the
samples were deformed in an Instron machine at a strain rate of $1.37 \times 10^{-4} \text{sec}^{-1}$
at room temperature. Thin foils for transmission electron microscopy were
obtained by electropolishing the samples in an electrolyte containing 35 parts
of perchloric acid, 165 parts of n-butanol and 300 parts of methyl alcohol,
using voltages in the range of 25-30V at temperature below -50°C.

### Results and Discussion

A brief description of the effect of solutionising temperature on the
microstructure of the quenched Ti-3Al-1Mo-1V alloy is given below. The salient
features of the results of the tensile tests and the X-ray diffraction studies
are also presented. An attempt has been made to rationalise the observed
tensile properties on the basis of the microstructural characterisation.

1. Microstructure of solution treated samples

On quenching the alloy from the beta phase field, a fully martensite
structure was obtained as shown in Fig.1(a). This martensite showed a lath
morphology (Fig.1(b)). Several near parallel laths were stacked in a colony,
the misorientation between the neighbouring laths being very small. The
majority of the laths had a dislocated substructure (Fig.1(c)) suggesting
that a slip mode of inhomogeneous shear accompanied the transformation. In
majority of these laths, some planar features similar to stacking faults were
observed. However, the incidence of faulting was found to be much more in
samples subjected to a tempering treatment subsequent to beta quenching
(Fig.1(d)). These faults were found to be of the basal type with $1/3 \langle 10\bar{1}0 \rangle$
type fault vectors. The faults observed in this study were very similar to
those noticed by Williams and Blackburn [7] in the same alloy. It is known
that addition of Al lowers the stacking fault energy of Ti [8] and the high
density of faults in tempered samples is suggestive of a preferential migration
of Al atoms to the faults resulting in a further reduction in the stacking
fault energy and in further growth of the faults.

X-ray and electron diffraction studies showed that no high temperature beta
phase was retained, suggesting that the transformation was complete. This is
expected in a Ti alloy containing Al, an alpha stabilising element, as the
major alloying addition. Diffraction studies also indicated that the martensite
produced in the Ti-3Al-1Mo-1V alloy, irrespective of the solutionising temperature, had the hcp structure and no evidence of the presence of an orthorhombic
martensite could be obtained.

When solutionising was carried out in the alpha + beta phase field, these
two phases constituted the alloy structure prior to quenching and the quenched
structure showed a mixture of primary alpha and either transformed or retained
beta depending on the temperature of solutionising (Fig.2). The volume
fractions of the constituent phases in these samples were determined by the
point count technique and it was noticed that the solutionisation temperature
had a pronounced effect on the relative amounts of the alpha and the beta
phases (Fig.3). This could be qualitatively seen from the micrographs

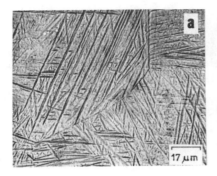

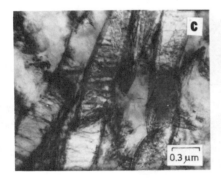
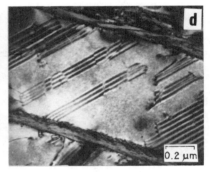

Fig.1 : Ti-8Al-1Mo-1V alloy, quenched from 1100°C; (a) Optical
microstructure; (b) lath morphology of the martensite
(c) dislocated lath substructure (d) stacking faults within
a lath on ageing at 600°C for 2 hours.

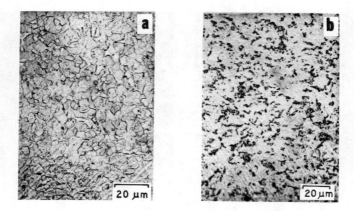

Fig.2 : Optical micrographs of duplex alpha plus beta structures
produced by solutionising for 1 hour  (a) at 950°C and
(b) at 800°C.

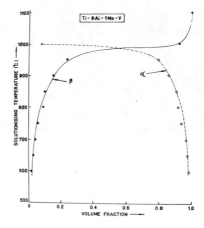

Fig.3 : Plots of the volume fractions of the alpha and the beta
phases against solutionisation temperature.

presented in Fig.2 also.

    TEM examination of thin foils of Ti-8Al-1Mo-1V samples solutionised at
900°C and at lower temperatures showed a distribution of globules of the beta
phase in the alpha matrix. It was also noticed that only a fraction of the
total number of the beta globules underwent the bcc to hcp martensitic transformation on quenching (Fig.4(a)). This observation suggested that the Ms
temperature of all the beta globules were not the same and that only those
which contained embryos that became operational during quenching underwent
martensitic transformation. In bulk samples, the successive autocatalytic
events of martensite nucleation are responsible for the propagation of the
transformation in the entire bulk. Such a process is not possible in a situation where the beta globules are widely separated from one another by intervening alpha regions. The present situation is analogous to the martensitic
transformation in small precipitates of Fe in a Cu matrix where only a few Fe
particles were seen to have undergone the transformation. In samples quenched
from 850°C and still lower temperatures, the presence of totally untransformed
beta globules was noticed, indicating that the degree of stabilization of the
beta phase in these samples was sufficient to depress the Ms temperature
below room temperature (Fig.4(b)). Selected area diffraction patterns obtained from such beta crystals showed diffraction effects associated with the
omega phase (Fig.4(c)). This observation suggested that eventhough the beta
phase was stabilized with respect to the bcc to hcp martensitic transformation
by Mo and V enrichment, the degree of stabilization was not adequate for
suppressing the athermal beta to omega displacive transformation.

    The equiaxed alpha grains were found, in general, to be dislocation free
as would be expected in well annealed structure. However, planar arrays of
dislocations lying on the $\{10\bar{1}0\}$ planes, which are the predominant slip planes
in this alloy [4] , were often noticed (Fig.5(a)). In samples quenched from
800°C and below, pairing of dislocations was occasionally observed though
SAD patterns did not reveal the presence of superlattice spots (Fig.5(b)).
However, SAD patterns obtained from samples aged at 600°C for 2 hrs. showed
the presence of $DO_{19}$ spots suggesting that the $Ti_3$ Al phase did form on ageing
(Fig.5(c)). It was possible that a state of short range order existed in the
quenched samples.

    The retained beta globules were very often found to be enveloped by a
layer which separated these from the surrounding alpha crystals (Fig.6(a)).
The existence of such transition regions sandwiched between the beta and the
alpha regions has been observed in many Ti alloys under certain conditions
and these interfacial features have been extensively investigated  [9, 10] .
Apart from the interfacial layer around the beta globules, plate shaped
products residing wholly within the alpha matrix regions were also noticed
(Fig.6(b)). Blackburn [4] also has observed similar features and has shown
that these plates constituted of the gamma hydride phase. In the present work,
it was not possible to associate such features with the hydride phase in all
instances and distinct diffraction spots associated with the hcp alpha phase
oriented differently from the matrix alpha crystals could be observed in SAD
patterns obtained from these regions. A detailed analysis of the interface
phase observed in the Ti-8Al-1Mo-1V will be presented elsewhere [11] .

    The results of the X-ray diffraction studies indicated that distinct types

1226   S.J. Vijayakar et al.

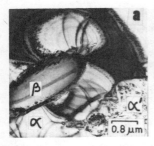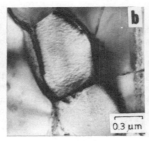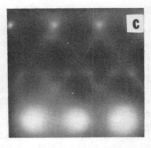

Fig.4(a) : Primary alpha, retained beta and transformed beta regions in sample
quenched from 950°C, (b) Untransformed beta grains in sample quenched from
850°C; (c) SAD pattern from untransformed beta indicating the presence of the
omega phase.   Zone axis : $\langle 133 \rangle_\beta$.

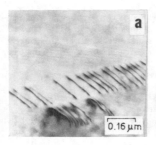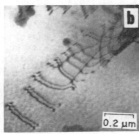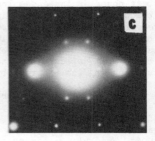

Fig.5(a) : Dislocations lying on a $\{10\overline{1}0\}$ type plane in a primary alpha
crystal; (b) Pairing of dislocations in an alpha grain in a sample solutionised at 750°C; (c) SAD pattern showing $DO_{19}$ superlattice reflections obtained
from an aged sample.

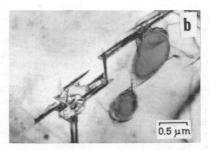

Fig.6(a) Typical appearance of the interface region separating contiguous
beta and alpha crystals; (b) plates emanating from the interphase interface
into the alpha regions.

of textures were produced in samples quenched from the beta phase field and those quenched from the two phase field samples quenched from 950°C and below showed a very strong basal texture and it was noticed that the ratio of the intensity of the basal reflection to that of the $\{10\bar{1}1\}$ reflection increased with decreasing solutionising temperature in the alpha + beta phase field. This intensity factor has been reported to increase with solute content in binary Ti-Al alloys: the basal texture becoming more perfect in high Al bearing alloys [12] . In the present case, decreasing solutionising temperature corresponded to higher Al enrichment of the alpha phase. In contrast to the texture observed in the samples quenched from the alpha + beta phase field, the beta quenched samples exhibited a pronounced $\{10\bar{1}0\}$ texture.

2. Tensile properties and microstructure of deformed samples

The true stress versus true plastic strain plots of specimens solutionised at different temperatures are shown in Figs.7(a) to (c). It could be seen clearly that the yield strength, the ultimate tensile strength and the elongation depended strongly on the solutionisation temperature. Samples quenched from the beta phase field appeared to strain harden at an appreciably higher rate as compared to samples solutionised in the high alpha + beta field. However, the work hardening rate appeared to be rather weakly dependent on the temperature of solutionisation in the low alpha + beta phase field. It was also found that the volume fraction of the martensite present in the structure had a pronounced effect on the yield strength of the alloy as could clearly be seen from Fig.8(a) where the yield strength has been fitted as a function of volume fraction of the beta phase (transformed or untransformed). A comparison of Fig.3 and Fig.8(a) shows that any reduction in the beta volume was associated with a corresponding drop in yield strength. This trend was noticed down to solutionisation temperatures of about 750°C (corresponding to a beta volume fraction of about 6-7%). On lowering the solutionisation temperature further, however, the yield strength registered an increasing trend. The enhanced strength and the reduced ductility of samples solutionised at temperatures below about 750°C and subsequently quenched could probably be attributed to the presence of a uniform dispersion of very fine particles of the strong but brittle, ordered $Ti_3Al$ phase in the alpha matrix.

The influence of solutionising temperature on yield strength and uniform elongation are reflected in the plots in Fig.8(b). Conforming to the observations made by Fopiano and Hickey [1] , a minimum in the yield strength versus solutionising temperature plot was found to occur in the region 750-800°C. In this temperature range, the microstructure consisted of a mixture of the alpha and the untransformed beta phases. During deformation one could expect the formation of stress induced martensite in the metastable beta phase. Under such circumstances, the two processes, the stress induced transformation of the beta phase and plastic deformation of the two phase alloy - would be competetive and the point of yielding would indicate the stress value at which the easier of the two processes would set in. If such were the case then in the low value of the yield stress in samples quenched from 750-800°C could be attributed to the formation of stress induced martensite in the metastable beta phase prior to the event of the "true" yielding of the sample. In order to verify whether such an explanation was valid, the structure developed in samples plastically deformed at room temperature was examined. Surprisingly, it was noticed that while the beta phase did not appear to have undergone any phase

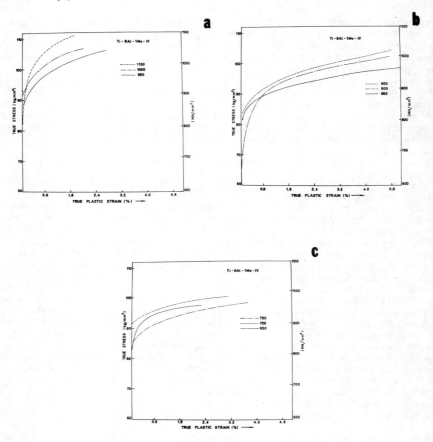

Fig.7(a)-(c) : True stress versus true plastic strain plots. The corresponding solutionising temperatures in  C are indicated.

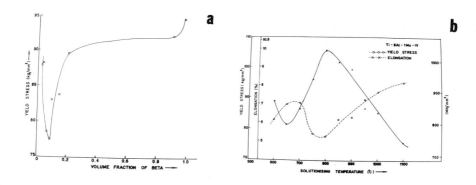

Fig.8(a) : Plot of yield strength against the beta (transformed and/or untransformed) volume fraction.

(b) : **Plot of yield strength and uniform elongation versus solutionising temperature.**

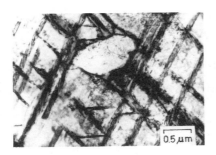

Fig.9 : Planar features fragmenting the alpha grains in sample quenched from 800°C and subsequently tensile tested at room temperature.

transition, the alpha grains were heavily fragmented by planar features(Fig.9). The deformed structure appeared to be very similar to the structure produced in austenitic stainless steels by stress induced transformation. X-ray diffraction could not detect the presence of any additional phase in the deformed samples. However, the textures of these samples were found to be significantly different : the strong basal texture was seen to have changed into a strong {11$\bar{2}$0} texture after plastic deformation. Though similar structures have been reported earlier in Ti-Al [13] , there is no general agreement on the mechanism of the formation of these structural features. It has not been reported whether such a structure could be produced by the application of stress.

In the present case, the absence of such a structural transformation within the alpha phase in the quenched samples suggested that the Al enrichment of this phase was inadequate for a transition similar to that observed in concentrated Ti-Al alloys [13] to occur. However, on the application of stress, the supersaturated alpha phase underwent this "transformation". Electron diffraction evidences pointed out that the planar features noticed were constituted of the same phase as the matrix phase, but oriented in a different manner. The nature and the formation mechanism of these reoriented alpha crystals would be discussed in detail elsewhere [11] .

## References

1. P.J. Fopiano and C.F. Hickey, Jr.: Army Materials and Mechanics Research Centre, Technical Report No.AMMRC TR 73-17 (1973).
2. P.J. Fopiano and C.F. Hickey, Jr: Journal of Testing and Evaluation, 1(1973), 514.
3. M.J. Blackburn : Trans. ASM, 59 (1966), 695.
4. M.J. Blackburn: Trans. ASM 59 (1966), 876.
5. C.F. Hickey, Jr. and P.J. Fopiano: AMMRC TR.70-4 (1970).
6. R.G. Sherman and H.D. Kessler: Trans. ASM, 48 (1956), 657.
7. J.C. Williams and M.J. Blackburn: Trans. ASM, 60 (1967), 373.
8. M.J. Blackburn; in "The Science, Technology and Application of Titanium", 633, Ed. R.I. Jaffee and N.E. Promisel, Pergamon Press Ltd., Oxford,1970.
9. C.G. Rhodes and N.E. Paton : Met. Trans. 10A (1979), 209.
10. C.G. Rhodes and J.C. Williams: Met. Trans. 6A (1975), 2103.
11. S.J. Vijayakar, E.S.K. Menon: P. Mukhopadhyay and S. Banerjee: Paper communicated to J. Mater. Sci.
12. H.W. Rosenberg : Met. Trans. 8A (1977), 451.
13. H. Sasano, T. Tsujimoto and H. Kimura: Titanium Science & Technology, 3 (1973), 1635, Plenum Press, New York.

# PLASTIC DEFORMATION OF TiAl AND Ti$_3$Al*

S. M. L. Sastry
McDonnell Douglas Research Laboratories
St. Louis, MO 63166, U.S.A.


H. A. Lipsitt
Metals and Ceramics Division
Air Force Wright Aeronautical Laboratories
Wright Patterson Air Force Base, OH 45433, U.S.A.

## 1.  Introduction

The slip character of titanium-aluminum alloys changes drastically at aluminum concentrations greater than 3-4 at.%. Pure titanium deforms by slip and twinning at low temperatures and predominantly by slip at high temperatures. In Ti-Al alloys, however, with increasing aluminum concentration, twinning is suppressed, and plastic deformation occurs by planar a-slip ($\underline{b}$ = a/3 <11$\bar{2}$0>) and c-component slip [1-3]. In alloys containing 10 to 20 at.% Al, the compound $\alpha_2$, based on the composition Ti$_3$Al and having the DO$_{19}$ type crystal structure, precipitates in the solid solution matrix. Deformation in this class of alloys occurs by the shearing of $\alpha_2$ precipitates by glide dislocations, leading to the formation of coarse, planar, slip bands. The stress concentrations at slip-band intersections and grain-boundary/slip-band intersections promote cleavage and reduce the ductility and toughness of the alloys. As the aluminum concentration is increased beyond 20 at.%, the directional character of the bonding between Ti and Al atoms increases, and this leads to the formation of the first intermetallic compound Ti$_3$Al in the Ti-Al system. The second intermetallic compound, which has the LI$_0$-type superlattice structure occurs near 50 at.% aluminum concentration. Because the ordered arrangement of atoms in Ti$_3$Al and TiAl persists to near the melting temperatures, these compounds have high-temperature strengths and oxidation resistance unequalled by commercial titanium alloys and thus are candidates for high-temperature applications.

The crystal structures of Ti$_3$Al and TiAl are shown in Figures 1a and 1b. In Ti$_3$Al, the a-dislocations lying on the basal plane are superlattice dislocations bound by four partial dislocations of the type a/6 <10$\bar{1}$0>. When slip occurs on prism planes, however, dislocations can travel as unit dislocations without disruption of the ordered arrangement of nearest neighbor atoms. The c+a type activity becomes important at higher temperatures. Although c-type dislocations are observed occasionally, they do not play a major role in the deformation process. In TiAl, the layered arrangement of Ti and Al atoms on successive {100} planes and the slight tetragonality result in nonequivalence of different 1/2 <110> slip vectors. The 1/2 [011] and 1/2 [101] dislocations must move in pairs if the ordered arrangement of atoms is

---

*This work was performed at the Air Force Wright Aeronautical Laboratories under a grant from the National Research Council.

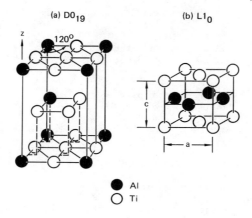

**Figure 1. Crystal structures of (a) Ti₃Al and (b) TiAl.**

Table 1   Nominal Compositions and Crystal Structures
of Titanium Aluminides

| Alloy | Crystal Structure |
|---|---|
| Ti-17Al<br>Ti-16Al-10Nb | Ordered hexagonal close packed<br>$DO_{19}$ |
| Ti-36Al<br>Ti-36Al-4Nb | Ordered face centered tetragonal<br>$L1_0$ |

not to be disrupted, whereas the 1/2 [110] dislocations can travel as ordinary unit dislocations without destroying order. Although the possible dislocation configurations in $L1_0$ and $DO_{19}$ superlattices have been known for some time [4,5], detailed investigations of slip systems in TiAl and Ti₃Al have been carried out only recently. The salient features of deformation substructures in Ti₃Al and TiAl at 25-900°C, the role of the unique dislocation configurations of $L1_0$ and $DO_{19}$ superlattices in the development of dislocation substructures, and the consequent effects of substructures on fracture processes are reported here.

## 2.  Experimental Procedure

The nominal compositions of the alloys studied are given in Table 1. The alloys were cast as stick ingots and converted by Nuclear Metals, Inc., to <500-μm size powders using a rotating electrode process. The powders were canned in Ti-6Al-4V cans and hot extruded at 1200-1400°C with an extrusion ratio of 26:1. Cylindrical test specimens having a reduced gauge diameter of 3.6 mm and flanges for clamping the strain measuring device were machined from the extruded rods. The fracture modes and deformation substructures of the specimens deformed at 25-900°C in tension, creep, and fatigue were determined by scanning and transmission electron microscopy.

## 3.  Results

3.1  Mechanical Properties and Deformation Substructures of Ti₃Al and Ti₃Al-Nb

The variations with temperature of tensile and fatigue properties of Ti₃Al are shown in Figure 2. Ti₃Al specimens deformed in uniaxial tension do not exhibit appreciable plastic deformation below 600°C. The high values of tensile and fatigue fracture stresses are retained up to 700°C, with a large decrease in strength occurring only above 800°C. The fatigue ductility (cumulative strain for fracture) is higher than the tensile ductility, and both fatigue and tensile ductilities increase monotonically with temperature up to 700°C. The difference between the tensile and fatigue fracture stresses is less than 8% at temperatures up to 700°C, but the saturation stress in fatigue at 800°C is considerably lower than the tensile fracture stress.

Figure 3 shows the dislocation arrangement on the basal plane of Ti₃Al fractured in tension at 25°C. All the dislocations in Figure 3 have a/3 <11$\bar{2}$0> Burgers vectors, and the paired dislocations (as at B) lie on the basal plane, whereas the unit dislocations (shown at P) lie on prism planes in agreement with the predicted dislocation configuration for $DO_{19}$ type superlattices. The c-component dislocations were not observed in specimens deformed at 25°C. Unstable shear containing long pile-ups of dislocations on basal and prism planes was observed in specimens deformed at 25-600°C.

Figure 4 is a comparison of dislocation substructures in Ti₃Al specimens deformed at 700°C in tension, creep, and fatigue. Slip activity occurs on both the basal and prism planes for each mode of deformation. The dominant slip vector in unidirectional deformation is the a-type (Figure 4a). The dislocations are straight, occurring for the most part in a screw orientation, and they are arranged in planar bands. In contrast, the dislocation distribution in specimens deformed in creep at 700°C (Figure 4b) is characterized by a much reduced planarity of slip, a tangled dislocation

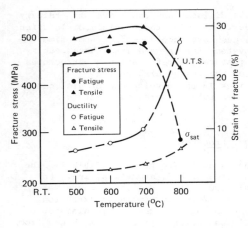

Figure 2.  Variation of tensile and fatigue properties with temperature.

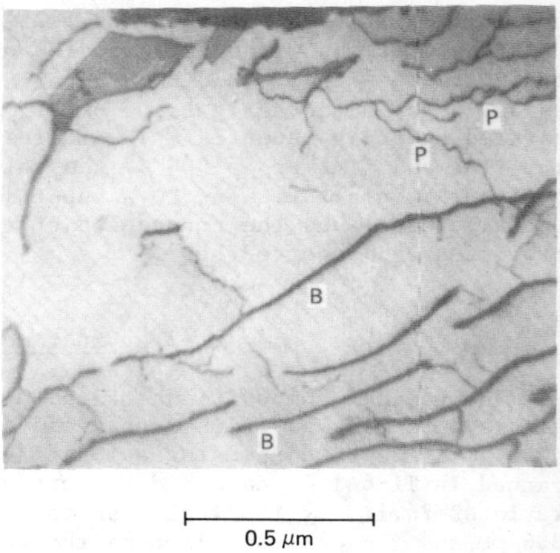

Figure 3.  a/3 <11$\bar{2}$0> {0001} superlattice dislocations and a/3 <11$\bar{2}$0> {10$\bar{1}$0} unit dislocations in Ti$_3$Al deformed in tension at 25°C.

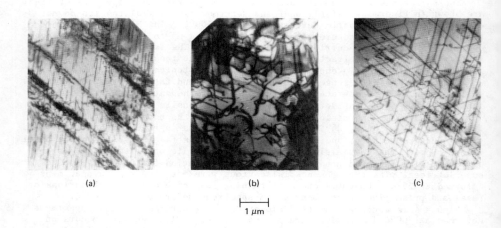

(a)                    (b)                    (c)

Figure 4.  Dislocation substructures on basal planes in Ti$_3$Al deformed at 700°C; (a) deformed in tension to 3% strain, (b) deformed in creep to 3% strain, and (c) fatigued at $\gamma_p$ = ±0.25%, and $\epsilon_{cum} \approx 0.03$.

arrangement instead of planar bands, jogged superlattice dislocations, dislocation loops, and an increased incidence of nonbasal slip vectors. Simple tilt boundaries consisting of one set of superlattice dislocations and mixed tilt-twist boundaries consisting of cross grids of superlattice dislocations were observed in $Ti_3Al$ deformed in creep. In specimens deformed in constant plastic-strain-amplitude fatigue, the type of dislocation structure produced depends on the strain amplitude and cumulative strain. An example of the dislocation structure observed in $Ti_3Al$ fatigued at $\gamma_p = \pm 0.25\%$ to a cumulative strain of 0.03 is shown in Figure 4c. The preferential screw orientation of dislocations is less pronounced in fatigued specimens than in unidirectionally deformed specimens, and many dislocations are arranged in dipole configurations. At higher strain amplitudes and cumulative strains, a dislocation braid structure is developed. The braids are aligned along the traces of the prism and basal planes (Figure 5). The spacing between the braids decreases with increasing strain amplitude.

The c-component dislocations in fatigued $Ti_3Al$ (Figure 6a) are arranged mainly in dipole configurations and are distributed both between the braids and within the braids. In specimens deformed in creep, the c-component dislocations are long and heavily jogged (Figure 6b).

The fracture surfaces of specimens deformed at $25°C$ and $800°C$ are shown in Figures 7a and 7b. At $25°C$, the fracture is characterized by extensive transgranular cleavage, but at $800°C$, the fracture is mixed cleavage, intergranular, and ductile-dimple.

The most important effects of niobium additions to $Ti_3Al$ are to slow the ordering kinetics, reduce the planarity of slip, and increase nonbasal slip activity, all of which result in increased ductility in the niobium-containing specimens. Figure 8b shows the increased c-component dislocation activity in $Ti_3Al$-Nb compared with $Ti_3Al$ (Figure 8a).

### 3.2 Mechanical Properties and Deformation Substructures of TiAl and TiAl-Nb Alloys

The temperature dependences of the yield stress and ductility of TiAl are shown in Figure 9. The ratio of fatigue strength (at $10^6$ cycles) to the ultimate tensile strength, shown in Figure 10, is significantly higher than that of several other high-temperature materials.

Figure 11 shows the dislocation distribution in a TiAl specimen deformed in tension at $25°C$. A detailed Burgers vector analyses of dislocations revealed the activity of predominantly a/2 [110] and a/2 [1̄10] dislocations and a/3 [111] faults. The occurrence of a/2 [101] and a/2 [011] dislocations was rare, and where they were observed occasionally, the superlattice dislocations were pinned by the trailing a/6 [112] partial dislocations.

The most important feature of the dislocation substructure of TiAl deformed in fatigue at room temperature is the presence of a large number of dislocation dipoles and prismatic loops (Figure 12a), which are rare in unidirectionally deformed specimens. The dislocation substructure in specimens fatigued at $800°C$ (Figure 12b) consists of a well-defined braid structure elongated in the [112] direction and containing edge dislocations and dipoles of all the a/2 <110> Burgers vectors. The regions between the braids contain mainly screw dislocations straddling the braids and edge dislocation dipoles and loops in the process of being swept into the braids. Whereas (111) twinning is the most common twinning mode in both tension and fatigue (Figure 13a), (211) twinning also occurs in fatigued specimens (Figure 13b).

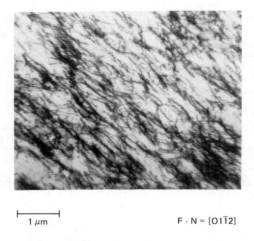

1 μm                                F · N = [O1$\bar{1}$2]

Figure 5.  Dislocation arrangement in Ti$_3$Al
            fatigued at 700°C at $\gamma_p$ = ± 0.125%.

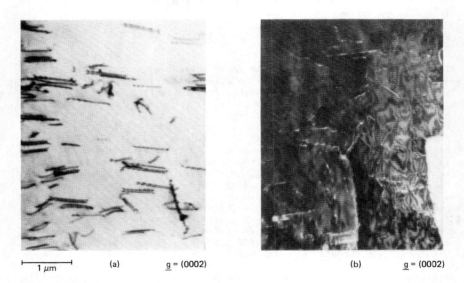

1 μm        (a)        g = (0002)                    (b)        g = (0002)

Figure 6.  C-component dislocation activity in Ti$_3$Al; (a) fatigued at 700°C at $\gamma_p$ = 0.125%,
            $\epsilon_{cum}$ = 0.05  and (b) deformed in creep at 700°C to ~5% strain.

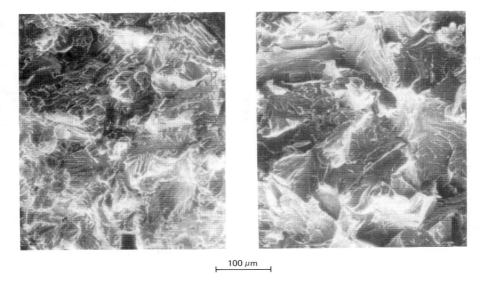

⊢ 100 μm ⊣

Figure 7.  Fracture surfaces of Ti₃Al deformed at (a) 25°C and (b) 800°C.

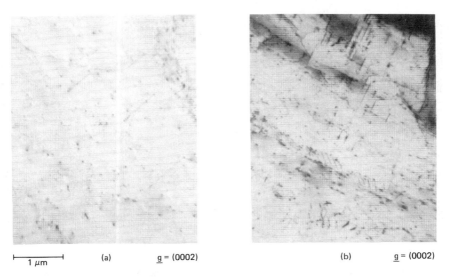

⊢ 1 μm ⊣     (a)          $\underline{g}$ = (0002)                    (b)          $\underline{g}$ = (0002)

Figure 8.  C-component dislocation activity in (a) Ti₃Al and (b) Ti₃Al-Nb deformed in tension at 700°C.

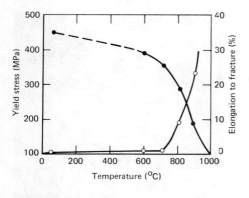

Figure 9. Variation with temperature of yield stress (•) and percentage elongation to fracture (○) of TiAl.

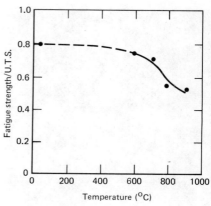

Figure 10. Variation with temperature of fatigue strength to ultimate-tensile-strength (U.T.S.) ratio.

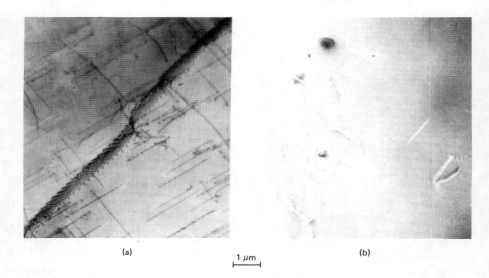

(a)

1 μm

(b)

Figure 11. Dislocation structure in TiAl deformed in tension at 25°C; (a) a/2 [110] and a/2 [1$\bar{1}$0] dislocations and (b) a/3 [111] stacking faults.

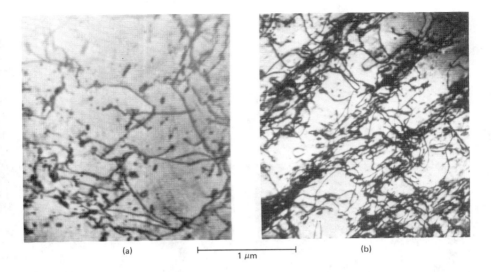

**Figure 12. Dislocation substructures in TiAl fatigued at (a) 25°C and (b) 800°C.**

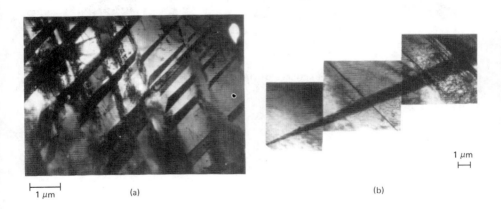

**Figure 13. Deformation twins in TiAl (a) {111} twins and (b) {211} twin.**

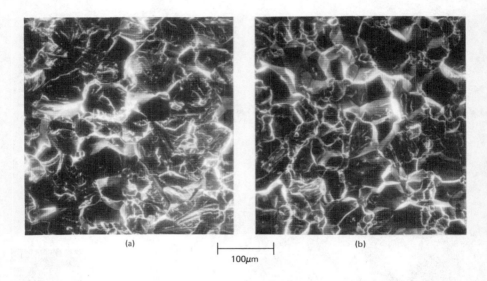

(a)

(b)

├─────┤
100μm

**Figure 14.  Fracture surfaces of TiAl deformed at (a) 25°C and (b) 800°C.**

The fracture surfaces of specimens deformed at 25°C and 800°C are shown in Figures 14a and 14b. At 25°C, fracture occurs by transgranular cleavage. With increasing temperature, the tendency towards intergranular fracture increases, and intergranular fracture dominates at 800°C.

The major influence of niobium additions to TiAl is a lowering of the temperature at which twinning becomes an important mode of deformation and thus a lowering of the ductile-brittle transition temperature of TiAl.

## 4. Discussion

The dominant mode of slip in Ti₃Al below 700°C is the a-type slip activity. Because of the absence of twinning and c-component dislocation activity, which are required to produce deformation normal to the basal plane, the compound has limited ductility in tension below 700°C. Above 700°C, the activity of nonbasal dislocations combined with the ease of cross slip and climb reduces the work hardening and increases the ductility both in tension and fatigue.

The most striking difference between the deformation substructures of Ti₃Al and other hexagonal close-packed metals and alloys is the absence of twinning in Ti₃Al. This is due to the effect of long range order on suppressing twinning because such deformation in Ti₃Al results in the generation of a net disorder. The dislocation substructure in fatigued Ti₃Al is also considerably different from that observed in fatigued Ti and Mg [6,7]. Whereas walls of edge dipoles are prominent in fatigued Ti and Mg, the dislocations in fatigued Ti₃Al are mostly in a screw orientation, which is characteristic of many ordered structures. Because of the high Peierls' stress in this type of compound, dislocations tend to take up screw orientaions, and the mobility of these dislocations is considerably reduced by directional bonding. Thus, in Ti₃Al, planar bands and dislocation walls, comprised predominantly of a/3 ⟨1120⟩ screw dislocations are observed up to 700°C. At higher temperatures the mobility of dislocations, is increased because of a reduction in the Peierls' stress and a simultaneous increase in thermally activated cross slip and dislocation climb. These dislocation processes are responsible for the increased ductility of Ti₃Al above 700°C.

In TiAl, the major constraints on plastic flow at low temperatures are the absence of twinning and [011] and [101] type superlattice dislocation activity. The antiphase boundary energy on {111} planes of TiAl is ≈ 0.45 $J \cdot m^{-2}$, and because of directional bonding the lattice friction stress in TiAl is significantly higher than in other face-centered-cubic (f.c.c.) metals and alloys. Furthermore, the effective Burgers vector of superlattice dislocation is twice that of unit dislocation. Consequently, the mobility of superlattice dislocations is greatly reduced because of higher stress required to move superlattice dislocations. The movement of [011] superlattice dislocations is further hindered because of the pinning of the constituent a/6 [112] partial dislocation [8]. The presence of the faulted ribbon bounded by the looped segment of the trailing a/6 [112] partial thus tends to immobilize the three leading partial dislocations. Above 700°C, the a/6 [112] partial is no longer pinned, and the activity of the [011] and [101] superlattice dislocations increases rapidly with increasing temperature. Furthermore, above 700°C, twinning assumes increasing importance as a deformation mode with the concommittant result of increased ductility of TiAl above 700°C.

The dislocation structure in TiAl that has been deformed in fatigue at 700-900°C closely resembles that observed in other f.c.c. metals and alloys, namely, a dislocation mat structure at small stress amplitudes changing into a closed cell structure at high stress amplitudes. The dislocation braid structure is fully developed near the grain boundaries because of enhanced primary and secondary slip at grain boundaries. This enhanced slip activity at grain boundaries increases the probability that the grain boundaries serve as potential sites for crack nucleation and propagation and hence contribute to intergranular fracture at high temperatures.

## 5.   Summary and Conclusions

The deformation behavior at 25–900°C of TiAl and Ti$_3$Al was investigated by tensile, creep, and fatigue testing, transmission electron microscopic examination of deformation substructures,and scanning electron microscopic examination of fracture modes.   In TiAl, a high Peierls' stress, planar slip, and sessile superlattice dislocation of a/2 [101] and a/2 [011] Burgers' vectors were identified as the sources of low-temperature brittleness.   The ductile brittle transition in TiAl occurs at 700°C.   The incidence of dislocations of all a/2 <110> slip vectors and {111} <112> twinning results in enhanced ductility above 700°C.   The dominant mode of fracture changes from cleavage at room temperature to intergranular above 700°C.   In Ti$_3$Al, the dominant deformation mode at low temperature is a-type [$\underline{b}$ = 1/3 <$1\bar{1}20$>] slip activity without any twinning, and fracture is by cleavage.   The compound exhibits limited ductility below 700°C.   However, in contrast with the abrupt ductile-brittle transition observed at 700°C in TiAl, the ductility of Ti$_3$Al increases gradually with temperature and remains below that of TiAl at high temperatures.   The c+a dislocation activity becomes pronounced above 700°C. Niobium additions to Ti$_3$Al retard the ordering kinetics, reduce the planarity of slip, and increase nonbasal slip activity.   The major influence of niobium additions to TiAl is to lower the temperature at which twinning becomes an important deformation mode and thus lower the ductile-brittle transition temperature of TiAl.

## Acknowledgements

The authors gratefully acknowledge the valuable contributions of Drs. M. J. Blackburn, D. Shechtman, R. Schafrik, M. G. Mendiratta, and W. J. Yang and Mr. P. L. Martin to the research described in this paper.

## References

1. J. D. Boyd and R. G. Hoagland, in Titanium Science and Technology, Proceedings of Second International Conference on Titanium, ed. by R. I. Jaffee and H. M. Burte, (Plenum Press, New York, 1973), p. 1071.
2. G. Lutjering and S. Weissmann, Acta Met., 18, (1970), p. 785.
3. M. G. Mendiratta, S. M. L. Sastry, and J. V. Smith, J. Mater. Sci., 11, (1976), p. 1835.
4. M. J. Marcinkowski, in Electron Microscopy and Strength of Crystals, ed. by G. Thomas and J. Washburn, (Interscience, New York, 1963), p. 333.
5. M. J. Marcinkowski, N. Brown, and R. M. Fisher, Acta Met., 9, (1961), p. 129.
6. P. G. Partridge, Phil. Mag., 12, (1965), p. 1043.

7.  R. Stevenson and J. F. Breedis, Acta Met., 23, (1975), p. 1079.
8.  D. Shechtman, M. J. Blackburn, and H. A. Lipsitt, Met. Trans., 5, (1974),
    p. 1373.

# THE EFFECTS OF ALLOYING ON THE MICROSTRUCTURE
## AND PROPERTIES OF Ti$_3$Al AND TiAl

P. L. Martin*, H. A. Lipsitt*, N. T. Nuhfer**, and J. C. Williams**

*AFML, Wright-Patterson AFB, OH, U.S.A.
**Carnegie-Mellon University, Pittsburgh, PA, U.S.A.

## Introduction

It has been known for sometime that the intermetallic phases based on Ti$_3$Al and TiAl have attractive strength and modulus values, especially when normalized by their relatively low density[1]. However, the inherent brittleness of these phases severely limits their usefulness. These limitations include the difficulty encountered in producing these materials by conventional ingot metallurgy methods. However, recent developments in processing techniques for brittle materials prompted further studies of alloys based on Ti$_3$Al and TiAl[2-4].

The structure of Ti$_3$Al, also known as $\alpha_2$, is DO$_{19}$, which is hexagonal. The structure of TiAl, also known as $\gamma$, is Ll$_0$, which is tetragonal. In addition, both $\alpha_2$ and $\gamma$ are ordered and exhibit planar slip. Thus the limited ductility of these phases is partly the result of lower symmetry and partly the result of planar slip[3,5]. It appears that some improvement in ductility can be accomplished by reducing the slip length[6]. Several possibilities for this exist. One is simply to refine the grain size by cold working. A second is to refine the slip length by using an allotropic transformation to give smaller microstructural units. A third is to precipitate second phase particles which disperse the slip. We have found that the latter two possibilities hold some promise, especially for $\alpha_2$ base alloys. In this paper we describe some of the results we have obtained.

## Experimental Procedures

All alloys studied were prepared by powder metallurgy techniques. The compositions to be discussed are listed in Table 1. Ingots of the desired composition were consumably arc melted, homogenized, and converted to powder using the Rotating Electrode Process (REP). The powders were sieved to -35 mesh and consolidated by hot extrusion as described elsewhere[3]. The temperature of the extrusion billets was 1200°C for Ti$_3$Al base alloys, and 1400°C for TiAl base alloys. Mechanical testing was performed in air on an Instron testing machine as also previously described[3]. Microstructural characterization of the material was performed on a JEOL JSM35U SEM, a Phillips 300 TEM, a JEOL 100C TEM and a JEOL 120CX AEM.

## Results and Discussion

1. Ti$_3$Al Base Alloys:

The binary Ti-Al phase diagram shown in Figure 1 is useful in under-

standing the structures attainable through heat treatment. Upon cooling
from the β-phase field the Ti$_3$Al composition forms the ordered α$_2$ structure.
If the cooling rate is rapid enough, the transformation occurs martensitically.   Figure 2 shows the martensite for both the stoichiometric α$_2$ and
the quaternary α$_2$ alloys when they are quenched from the β-phase field into
boiling water.   Each is essentially single phase α$_2$, however, the platelet
size is dramatically smaller in the quaternary α$_2$ alloy.   This is because
Nb and W reduce the M$_s$ temperature causing the transformation to occur at
lower temperatures where the β-phase is stronger and growth of the individual martensite plates is more difficult.   The differences in martensite
morphology also change the microstructural stability as will be discussed
below.   Reheating α$_2$ to 900°C does not significantly alter its microstructure.   Reheating the quaternary α$_2$ to 900°C causes the heterogeneous nucleation of the W and Nb stabilized β-phase.   Figure 3 shows these precipitates
and qualitatively shows their composition relative to the matrix as determined by AEM.

The ordering kinetics in both of these alloys are very rapid and thus
the individual martensite plates in both of the quenched alloys contain
ordered domains.   The Nb and W additions do retard the ordering kinetics
somewhat because the domain size is smaller in the quaternary α$_2$ alloy.

The effect of these microstructural changes on tensile properties can
be seen through the data in Table 2.   Slip length refinement is seen to not
only increase the strength but also to increase the ductility.   This is
shown through comparison of the room temperature ductilities.   The attainment of sufficient ductility to exhibit a 0.2% yield stress at room temperature, as shown for the quaternary α$_2$ alloy, is a major accomplishment for
Ti$_3$Al base alloys or any other brittle material.   The tensile fracture surfaces of the two alloys are shown in Figure 4.   Although the principal
fracture mode of the quaternary α$_2$ is still cleavage, refinement of the
cleavage facet size and the presence of numerous tear ridges where the crack
changes orientation is consistent with the observed ductilizing effect of
finer structure and the presence of β-precipitates.

The rational limit to structure refinement is high temperature stability during use.   The quaternary α$_2$ alloy in the structure discussed above
showed a tendency to recrystallize along prior β-grain boundaries during low
temperature* tensile deformation.   This is shown optically in Figure 5 for a
sample tested at 600°C.   This reaction appears to be a recrystallization
analogue of the cellular or discontinuous reaction seen in many alloys.   It
is initiated at prior β-grain boundaries because these sites provide the dis-
ordered interface which bounds the recrystallized regions.   The driving
force for this reaction appears to be largely surface energy.   That is, the
small size of the martensite plates produces a large driving force for recrystallization.   The equiaxed nature of the recrystallized grains is shown
in Figure 6.   The fine precipitates in the equiaxed grains are fine β-phase.
The smaller of these appear to be coherent.   The alignment of these precipitates (Fig. 6b) is reminiscent of the precipitation effects which occur at
moving interfaces in Fe-base alloys[7].

---

* at temperatures less than aging temperature for β-phase precipitation

Table 1   Alloy compositions in atomic percent

| Alloy Designation | Ti | Al. | Nb | W |
|---|---|---|---|---|
| $\alpha_2$ | bal. | 27 | – | – |
| Quaternary $\alpha_2$ | bal. | 25 | 5 | 1 |
| $\gamma$ | bal. | 50 | – | – |
| Ternary $\gamma$ | bal. | 49 | – | 2 |

Table 2   As extruded mechanical properties

Tensile

| Alloy Designation | Temperature (°C) | 0.2% Yield (MPa) | Ultimate (MPa) | Plastic Elongation (%) | R.A. (%) |
|---|---|---|---|---|---|
| $\alpha_2$ | 25 | – | 552 | – | 0.1 |
| Quaternary $\alpha_2$ | 25 | 796 | 920 | 0.4 | 1.1 |
| $\alpha_2$ | 700 | 413 | 521 | 2.7 | 4.0 |
| Quaternary $\alpha_2$ | 700 | 544 | 696 | 4.4 | 7.0 |
| $\gamma$ | 25 | 335 | 450 | 0.9 | 1.6 |
| Ternary $\gamma$ | 25 | 642 | 790 | 0.8 | 1.3 |
| $\gamma$ | 700 | 300 | 360 | 14.0 | 22.0 |
| Ternary $\gamma$ | 700 | 537 | 803 | 2.9 | 4.7 |

Creep

| Alloy Designation | Temperature (°C) | Initial Stress (MPa) | Time to $1\%\varepsilon_p$ (hr) | $\dot{\varepsilon}_{s.s.}$ (hr$^{-1}$) |
|---|---|---|---|---|
| $\gamma$ | 760 | 279 | 5 | $2 \times 10^{-3}$ |
| Ternary $\gamma$ | 760 | 279 | 260 | $2 \times 10^{-5}$ |

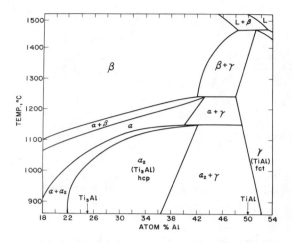

Fig. 1 - Binary Ti-Al phase diagram.

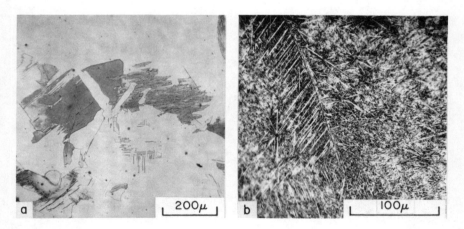

Fig. 2 - Optical micrographs of Ti$_3$Al base alloys quenched from 1200°C into boiling water, a) $\alpha_2$ alloy, b) quaternary $\alpha_2$ alloy.

Fig. 3 – Transmission electron micrographs of the quaternary $\alpha_2$ alloy quenched from 1200°C and aged at 900°C a) bright field image of the martensitic platelets, b) dark field image of the $\beta$ precipitates, c) STEM comparison of the compositions of the matrix $\alpha_2$ (background) and $\beta$ precipitates (overlayed dots).

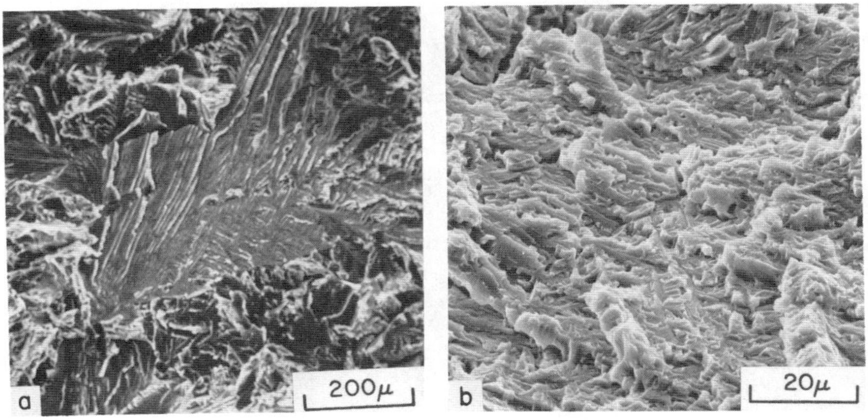

Fig. 4 – SEM fractographs of room temperature tensile tests, a) $\alpha_2$ alloy b) quaternary $\alpha_2$ alloy.

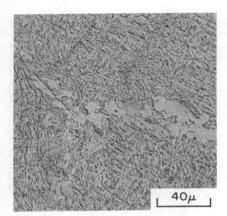

Fig. 5 - Prior β grain boundary in
the quaternary α₂ alloy following:
1200°C boiling water quench +
aging at 900°C for 8 hrs + tensile
testing at 600°C.

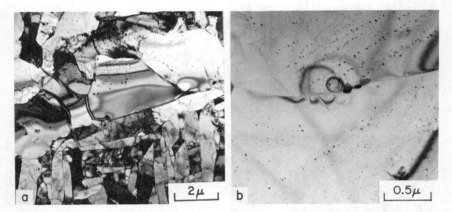

Fig. 6 - Transmission electron micrograph of a prior β grain boundary in the
quaternary α₂ alloy tensile tested at 500°C; a) interface between the martensite
platelets and the recrystallized grains, b) β precipitates within the equiaxed
grains indicative of solute dumping during interface motion.

## 2. TiAl Base Alloys:

The optical microstructures of the stoichiometric $\gamma$ and the ternary $\gamma$ alloys are shown in Figure 7. The thermo-mechanical history of each alloy is identical, yet the ternary $\gamma$ shows a much finer grain size than stoichiometric $\gamma$. Again, referring to Figure 1, the alloys were extruded at 1400°C where the stoichiometric $\gamma$ alloy is primarily $\gamma$ and the ternary $\gamma$ alloy (due to the W stabilization) is mostly $\beta$. The W content of the ternary $\gamma$ stabilizes a higher volume fraction of $\beta$ and this $\beta$ should persist to a lower temperature than in the $\gamma$ alloy. These two effects would lead to a grain size refinement through the $\beta \rightarrow \alpha_2$ transformation as it did for the quaternary $\alpha_2$ alloy. In this case, however, the $\beta$-phase decomposes to a mixture of $\alpha_2$ and $\gamma$ producing the lamellar structure shown in Figure 8. These lamellae have the orientation relationship; $\{111\}_\gamma || \{0001\}_{\alpha_2}$ and $<110>_\gamma || <11\bar{2}0>_{\alpha_2}$. Neighboring $\gamma$ lamellae are twin related across $\{111\}$ planes. The effect of the mixed $\alpha_2$ and $\gamma$ lamellae is to somewhat limit slip length, especially along their thickness direction. Further, the lamellae interfaces can lead to crack bifurcation during fracture, thereby enhancing ductility.

As in the quaternary $\alpha_2$ alloy, the presence of W also stabilizes the $\beta$-phase to lower temperatures. After air cooling from 1300°C, the ternary $\gamma$ alloy consists of equiaxed $\gamma$ grains surrounded by $\alpha_2+\gamma$ lamellae as in Figure 8. Aging at 950°C causes small W rich $\beta$ precipitates to nucleate heterogeneously on internal boundaries. Figure 9 shows these precipitates and their composition relative to the matrix $\gamma$ as determined by AEM.

The increased strength levels due to the refined microstructure of the ternary $\gamma$ relative to $\gamma$ are shown in Table 2. Figure 9 shows that there are precipitates within the grains in the ternary $\gamma$ alloy. However, upon measuring the spacing of these precipitates, a calculation of the increase in yield stress due to them is insignificant. Therefore, the strength increases shown in Table 2 must arise largely from a decreased slip length (Hall-Petch hardening) with a small contribution from solid solution strengthening. In this regard, Figure 9 also shows that some W remains in solid solution in the $\gamma$ phase. The ductility differences are not as dramatic as for the Ti₃Al base alloys. The ability to significantly strengthen the material without further losses in ductility is thought to be due at least in part to the structural refinement. The most significant improvement of the ternary $\gamma$ over the $\gamma$ alloy is seen to be the creep resistance. There are three possible sources for this improved creep resistance. First, the relatively high density of precipitates at grain boundaries can retard the extent of grain boundary sliding. Second, the presence of a low diffusivity solute such as W retards diffusion assisted dislocation motion. Finally, at the reduced stresses pertinent to creep, the interaction of dislocations with the secondary $\beta$ precipitation may also reduce the creep rate of the ternary $\gamma$ alloy.

## Conclusions

1. The refinement of slip length in Ti₃Al base alloys causes a modest increase in ductility and a significant increase in strength.

2. Structural refinement in TiAl base alloys causes appreciable strengthening to occur without significantly decreasing the ductility. The high strength conditions also have significantly improved creep resistance.

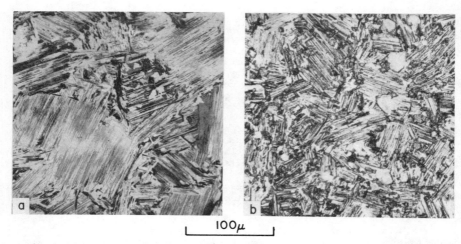

$\vdash$ 100μ $\dashv$

Fig. 7 - Optical micrographs of TiAl base alloys, a) γ alloy b) ternary γ alloy.

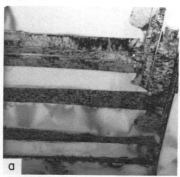

Fig. 8 - Transmission electron micrographs of the lamellar regions of the ternary γ alloy quenched from 1300°C a) bright field image, b) dark field image of the $Ll_0$ (TiAl) phase, c) dark field image of the $DO_{19}$ ($Ti_3Al$) phase showing APB contrast.

$\vdash$ 1μ $\dashv$

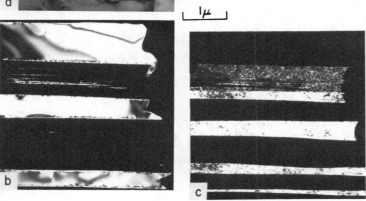

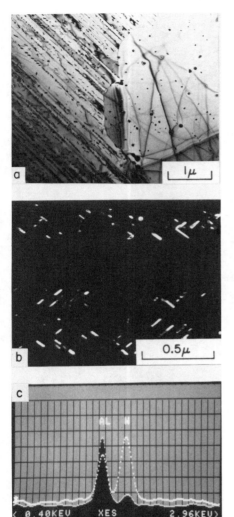

Fig. 9 - Transmission electron micrographs of the ternary γ aged at 950°C, a) bright field image showing β precipitates within both the lamellae and the equiaxed grains, b) dark field image of β precipitates along lamellae surfaces, c) STEM comparison of the compositions of the matrix γ (background) and the β precipitates (overlayed dots).

3.  Structural refinement can be accomplished through the control of the martensitic $\beta \rightarrow \alpha$ phase transformation both by alloying and cooling rate control.

4.  Additional slip length refinement can be attained through the secondary precipitation of $\beta$-phase stabilized by the presence of Nb and/or W.

### Acknowledgments

The authors would like to acknowledge the assistance of A. M. Adair for extruding the powders and D. B. Miracle for assistance with tensile testing. The Garrett AiResearch Corporation supplied the ingot of the quaternary $\alpha_2$ alloy. Use of facilities provided by the Center for the Joining of Materials is also gratefully acknowledged.

### References

1.  J. B. McAndrew and H. D. Kessler:  J. Metals, vol. 8, (1956), 1348.
2.  D. Shechtman, M. J. Blackburn, and H. A. Lipsitt:  Met. Trans., vol. 5, (1974), 1373.
3.  H. A. Lipsitt, D. Shechtman, and R. E. Schafrik:  Met. Trans., vol. 6A, (1975), 1991.
4.  S.M.L. Sastry and H. A. Lipsitt:  Met. Trans., vol. 8A, (1977), 1543.
5.  H. A. Lipsitt, D. Shechtman, and R. E. Schafrik:  Met. Trans., in press.
6.  A. N. Stroh:  Proc. Roy. Soc., London, vol A232, (1955), 548.
7.  K. Campbell and R.W.K. Honeycombe:  Met. Sci. J., vol. 8, (1974), 197.

# MECHANICAL PROPERTIES OF FeTi, CoTi, AND NiTi
## AT ELEVATED TEMPERATURES

Toshiyuki Suzuki and Shigeaki Uehara

National Research Institute for Metals
Tokyo 153, Japan

## Introduction

It is well known that intermetallic compounds are brittle at room temperatures but fail with sufficient plastic deformation above 0.5 Tm and that the strength of intermetallic compounds is relatively high because of these high elastic modulus. Since the strength of intermetallic compounds is essentially related to the bonding force and is not involved in the strengthening mechanisms such as martensitic transformation and precipitation, intermetallic compounds seem to be useful for the high temperature materials.

The compounds having CsCl structure are one of the groups of the intermetallic compound family which consist of huge number of compounds. About 200 of intermetallic compounds have been identified to have CsCl structure(1). It is clear from the phase diagrams determined that almost all the compounds having CsCl structure have homogeneity range at elevated temperatures(2)(3)(4).

The intermetallic compounds which have homogeneity range contain structure defects at non-stoichiometric compositions to achive the required composition with ordered structure(5). According to a simple calculation, the concentration of the defects attains $1 \times 10^{-2}$ for substitutional atoms and $2 \times 10^{-2}$ for vacancies at the composition of 1 at% off from the stoichiometry. It is, therefore, obvious that the mechanical properties of non-stoichiometric compounds are largely affected by the defects.

This investigation was undertaken by means of compression test and hot hardness measurement in order to evaluate the effect of the structure defects on the strength of FeTi, CoTi, and NiTi. They have CsCl structure and have wide homogeneity range. The type of the structure defects introduced at non-stoichiometric compositions was already determined by density measurements and was confirmed to be substitutional wrong atoms at the Ti deficit side of the stoichiometry for each compounds(6).

## Experimental

Samples were prepared by non-consumable arc melting under gettered argon using sponge titanium of ELI grade, electrolytic iron, electrolytic cobalt, and electrolytic nickel as starting materials. These elements were weighed and mixed to have 100 g of charge weight and then melted. Ingots were remelted more than three times to minimize segregation. All ingots obtained were weighed again to detect the weight loss during the melting. The specimens used in this investigation were prepared from the ingots which showed less than 0.1 % of weight loss. The ingots were annealed in dynamic vacuum for more than 25 hr at 1000°C to insure homogeneization. The ingots were then canned in mild steel and forged on a conventional forging press at temperatures from 850 to 950°C. The specimens having 3 x 3 x 9 mm in size were cut by multi-wire saw from the forged plate which was taken out from the can. In order to normalize, all specimens were sealed in evacuated quartz ampoules and annealed for 1 hr at 1050 °C for FeTi and at 1000°C for CoTi and NiTi prior to further heat treatments.

All heat treatment were performed under vacuum of more than $2 \times 10^{-5}$ mmHg. When rapid cooling is necessary the specimens were quenched by breaking the quartz ampoules under water.

Hot hardness measurements were carried out by means of Vickers hardness tester which is equipped with evacuated specimen chamber in a load of 300 g. Compressive strength measurements were carried out on an Instron type testing machine at temperatures from R.T. to 900°C at a strain rate of $9.3 \times 10^{-4}$ sec$^{-1}$ installing evacuated chamber with tantalum heater.

## Results

### 1.  FeTi

The change in hardness of FeTi with temperature is shown in Fig. 1.  The hardness of FeTi remains constant up to 300°C then decreases with increasing temperature.  Hardness level of the non-stoichiometric compound is higher than that of stoichiometric compound.  The relation between temperature and yield stress of FeTi is shown in Fig. 2.  Though FeTi behaves in brittle manner below 500°C the compound is completely ductile above 600°C.  Therefore, yield stress measurements could be performed without any difficulty above the temperature. The yield stress of FeTi decreases steeply with increasing temperature in the temperature range from 600 to 900°C.  The yield stress of non-stoichiometric FeTi is higher than that of stoichiometric compound in the ductile fracture range.

### 2.  CoTi

The change in hardness of CoTi with temperature is shown in Fig. 3.  The hardness of non-stoichiometric CoTi decreases gradually with temperature up to 400°C and then decreases rapidly above the temperature.  The hardness of CoTi having stoichiometric and neary stoichiometric compositions remains constant up to 400°C.  It is also clearly observed that the hardness of CoTi increases as the composition deviates from the stiochiometry.  This tendency, however, becomes small at higher temperatures.  The relation between temperature and yield stress of CoTi is shown in Fig. 4.  CoTi compounds fail without plastic deformation below the specific temperature.  However, above that temperature they show completely ductile behavior.  The brittle to ductile transition temperature varies with deviating composition from the stoichiometry.  The more the composition deviates from the stoichiometry the higher the transition temperature becomes.  In the ductile fracture range yield stress of the compound decreases rapidly with increasing temperature.

### 3.  NiTi

In this investigation, hardness and yield stress measurements for NiTi were carried out on the specimens which were quenched from 1000°C because NiTi phase has wide homogeneity range at high temperature but restricted around room temperature(2).  Therefore, it should be minded that the data obtained below 600°C are irrelevant for discussing the effect of structure defects on the mechanical properties precisely.

The change in hardness of NiTi with temperature is shown in Fig. 5.  The hardness of quenched NiTi keeps initial value up to 400°C and then decreases with increasing temperature.  At room temperature, the hardness of NiTi increases as the composition deviates from the stoichiometry.  However, only a little composition dependence was observed above 600°C.  The relation between temperature and yield stress of NiTi is shown in Fig. 6.  It is obvious that the plastic behavior of stoichiometric NiTi is completely different from the

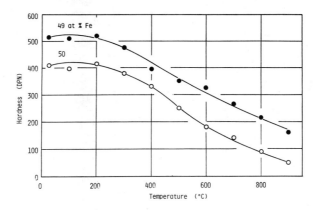

Fig. 1   Temperature dependence of Hardness of FeTi

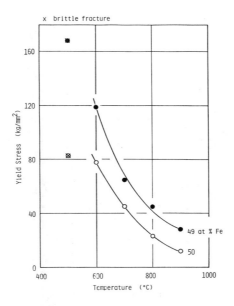

Fig. 2   Temperature dependence of yield stress of FeTi

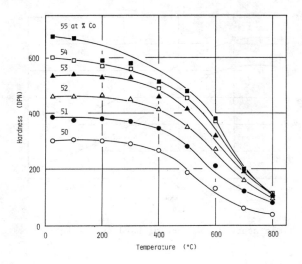

Fig. 3   Temperature dependence of Hardness of CoTi

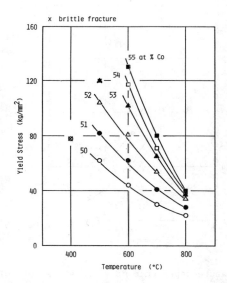

Fig. 4   Temperature dependence of yield stress of CoTi

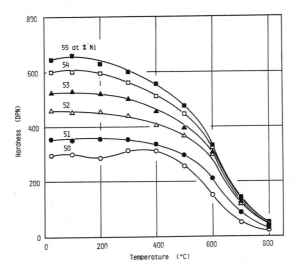

Fig. 5    Temperature dependence of hardness of NiTi

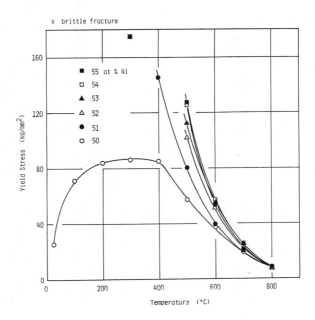

Fig. 6    Temperature dependence of yield stress of NiTi

non-stoichiometric compound.  The stoichiometric NiTi is ductile at all temperatures tested and also shows anomalous behavior in the temperature range between R.T. and 400°C.  Non-stoichiometric NiTi shows the same plastic behavior as was observed in FeTi and CoTi.  In the NiTi, brittle to ductile transition was observed around 400°C.  The relatively small composition dependence of yield stress was also observed.

## Discussion

1.  Temperature dependence of yield stress in stoichiometric NiTi

In the measurements of temperature dependence of yield stress, brittle to ductile transition was clearly observed around 0.5 Tm in FeTi, CoTi, and non-stoichiometric NiTi.  Only the stoichiometric NiTi was ductile at all temperatures tested and showed to have anomalous temperature dependence of yield stress in the temperature range from R.T. to 200°C followed by temperature independent range between 200 and 400°C.  The same behavior has been reported by Rozner and Wasilewski(7) in tensile testing.  The cause of this anomalous behavior will be discussing fully now.

The temperature dependence of yield stress of intermetallic compounds having CsCl structure has been measured on AgMg(8) and NiAl(7).  The general trend obtained is that the temperature dependence of yield stress is small at the temperatures below 0.5 Tm but large above that temperature and no anomaly exists.  It is well known that in some intermetallic compounds which have $L1_2$ structure yield stress increases with increasing temperature at specific temperature range.  In order to explain the anomaly, several theories have been proposed(9).  But no successful explanation has been given.

In stoichiometric NiTi there are two distinguished phenomena affecting the plastic behavior around room temperatures.  One is stress induced martensitic transformation and the other is elastic modulus anomaly which was first observed by Wasilewski(10).  When the stress is applied at the temperatures between Mf and Md mother phase must transform into martensite.  This causes stress relaxation at relatively low stress level and consequently yield like profile is observed in a stress-strain curve.  However, Md temperature of stoichiometric NiTi was reported to be 130°C by Sastri and Marcinkowski(11).  So that above 130°C the elastic modulus anomaly is considered to be a main factor affecting the yield stress.  As a matter of fact, elastic modulus of stoichiometric NiTi increases with increasing temperature from Ms to 600°C, and as shown in Fig. 6, the yield stress changes with temperature in the same manner as the elastic modulus does.  Therefore, it may be concluded that the yield stress anomaly observed in stoichiometric NiTi is caused by both stress induced martensitic transformation and the elastic modulus anomaly.

2.  Effect of structure defects on hardness and yield stress

Fig. 7 illustrates the relation between hardness and composition for CoTi which has wide homogeneity range.  It is clear from the figure that hardness minima appear at the stoichiometric composition throughout the temperatures tested, that the hardness increases as the composition deviates from the stoichiometry below 0.5 Tm, that above 0.5 Tm the increase in hardness with composition becomes small and that the hardness maxima appear at specific composition in the off stoichiometry.  In other words, the introduction of structure defects is effective to increase the hardness below 0.5 Tm but the introduction of large amount of the defects causes the decrese in hardness for off-stoichiometric CoTi above 0.5 Tm.  The same tendency has also been observed in several compounds having CsCl structure(12)(13)(14).  In these compounds hardness maxima were clearly observed above 0.6 Tm.  This behavior can be

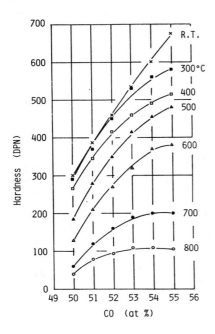

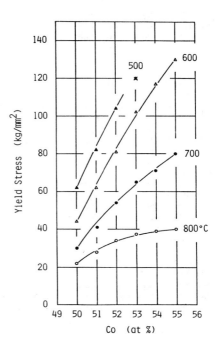

Fig. 7 Hardness isotherms as a function of composition for CoTi

Fig. 8 Yield stress isotherms as a function of composition for CoTi

explained by the following idea that though structure defects act as barriers against the motion of dislocations at lower temperatures where deformation is controlled by slip, the defects even substitutional atoms act as prefered jump site at higher temperatures where the deformation is mainly controlled by diffusion(15). This idea is considered to be strongly supported by the fact that diffusion coefficient increases as the composition deviates from the stoichiometry in AgMg, AuCd, AuZn, Ni rich side of NiAl, Co rich side of CoAl, and Ni rich side of NiTi(16)(17)(18)(19)(20) where substitutional wrong atoms exist as structure defects(6)(21). Domian et al(19) theoretically gave the following concept by using the random walk theory and six jump mechanism that the activation energy of diffusion should decrease as the composition deviates from the stoichiometry. If the maximum of the activation energy of diffusion means the minimum of the diffusion coefficient this may be the strong evidence to support the fact that the diffusion coefficient of the compounds increases as the composition deviates from the stoichiometry.

In this investigation, the activation energies of diffusion for FeTi, CoTi, and NiTi were obtained from the relation between hardness and temperature, such as H = A exp 2Q/RT, by using the procedure proposed by Sherby and Armstrong(22). However, no clear relation was obtained.

As shown in Fig. 8, the introduction of substitutional atoms brings increase in yield stress at relatively lower temperatures of the ductile fracture range but little increase in yield stress was given at higher temperatures. This behavior is completely identical with the hardness does and may be attributed to the action of substitutional atoms. In fact, the same tendency has been obtained in the measurements of flow stress of AgMg. It also has CsCl structure(10). In the AgMg, it was found that the introduction of structure defects causes the decrese in flow stress at temperatures near Tm.


## Conclusion

In order to clarify the effect of structure defects on the mechanical properties of FeTi, CoTi, and NiTi which have CsCl structure and wide homogeneity range, the change in hardness and yield stress with temperature were measured.

Stoichiometric NiTi showed sufficient ductility at all temperatures tested. However, the relation between yield stress and temperature is rather complicated. The NiTi showed yield stress anomaly between R.T. and 200°C where yield stress increases with increasing temperature. The cause of the anomalous behavior is considered to be due to both stress induced martensitic transformation and elastic anomaly.

The effect of structure defects on yield stress and hardness was clearly observed in CoTi because CoTi has relatively wide homogeneity range. The hardness is largely affected by the structure defects at lower temperatures. However, the effect becomes small as temperature increases. The hardness maxima were observed above 0.6 Tm. In yield stress measurements, brittle to ductile transition was observed around 0.5 Tm. The transition temperature increases as the composition deviates from the stoichiometry. The relation between yield stress and composition in the ductile fracture range is almost the same as in hardness.

The decrease in hardness and yield stress with increasing amount of structure defects observed at the compositions where large amount of structure defects exist can be explained by the fact that in the intermetallic compounds having CsCl structure, diffusion coefficient increases as not only vacancies are introduced but also substitutional atoms are introduced. Therefore, stoichiometric compounds should be chosen for high temperature materials.

## References

1.  M. Ettenberg, K. L. Komarek, and E. Miller: *Ordered Alloys; Structural Application and Physical Metallurgy*, Ed. by B. H. Kear, C. T. Sims, N. S. Stoloff and J. H. Westbrook, Claitor's Publishing Division, Baton Rouge, (1970), 49.
2.  M. Hansen and K. Anderko: *Constitution of Binary Alloys*, McGraw Hill, New York, (1958).
3.  R. P. Elliott: *Constitution of Binary Alloys, First Supplement*, McGraw Hill, New York, (1965).
4.  F. A. Shunk: *Constitution of Binary Alloys, Second Supplement*, McGraw Hill, New York, (1968).
5.  R. J. Wasilewski: J. Phys. Chem. Solids, 29(1968), 39.
6.  T. Suzuki and K. Masumoto: Met. Trans., 3(1972), 2009.
7.  A. G. Rozner and R. J. Wasilewski: J. Inst. Metals, 94(1966), 169.
8.  D. L. Wood and J. H. Wesrbrook: Trans. Met. Soc. AIME, 224(1962), 1024.
9.  A. Kuramoto: Bulletin Japan Inst. Metals, 14(1975), 567.
10. R. J. Wasilewski: Trans. Met. Soc. AIME, 233(1965), 1691.
11. A. S. Sastri and M. J. Marcinkowski: Trans. Met. Soc. AIME, 242(1968) 2393.
12. J. H. Westbrook: J. Electrochem. Soc., 103(1956), 54.
13. J. H. Westbrook: J. Electrochem. Soc., 104(1957), 369.
14. R. W. Guard and J. H. Westbrook: Trans. Met. Soc. AIME, 215(1959), 807.
15. J. H. Westbrook: *Mechanical Properties of Intermetallic Compounds*, Ed. by J. H. Westbrook, John Wiley & Sons, New York, (1960), 1.
16. G. F. Bastin and G. D. Rieck: Met. Trans., 5(1974), 1827.
17. W. C. Hagel: *Intermetallic Compounds*, Ed. by J. H. Westbrook, John Wiley & Sons, New York, (1967), 377.
18. G. F. Hancock and B. R. McDonnell: phys. stat. sol. a, 4(1971), 143.
19. H. A. Domian and H. I. Arronson: Trans. Met. Soc. AIME, 230(1967), 44.
20. D. Gupta, D. Lazarus, and D. S. Lieberman: Phys. Rev., 153(1967), 863.
21. G. G. Libowitz: Met. Trans., 2(1971), 85.
22. O. D. Sherby and P. E. Armstrong: Met. Trans., 2(1971), 3479.

# ADVANCED TRENDS IN DEVELOPING
# HIGH-STRENGTH TITANIUM ALLOYS

R. E. Shalin, S. G. Glasunov, A. I. Khorev

All-Union Institute of Aviation Materials
USSR

Study of complex alloying, microalloying, heat and thermomechanical treatment effect on mechanical properties of titanium
alloys permitted to make general conclusions presented below.

## 1. Complex Alloying

Systematic study of alloying elements effect during separate
and complex alloying of titanium alloys was conducted on the ini-
tiative and under the leadership of academician S. T. Kischkin.

These investigations permitted to determine main principles
of complex alloying high-strength weldable titanium alloys intended for use in the annealed and thermally strengthened conditions:
— simultaneous alloying by $\beta$-stabilizers of opposite distribution on dendritic segregation assuring macrouniformity of different alloy volumes strengthening, minimizing grain size difference, stability of polymorphic transformation temperature in the
alloy total volume, the uniform conditions of new phase initiation, decrease of phase constituents size and relationship between
the particle length and width, decrease of parallel plate bundles
size, increase of phase constituents disorientation, constant degree of solution metastability in different alloy zones and consequently more uniform and dispersed decomposition;
— alloying for effective strengthening of $\alpha$ and $\beta$-solid solutions and decrease of their strength difference giving rise to
strengthening microuniformity, decrease of the phase boundary
stress gradient, uniform $\alpha$-$\beta$-phases participation in the loading

process; uniform plastic deformation and hence decrease of structure banding and heterogeneity, increase of metastable phase decomposition products strength;

   – alloying with the relation of isomorphous and eutectoid-
forming alloying elements, expressed in terms of equivalent to
molybdenum values from 2:1 to 1:1, resulting in constant quantity
of β-stabilizers in different zones and suppression of eutectoid
decomposition. Developed principles of titanium alloying formed
the basis for developing heat treatable weldable BT23 and BT19
alloys [1] .


## 1.1 Complex-alloyed BT23 alloy

   BT23 alloy with the composition Ti-(4–6.2)Al-4(4–5)V-(1.5–
–2.5)Mo-(0.8–1.4)Cr-(0.4–0.8)Fe is alloyed by β-stabilizers with
distribution coefficients higher and lower than one. Total content of isomorphous β-stabilizers (equivalent to 4.2–5.8% Mo) in
the alloy is 1.5 times higher compared to that of eutectoid-for-
ming elements (equiv. to 2–4.2% Mo). The alloy strength varies
from 105 to 190 $\text{kgf/mm}^2$ depending on annealing and strengthening
heat treatment conditions. BT23 alloy structure in the near equilibrium state consists of 30% β-phase (Ti-2Al-5Mo-7.2V-2Cr-2.2Fe)
with the cell parameter $a_\beta$= 3.21 Å and 70% $\alpha$-phase (Ti-5.6Al-
–0.7Mo-2.5V-0.5Cr-0.3Fe). Total content of β-stabilizers in $\alpha$-
phase is equivalent to 3.7% Mo, while on alloying by each element
separately their content in $\alpha$-phase is 2–3 times less, which
determines higher strengthening of $\alpha$-phase on complex alloying.
Total content of four β-stabilizers in β-phase is practically
the same as in case of alloying by one β-stabilizer. Ultimate
strength of BT23 alloy phase constituents are close: $\sigma_{b\alpha}$= 88 $\text{kgf/mm}^2$
($H_{20}$ = 480 $\text{kgf/mm}^2$) and $\sigma_{b\beta}$= 90 $\text{kgf/mm}^2$ ($H_{20}$ = 530 $\text{kgf/mm}^2$).

   When quenching temperature rises to 800°C β-phase quantity
increases to 75%, on quenching from 900°C ($t_n$ =920°C) $\alpha''$-phase
is fixed and on quenching from 1000°C $\alpha'$-phase is fixed.

   The analysis of alloying elements distribution in automatic
argon-arc welded joint showed that total quantity of β-stabilizers
in the dentrite central volumes is equivalent to 7.9% Mo (4.1 V;

1.9 Mo; 0.97 Cr; 0.78 Fe), that is practically the same. BT23$\alpha$+
+ $\beta$ alloy specific strength ( $\sigma_{b/d}$ =32 km; $\delta$=6%) is considerably higher than that of BT14 alloy. ( $\sigma_{b/d}$ = 25 km; $\delta$=7%) of
the same class. Welded joint strength of BT23 alloy in the annea-
led ($\sigma_\beta$ =100-106 kgf/mm$^2$; $\alpha \geqslant$ 50$^\circ$) and heat treated conditions
($\sigma_\beta$ =125-130 kgf/mm$^2$; $\alpha$ =25-30$^\circ$) is 20 kgf/mm$^2$ higher compared to
that of BT14 alloy. BT23 alloy large-sized plates with $\sigma_\beta$=100 kgf/
mm$^3$ have $K_{IC} \geqslant$ 420 kgf/mm$^{3/2}$. The most promising US «Corona-5»
alloy [2]  Ti-4.5Al-5Mo-1.5Cr ($K_{IC} \geqslant$ 350 kgf/mm$^{3/2}$ with $\sigma_\beta$=95 kgf/
mm$^2$) phase composition is close to that of BT23 alloy for which
complex alloying by two isomorphous 2 Mo+ 4.5 V (equivalent to
5% Mo) and 0.8 Cr+ 0.6 Fe (equivalent to 1.7 Cr) elements is used.


## 1.2  Complex Alloyed BT19 alloy


   BT19 alloy is alloyed in accordance with the developed principles. Central dendritic zones of automatic argon-arc weld metal
contain $\beta$-stabilizers in the quantity equivalent to 16.6% Mo (6.7
Mo; 3.2 V; 4.8 Cr) and interdendritic zones — to 17.8 Mo (5.0 Mo;
3.4 V; 6.4 Cr). Ultimate strength of $\beta$-(87 kgf/mm$^2$) and $\alpha$-phases
(84 kgf/mm$^2$) are close together. In the alloy quenched from $\alpha$+ $\beta$
region temperature metastable phase decomposition during ageing
takes place according to the following scheme $\beta_{met.} \rightarrow \beta_{depl.}$ +
$\alpha_{enrich.} \rightarrow \alpha$ equiax. + $\beta_{equiax.}$ and in the alloy quenched
from the $\beta$-region temperature — according to the scheme $\beta_{met.} \rightarrow$
$\beta_{depl.}$ + $\omega \rightarrow \beta_{depl.}$ + $\alpha$ enrich.$\rightarrow \alpha$ equiax. + $\beta$equiax. with
is confirmed by X-ray, chemical and thermophysical analysis. BT19
$\beta$-alloy structural strength is 30 kgf/mm$^2$ higher than that one of
BT15 $\beta$-alloy. BT19 $\beta$-alloy specific strength ( $\sigma_{b/d}$ = 33 km;
$\delta$ =3.5%) is also higher than that of BT15 $\beta$-alloy (26.5 km; $\delta$=3%).
After strengthening heat treatment BT19 alloy welded joints with
$\sigma_\beta$ =115 kgf/mm$^2$ have high characteristics a$_H$ =3.5 kgf/m/cm$^2$;
a$_{TY}$ =3.0 kgfm/cm$^2$; a$_{TC}$ =2.8 kgfm/cm$^2$, while BT15 alloy welded
joints are brittle. BT19 alloy while having high tensile strength
$\sigma_\beta$=155 kgf/mm$^2$ has moderate value of $K_{IC}$= 210-225 kgf/mm$^{3/2}$.

## 2. Microalloying of alloys

Microalloying of $\alpha + \beta$ - and - $\beta$-structure alloys by III (Y, Gd, Dy), IV (Zr, Hf) and VII (Re) group elements participating in structure formation at all stages of semifinished and final product fabrication, which, in its turn, greatly effects alloy mechanical properties and structural strength was studied.

Positive effect of microalloying by horophobic (Re), horophilic elements (rare metals) and also by elements having greater affinity to oxygen than titanium (rare metals, Zr, Hf) is shown. These elements accomplish refractory seed function or decrease the surface energy and serve as deoxidants of near-boundary alloy volumes.

## 2.1 $\alpha + \beta$-alloy microalloying

The introduction of up to 0.02% Re and 0.1% Zr into $\alpha + \beta$-alloys changes the weld metal solidification conditions of $\alpha \to \beta$ polymorphic and $\beta \to \alpha'$ martensite transformations which is accompanied by welded joint ductility increase by 1.5-2 times. As X-ray analysis showed, more intensive diffusion processes over post-martensite temperature range give rise to the formation of greater amount of primary $\alpha$ and $\beta$-phases and martensite content decrease. In addition finer martensitic structure with smaller size packets of parallel martensitic needles and their greater disorientation is formed. Microalloying by zirconium is likely to increase the surface tension, decrease martensitic needle acuity and, hence diminishes stress concentration on its tip. Heat treated $\alpha + \beta$-alloy welded joint strength increases by 10-15 kgf/mm$^2$ with the increase of zirconium content up to 0.2% which is due to the change of metastable phase decomposition character [3].

The introduction of 0.02% rare metals into BT23 $\alpha + \beta$-alloy with overheated at 1000°C structure (is characteristic of heat-affected zone during welding) diminishes grain (by 2-3 times) and intragranular platelet size. Rare metals segregate over $\beta$-phase

and precipitating $\alpha$-phase matrix boundary lowering surface
energy, which in its turn results in the decrease of $\alpha$-phase intragranular platelet critical nucleus. Finer boundaries of primary β-grains are indicative of rare metals deoxidation function
and impurity content reduction on the boundary. Electron microscopic analysis showed that primary $\alpha$-phase in BT23 alloy has
a laminated form and in the alloy containing rare metals a polyhedral one. Rare metals reduce the stress gradient on the interphase by changing $\alpha$-phase morphology. BT23 alloy with 0.02–0.1%
rare metals heated at 1000°C followed by strengthening heat treat-
ment has twice as much ductility ( $\delta$=6%; $\psi$=10% with $\mathcal{G}_6$=135 кgf/
mm$^2$) compared to that of BT23 alloy after the same treatment.
Structure of the alloy with rare metals is characterized by more
uniform and more dispersed decomposition. Electron microscopic
analysis of specimen fractures showed that the introduction of
rare metals greatly increases the cup size around the strengthening phase particles. This indicates greater local ductility
(Fig. 1).

Fig. 1 Electron fractographs of BT23 (a) and BT23+ 0.05% Dy
(b) alloy fractured specimens.
Heating at 1000°C – 1 h, Water quenching from 800°C
(30 min.). Ageing at 500°C – 10 h. x 10,000.

## 2.2 Microalloying of β-alloys

The introduction of up to 0.5–1% Zr into β-alloys (BT15, BT19)
doubles base metal ductility (in thermally strengthening condition).

Auger-spectroscopy revealed that BT19 alloy containing 1% Zr has
two times less oxygen concentration along grain boundaries and
three times less oxygen enriched near-boundary zone compared to
those of ß-alloy without zirconium. Zirconium function as deoxidant of near-boundary volumes is the most effective in ß-alloys
having coarse boundaries of ß-grains. To improve their condition
greater zirconium content is required than for $\alpha$+ ß-alloys. Zirconium increases metastable ß-phase decomposition uniformity during ageing, diminishes the detrimental effect of alloying elements segregation on high-alloyed ß-alloys structure thereby contributing to more uniform participation of structural elements
in loading plastic strain.

While BT15 alloy thermally strengthened welded joints frac-
ture is brittle at $\sigma_\ell$ =80-100 кgf/mm$^2$ those of BT15-1 alloy containing 0.5-1.5% Zr at $\sigma_\ell$=125 кgf/mm$^2$ posssess a satisfactory ductility (Fig. 2). Zirconium positive effect is greater in ß-alloy
with higher oxygen content, which confirms once more its deoxidant
function. Addition of 0.5-1.0% Hf produces the same effect on ß-
alloys (Fig. 2). When introducing 0.02% rare metals ductility of
ß-alloy increases by 1.3 times in annealed condition and by 1.5-
-2 times in thermally strengthened condition. Microalloying produces greater effect on ß-alloys with coarse-grained overheated
structureᵉˡ, differing in mechanical properties, have considerable sizes. While in BT15 alloy with this structure metastable
ß-phase decomposition is heterogeneous, near-boundary regions de-
composing particularly intensively, in BT15 alloy with rare metals
decomposition is homogeneous over the whole section field. BT15
alloy microhardness variation range is $\Delta H_{50}$= 532-354= 178 кgf/mm$^2$,
for BT15 alloy with rare metals $\Delta H_{50}$ decreases down to 75 кgf/mm$^2$.
Rare metals concentrate along grain boundaries of ß-alloy, decrease surface energy and probability of $\alpha$-phase formation along the
boundaries. One may suppose, that rare metals, which atomic radius
is much greater than that of titanium, precipitate preferably on
vacancies decreasing alloying elements diffusion rate in the near-
boundary volumes making it equal to that of grain body, which also
contributes to greater ß-phase decomposition homogeneity. Electron
fractographic analysis is indicative of greater cup relief size
and depth in the fracture of alloys with rare metals, and also of

greater and more uniform plastic strain. Microalloying by elements possessing greater affinity to oxygen than titanium (219
кcal/mol), such as zirconium (259 кcal/mol), hafnium (272 кcal/
mol), rare metals (420-437 кcal/mol) increases mechanical properties of base metal and particularly of welded joints in accordance with rising negative values of these elements oxides formation heat. The microalloying has positive effect on structure:
reduces grain size and impurity content along its boundaries,
assures more dispersed and uniform decomposition of metastable
phases on ageing contributes to the formation of weld finer martensitic structure after welding with greater martensitic plates
disorientation. Structural variations are accompanied by internal
stress concentration decrease and contribute to more uniform participation of structure elements in plastic strain which assures
base metal and welded joints mechanical properties increase.

The effect of improving mechanical properties on introducing
microalloying elements is greater in thermally strengthened condition than in annealed condition. In the alloy with overheated
in β-region structure or with cast structure (which is characteristic of heat affected zone and weld metal respectively) the
effect of properties improvement is considerably greater than in
the alloy with equiaxial fine-grained structure. Microalloying
does not practically change phase constituents mechanical and
physical properties, but changes their morphology; reducing structural stress concentrators negative effect and increasing polycrystalline material strengthening uniformity.

## 3. Heat and thermomechanical
### strengthening treatment

To improve titanium alloys structure and mechanical properties in service thermomechanical treatment (TMT) methods are proposed which are classified depending on deformation temperature
and treatment thermal cycle sequence.

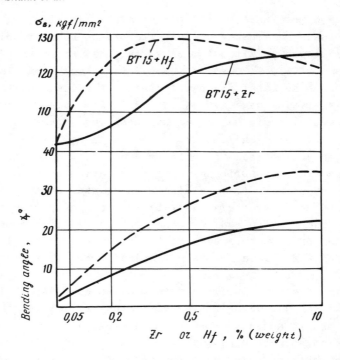

Fig. 2 Zirconium and hafnium effect on BT15 alloy welded
joints mechanical properties.

### 3.1 Welded joints deformation
### and heat treatment

On thermoplastic treatment of weld metal primary β-grain
boundaries are destroyed, intragranular precipitations size reduces, their disorientation increases, length to width ratio  of
laminated $\alpha$-phase, which aquires a polyhedral form, reduces. Deformation before heat treatment, while refining the structure increases homogeneity of metastable phase decomposition process
during ageing, rises appreciably the dispersion hardening degree,
assures more uniform structural constituents participation in
loading and thereby enhances strengthening heat treatment efficiency. Plastic strain (up to 30-40%) of welded joints at 750-
800°C before strengthening heat treatment increases its tensile
strength by 15-35 kgf/mm² for $\alpha$+ β-alloys (BT14, BT16, BT23) and

by 40–50 $kgf/mm^2$ for ß-alloys (BT15), which is accompanied by
simultaneous weld metal ductility and welded tanks structural
strength improvement. X-ray analysis of BT23 $\alpha$ + ß-alloy weld metal showed that deformation before quenching changes matrix ß-
phase cold working degree, increasing it up to 30 $kgf/mm^2$ and
disperses substructure up to 700–1000 Å. After ageing at 450°C
stresses due to phase cold working increase up to 40 $kgf/mm^2$ and
subgrain sizes decrease down to 100 Å. After ageing at 500°C
stress relaxation down to 15 $kgf/mm^2$ is observed and noticeable
subgrain growth is not revealed. Thus ageing at 500°C results in
lower strengthening due to phase cold working with equal dispersion
hardening, which is accompanied by welded joint ductility increase
while retaining strength, compared to ageing at 450°C.

### 3.2 Thermomechanical treatment conditions.
### Optimization and efficiency of its
### utilization

Fundamentals of mechanical properties variation in the alloys, treated according to different TMT regimes, due to ß-sta-
bilizer alloying over wide concentration range in Ti–3Al–ß systems
and phase composition change. Maximum strength and mechanical
properties improvement effect after TMT are observed in $\alpha$ + ß-alloys of undercritical concentration Ti–3Al–5Cr, Ti–3Al–5Mo and
Ti–3Al–3Fe. ß-stabilizers concentration in these alloys is equivalent to 5–8% Mo (Fig.3 ). Maximum strength after high-temperature thermomechanical treatment (HTTMT) is obtained in BT23
$\alpha$ + ß alloy ($\sigma_b$ =160 $kgf/mm^2$). Study of low-temperature thermomechanical treatment (LTTMT) effect on the alloy mechanical properties showed that with the increase of alloying by ß-stabilizers
maximum strengthening due to LTTMT (compared to strengthening
heat treatment) is attained for Ti–3Al–15Mo and BT19 alloys (Fig.3).

It is experimentally found that during HTTMT it is necessary
to accomplish deformation at ß-region temperature for BT1–0, BT5–1
$\alpha$ -alloys and BT15, BT19 ß-alloys and at $\alpha$ + ß-region temperature
for BT6, BT14, BT3–1, BT23 $\alpha$ + ß-alloys. BT16 and BT22 alloys al-

loyed by ß-stabilizers in the quantity close to critical concentration should be subjected to preliminary thermomechanical treatment (HTTMT plus quenching). HTTMT-2 utilization is promising. This strengthening method is energetically efficient and does not require the use of quenching equipment, it permits only low-temperature furnaces ageing without protective atmosphere. Its use for BT23 alloy bars, forgings and die forgings fabrication has assured high combination of mechanical properties ($\sigma_{\ell} \geqslant 120$ kgf/mm$^2$, $K_{IC} \geqslant 260$ kgf/mm$^2$) and reduction of power consumption by 350 кw. h/t.

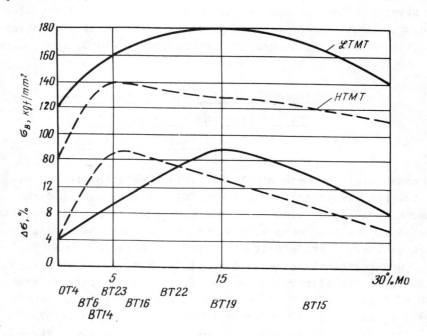

Fig. 3 Titanium alloys strengthening: ———— LTTMT; – – – HTTMT.

According to the efficiency of increasing ß-alloy tanks mechanical properties and structural strength TMT methods may be placed as follows: PTMT, HTTMT-2, HTTMT-1, HTTMT, MTT, HLTTMT (HTTMT$^+$ LTTMT), HLTTMT utilization allowed to increase structural strength of cylindrical tanks up to 165 кgf/mm$^2$.

It is found that mechanical property improvement effect is due to primary β-grain size reduction; boundary improvement; intragranular structure considerable refinement; more dispersed and more uniform decomposition during ageing; impurities and alloying elements distribution character change; dislocation density increase and coherent scattering region reduction.

## Conclusions

Uniform participation in resistance to service stresses of titanium alloys phase constituents and structure elements is attained by simultaneous effect of metallurgical and technological factors (complex alloying, microalloying, heat and thermomechanical treatment) on metal and structural concentrators ($\alpha$ -phase acute tips, etc.) effect is minimized.

## References

1. A. I. Khorev. Main Principles of High-Strength Weldable Titanium Alloys Development. J. Svarochnoe proizvodstvo, 1975, N 10, pp. 4-7 (in Russ.).
2. F. H. Froes, R. G. Berryman, J. C. Williams. Proc. of the 3rd International Conference on Titanium. M., VILS, 1978, v. 3, pp. 323-327.
3. A. I. Khorev, S. G. Glasunov, L. G. Mukhina. Effect of Modifying Additions on Titanium Alloy Properties. Tsvetnye metalli, 1966, N 7, pp. 86-88.
4. S. T. Kishkin, S. G. Glasunov, A. I. Khorev. Use of High-Temperature Thermomechanical Treatment in Producing Extruded Tubes of BT15 Titanium Alloy. Isvestiya AN SSSR, Metalli, 1966, N 3, pp. 125-129.

# LOW MELTING HYPEREUTECTOID TITANIUM-COPPER ALLOYS

H. B. Bomberger*, G. S. Hall** and S. R. Seagle*

*RMI Company, Niles, Ohio, and **American Welding & Mfg. Co., Warren, Ohio

## Introduction

Most alloys employed by the titanium castings industries were developed originally as wrought products with little or no attention given to their cast properties and castability. In the present study a casting system is proposed which permits use of simpler and less costly methods than those currently employed.

Many of the problems and high costs involved in skull casting titanium are related directly to the high reactivity of the molten metal. At normal melting and casting temperatures titanium combines rapidly with most crucible and mold refractories. Since reactivity is a direct function of temperature, it seems apparent that some of the problems could be alleviated if a useful alloy with a significantly lower casting temperature were available.

With this concept, a number of low melting point alloys were prepared and evaluated. The most promising of these involved large additions of beta eutectoid alloying elements. Among the most interesting were binary and ternary alloys containing substantial amounts of copper. In addition to low melting points, several compositions in this series were found to have surprisingly good mechanical properties, including strength, ductility, toughness, stability, creep and fatigue resistance.

Copper and other beta eutectoid alloying elements are well known to depress both the melting points and the alpha-beta transformation temperatures[1]. Copper as well as cobalt, nickel, silicon and certain other elements serve as active eutectoid beta phase stabilizers in titanium. On cooling such compositions through the transformation temperature, the beta phase transforms rapidly to alpha and an intermetallic compound. Concern for instability and a loss of ductility generally associated with this reaction has heretofore limited the use of active eutectoid elements to not much more than the small amount soluble in the alpha phase. In the case of copper, the maximum solubility in alpha is about 2.1 weight percent and in beta about 17 weight percent[1]. Published studies on the Ti-Cu system indicated little or no ductility in compositions containing 10 percent copper or more[2,3,4]. High cooling rates and contaminants may have affected the ductility in the earlier work.

## Materials and Testing

All alloys were double consumably arc melted. The alloying metals were used in elemental forms and the titanium sponge had an oxygen level of 0.05 percent. Most ingots weighed eight Kg (17 lbs) and were 11.4 cm (4.5 in.) in diameter. A few 20.3 cm by 50 Kg (8-in. by 110 lb) ingots were also evaluated.

Most alloys were tested for cast properties using a 1.6 cm (0.625-in.) thick slice cut transversely through the ingot centers.  Standard ASTM mechanical test procedures were employed except where indicated otherwise. Liquidus temperatures were determined in vacuum and are reported as melting points.  Chemical analyses and details on these and other tests were reported earlier[5].


## Results

Compositions, melting points and tensile properties of the most interesting titanium-copper alloys are given in Table 1 and some of these data are illustrated in Figures 1 and 2.  This information shows that titanium alloys can be formulated to have not only attractive cast properties but also melting points several hundred degrees below the temperature (about 1600°C or 3020°F) at which unalloyed titanium melts.

Table 1  Alloy compositions, melting points and tensile properties

| Alloy, wt% | Melting[a] Points,°C | As-Cast[b] | | | Annealed[b,c] | | |
|---|---|---|---|---|---|---|---|
| | | UTS | YS | % El | UTS | YS | % El |
| Ti-8Cu-6.5Ni | 1380 | 786 | 579 | 6.0 | 669 | 448 | 5.2 |
| Ti-10Cu | 1496 | 924 | 696 | 6.0 | 565 | 421 | 23.0 |
| Ti-10Cu-2.5Fe | 1471 | 1048 | 896 | 3.5 | 772 | 545 | 7.0 |
| Ti-10Cu-2.5Fe-2Al | 1463 | 1186 | 1131 | 2.0 | 903 | 731 | 7.0 |
| Ti-13Cu | 1434 | 717 | 565 | 10.0 | 579 | 414 | 14.0 |
| Ti-13Cu-0.2 O | 1434 | 869 | 724 | 3.0 | 655 | 496 | 4.7 |
| Ti-13Cu-0.25Y$_2$O$_3$ | 1434 | 710 | 524 | 8.0 | 600 | 379 | 11.0 |
| Ti-13Cu-1.5Al | 1430 | 758 | 621 | 3.0 | 676 | 552 | 8.0 |
| Ti-13Cu-2Al | 1430 | 1054 | 986 | 1.0 | 724 | 634 | 12.0 |
| Ti-13Cu-4Sn | 1425 | 807 | 690 | 8.0 | 655 | 538 | 8.5 |
| Ti-13Cu-1Co | 1425 | 745 | 600 | 6.5 | 621 | 428 | 10.2 |
| Ti-13Cu-3Al | 1425 | 993 | - | - | 827 | 724 | 10.0 |
| Ti-13Cu-2.5Fe | 1410 | 855 | 676 | 4.0 | 676 | 510 | 5.0 |
| Ti-13Cu-1Be | 1375 | 496 | - | - | 386 | 359 | 1.0 |
| Ti-13Cu-4.5Ni | 1340 | 690 | 524 | 6.0 | 607 | 441 | 5.0 |
| Ti-13Cu-4Ni-1Al | 1330 | 710 | 593 | 5.0 | 648 | 517 | 5.0 |
| Ti-16Cu | 1380 | 669 | 524 | 9.0 | 572 | 393 | 9.0 |
| Ti-16Cu-1.5Al | 1370 | 786 | 669 | 4.0 | 648 | 545 | 5.2 |
| Ti-19Cu | 1325 | 600 | 448 | 4.0 | 469 | 359 | 4.0 |
| Ti-19Cu-1Al | 1320 | 669 | 531 | 3.0 | 607 | 462 | 5.0 |

a.  Liquidus temperature
b.  Tensile tested at 21°C, average of two values, MPa units used
c.  Annealed one hour at 850°C, cooled 55°C/hr to 480°C and air cooled

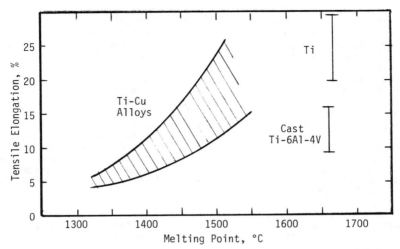

Fig. 1  Melting points and ductilities of cast Ti-Cu
and Ti-Cu-X alloys in comparison to Ti and
Ti-6Al-4V

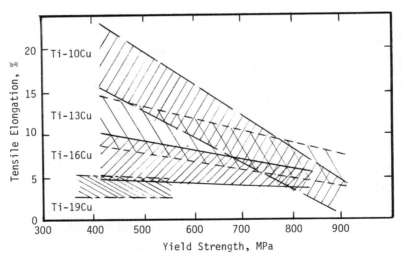

Fig. 2  Yield strengths and ductilities of Ti-Cu
and Ti-Cu-X alloys

Over the range of interest, copper additions were found to lower the melting point of titanium about 17°C (30°F) and the strength was decreased about 165 MPa (24 Ksi) for each weight percent of copper added. This change was accompanied with a loss in ductility. Other additions, including aluminum, iron, nickel, and tin also lowered the melting temperature as expected. Nickel in combination with copper had the greatest effect by depressing the melting point about 22°C (40°F) with each percent added. Iron had a similar effect in the Ti-Cu system and the melting point was depressed only about 10°C (18°F) for each percent added. Aluminum was especially effective as a strengthening agent but it had only a small effect on the melting point. Additions of aluminum, tin, iron and oxygen generally gave fine transformation products in the Ti-13Cu base whereas nickel and cobalt gave coarser products. Alloys containing 10 to 13 percent copper and in combination with a few percent aluminum, iron, nickel or tin were found to offer an interesting combination of cast properties. Additions of oxygen and yttria to the Ti-13Cu base show these materials can be tolerated.

Annealing was useful for lowering strength and improving ductility. Duplex heat treatments, listed in Table 2, show the cast alloys respond and offer opportunities for further property enhancement.

Table 2   Tensile properties and stability of heat treated alloys

| Alloy, wt% | Final Heat[a] Treatment | Test Temp.°C | As Heat Treated[b] | | | After 200hr at 427°C[b] | | |
|---|---|---|---|---|---|---|---|---|
| | | | UTS | YS | % El | UTS | YS | % El |
| Ti-8Cu-6.5Ni | 705°C-1hr-AC | 21 | 841 | 593 | 5.0 | | | |
| Ti-8Cu-6.5Ni | 760°C-8hr-AC | 21 | 800 | 559 | 7.0 | | | |
| Ti-10Cu-2.5Fe | 760°C-8hr-AC | 21 | 910 | 683 | 5.0 | | | |
| Ti-10Cu-2.5Fe | 760°C-16hr-AC | 21 | 896 | 703 | 5.0 | | | |
| Ti-13Cu | 705°C-8hr-AC | 21 | 814 | 607 | 10.0 | 786 | 586 | 5.5 |
| Ti-13Cu | 790°C-8hr-AC | 21 | 731 | 565 | 10.0 | 696 | 559 | 10.5 |
| Ti-13Cu | 790°C-8hr-AC | 316 | 517 | 324 | 14.5 | | | |
| Ti-13Cu-1.5Al | 790°C-16hr-AC | 21 | 793 | 621 | 10.0 | | | |
| Ti-13Cu-1.5Al | 790°C-16hr-AC | 316 | 621 | 428 | 19.0 | | | |
| Ti-13Cu-2Al | 790°C-8hr-AC | 21 | 869 | 724 | 6.5 | | | |
| Ti-13Cu-2Al | 850°C-1hr-AC[c] | 21 | 724 | 634 | 12.0 | 737 | 613 | 13.5 |
| Ti-13Cu-3Al | 705°C-8hr-AC | 21 | 1076 | 938 | 2.0 | | | |
| Ti-13Cu-3Al | 815°C-8hr-AC | 21 | 945 | 855 | 6.0 | | | |
| Ti-13Cu-4.5Al | 790°C-8hr-AC | 21 | 683 | 496 | 5.5 | | | |
| Ti-13Cu-4.5Al | 790°C-8hr-AC | 316 | 579 | 379 | 14.0 | | | |
| Ti-13Cu-2.5Fe | 705°C-8hr-AC | 21 | 896 | 807 | 3.5 | | | |
| Ti-13Cu-2.5Fe | 760°C-8hr-AC | 21 | 876 | 703 | 3.5 | 889 | 676 | 5.0 |
| Ti-13Cu-4.5Ni | 745°C-8hr-AC | 21 | 821 | 717 | 4.0 | | | |
| Ti-13Cu-4.5Ni | 760°C-8hr-AC | 21 | 745 | 565 | 5.0 | 731 | 614 | 6.5 |

a. Heat treated one hour at 960°C, air cooled and followed by treatment shown
b. Average of two tensile values, MPa units used
c. Single heat treatment only, as shown

As-cast and annealed tensile strengths of binary Ti-Cu alloys are illustrated in Figure 3 as a function of copper content.

Alloys containing 13Cu, 13Cu-1.5Al and 13Cu-4.5Al were tensile tested at 316°C (600°F) in a heat treated condition and found to have useful properties at this temperature. Furthermore, Ti-13Cu, Ti-13Cu-2Al, Ti-13Cu-2.5Fe and Ti-13Cu-4.5Ni showed good thermal stability after 200 hours at 427°C (800°F) as indicated in Table 2.

Smooth rotating-beam fatigue endurance limits for heat treated Ti-13Cu and Ti-13Cu-4.5Ni alloys are about 207 MPa (30 Ksi) as shown in Figure 4. Interestingly, the limit for Ti-13Cu-1.5Al appears to be significantly higher at about 276 MPa (40 Ksi). The reason for this difference is not clear at this time but suggests improvement may be possible by optimizing heat treatments and compositions. In any case, the ratio of endurance limit to ultimate tensile strength of these alloys is favorable. In the order given above, the ratios are about 0.30, 0.32 and 0.35, respectively. Cast low-oxygen Ti-6Al-4V was reported to have an endurance limit of 241 MPa (35 Ksi) and a ratio of 0.32[6].

A brief study indicated the toughness of as-cast Ti-13Cu (at a UTS and YS of 949 and 707 MPa and 6% El) is about 55 MPa $\sqrt{m}$ (50 Ksi $\sqrt{in}$) and the $K_{Iscc}$ in a 3.5% NaCl solution is between 20.5 and 27.4 MPa $\sqrt{m}$ (18.6 and 24.9 Ksi $\sqrt{in}$). However, when this material was annealed (850°C-1hr-FC 38°C/hr to 482°C-AC), the properties changed to an ultimate and yield strength of 570 and 427 MPa, 23% El and a $K_{Ic}$ of 114.6 MPa $\sqrt{m}$ (104.2 Ksi $\sqrt{in}$). The $K_{Iscc}$ also increased to between 103.6 and 114.6 MPa $\sqrt{m}$ (94.2 to 104.2 Ksi $\sqrt{in}$). This information suggests reasonably good toughness can be realized with the Ti-Cu alloy system although the values will be determined by the alloy selected and heat treatments employed.

Recent work has shown that cast Ti-13Cu-1.5Al has creep and creep-rupture resistance superior to unalloyed titanium and slightly below the values for cast Ti-6Al-4V[7]. Alloys of this type have densities of about 4.9 gm/cc (0.175 lb/cu in.) and elastic moduli of about 114,000 MPa ($16.5 \times 10^6$ psi).

Preliminary studies also indicate the Ti-Cu binary alloys have good hot workability but more limited cold rollability. The alloys also exhibited good weldability and the weld properties were similar to the cast properties.

Microstructural examinations of as-cast material from the center of 11.4 cm diameter ingots showed large grained structure but it contains a fine substructure of alpha and $Ti_2Cu$ as shown in Figure 5a. The substructure of $Ti_2Cu$ coarsens by annealing at 848°C (1560°F) which is in the beta-$Ti_2Cu$ phase region. Annealing at 960°C (1760°F) and air cooling (Fig. 5c) resulted in a beta matrix and fine particles of $Ti_2Cu$ showing the material was heated close to the hypereutectoid beta transus temperature. Reheating this material to 788°C (1450°F) (which is below the eutectoid temperature) produced a very fine structure of alpha and $Ti_2Cu$ (Fig. 5d). Thus, alloys of this type can be heated high into the beta-$Ti_2Cu$ or all-beta fields to produce very fine structures.

The as-cast structures were found to become coarser on going to larger ingot diameters (from 11.4 to 20.3 cm). The as-cast strengths were lower

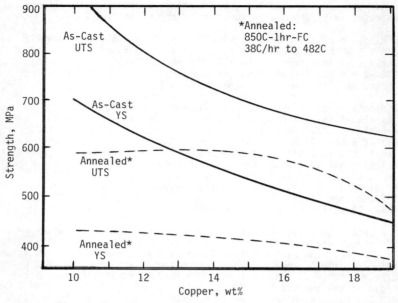

Fig. 3   Tensile strengths vs copper in Ti-Cu binary alloys

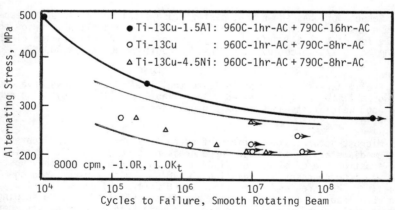

Fig. 4   Fatigue resistance of heat-treated alloys

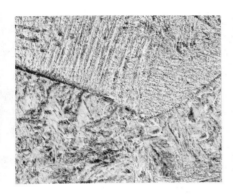

a. As-cast

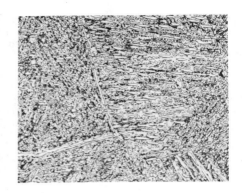

b. 848°C-1hr-FC at
   38°C/hr to 480°C, AC

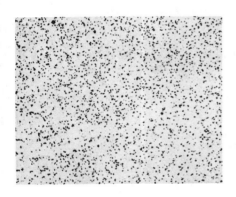

c. 960°C-1hr-AC

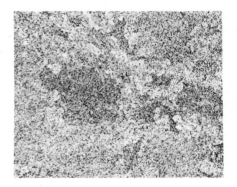

d. 960°C-1hr-AC
   + 788°C-16hr-AC

Fig. 5 Microstructures of cast Ti-13Cu-1.5Al alloy (250X)

but this effect seems to be related to cooling rates since comparable values were achieved by annealing pieces of similar size. Smaller cross sections would, of course, be expected to have smaller grains and correspondingly better properties.

## Discussion

The proposed casting system was demonstrated and reported elsewhere[5,8]. Ti-13Cu, Ti-13Cu-1.5Al and Ti-13Cu-4.5Ni alloys were induction melted and cast to produce precision aircraft parts using existing superalloy vacuum investment casting facilities. Titanium modified yttria ($Y_2O_3 \cdot 15Ti$) was selected as the most promising of several primary crucible materials tested and a rare earth oxide mixture was preferred for mold face coats. These ceramics had reasonably good resistance to the alloys and they in turn displayed good tolerance for contaminants from the ceramics.

Tensile properties and a high-cycle fatigue endurance limit of 276 MPa (40 Ksi) of the Ti-13Cu-1.5Al castings compared favorably with values reported for cast Ti-6Al-4V.

An economic evaluation indicated the suggested casting systems using low-melting Ti-Cu alloys has a potential cost savings of 20 to 25 percent in producing small (up to 2 Kg) aerospace castings and more in larger parts.

## Conclusions

A casting system was proposed and demonstrated which involved the development and use of a series of low-melting hypereutectoid titanium-copper alloys. These alloys permit the use of casting technology familiar to the superalloy vacuum casting industries. Ti-Cu alloys, including Ti-13Cu-1.5Al and Ti-13Cu-4.5Ni were found to have excellent castability and melt at temperatures at which there was very little damage to the crucible and mold refractories employed. Alloys in the series were found to have a good combination of properties which approached those of cast Ti-6Al-4V. These alloys appear to have potential usefulness in aircraft, dental and a variety of industrial applications. An economic study indicates the proposed system could result in significant cost savings.

## Acknowledgements

The authors wish to thank Messrs. D. R. Schuyler and J. A. Petrusha of the AiResearch Manufacturing Company for the successful completion of their work on ceramic materials and investment casting. Appreciation is also extended to the U.S. Air Force Materials Laboratory for the portion of this work supported under Contract No. F33615-74-C-5055.

## References

1. F. A. Shunk, *Constitution of Binary Alloys*, (1969).  McGraw-Hill.
2. F. C. Holden, et al, Trans. AIME, 203(1955), 117-125.
3. K. M. Buchanan, British Non-Ferrous Metals Res. Assoc., Final Report S295111(1973).
4. R. F. Bunshah and H. Margolin, Trans. ASM, 51(1959), 961-980.
5. D. R. Schuyler, et al, "Development of Titanium Alloy Casting Technology, WPAFB, AFML-TR-76-80 (Aug. 1976).
6. R. A. Brown, "Precision Casting of Titanium", Lecture 11-A, NY Univ. Titanium Course (Sept. 8-11, 1969).
7. Unpublished data, AiResearch Manufacturing Co., Phoenix, Arizona.
8. D. R. Schuyler and J. A. Petrusha, "Investment Casting of Low-Melting Titanium Alloys". *Proc., Vacuum Metallurgy Conference,* Am. Vac. Soc. (1977).

# EFFECT OF HEAT TREATMENT ON STRUCTURE AND PROPERTIES OF IMI 829

D.F. Neal and P.A. Blenkinsop
IMI Titanium, IMI Kynoch Ltd., Birmingham, England

## Introduction

IMI 829 has been designed as a creep resistant, beta heat treated titanium alloy for gas turbine use up to around 550°C. Its nominal composition is Ti-5.5%Al-3.5%Sn-3%Zr-1%Nb-0.25%Mo-0.3%Si, and as such is regarded as a near-α alloy with the capability of being welded without significant loss of mechanical properties.

The alloy's standard heat treatment is a solution treatment in the beta phase field at 1050°C (β transus 1015°C ± 10°C), followed by ageing/stress relieving at 625°C.

Changes in composition, and advances in thermomechanical processing knowledge compared with previously known alloys have given significant improvements in structure and creep resistance of IMI 829 and a tolerance of higher stress relieving temperatures. This paper will describe some of the effects of heat treatment on the structure and mechanical properties of IMI 829.

## Results and Discussion

To date, IMI 829 has been produced both as small ingots and as three tonne production melts, and the work to be described has been taken from both sources.

### Effect of Solution Treatment Variables

Two of the major structural factors in the control of basic properties of β heat treated titanium alloys are :

a)   β grain size and
b)   β transformation product (acicular α)

### a) β grain size

This factor is controlled initially by composition and processing variables and in this regard IMI 829 has been designed to improve on previously known alloys, particularly in the case of larger (~ 300mm Ø) billet sections. Time and temperature of β solution treatment then determines β grain size as shown in Fig.1.

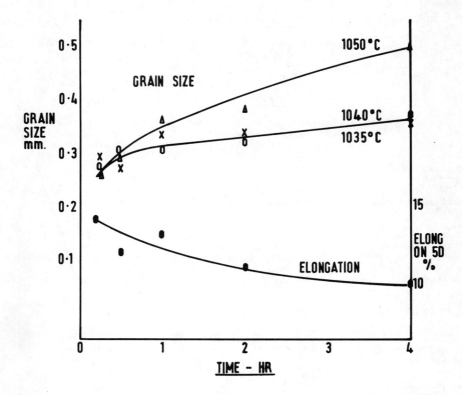

## FIG. 1. EFFECT OF TIME AND TEMPERATURE OF SOLUTION TREATMENT ON $\beta$ GRAIN SIZE AND TENSILE DUCTILITY OF IMI 829

A temperature of 1035°C – 1040°C would appear to be 'adequate' for IMI 829, but in practice 1050°C solution treatment is chosen to ensure complete "solutioning" and allow extra leeway under production conditions.

The effect of time and temperature on tensile elongation is also illustrated in Fig.1 (similar results were achieved for all solution treatment temperatures) and the inverse relationship of grain size and elongation is clearly shown. A stable situation is achieved in practice after approximately one hour at temperature.

b) $\beta$ transformation product (acicular $\alpha$)

The second major influence in solution treatment is that of quenching rate and its effect on $\beta$ transformation product and phase composition. Fig.2 and Table 1 show the effect of slow (air cool) and 'typical' (oil quench) cooling rates on structure, tensile, creep and post creep properties of IMI 829. This data is taken from typical gas turbine engine disc cut-up trials and illustrates several of the effects of cooling rate.

## TABLE 1
### EFFECT OF COOLING RATE FROM BETA HEAT TREATMENT ON TENSILE, CREEP AND POST CREEP TENSILE PROPERTIES OF IMI 829 DISCS.

| CONDITION | CREEP TOTAL PLASTIC STRAIN 540°C/300 N.mm⁻² 100 HR% | 300 HR% | 0·2%PS N.mm⁻² | U.T.S. N.mm⁻² | EL 5D % | R in A % |
|---|---|---|---|---|---|---|
| AIR COOLED FROM β S.H.T (+ AGE) | NO EXPOSURE | | 815 | 931 | 9 | 17 |
|  | 0·127 | 0·220 | 859 | 942 | 6 | 14 * |
| OIL QUENCHED FROM β S.H.T (+ AGE) | NO EXPOSURE | | 843 | 956 | 12 | 22·5 |
|  | 0·056 | 0·147 | 892 | 981 | 5·5 | 7·5 * |

✻ POST CREEP TENSILE AFTER 300 HR/540°C/300 N.mm⁻² SURFACE RETAINED

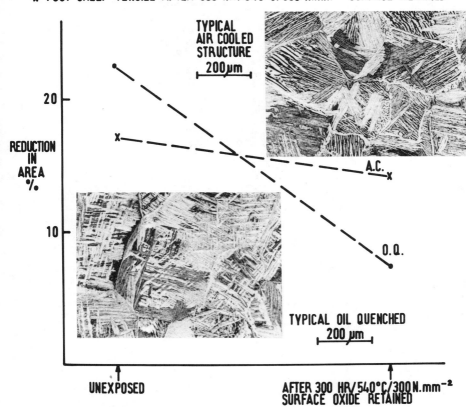

TYPICAL AIR COOLED STRUCTURE
⊢ 200μm ⊣

REDUCTION IN AREA %

A.C.

O.Q.

TYPICAL OIL QUENCHED
⊢ 200 μm ⊣

UNEXPOSED

AFTER 300 HR/540°C/300 N.mm⁻²
SURFACE OXIDE RETAINED

### FIG.2. EFFECT OF COOLING RATE FROM β SOLUTION HEAT TREATMENT, AND CREEP EXPOSURE ON THE DUCTILITY OF IMI 829

Firstly, structure is controlled principally by cooling rate, with air cooling of thick (> ~ 30mm) sections producing a thickening of α at β grain boundaries, and large colonies of aligned α within the β grains. Oil quenching which gives cooling rates around $1 - 10°C$ sec$^{-1}$ produces a basic basketweave α structure with minimal amounts of grain boundary α or aligned α.

The volume fraction of retained β phase after solution treatment is also influenced by cooling rate; faster rates giving less retained β, but compositions of the phases are also important and this latter feature particularly, has an influence on creep properties.

The other well known effect of cooling rate (on strength) is shown in Table 1, which also reveals the less obvious effects on ductility, creep resistance and stability. Faster cooling rates improve ductility and creep resistance (generally but not universally) whereas they tend to reduce stability - Fig.2. Ductility improves because of the greater availability of slip systems in a less aligned structure, whilst creep resistance achieves an optimum at cooling rates around $1 - 10°C$ sec$^{-1}$ and reduces thereafter because of phase compositional factors. The reduction in stability with faster cooling rates is believed to be related to the increased propensity for ordering to occur, again due to phase composition changes with cooling rate.

## Effect of Ageing Variables

The purpose of 'ageing' near-α alloys, which are by definition lean in the 'strengthening' β stabilising elements, is to stabilise and stress relieve although some strengthening does occur. A careful control of ageing temperature is needed because of the loss of tensile ductility and creep resistance which occurs if temperatures are too high or times too long. Fig.3 shows the effects of second heat treatments (after oil quenching 50mm Ø bar from β solution treatment) on structure, tensile strength, ductility and stability.

## Microstructural Changes After 'Ageing'

Included in Fig.3 are five transmission electron micrographs which illustrate some of the complex changes that occur over the temperature range $500 - 950°C$ after β solution treatment.

In the unaged condition, retained β is present at α lath boundaries and has a thickness of around 50nm. Ageing at temperatures up to around 550°C for 24 hours has no significant effect on microstructure, but at temperatures above this, the width of β phase reduces and discrete particulate 'precipitates' begin to form within it. The temperature range from ~575°C to ~650°C is very critical, in that α lath boundary morphological changes are relatively rapid. At 625°C for example, a treatment of two hours produces a slight reduction in β boundary width, whilst after 24 hours discrete (TiZr) Si type precipitates have formed.

The changes occur more rapidly at temperatures in the range 700 to ~920°C where coarser 'precipitates' are formed, but above ~ 920°C an α/β structure reforms and true β phase is again obtained.

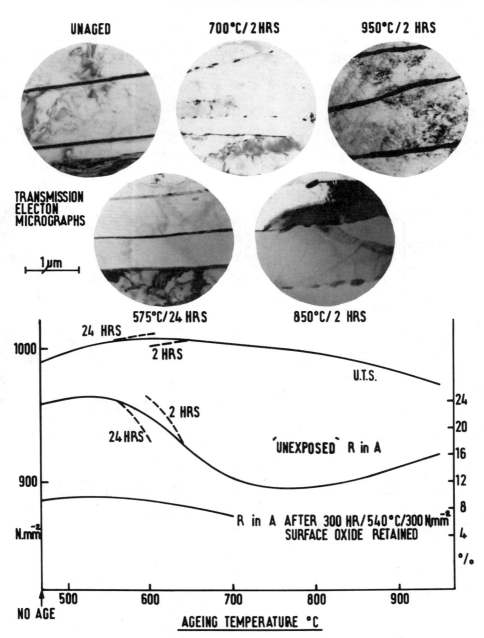

FIG. 3. EFFECT OF AGEING TEMPERATURE ON STRUCTURE AND TENSILE PROPERTIES OF IMI 829

Tensile Strength and Ductility Changes After Ageing

Although ageing upto around 625 °C produces some strengthening, followed by softening as ageing temperature is increased, the major change in tensile properties with ageing is in ductility. Losses in ductility over the temperature range 575 to 800 °C can be related to structural factors, particularly the appearance of discrete silicide particles (although the influence of phase compositional changes cannot be completely divorced from this). These particles are relatively brittle compared with the matrix, and early voiding occurs in tensile testing when planar dislocation arrays meet silicide particles. This affects "reduction in area" more than "elongation".

A minimum in ductility is reached after treatment around 850 °C whilst higher temperatures induce β formation which is considerably more ductile.

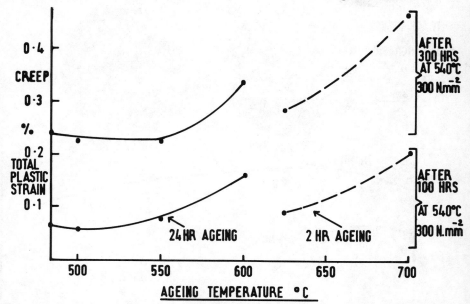

FIG 4. EFFECT OF AGEING TEMPERATURE ON CREEP BEHAVIOUR OF IMI 829

Despite the formation of β phase at around 900 °C the acicular nature of the structure still has the major influence. Treatments at 900 °C without a prior β solution treatment give "reduction in area" values approximately twice that seen after β + 900 °C treatments.

Creep and Stability Changes After Ageing

The structural and phase changes which occur above ~575 °C and which influence ductility, are accompanied by phase compositional changes which affect creep resistance. Fig.4 shows the effects of two times — 2 hours and 24 hours — on creep resistance of IMI 829 and again highlights the critical 575 - 650 °C range. In practice a compromise is achieved between adequate stress relief and good creep resistance by ageing IMI 829 at 625 °C for two hours.

Stability is defined as the retention of tensile ductility after creep exposure, (typically tests are for 300 hours at 540°C at a stress of 300N.mm$^{-2}$) with the surface oxide retained. The major portion of instability is due to metallurgical changes rather than to surface changes (at least up to exposure conditions of 600°C/300 hours in IMI 829 – Fig.5),

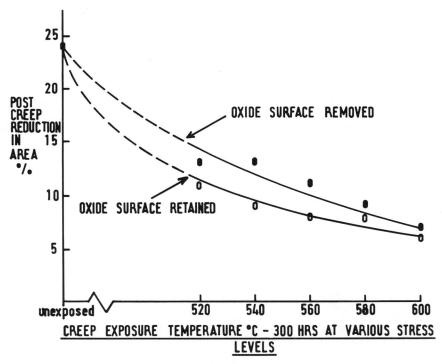

FIG. 5. STABILITY OF IMI 829

and the effects of prior ageing treatment are overriden during creep exposure. Hence, although a greater proportion of ductility is lost in material aged at low temperatures, the actual ductility value in any aged condition after creep exposure is little different – Fig.3. This is a consequence of the slow ordering reactions which occur during creep exposure, and which benefit creep resistance but which lower ductility.

Conclusion

The improvement in creep resistance, fracture toughness and crack propagation resistance of beta heat treated near-α alloys over α/β heat treated alloys has been demonstrated many times. Whilst slight loss of strength and fatigue resistance may be seen in near-α alloys e.g. IMI 685 v Ti-6-2-4-2-S, high temperature capability is improved. A further improvement in high temperature capability has been achieved in IMI 829 which is designed for sustained running at up to 550°C in aircraft gas turbine environments. Short term, low stress excursions e.g. up to 600°C for several hundred hours, are also acceptable, and the ability of this type of alloy to be welded is also a major advantage.

Properties of near-α alloys such as IMI 829 can be markedly improved by correct choice of heat treatment both during the β solutioning stage and during 'ageing'. This paper has attempted to show that by correct control of processing and heat treatment, an exceptionally good balance of properties can be achieved in IMI 829 and maintained up to the high temperatures of modern gas turbine compressor and turbine stages.

# PHASE TRANSFORMATION AND HEAT TREATMENTS

# A. Precipitation

# THE ω-PHASE REACTION IN TITANIUM ALLOYS

T. W. Duerig, G. T. Terlinde, and J. C. Williams

Carnegie-Mellon University, Pittsburgh, PA, U.S.A.

## Introduction

The $\beta \rightleftharpoons \omega$ transformation represents a somewhat unusual problem in that it has received significant attention, both theoretically and experimentally, from materials scientists and from materials engineers. The former group view the transformation as a fully reversible, displacement controlled transformation and continue to seek new ways to study the phenomenon. The engineers, on the other hand, view ω-phase formation as the source of embrittlement and, therefore, search for ways to avoid it. As a result of these variour motivations for studying it, the $\beta \rightleftharpoons \omega$ transformation is now much better, but still not completely, understood.

In this paper we will first briefly review the general characteristics of the ω-phase. Then we will discuss the current status of the athermal $\beta \rightleftharpoons \omega$ transformation and comment on the relationship between this transformation and isothermal ω formation. Finally, we will present some new observations regarding the morphological changes during isothermal ω-phase formation including observations on the relationship between the ω and α phases.

## Characteristics of the ω-Phase

The ideal omega structure can be viewed as having a hexagonal (P6/mmm) unit cell, with 3 atoms per cell, positioned at: $(0,0,0)$, $(1/3,2/3,1/2)$, and $(2/3,1/3,1/2)$. The resulting c/a ratio is 0.613. In Ti alloy systems, this ideal structure has been confirmed both by X-ray and electron diffraction [1-3]. In other ω forming systems (based on Zr or Hf), an axial shift of the cell internal atoms has been reported, which results in a (c/a) ratio slightly greater than .613 [4].

The ω-phase can form in either of two distinct modes. In relatively dilute alloy compositions, the ω-phase appears during the rapid quenching from the β-phase field to temperatures below the ω start temperature $(T_\omega)$. This type of ω is referred to as athermal ω $(\omega_{ath})$. Isothermal aging of these alloys, as well as alloys richer in β stabilizing alloying additions, produces isothermal ω $(\omega_{iso})$.

Isothermal ω is generally reported as having either an ellipsoidal or cuboidal morphology. As will be discussed in this paper, the primary shape controlling factor is the β/ω misfit which in turn is controlled by the solute misfit in the β-phase.

Another aspect of the $\beta \rightarrow \omega$ transformation is observed after quenching alloys slightly too rich to form $\omega_{ath}$. Specifically, a "pre-transition" phenomenon is observed. Although no discrete particles can be visibly associated with this region, streaking in reciprocal space has been observed by both electron and X-ray diffraction. This phenomenon has been the subject of many theoretical and experimental investigations.  However, discussions of the "pre-transition" streaking are beyond the scope of this paper[5,6,7].

Although the emphasis of this paper will be on the isothermal formation of $\omega$, we feel that $\omega_{iso}$ and $\omega_{ath}$ are more closely related than has been previously recognized.  Thus, it will first be necessary to review the displacement controlled reaction which is attributed to $\omega_{ath}$ formation before presenting our views and evidence which relate the two reaction products, $\omega_{iso}$ and $\omega_{ath}$.

## The Displacement Controlled Reaction - $\omega_{ath}$

In Figure 1, free energy versus composition curves are represented for the $\alpha$, $\omega$, and $\beta$ phases in a hypothetical $\omega$ forming alloy system (viz-Ti-V).  Although the $\beta+\omega$ common tangent is metastable relative to the $\alpha+\beta$ tangent, in practice the $\alpha$-phase precipitates very slowly compared to $\omega$.  For the purposes of following discussions, we will focus our attention on $\omega$-formation and ignore the subsequent precipitation of $\alpha$.

Between compositions A and C of Figure 1, $\beta$ is energetically unstable with respect to decomposition into the $\omega$ and $\beta$ phases.  We should expect that the $\beta$-phase will decompose into a $\beta$ matrix at composition A, with $\omega$ particles of composition C.  Since this reaction requires diffusion, it can be suppressed by quenching.  If the alloy composition falls between A and B, however, a compositionally invariant transformation from $\beta$ to $\omega$ is still energetically favorable.  But for this to occur, a diffusionless mechanism for changing the $\beta$ structure into $\omega$ is required.

Experimental evidence for the existence of such a compositionally invariant mechanism is abundant, but the details of the mechanism are much more difficult to determine.  The observation that the $\omega_{ath}$ reaction cannot be suppressed by quenching rates as high as 11,000°C/sec[8] certainly indicates that a composition change is unlikely.  Reversibility studies performed below 0°C with cold stage TEM[9] provide further evidence that the reaction is compositionally invariant.  Evidence that the $\beta \to \omega$ reaction does not involve high diffusivity paths is evidenced by the consistently reported uniform nature of $\omega_{ath}$, as well as the extremely high particle density ($\sim 10^{18}/cm^3$) (Fig. 2).  The small size and the uniformity of $\omega$, the lack of surface effects, and the short range atom movements associated with the crystal structure change are among those factors which separate the $\omega$ mechanism from the classical martensitic reactions which are dominated by lattice shear.  The damping associated with the $\beta \rightleftarrows \omega$ reaction also is consistent with an atomic shuffle transformation, rather than a diffusional one.

De Fontaine[10] first pointed out that a $\underset{\sim}{k} = \frac{1}{3}$ <112> transverse bcc lattice displacement wave of the proper amplitude would convert the bcc $\beta$-phase to the ideal $\omega$ structure.  Further, he showed that such a mechanism would result in the observed orientation relation:

$$[111]_\beta || [0001]_\omega$$
$$(1\bar{1}0)_\beta || (11\bar{2}0)_\omega$$

Figure 3 demonstrates this by viewing the (1$\bar{1}$0) plane of the β matrix (closed circles), during passage of a $\underset{\rightarrow}{k} = \frac{1}{3}$ [11$\bar{2}$] sinusoidal transverse displacement wave. The displaced atomic positions resulting from such a wave are shown by the open circles. This displaced, or periodically disturbed, structure is then the (11$\bar{2}$0) plane of the ω structure.

De Fontaine was able to show by harmonic lattice theory that minimum stability contours in the bcc structure fell along the {111} octahedral planes in reciprocal space. Further, that the stability minimums corresponded to $\underset{\rightarrow}{k} = \frac{1}{3}$ <112> transverse waves, which can be equivalently viewed as 2/3 <111> longitudinal waves.

Harmonic theory was not, however, able to account for several aspects of the β→ω transformation. Specifically, it provided no activation barrier for the lattice displacement. The implications of this were that nothing would prevent the ω reaction from going to completion. Also, there was no preference provided for phonons of the specific wave amplitude required to form ω from β. This problem was corrected by Cook[11,13], who used an anharmonic approximation of lattice energy. Cook's approximation of lattice energy took the form:

$$F \sim \frac{1}{2} \gamma_{ij} U_i U_j + \frac{1}{3!} \gamma_{ijk} U_i U_j U_k$$

where $\gamma_{ij}$ and $\gamma_{ijk}$ are coupling parameters, and are physically the second and third derivatives of the lattice free energy-F with respect to the planar displacements - $U_i$, $U_j$ and $U_k$. It is the second term, the anharmonic term, of this expression that drives the ω reaction. This third order coupling term appears to be important only at low temperatures. (Graphically we can visualize this by studying Figure 4). Plotted are the lattice energy and displacement amplitude for a transverse wave of $\underset{\rightarrow}{k} = \frac{1}{3}$ <112>. Below the omega start temperature - $T_\omega$, the energy minimum corresponding to the ω structure must lie below that for the β structure. Separating the β and ω energy wells, there must be an activation energy barrier and an activated complex state - β*. The reversibility and quenching studies discussed above are evidence that F* must be very small. Above $T_\omega$, the energy minimum associated with the ω structure may still exist, but is now metastable to the β. From Figure 4, we can see that a wave of amplitude greater than A* is required to form ω. To achieve this, Cook[11] has proposed that the amplitude of the above displacement wave be thermally modulated. The result is pictured in Figure 5, as regularly spaced wave packets. Since the amplitude of the modulated displacement wave alternates polarity from packet to packet, every other wave packet is out of phase by 180° with the required ω displacement. Thus every other packet is a potential ω former. If the envelope amplitude is above A*, ω will form. A phonon flipping mechanism[11] could subsequently transform the remaining packets to ω. The end result, as predicted by this model, is a field of ω particles spaced periodically along the <111> directions (Fig. 5b), with a spacing controlled by the envelope frequency or, in this case, by the temperature.

Finally, it is useful to examine the sequence of events occurring as an ω forming alloy is rapidly heated to an aging temperature - $T_a > T_\omega$. During heating, the ω free energy curve of Figure 1 will rise relative to the β curve. This is qualitatively equivalent to shifting our alloy composition to the right in Figure 1. Above $T_\omega$, our alloy composition must lie to the right of composition B. Between compositions B and C, the β structure is energetically preferable to ω, assuming, of course, local composition changes by diffusion are suppressed.

From both Figure 1 and Figure 4, we do not then expect to observe the athermal ω reaction above $T_\omega$. Cold stage microscopy has revealed, however, that the reversion of ω to β occurs over a range of temperatures[9]. Diffuse neutron scattering work[14] has shown that ω particles are present well above $T_\omega$, even though no particles were visible in TEM. Cook[13] has stated that this presents no theoretical difficulties. The apparent contradiction between microscopy and neutron scattering can be understood in terms of particle lifetime. Fluctuations between the β and the quasi-static ω structures could be too rapid for detection using microscopy, but could be readily found by neutron scattering[14]. Moreover, these heterophase fluctuations above $T_\omega$ should be expected, since an ω energy well in Figure 4 still exists, and because β-$\omega_{ath}$ strain fields are likely to be extremely small. For the purposes of the following discussion, it will be necessary to assume that the ω structure can persist above $T_\omega$ in a steady state equilibrium, but that the number of ω particles decreases as $(T-T_\omega)$ increases.

### Isothermal ω

The nucleation of ω during isothermal aging appears to progress independently of both grain boundaries and dislocations. Further, recent work[15] has demonstrated that residual solute depleted zones (resulting from a phase separation reaction in the Ti-Cr system) do not seem to effect ω nucleation. We suggest that at aging temperatures near $T_\omega$, growth may continue from the quasi-static $\omega_{ath}$ particles which remain after heating to $T_a$. Figure 6 supports this view. The size consistency of the isothermal ω indicates that all nucleation events were essentially simultaneous. Also shown in Figure 6 is hyperfine ω; $\omega_{ath}$ that has reprecipitated during the requench of $T_a$. If this nucleation model is correct, re-aging at a lower temperature would mean that fewer $\omega_{ath}$ particles revert to β, and more would become available as nuclei for $\omega_{iso}$ growth. Thus, a bimodal size distribution should be expected and, in fact, is found (Fib. 7). If the original $T_a$ is lowered, we should expect and find a higher number density of particles (Fig. 8).

A mechanism of isothermal growth is now required. The apparent anomalously rapid growth of ω after only 1 minute at 400°C (Fig. 6) cannot be analyzed using diffusional growth. That is, simple calculations of diffusion at such low temperatures indicates that precipitation by solute segregation to the β+ω tie line compositions of Figure 1 is impossible. Furthermore, the displacement controlled mechanism applied to athermal ω formation is energetically unfavorable in this regime.

To surmount these difficulties, we visualize a β compositional fluctuation in the vicinity of a quasi-static $\omega_{ath}$ particle. This would reduce the ω free energy with respect to β. In terms of Figure 1, the local β composition would enter the A-B regime. Following this event, growth by the displacive mechanism could then proceed as already discussed. We should thus observe a rapid physical growth of the ω-phase, followed by a gradual chemical equilibration to the β-ω tie line compositions A and C. Such a process was alluded to by De Fontaine et al[9].

Note also, that in alloy compositions just to the right of 'B' in Figure 1, particle growth will be far more rapid than chemical equilibration, while in heavily β-stabilized compositions (near composition C of Figure 1), particle growth and chemical equilibration should correlate reasonably well. In fact, precipitation in these latter solute rich compositions should begin to approach the classical diffusion controlled nucleation and growth.

There is experimental evidence for this in the literature. Hickman[16] measured changes in β lattice parameters during ω precipitation in Ti-V alloys of several compositions. Since V markedly contracts the Ti lattice, he was able to use lattice parameter and volume fraction data obtained from X-ray diffraction to follow changes in β and ω compositions during aging. The physical growth of ω was, in fact, observed to occur over a shorter time scale than the solute rejection of ω. Recent Mössbauer work[17] in Ti-7.1wt%Fe shows a similar effect.

Perhaps the most definitive support for the diffusionally assisted, displacement controlled growth of ω can be found in Figures 6 and 7. It has been a generally accepted observation that low misfit ω forming systems such as Ti-Mo and Ti-Nb form ellipsoidal $\omega_{iso}$ with the major axis of the ellipsoidal lying along the "elastically soft" $<111>_\beta$ direction[18,19]. High misfit systems such as Ti-Fe, Ti-V, Ti-Cr, Zr-Nb, Ti-Mn form cuboidal ω with the cuboid faces lying parallel to $\{100\}_\beta$ planes[19]. The generally accepted explanation of this phenomenon is that high misfit systems must assume a shape minimizing strain energy, while low misfit systems take on a shape minimizing surface energy. The observation of ellipsoidal ω in the high misfit Ti-10V-2Fe-3Al alloy (Fig. 6) can only be explained by the incomplete solute segregation between β and ω. Continued aging (and continued solute segregation) then converted the sllipsoids to the anticipated cuboids (Fig. 7). It was necessary to lower $T_a$ from 400°C to 300°C during the second aging treatment in order to suppress α formation.

The hyperfine particles of Figure 6 demonstrate still more support for the diffusionally assisted displacement controlled growth process. During requenching, a uniform distribution of $\omega_{ath}$ appeared throughout the untransformed β matrix. The lack of an $\omega_{ath}$ free zone around $\omega_{iso}$ particles demonstrates solute rejection could not have been extensive during $\omega_{iso}$ growth. Note that the $\omega_{ath}$ is absent in Figure 7 because the β matrix is now enriched in solute due to diffusional partitioning.

The $\omega_{iso}$ final particle size appears to be dependent on temperature more than time. We can visualize four possible reasons that an isothermal ω particie could stop growing.

1.  Chemical stabilization of the β matrix
2.  Coherency strains
3.  Dilatation strain fields overlap

The first of these is likely to be controlling when the aging times are very long, or when the alloy composition is far to the right of composition 'B' in Figure 1. Coherency strains,on the other hand, will tend to limit growth at high temperatures and short times when the alloy composition is near composition 'B' of Figure 1. In this situation, the β matrix remain chemically unstable relative to the ω structure, but the increase in strain energy that would result from further growth outweighs the chemical energy gains of transforming the β structure to ω. It is yet unclear what magnitude or even what sense the dilatational strain fields surrounding a growing ω particle might have. It may be that when the particle density is high (Fig. 8), growth is limited by the interaction of such strain fields. Perhaps it is this interaction which prevents high ω dispersion densities from adopting a specific particle shape.

## Conclusions

A model for the isothermal precipitation of ω in Ti alloys has long been presented, which utilizes chance compositional fluctuations to set the stage for a compositionally invariant growth mechanism. In heavily β-stabilized alloys, this type of growth does not differ from "classical" diffusion controlled growth. As the alloy content becomes leaner (or the β structure less stable), the compositionally invariant growth mechanism begins to dominate, and certain "unusual" growth phenomenon is observed, which distinguishes the β→ω reaction from most others. For example, a morphology transition from ellipsoids to cuboids was noted in Ti-10V-2Fe-3Al at a constant size. It was proposed that this could result from a rapid, chemically invariant ω particle growth, followed by a slow composition equilibration.

## Acknowledgments

This work has been supported by the Office of Naval Research. Experimental work has been conducted using facilities provided by The Center for the Joining of Materials. The authors gratefully acknowledge the experimental assistance of M. Glatz and the secretarial help of Mrs. A. M. Crelli.

## References

1.  J. M. Silcock, M. H. Davies, and H. K. Hardy:  in "The Mechanism of Phase Transformations in Metals", Inst. of Met. London, (1956), 93.
2.  B. S. Hickman:  Trans. AIME, 245, (1969), 1329.
3.  M. J. Blackburn and J. C. Williams:  Trans. AIME, vol. 242, (1968), 813.
4.  B. A. Hatt and J. A. Roberts:  Acta Met., 8, (1960), 575.
5.  B. Borie, S. L. Sass, and A. Andreassen:  Acta Cryst., A29, (1973), 585.
6.  J. M. Sanchez and D. De Fontaine:  J. Appl. Cryst., 10, (1977), 220.
7.  S. L. Sass:  Acta Met., 17, (1969), 813.
8.  Y. A. Bagaryatskiy and G. I. Nosove:  Physics of Met. and Metellog., 13, (1962), 92.
9.  D. De Fontaine, N. E. Paton, and J. C. Williams:  Acta Met., 19, (1971), 1153.
10. D. De Fontaine:  Acta Met., 18, (1970), 275.

11. H. E. Cook: Acta Met., 22, (1974), 239.
12. H. E. Cook: Acta Met., 23, (1975), 1041.
13. H. E. Cook and W. J. Pardee: Acta Met., 25, (1977), 1403.
14. S. C. Moss, D. T. Keating, and J. D. Axe: Phase Transformations 1973, (ed. L. E. Cross) Pergamon Press, Oxford, (1973), 179.
15. V. Chandrasekarant, R. Taggert, and D. H. Polonis: Metallography, 6, (1973), 313.
16. B. S. Hickman: J. Inst. Met., 96, (1968), 330.
17. M. M. Stupel, M. Ron, and B. Z. Weiss: Met. Trans., 9A, (1978), 249.
18. B. S. Hickman: J. Mat. Sci., 4, (1969), 554.
19. H. E. Cook: Acta Met., 21, (1973), 1445.

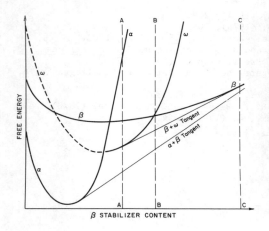

Fig. 1 – Schematic free energy curves for the α, β, and ω phases in a β stabilized Ti alloy.

Fig. 2 – Athermal ω in β–ST and water quenched Ti–10V–2Fe–3Al shown by dark field TEM.

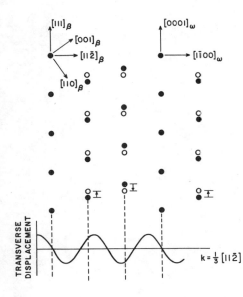

Fig. 3 – Representation of the atomic "shuffles" involved in the β–ω transformation. Closed circles represent ω lattice positions, and open circles represent the displaced, or ω lattice sites. The $(1\bar{1}0)_\beta$ plane is shown with the atoms displaced by a

$$\underline{k} = \frac{1}{3} [11\bar{2}] \text{ wave.}$$

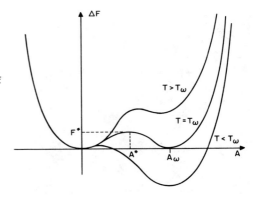

Fig. 4 - Schematic representation of a free energy versus wave amplitude curve for a $\underset{\sim}{k} = \frac{1}{3} <11\bar{2}>$ transverse wave.

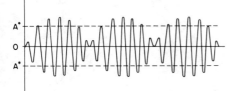

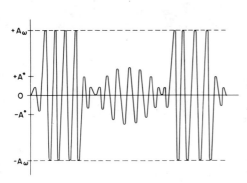

Fig. 5 - (a) Modulated sinusoidal wave, forming wave "packets" of alternating "sense", and (b) the growth of discrete ω particles from the above packets. The displacements of the center packet are 180° out of phase with the "proper" ω displacements, and are therefore unstable.

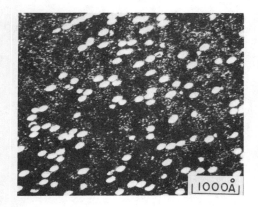

Fig. 6 – Ellipsoidal isothermal $\omega$ in $\beta$-ST Ti-10V-2Fe-3Al aged at 400°C for 1 minute.

Fig. 7 – Cuboidal isothermal $\omega$, in $\beta$-ST Ti-10V-2Fe-3Al duplex aged at 400°C for 1 minute, and 300°C for 45 minutes.

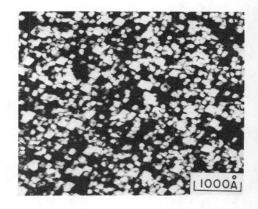

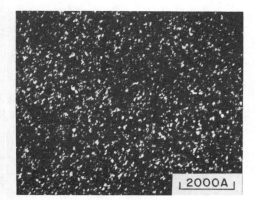

Fig. 8 – Isothermal $\omega$ of non-descript form, in $\beta$-ST Ti-10V-2Fe-3Al aged at 260°C for 10,000 mins.

# ω PHASE FORMATION IN Ti-Fe ALLOYS

Toshimi Yamane and Akihiko Miyakubi[*]

Department of Materials Science and Engineering,
Osaka University, Suita, Osaka 565, Japan
* Formerly Graduate Student, now at Mitsubishi Heavy
Industry Co., Ltd. Nagasaki, Japan

## Introduction

Metastable ω phase is formed in titanium alloys with iron, cobalt, vanadium and so on as solute elements which make stable the β phase, during aging at about 250~500°C. The crystal structure of the ω phase has been investigated by many authors.[1]-[6] But the researches on the precipitation of the ω phase from the point of the view of kinetics are not so many.[7],[8]

In this paper, we aim to analyse the precipitation of the phase by the kinetics, and to know the origin of the formation of the ω phase by the measurement of internal friction.

## Experimentals

Alloy compositions were decided by Luke's report[9] in which the ω phase was formed in the composition range of the valency electron number of 4.06~4.14. The chemical compositions of the titanium-iron alloys used in this investigation are shown in Table 1.

Table 1 Chemical compositions of Ti-Fe alloys (at%)

| Symbols | Compositions |
|---------|--------------|
| F - 3 | Ti - 2.59at%Fe |
| F - 5 | Ti - 5.05at%Fe |
| F - 6 | Ti - 5.96at%Fe |
| F - 7 | Ti - 6.67at%Fe |

Pure titanium (99.8%) and electrolysis iron were melted in an arc melting furnace in argon atmospher. Ingots were heated at 800°C for 2 days for homogenization, then rolled to 1 mm thickness. Specimens were heated at at 1000°C for 1 hour in vacuum of $10^{-5}$ Torr., then quenched in ice water. The aging for the $\omega$ phase formation in the quenched alloys were performed isothermally below 450°C in a $NaNO_3(1) + KNO_3(1)$ salt bath.

## Results and Discussion

Isothermal aging was performed at 300, 350, 400 and 450°C. The changes in the hardness of the alloys are shown in Figs.1~4 where the hardness increases with aging time and reaches a maximum. The specimens which contain higher iron concentration, have a longer time to reach the maximum hardness. The retained $\beta$ is not transformed into $\alpha'$ or $\alpha$ during aging at the temperature below 450 °C.[8] If the $\omega$ phase formation has only single mechanism and obeys an activation process, the time ($t_{max}$) when the hardness reaches the maximum, has the following relation between the aging temperature (T) and the activation energy (Q),

$$\log t_{max} = Q/(2.3RT) - C \text{ -----------(1)}$$

where R is the gas constant and C is a constant.

Fig.5 shows the relation between $\log t_{max}$ and 1/T. The activation energy obtained from the linear line slopes are 15.9, 31.8, 37.6 and 41.2 Kcal/mol for F-3, F-5, F-6 and F-7 respectively. These values have good agreement with those of the tracer diffusion of iron in the $\beta$ titanium-iron alloys, which are 39.6 Kcal/mol for Ti-5wt%Fe and 48.6 Kcal/mol for Ti-10wt%Fe alloys.[10] The $\omega$ phase formation in the titanium-iron alloys is considered to be controlled by the diffusion of iron in the $\beta$ phase.

Fig. 6 shows microstructures of the F-3, F-5 and F-7 alloys quenched from 1000°C in ice water. $\alpha'$ can be seen in Fig.6(a),on the other hand, only single $\beta$ phase is boserved in Fig.6(b) and (c) as reported previously.[11] We can consider that the $\omega$ phase precipitates from the $\beta$ phase.

It has been reported by Miller[12] that Ti-0.3at%Fe alloys had a Zener type relaxation. Internal friction measurements

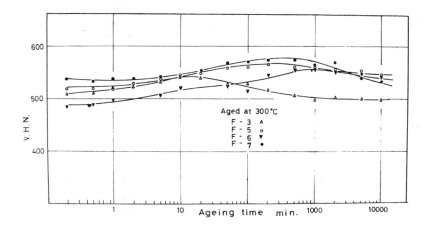

Fig.1   Micro-Vickers hardness (Load 500g) of the
specimens aged at 300°C after quenching in
ice water from 1000°C.

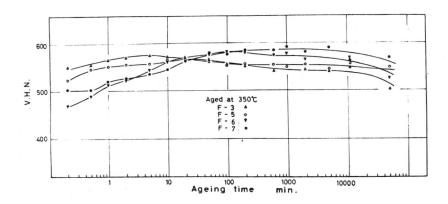

Fig.2   Micro-Vickers hardness (Load 500g) of the
specimens aged at 350°C after quenching in
ice water from 1000°C.

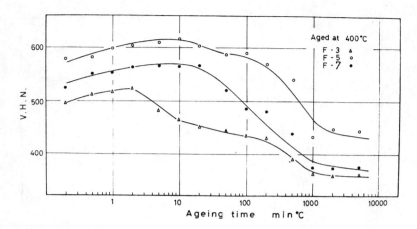

Fig.3   Micro-Vickers hardness (Load 500g) of the
specimens aged at 400°C after quenching in
ice water from 1000°C.

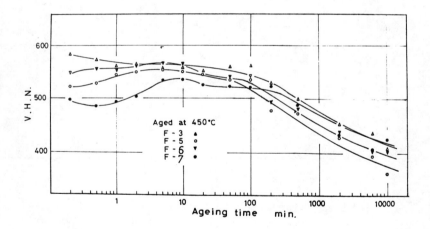

Fig.4   Micro- Vickers hardness (Load 500g) of the
specimens aged at 450°C after Quenching in
ice water from 1000°C.

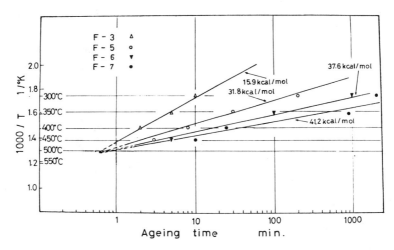

Fig.5 Relations between reciprocals of aging temperatures
(1/T) and aging time when hardness reached maxima.

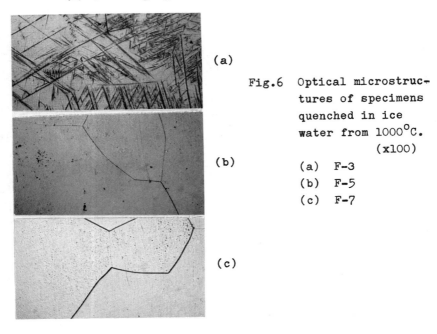

Fig.6 Optical microstruc-
tures of specimens
quenched in ice
water from 1000°C.
(x100)

(a) F-3
(b) F-5
(c) F-7

were performed by Ke's torsional pendulum method. The results
are shown in Figs. 7 and 8 as the function of temperatures. The
aged specimen F-3 has no internal friction peak but as quenched
one has peaks near 350 and 250°C (Fig.7), on the other hand, the
aged specimen F-5 has only the 250°C peak and no 350°C peak. The
peak at 350°C is considered to correspond to the α' phase for this
peak is observed only in the specimen F-3 which contains the α'
phase as seen in Fig.6(a). The peak at 250°C may owe to the quenched β phase for it is observed in both F-3 and F-5 alloys.

The activation energy of the 350°C peak obtained by Wert-
Marx's method[13) is about 42 Kcal/mol which is not compared with
no previous data on the diffusion of iron in α titanium. The activation energy of the 250°C peak is about 35 Kcal/mol which is near
that of the diffusion of iron in the β Ti-5at%Fe alloy as mentioned above.

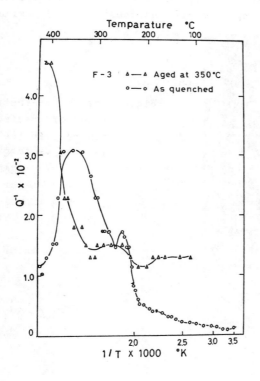

Fig.7 Internal friction
of the specimen F-3
during heating at the
heating rate of 1.5°C/
min., The specimen was
a wire of 1 mm diame-
ter and quenched in
ice water from 1000°C.
Aging was performed
at 350°C for 10 min.,
Resonant vibration
frequency was about
1.2 Hz.

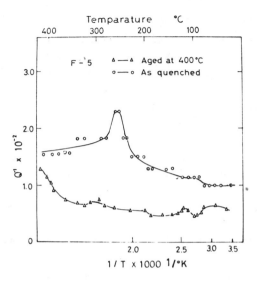

Fig.8 Internal friction of the specimen F-5 during heating at the heating rate of 1.5°C/min., The specimen was a wire of 1 mm diameter and Quenched in ice water from 1000°C. Aging was performed at 400°C for 10 min., Resonant vibration frequency was about 1.2 Hz.

## Summary

The process of the ω phase precipitation is controlled by the diffusion of the solute iron atoms. The activation energy obtained from the changes in the hardness of the titanium-iron alloys aged isothermally are 15.9, 31.7, 37.6 and 41.2 Kcal/mol for the F-3, F-5, F-6 and F-7 alloys respectively. These values have good agreement with the activation energy of the tracer diffusion of iron atoms in the β titanium-iron alloys in which they are 39.6 Kcal/mol for Ti-5wt%Fe and 48.6 Kcal/mol for Ti-10wt%Fe alloys.

Moreover, it is considered that the 250°C peak of internal friction is caused from the quenched β phase and the 350°C peak is done from the α′ phase.

## References

1.  S. A. Spachner: Trans. AIME., 212(1958), 57.
2.  H. Yoshida: J. Jap. Inst. Metals, 20(1956), 292.
3.  J. M. Silcock, M. H. Davies and H. K. Hardy: "The Mechanism of Phase Transformation in Solid" Inst. of Metals, London, (1958), 93.
4.  J. M. Silcock: Act. Met., 6(1958), 481.
5.  M. Miyagi, N. Shin and H. Ikawa: J. Jap. Inst. Metals, 33 (1969), 147.
6.          ibid                :     ibid   33(1969), 152.
7.  P. D. Frost, W. M. Pariris and L. L. Hirsch: Trans. Amer. Soc. Metals, 46(1954), 1056.
8.  F. R. Brotzen, E. L. Harman and A. R. Troiano: Trans. AIME., 203(1955), 413.
9.  C. A. Luke, R. Taggart and H. Polonis: Trans. Amer. Soc. Meta Metals, 57(1964), 142.
10. R. F. Peart and D. H. Tomlin: Act. Met., 10(1962), 123.
11. B. S. Hickman: Trans. AIME., 245(1969), 1329.
12. D. R. Miller: Trans. AIME., 224(1962), 275.
13. C. Wert and J. Marx: Act. Met., 1(1953), 113.

# ON THE STRENGTHENING OF A METASTABLE β-TITANIUM ALLOY BY ω- AND α-PRECIPITATION

A W Bowen

Materials Dept, Royal Aircraft Establishment,
Farnborough, Hants GU14 6TD, UK

## Introduction

Some years ago it was established that a metastable β-Ti-Mo alloy containing small ω-phase particles did not exhibit the severe embrittlement normally associated with the presence of high volume fractions of this phase [1]. The unusual ductile characteristics of the aged alloy prompted a subsequent general survey of its tensile properties, the results of which are presented in this paper. Strength, modulus and ductility parameters of the alloy, Ti-15% Mo*, are given as a function of ageing temperature and time and are related to the size and volume fraction of ω- and α- phases in the β- matrix. The data are interpreted in terms of the roles played by the ω-and α- phases in the strengthening and fracture of metastable β-titanium alloys.

## Experimental

Tensile tests were carried out on 1.6 mm thick Ti-15% Mo sheet (composition in wt%: 16.25 Mo; 0.05 Fe; 0.03 C; 0.08 $O_2$; 0.06 $N_2$ and 0.003 $H_2$). Test piece blanks were cut in the longitudinal direction, solution treated at 800°C for 20 minutes in a flowing argon atmosphere, and water quenched. The average β-grain size was 23 μm and texture analysis showed strong concentrations of (111) poles normal to the sheet surface and tilted 54° towards the rolling direction. Ageing treatments were carried out subsequently at temperatures between 250° and 560°C in air ovens or salt baths. The blanks were then machined to size (28.5 mm gauge length, 4.8 mm wide and 1.3 mm thick), electropolished, and tested as described previously [1]. Hardness tests were also carried out. Optical and electron microscopy and X-ray diffraction studies were made on selected heat treatments.

## Results

### Tensile Properties

The extent of hardening due to precipitation in the Ti-15% Mo alloy is shown by hardness changes in Fig 1 and by tensile property changes in Figs 2 and 3. In almost all cases macroscopic yielding took place at a stress which equated to the tensile strength, and only in a few cases, when the uniform elongation was not zero, was it possible to measure a true yield stress ie 0.2% proof stress. These are shown by the filled-in symbols in Fig 2.

---

*All compositions in this paper are given in wt%.

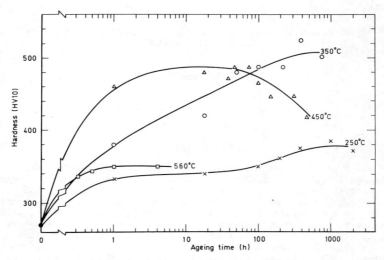

Fig 1.    Variation in hardness of the Ti-15% Mo alloy when aged at the temperatures shown (X-250°C; O-350°C; Δ-450°C; □-560°C)

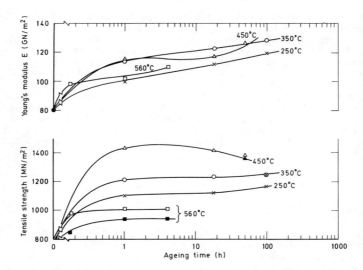

Fig 2.    Variations in Young's modulus and tensile strength of the Ti-15% Mo alloy when aged at the temperatures shown (X-250°C; O-350°C; Δ-450°C; □-560°C). Note that yield strength values are given only for the ageing temperatures of 450°C (▲) and 560°C (■).

Values of yield and tensile strength σ followed hardness variations very closely, whereas the changes in Young's modulus E were quite different (cf Figs 1 and 2). This difference can probably be attributed to σ reflecting both the size and volume fraction of precipitate, while E reflects only the volume fraction. The poorest agreement between tensile strength and hardness occurred after ageing in excess of 18 hours at 350°C. Here the ductility was zero (Fig 3) and compression tests would probably be required for more accurate

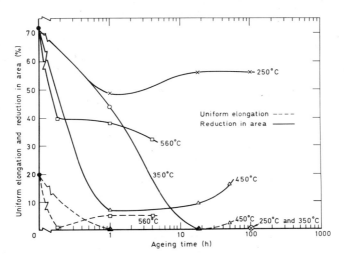

Fig 3. Variations in uniform elongation and reduction in area of the Ti-15% Mo alloy when aged at the temperatures shown (X-250°C; O-350°C; Δ-450°C; □-560°C).

strength values. Nevertheless, in view of the suggestion that the ratio σ/E is a reflection of strengthening efficiency [2], it is unlikely that the measured tensile strength after ageing for 100 hours at 350°C (Fig 2) is lower than the true value by more than ~10%. Hence assuming that σ/E should be ~1.1 x 10$^{-2}$ (as it is for all other times at 350°C) the tensile strength value after 100 hours at 350°C should perhaps be increased from the measured value of 1237 MN/m$^2$ to ~1420 MN/m$^2$. The tensile strength curve would then follow the hardness curve much more closely. All σ/E values were close to 1.0 or 1.1 x 10$^{-2}$, with the exception of the ageing times of 1 and 18 hours at 450°C, where the ratios were 1.25 and 1.21 x 10$^{-2}$ respectively. The significance of this increase in σ/E has been discussed elsewhere [2].

Values of uniform elongation (e$_n$) fell to zero with ageing time at all temperatures except 560°C (Fig 3). As ageing progressed there was a further increase in e$_n$ at 560°C and a very small recovery after 50 hours at 450°C. Reduction in area values showed a much more gradual decrease with ageing time at all temperatures and a recovery at long ageing times at 250° and 450°C (Fig 3).

## Microstructural Examination

In contrast to earlier work [1], water quenching the Ti-15% Mo alloy from 800°C (in the β-field) did not completely retain the bcc β-phase.   X-ray diffraction and transmission electron microscopy (TEM) evidence indicated the presence  of a very small amount of ω-phase.   On ageing in the temperature range 250° to 560°C this ω-phase grew and the α-phase nucleated and grew. The α-phase was detected only after ageing at 560°C and after long ageing times at 450°C.   No significantly different observations were recorded over those described in the literature (eg see reviews [3] and [4]).   Examples of some of the microstructures formed in this composition alloy can be found in [2]. The lattice parameter of the β-phase ($a_β$) in the water quenched condition was 3.260Å, in excellent agreement with the data of Hake et al [5].   On ageing $a_β$ decreased, indicating an enrichment of molybdenum in the β-phase, at a rate which increased with increasing temperature.   Thus after 100 hours at 250°C and 18 hours at 350°C $a_β$ = 3.255Å ( ≡21% Mo), while after 24 hours at 450°C $a_β$ = 3.253Å (≡ 22.2% Mo).   After 4 hours at 560°C $a_β$ = 3.250Å (≡ 24.6% Mo).

Analysis of the volume fraction of ω-phase showed reasonable agreement with the trends exhibited by the data of Hickman [6].   Although the absolute values were somewhat lower than those of Hickman the discrepancies can probably be attributed to the inaccuracies involved with measuring volume fractions in material exhibiting preferred orientation and to impurity elements present in the Ti-15% Mo alloy, particularly oxygen, which act to reduce the volume fraction of ω-phase [7].   It was found that a volume fraction (f) of ~0.5 was achieved in 100 hours at 250°C, 18 hours at 350°C and 1 hour at 450°C, for which the approximate particle sizes were 150, 300, and 1000Å respectively.   These are the maximum dimensions of the ellipsoidal [3,4] particles, the major/minor axis ratio varying from ~1.5 at 150Å to ~2.5 at 1000Å.

The volume fraction of α-phase after 4 hours at 560°C was ~0.55, in reasonable accord with other work on metastable β-titanium alloys [8].

## Deformation and Fracture Observations

As expected, the wide range of strength and ductility of the Ti-15% Mo alloy (Figs 2 & 3) was reflected in the deformation behaviour, which extended from profuse slip and twinning to very sparse evidence of slip adjacent to fracture surfaces.   A reasonably accurate impression of this range of deformation is exemplified by:

a.   the β-quenched condition, which deformed by twinning and slip (Fig 4(a)). This type of deformation has been discussed by Blackburn & Feeney [9] and Carter et al [10].

b.   the β+ (ductile and brittle) ω conditions, which deformed by planar and wavy slip, the degree of slip varying markedly between small and large ω-particles (Fig 4(b) & (c)).   These slip lines have recently been correlated with dislocation configurations and the destruction of ω-particles in these slip bands [11,12], as suggested earlier [1].

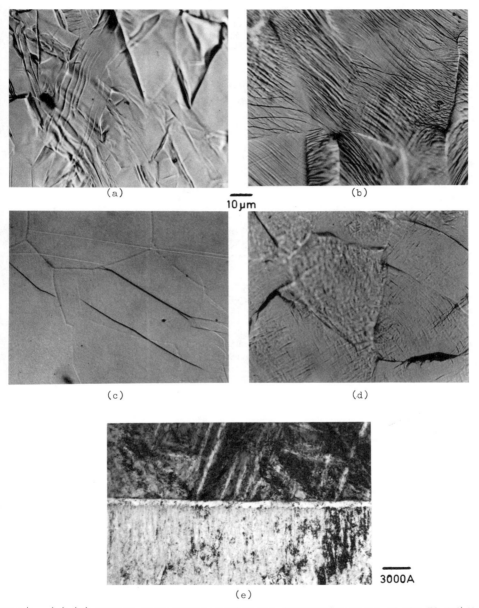

Fig 4.  (a)-(d) Optical micrographs showing the deformation modes of the Ti-15% Mo alloy in the (a) β-quenched condition (to maximum load only), (b) β + ductile ω conditions (aged 100 hours at 250°C), (c) β + brittle ω conditions (aged 18 hours at 350°C) and (d) β+α conditions (aged 4 hours at 560°C.)   (e) transmission electron micrograph of the β+α microstructure (aged 4 hours at 560°C) showing precipitation of acicular α at the boundaries of two β-grains.

c.    the β + α conditions, which deformed by slip with evidence of
considerable cracking in grain interiors and, in particular, at grain
boundaries (Fig 4(d)).    TEM analyses ([13] and Fig 4(e)) suggest that these
crack initiation sites are along the interfaces of the acicular α , which
may itself be preferentially nucleated on an impurity phase [14].
Confirmatory evidence for such nucleation is lacking at present.    The
deleterious effect of this type of precipitation may be minimised by cold
rolling prior to ageing [15].

Fracture surfaces were examined by scanning electron microscopy and examples
may be found in [16].

## Effect of Cold Rolling

In view of the unusual ductile nature of the Ti-15% Mo alloy after
ageing for 100 hours at $250^{\circ}$C, further experiments were carried out on
deformed material (ie cold rolled (CR) to 50% of original thickness), both
before and after ageing.    The results are given in Table 1, the undeformed β-
quenched and β-quenched-and-aged conditions being included for comparison.
After cold rolling the β-quenched condition showed a 50% increase in strength
but E did not change.

Cold rolling prior to ageing produced slightly better results when compared
with cold rolling subsequent to ageing.    The $\sigma/E$ ratio was high in both of these
rolled-and-aged conditions.    Comparisons of the $\sigma/E$ values indicates that the
ratio is increased from $0.7 \times 10^{-2}$ in the β-quenched condition to $1.0-1.1 \times 10^{-2}$
when aged to contain a single type of precipitate (see also [2]).    Cold rolling
seemed to increase this ratio a little ($1.15 \times 10^{-2}$) while the parameter can
be as high as 1.2 or $1.3 \times 10^{-2}$ for combinations of strengthening modes
(see Table 1 and [2]).

TABLE 1

Some Tensile Properties of the Heat Treated and Cold Rolled Ti-15% Mo Alloy

| Condition | 0.2% Proof Stress $MN/m^2$ | Tensile Strength $MN/m^2$ | E $GN/m^2$ | Uniform Elongation % | Reduction in area % |
|---|---|---|---|---|---|
| β-quenched* | 565 | 771 | 81 | 20.4 | 72.0 |
| β-quenched + 100h $250^{\circ}$C* | - | 1169 | 120 | 0 | 55.8 |
| β-quenched + CR 50% | 952 | 1195 | 83 | 1.5 | 43.9 |
| β-quenched + CR 50% + 100h $250^{\circ}$C | 1353 | 1379 | 102 | 1.0 | 35.1 |
| β-quenched + 100h $250^{\circ}$C + CR 50% | 1282 | 1366 | 97 | 1.2 | 25.7 |

*from [1]

An important point to note from Table 1 is that on cold rolling, before or after ageing, the Ti-15% Mo alloy reverts to normal tensile behaviour for high strength titanium alloys [17] ie $e_n \neq 0$.

X-ray examination of the cold rolled Ti-15% Mo alloy showed features which were initially identified as stacking faults.   These features are the subject of continuing investigation and will be reported upon in another publication.

Cold rolling before or after ageing decreased both the size and volume fraction of the $\omega$-phase.   This effect of deformation prior to ageing agrees with other work [9,18,19], while the effect of cold rolling subsequent to ageing would seem to offer additional evidence that the $\omega$-phase is indeed destroyed by plastic deformation [11,12].   The lower modulus in these cold rolled conditions is also indicative of a smaller volume fraction of $\omega$-phase.

## Discussion

The trends in strength, modulus and ductility presented in this paper are in agreement with much of the published results on metastable $\beta$-titanium alloys [eg 1,11,20-27].   Discussion will therefore be restricted to brief comments on some aspects of the mechanisms of strengthening and fracture of these alloys when hardened by $\omega$-and $\alpha$-precipitates.

Ageing the Ti-15% Mo alloy below $450^\circ C$, or at short times at $450^\circ C*$, produced only $\omega$-precipitation, the particle size increasing markedly with higher ageing temperatures.   For the heat treated conditions which gave $\omega$-phase volume fractions of ~0.5 (100 hours $250^\circ C$, 18 hours $350^\circ C$, 1 hour $450^\circ C$) tensile strength values (Fig 2) increased as the particle size increased from 150 to $1000\overset{\circ}{A}$.   This type of behaviour is characteristic of an alloy system containing deformable particles which are sheared by moving dislocations [eg 28].   Such behaviour has already been proposed for a metastable $\beta$-titanium alloy hardened by $\omega$-phase [11], but an increase in strength with increasing particle size, at constant volume fraction, does not appear to have been established previously.   For larger particles one might expect a change in deformation behaviour to that of an alloy system containing non-deformable particles, ie Orowan hardening, when dislocations are forced to by-pass particles [eg 28].   The strength in this case decreases with increasing particle size, again at constant volume fraction.   The critical particle size for the change in deformation behaviour has not been established;   it may be in the range 1-$2000\overset{\circ}{A}$ since Williams et al [20] have observed dislocation by-pass for $2000\overset{\circ}{A}$ size $\omega$- particles in a Ti-20%V - 10%Zr alloy.   Unfortunately it has not been possible to test for this by-pass mechanism in the work reported here because particles of sufficient size have not been obtained.

Little attention has been paid to modelling the strengthening of $\beta$ by $\alpha$-precipitates.   Rhodes and Paton [23] have shown recently that if types 1-and 2-$\alpha$ [29] deform by a dislocation by-pass process, then the theoretical strength increment underestimates the measured strengthening;   an alternative

---

*After 1 hour at $450^\circ C$ no $\alpha$-phase was detected by X-ray diffraction but in view of the high $\sigma/_E$ ratio this phase may be present as very small particles [2].   This will be studied further.

coherency model of Brown and Ham [30] gave better agreement for type 1-α.
From results on thermo-mechanically worked β-III alloy (Ti-11.5% Mo-6%Zr-
4.5%Sn) Rosales et al [27] concluded that dislocation by-pass was an
appropriate mechanism for β+α microstructures.

A complication in the application of strengthening models to
metastable β-titanium alloys hardened by either ω- or α-precipitates is
that the volume fraction of precipitates is probably considerably higher
than that assumed in any model (although it is not clear by how much). In
addition, interparticle distances (λ) are of the same magnitude as the
particle sizes (d). For f ≤ 0.18 and λ/d ≥ 2.3 Gysler et al [11] observed
very good agreement between their experimental data and the values predicted
by the Gleiter-Hornbogen order strengthening model for misfit-free
particles [31], when modified for a dis-ordered system. But for larger
values of f (and smaller λ/d values) it is almost certain that the proportion
of each phase cannot be considered to be uninfluenced by each other and elastic
stresses at the interface may be of great importance. Moreover, the stress
fields around closely spaced particles are almost certain to interact with one
another, complicating the analysis of dislocation motion and interaction.
In these cases strengthening mechanisms, such as the above mentioned order
strengthening, may need to incorporate additional considerations, for example,
the modulus difference between precipitate and matrix [28]. Such an effect
may only be a short range one ie it will be most effective while the
dislocations are inside the particles. A test for this mechanism may be
difficult to carry out because precipitate and matrix atoms need to be of
identical size and the crystal structure of precipitate and matrix need to be
the same, in order to eliminate other strengthening mechanisms. It is
unfortunate that little attention has been paid to ways of combining
strengthening mechanisms, not only those described above, and this could be
an area for future work.

The increase in modulus when either the ω- or α-phases are precipitated
in the β-matrix suggests that from a knowledge of the modulii of both
phases in a two phase mixture it should be possible to obtain an indication
of the volume fraction of the precipitate, knowing the increase in modulus
[32] It was proposed some years ago that $E_\omega \sim 2E_\beta$ [1]. It is now clear
that some ω was, in fact, present in the quenched alloy, hence the true
modulus of the ω-phase will be slightly less than the quoted value of
82 GN/m$^2$. However, for the purposes of this general discussion it will
continue to be assumed that $E_\omega \sim 2E_\beta$; particularly in view of the dis-
agreement on the absolute value of $E_\omega$ [1,11,33,34]. The variation in the
modulus of α is well documented [35] and for precipitated α we shall assume
$E_\alpha \sim 1.5E_\beta$. This is somewhat higher than a random value [35] but is in keeping
with the experimental data (Fig 2). To be able to predict f (or E) of the
precipitating phases one must assume an appropriate equation. The assumption
to date has been that for ω-precipitation an equation of the type

$$E = E_1 f_1 + E_2 f_2 \tag{1}$$

is applicable [1,11,32,34], where subscripts 1 and 2 denote matrix and
precipitate resp. However, the problem with equation (1) is that it is
only one of two limiting equations, equation (1) being the upper bound
(provided the Poisson's ratios of both phases are the same [36]) while
the lower bound is approximated by:

$$\frac{1}{E} = \frac{f_1}{E_1} + \frac{f_2}{E_2} \tag{2}$$

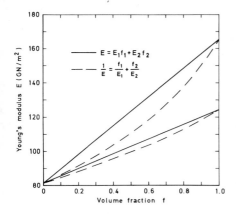

Fig 5.   Variations of Young's modulus for a two phase mixture, according to equations (1) and (2), taking as examples $E_\beta = 82$, $E_\alpha = 124$ and $E_\omega = 165$ GN/m$^2$. The upper set of curves is for $\beta + \omega$ mixtures and the lower set of curves for $\beta + \alpha$ mixtures.

These two equations are shown in Fig 5 for $\beta$ hardened by $\omega$-and $\alpha$-precipitates taking as examples $E_\beta = 82$, $E_\alpha = 124$ and $E_\omega = 165$ GN/m$^2$. A maximum divergence is indicated at $f = 0.5$, with this difference being considerably greater for $\omega$-than for $\alpha$-precipitates. The significance of these equations is that they assume different conditions, namely that both matrix and precipitate are either strained equally (eqn (1)), or stressed equally (eqn (2)).

Real materials are claimed to fall between these extremes [36] and recent data of Jinoch et al [37] would seem to confirm that this is indeed the case for a $\beta + \alpha$ Ti-Mn alloy. Jinoch et al also showed that the stresses are higher but the strains are lower in the $\beta$-phase compared with the $\alpha$-phase. After ageing to precipitate $\alpha$ the $\beta$ will be solute rich ($\equiv 24.6\%$ Mo) while the $\alpha$ will be solute lean ($\sim 1\%$ Mo). From experimental evidence [37-39], such differences in chemical composition suggest that the $\beta$-phase will be very much stronger than the $\alpha$-phase [37,38] although possessing a smaller $e_n$ [38] and slightly lower E [39]. Taking these points into consideration it is probable that on stressing, the $\alpha$-phase will yield and begin to neck before the $\beta$-phase. Thus voids can nucleate at the $\alpha/\beta$ interfaces, as shown in Fig 4(e) and elsewhere [13,15]. (This, of course, assumes that precipitated $\alpha$ in $\beta$ behaves in the same manner as discrete grains of $\alpha$ in $\beta$, and that constraint around the acicular $\alpha$ does not alter markedly the properties of the $\alpha$-or $\beta$- phases.)

No comparable data are available for $\beta + \omega$ microstructures but from the results of Ganesan et al [32] and Vigier et al [33,34], it is not unreasonable to conclude that equation (1) is observed in practice. If both phases are strained equally under load the stress will be considerably higher in the $\omega$-phase. As the yield strength of the $\omega$-phase is approached the $\omega$-particles will be destroyed in an increasing number of slip bands. These bands are therefore likely to change to thin zones of lower strength and modulus (ie equation (1) will no longer be obeyed locally). Thus from an initial uniform loading system of high stress and low strain, there may be load shedding (depending on strain rate etc) on to the slip bands because these will now be subjected to a loading system of lower stress and higher strain. Deformation will thus be concentrated in these bands, and while the ductility in each slip band may be quite high, the overall ductility will be governed by the density of slip bands. This, then, qualitatively explains the ductility difference between the conditions shown in Figs 4(b) and (c). Furthermore, the highly concentrated slip in a slip band will be conducive to dislocations moving at high velocity, causing pile-ups and the eventual opening up of voids at slip band intersections and grain boundaries [11,12]. The result will almost certainly be rapid local ductile fracture accommodated by shallow dimples, the depths of which will be limited because of the marked difference in properties between the narrow softened slip bands and the surrounding material which continues to be hardened by the $\omega$-phase.

## References

1   A. W. Bowen:  Scripta Met., 5 (1971), 709
2   A. W. Bowen:  J. Mater. Sci., 12 (1977), 1355
3   B. S. Hickman:  J. Mater. Sci., 4 (1969), 554
4   J. C. Williams:  Titanium Science and Technology, 3 (1973), 1433,
    Plenum, New York
5   R. R. Hake et al:  J. Phys. Chem. Solids, 20 (1961), 177
6   B. S. Hickman:  Trans. TMS-AIME, 245 (1969), 1329
7   J. C. Williams et al:  Met. Trans., 2 (1971), 477
8   D. E. Gordon and J. W. Hagemeyer:  J. Mater. Sci., 10 (1975), 1725
9   M. J. Blackburn and J. A. Feeney:  J. Inst. Metals, 99 (1971), 132
10  G. Carter et al:  J. Mater. Sci., 12 (1977), 2149
11  A. Gysler et al:  Acta. Met., 22 (1974), 901
12  E. Levine et al:  ibid, 1443
13  R. Chait:  Met. Trans. A, 6A (1975), 2301
14  A. W. Bowen and C. A. Stubbington:  J. Less-Comm. Metals, 20 (1970), 367
15  J. C. Chesnutt and F. H. Froes:  Met. Trans. A, 8A (1977), 1013
16  A. W. Bowen:  Met. Technol., 5 (1978), 17
17  A. W. Bowen and P. G. Partridge:  ref [4], 2, 1021
18  J. M. Silcock:  Acta. Met., 6 (1958), 481
19  M. J. Blackburn and J. C. Williams:  Trans. TMS-AIME, 242 (1968), 2461
20  J. C. Williams et al:  Met. Trans., 2 (1971), 1913
21  G. A. Sargent et al:  Microstructure and Design of Alloys, 1 (1973), 114
    Inst. of Metals, London
22  P. Ganesan et al:  J. Less-Comm. Metals, 34 (1974), 209
23  C. G. Rhodes and N. E. Paton:  Met. Trans. A, 8A (1977), 1749
24  M. K. Koul and J. F. Breedis:  Met. Trans., 1 (1970), 1451
25  D. Kalish and H. J. Rack:  Met. Trans., 3 (1972), 1885
26  K. Ono et al:  Mech. Behaviour of Materials, 1 (1972), 66,
    Soc. of Mat. Sci., Japan
27  L. A. Rosales et al:  ref [4], 3, 1813
28  P. M. Kelly:  J. Aust. Inst. Metals, 16 (1971), 104
29  C. G. Rhodes and J. C. Williams:  Met. Trans. A., 6A (1975), 2103
30  L. M. Brown and R. K. Ham:  in Strengthening Methods in Crystals, (1971), 9
    Wiley, New York
31  H. Gleiter and E. Hornbogen:  Phys. Stat. Solidi., 12 (1965), 235
32  P. Ganesan et al:  Metall, 10 (1977), 399
33  G. Vigier et al:  Mém. Sci. Rev. Mét., 72 (1975), 677
34  G. Vigier and J. Merlin:  Metall, 12 (1979), 113
35  F. R. Larson and A. Zarkades:  MCIC Report - 74-20 (1974)
36  I. L. Mogford:  Met. Rev., 12 (1967), 49
37  J. Jinoch et al:  Mat Sci. and Eng., 34 (1978), 203
38  F. C. Holden et al:  J. of Metals, Trans-AIME, 8 (1956), 1388
39  S. G. Fedotov:  ref [4], 2, 871

## Acknowledgements

Crown copyright (C) Controller HMSO, London 1980.

# THERMAL INSTABILITY AND MECHANICAL PROPERTIES
# OF BETA TITANIUM-MOLYBDENUM ALLOYS

Moritaka Hida 1), Eiichi Sukedai 2),
Yukio Yokohari 3) and Akihiro Nagakawa 4)

1) School of Engineering, Okayama University, Okayama,
      Japan
2) Dept. of Mechanical Science, Okayama University of
      Science, Okayama, Japan
3) Okayama Technical High School, Okayama, Japan
4) Okayama University of Science Senior High School,
      Okayama, Japan

## Introduction

It has been reported that the ductility of beta Ti-Mo alloys is remarkably influenced by the chemical composition of molybdenum tending to stabilize beta phase[1].    It is important to investigate the source of the ductility in order to improve the workability of    beta titanium alloy.    The ductility seems to depend strongly upon the instability of beta phase, for deformation twin and athermal omega phase have been taken to be the products formed when instable or metastable Ti-Mo alloys are distorted[2,3].    The present investigation was carried out to elucidate the correlation of the ductility whith the thermal instability through surveying the mechanical properties, the deformation mode and the microstructure of beta Ti-Mo alloy, and to obtain the fundamental knowledge concerning the improvement of the ductility for stable beta titanium alloys.  Two kinds of Ti-Mo alloys were prepared for the present examination. One of them is Ti-14wt pct Mo alloy being thermally instable beta phase alloy. The other is Ti-20wt pct Mo alloy being comparatively stable beta phase alloy.

## Experimental Procedure

Button shaped ingots of Ti-14 and -20wt pct Mo(abbreviated to 14Mo and 20Mo later, respectively) alloys were prepared by a non-consumptive arc-melting practice and from materials, titanium and molybdenum as shown in Table 1  and Table 2, respectively.  The ingots were homogenized for 3456ks at 1173K in

Table 1  Chemical composition of Ti (wt pct)

| C | Fe | N | H | O | Ti |
|---|---|---|---|---|---|
| 0.004 | 0.030 | 0.030 | 0.0014 | 0.0036 | bal |

vacuum(about 3mPa).    The specimens used for this work were machined from plate

Table 2  Chemical composition of  Mo (wt pct)

| Mo | Fe | P | S | C |
|---|---|---|---|---|
| 99.95 | 0.01 | 0.005 | 0.005 | 0.003 |

(0.6mm in thickness) obtained by conventional procedure of forging and rolling for ingots of titanium alloys.    The shape and size of the specimens are shown in figure 1.

Silica capsules(vacuum, 0.13mPa) containing specimens of alloys were heated for 4.5ks at 1223K, quenched into iced water and then mechanically fractured instantly after immersion.    The quenched specimens were confirmed to be beta mono-phase by inspecting with a conventional X-ray diffractometer.    All of the quenched specimens was metallographically polished, and was subsequently electro-polished at nearly 220K in a solution of the volume ratio, 1 for 70wt pct perchrolic acid/  6 for n-butyl alcohol/  10 for metyl alcohol.

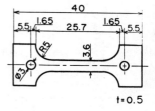

The hardness testing was performed within the temperature range 77 to 823K, and carried out at room temperature for isochronal aged alloys.    The tensile testing was performed with an Instron type testing machine under the condition of strain rate, 0.53/ks and of temperature range, 77 to 300K.

Fig.1 Shape and dimenion ( in mm ) of specimen for tensile test.

Thin foils for observing microstructure of alloys were prepared with using standard techniques and the above mentioned solution . The used electron microscope was Hitachi 200D, which was operated with using accelerating voltage at 200kV.

## Results and Discussion

### 1.    Thermal instability reflected on hardness changes

As-quenched beta Ti-Mo alloys are in a thermal instable state, and that is in the state of pre-transition to nucleate omega phase.    The instability will depend upon the concentration of molybdenum and internal elastic stress. In the case of too much beta stabilizing element, beta phase has been considered to decompose to beta(1) and beta(2), before omega phase nucleates[4].

In order to survey the relationship between the thermal instability and the mechanical properties of the alloys, the dependence of the hardness upon the temperature and upon the isochronal(1.2ks ) ageing are investigated within the temperature range 77 to 823K.    As shown in figure 2, increase in hardness due to omega precipitation[5] appears to be very marked, e.g. in 14Mo alloy the highest hardness peak of up to 540 VHN are obtained at about 723K in the isochronal ageing curves, and the remarkable hardening of 14Mo alloy initiates at lower temperature(400K) than that(650

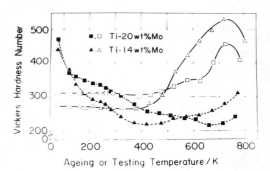

Fig.2 Hardness changes of Ti-14 and -20 pct Mo alloys; solid lines show isochronal(1.2ks) ageing curves and dotted lines show dependence of the hardness upon the temperature.

K) of 20Mo alloy.   The isochronal ageing curve of 20Mo alloy has two stages, the first one shows a plateau at about 573K and the amount of hardening is about one third of the second stage belonging to the peak at 725K due to the precipitation of omega phase.   In the curve of the dependence of the hardness upon the temperature, the increases of hardening due to precipitation of omega phase in 14Mo and 20Mo alloys are found upper above 400K and 700K, respectively, but the increase in hardness corresponding to the plateau shown on isochronal ageing curve of 20Mo alloy appears to be suppressed.   There should be some differences in hardening mechanism between the first stage and the second one. The plateau of 20Mo alloy seems to be corresponded to the pretransition of omega phase, i.e. two phase decomposition( beta $\longrightarrow$ beta(1) + beta(2) )[4].   The beta phase of 14Mo alloy is presumed nearly to be beta(1) phase, where molybdenum concentration of the beta(1) is supposed to be lower than that of the beta (2).   It is considered from the results of hardness testing that in 14Mo alloy, the thermal instability is the largest and thermal omega phase is easily precipitated.

Many investigators have clarified that stress-induced products are formed in beta titanium alloys[2,3,6,7,8].   It must be considered that internal stress caused by lattice distortion makes metastable beta titanium alloys increase the instability.   If this is true, the increment of instability will be also reflected on promoting precipitation of thermal omega phase in cold worked beta titanium alloys.

The effect of cold work( 10 and 30 pct in reduction ) on the thermal inatbility of 14Mo alloy is shown in figure 3.   By cold work, age-hardening initiates more steeply at lower ageing temperature, i.e. the initiation temperature of hardening varies from 400K to 300K. Such as effect,however, was not apprently observed for 20Mo alloy.   It is therefore assumed that the athermally stress-induced products enhance to nucleate thermal omega phase in metastable 14Mo alloy.   And also the increment of hardness by only the cold work was greater on 14Mo than on 20Mo alloy. It is interesting to study whether the high work hardening rate of 14Mo alloy indicates or not the formation of athermal omega phase.   It is natural to be expected from the differences in work hardening rate and in the effect of cold work on thermal instability that there are notable difference in their deformation mode between 14Mo and 20Mo alloys.

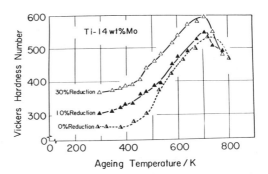

Fig. 3   Hardness changes of Ti-14 pct Mo alloy rolled to 0, 10, 30 pct in reduction by rolling and isochronal-aged for 1.2ks.

2.   Relationship between ductility and deformation mode

Ductility of titanium alloys should depend upon deformation mode(designated as slipping or twinning),propagation of deformation region and density of substructure in alloy.   The dependence of concentration of molybdenum on the mode and the substructure[1,3,8] has not been studied enough about beta Ti-Mo alloys.   It is reasonable  to presume that the tendency to form omega phase in metastable or instable beta phase becomes a driving force to lead a mode of twinning,as far as we allow for the formation of omega phase being a displacive (acoustic) transformation[9,10].

Figure 4   shows a part of typical stress-strain curves obtained in temperature range, 77K to 300K.   It appears that 14Mo alloy has excellent mechanical properties even at 77K, which are the homogeneous deformation with remarkable e-longation and linear work hardening rate, and that 20Mo alloy has poor uniform elongation and local shear strain.   Interesting to us, the fracture strain and work hardening of 20Mo alloy are slightly increase(see dotted line in Fig.4) by ageing for 1.2ks at 573K.   This ageing temperature corresponds to the first stage , i.e. beta $\longrightarrow$ beta(1) + beta(2) decomposition, on age-hardening curve of 20Mo alloy in figure 2. The proof stress at 0.2 pct strain and fracture stress on both alloy in-

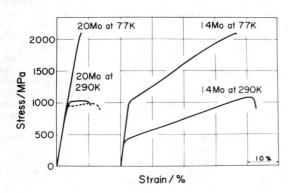

Fig. 4   Stress-strain curves of Ti-14 and -20Mo alloys deformed in tension at at 77K and 290K.   Dotted line shows the curve of Ti-20 pct Mo alloy aged for 1.2ks at 573K.   This ageing treatment corresponds nearly to the occurrence of the plateau on age-hardening curve in Fig. 2.

creased monotonically with decrease of deformation temperature.   The dependence of the fracture strain upon the deformation temperature(77 to 300K) is shown in figure 5.   Ductility of 14Mo alloy is very large all over the range of temperature as compared with that of 20Mo alloy.   Furthermore, it seems that fracture strain of both alloys tends to be increased by some softening effect under 77K.

The fine and homogeneous dimples on the fracture surface of 14Mo alloy deformed at room temperature was observed.   The feature  of the dimples fractured at 77K seemed to be almost the same as the former. The coase dimples was observed on the fracture surface of 20Mo alloy at room temperature, however cleavage-like fracture was found on some places.

The difference of the ductility on these alloys is disclosed on the surface fine-structure of their specimens, as shown in Photo. 1-(1) .
It is observed that there are twin-like straight bands on the surface of 14Mo alloy, and straight and wavy slip lines probably caused by cross slip on the surface of 20Mo alloy.

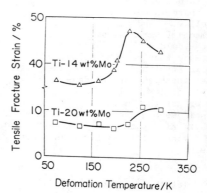

Fig. 5 Dependence of tensile fracture strain of Ti-14 and -20 pct Mo alloys upon the deformation temperature.

These contrasts on the surface were made sure to be some stress-induced products , not to be ordinary slip lines, by the techniques of electrical polishing and of chemical etching. The microstructures in both 14Mo and 20Mo alloys are shown in transmission electron micrograpghs, Photo. 1-(2). These microstruc-tures seem to correspond enough to the surface fine structure in Photo. 1-(1). The substructures of the stress-induced products in Photo. 1-(1) have very com-plex microstruture. It seems that remarkable difference of these substructure cannot be recognized even in case of deformation under room temperature.

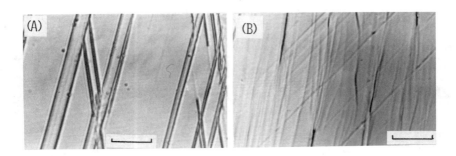

Photo.1-(1)  Fine structures of Ti-Mo alloys chemically etched after deformed in tension to a few pct strain and electroytically polished; (A) In Ti-14 pct Mo alloy, twin-like straight bands are crosscut with each other. (B) In Ti-20 pct Mo, the contrasts are not ordinary straight and wavy slip lines, but some stress-induced products.          ( Scale = 10 μm )

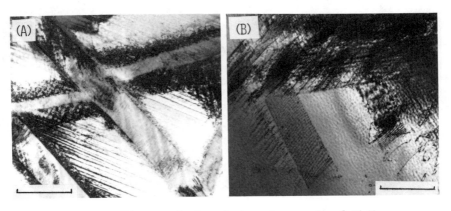

Photo.1-(2) Transmission electronmicrographs of Ti-Mo alloys deformed in tension to a few pct strain; (A) In Ti-14 pct Mo alloy, twin bands are crosscut with each other.  (B) In Ti-20 pct Mo alloy, cross slipped and piled up dislocations are observed. ( Scale = 1.0 μm )

14Mo alloy is mainly deformed by formation of twins involving high density of
contrast  like stacking faults shown in Photo. 2(A) and 2(B).   The plane of
twin is confirmed to be {332}β from the selected area diffraction patterns taken
from matrix ( zone axis is <111 > , see Photo. 2(B) and 2(C) ) and from deformation twin band (zone axis is <311> , see Photo. 2(B) and 2(D) ) regarded as
the stress-induced products.   Secondary twins could not been observed for 14Mo
alloy deformed in tension to a few pct strain .   The extra-spots of omega phase
are detected in the diffraction pattern taken from the twin band (2(A) and 2(D)).
But we could not accomplish to observe some clear contrasts of plate shaped
omega phase in dark field image taken from the extra-spots.   It is therefore
considered that athermal omega phase is constructed from stacking faults in
deformation twin.   20Mo alloy in tension to a few pct strain  is mainly de-

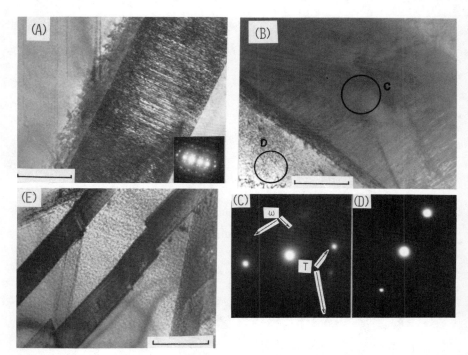

Photo. 2  Transmission electron micrographs of Ti-14 pct Mo
alloy  deformed in tension to a few pct strain show  many
contrasts like stacking fault in twin band in (A) and (B).
Selected area diffraction patterns show <113 > zone axis in
(C) and <111> zone axis in (D), which are taken from a twin
band and matrix in (B), respectively.   These patterns correspond with (233̄)β twinning.  And the extra-spots,labeled as
ω, according with omega phase around twin spots, T, are shown
in (C).   Intersecting of bands like stacking faults is observed in (E).   These faults may be early stage of twin
formation.    ( Scale = 1.0 μm )

formed  by slip and cross slip (see Photo. 1-2(B) and Photo. 3 (A) ).   Piled-up

dislocations (see Photo. 1-2(B) and Photo. 3(A) ), dislocation bundles (see
Photo. 3(B) ) and stacking fault fringes (see Photo. 3(C) ) are observed.
Photo. 3(C) seems to be an evivence of the formation of stacking fault in B.C.C.
structure.   The possibility of formation of stacking faults in B.C.C. phase
has been pointed out by Sleeswyk[11].   In this study, any twin including {112}β
twin[3] and any extra-spot due to omega phase could not been detected.   It is
observed that the bundles, which had presumably been before, have partialy
been annihilated by reaction with moving dislocations (see the arrow in Photo.
3(B) ).   Stacking faults and dislocation bundles are considered to be stress-
induced products in 20Mo alloy deformed in tension to a few pct strain.

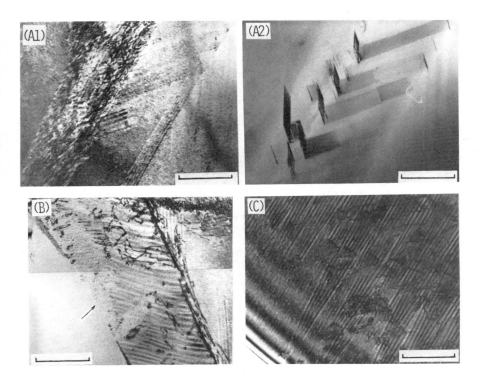

Photo. 3   Transmission electron micrographs of
Ti-20 pct Mo alloy deformed in tension to a few pct
strain show  piled-up and cross-sliped dislocations
in (A1) and (A2) which are taken from same area
except the difference in diffraction condition.
Dislocation bundles can be imagined to be annihilated
partially (see the arrow in (B) ).   Many stacking
fault fringes are observed in (C).   Photo. 3(C)
seems to be an evidence of the formation of stacking
fault in B.C.C. structure.   Moiré fringes constructed from stacking fault fringes are observed in (B).
( Scale = 1.0 μm )

## Conclusion

From the facts mentioned above, the most important matters are that in thermally instable beta phase such as 14Mo alloy, there is the definite relation between the twinning and the easiness of the formation of athermal omega phase, and that the remarkable ductility of 14Mo alloy correlates to the formation of {332}β twin .   As the twins contain a lot of athermal omega phase, i.e. stacking faults, they must be hardened rapidly with proceeding of cold work, and many twins will be newly created in undeformed regions.   There is no doubt that the ductility and the linear work hardening rate of 14Mo alloy are caused by growth,propagation and intersection of these twins, while these twins are storing up athermal omega phase.

Decomposition of beta , beta ⟶ beta(1) + beta(2), could be applied to improve the ductility of stable titanium alloy such as 20Mo alloy, as expected from figure 2.   This decomposition is performed by following heat treatments;(1) beta phase  is annealed at lower temperature than that omega phase cannot be formed at, (2) beta phase is annealed at higher temperature than that monotectoid reaction [12] is found in Ti-Mo constitution at.

## References

1.   F.C. Holden, H.R. Ogden and R.I. Jaffe: J. Metals, 8(1956),1388.
2.   R.M. Wood: Acta Met., 11(1963),907.
3.   M. Oka and Y. Taniguchi: J. Japan Inst. Metals, 43(1978),814.
4.   E.L. Harmon and A.R. Troiano: Trans. ASM., 53(1961),43
5.   B.S. Hickman: J. Inst. Metals, 96(1968),330.
6.   J.A. Roberson, S. Fujishiro, V.S. Arunachalum and C.M. Sargent: Met. Trans. , 5(1974),2317.
7.   M.K. Koul and L.F. Breedis: Acta Met., 18(1970),579.
8.   M.J. Blackburn and J.A. Feeny: J. Inst. Metals, 99(1971),132.
9.   B.A. Hutt and J.A. Roberts: Acta Met., 8(1960),575.
10.   D. de Fontaine and D. Buck: Phil. Mag., 27(1973),967.
11.   A.W. Sleeswyk: Phil. Mag., 8(1963),1467.
12.   S. Terauchi, H. Matsumoto, T. Sugimoto and K. Kamei: J. Japan Inst. Metal, 41(1977),632.

# OMEGA AND ALPHA PRECIPITATION IN Ti-15 Mo ALLOY

Uma M. Naik and R. Krishnan

Metallurgy Division, Bhabha Atomic Research Centre,
Trombay, Bombay-400 085, India.

## Introduction

It is known that in beta titanium alloys the omega phase precipitates as a transition phase during the decomposition of the high temperature beta phase into the equilibrium mixture of the alpha and the beta phases. Omega precipitation can occur either during quenching or during ageing after quenching. The precipitation process in the Ti-Mo system has been investigated by several workers earlier [1-5] . It has been established that the beta → alpha reaction occurs less readily than the beta → omega → alpha reaction and that it occurs preferentially at grain boundaries. It has been found that omega precipitation is most active during ageing around 400°C. The influence of ageing temperature on the tensile strength and the elongation of the Ti-15Mo alloy has also been reported [4,6,7] . When this alloy is aged at temperatures between 350 and 450°C, it becomes very brittle and the reduction in area and the elongation are negligible.

The present work was undertaken to study the omega and the alpha precipitation reactions by transmission electron microscopy and to carry out a fractographic study of the fractured surfaces of quenched and aged samples after tensile testing, so that a correlation between the microstructure, the mechanical properties and the fracture topography could be established.

## Experimental Procedure

The material used for investigation was a commercial IMI-205 alloy which contained 2000 ppm oxygen and 80 ppm nitrogen. The molybdenum content analysed to 15wt%. The solutionisation treatment was carried out in sealed silica tubes, initially evacuated and subsequently filled with helium at a pressure of 175 mm of mercury. The solutionisation temperature was 900°C and the duration of the treatment was 1 hour.

After solutionisation, the samples were quenched in water and simultaneously the silica tubes were broken. Subsequent ageing was carried out in evacuated and sealed silica capsules. Ageing was carried out for different durations in the temperature range 300 to 700°C.

TEM samples were prepared by electropolishing at -50°C in an electrolyte containing 30 parts perchloric acid, 175 parts n-butanol and 300 parts methyl alcohol. Thinning was carried out at 20 volts and the thin foils were examined in a Siemens Elmiskop 102 microscope operating at 125 KV.

Tensile testing of the as-quenched and the aged samples was carried out in an Instron tensile testing machine at room temperature. Half an inch gauge length specimens were used. The fractured surfaces were then examined in a

ETEC Autoscan Scanning Electron Microscope.

## Results and Discussion

1. Precipitation of Omega and Alpha Phases

The precipitation of the omega and alpha phases in the quenched and aged Ti-15 Mo alloy was examined by transmission electron microscopy (TEM). The typical as-quenched structure is shown in Fig.1. It can be seen that the matrix is essentially beta with a fine dispersion of athermal omega, the presence of which is evident from the corresponding selected area diffraction pattern. One could also see that the beta grains were broken up into plates. These platelets could be the products of the reported thinning transformation, which leads to the production of an fct phase and its subsequent reversion to the bcc structure, as suggested by Pennock et al [8] . On ageing, the quenched sample at 300°C for 1 hr., one could observe a high density of fine and uniformly distributed omega particles (Fig.2). Here again, the banded structure was apparently due to the thinning transformation. On increasing the ageing time from 1 hr. to 24 hours at 300°C, it was found that the average size of the omega precipitates did not increase very significantly. Samples aged at 400°C showed essentially similar features except that a substantial amount of coarsening of the precipitates could be brought about by ageing for periods upto 1 week (Fig.3). Even at 450°C, ageing for periods as long as 17 hrs. did not result in the precipitation of any appreciable quantity of the alpha phase (Fig.4). In fact, the omega phase continued to persist even at 500°C, when the ageing was carried out for brief intervals of time. At still higher temperatures, however, the alpha phase rapidly precipitated as Widmanstatten plates, as can be seen from (Fig.5), a typical micrograph obtained from a sample aged at 600°C for 15 mts. Ageing at 700°C for 15 mts. showed substantial amounts of alpha (Fig.6). In addition, it could be seen that the peripheral regions of these plates often contained agglomerations of non-Burgers type 2 alpha crystallites.

2. Fractography

Tensile testing of the Ti-15 Mo alloy was carried out at room temperature in the as-quenched and in the quenched and aged conditions in an Instron tensile testing machine, using a strain rate of $2 \times 10^{-3}$/sec. The as-quenched material gave an yield strength of 71.56 Kg/mm$^2$ and an uniform elongation of 33%. Ageing at 300°C for 1 hour brought down the elongation to 5% with only a marginal increase in yield strength, 75.3 Kg/mm$^2$. Specimens which were aged at 400°C for 1 hour showed very little elongation of the order of 3% only. However, samples aged at 600°C for 15 mts. gave an yield strength of 148.0Kg/mm$^2$ with an uniform elongation of 7%. On increasing the ageing time or temperature, the yield strength dropped down to a value of the order of 88 Kg/mm$^2$ but with no appreciable increase in uniform elongation.

The fractured surfaces of these tensile tested samples were examined in a Scanning Electron Microscope. Fig.7 shows the fracture morphology observed in the as-quenched sample. A dimple structure typical of a ductile mode of failure could be seen. Thus, eventhough the quenched alloy contained athermal omega particles, the mode of fracture was ductile, confirming the earlier observations in Ti-Mo [9] and Ti-Cr [10] systems. The athermal omega particles

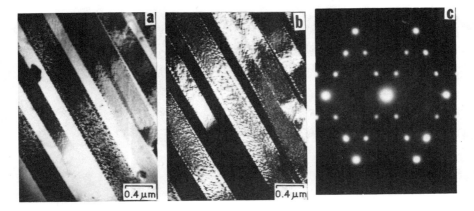

Fig.1        : Ti-15 Mo quenched from 900°C showing athermal
                omega (a) Bright field, (b) Dark field, (c) SAD

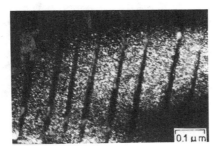

Fig.2 : Omega particles in sample aged at 300°C for 1 hour.

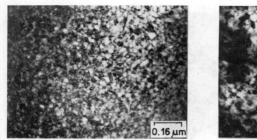 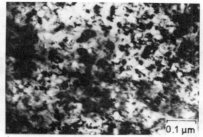

Fig. 3                              Fig.4

Fig.3 : Omega particles after ageing at 400°C for 7 days.

Fig.4 : Omega particles in sample aged at 450°C for 17 hours.

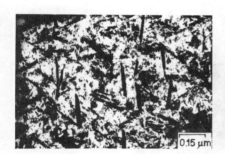 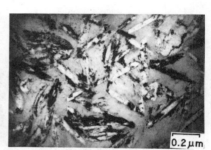

Fig. 5                              Fig. 6

Fig.5 : Alpha plates in the alloy on ageing at 600°C for 15 mts.

Fig.6 : Ageing at 700°C for 15 minutes showing type 2 crystallites
in alpha plates.

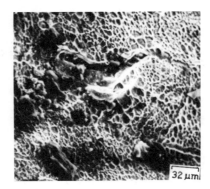

Fig.7 : Fractograph obtained from as-quenched alloy showing
        dimple structure.

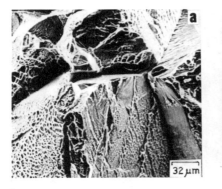
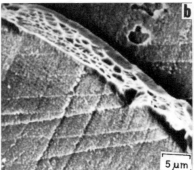

Fig.8(a): Fractograph of a sample aged at 300°C for 1 hour.
    (b): Fractograph of a sample aged at 300°C for 1 hour
         showing step like fracture through the matrix.

Fig.9 : Fractograph obtained from a sample aged at 400°C for 1 hour.

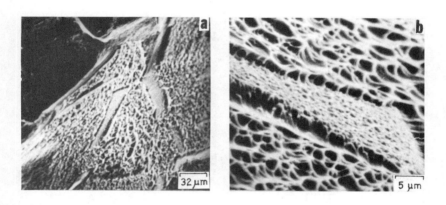

Fig.10 : Fractographs obtained from the alloy on ageing at 600°C
for 15 minutes.

(a) showing planar features in the general fracture and
(b) showing the ductile nature of the planar features.

in the quenched alloy were very small in size (of the order of 30 to 40A°) and it was difficult to resolve them clearly during TEM examination. Therefore, no correlation could be obtained between the size and the spacing of the omega particles and the dimensions of the net work of fine dimples obtained on the fracture face. If one assumes the omega particles to be non-deformable, then the interfaces between beta and omega could act as stress raisers and lead to crack nucleation. Then the observed dimple morphology can be easily explained. Fig.8 corresponds to a sample that was aged at 300°C for 1 hr. The intergranular nature of failure is very clearly visible. In addition, it may be noted that there are two distinct types of fracture morphology - one corresponding to a ductile dimple structure and the other exhibiting a step like fracture through the matrix. The observation was similar to that of Gysler et al [9] who have proposed that the step like features arise from inhomogeneous slip bands. They have also concluded that dislocations are able to move through the omega particles. This is in contradiction to the observations made by Williams et al [11] . According to Chandrasekaran et al [10] , the metastable beta phase in a quenched Ti-11%Cr alloy undergoes a strain induced martensitic transformation and the step like features observed by them have been explained as the separation along matrix martensite interfaces which constitute planes that are susceptible to crack propagation. Silcock [2] has shown that out of the four possible omega variants, only one is suitably oriented with respect to the matrix for slip to take place and as such omega particles should be effective barriers for the motion of dislocations. In view of these facts and of the observation made in the present work that during tensile testing, a strain induced product did form, it is likely that the step like features observed on the fractured surfaces were formed as a result of the strain induced transformation rather than due to inhomogeneous slip operating through the matrix and omega precipitates as suggested by Gysler et al.

Fig.9 shows the fracture morphology of a sample which was aged at 400°C for 1 hour. It could be seen that the structure is typical of a cleavage fracture. The step like features showed on a microscopic scale, ductile dimples which were visible on both sides of the individual steps.

Samples that had been aged at 600°C for 15 mts. and showed the maximum strength were associated with a fracture morphology illustrated in Fig.10. One could see that there were some planar features in the general fracture pattern. It could be seen at high magnification, that in this case also, the planar features contained a large number of dimples, indicating that the mode of failure was ductile.

## References

1. B.S. Hickman : Journal of Inst. of Metals, 96 (1968), 330.
2. J.M. Silcock; Acta Met., 6 (1958), 481.
3. R.M. Wood : Acta Met., 11 (1963), 907.
4. T. Yukawa et al : The Science, Technology and Application of Titanium, Ed. R.I. Jaffee and N.E. Promisel, Pergamon Press (1970), 699.
5. H. Ikawa et al. Titanium, Science and Tech.3 (1973), 1545 Plenum Press, N.Y.
6. A.W. Bowen : Scripta Met., 5 (1971), 709.
7. M.J. Blackburn and J.C. Williams : Trans. Met. Soc. AIME,242 (1968), 2461.
8. G.M. Pennock, H.M. Flower and D.R.F. West : Metallography,10 (1977), 45.
9. A. Gysler, G. Lutzering and Gerold: Acta Met.,22 (1974), 901.

10. V. Chandrasekaran, R. Taggart and D.R. Polonis, Metallography, 5 (1972) 235.
11. J.C. Williams, B.S. Hickman and H.L. Marcus, Met.Trans. 2 (1971) 1913.

# THE CONTROL OF α PRECIPITATION BY TWO STEP AGEING IN β Ti-15%Mo

G.M. Pennock, H.M. Flower and D.R.F. West

Department of Metallurgy and Materials Science
Imperial College, London, SW7 2BP, England

## Introduction

In many metallurgical systems the microstructural development of a precipitate phase can be modified by the prior precipitation of a less stable phase. This may occur as a result of metastable precipitation processes taking place during heating to the ageing temperature or via two step ageing treatments in which the metastable phase is produced in a controlled manner during the first ageing step at a relatively low temperature followed by ageing at a higher temperature. Such processes are well established in the heat treatment of aluminium alloys (1) and in β titanium systems ω precipitates have been observed to act as α phase nucleation sites (2). The present work aimed therefore, to investigate the role of ω precipitation upon α phase formation in a Ti 15wt% Mo alloy and to determine if suitable two step ageing treatments could be formed to improve the tensile properties of the material.

## Experimental Procedure

The Ti 15wt% Mo alloy (oxygen content 2050 ppm) was supplied, by IMI Ltd. in the form of as-rolled plate 7.5 mm thick. Specimens for hardness and microstructural studies were prepared by rolling part of the plate at 650°C down to 1.1 mm and finish rolling to 0.8 mm at room temperature. The remainder of the plate was machined to produce standard round bar tensile specimens of gauge length 16 mm and cross-sectional area 16.1 mm². The gauge length was parallel to the original rolling direction. The strip material was solution treated at 800°C for 60 minutes and water quenched whereas the tensile samples were treated at 890°C and water quenched. All heat treatment was carried out with the specimens sealed in argon filled silica capsules. Samples cut from the strip material were aged in the range 300-550°C for times up to 1000 minutes to investigate the ageing response of the alloy. Both salt bath and furnace heat treatments were employed to determine the role of heating rate on the phase transformations and hardening response. Both light and electron microscopy were used in the microstructural studies and X-ray diffraction complemented electron diffraction in the determination of the phases present.

On the basis of the results of these heat treatments the tensile specimens were given selected ageing treatments and the tensile properties were determined. Fractography, using scanning electron microscopy, was employed to determine the correlation between the fracture characteristics and the microstructure.

## Results

As solution treated, the strip material had a grain size of 35 μm as

compared with 70 μm in the round bar tensile specimens.   The difference in
grain size resulted in the hardness of strip being in general about 10 VHN
greater than that of the tensile specimens.   Therefore, throughout the
results material is identified as 'strip' or 'tensile' and any significant
differences between the two are noted.

1.   Mechanical Properties.

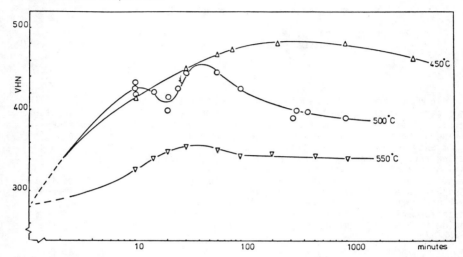

Figure 1.   Ti-15wt%Mo alloy.   Hardness changes on ageing of furnace treated
samples.

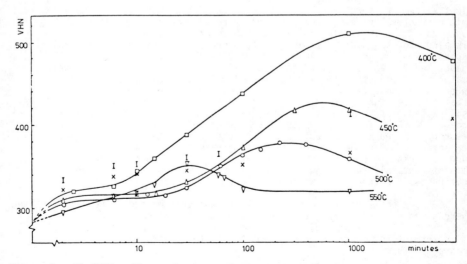

Figure 2.   Ti-15%Mo alloy.   Hardness changes on ageing of salt bath treated
samples (curves for 300°C = x and 350°C = I have been omitted for
clarity).

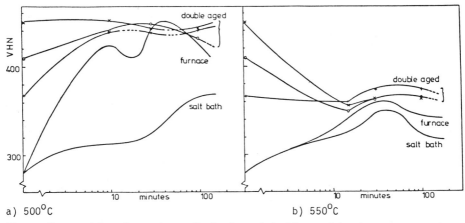

a) 500°C                                    b) 550°C

Figure 3. a and b.   Comparison of single and double aged hardness curves at
                     500 and 550°C.  (Pre-aged at 400°C for 10 mins = +;
                     30 mins = 0; 100 mins = x)

The hardening responses on ageing are shown in Figure 1 for furnace heated
specimens and in Figure 2 for specimens heated in a salt bath; a comparison of
furnace, salt bath and double aged specimens at 500 and 550°C is shown in
Figure 3.   In all cases the results refer to strip material.

Treatments at very low ageing temperatures (300 and 350°C) were carried
out with salt bath heating only.   At these temperatures the hardness increased
by ∿100 HV during a period of ∿1000 minutes.   On ageing at 400 and 450°C
larger hardness changes occurred (∿200 VHN) and overageing occurred beyond
∿1000 minutes.   At higher ageing temperatures the hardening effect decreased,
becoming negligible at 600°C and above.

The initial heating rate had a significant effect on subsequent ageing,
particularly in the temperature range 450-500°C.   For the same ageing temperature, the slower heating rate of the furnace aged specimens caused a
greater age hardening effect and the peak hardness was attained sooner,
Figure 3.

After pre-ageing at 400°C the hardness values after a second age of 10
minutes duration were seen to increase at 500°C and to decrease at 550°C,
see Figure 3.   The hardness difference between specimens pre-aged at 400°C
for 10 minutes, and those pre-aged for longer times, was small after 10
minutes ageing.   The double aged hardnesses were all higher than the corresponding single aged values.

The tensile properties for the aged specimens are presented in Table 1;
where duplicate specimens were tested the scatter was small.   (The abbrev-
iations SB and F are used to represent salt bath and furnace methods of
treatment respectively.)

As the temperature of ageing was raised from 450 to 550°C the tensile
strength decreased from a maximum of 1526 MNm$^{-2}$ to a minimum of 1067 MNm$^{-2}$,
with an attendant increase in total elongation from 1.5 to 10.3%.   Salt bath
treated specimens were more ductile than those treated in the furnace at all
ageing temperatures, furnace ageing at 450 and 500°C causing brittle fracture
at the shoulder of the test piece.   After ageing at 550°C, the furnace
treated specimens had higher tensile strengths.   The tensile properties of

specimens aged at 500°C were not improved by pre-ageing at 400°C, the double aged samples all fracturing in a brittle manner.   However, specimens aged at 550°C after pre-ageing showed an increase in tensile strength and still maintained elongation values of ~8-10%.

Table 1.   Tensile Properties of Ti-15Mo solution treated at 890°C and aged.

| AGEING TREATMENT | | | | TENSILE PROPERTIES | | | |
|---|---|---|---|---|---|---|---|
| F or SB | Temp. °C | Time mins. | Hardness Position | 0.2% Proof Stress MNm$^{-2}$ | Tensile Stress MNm$^{-2}$ | Elongation | |
| | | | | | | Uniform % | Total % |
| - | - | - | As quenched | 660 | 821 | 30 | 44 |
| SB | 400 | 340 | peak | -* | - | - | - |
| F | 450 | 200 | peak | - | - | - | - |
| SB | 450 | 340 | peak | 1454 | 1526 | - | 1.5 |
| F | 500 | 100 | overaged | - | - | - | - |
| F | 500 | 35 | peak | - | - | - | - |
| SB | 500 | 210 | peak | 1256 | 1425 | - | 7.4 |
| SB | 500 | 100 | ½ peak | 1234 | 1359 | - | 4.6 |
| | | | | 1219 | 1347 | 4 | 4.6 |
| F | 550 | 35 | peak | 1083 | 1210 | - | 4.0 |
| F | 550 | 240 | overaged | 1213 | 1289 | - | 3.4 |
| SB | 550 | 30 | peak | 1024 | 1161 | 7.8 | 10.0 |
| SB | 550 | 240 | overaged | 964 | 1067 | 9.8 | 10.3 |
| SB | 400 | 10,30 or 100 | } DOUBLE | - | - | - | - |
| | 500 | 100 | } AGED | | | | |
| SB | 400 | 10 | } DOUBLE | 1216 | 1337 | 8.0 | 10.2 |
| | 550 | 30 | } AGED | 1246 | 1319 | 5.0 | 7.3 |
| SB | 400 | 30 | } DOUBLE | 1216 | 1295 | 8.7 | 9.5 |
| | 550 | 30 | } AGED | 1216 | 1278 | 7.7 | 8.8 |
| SB | 400 | 100 | } DOUBLE | 1106 | 1189 | 7.8 | 8.2 |
| | 550 | 30 | } AGED | 1131 | 1188 | 7.4 | 8.0 |

*A dash indicates brittle behaviour.

2.  Fractography.

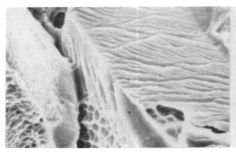 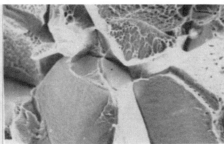

Figure 4. Mixed intergranular fracture features observed with sample aged at 450°C for 340 mins in the salt bath.  X 1940.

Figure 5.  Smooth and dimpled intergranular fracture facets. Specimen was aged at 550°C for 30 mins in the salt bath.  X 490

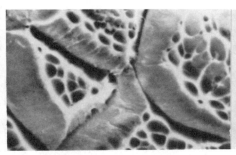 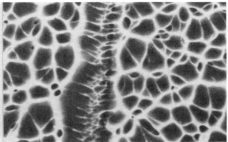

Figure 6. Grain boundary triple point region showing local brittle fracture.  Sample aged at 550°C for 30 min in the salt bath.  X 1950. (shear lip region of fracture.)

Figure 7. Grain boundary triple point region showing a fine dimpled fracture surface.  Sample aged at 400°C for 30 min + 550°C for 30 min in the salt bath.  X 1940.(shear lip)

Scanning electron fractographs of selected aged tensile specimens are shown in Figures 4-7.    Figure 4 was taken from the fibrous region of a specimen aged in the salt bath at 450°C, and shows a mixture of cleavage and ductile intergranular fracture.    Fracture surfaces of samples aged at 550°C in the furnace and at 500 and 550°C in the salt bath showed large areas of extensive dimpling together with smooth intergranular fracture facets, Figure 5.    Pre-ageing at 400°C increased the amount of dimpled fracture surface. Figures 6 and 7 are high magnification fractographs of the grain boundary region, observed in single and double aged specimens at 550°C.    Both specimens showed a change in fracture features in the region adjacent to the grain boundary:   the single aged specimens showed a brittle fracture, whereas double aged specimens showed a dimpled fracture surface, which was on a finer scale than that observed in the centre of the grain.

## 3. Microstructural Observations.

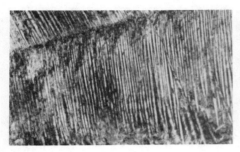

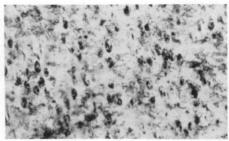

Figure 8. Extensive plate-like α
occurring at the grain
boundary. Sample aged at 500°C
for 100 mins in the salt bath.
This microstructure is representative of the small grain
size strip material only. The
larger grain size tensile
samples showed much less well
developed grain boundary
precipitation. X 27,000

Figure 9. Ellipsoidal α observed
in furnace and double
aged samples at 500°C;
sample was pre-aged at
400°C for 40 mins plus
500°C for 100 mins, salt
bath. X 87,000

Figure 10. Lenticular regions of α,
showing a complex internal
structure; typical of higher
ageing temperatures. 500°C
100 mins salt bath. X 87,000

Figure 11. Reduced amount of plate-
like α at the grain
boundary; sample aged
at 550°C for 30 mins
in the salt bath.
X 27,000

Light and electron microscopy showed that in the temperature range at and
above 500°C α precipitation occurred preferentially on the β grain boundaries
and upon sub-grain boundaries. This tendency was more obvious in salt bath
aged material than furnace aged samples (Figure 8). It should also be
emphasised that the extent of grain boundary α formation was considerably
reduced in the coarser grained tensile material which had undergone higher
temperature solution treatment. Electron microscopy and X-ray diffraction
showed that, in salt bath treated samples, no ω formed in this temperature range.
Ageing at 400°C, however, produced only ω phase in the form of uniformly

distributed ellipsoidal particles.   At 450°C and above the ω was replaced
by α, as ageing progressed but the ω persisted for longer in furnace aged
samples, than in salt bath treated specimens.   The α precipitation in
furnace heated samples was slower and produced finer precipitate dispersions.

Figure 12. Double aged sample at
550°C showing no plate-
like α at the grain boundary; sample aged at 400°C
30 mins + 550°C 30 mins,
salt bath.  X 27,000

Pre-ageing within the temperature range of ω formation accentuated this
difference on subsequent ageing at 500°C; the double aged specimens contained
ω and initially ellipsoidal α precipitates, which were uniformly distributed
through the grains up to the grain boundaries (Figure 9).   Material aged
only at 500°C contained a Widmanstatten array of coarse lenticular α phase
precipitates (Figure 10) which changed in morphology to a more ellipsoidal
shape towards the grain boundaries where α nucleation also occurred.   At
550°C the different heating rates in furnace and salt bath did not produce
such significant differences and the main effect of pre-ageing at 400°C was
to reduce the amount of α preferentially nucleated at, and growing from, the
grain boundaries although some refinement of the lenticular α precipitation
was still obtained (Figures 11 and 12).   No ellipsoidal α was observed in
this case.   Ageing above 550°C produced coarse α dispersions and no significant age hardening.

## Discussion

     The large ageing response obtained at 400°C resulted from ω formation.
In the range 450-550°C the ageing responses after furnace and salt bath
heating involved both ω and α precipitation.   The greater and more rapid
ageing response of the furnace heated material as compared to the rapidly
(salt bath) heated material can be interpreted in relation to the kinetics,
and the structural aspects of ω and α formation.   The ω phase remained in
the structure to longer times in the case of the furnace heated material.
This may be interpreted as resulting from a greater degree of ω formation
occurring during slow heating from room temperature up to ∿450-550°C.   This
ω formed as a fine dispersion throughout the β phase, being subsequently
replaced by α.   At 450°C the peak hardness was mainly associated with ω, but
at 500°C α was also present at the peak.   The slight decrease in hardness
observed after ∿10 minutes ageing at 500°C can be attributed to the dissolution of the ω phase which accompanies the onset of α formation since weaker ω
reflections were observed together with distorted α reflections in diffraction
patterns obtained from thin foils.

     With salt bath heating, less ω phase was formed during the heating period,
and ω phase was not detected after prolonged ageing.   The peak and overaged

hardness values were lower than those for the furnace heated specimens and were associated with α phase, present as lenticular particles (e.g. of the type shown in Figure 10).

Pre-ageing at 400°C produced much more ω than was formed during furnace heating and therefore the ω had a greater influence on α formation during subsequent, higher temperature, ageing.   In particular the α distribution was rendered much more uniform.   Thus a high hardness was retained as ω dissolved and α grew (Figure 3a).   At the highest ageing temperature the insensitivity of the ageing response and α precipitation to heating rate can be attributed to the complete dissolution of ω prior to α formation.   Even ω grown at 400°C rapidly dissolved at 550°C, as indicated by the rapid fall in hardness over the first ten minutes of ageing (Figure 3b)in material pre-aged for 30 or 100 minutes at 400°C.   However, even in material pre-aged for only 10 minutes at 400°C the ω persisted sufficiently long to produce a refined lenticular α dispersion with a consequently greater hardness and to inhibit grain boundary α precipitation.

The nature of the precipitating phase, particle size and morphology are critical in obtaining good combinations of strength and ductility.   The correlation of microstructure with the tensile test data of Table 1 shows that all heat treatments which produced ω or fine ellipsoidal α were fully brittle and no tensile elongation could be determined.   As the α grows ductility is restored (e.g. 450°C/340 min in salt bath) to measurable values and the mixed mode of fracture shown in Figure 4 is obtained.   At higher temperatures and/ or longer times the α assumes a lenticular Widmanstatten morphology and gives rise to good combinations of strength and ductility.   The superiority of two step ageing lies in the refinement of this lenticular morphology.   Treatments producing ellipsoidal α (e.g. Figure 9, salt bath, 400°C/40 mins + 500°C/100 mins) still produce a fully brittle material.   The brittle fracture facets at grain boundary triple points (Figure 6) obtained at 500°C and 550°C corres-pond to regions (a) where a more ellipsoidal α morphology persists in the matrix adjacent to the boundary, and (b) to regions of grain boundary α precipitation.   Double ageing treatments which render the dispersion of lenticular α particles more uniform and inhibit grain boundary precipitation (e.g. 400°C/30 mins + 550°C/30 mins) remove the brittle fracture zone at triple points and result in high strength and good ductility with relatively short ageing times.

## Conclusions

1. The strength and ductility of aged Ti-15%Mo alloy are controlled by the amounts and distribution of ω and α phases.

2. Microstructures containing ω and/or ellipsoidal α precipitates are brittle.

3. A fine Widmanstatten dispersion of lenticular α particles is required to obtain good combinations of strength and ductility; grain boundary precipitation of α lowers the ductility.

4. The morphology and distribution of α can be controlled by prior precipitat-ion of ω phase and therefore two step heat treatments can be employed to produce desirable α phase dispersions.

## Acknowledgements

Thanks are due to Imperial Metal Industries for the provision of the alloy and to Professor J.G. Ball for laboratory facilities.    The receipt of a grant by one of the authors (GMP) from the Procurement Executive, Ministry of Defence during the course of part of this work is also acknowledged.

## References

1.  G.W. Lorimer and R.B. Nicholson: *The Mechanism of Phase Transformations in Crystalline Solids*, Institute of Metals (1969) 36.
2.  J.C. Williams: *Titanium Science and Technology*, 3 (1973) 1433, Plenum Press, New York.

# AGING BEHAVIOR OF THE Ti-15Mo-5Zr ALLOY

Shinya Komatsu, Takashi Sugimoto, Kiyoshi Kamei

Department of Metallurgical Engineering, Faculty of Engineering, Kansai University.   Suita, 564, Japan

and

Osamu Maesaki

Graduate School of Kansai University, presently with Azumi Co. Ltd., Asahi, Osaka, 535, Japan

## Introduction

Ti-15Mo-5Zr alloy developed as a corrosion resistant material has also both excellent formability and ability of age-hardening [1]. The ω phase in this alloy as well as in other meta-stable β-Ti alloys has been undesirable as the cause of embrittlement. It has been reported that the tensile strength of Ti-15Mo-5Zr alloy is improved not to cause the embrittlement by a duplex aging within the aging temperature range of ω formation [2]. There have been some evidences about the acceleration of α precipitation accompanied with a slightly high density of dislocations [3]. However, detailed investigation about the effects of cold work and pre-aging on the aging behavior of this alloy have not been published.

The resistivity change often gives important informations about the aging of alloys. The specific electrical resistivity of meta-stable β-Ti alloys has two outstanding characteristics. The one is the decrease with increasing concentration of β stabilizing elements [4], and the other is the negative temperature dependence in the instable β single phase state [5]. The latter has been pointed out to be quite useful to investigate the formation and the reversion or retrogression of ω phase in Ti-20V alloy [6].

In the present work, a detailed study was carried out employing mainly the resistivity measurement to clarify the aging behavior and effects of cold work and pre-aging on isothermal aging of this alloy.

## Experimental procedure

All specimens were cut from a sheet of the alloy of 1 mm thick made by Kobe Steel Ltd. In as-received state, the sheet had been 90% cold rolled after hot rolling and final annealing at the temperature above the β transus. Chemical composition of the alloy (in wt%) is 14.74 Mo, 4.84 Zr, 0.046 Fe, 0.17 O, 0.0175 H, 0.061 N with Ti making up the balance.

Solution treatment was carried out for 3.6 ks at 1173K in quartz ampoule evacuated to 1.3 mPa. Specimens were quenched into iced water by crashing the ampoule. Neither ω nor α reflection was detected by X-ray diffraction in both the as-received and the as-quenched states. Aging was carried out in the temperature range between 623 and 873K with intervals of 50K. Specimens were encapsulated when they were aged at 873K and when they were aged for a long period above 773K. For the duplex aging, 673K was chosen as the pre-aging temperature (at which only the ω formation occurs) and as the final aging temperature 773K (at which the α precipitation occurs) was selected. It is sometimes experienced that the specimen aged without interruption differs in some properties from that intermittently aged. Then, the time interval of isothermal aging was selected to be nearly same for all specimens used in various techniques.

Resistivity was measured with specimens of 1×2×100 mm at 77K and room temperature (300K). Specimens for electron microscopic observation were ground by wet emery paper from 1 mm to 0.2 mm in thickness after all heat treatments, then thinned by electropolishing at 200K in a methanol-sulfuric acid electrolyte. Micro-Knoop hardness at 0.5 kg load was measured with 20 indentations lined in the short transverse direction on the cross section of the sheet specimen. X-ray diffraction profile was taken with Cu-Kα radiation filtered by Ni.

Results

The change of electrical resistivity at 77K during isothermal aging is shown in Fig. 1. Decrease of resistivity is observed without incubation period below 723K, corresponding to the ω formation. The final value of resistiv-

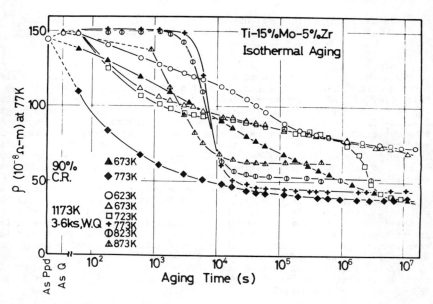

Fig. 1 Resistivity change during isothermal aging. As Ppd stands for as-prepared.

ity decreased by the ω formation is ca. $70×10^{-8}\Omega$m. Above 773K, the resistivity starts to decrease after remarkable incution periods. Saturated values of the resistivity decrease are lower than those by the ω formation and lessened with lowering aging temperature, as shown in Fig. 2. It has been reported by Nishimura et al. [2] that the α precipitation becomes dominant above 773K in this alloy. These results may indicate that the resistivity decrease appearing above 773K with the incubation period is attributed to the α precipitation and that the amount of precipitated α phase decreases with elevating aging temperature.

Figure 3 shows isothermal age-hardening curves. Below 723K, marked age-

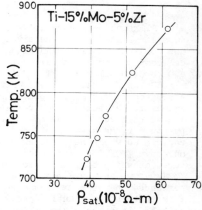

Fig. 2 Relation between the final value of resistivity decrease and the aging temperature.

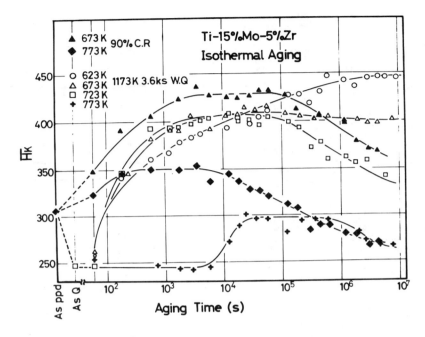

Fig. 3  Isothermal age-hardening curves.

hardening due to the ω phase occurs without the incubation period, in good agreement with the resistivity change. At 773K, a small amount of hardening appears after the incubation period of ca. 3 ks. This may be attributed to the nucleation and growth of the α precipitates. The resistivity change at 723K shows a second stage decrease after the initiation of over-aging, as shown in Figs. 1 and 3. It is considered that these results indicate an isothermal transition from β+ω to β+α through β+ω+α and that the start of the growth of α precipitates results in the gradual annihilation of ω particles which mainly contribute to the hardness. The concentration of Mo in the β matrix will remain constant during the growth of α precipitates at expense of the ω particles. The second stage decrease of resistivity

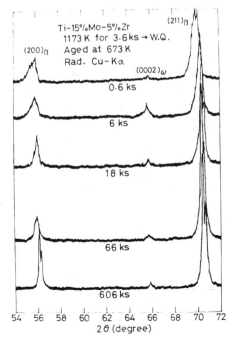

Fig. 4  X-ray diffraction profiles of a specimen aged at 673K.

will indicate that all of the ω
particles are annihilated and the
Mo concentration in the β matrix
starts to increse.

Figure 4 shows the change in
the X-ray diffraction profile by
aging at 673K. The reflection of
$(0002)_\omega$ appears from the aging
for 0.6 ks, and the $(211)_\beta$ reflection shifts to higher angle with
aging time, indicating the increase of Mo concentration of the
β phase. In the case of aging at
773K, no change is detected during the incubation period, as shown
in Fig. 5. The shift of β peaks
to higher angle occurs with the
appearance of α peaks.

Photos. 1 and 2 show the microstructure of the specimens aged at
673 and 773K. Ostwald ripening
seems to occur during the growth of
ellipsoidal ω particles, as shown
in Photo. 1. No precipitate is observed at the end the incubation
period, as shown in Photo. 2(a).
Plate shaped α precipitates are observed in Photos. 2(b) and (c).

The resistivity of the rolled
specimen aged at 673K decreases to
the value corresponding to that after the α precipitation, as shown
in Fig. 1. The two stage decrease
of resistivity appeared in the quenched specimen aged at 723K is not observed in the case of rolled specimen
aged at 673K. Figure 6 shows the X-
ray diffraction profiled of a rolled
specimen aged at 673K. The $(0002)_\omega$
peak at about 66° is hardly recognized. Reflections of the α phase
appear in the aging for 120 ks, where
the over-aging occurs as shown in
Fig. 3. The rolled specimen aged at
773K shows no incubation period both
in the resistivity change and the
age-hardening.

Figure 7 shows the change of
resistivity and hardness during the
final aging after pre-aging at 673K.
Curves of the resistivity measured
at room temperature (300K) are also
shown in the figure. In the specimen
not pre-aged, essentially no change
occurs during the incubation period,
and then its resistivity and hardness
are the same as those of as-quenched
state. The resistivity of as-quench-

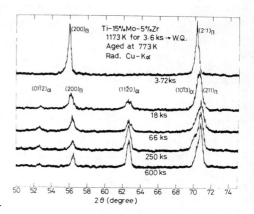

Fig. 5 X-ray diffraction profiles
of a specimen aged at 773K.

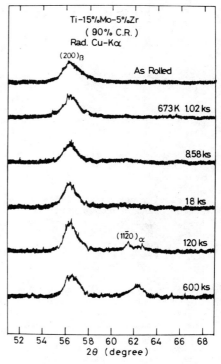

Fig. 6 X-ray diffraction profiles
of a 90% cold rolled specimen aged
at 673K.

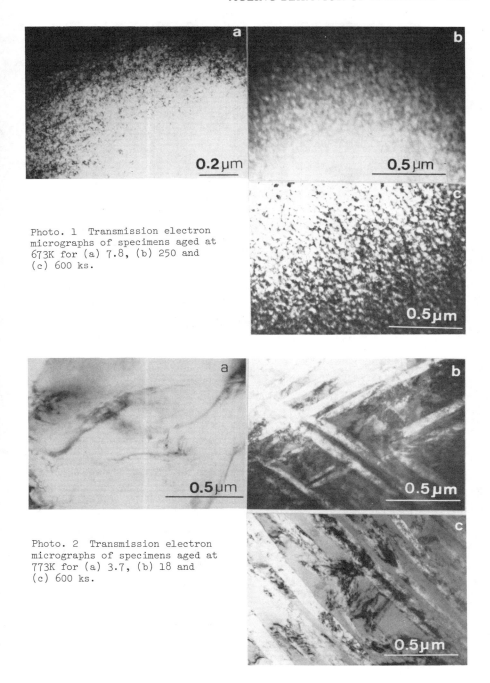

Photo. 1   Transmission electron micrographs of specimens aged at 673K for (a) 7.8, (b) 250 and (c) 600 ks.

Photo. 2   Transmission electron micrographs of specimens aged at 773K for (a) 3.7, (b) 18 and (c) 600 ks.

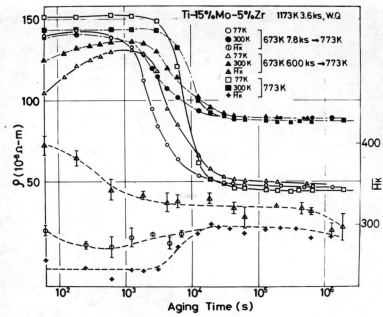

Fig. 7 Changes of resistivity and hardness during the final aging at 773K after pre-agings at 673K.

ed specimen measured at 300K is lower than that measured at 77K; namely the as-quenched specimen shows the negative temperature dependence of resistivity (NTDR). In the specimen pre-aged for 7.8 ks at 673K, which shows the positive temperature dependence, the resistivity is increased by the final aging up to 0.18 ks to the value slightly lower than that of the as-quenched state and returns to the state of NTDR. The initiation of the rapid decrease of resistivity in the pre-aged specimen occurs earlier than that in the not pre-aged specimen. These observations suggest that the great majority of ω particles annihilates rapidly in the final aging and the nucleation of the α phase is accelerated by the pre-aging for 7.8 ks at 673K. When the specimen is pre-aged for 600 ks, the resistivity in-

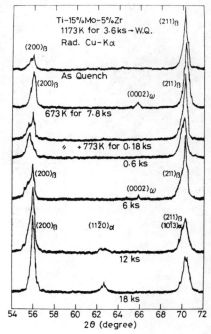

Fig. 8 X-ray diffraction profiles of a specimen pre-aged for 7.8 ks at 673K and finally aged at 773K.

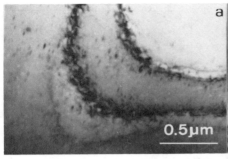
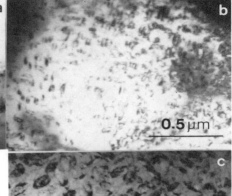
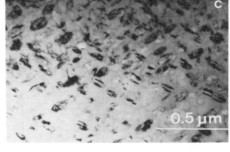

Photo. 3  Transmission electron micrographs of specimens aged at 773K for (a) 0.18, (b) 0.6 and (c) 6 ks after the pre-aging for 7.8 ks at 673K.

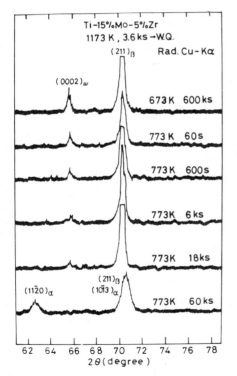

Fig. 9  X-ray diffraction profiles of a specimen pre-aged for 600 ks at 673K and finally aged at 773K.

crease due to the reversion of $\omega$ particles is retarded by the larger particle size than that in the case of the pre-aging for 7.8 ks, and then the rapid decrease of resistivity due to the growth of $\alpha$ precipitates is also retarded by the high Mo concentration in the $\beta$ matrix due to the remaining $\omega$ particles. In good agreement with the resistivity changes, an early softening occurs in the specimen pre-aged for short period. Changes in the X-ray diffraction profile shown in Figs. 8 and 9 substantiate the retardation in the reversion of $\omega$ particles in the specimen pre-aged for 600 ks at 673K.

Photo. 3 shows the microstructure of specimens pre-aged for 7.8 ks at 673K and finally aged at 773K. Comparing with Photo 1(a), Photo 3(a) shows that the great majority of $\omega$ particles is annihilated by the final aging for 0.18 ks. Though it is not clear whether small particles in Photo. 3(a) are $\omega$ particles or small $\alpha$ particles newly precipitated, it is clear that such small particles grow with increasing final aging time. If one sets importance on the diffraction profile of the specimen finally aged for 6 ks (Fig. 8) and the morphology of the

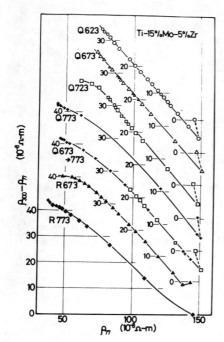

Fig. 10   Relations between $\rho_{300}-\rho_{77}$ and $\rho_{77}$ for various isothermal agings. Q and R represents the quenched and the 90% cold rolled specimen, respectively.

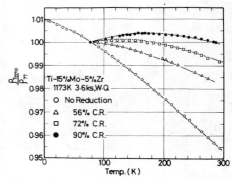

Fig. 11   Temperature dependence of the resistivity in the as-quenched and several cold rolled specimens.

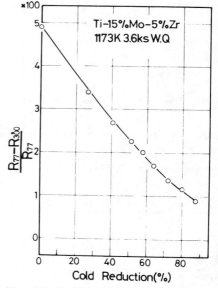

Fig. 12   The change of the temperature dependence of resistivity with the degree of cold reduction.

particles (Photo. 3(c)), there is a possibility that lenticular particles in Photo. 3(c) would be ω particles. However, it is difficult to consider that the phase once annihilated grows further by the aging even at the same temperature.  Furthermore, both the decrease of resistivity and the temperature dependence of resistivity suggest that the α phase is precipitated by the final aging for 6 ks.

The difference of the resistivity at 300K from that at 77K is plotted versus the resistivity at 77K in Fig. 10 to show the change in the temperature dependence of resistivity during isothermal aging.  Well defined straight lines appear in the aging temperature range in which the ω phase forms.  In the temperature range of the α precipitation, the plots give curves.  When the specimen is aged at 723K, the plots lie on a straight line during the ω formation and on a curve after the beginning of the α precipitation.  A curve of the specimen aged at 773K after the pre-aging for 7.8 ks at 673K resembles to that of aged at 773K immediately after quenching; compare Q773 with Q673-773 in Fig. 10.  Figure 11 shows the change of resistivity with the measurement temperature for the as-quenched specimen and several cold rolled specimens.  In the as-quenched specimen without any reduction, the resistivity decreases with increasing measurement temperature from 4.2 to 300K.  The NTDR decreases with increasing degree of the cold reduction.  A resistivity peak appears in the 72% cold rolled specimen at 130K but in the 90% cold rolled specimen at 190K.  As the resistivity of the cold rolled specimens shows the parabollic change with the measurement temperature, the value of $\rho_{300} - \rho_{77}$ shown in Fig. 10 is proportional to the temperature coefficient averaged over the temperature range between 300 and 77K.  Figure 12 shows the change of NTDR with the reduction of thickness.  The specimen is cumulatively rolled at room temperature using a roll driven by hand.  The averaged NTDR decreases with increasing degree of cold rolling, but does not drop to the value of the 90% cold rolled specimen shown in Figs. 10 and 11.  This difference may be caused by the difference in dislocation density due to that in the rolling speed.

## Discussion

Nishimura et al.[2] reported the formation of the ω phase in this alloy aged at 798K and the α precipitation above 698K, whereas the present results give both the upper temperature limit for the ω formation and the lower limit for the α precipitation to be 723K.  The discrepancy may be due to differences in the specimen size and in the solution treatment temperature.  As the present specimens are smaller than those of Nishimura et al., the quenching rate is higher in the present work.  The low cooling rate will increase the amount of the ω phase formed during cooling.  Moreover, the larger specimen size decreases the rate of up-quenching to the aging temperature, resulting to stabilize ω particles and to elevate the upper temperature limit for the ω formation. High solutionizing temperature and severe cold work in the present specimen give the more perfect recrystallization which deprives α precipitates of their nucleation sites thus will elevate the lower temperature limit of the α precipitation.

According to the result of X-ray small angle scattering by Gysler et al. [7], the ω particles in Ti-15Mo alloy show the Ostwald ripening.  Larger ω particle will require longer time to dissolve at the final aging temperature when the stability of ω particle is dominantly determined by the particle size. They also reported the partial suppression of the ω formation in Ti-25Mo alloy by the cold rolling of 75% which will facilitate the α precipitation.  Williams et al.[8] reported the concentration separation of the β phase in Ti-Mo-Al alloys and the preferred nucleation of the α phase in the Mo lean $\beta_2$ region.  The present specimen does not contain α stabilizing elements except oxygen as an impurity, and in the binary Ti-Mo alloy the β phase separation seems not to occur [8,9].  However, in the case of duplex aging, the Mo concentration may be lowered after the re-solution of the ω particles in the sites where the ω particles once existed.  This will be the case when the ω→β transition is faster than the homogenizing of Mo concentration by diffusion.  The existence of the solute lean β region remained after the re-solution of the ω particle has been substantiated by Jones et al.[10] in the reversion of the cuboidal ω particles in a Zr-Ti-Nb alloy.  Thus, the β phase separation may occur even in the present alloy in the initial stage of the final aging after pre-aging, as indicated by the variation in line profiles of $(200)_\beta$ in Fig. 8.

The NTDR was interpreted in terms of the energy overlap across the second Brillouin zone [6].  Recently, Williams et al.[8] proposed that the NTDR can also be explained by the reversible formation of the athermal ω phase.  The present result shown in Fig. 10 is in good agreement with the previous results on Ti-20V aged at 626K [6].  The NTDR in the as-quenched specimen changes to the positive one after a short time aging at the temperature within the range of the ω formation.  By the reversion of the isothermal ω phase, the NTDR is recovered or, at least, the positive temperature coefficient is decreased. Though there are still some ambiguities about the cause of the NTDR, the temperature dependence of resistivity is quite useful to investigate the aging behavior of the meta-stable β-Ti alloys.

It is necessary to know about the effect of cold work on the NTDR for the investigation of that effect on the aging behavior of the meta-stable β-Ti alloys by the resistometry. The effect of cold rolling on the NTDR has been reported by Zwicker in Ti-33Nb alloy [11]. The NTDR in the solution treated Ti-33Nb alloy is changed to the positive one by the 5% cold rolling ; namely the resistivity measured at 300K is larger than that measured at 77K in the 5% cold rolled specimen. In contrast with the Ti-33Nb alloy, the present Ti-15Mo-5Zr alloy shows the NTDR even after the 87% cold rolling, as shown in Fig. 12. Though the effects of cold work on the NTDR in both alloys have a similar tendency that the NTDR in the solution treated specimen is changed (or brought close) to the positive one by the cold work, the effect in the Ti-15Mo-5Zr alloy is smaller than that in the Ti-33Nb alloy.

## Conclusions

1. The upper temperature limit for the ω formation and the lower limit for the α precipitation by isothermal aging are 723K in the Ti-15Mo-5Zr alloy sheet quenched from 1173K.
2. The isothermal ω formation is clearly separated from the α precipitation by the kinetics of the resistivity change and the decrement of resistivity.
3. The cold reduction of 90% extremely accelerates the α precipitation and also lowers the lower temperature limit for the α precipitation.
4. The pre-aging slightly accelerates the α precipitation by the final aging.
5. The temperature dependence of resistivity is quite useful to investigate the aging behavior of the meta-stable β-Ti alloys.

## Acknowledgement

Authors wish to express their thanks to Kobe Steel Ltd. for supply of the alloy sheet. This work was partially supported by Grants-in-Aid for Fundamental Scientific Research from Ministry of Education.

## References

1. S.Ohtani and M.Nishigaki: J. of Japan Inst. Met.,36(1972), 346.
2. T.Nishimura, M.Nishigaki and S.Ohtani: J. of Japan Inst. Met.,40(1976), 219.
3. S.Ohtani, T.Nishimura and M.Nishigaki: J. of Japan Inst. Met.,36(1972), 1105.
4. R.R.Hake, D.H.Leslie and T.G.Berlincourt: J.Phys.Chem.Solids, 20(1961), 177.
5. U.Zwicker: Ti und Ti Legierungen, Springer, Berlin, (1974), p.73.
6. F.R.Brotzen, E.L.Harmon Jr. and A.R.Troiano: Trans. AIME., 203(1955), 413.
7. A.Gysler, W.Bunk and V.Gerold: Z.Metallkde.,65(1974), 411.
8. J.C.Williams, B.S.Hickman and D.H.Leslie: Met. Trans., 2(1971), 477.
9. J.C.Williams, D.de Fontaine and N.E.Paton: Met. Trans., 4(1973), 2701.
10. W.B.Jones, R.Taggart and D.H.Polonis: Met. Sci., 13(1979). 60.
11. U.Zwicker: Metal, 18(1964), 941.

# THE POTENTIOKINETIC METHOD : A USEFUL TOOL TO
# STUDY PHASE TRANSFORMATIONS IN BETA III TITANIUM ALLOY

J.A. Petit, D. Delaunay, D. Leroy and G. Chatainier

Laboratoire de Métallurgie Physique. E.R.A. n° 263 du C.N.R.S.
Institut National Polytechnique
118 Route de Narbonne - 31077 Toulouse Cédex, France

## Introduction

Phase transformations in alloys are generally studied with classical experimental techniques including tensile and hardness tests, dilatometric and resistometric experiments and metallographic examinations. Titanium alloys aging behavior has been established by this way. In particular the phase transformations occuring in the metastable Beta III alloy have been thoroughly studied (1-5).

However several authors have shown that the systematic investigation of potentiokinetic anodic polarization curves is a very effective means to study phase transformations in various steels (6-8). As we stated recently (9), works to date on the effect of microstructure on the electrochemical behavior of titanium alloys are rather limited. Moreover, either the considered media in these works are often not agressive enough or the galvanic effects between the phases are too small to produce important variations allowing to follow phase transformations. In titanium alloys, indeed, the galvanic effects mainly come from the greater solubility of passivation elements like Mo in the beta phase than in the alpha phase which then is selectively dissolved. Molybdenum exhibits a transpassive behavior, not shown by titanium, so the Mo rich beta phase presents a secondary activity, which may supply supplementary informations. Therefore Beta III alloy with 11.5 wt % Mo was chosen. The main purpose of this work is to show that the potentiokinetic method may be, under some circumstances, a complementary tool to investigate aging behavior of titanium alloys.

## Experimental procedure

The Beta III alloy was supplied by the Crucible Steel Corp : the chemical composition of the material is shown in Table 1.

Table 1 - Chemical composition of Beta III (wt. %)

| Mo | Zr | Sn | Fe | C | N | O | H | Ti |
|------|------|------|------|------|-------|------|--------|------|
| 10.2 | 7.4 | 4.5 | 0.11 | 0.01 | 0.012 | 0.10 | 0.0091 | Bal. |

Sheet material was used to study phase transformations. All specimens were solution treated above the beta transus at 815 C for 1 hr., then water quenched. The temperature ranges of stability of the different phases formed upon aging the as-quenched Beta III alloy are well defined. The alpha phase precipitation occurs only above 400 C and that of the omega phase below this temperature. In a very narrow temperature range near to 400 C, the beta, alpha and omega phases coexist. Therefore aging was carried out under vacuum at three characteristic temperatures : 350, 400 and 550 C, corresponding to these three temperature ranges. Aging time extended from 1 min. to 150 hr.

Potentiokinetic experiments were performed at a $4V.h^{-1}$ polarization rate in deaerated 40 % (~ 10 N) sulphuric solutions at 70 C. Experimental devices and procedures have been already described (10). Rockwell C hardness tests were performed on a standard Frank machine. Phase identification by X-ray diffraction was carried out on a Siemens Kristalloflex IV instrument using a copper tube. Experiments were completed by microscopic examinations and measurements of the amounts of Ti and Mo in solution after potentiostatic polarization.

### Results and discussion

Molybdenum in Beta III titanium alloy mainly acts as a passivator element, like chromium in stainless steels. The main effects are the following : corrosion potentials and primary passivation potentials rised towards noble values and the active and passive current densities decreased by comparison with unalloyed titanium. But as shown in figure 1 the passivity range of Beta III alloy in sulphuric solutions is more or less disturbed by a secondary activity due to the transpassivity of molybdenum.

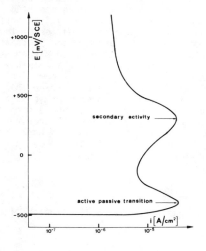

The as-quenched metastable beta phase alloy exhibits a very low corrosion rate in active and passive states ; the ratio between Ti and Mo found in solution after anodic polarization in the active state for 5 hr. is about the same as in the alloy. The single phase alloy has also a high stability in the range of potentials of molybdenum transpassivity. As a result of aging at various temperatures leading to the precipitation of the alpha phase and (or) the omega phase the dissolution rates both in the active and passive states markedly change (figure 2). The corrosion in the range of potentials of transpassivation for molybdenum is also affected (figure 2).

Figure 1 - Anodic polarization curve of Beta III alloy, solution-treated and aged at 550 C for 20 hr.

### 1. Aging at 550 C.

For a 550 C aged alloy, after a 5 min. incubation period an important increase of the maximum active dissolution rate, due to alpha phase precipitation, is observed. The dissolution rate is maximum after 2 hr. aging. This aging time

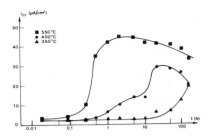

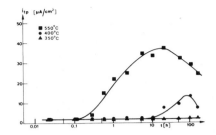

Figure 2 - Variation of the critical current density with aging time
for specimens of Beta III aged at
350, 400 and 550 C.

Figure 3 - Variation of the maximum
current density in the range of potentials of transpassivation for
molybdenum with aging time for specimens of Beta III aged at 350, 400 and
550 C.

also corresponds to the maximum strengthening of the alloy (figure 4). Then the
Widmanstätten fine alpha phase precipitation (figure 5) seems to the ended,
which gives rise to the greater galvanic effects between the Mo depleted alpha
phase and the Mo enriched beta phase. Moreover, after anodic polarization, a
powdery dark layer, identified as pure beta phase by X-ray diffraction, is accumulated on the surface of the sample aged at 550 C for 5 min. to 2 hr. The ratio
Mo/Ti in the solutions are considerably less than in the alloy, the Mo content
in solution being less than 4 % of the amount of alloy dissolved. Therefore,
owing to its lower hydrogen overvoltage, the beta phase accumulated on the surface facilitates the hydrogen evolution reaction, but not enough to shift the
corrosion potential of the alloy in the passive range. So an increase of the
active dissolution rate ought to be observed as it appears in figure 2. Owing
to these important effects the potentiokinetic method behaves a very sensitive
technique to study the alpha phase precipitation whereas hardness experiments
are less interesting because this phenomenon leads to a limited strengthening of
the alloy.

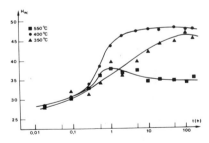

Figure 4 - Variation of hardness
with aging time for specimens of
Beta III aged at 350, 400 and
550 C.

After aging for 2 hr. the overaging visible on the hardness versus
aging time curve also appears on the
dissolution rate-aging time curve.
This overaging can be connected to
subsequent coarsening of the Widmanstätten alpha phase as shown in figure
6. The amount of the alpha phase remaining almost constant (~ 50 vol. %),
after aging for 2 hr. the influence of
the hydrogen depolarization is about
the same as that previously involved
for the strengthening period. However
the galvanic effects between the thickened alpha phase and the beta phase
are smaller than between the fine dispersed alpha phase and the beta phase.
Therefore the active dissolution rate
slightly decreases for the overaging
period. Moreover, as stated by Froes et al. (11), the Mo content in the beta
phase reaches a maximum between 10 to 15 hr. Correspondingly the dissolution

rate in the range of potentials of transpassivation for molybdenum is maximum after aging for about 10 hr. But the variation of this dissolution rate with aging time due to the precipitation of the alpha phase is less sharp than the variation of the active dissolution rate because the Mo enrichment of the beta phase leads to smaller effects than the precipitation of the alpha phase. As a result of anodic polarization in the potentials range of molybdenum transpassivity the alloy surface is enriched with alpha phase while the relative amount of molybdenum in solution is larger than in alloy. As the percentage of the beta phase decreases when the aging period increases, even if the beta phase is enriched with molybdenum, the galvanic effects have again a great influence which should account for the shape of the dissolution rate-aging time curve in figure 3.

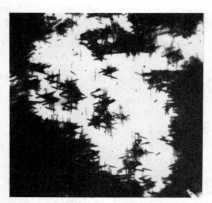 

Figure 5 - Optical micrograph of fine Widmanstätten alpha phase formed in Beta III aged 30 min. at 550 C (x 1300).

Figure 6 - Optical micrograph of thickened Widmanstätten alpha phase formed in Beta III aged 5 hr. at 550 C (x 1300).

## 2. Aging at 350 C

As a result of aging at 350 C a significant increase of the active dissolution rate can be seen after 10 hr. only. From this time to 150 hr. the dissolution rate continuously increases without reaching a maximum. The shift of the dissolution rate remains small by comparison with that due to alpha precipitation. As shown in figure 4, at 350 C aging starts without an incubation period and an important rise in hardness takes place. The highest hardness is obtained after 100 hr. The precipitation of the omega phase is responsible for this increase in hardness. Vigier et al. (3) have shown that 80% of the total amount of the omega phase able to precipitate is obtained after 5 hr. It seems from our results that the total amount ($\sim$ 65 vol. %) of the omega phase is precipited after 10 hr.

From an electrochemical point of view, the omega and beta phases do not exhibit very different behavior, since they are both molybdenum rich phases. However the amount of titanium in the omega phase ($\sim$ 95 %) is more important than in the beta phase ($\sim$ 90 %). After anodic polarization in the active state of samples aged for 100 hr. at 350 C, only about 6 % of molybdenum are found in solution. This amount is lower than the molybdenum content of the alloy which may indicate that the increase of the active dissolution rate comes from selective dissolution of the titanium enriched omega phase. The galvanic effects between the omega phase and the beta phase are only noticeable after some growth of omega phase particles, which maximum size may be about 900 Å (1), and a corresponding enrichment of the particles with titanium. In the potentials range of molybdenum transpassivation a 100 hr. aging is required to find a very slight variation of the dissolution rate.

These results show that the potentiokinetic method is little sensitive to study the beta phase decomposition as a result of aging at 350 C because the electrochemical behaviors of the omega phase and of the beta phase are too close. However the low variation of the dissolution rates observed allows to show up the weakness of the hypothesis of alpha phase precipitation as a result of overaging. Feeney and Blackburn (1), indeed, have shown that overaging at 370 C is not associated with the initiation of the precipitation of the alpha phase from an oversaturated beta phase since the omega phase is the only precipitate present after aging for 1000 hr.

3. Aging at 400 C

From the experimental results (figures 2 to 4) the effects of aging at 400 C seem to be a compromise between those at the two other aging temperatures. At this aging temperature the alpha phase and the omega phase can grow simultaneously (1-3). The shape of the active dissolution rate versus aging time curve indicates that the alpha phase precipitation does not control alone the aging kinetics. If it occurs the incubation period would be much longer at 400 C than at 550 C and the slope of the curve in the incubation period would not be positive. Besides it seems from electrochemical experiments that up to 10 hr. aging only a small amount of alpha phase is precipitated together with omega phase. Variation of the dissolution rate in the range of potentials of molybdenum transpassivation appears after 10 hr. only (figure 3). This result agrees very well with the two types of variation of the active dissolution rate. After 10 hr. an important increase of this dissolution rate is observed and a maximum is reached after aging for 30 hr. During the first period of aging simultaneous precipitation of the alpha phase in small amount, as shown on the optical micrograph in figure 7, and of the omega phase is responsible for the slight increase in the active dissolution rate. The enrichment with molybdenum of the beta phase is too low to cause a variation of the dissolution rate in the range of potentials of molybdenum transpassivation.

On the other hand, after aging for 10 hr. the amount of alpha phase precipitated increases owing in part to some transformation of the omega phase as pointed out by Feeney and Blackburn (1). This leads to some increase of the active dissolution rate since the galvanic effects between the Mo poor alpha phase and the Mo rich beta and omega phases are more important. During the same period the enrichment of the beta phase with molybdenum may account for the increase of the secondary activity corrosion rate. Some thickening of the alpha phase after aging for 30 hr. (figure 8) reduces the galvanic effects, so the dissolution rates slow down. However the hardness remains maximum (figure 4) owing to the simultaneous presence of the alpha and the omega phases which gives the alloy a maximum strengthening. Then the aging behavior at 400 C of as-quenched Beta III titanium alloy provides a good exemple of the advantage of the potentiokinetic method as a means of investigation. In this case, the method allows to characterize some aspects of the aging kinetics that cannot be displayed so accurately by other methods.

Figure 7 - Optical micrograph showing alpha phase precipitation in
Beta III aged 5 hr. at 400 C
(x 250).

Figure 8 - Optical micrograph of
overaged Beta III at 400 C for
150 hr. (x 250).

## Conclusion

Under some circumstances, the electrochemical techniques, sensitive to slight variations of alloys composition, seem very useful complementary methods to investigate the aging kinetics of titanium alloys. It is possible if galvanic effects between mainly the precipitated alpha phase and the beta phase are important enough to induce measurable variations of the dissolution rates. Such circumstances are met for aged Beta III alloy, owing to the passivating power of molybdenum, to its lower hydrogen overvoltage than that of titanium and to its transpassivation in the passive range of potentials for titanium.

## References

1.   J.A. Feeney and M.J. Blackburn : Met. Trans., 1 (1970), 3309.
2.   F. Vial, B. Hocheid and C. Beauvais : C.R. Acad. Sci. Paris, 276 C (1973), 1441.
3.   G. Vigier, J. Merlin and P.F. Gobin : Mem. Sci. Rev. Met., 72 (1975), 678.
4.   D. Kalish and H.J. Rack : Met. Trans., 3 (1972), 1885.
5.   R. Molinier, L. Seraphin, R. Tricot and R. Castro : Rev. Metallurgie, 71 (1974), 63.
6.   L. Priester and M. Aucouturier : Metallography, 6 (1973), 195.
7.   E. Jolles, L. Priester, M. Aucouturier and P. Lacombe : Mem. Sci. Rev. Met., 67 (1970), 261.
8.   N. Bui, B. Pieraggi and F. Dabosi : Mem. Sci. Rev. Met., 68 (1971), 223.
9.   J.A. Petit, P. Lafargue and F. Dabosi : *Selective phase dissolution in titanium alloys*. Journées d'Automne Soc. Française Métallurgie, Paris Oct. 1979.
10.  J.A. Petit and F. Dabosi : Mem. Sci. Rev. Met., 68 (1971), 595.
11.  F.H. Froes, J.M. Capenos and M.G.H. Wells : *Titanium Science and Technology*, 2 (1973), 1621. Plenum Press, New York.

# EFFECTS OF COLD-WORKING AND OXYGEN ADDITION
## ON PRECIPITATION BEHAVIOR IN SUPERCONDUCTING TI-NB ALLOY

Kozo Osamura, Eiichiro Matsubara, Takashi Miyatani and Yotaro Murakami

Department of Metallurgy, Kyoto University, Kyoto 606, Japan

and
Takefumi Horiuchi and Yoshiyuki Monju

Asada Research Laboratory, Kobe Steel Ltd., Kobe 657, Japan

## Introduction

A type II superconductor is characterized by a mixed state over a certain range of magnetic field strength, where magnetic flux is quantized into units known as fluxoids. When a current flow along the conductor, the Lorentz force attempts to move the fluxoids through the material. This movement of fluxoids causes resistance to current flow. Thus when fluxoids can be pinned against the action of the Lorentz force by structural inhomogeneities, a greater critical current density can be achieved. Several types of defects have been suggested for the flux pinning center as secondary defects of vacancies, dislocation, grain boundary, precipitates and nonmetallic inclusion.

It is reported that dislocation networks have a direct role for flux pinning in Ti-40 at%Nb [1]. On the other hand, ina Ti-20.7 at%Nb alloy, fluxoids are pinned by $\alpha$-Ti precipitates[2], or possibly by $\omega$ phase [3]. When the Ti-36 at%Nb alloy was heavily cold-drawn and aged at 653 K for long time, the enhancement of critical current density [4] is attributed to the precipitation of $\alpha$ phase as reported in our recent work [5]. In order to precipitate more easily and stabilize $\alpha$ phase, it seems an effective way that small amount of oxygen is added in the alloy. In the present report, the effects of cold-working and oxygen addition on precipitation of $\alpha$ phase have been investigated in a Ti-36 at%Nb alloy by means of small angle X-ray scattering (SAXS) measurement as well as transmission electron microscopy (TEM).

## Experimental Methods

Two types of specimens were used, that is, foils and wires depending on the rate of cold working. Their arc-melted Ti-36 at%Nb alloy was supplied by Kobe Steel Co. Ltd.. The principal impurities were 0.025%Fe, 0.018%C, 400ppmO and 6ppmH. Thin foils with different reduction rates were prepared by cold-rolling at room temperature after the alloy was solution-treated 1 hr at 1073 K and was slow-cooled under vacuum. As listed in Table 1, the reduction rate was selected as 9, 50 and 90%. Final thickness of all specimens was adjusted to about 100 $\mu$m. The ordinary ageing was performed at 653 K. In order to investigate the oxygen addition on precipitation, small amount of oxygen was added during ageing. The solution treated thin foil was put into the measured silica ampule. After the constant pressure of oxygen gas was introduced using the manometer, the ampule was sealed. All amount of oxygen was assumed to dessolve into the alloy during ageing.

The wire specimens were got from the copper clad Ti-Nb alloy. The compo-

site was heavily drawn and then copper was dissolved in the nitric acid bath.
Two kinds of wires with diameters of 60 and 110 $\mu$m had a reduction rate of
area of 90 %.  The other with diameter of 40 $\mu$m had a reduction rate of 99.995
%.  A quaternary alloy of Ti-27at%Nb-5at%Zr-5at%Ta which is commercially used
for the practical magnet, was also examined.  The diameter of its wires and
the reduction rate of area were 40 $\mu$m and 99.994%, respectively.  These wires
were aged under vacumm at 653 K.  For SAXS measurement, the short pieces of the
wires were closely arranged between the frames.

Table 1  Chemical composition, shape and reduction rate of specimens

| Sample No | Composition | Shape | Thickness/ Diameter [ $\mu$ m] | Reduction Rate [ %] |
|---|---|---|---|---|
| NT-36,HO | Ti-36at%Nb | plate | 100 | 0 |
| NT-36,H9 | " | " | " | 9 |
| NT-36,H50 | " | " | " | 50 |
| NT-36,H90 | " | " | " | 90 |
| NT-36,110$\mu$ | " | wire | 110 | 90 |
| NT-36,60$\mu$ | " | " | 60 | 90 |
| NT-36,40$\mu$ | " | " | 40 | 99.995 |
| NZT-2,40 $\mu$ | Ti-27at%Nb -5at%Zr-5at%Ta | " | 40 | 99.994 |

The SAXS intensity was measured between 0.05 and 5 degree using the MoK$\alpha$
line monochromatized by Zr filter.  After the scattering intensity was corrected for the background and the absorption, the integrated intensity and Guinier
radius were obtained.  The thin foil for the TEM observation was prepared by
chemical etching.  The etching solution was an equimolar mixture of HNO3, HF
and H2SO4 cooled at 0 to -10°C.  The TEM observation was performed by using
JEOL-120 accelerated to 120 kV.

## Experimental Results and Discussion

Fig.1 shows the integrated intensity for various thin foil specimens aged
at 653 K.  The remarkable increase occured above 1000 min ageing, corresponding
to the onset of precipitation, because the integrated intensity is in general
proportional to the total volume of precipitates.  It was also found out that
the increasing rate of cold working accelerates     precipitation.  The change
of Guinier radius as a function of ageing time is shown in Fig. 2.  The radius
was greatest in the specimens without cold working and became smallest in the
specimens with the 9%  cold-rolling over the whole region up to 30000min.
Mendiratta et al [6] reported the Guinier radius of about 10 nm in the Ti-41 to
45 wt%Nb alloys aged at 648 to673 K for 6000 min.  Fig. 3 shows the change of
both integrated intensity and Guinier radius as a function of rate of cold-
working for the specimens aged at 653 K for 10000 min.  It should be noted that
the precipitation enhanced remarkably by a 9 % cold rolling and the average
size of precipitates became finer.

The relative integrated intensity is related to the volume fraction, V
defined by the equation,

$$Q \propto V = \frac{4}{3}\pi \langle R^3 \rangle N. \tag{1}$$

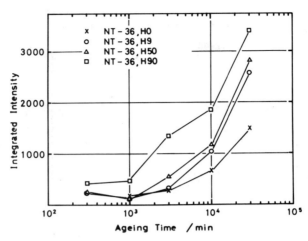

Fig.1   Ageing-time dependence of relative integrated intensity
in Ti-36at%Nb alloys aged at 653 K.  [5]

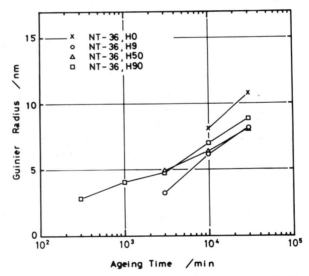

Fig.2   Ageing-time dependence of Guinier radius in Ti-36at%Nb
alloys aged at 653 K.  [5]

where $\langle R^3 \rangle$ means the average of $R^3$ with respect to the size distribution and
N is the number of particles per unit volume.  When the particles are arranged
in a close-packed manner, the interparticle distance L might be written in
terms of the Guinier radius as follows,

$$L = (\sqrt{2} / N)^{1/3} \propto R_G / Q^{1/3} \qquad (2)$$

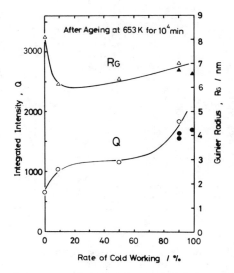

Fig.3   Changes of integrated intensity and Guinier radius as a function
of rate of cold working, where O and Δ are of foil specimens,
and ● and ▲ are of wires, respectively.

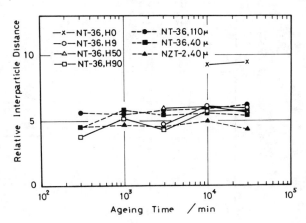

Fig.4   Relative interparticle distance among precipitates as a function
of ageing time. [5]

Fig.4 shows the change of this quantity as a function of ageing time.   The
interparticle distance was found out to decrease by a factor two by the cold-
working.   At 30000 min ageing, the distance was nearly same in all specimens
with the different rate of cold-working.   Therefore the increase of integrated
intensity with increasing cold-working was related to the growth of each particle during ageing.

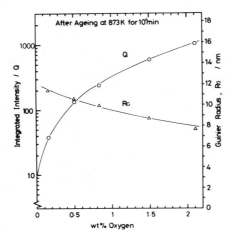

Fig.5    Changes of integrated intensity and Guinier radius
as a function of oxygen content.

When the alloy was aged under oxygen atomosphere, the precipitation of $\alpha$ phase was accelerated. Fig.5 shows the effect of oxygen addition. On increasing oxygen content, the integrated intensity increased rapidly, but the Guinier radius became smaller.

The TEM observation was performed in order to determine the structure of precipitates generating the SAXS intensity. The same specimens were examined after the SAXS measurement. The observation was very difficult for the 90% cold-rolled thin foil and all the wire specimens because of the high density of dislocation. Fig. 6(A) shows a set of electron micrographs for the alloy aged at 653 K for 10000 min after the 50% cold-rolling. From the analysis of selected area diffraction pattern, the incident beam was perpendicular to (001) plane of the $\beta$ matrix and the other extraspots were assigned to $\alpha$ phase as shown in the key diagram of Fig.6(C). In the bright field, there exists the homogeneously distributed dislocation structure, but it was difficult to see any precipitates directly. Dark field indicates plate-like $\alpha$ precipitates. The crystallographical relation between $\alpha$ precipitates and $\beta$ matrix obeys exactly the Burgers relationships [7]. Here it should be noted that both zone axes of $(001)_\beta$ and $\{11\bar{2}0\}_\alpha$ incline mutually by about 5 degree as shown in Fig.6(C). Accordingly the diffraction pattern of precipitates does not appear when the incident beam is strictly parallel to the direction of $[001]_\beta$. Fig. 6(B) shows the large plate-like $\alpha$ precipitates in the alloy containing 1 wt%oxygen aged at 1173K for 10 min. Their structure and crystallographical relationship were completely same to those for the specimens without oxygen aged at lower temperature.

From the comparison of both experiments of TEM and SAXS measurements, the precipitation behavior of $\alpha$ phase was discussed as follows. These precipitates seemed to nucleate on the homogeneously distributed dislocation networks, because the integrated intensity increases with cold-working. As precipitates appeared thread-like at first along dislocation lines and became plate-like after long period ageing, Guinier radius, which is a sort of radius of gyration, was interpreted to bring us major information on width and/or thickness of precipitates. The number of precipitates did not increase apparently during ageing.

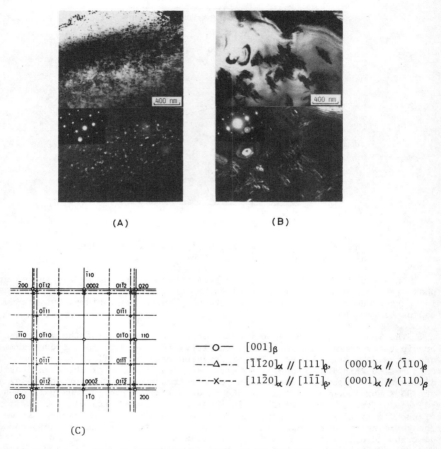

Fig.6   Transmission electron micrographs in Ti-36at%Nb alloys. (A); after
ageing at 653 K for 10,000 min after cold-rolling, (B); after ageing
under atomosphere of 1 wt% oxygen at 1173 K for 10 min, (C); schematic
illustration of superposed diffraction patterns of β matrix and two
variants of α precipitates for $\mathbb{B} \sim [001]_\beta$. It should be noted that
all kinds of spots do not appear simultaneously.

Brammer and Rhodes [8] mentioned that α phase does not precipitates so easy
in the non-deformed Ti-Nb alloy and nucleates only along the grain boundary in
the aged alloy at 1073 K. As shown in the present report, the precipitation
of α phase occurred at 653 K when the alloy was cold-rolled. Accordingly the
precipitation was concluded to be accelerated by the cold-work, because the
introduced dislocation offers nucleation sites.

The ω phase was also observed partly in the aged alloy at 653 K for 30000
min, but did not appear up to 10000min ageing. The fine ω precipitates were
ellipsoid as reported previously[8] and their crystal orientation was confirmed
to obey the Silcock's relationship[9]. The axial ratio of ellipsoid was about 3

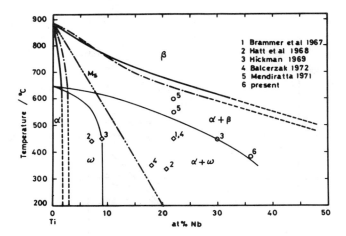

Fig.7  A proposed phase relation in Ti-Nb binary system, where references
are as follows; Brammer et al, 1967[8], Hatt et al, 1968[10], Hickman, 1969[11], Balcerzak, 1972[12] and Meandiratta, 1971[6].

and the diameter of short axis distributed in the region of 5 to 20 nm.  Both
$\alpha$ and $\omega$ phases sometimes coexisted in the same specimen, and the $\omega$ phase was
recognized to be the minor precipitation at that ageing condition.  According
to Hickmann[11], the $\omega$ pahse precipitated in the Ti richer region, but was not
observed in the Ti- 30 to 35 at%Nb alloys aged above 473 K.  He obtained also
the similar result in the Ti-Mo alloy system, where the aged $\omega$ phase did not
precipitates in the Ti-14 at%Mo alloy.  On the other hand, Gysler, Bunk and
Gerold[13] observed the precipitation of $\omega$ phase in the Ti-Mo alloys with the
same composition which was aged at the same temperature for long period.  The
present result for the Ti-Nb alloy did not agree with Hickman[11], but suggests
that the precipitation of $\omega$ phase is retarded with increasing the solute element.
Here a temporary phase relation for various phases is proposed as shown in Fig.7
, where the $\alpha+\omega$ phase field was assumed to have the similar shape as that in a
Ti-Mo system[13].

It has been recognized that the superconducting property is improved when
the wire of Ti-36at%Nb alloy is heavily drawn and aged at 653 K for long period
[ 4, 14].  Its improvement seemed to correlate to the existence of fine precipitates.  Fig.8 shows the change of integrated intensity during ageing for wire
specimens.  The ageing time dependence for two kinds of specimens with diameters
of 60 and 110 $\mu$m was similar to that for the thin foil specimens.  On the other
hand, the intensity for the specimen with diameter of 40 $\mu$m became large at
short time ageing and tended to saturate beyond 10000 min ageing.  Its tendency
was clearly appeared in the change of Guinier radius during ageing as shown in
Fig.9.  The Guinier radius for the 40 $\mu$m wire specimen increased more rapidly
in the early state of ageing than for the other two kinds of wire specimens,
which had almost the same size each other during ageing.  The relative interparticle distance is shown in Fig.4.  It was found out that its dimension for
all kinds of wire specimens is almost constant and coincides with the result
for thin foil specimens.  Accordingly the ageing behavior is independent on the
external shape of specimen, but depends only on the rate of cold working.  The
quaternary alloy had the largest integrated intensity and Guinier radius.  Its

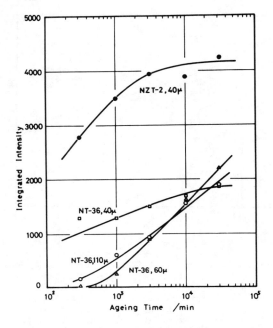

Fig.8  Ageing time dependence of relative integrated intensity
in wire specimens aged at 653 K. [5]

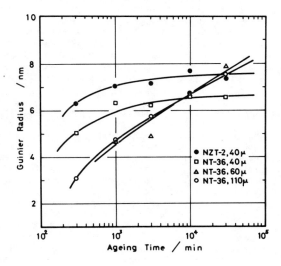

Fig.9  Ageing time dependnece of Guinier radius in wire specimens
aged at 653 K. [5]

interparticle distance was almost same as those of binary alloys.  Its commercial quaternary alloy was reported to have a high critical current density in
the medium magnetic field, for instance, 2.4 x10$^5$ A/cm$^2$ at 5 T, in comparison
with the binary Ti-Nb alloys[15].

## Summary

When the thin foil specimens of Ti-36at%Nb alloy were aged at 653 K, the
SAXS intensity was observed.  The integrated intensity increased with increasing
ageing time as well as with increasing rate of cold-rolling before ageing.  The
Guinier radius increased with ageing time, but decreased with increasing rate
of cold-rolling.  The relative interparticle distance became constant after
the short time ageing for all specimens with various rates of cold-rolling.
From the TEM observation, those precipitates generating SAXS intensity were
identified to be $\alpha$ phase and nucleated heterogeneously on dislocations.  The
crystallographical relationship between the $\alpha$ precipitates and the $\beta$ matrix
obeyed the relation proposed by Burgers.  After the long period ageing, the
precipitation of $\omega$ phase was also observed in a small amount.  By adding oxygen
and ageing, the precipitation of $\alpha$ phase was confirmed to be enhanced.

The major precipitates in the wire speciemns were $\alpha$ phase and the tendency
on the ageing time and the rate of cold working was completely the same as in
the case of thin foils.  In the heavily cold drawn quaternary alloy  applied to
the practical use, the largest integrated intensity was observed.  Accordingly
a large number of fine precipitates of $\alpha$ phase was concluded to improve the
superconducting property.

## Acknowledgement

The authors wish to express their gratitude to Dr. A. Terauchi for his
valuable discussion and also to Mr. H. Tsunekawa  for his  help.

## References

1.  D.F.Neal, A.C.Barber, A.Woolcock and J.A.F.Gidley: Acta Met., 19(1971),143.
2.  J.B.Vetrano and R.W.Boom: J. Appl. Phys., 36(1965),1179.
3.  D.Kramer and C.G.Rhodes: Trans. Met. Soc., AIME, 236(1967),1612.
4.  S.Katoh, H.Tsubakihara, T.Okada and T.Suita: J. Nucl. Sci. Technol., 72
    (1975),193.
5.  K.Osamura, E.Matsubara, T.Miyatani, Y.Murakami, T.Horiuchi and Y.Monju:
    submitted to Phil. Mag.
6.  M.G.Mendiratta, G.Lütjering and S.Weissman: Met. Trans., 2(1971),2599.
7.  W.G.Burgers: Physica, 1(1934),561.
8.  W.G.Brammer, Jr and C.G.Rhodes: Phil. Mag., 16(1967),477.
9.  J.M.Silcock: Acta Met., 6(1958),481.
10. B.H.Hatt and J.A.Roberts: Acta Met., 8(1960),575.
11. B.S.Hickman: Trans. TMS-AIME, 245(1969),1329.
12. A.T.Balcerzak and S.L.Sass: Met. Trans., 3(1972),1601.
13. A.Gysler, W.Bunk and V.Gerold: Z. Metallkde., 65(1974), 411.
14. T.Horiuchi and T.Takashima: Kobe Steel Eng. Reports, 24(1974),14.
15. T.Horiuchi, Y. Monju and I.Tatara: Kobe Steel Eng. Reports, 23(1973),51.

# Al$_3$Ti PRECIPITATION IN ALUMINIUM-TITANIUM ALLOYS

H.A.F. El Halfawy
Metallurgy Division, Atomic Energy Establishment, Cairo, Egypt
and
E.S.K. Menon, M. Sundararaman and P. Mukhopadhyay
Metallurgy Division, Bhabha Atomic Research Centre, Bombay, India

## Introduction

Small additions of transition metals like titanium and zirconium to aluminium base alloys can bring about substantial grain refinement and titanium is very efficient in this respect [1, 2] . In binary aluminium-titanium or aluminium-zirconium alloys, the grain refinement is associated with the presence of the intermetallic phases Al$_3$Ti and Al$_3$Zr respectively. The accepted phase diagram [3] for the aluminium-titanium system shows a peritectic reaction at 665°C, in which Al$_3$Ti and a liquid solution of aluminium containing 0.14 wt% titanium react to form a solid solution based on aluminium and containing 1.15 wt% titanium. The mechanism of grain refinement involves heterogeneous nucleation of aluminium grains on properitectic particles of the Al$_3$Ti [4] . The equilibrium solid solubility of titanium in aluminium is strongly temperature dependent and falls rapidly with decreasing temperature - from 1.15 wt% at 665°C to only about 0.24 wt% at 510°C [5] .

For the peritectic reaction to occur at the equilibrium peritectic temperature, the cooling rate has to be very slow. Under equilibrium conditions, the free energy associated with a hyperperitectic liquid would be lowered by the formation of Al$_3$Ti and a simultaneous change in the composition of the liquid. However, the formation of Al$_3$Ti needs major compositional fluctuations. Rapid cooling tends to reduce the possibility of the occurrence of such fluctuations and thus inhibits Al$_3$Ti formation. Even if nuclei of this phase form, their diffusion controlled growth would be severely restricted owing to the insufficient time available during rapid cooling. If the rate of cooling is not slow enough to allow the composition of the liquid to follow the equilibrium liquidus closely, the free energy of the system can still be lowered by the direct formation of a supersaturated Al (Ti) solid solution. As Kerr et al [6] have pointed out, though the driving force in such a situation would be lower than that associated with the liquid→Al$_3$Ti reaction, the reduced requirement of compositional changes would be favourable kinetically. When the cooling rate is sufficiently rapid, it may be possible to suppress Al$_3$Ti formation completely. At intermediate rates of cooling, the possibility of the emergence of metastable phases Al Ti would also exist [6, 7] . The critical cooling rate required to inhibit the precipitation of Al$_3$Ti or of any other metastable phase would increase with increasing titanium content of the alloy.

The Al$_3$Ti phase has the body centred tetragonal DO$_{22}$ structure, with lattice parameters : a = 3.85 A° and C = 8.596A° [8] . As shown in Fig.1(a), the unit cell contains six aluminium and two titanium atoms and is composed of two distorted simple cubic L1$_2$ subcells stacked one above the other along the [001] direction, with a step shift at the interface between them. Ideally, the DO$_{22}$ structure(associated with the stoichiometry A$_3$B) is built up of close packed atomic layers of composition A$_3$B which lie parallel to the (112) planes (Fig.1(b)). There are no B-B nearest neighbours. The ratio of the edges of the

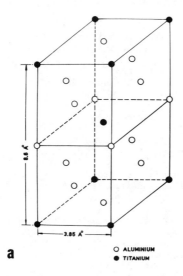

**a**

○ ALUMINIUM
● TITANIUM

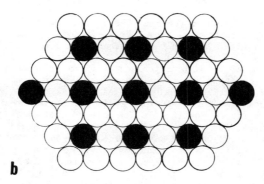

**b**

Fig.1 (a) : Unit cell of body centred tetragonal
            $DO_{22}$ structure

      (b) : Disposition of the two different atomic species
            on the close packed (112) plane   in an ideal
            $DO_{22}$ structure.

rectangles formed by the B atoms in the (112) plane has the value $2/\sqrt{3}$ or 1.155 in this ideal case where $c/a = 2.0$. In the case of Al₃Ti, for which $c/a = 2.233$, this ratio is about 1.030 [9].

The present work was undertaken with a view to examining the distribution, the morphology and the substructure of Al₃Ti precipitates in two rapidly cooled (chill cast) aluminium-titanium alloys, one dilute (Al-1wt%Ti) and the other concentrated (Al-14wt%Ti).

## Experimental Procedure

The alloys were prepared by melting appropriate amounts of 99.99% pure aluminium and iodide pure titanium in a non-consumable arc furnace, on a water cooled copper hearth, under a protective argon atmosphere. For each alloy, the melting operation was carried out several times in order to ensure homogeneity. The weight changes in melting were found to be negligible. Thin slices were cut from the 'fingers' obtained, given the desired heat treatments (if any) and subsequently subjected to metallographic examination. Samples for heat treatment were encapsulated in silica capsules under a helium atmosphere. Thin foils for transmission electron microscopy were obtained by mechanically grinding the slices and finally electropolishing them in a 20% perchloric acid - 80% ethanol solution, using the 'window' technique.

## Results and Discussion

X-ray diffraction from both the alloys (in the as-cast condition) showed only two sets of peaks - one belonging to the Al(Ti) solid solution and the other to the Al₃Ti phase. This implied that the cooling rate involved in the alloy making process was insufficient to suppress the formation of this intermetallic phase. As expected, the relative intensities of the Al₃Ti peaks were higher in the concentrated alloy than in the dilute one. The absence of reflections from any other phase suggested that no metastable phase was present in any appreciable amount in either of the alloys. This observation was consistent with that made by Maxwell and Hellawell [10] who found that in chill cast aluminium-titanium alloys (cooling rates ∼800°C/sec.) containing upto 5wt% titanium, no intermetallic phase other than Al₃Ti could be detected. In the present case, the formation of the Al₃Ti phase during the rapid solidification of the alloy could occur only to a limited extent because of the extremely short time available for precipitation to occur. This was verified by ageing the alloys at 450°C for different periods of time. It was found that with the progress of ageing, the intensity, $I_P$, of the strongest reflection from the Al₃Ti phase ( {112} type) gradually increased in comparison with the intensity, $I_M$, of the strongest reflection from aluminium ( {111} type) in both the alloys (Table 1), indicating a steady increase in the volume fraction of the former phase with ageing. This trend was also detected by optical microscopy. The lattice parameters of the two phases, obtained from samples of the Al-14wt%Ti alloy, aged for 240 hours at 450°C, were found to be : a = 4.051A° for aluminium; a = 3.845A° and c = 8.582A° for Al₃Ti.

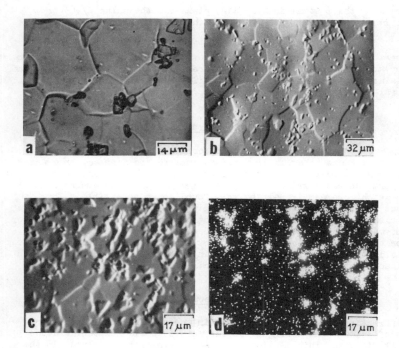

Fig.2 : Al-1 wt%Ti alloy, aged at 400°C for 24 hours

   (a) Optical micrograph showing the presence of the
   precipitate phase at the grain boundaries as well as
   within the grain interior.

   (b) and (c) Back scattered electron image showing a
   non-distribution of the precipitate phase in an equiaxed
   matrix.

   (d) X-ray image corresponding to figure 2(c) obtained
   with the titanium K$_\alpha$   radiation.

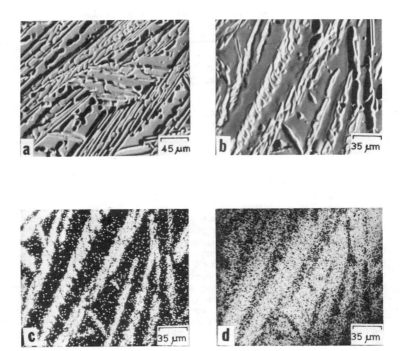

Fig.3 : Al-14 wt% Ti alloy; Aged at 450°C for 240 hours :
(a) and (b) Scanning electron micrograph showing
large plate shaped precipitates in the Al matrix.

(c) X-ray imaged obtained with TiK$_\alpha$   corresponding
to region in (b)

(d) X-ray image obtained with AlK$_\alpha$   corresponding to
region in (b).

Table 1.    Relative Diffracted Intensities

| Alloy | Treatment | $I_P / I_M$ |
|---|---|---|
| Al-1wt%Ti | Aged at 450°C for 1 hour | 0.16 |
| | Aged at 450°C for 24 hours | 0.50 |
| Al-14wt%Ti | As cast | 0.30 |
| | Aged at 450°C for 20 hours | 0.63 |
| | Aged at 450°C for 240 hours | 1.49 |

Optical and scanning electron microscopy showed that in the Al-1wt%Ti alloy, both in the as-cast and in the aged conditions, the majority of the $Al_3Ti$ particles were either globular or in the form of irregularly faceted platelets, with a tendency to be located at or near the aluminium grain boundaries. The distribution of the precipitates appeared to be quite inhomogeneous. X-ray imaging with the titanium $K_\alpha$ radiation demonstrated the higher titanium content of the precipitates with respect to that of the surrounding regions. These features are illustrated in Fig.2. In the cast Al-14wt%Ti alloy, the major fraction of the $Al_3Ti$ phase appeared to be distributed in the form of plate shaped precipitates (Fig.3(a)). The morphology and the distribution of these precipitates did not undergo any drastic change on prolonged ageing. The ageing treatment, however, helped in relieving the titanium supersaturation of the aluminium grains and in bringing about a redistribution of the two atomic species between the two constituent phases. As a result, in scanning electron microscopy examination, the X-ray images followed the contours of the back scattered electron images very closely (Fig.3(b)-(d)).

Transmission electron microscopy of aged specimens of both the alloys revealed several interesting microstructural features. The precipitate-matrix interfaces contained arrays of interfacial dislocations (Fig.4). Profuse faulting was observed in the precipitates, more so in the concentrated alloy (Fig.5). A limited amount of single surface trace analysis indicated that the fault planes were the close packed (112) type planes of $Al_3Ti$. The presence of faults suggested that the stacking fault energy of this phase is low. Regions within the $Al_3Ti$ particles, adjacent to the precipitate - matrix interfaces, were often noticed to be heavily dislocated (Fig.6). The aluminium grains, even after prolonged ageing, were found to contain a large population of dislocations which formed dense tangles in some regions and cells in some others (Fig.7). The observed high density of dislocations in samples aged for as long as 240 hours at 450°C was somewhat unusual. Selected area diffraction patterns obtained from the two phases indicated that there existed a wide range of orientation relationships between these phases. This implied that the two lattices were not uniquely oriented with respect to each other. The two orientation relationships encountered most frequently were :

$$(111)_M \parallel (112)_P \; ; \; [\bar{1}10]_M \parallel [\bar{1}10]_P$$

$$\text{and } (011)_M \parallel (\bar{1}15)_P \; ; \; [3\bar{1}1]_M \parallel [110]_P$$

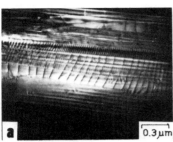
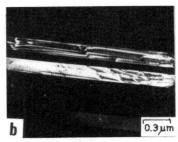

Fig.4 : Dark field image taken with a precipitate reflection showing a regular array of interfacial dislocations decorating the Al/Al₃Ti interface: (a)Al-14% Tialloy aged at 400°C for 24 hours (b) Al-1% Ti  alloy, aged at 450°C for 240 hours.

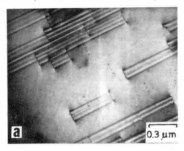

Fig.5 : Fringe contrast associated with stacking faults within the Al₃Ti phase in (a) Al-1%Ti alloy, aged at 400°C for 24 hours and (b) Al-14%Ti alloy, aged at 450°C for 240 hrs.

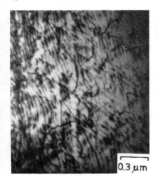
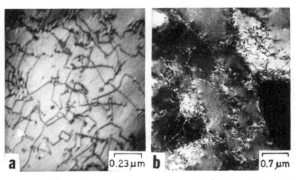

Fig.6: Al-14%Ti alloy:
High density of disloca-
tions within the precipi-
tate phase near the matrix-
precipitate interface.

Fig.7: Al-14%Ti  alloy, aged at 450°C for 240 hrs:
Substructural features associated with the aluminium
matrix showing (a) dislocation tangles and (b) the
cellular arrangement of dislocations.

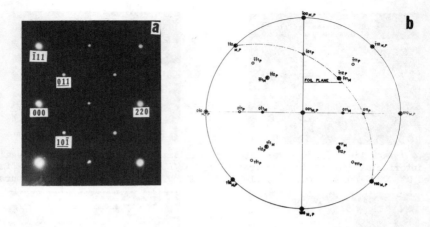

Fig.8(a) : Superimposed reciprocal lattice sections of the matrix and the
precipitate phases : $[11\bar{2}]_{Al}$ and $[1\bar{1}1]_{Al_3Ti}$ respectively.
(b) Stereogram illustrating the orientation relationships
between the two phases indicating the parallelity of all the
three orthogonal axes.

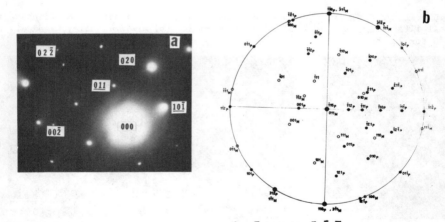

Fig.9(a) : SAD pattern, zone normals : $[100]_{Al}$ and $[1\bar{1}1]_{Al_3Ti}$
(b) Stereogram indicating the orientation relationship
between the two phases.

and equivalent variants of these, the subscripts M and P referring respectively to the Al(Ti) and the Al₂Ti phases. These relationships are illustrated in Fig.8 and Fig.9.  It is pertinent to mention here that multiple orientation relationships between these two phases have been observed by other workers too [11, 12] .

## References

1. A. Cibula : J. Inst. Metals, 76(1949), 321.
2. G.W. Delamore and R.W. Smith : Met. Trans., 2 (1971), 1733.
3. R.P. Elliott : Constitution of Binary Alloys, First Supplement (1965), McGraw Hill, New York.
4. D.H.St. John and L.M. Hogan : J. Australian Inst. Metals, 22 (1977), 160.
5. W.L. Fink : Physical Metallurgy of Aluminium Alloys, ASM (1949), Cleveland, Ohio.
6. H.W. Kerr, J. Cisse and G.F. Bolling : Acta Met., 22 (1974), 677.
7. J. Cisse, H.W. Kerr and G.F. Bolling : Met. Trans., 5 (1974), 633.
8. W.B. Pearson : A Handbook of Lattice Spacings and Structures of Metals and Alloys, 2 (1967), 133. Pergamon Press, Oxford.
9. W.B. Pearson : The Crystal Chemistry and Physics of Metals and Alloys, (1972). Wiley Interscience, New York.
10. I. Maxwell and A. Hellawell : Acta Met., 23 (1975), 895.
11. J.A. Marcantonio and L.F. Mondolfo : J. Inst. Metals, 98 (1970), 23.
12. K. Asboll and N. Ryum : J. Inst. Metals, 101 (1973), 212.

# PHASE TRANSFORMATIONS KINETICS AND MECHANICAL CHARACTERISTICS OF TWO-PHASE TITANIUM ALLOYS

M. A. D'yakova, Z. F. Zvereva, I. I. Kaganovich,
E. A. L'vova, E. S. Makhnyov, S. S. Meshchaninova

All-Union Institute of Light Alloys
USSR

The study of phase and structural transformations in titanium alloys under heating followed by quenching at various rates is of interest for a competent approach to the choice of thermal treatment conditions for titanium alloys with the purpose of providing the necessary level and stability of mechanical and maintenance characteristics. With that aim in veiw knowledge of decomposition kinetics of β-solid solution is necessary [1-5].

This paper deals with kinetics of phase transformations in industrial titanium alloys of the following systems: Ti-Al-V, Ti-Al-Mo-Zr-Si, Ti-Al-Mo-Cr-Fe-Si, quenched from temperatures, corresponding to $\alpha + \beta$ area, and for these alloys diagrams of isothermal decomposition of metastable β-solid solution were plotted.

These alloys samples 20x3 mm of size were quenched from the heating temperature $(T_{pt} - 70^{\circ}C)$ in salt baths at temperatures of 100-700$^{\circ}$C, held in them for the period from 10 seconds up to 2 hours, and then quenched in water. For alloys Ti-Al-V and Ti-Al-Mo-Cr-Fe-Si the heating temperature was 900$^{\circ}$C, and for Ti-Al-Mo-Zr-Si it was 920$^{\circ}$C.

An X-ray diffraction study was carried out by means of a difractometer DRON-1.5 in CuK$_{\alpha}$ radiation.

On the basis of the X-ray structural study and the total
data about the change of relative intensity of $\alpha$, $\beta$, $\alpha''$ phases
diffractional lines, depending on temperature and time of holding,
diagrams of isothermal decomposition of metastable $\beta$-solid solution of the investigated alloys were plotted (Fig. 1).

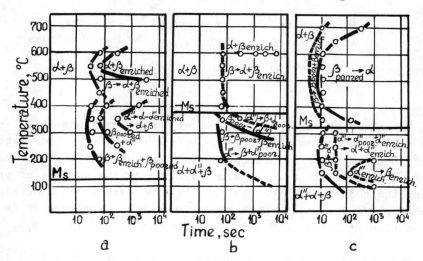

Fig. 1 Diagrams of isothermal decomposition of $\beta$-phase,
quenched from temperature $T_{pt}$-70°C for Ti-Al-Mo-Si
(c), Ti-Al-V (b) and Ti-Al-Mo-Cr-Fe-Si (a) alloys.

In Ti-Al-Mo-Cr-Fe-Si alloy quenched from 900°C the tempera-
ture of the beginning of martensite transformation $\beta \rightarrow \alpha''$ is in
the range of 150-100° (see Fig. 1a). Quenching and holding at
the martensite range of 100° results in formation of isothermal
martensite. In that case with the increase of holding period up
to a certain length the increase of martensite amount and the
decrease of $\beta$-phase amount take place without any change of their
crystal lattice parameters, i. e. the transformation proceeds
without any change of phase composition. Further increase of the
holding period at 100°C results in stabilization of $\beta$-phase martensite $\beta \rightarrow \alpha''$-transformation stops (Fig. 2).

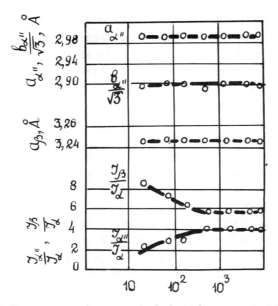

Fig. 2 The change of crystal lattices parameters and of
relative intensity of difraction lines of phases
$\alpha$, $\beta$, and $\alpha''$ during the holding at 100°C for
Ti–Al–Mo–Cr–Fe–Si alloy, quenched from $T_{pt}$–70°C.

Quenching in temperature range of 200–400° makes for the
decomposition of metastable β-phase according to the intermediate
mechanism. The process starts with the formation of enriched and
poor areas of β-solid solution; in the poor areas $\alpha''$-phase occurs through shifting. (X-ray examination showed that at an increase of $\alpha''$-phase amount its composition does not change). Such
a mechanism of β-solid solution decomposition was observed while
tempering at temperature range of 300–400° Ti–Al–Mo–Zr–Si and
Ti–Al–Mo–Cr–Fe–Si alloys hardened from 860° [6] . With the increase of holding period at temperatures of 200–400° the formed
$\alpha''$-phase decomposes according to the following reaction:

$$\alpha'' \to \alpha + \alpha''_{\text{enriched}} \to \alpha + \beta$$

In this paper the possibility of $\alpha''$-phase formation both
at isothermal holding and at water quenching (athermal formation

of $\alpha''$-martensite of hot decomposed $\beta$-phase) was considered. As the criterion of separation of those types of $\alpha''$ phases the data of changing the crystal lattice periods and the relative intensity of $\beta$ and $\alpha''$-phases difractional lines were taken.

At the temperatures of quenching of $450^\circ$ and higher the metastable $\beta$-phase in the Ti-Al-Mo-Cr-Fe-Si alloy decomposes according to the reaction: $\beta \rightarrow \beta_{poor} + \beta_{enriched} \rightarrow \alpha + \beta_{enriched}$. If the decomposition is not fully completed, then under cooling martensite transformation occurs $\beta_{poor} \rightarrow \alpha''$.

In the Ti-Al-V alloy quenching from $900^\circ$, the temperature of the beginning of martensite transformation is in the range between $300^\circ$ and $350^\circ$ and in Ti-Al-Mo-Zr-Si alloy, quenched from $920^\circ$, $M_s$ is somewhat higher than $350^\circ$ (see Fig. 1b, c). At quenching Ti-Al-V and Ti-Al-Mo-Zr-Si alloys down to temperatures lower than $M_s$ $\alpha$ and $\beta$ phases are fixed and martensite $\alpha''$ is formed. $\beta$ and $\alpha''$ phases are metastable ones, and they decompose during the isothermal holdings. In the Ti-Al-Mo-Zr-Si the decomposition proceeds according to the following reactions:

$$\beta \rightarrow \beta_{poor} + \beta_{enrich.} \rightarrow \alpha'' + \beta_{enrich.} \rightarrow \alpha + \beta_{enrich.} \quad (1)$$

$$\alpha'' \rightarrow \alpha''_{enrich.} + \alpha''_{poor} \rightarrow \beta + \alpha''_{poor} \rightarrow \alpha + \beta \quad (2)$$

The higher the quenching temperature in the martensite range, the more intensive the composition proceeds.

The decomposition of martensite $\alpha''$ in the Ti-Al-V alloy at temperatures lower than $M_s$ proceeds as follows: $\alpha'' + \alpha''_{poor} + \alpha''_{enriched} \rightarrow \alpha + \alpha''_{enriched} \rightarrow \alpha + \beta_{enriched}$, i. e. unlike Ti-Al--Mo-Zr-Si alloy first of all $\alpha$-phase is released, then $\beta$-phase nuclei are formed in the zones of $\alpha''$ martensite enriched with alloying elements.

Short holdings at temperatures, lower than $M_s$, of Ti-Al-V alloy result in $\beta$-phase stabilization, as a result of which athermal transformation $\beta \rightarrow \alpha''$ becomes slower. During the process of isothermal treatment at the temperature of $100-250^\circ C$ transformation of $\beta \rightarrow \alpha''$ is most probable of isothermal nature, at higher temperatures ($300^\circ$C) $\alpha''$-phase formation develops according to the type of an intermediate transformation (reaction 1) [7].

In Ti-Al-V and Ti-Al-Mo-Zr-Si alloys quenched in the temperature range higher than $M_s$ the decomposition of metastable $\beta$-solid solution starts with redistribution of alloying elements. In zones poor of alloying elements particles of $\alpha$-phase are formed, the enriched $\beta$-solid solution is fixed at cooling down to room temperature, and in the parts of poor $\beta$-phase, where the $\alpha$-phase formation has not yet finished, at cooling athermal martensite transformation $\beta_{poor} \rightarrow \alpha''$ occurs like in Ti-Al-Mo-Cr-Fe-Si alloy. In Ti-Al-V alloy at all the temperatures of quenching higher than $M_s$ and a definite time of holding one can observe a thermal stabilization of poor zones of $\beta$-phases [7]. In Ti-Al--Mo-Zr-Si alloy thermal stabilization of the poor $\beta$-phase occurs at temperature range of 400-500$^\circ$C.

The change of phase composition of the investigated alloys after isothermal treatment essentially influences mechanical properties. So the decomposition of a quenched from 900$^\circ$C $\beta$-solid solution of Ti-Al-Mo-Cr-Fe-Si alloy at temperatures of 200-400$^\circ$C, according to the type of intermediate transformation, provides an increase of strength characteristics and a decrease of plasticity. A similar change of properties is also characteristic of Ti-Al-V and Ti-Al-Mo-Zr-Si alloys following an isothermal treatment at temperatures lower than $M_s$. Besides for Ti-Al-V and Ti--Al-Mo-Zr-Si alloys after such a treatment an increased difference is observed between the ultimate strength and the yield point (about 15-20 kg/mm$^2$) which is due to the presence of a mechanical nonstable $\beta$-phase in the alloys [8].

The full decomposition of $\beta$-solid solution of the investigated alloys according to the first step at the temperatures of quenching of 450$^\circ$C and higher results in achieving high strength properties with a sufficient level of plastic characteristics (Fig. 3). The most stable mechanical properties result from quenching the investigated alloys down to 550$^\circ$C, where the maximum rate of $\beta$-solid solution phase decomposition takes place accompa-

nied with the formation of $\alpha + \beta$ structure. In case of quenching
the alloys down to temperatures lower or higher than 550°C a complicated change of properties is observed due to, firstly, not
full decomposition of poor β-solid solution with $\alpha$-phase formation at short holdings, secondly, to thermal stabilization of
β-phase, and thirdly, to formation of poor $\alpha''$ cooling martensite.

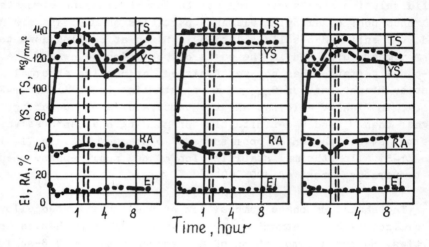

Fig. 3 The dependence of mechanical characteristics of the
       alloy Ti-Al-Mo-Cr-Fe-Si quenched from $T_{pt}$-70°C on
       the isothermal holding temperature: a) 500; b) 550;
       c) 600°C.

The results of the investigation also showed that the isothermal treatment of two-phase titanium alloys quenched from $\alpha + \beta$
zone down to temperatures corresponding to maximum rate of decomposition according to the first step of metastable β-phase provides the best combination of strength and plasticity characteristics, impact strength (KCU) and impact strength of samples
with a fatigue crack (KCT) of half-finished products of Ti-Al-
-Mo-Cr-Fe-Si depending on the temperature of isothermal holding.
At the temperature of isothermal holding of 550°C the above-mentioned characteristics are much higher than in the case of 650°C.
At testing rupture and fatigue strength the same dependence is
observed (Fig. 5).

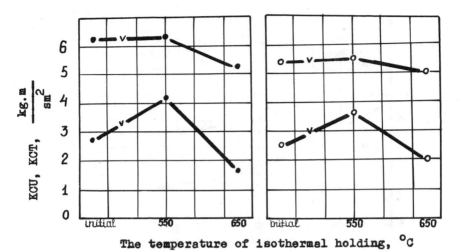

The temperature of isothermal holding, °C

Fig. 4 The effect of thermal treatment conditions on impact strength (KCU) and on impact strength with a
fatigue crack (KCT):
a) a rod 65 mm dia.; b) a stamping of round section.

* The temperature of the first annealing step 920°C,
60 min. holding, transfer into the furnace at
550°C or (650°C), holding in the furnace 2 hrs.,
air cooling.

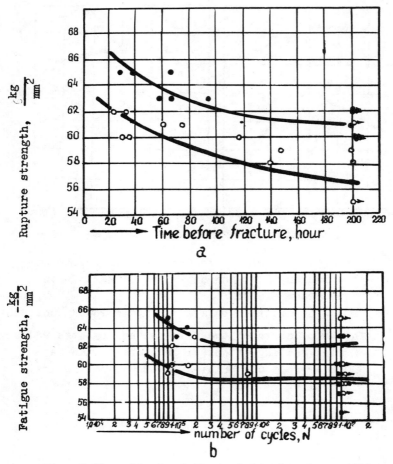

Fig. 5 The effect of thermal treatment conditions on rupture
strength (T)-(a), and on fatigue strength-(b) of the
alloy Ti-Al-Mo-Cr-Fe-Si:
    - - - annealing 920°C, holding 60 min., transfer
        into the furnace at 550°C for 2 hrs, air
        cooling;
    o-o-o - annealing 920°C, holding 60 min., transfer
        into the furnace at 650°C for 2 hrs, air
        cooling.

## References

1. V. S. Lyasotskaya. Investigation of Transformations in Titanium Alloys. Autoreferat diss. for bachelor's degree in technical sciences, M., 1966 (MATI).
2. I. S. Pol'kin. Thermal and Thermomechanical Treatment of the Alloy BT15. Autoref. diss. for bachelor's degree in technical sciences, M., 1966 (MISA).
3. Thomas Andersson. Scand. f. of Metallurgy, 1973, 2, N 5, 251–256.
4. Majda Majdic, Günter Ziegler. Z. Metallkunde, 1973, 64, N 11, 751–758.
5. F. L. Lokshin, V. S. Lyasotskaya, P. G. Kocknaev, L. S. Krasnoyartseva. Isvestiya Vusov. Tsvetnaya Metallurgia, 1976, N 5, 115–119.
6. M. A. D'yakova et al. Fizika metallov i metallovedeniye, 1976, v. 42, issue 2, 333–339.
7. M. A. D'yakova, T. G. Potyomkina, N. A. Krasilnikova. Fizika metallov i metallovedeniye, 1977, v. 44, issue 1, 142–145.
8. M. A. D'yakova et al. Izvestiya AN SSSR. Metalli, 1977, N 3, 142–147.

EFFECT OF DISLOCATION STRUCTURE UPON REGULARITIES
OF DECOMPOSITION AND CHARACTER OF FRACTURE OF
HIGH-ALLOY TITANIUM ALLOYS

I. S. Pol'kin, A. B. Notkin, O. S. Korobov,
N. M. Semenova, V. G. Kudryashov

All-Union Institute of Light Alloys, USSR

It has been previously shown [1-3], that hot deformation of high-alloy
titanium alloys results in formation of a variety of structures in β-solid
solution, such as: recrystallized, polygonized and mixed ones. Yhe polygonized structure parameters (dislocation density within the subgrains, their shape
and distribution, etc.) change depending upon the deformation conditions.
    The mechanical properties of alloys in a thermally hardened state are de-
termined either by parameters of hot deformation or temperature-time conditions of heat- treatment. Effect of the hot deformation parameters upon properties is caused by that the ageing on heat-treatment is strongly affected by
the state of initial structure of alloy (polygonized or recrystallized)[3].
    In connection with the results given above, it was the aim of the present
work to study the ageing processes in high-alloy titanium alloys possessing
various structures of β-solid solution, formed after hot deformation.

## Materials and Experimental Procedure

    The study has been carried out with the specimens of Ti-Al-Mo-V-Cr-Fe al-
loy. The specimens with dimensions of 5×10 mm were water quenched (their initial structure being recrystallized one as well as polygonized of various
kinds); this was followed by their simultaneous anneal in the nitre and lead
baths at 200-850°C for 1 min to 10 hrs of holding time. To obtain the X-ray
diffuse scattering patterns the grains of about 2-3 mm in dia. were grown in
the specimens using the thermocycling technique. The X-ray analysis was carried out in a monochromatic MoK-α radiation using the camera for monocrystals
analysis. The specimens destined for electron microscopy were prepared from
the bulk specimens, which also served for determination of mechanical properties and X-ray analysis. The foils were thinned in the solution as that given
in [4].
    The character of fracture was investigated using the electron scanning
microscope of the MINI-SEM type.

## Experimental Results

    When studying the ageing process using the light microscopy, the polygo-
nized structure was shown to promote the decomposition of β-solid solution.
In this case the particles of precipitating phases are revealed at elevated
temperatures and lesser holding times. So, for example, no precipitations of
α-phase sere revealed after heating at 850°C for 10 hrs in recrystallized specimens (Fig. 1a) while they are clearly seen in the specimens with polygonized
structure (Fig. 1b). These results are confirmed by the X-ray analysis. When
heating up to 750°C, the decomposition takes place in both recrystallized and
polygonized specimens. However, in the first case no α-phase is seen after 1
min holding, and 5 min of holding are required for the first particles to
appear, while in polygonized specimens decomposition of β-phase rapidly pro-

ceeds after only 1 min of holding (Fig. 1c, d).  The presence of the polygonized structure makes the decomposition phases much more disperse.  This is
clearly seen in Fig. 1d, where separate recrystallezed grains may be observed
in the polygonized matrix.  At lower ageing temperatures in the specimens
possessing both types of original structure the size of precipitates is very
small thus reducing the effectiveness of the light microscopy.

The electron microscopy and X-ray data indicate, that even in the quenched state a large number of diffuse maxima are observed on the selected area
diffraction and Bragg patterns.  The position of these maxima corresponds to
possible reflections from α (α") and ω-phases.  Fig. 2 presents the SAD (a)
and Bragg patterns (b) for quenched specimens with {100} -orientations.

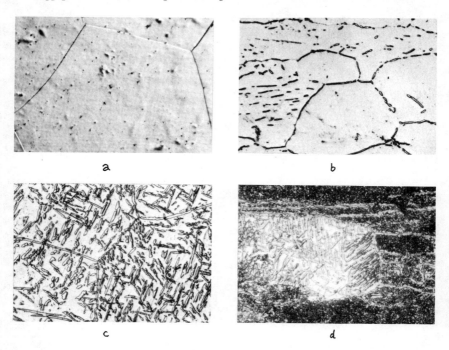

Fig. 1 Effect of the annealing conditions and initial structure of the
        -solid solution on the alloy microstructure (x500):
        a, c -recrystallized structure; b, d -polygonized structure;
        a, b -heating up to 850°C, 10 hrs, water cooling;
        c, d -heating up to 750°C, 1 hr, water cooling.

When heating up to 850°C as well as slow cooling from the β-phase area
down to this temperature no qualitative change of the diffuse scattering pattern is observed in the specimens with recrystallized structure.  The particles of α-phase are observed in the specimens with polygonized structure and
respective reflections are seen in the SAD patterns.  The position of the α-
phase particles is determined by the character of polygonized structure.
This is illustrated by Fig. 3, where the electron micrograph is given of the
specimen with polygonized structure aged at 850°C.

Using the dark-field technique [5] the net disorientation was measured
between the subgrains 1-2, 2-3, 3-5, 1-5, 4-5, 3-4.  The disorientation bet-

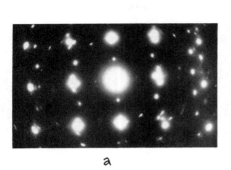

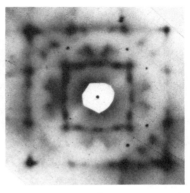

Fig. 2 SAD pattern (a) and Bragg diffraction pattern (b) of the quenched alloy, zone normal <100> β.

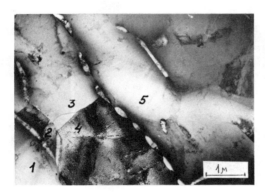

Fig. 3 Electron micrograph of the alloy with polygonized structure after heating to 850°C (1 hr, water cooling).

ween the subgrains 1-2, 3-5, 4-5 was equal to 3° and over (those were the subgrain boundaries where the particles of α-phase precipitated). For the other subgrain boundaries, including those revealed by inclining the goniometer the disorientation did not exceed one degree.

The analysis of this case and analogous ones, makes it possible to assure that subgrain boundaries, possessing higher disorientation are the sites where the particles of α-phase chiefly precipitate. When disorientation between the subgrain is small (~1°) and dislocation density is rather high the α-phase precipitates either at the subgrain boundaries or at the separate dislocations.

The average thickness of the α-phase particles after ageing at temperatures given above is about 0.5-1 μm. separate dislocations and their nets are seen inside the crystals (Fig. 4a).

When reducing the ageing temperature down to 750°C in a recrystallized structure after several minutes of holding separate precipitates of the α-particles are seen of 0.2-.03x1 μm in dimension (Fig. 4b).

With increasing ageing time the particles coarsen, reaching 15   m in length. This is accompanied by formation of the fine structure in the form of thin plates of various directions (Fig. 4b). It was earlier shown by the authors [6], that plates inside the α-crystals can be considered as microtwins along various planes {101}. When heating the polygonized specimens, an intensive decomposition of β-solid solution starts after only one minute of holding time. The β-phase precipitates in a l rge number of nucleating sites (subgrain boundaries and dislocations). T e α-crystals dimensions are one order less than those in the specimens with ecrystallized structure at respective ageing temperatures (0.2x 1.5 μ m); th  par icles of the α-phase possess rather simple structure (without twins).

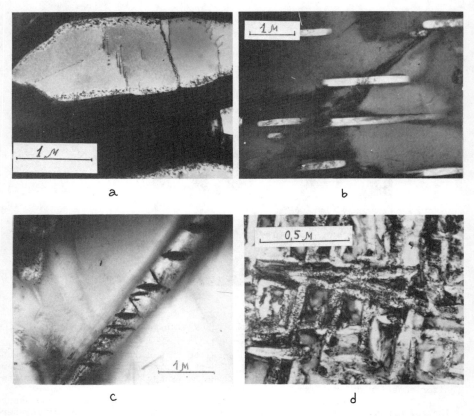

Fig. 4 Electron micrographs of α and α' crystals after various
       regimes of ageing:
       a- polygonized state, 850°C, 5 hrs, dark-field image
          from the α-phase reflection;
       b, c- recrystallezed state, 750°C, 1 and 60 minutes respectively;
       d- recrystallized state, 600°C, 4 hrs.

When the ageing temperature range over 350-600°C is concerned then initial structure exerts its effect only upon dispersion of precipitating phase. The α-phase is twinned along the {101} planes for all the temperature indicated.   Fig. 4 shows the structure of α(α") crystals after various ageing regimes.   The analysis of SAD patterns as well as that of dark-field image from various reflections of the α-phase, has revealed it to precipitate in the form of the plates, whose orientations obey the orientation relationship of Buergers.   According to the X-ray data this is the orthorhombic martensitic phase α" which precipitates from the β-solid solution at temperatures ranging from 350-600°C rather than hexagonal α(α") phase.   The extent to which the α" phase is <<rthmbic>> increases with ageing temperature.   Fig. 5 illustrates the data of X-ray analysis dealing with the distance between the (100) and (020) peaks of the α" phase, which appear instead of (100) reflection from hexagonal close-packed lattice when it is rhombically distorted.   The extent to what the lattice is rhombic is seen to decrease with increasing temperature according to linear law.   The X-ray data indicata the presence of α' phase at 600°C

The effect of initial dislocation structure on the ageing processes was not revealed at low temperatures (200-300°C).   The patterns of diffuse scattering for the specimens with both polygonized and recrystallized structure in the latent stage of precipitation correspond to that of quenched condition. No dark-field image was obtained of ω or α (α") phases up to the ageing temperatures of about 200°C.   At 250-300°C (for 30 and 15 min holding respectively) clear reflections from ω-phase appear in those sites of the electron diffraction patterns of the quenched specimens, where diffuse scattering was more dense.   Fig. 6 shows the sections of reciprocal lattice of the β-phase and dark-field image in the ω-phase reflection.   The ω-phase particles are seen to have the round shape and 50-100 A° in dimension.   There were no evidence obtained concerning the precipitation of the ω-phase on dislocations.

Precipitation of all the phases, detected by X-ray and transmission electron microscopy is accompanied by strengthening.   The strengthening effect depends upon the temperature-time conditions of ageing and dislocation structure. Fig. 7 shows the data dealing with the effect of temperature and time of ageing as well as dislocation structure of the β-soled solution on the alloy hardness.   The maximum strengthening is achieved on ageing at 500-600°C.   The magnitude and rate of strengthening for polygonized specimens are more than those for recrystallized ones.   The maximum difference in strengthening for those structures is observed over the temperature range of 550-700°C.   At low (300°C) and high (800-850°C) temperatures the hardness of alloy is slightly affected by precipitating phases.

It was shown by the study of the character of fracture of tensile specimens possessing different initial dislocation structure, that those having polygonized structure break mainly through the grain, while specimens with recrystallized structure do it in an intergranular manner.

The character of fracture, indicated above for both types of the structure is a common one on ageing over different temperature intervals.   At 250-350°C ageing temperatures a slightly pronounced relief may be observed in both cases; for those specimens possessing low plasticity some separate areas of the grains can be met, having practically no relief.   At ageing temperatures of 500-750°C the relief not pronounced and this is to a lesser degree for recrystallized specimens; however, increasing temperature and holding time result in appearance of more grain parts, having clear traces of local strain in each grain.

At 800-850°C the grain areas possessing weak traces of local strain may be seen in recrystallized specimens, while clearly shaped cells may be observed for the polygonized structure, indicating considerable deformation.

## Discussion

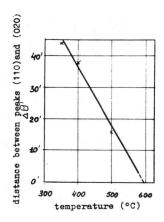

Fig. 5 Effect of heating temperature upon <<rhombicity>>
of the α" -phase precipitating on ageing.

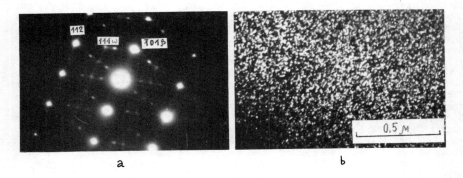

a                           b

Fig. 6 Precipitation of ω-phase after ageing at 300°C:
a - SAD pattern, [Ī13] zone normal;
b - dark field image from phase reflection.

The investigation of structural changes on ageing show the existence of
various temperature intervals in  the alloy under study within which various
schemes of α-solid solution decomposition take place and initial dislocation
structure influences the decomposition kinetics in a different manner.   The
study of X-ray and electron diffuse scattering patterns obtained for the latent stage of precipitation makes it possible to assume, that even in the
quenched condition the atom configuration exists in alloy, whose symmetry is
close to that of phases, precipitating on ageing.

This taken into account, one can suggest, that there is no individual
decomposition stage on ageing of this alloy and nucleation of the phase occurs
due to fluctuation.

At 200-300°C β-phase decomposes according to the scheme: β⟶ β+ω .  The
shape of the ω-particles is, however, rounded one and strengthening of the β-
solid solution by ω-particles is not high.  These results do not correspond to
those of [7], obtained for binary alloys Ti-Mo, Ti-V, where low strengthening

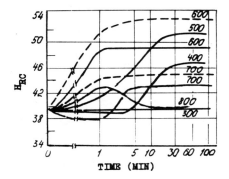

Fig. 7 Change of the alloy hardness after ageing at 300-800°C:
—— -specimens with initial recrystallized structure;
--- -specimens with initial polygonized

of alloy was indicated for cubic ω-phase, while considerable strengthening was observed in the case of formation ellipsoidal ω-particles.

At 350-600°C the martensitic α" phase precipitates from the β-solid solution. The martensitic mechanism of transformation at these temperatures is confirmed by the structure of precipitating phases: the plate-shaped crystals, internally twinned along the {101} planes are characteristic of the α-phase. The internal twinning on martensitic transformation is known to serve as an additional deformation, providing for the minimum elastic energy of transformation [8].

The phenomenological consideration of the martensitic transformation of titanium alloys [9] when taking internal twinning along the {101} planes into account, predicts the crystal growth along the {334} habit planes. This is fully confirmed by experiments.

At temperatures above 600°C the rhombic distortions in lattice decrease down to zero and formation of hexagonal phase is detected by the X-ray method. The hexagonal α and α'-phases differ from each other only in the broadening of diffraction lines, and therefore their identification is rather uncertain. The transmission electron microscopy indicates, that over the 600-700°C temperature range the crystals of hexagonal phase inherit the features of fine structure which are characteristic of martensitic phase. On further temperature increase fine structure of crystals becomes more plane- the microtwins disappear. Some separate twins are seen only in very coarse plates, precipitated at ~750°C. At 800-850°C they completely disappear, the particles aquire nonregular shape. The structure of the α-phase particles, observed at these temperatures may indicate the diffusion mechanism of transformation. Hence, conclusion can be made from the data presented, that increasing temperature results in the change of the mechanism of the (α", α and α') phases formation: from martensitic mechanism through intermediate to diffusion one. Since the difference between the α' and α-phase is rather conventional, it is difficult to determine the exact temperature of their formation.

As shown in experimental part, α", α' and α-precipitations strengthen the alloy due to differing mechanisms. Decreasing strengthening with increasing ageing temperature is due to coarsening the particles and their internal structure becoming more plain. Some additional strengthening on ageing may be achieved by producing high dislocation density in β-solid solution, using, for example, cold deformation. Thereby, very high strength and low plasticity were obtained for a number of high-alloy alloys due to decomposition phases becoming more disperse [10]. A set of high mechanical properties may be achieved when forming the regular polygonal structure in β-solid solution. This kind of structure in high alloys may be formed when certain thermo-mechanical conditions of hot deformation are fulfilled [1,2], in particular low temperatures (not more than 50-100°C above the α→β transformation temperature ) and high reduction (ε>40%). Some slight increase of plasticity may be obtained for polygonized specimens using the one-stage ageing [1]. However, the step-by-step heat-teratment is more effective, which enables to produce the structure, resembleng the <<composite>> one, consisting of the plate-like framework of the twin-free α-phase, situated along the grain and subgrain boundaries as well as strong matrix, formed by precipitation of the α" and α-phase inside the subgrains. This structure, in particular, may be obtained using the treatment consisting in quenching or anneal, or slow cooling from the β-phase region down to 800-850°C and subsequent ageing at 550-600°C. During the first stage of such treatment coarse α-phase particles precipitate along the grain boundaries and those of subgrains disoriented by 3° and more, since these are the subgrain boundaries which serve as the sites of preferential nucleation of the particles. The β-solid solution is fixed inside the subgrains on cooling. It is of importance to note, that at temperatures mentioned, the α-particles are in main free from internal twins and therefore rather plastic. The β-solid solution inclusions inside the subgrains decompose on further ageing, forming fine internally twinning crystals of martensite α" (α), which considerably strengthen the alloy.

Fig. 8a, b, d, e schematically show the structure, which form on step-by-step treatment of the speciments with polygonized and recrystallized structure, while Fig. 8c and present the corresponding electron micrographs taken from the specimens of the alloy under study. Besides, characteristic values of mechanical properties are given.

When badly formed polygonal structure results from hot deformation, the plate-like α-phase particles situate randomly either at the subgrain boundaries or inside the subgrains at separate dislocations. Plasticity of these specimens is less than that for perfect subgrain structure.

The step-by-step treatment of those specimens having recrystallized structure does not result in high mechanical properties. This is due to very small amount of α-phase formed during anneal (or quenching) at 800-850°C, since the process of α-phase precipitation in recrystallized specimens is strongly retarded and its amount is not enough to provide for plasticity.

When reducing the temperature below 800°C then twinned phase precipitates at the boundaries and inside the grains.

Hence, <<composite>> structure is formed due to complete step-by-step treatment in both recrystallized and polygonized specimens, however, thereby either primary or secondary α-phase being of martensitic origin; it is considerably strengthened and its high strength is not accompanied by high plasticity.

The fractographic analysis of the specimens with polygonized structure shows, that formation of the plastic framework from the α-phase eliminates the embrittlement caused by the grain boundaries, this resulting in the failure through the grain accompanied by considerable plastic deformation.

Coarse precipitations of the twinned α-phase along the grain boundaries in the specimens with recrystallized structure result in the fracture with a very small plastic deformation.


## Conclusions

A systematic study of fine structure of the phases, precipitating on ageing of high-alloy titanium alloys depending upon the conditions of heat-treatment and dislocation structure of the β-solid solution, made it possible to determine the optimum structure providing for the set of high mechanical pro-

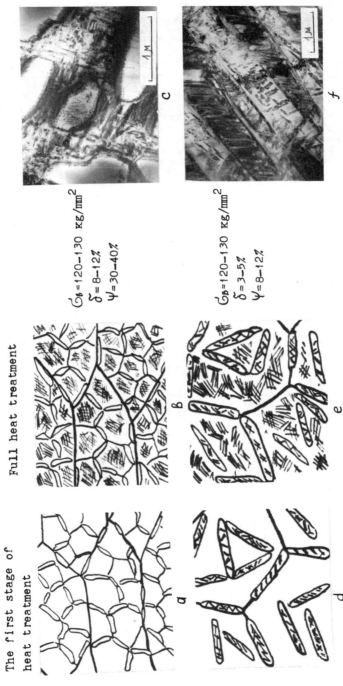

The first stage of heat treatment

Full heat treatment

$\sigma_b = 120-130 \text{ кg/mm}^2$
$\delta = 8-12\%$
$\psi = 30-40\%$

$\sigma_b = 120-130 \text{ кg/mm}^2$
$\delta = 3-5\%$
$\psi = 8-12\%$

Fig. 8 The scheme and electron photomicrographs of the structures, formed on step-by-step treatment of the specimens with polygonized (a-c) and recrystallized (d-f) structures. Near the schemes typical mechanical properties are given:

perties. This structure is analogous to a <<composite>> structure and consists of the mixture of the plastic $\alpha$-phase particles, situated at the grain and subgrain boundaries as well as fine-grain strengthened plates of $\alpha''$ and $\alpha$ martensite inside the subgrains. A perfect polygonized structure of the $\beta$-solid solution is required which can be obtained by means of certain conditions of hot deformation and step-by-step heat-treatment is to be carried out in order to obtain the <<composite>> structure.

## References

1.  I. S. Pol'kin, N. M. Semenova, A. B. Notkin. Proceedings of the 3rd Intercational Conference on Titanium. M., VILS, 1978, v. 3, p. 37.
2.  A. B. Notkin, A. A. Gelman, N. Z. Pertsovsky, N. M. Semenova, G. V. Golubeva. Fizika Metallov i Metallovedeniye, 1976, 42, p. 1257.
3   R. M. Lerinman, G. V. Murazaeva. In; Structure and Mechanical Properties of Metals and Alloys. Trans. of the Institute of Physics of Metals, Academy of Sciences of the USSR, 1975, 30, p. 30.
4.  M. I. Blackburn, I. L. Williams. Trans. Met. Soc. AIME, 1967, 236, No. 2, p. 287.
5.  L. M. Utevsky. Diffraction electron microscopy in physical metallurgy. M., published by <<Metallurgiya>>, 1972.
6.  L. M. Utevsky, A. B. Notkin, N. M. Semenova, N. Z. Pertsovsky, A. S. Fainbron. Proceeding of the 3rd International Conference of Titanium, M., VILS, 1978, v. 2, p. 547.
7.  M. I. Blackburn, I. C. Williams. Trans. Met. Soc. AIME, 1968, v. 242, p. 2462
8.  G. V. Kurdyumov, L. M. Utevsky, R. I. Entin. Transformation in Iron and Steel. M., <<Nauka>>, 1977.
9.  C. Hammond, P. M. Kelly. Acta Metallurgica, 1969, v. 17. p. 869.
10. In: Thermoplastic Strengthening of Martensitic Steels and Titanium Alloys. M., <<Nauka>>, 1971.

# ISOTHERMAL DECOMPOSITION OF SOLID SOLUTIONS IN A
## Ti-2,5 wt % Cu ALLOY.

J.M. Pelletier, G. Vigier, R. Borrelly and J. Merlin

Groupe d'Etudes de Métallurgie Physique et de Physique des Matériaux (ERA 463) - INSA de LYON - Bât. 502 - 69621 VILLEURBANNE CEDEX - FRANCE

## Introduction

Ti-Cu alloys are especially attractive because structural hardening may occur (however this possibility is limited because the formation of hardening precipitates is very slow). It is interesting to note that precipitation phenomena can be observed starting from 2 different solid solutions :
- the supersaturated $\alpha$ solid solution, obtained by quenching from the $\alpha_0$ state (h.c.p. structure).
- the supersaturated $\alpha'$ solid solution, obtained by martensitic quenching from the $\beta_0$ state (b.c.c. structure).
The purpose of the present work is to compare the decomposition of the 2 different solid solutions, in a TU2 alloy (titanium with about 2,5 wt % Cu) and then to determine the time-temperature-precipitation (T.T.P.) diagram ; this study will be continued on the conditions of heat treatments which allow an acceleration of the formation of precipitates.
The microstructure evolution is observed by electron microscopy. The nature and morphology of the different precipitates have already been studied by several authors (1, 2, 3) ; however precipitation kinetics have not been detailed. Then we have chosen the T.E.P. measurements to follow these decomposition kinetics ; in effect this technic is very sensitive to the solute content in solid solution in the matrix and almost insensitive to the existence of precipitates, as we have shown in previous works (4, 5, 6) ; these properties are especially attractive, because scarcely met by using other more classical technics (hardness or electrical resistivity measurements for example).

## Experimental Procedure

The chemical composition of the Ti-Cu alloy is as follows : (in wt %) Cu : 2,40 % ; N = 0,012 ; C : 0,028 ; O : 0,100 ; H : 0,0011 ; Si < 0,01.
The alloy was prepared and analyzed by the Ugine-Aciers Society. Specimens were in the form of sheets with dimensions 75 x 6 x 1 mm$^3$. In order to avoid anisotropy problems (7), the samples are cut perpendicularly to the rolling direction and apreliminary recrystallisation treatment (15 min. at 1025°C) is performed in a pure argon atmosphere furnace. Salt baths were used for the aging treatments.
The thermoelectric power measurements were carried out at room temperature with an apparatus described in a previous work (8). Before T.E.P. measurements, the samples were chemically

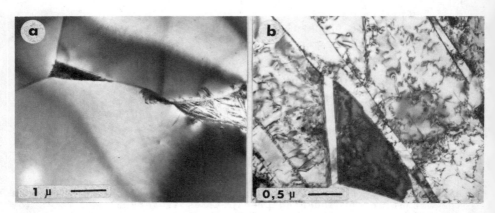

Fig. 1 : transmission electron micrographs of TU2 alloy water quenched
a) from 780°C,  b) from 950°C .

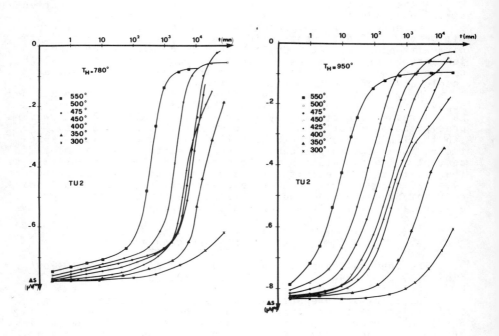

Fig. 2 : thermoelectric power change during isothermal agings at various
temperature, after quenching :  a) from 780°C .
b) from 950°C .

polished in a solution of one part HNO3 and one part HF ; $\Delta$S is the T.E.P. difference between the alloy and titanium recrystal-lized in β phase.

For transmission electron microscope (JEOL 100C) investiga-tions the specimens were electrolytically thinned and polished following the procedure described by Blackburn and Williams (9).

In a previous work (10) the phase diagram has been determi-nated in the pertinent portion and it was shown that the best homogeneization temperatures were as follows : 950°C for the $\beta_0$ state and 780° for the $\alpha_0$ state. Upon quenching from 950°C the β(b.c.c.)→α' (h.c.p.) transformation gives rise to a martensitic reaction and upon quenching from 780°C $\alpha_0 \rightarrow \alpha$. α and α' are both supersaturated solid solutions ; however the two states are dis-tinct ; in effect :

- the copper supersaturation is 2,4 % in α' and 1,8 % in α ; this last value is the maximum copper solubility at 780°C and then a part of the copper content is precipitated in grain boun-daries, in the shape of large $Ti_2$ Cu particles (see micrograph fig. 1a.)

- the structure defects density is very different : only a small number of dislocations may be observed in the α solid solu-tion, while martensitic plates observed in α' solid solution contain a very high dislocation density and twin boundaries (see micrograph fig. 1b.)

Decomposition of α and α' solid solutions.

The T.E.P. evolution ($\Delta$S) has been studied during the two solid solutions decompositions at aging temperatures $T_A$ between 300°C and 550°C (fig. 2). Sigmoïdal curves are observed ; the initial value is more negative for the α' state than for the α state ; this is due to the difference of copper content in solid solution. The final value is the smaller as $T_A$ is smaller ; this is connected with the variation of the copper solubility in titanium

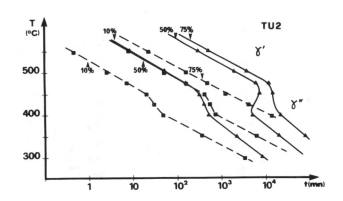

Fig. 3 : T.T.P. curves for the TU2 alloy :
—— after quenching from 780°C
--- after quenching from 950°C

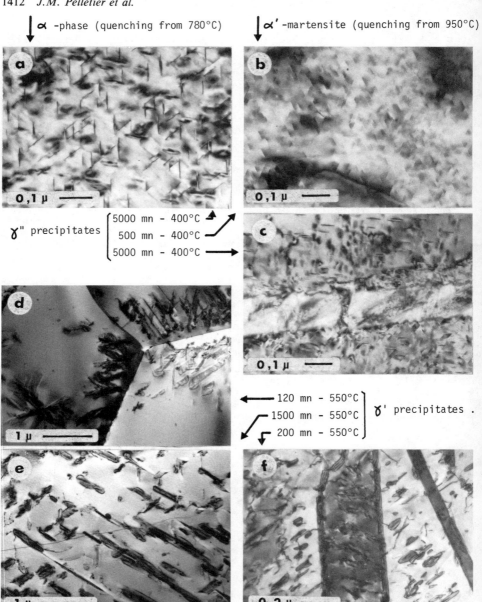

↓ α -phase (quenching from 780°C)    ↓ α′ -martensite (quenching from 950°C)

γ″ precipitates { 5000 mn - 400°C
                  500 mn - 400°C
                  5000 mn - 400°C

← 120 mn - 550°C
  1500 mn - 550°C } γ′ precipitates .
  200 mn - 550°C

Fig. 4 : transmission electron micrographs of TU2 alloy quenched from 780°C or 950°C, and aged at 400°C and 550°C .

with temperature (10). Kinetics are the faster as $T_A$ is higher ; then one or more thermally activated phenomena occur. By analyzing the curves it appears that two different phenomena exist, with a transition temperature at about 450°C (and that for the two different starting states) ; from these curves the T.T.P. diagram has been deduced (fig. 3) ; it was assumed that T.E.P. variations are directly connected to the precipitation rate. The existence of the two distinct temperature ranges may be clearly observed below and above 450°C. Electron micrographs are then carried out for characteristic states ; results are collected on fig. 4 ; the existence of the two different kinds of precipitates is shematically observed :

   - below 450°C : generalized precipitation of coherent disk-like precipitates (Υ")(about 15 Å thick).

   - above 450°C : localized precipitation of semi-coherent, plate-like precipitates (Υ' - $Ti_2$ Cu). The Υ' precipitation is heterogeneous, initiated on the dislocations (not on grain boundaries) and progressively spreads over the matrix.

   This precipitation is faster within martensitic microstructure than within α microstructure ; this fact may be explained by the existence of a high dislocations density within the martensitic plates ; these lattice defects act as nucleation sites for Υ' precipitates. Then such precipitates may be observed even below 450°C in this martensitic state.

   From the T.T.P. curves, the apparent activation energies of Υ' and Υ" formation are determinated ; for Υ' formation $E_A$ (Υ') = 1,9 eV, whatever the initial state may be ; for Υ" formation $E_A$ (Υ") values are 1,4 eV and 0,9 eV starting from α' and α respectively. The difference between these two last values may be due to the fact that during aging below 450°C after quenching from $β_0$ the two different precipitates Υ" and Υ' are observed ; 1,4 eV is effectively between 1,9 and 0,9 eV and takes into account the apparent activation energies of the two phases.

## Change of solid solutions decompositions
## by thermal or mechanical treatments

   The formation of Υ' and Υ" phases may be accelerated by heat treatments (duplex treatments) or by mechanical treatments (cold working). This acceleration is especially attractive because precipitation phenomena are very slow.

### 1. Υ' acceleration

   It has been already shown that dislocations act as nucleation sites for Υ' ; then a preliminary cold working of the solid solution will accelerate the Υ' formation. This can be observed effectively on fig. 5 : a drastic rolling (200 %) carried out on specimens quenched from 780°C leads to an important increase of the decomposition rate, during an aging at 475°C. This decomposition is even faster than that realized after quenching from 950°C. However the hardening due to this Υ' phase is not very important ($H_V$ max ≃ 220).

### 2. Υ" acceleration

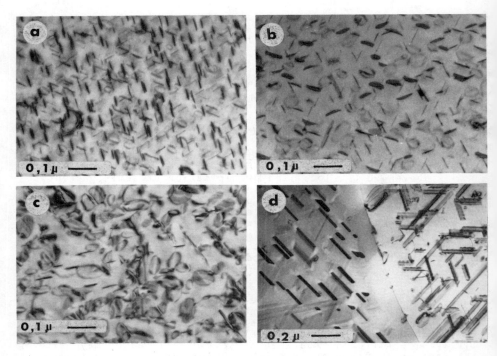

Fig. 7 : electron micrographs after quenching from 780°C, 5000 mn at 400°C
and :    a) 2 mn at 550°C
         b) 15 mn at 550°C
         c) 100 mn at 550°C
         d) 1000 mn at 550°C.

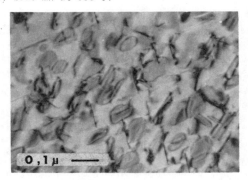

Fig. 9 : electron micrograph of TU2 alloy quenched
from 780°C, aged at 400°C for 1500 mn, and at 475°C
for 5000 mn.

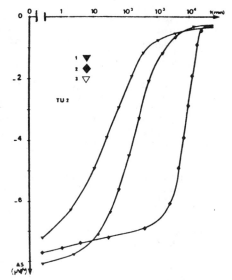

Figure 5 : ΔS variations during aging
          at 475°C :
1. after quenching from 780°C and
   cold rolling (200%)
2. after quenching from 780°C
3. after quenching from 950°C

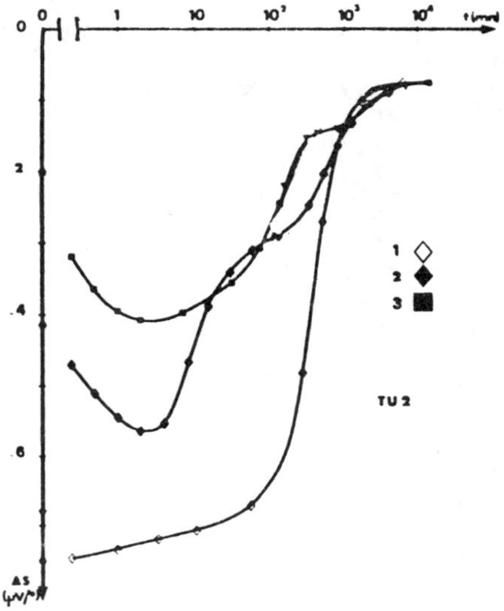

Figure 6 : **ΔS** variations during iso-
          thermal aging at 550°C :
1. **after** quenching from 780°C
2. **after** quenching from 780°C and
   aging of 1500 min. at 400°C
3. after quenching from 780°C and
   aging of 5000 min. at 400°C.

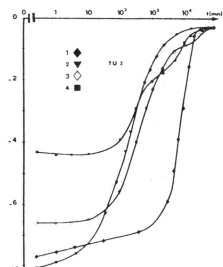

**Figure 8** : ΔS variations during aging
          at 475°C :
1. after quenching from 780°C
2. after quenching from 950°C
3. after quenching from 780°C and
   aging of 1500 min. at 400°C
4. after quenching from 780°C and
   aging of 5000 min. at 400°C.

The hardening due to the $\gamma''$ phase is more important (Hv ≃ 270) but very long aging times are necessary (several days or even several weeks). Then duplex treatments are especially attractive ; they are based on the following principle : nucleation of the $\gamma''$ phase below $T_C$ = 450°C and faster growth of these nuclei above $T_C$. The quench is always performed from 780°C in order to avoid any $\gamma'$ formation.

The first aging in then carried out at 400°C for various times (1500 mn and 5000 mn) and the second aging is carried out either at 475°C or at 550°C. The results obtained during the aging at 550°C are plotted on the fig. 6. Whatever the first aging time may be, all the curves have the same aspect and can be schematically decomposed in four stages. Electron micrographs on the fig. 7 illustrate these four stages in the case of a preliminary aging at 400°C for 5000 mn :

Stage 1 : T.E.P. decreases, this decrease may be related to a dissolution of the smallest $\gamma''$ precipitates. By comparison between the fig. 4b. and 7a. it is observed that the mean precipitates diameter is unchanged (∅ ≃ 500 Å), but the smallest precipitates have probably disappeared.

Stage 2 : T.E.P. increases quickly, this increase may be related to the growth of $\gamma''$ precipitates : ∅ goes from 500 Å to 700 Å (fig. 7b.)

Stage 3 : The T.E.P. increase is slowed down : the examination of the fig. 7c. indicates a growth and a loss of coherency of precipitates (dislocation loops around precipitates appear).

Stage 4 : a T.E.P. new increase makes this curve to join that corresponding to the direct formation of $\gamma'$ at 550°C. Plate-like semi-coherent precipitates may be seen on the fig. 7b. The coarsening has already begun.

During aging at 550°C, the T.E.P. values is never as negative as that corresponding to the solid solution, then the $\gamma''$ precipitates reversion is only partial. It may be assumed that a complete reversion should occur during aging at higher temperature. However $\gamma'$ precipitates should appear before the complete $\gamma''$ reversion. By contrast, during aging at lower temperature (475°C) no or little $\gamma''$ precipitates dissolution is observed (fig. 8). Then the density of $\gamma''$ nuclei, which are going to grown, is much greater and the precipitates will be finely dispersed (fig. 9). Consequently an important hardening will occur. The total aging time to reach the maximum hardness increase is much shorter than that related to a direct aging at 400°C.

## Conclusion

By using T.E.P. measurements and electron micrographs the kinetics of $\alpha$ and $\alpha'$ solid solutions decompositions and the morphology of precipitates phases have been determinated. These decompositions are similar and two different ranges of aging temperatures have to be distinguished : below 450°C a formation of coherent $\gamma''$ precipitates is observed, while above 450°C semi-coherent $\gamma'$ precipitates nucleate on structural defects and then spread over the matrix.

The influence of cold working and duplex aging on the formation of $\gamma'$ and $\gamma''$ respectively, have also been examined :
- $\gamma'$ precipitation is accelerated by the introduction of linear defects.
- after nucleation at low temperature some small $\gamma''$ precipitates

can be dissolved during aging at higher temperature, but a complete reversion cannot be achieved ; by contrast some larger " precipitates grow quickly before the ' formation ; the proportion of " dissolution depends on second aging temperature.

Thus thermoelectric power measurements appear to be a useful tool for the study of the alloys decomposition. The easiness of this technic is especially attractive and then T.E.P. measurements should be more used for T.T.P. diagram determination and in a general way for phase transformation study.

## References

1. J.C. Williams, R. Taggart and D.H. Polonis : Met. Trans., 2 (1971), 1139.
2. G. Lütjering and S. Weissmann : Met. Trans., 1 (1970), 1641.
3. A. Zangvil, S. Yamamoto and Y. Murakami : Met. Trans., 4 (1973), 467.
4. J.M. Pelletier, R. Borrelly and P.F. Gobin : Scripta. Met., 11 (1977), 553.
5. J.M. Pelletier, J. Merlin and R. Borrelly : Mat. Sci. Eng., 33 (1978), 95.
6. R. Borrelly, J.M. Pelletier and P.F. Gobin : Acta Met., 26 (1978), 1863.
7. R. Borrelly, J. Merlin, J.M. Pelletier and G. Vigier : J. Less -Common Metals, 69 (1980), 49.
8. J.M. Pelletier, R. Borrelly and E. Pernoux : Phys. Status Sol. A 39 (1977), 525.
9. M.J. Blackburn and J.C. Williams : Trans. T.S.M.-A.I.M.E., 239 (1967), 287.
10. G. Vigier, J.M. Pelletier and J. Merlin : J. Less-Common Metals, 64 (1979), 175.

# IMPROVEMENT OF MECHANICAL PROPERTIES
## OF Cu-Ti ALLOYS BY NEW PROCESS

Y.Nishi[*], H.Aoyagi[**], T.Suzuki[***]
and E.Yajima

[*]   Department of Material Science, Faculty of
      Engineering, Tokai University, Hiratsuka, Japan
[**]  Graduate School of Material Science, Tokai
      University, Hiratsuka, Japan
[***] Department of Physics, Faculty of Science,
      Tokai University, Hiratsuka, Japan

## Introduction

Aging of the supersaturated Cu-Ti alloys was investigated by many workers[(1)-(5)] and this alloys is used commercially as spring. In commercial process, the procedure is as follows:

(1)   melting
(2)   casting and forging
(3)   solution treatment and quenching
(4)   cold-working
(5)   prolonged aging

This procedure consists of high energy consumption processes.

By new process (quenching from liquid process), supersatulated crystal solid solution is directly produced.  The procedure is as follows.

(1)   melting
(2)   quenching from liquid
(3)   shortened aging

By the new process, Cu-Ti spring alloys are obtained without many pre-treatments (solution treatment, cold-working and prolonged aging) compared with the commercial process.  Furthermore, a high concentrated ductile crystal (Cu-10 at%Ti) is produced by the new process.  This crystal is more super-saturated,though the solubility limit of titanium is about 5 at% as shown in Fig.1.  The effect of the new process on aging is investigated for Cu-Ti alloys by means of hardness, tensile test and transmission electron microscopy.

## Experimental

Cu-Ti alloys were prepared from 99.99% copper and 99.8% sponge titanium in an argon arc furnace.   Specimens were prepared by quenching from liquid, which was melted about 1600 K for 100 s under argon atmosphere in an infrared furnace, and the cooling rate was over 10000 K/s to avoid precipitated phase formation.   The thickness of specimens was approximately 2 X $10^{-4}$m.   These specimens were aged in a salt bath or in vacuum.

Mechanical properties were investigated by the Vicker's hardness (100 g Lord) and tensile ( 5.5 X $10^{-2}$ $s^{-1}$strain rate) tests.   The changes in microstructure were investigated ʳˡᵉᶜtron microscopy and diffraction.

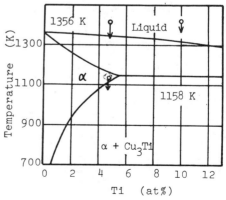

Fig.1  Cu-Ti phase diagram

Results

Hardness changes for the Cu-Ti alloys which are quenched from liquid and
then aged at 673 K are shown in Fig.2 (o is Cu-5 at% Ti alloy and ● is Cu-10 at%
Ti alloy.).   The result of Cu-5 at% Ti alloys which are produced by commercial
process is shown in this figur too (▲ is as solution treatment, Δ is cold-working
(20%) and X is cold-working (50%).).

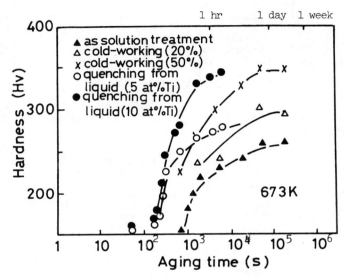

Fig.2   Typical changes in micro-Vicker's hardness of
Cu-Ti alloy aged at 673 K.

In the case of commercial process, maximum hardness  and rate of hardening
are greatly enhanced by cold-working.   This enhancement depends on the density
of dislocations.   This fact was investigated by transmission electron microscopy.
Aging-time of maximum hardness is slightly decreased with the increase of cold-
work.   The main effect of cold-work is enhancement of hardness.

Maximum hardness and rate of hardening are enlarged  and aging-time of
maximum hardness is shortened by the new process in comparison with the commercial
process.   Maximum hardness of Cu-5 at% Ti alloy quenched from liquid is equal to
that of cold-worked (20%) alloy.   Aging time for maximum hardness of the cold-
worked (20%) alloy is about 20 times that of quenching from liquid.  Consequently,
an important feature of the new  process is shortened aging of maximum
hardness.   Furthermore, more supersaturate crystal phase (over solubility limit
of Ti) is obtained by the process of quenching from liquid, though the equilibrium
solubility limit of titanium concentration is 5 at%, as shown in Fig.1.   Maximum
hardness of this crystal phase is equal to that of cold-worked (50%) Cu-5 at% Ti
alloy, as shown in Fig.2 and Table 1.   Excess addition of 10 at% Ti induces a
large maximum hardness that is equal to cold-work (50%).

(6)

Table 1  Maximum hardness and aging-
time of maximum hardness

|  | maximum hardness | aging-time |
|---|---|---|
| as solution treatment | 250 | 1 day |
| cold-working 20% | 300 | 20 hr |
| cold-working 50% | 350 | 16 hr |
| quenching from liquid (5 at%Ti) | 280 | 1 hr |
| quenching from liquid(10 at%Ti) | 350 | 1 hr |

One of the most important mechanical properties of the spring is modulus of elasticity. This is obtained by tensile test in this work. The typical type of Cu-10 at% Ti alloys quenched from liquid is shown in Fig.3. The modulus of elasticity of specimens after aging (1 hr) is about 1.5 times that of as casting specimens. Futhermore, the elongation after aging is about twice as large as before aging.

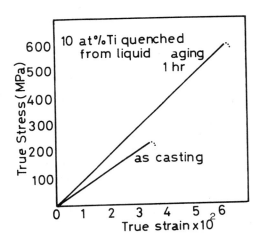

Fig.3  Stress-strain curves for Cu-10 at% Ti
quenched from liquid.

### Discussion

Johnson and Mehl have proposed a quantitative kinetic treatment. This has provided to be a very powerful tool in elucidation of many transformations. This equation is as follows

$$X = 1 - \exp\left(-kt^{n}\right),$$

where X is the fraction of transformation, k is a kinetic constant, t is the time and the exponent (n) is determined by the transformation mode. The parameter k and n are determined by plotting $\log(-\ln(1-X))$ versus $\log t$. X is calculated from hardness in this work. All specimens which are investigated in this work have same exponential value (n; 1.7<n<2) from Fig.4. This result shows that all specimens transformed in the same way and is reconfirmed by Photo.1, Photo.2 and Photo.3.

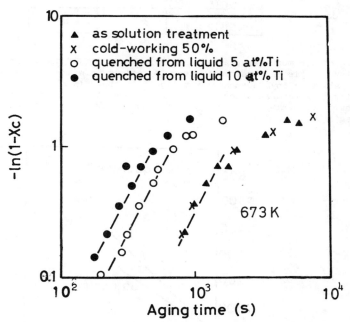

Fig.4   Relation between $\log(-\ln(1-X))$ and $\log t$.

First we discussed the effect of cold-working in the commercial process. Fig.4 shows the agreement of the logarithmic kinetic constants (log k) between the as solution treated and the cold-worked (50%) alloy. Same results are obtained by transmission electron microscopy, as shown in Photo.1 and Photo.2. Photo.1 shows typical modulated structure of Cu-5 at% Ti alloys aged for $6 \times 10^{4}$ s at 673 K. Photo.2 shows modulated structure of Cu-5 at% Ti alloys which is deformed (50%) and aged for $6 \times 10^{4}$s at 673 K. These photographs show that the wavelength of deformed (50%) alloy is equal to that of non-deformed alloy.

This result reconfirms the agreement of the rate of transformation between as soltution treatment and cold-working (50%).    The difference of Photo1 and Photo.2 is the form of the wavelength,    The cold-working (Photo.2) induces irregularity nd heterogenity in modulated structure.    These results show that the cold-working before aging dose not affect   the logarithmic kinetic constants.

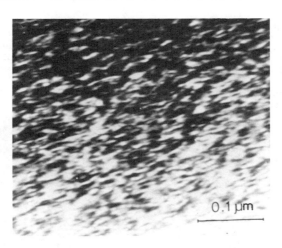

Photo.1    Cu-5 at% Ti alloy aged for $6 \times 10^4$s
          at 673 K.

Photo.2    Cu-5 at% Ti alloy deformed 50%
          aged for $6 \times 10^4$s at 673 K.

In the case of the new process, Photo.3 shows the structure of Cu-5 at% Ti alloy, which is quenched from liquid and then aged for 900 s at 673 K.   We can find fine modulated structure and this structure shows the satellite in the electron diffraction pattern.   The kinetic constant of the new process is 10 times larger than that of commercial process.   This feature is, mainly, induced from difference of mass transport.   We suggest that excess vacancy and secondary lattice defects induce the increase of mass transport.

Making a comparison between Cu-5 at% Ti and Cu-10 at% Ti, we understand that the logarithmic kinetic constants depend on titanium concentration.   Exponent (n) is about 2, as shown in Fig.4.   The mechanism of transformation for Cu-10 at% Ti is as same as that for Cu-5 at% Ti.   But the logarithmic kinetic constant increases with concentration of titanium.   Calculated temperature of spinodal decomposition ($T_c$) rises with concentration of titanium.   The rise of $T_c$ yields the enlargement of driving force of spinodal decomposition and then the increase of titanium accelerates the spinodal decomposition.

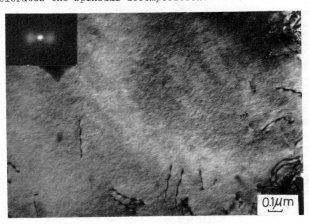

Photo.3   Cu-5 at% Ti alloy quenched from liquid
and aged for 900 s at 673 K.

### Conclusion

By the new process (quenching from liquid process), supersaturated Cu-Ti crystal solid solution is directly produced without many high energy consumption pre-treatments (solution treatment, cold-working and prolonged aging).

Maximum hardness of the new process is equal to that of cold-work (20%) in Cu-5 at% Ti alloys.   Excess addition of 10 at% Ti induces a larger maximum hardness that is equal to that of the cold-worked (50%) alloy.   The most important feature of the new process is shortened aging for maximum hardness. The modulus of elasticity of specimens after aging (1 hr) is about 1.5 times that of as casting specimens.

The mechanism of transformation on aging is same spinodal decomposition, but the decomposition rate of the quenched alloys from liquid is larger than that of  the commercial produced alloys and also depends on Ti content.

### Acknowledgement

The author wish to thank Prof. T.Miyazaki and Dr. H.Mori of Nagoya Institute of Technology for their helps in a part of Electron Microscopy.

## References

(1)   T.Miyazaki, E.Yajima and H.Suga: Trans.Japan Inst.Metals, 12(1971),119.

(2)   S.Saji, S.Ikeno and S.Hori :J.Japan Inst.Metals,42(1978),63.

(3)   K.Saitou, K.Iida.and K.Watanabe:J.Japan Inst.Metals,31(1967),641.

(4)   K.Osada, K.Umezu and S.Nishikawa:J.Japan Inst. Metals, 34(1970),1253.

(5)   S.Ikeno, S.Saji and S.Hori:J.Japan Inst. Metals,38(1974),1186.

(6)   M.Nakagawa, T.Ariga and E.Yajima:J.Japan Inst.Metals,43(1979),583.

# ON THE AGING PROPERTY OF Cu-BASE Cu-Ni-Ti ALLOYS

Masao Yakushiji, Yoshiyuki Kondo and Kiyoshi Kamei

Department of Metal Engineering, Faculty of Engineering,
Kansai University, Suita, Japan

## Introduction

Aging characteristics of copper base Cu-Ni-Ti alloys containing nickel three times as much as titanium have been studied by Maeda et alius[1], and they have attributed the age-hardening to the precipitation of $Ni_3Ti$.

On the other hand, in the previous work[2][3] the authors have investigated the equilibrium relationship of phases in Cu-Ni-Ti ternary system. As a result, it has been found that the T phase based on the ternary compound CuNiTi existed as the only compound phase in Cu-Ni-Ti system, and that the primary copper solid solution α was in equilibrium with the compound T and they constitute a quasi-binary system. As seen in Fig.1, Cu-T quasi-binary diagram shows that the solid solubility of α for T decreases markedly with falling temperature to below 2%T at room temperature, while the α holds about 10%T in solution at 1343K.

On the basis of the results of investigation on Cu-Ni-Ti phase diagram described above, the age hardening of Cu-T quasi-binary alloys containing 2 to 6%T was examined. Moreover the measurements of tensile strength and electric conductivity of the aged alloys were carried out, and the effect of thermo-mechanical treatment on Cu-4%T alloy was also examined.

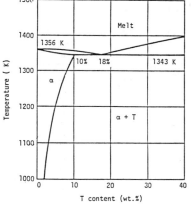

Fig.1 The Cu rich end of Cu-T quasi binary diagram.

## Experimental

### 1. Preparation of alloys

The chemical compositions of the alloys used in this work are shown in Table 1. The alloys were prepared by a conventional melting procedure using graphite crucible in argon atomosphere. The starting materials were electrolytic copper, cobalt free electrolytic nickel and ELI grade sponge titanium. About 1.2Kg electrolytic copper was melted and then

Table 1 Chemical composition of alloys(%)

| alloy | Ni | Ti | Cu |
|-------|------|------|------|
| Cu-2%T | 0.69 | 0.56 | bal. |
| Cu-3%T | 1.03 | 0.84 | bal. |
| Cu-4%T | 1.36 | 1.14 | bal. |
| Cu-5%T | 1.74 | 1.40 | bal. |
| Cu-6%T | 2.07 | 1.65 | bal. |

the desired amount of T which had been prepared by arc melting, was added to the melt. After stirring up and being kept at about 1423K for 10 min, the melt was cast into a permanent mold and an ingot of 50 mm in diameter was obtained. Chemical analysis showed the contamination by carbon from the melting process to be neglisible.

## 2.  Experimental procedure

The ingots were homogenized for 24hr at 1273K and forged to square bars of approximately 10mm x 10mm cross section. They were reduced further to strips of 1mm thickness by hot and cold rolling for the hardness test and observation by electron microscope. They were also drawn into wire of 1mm in diameter, for the measurement of tensile strength and electric conductivity.

The solution treatment was made by heating the specimen for 1hr at 1273K, followed by quenching into water. The aging and solution treatment were carried out both in vacuum-shield quartz ampoules.

Hardness was measured using a micro-Vickers hardness tester with a 200g load. Thin foils for the examination under electron microscope were prepared using an electro-polishing solution of 4 parts of ethyl alcohol and 1 part of nitric acid kept at 278K. Transmission electron micrography was carried out by using a JSEM-200 electron microscope. Tensile tests were made with a TENSILON machine at a constant cross head speed of 1mm/min, and electric conductivities were measured with a Kelvin double bridge. In both tensile test and electric conductivity measurement, at least two specimens for each treatment were tested to ensure reproducibility.

## Experimental Results and Discussion

### 1.  Isothermal aging

Isothermal age-hardning curves for the quasi-binary Cu-T alloys are shown in Fig.2 to Fig.6. As quenched hardness of the alloy increases as T content is raised. The appreciable hardening is observed in all the alloys aged at 573K. On aging at 723K, hardening appears to    occur in two steps, and the hardness does not reach its maximum even after aging for 2X10⁴min. It is considered,therefore, that the aging process of these alloys occurs in two stages,  whereas at other temperatures the hardening in two steps is not observed. When the alloys are aged at temperatures higher than 823K, the hardness increases initially and then decreases through its maximum.

It is found that both the maximum hardness value and the time required to reach the maximum hardness decrease with increasing temperature. Semi-logarithmic plots of the time required to reach maximum hardness against the

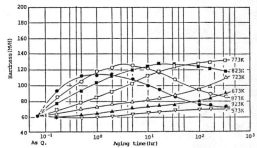

Fig.2 Isothermal age-hardning curves of Cu-2%T alloy.

reciprocals of aging temperatures yield a straight line
as seen in Fig.7. From the
slopes of these straight
lines, the activation energy
for aging process of Cu-T
quasi-binary alloys was calculated to be 39, 000cal/mol.
This value is     between
37, 000cal/mol and 41, 000
cal/mol which are the activation energies for growth
of TiCu and Ti$_2$Cu$_7$ respectively[4].

The structural change
associated with aging of Cu-
4%T alloy was examined by
transmission electron micrography.

Photo.1 shows an elec-
tron micrograph of the as
quenched alloy. No precipitate is apparent.

Transmission electron
micrograph of this alloy
aged for 1hr at 773K  after
quenching is shown in Photo.
2. Extremely fine precipitates responsible for harden
ing are observed. The size
of such precipitates are
estimated to be about 5nm.

Photo.3 shows a micro-
structure of the alloy aged for 10hr at
823K, which corresponds the
hardness maximum. The fine
precipitates are uniformly
dispersed in the matrix.
Moreover, preferential precipitation on grain boundary and precipitation free
zone along the grain boundary are observed, suggesting
the participation of vacancies in the nucleation of
precipitate.

Photo.4 shows a micro-
structure of the alloy aged
for 10hr at 873K,which corresponds to the begining of softening. Further growth of
both precipitates on grain
boundary and in matrix are
apparent.

The microstructures of
over-aged alloy are shown in
Photo.5 and Photo.6. In Photo.
5 the precipitates grown to
large acicular shape and dislocations around them are

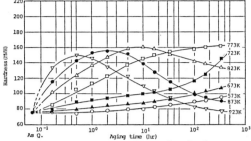

Fig.3 Isothermal age-hardening
curves of Cu-3%T alloy.

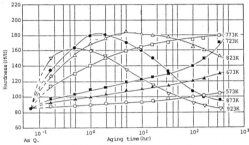

Fig.4 Isothermal age-hardening
curves of Cu-4%T alloy.

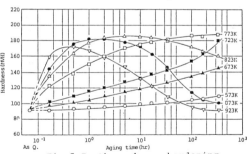

Fig.5 Isothermal age-hardening
curves of Cu-5%T alloy.

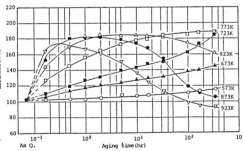

Fig.6 isothermal age-hardening
curves of Cu-6%T alloy.

observed. The contrast observed around
the precipitates are due probably to
coherency strain.

Photo.6 shows a microstructure of
the alloy aged for 1000hr at 873k.
Large rod shaped precipitates are apparently arranged in the particular directions. Such precipitates grown to
large dimmension as shown in Photo.5
and Photo.6 do not give any hardening
effect on the alloy.

The electron diffraction pattern
of Photo.6 is shown in Photo.7 (a),
together with its key diagram (b). In
Photo.7 (a), the (110) plane of matrix
α is nearly perpendicular to the electron beam, and some weak spots are
seen in addition to the strong spots
reflected from matrix α. It is considered that these additional spots are
due to the T phase, i.e., the spots

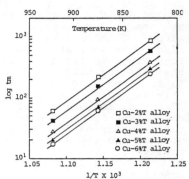

Fig.7 Plots of log tm vs 1/T; tm
being the time required to reach
the maximum hardness, and T the
aging temperature.

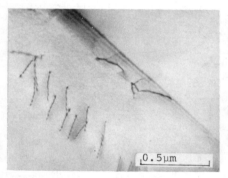

Photo.1 Transmission electron micrograph of Cu-4%T alloy solution-
treated for 1hr at 1273K.

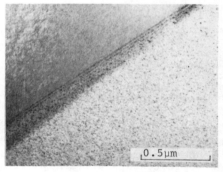

Photo.2 Transmission electron
micrograph of Cu-4%T alloy
aged for 1hr at 773K.

Photo.3 Transmission electron
micrograph of Cu-4%T alloy
aged for 10hr at 823K.

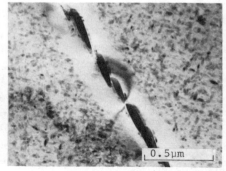

Photo.4 Transmission electron
micrograph of Cu-4%T alloy
aged for 10hr at 873K.

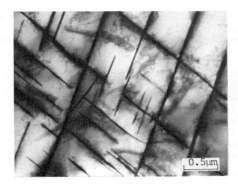

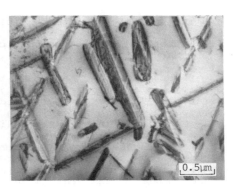

Photo. 5 Transmission electron
micrograph of Cu-4%T alloy
aged for 100hr at 873K.

Photo. 6 Transmission electron
micrograph of Cu-4%T alloy
aged for 1000hr at 873K.

aa'a''a''', bb' and cc'c''c''' can be explained as {110}, {002} and {121} reflection of T respectively.

These results suggest that the supersaturated α solid solution decomposes during
aging to the equilibrium α and T phases

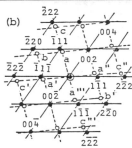

Photo.7 (a) Electron dif-
fraction pattern of photo.
6, and (b) the key diagram
of (a).

through the intermedi-
ate precipitate, and
that the hardening at
initial stage is due
to the intermediate
precipitates, and the
marked hardening in
second stage is due to
the finely dispersed
precipitation of T.
However, attempt to

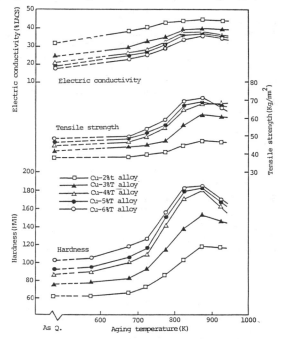

Fig.8 Effect of aging temperature on hardness,
tensile strength and electric conductivity of
quasi binary Cu-T alloys, solution-treated for
1hr at 1273K and aged for 1hr at each temperature.

examine the intermediate precipitate was not successfull.

2.  Effect of isochronal aging on the properties of Cu-T quasi-
    binary alloys

Fig.8 shows the hardness, tensile strength and electric conductivity of Cu-T quasi-binary alloys aged for 1hr at various temperatures. The increase  in hardness occurs gradually up to 723K
followed by marked hardening in a temperature range from 723K to
873K especially in alloys containing T more than 4%, leading to
the maximum at 873K. Above 873K softening takes place in all alloys. The changes of tensile strength and electric conductivity
of the alloys accompanying with increase of aging temperature are
similar in behavior to that of hardness. As seen in Fig.8, with
increasing T content the hardness and tensile strength are increased and electric conductivity is decreased. However the
content of T more than 4% is not so effective. The hardness, tensile strength and electric conductivity of Cu-4%T alloy isochro-
nally aged for 1hr at 873K were 178MVH ,68Kg/mm$^2$ and 36%IACS respectively.

3.  Effect of thermo-mechanical treatment on age-hardening of Cu-
    4%T alloy

Cold working has been found to accelerate the rate of aging
depending on the amount of working given. Therefore the effect of
cold rolling on the aging process of Cu-4%T alloy was investigated.
The quenched specimens of Cu-4%T alloy were rolled by 20%, 50% and
90% at room temperature prior to aging. The hardness of as-rolled
specimen  is raised ⋅remarkably with increasing reduction as seen
in Fig.9. When these specimens were aged isothermally at various
temperature, the rates of aging were accelerated as expected. It
is evident from Fig.9 and Fig.4 that the maximum hardness is increased and the time to reach maximum is decreased, as compared
with the case of no reduction.
Fig.10 shows the hardness obtained by aging for 1hr at

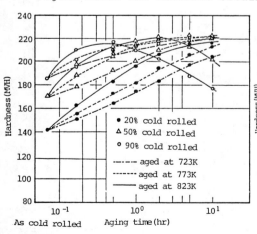

Fig.9 The effect of cold rolling on
isothermal aging of Cu-4%T alloy.

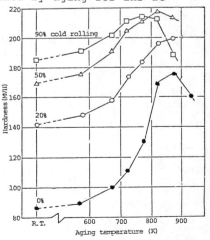

Fig.10 The effect of cold rolling
on the hardness of Cu-4%T alloy,
aged for 1hr at each temperature.

various temperature after rolling at room temperature. The hardness maximum moves to lower aging temperature with increasing reduction, while the maximum hardness is obtained at 50% reduction. The specimen of Cu-4%T alloy which was 50% cold rolled and aged for 1hr at 823K after quenching, showed the hardness of 218MVH. This value is considerably higher as compared to 176MVH, which is the maximum value obtained by isochronal aging of 1hr immediately after the quenching.

Fig.11 shows the effect of preaging for 1hr at 673K and 723K on the hardness of Cu-4%T alloy rolled at room temperature after preaging and aged for 1hr at 773K. Very little amount of hardness increase with preaging is observable at any reduction. It is concluded, therefore, that the preaging prior to aging treatment at higher temperature is not effective.

The hardness change of Cu-4%T alloy accompanying with reiteration of the treatment consisting of 20% cold rolling and aging for 1hr at 723K was investigated. As seen in Fig.12, the hardness was raised in a parabolic manner with increasing reiteration and the hardness 212 MVH was obtained after five times of reiteration. This hardness value is nearly equal to that of the specimen 50% cold rolled and aged for 1hr at 823K.

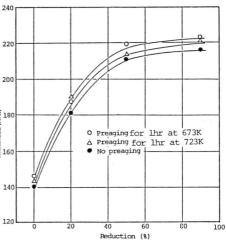

Fig.11 The effect of preagiing on the hardeness of Cu-4%T alloy deformed at room temperature after preaging and aged for 1hr at 773K.

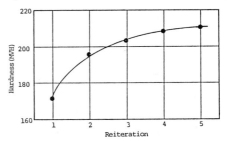

Fig.12 The hardness change accompanying with reiteration of the treatment consisting of 20% cold rolling and aging for 1hr at 723K.

## Summary

The aging characteristics of Cu-T quasi-binary alloys containing 2 to 6%T were investigated. The results obtained are summarized as follows.

(1) On the isothermal aging the beginning of hardening is observable in all the alloys aged at 573K. The hardening is accelerated with increasing aging temperature, and at temperatures higher than 773K the hardness increases with aging time, reaches a maximum, and then decreases.

(2) An activation energy for the age-hardening of Cu-T quasi-binary alloys was calculated to be 39,000 cal/mol.

(3) It is found from the hardening behavior of Cu-T alloys aged at 723K, that the aging processes of these alloys occur in two stages. Transmission electron microstructure observation suggests that the slight hardening at initial stage is due to the intermediate precipitates and the finely dispersed precipitation of T

contributes to the marked hardening in the second stage.

(4) The hardness and tensile strength of the alloys aged iso-chronally at various temperatures were increased and electric con-ductivity was decreased with increasing T content. However the ad-dition of T more than 4% was not so effective. On the other hand, both the  maxima  of hardness and tensile strength, and the mini-mum of electric conductivity occured at 873K. The hardness, ten-sile strength and electric conductivity of Cu-4%T alloy aged for 1hr at 873K were 178MVH, 68Kg/mm$^2$ and 36%IACS respectively.

(5) The hardness of Cu-4%T alloy deformed at room temperature after quenching is raised remarkably with increasing deformation. When these deformed alloys were aged the aging process was acceler-ated, and the maximum hardness of 218MVH was obtained in the spec-imen deformed by 50% and aged at 823K.

(6) Thermo-mechanical treatment such as preaging prior to the deformation at room temperature and aging at higher temperature, and as reiteration of the treatment consisting of 20% deformation and aging for 1hr at 723K, proved to be not so effective on this alloy.

## Acknowledgements

The authors wish to express their gratitude to Professor Dr. Y.Murakami,Kyoto University, for his encouragement and helpful discussion. Thanks are also due to Mr. K.Sibatomi, Japan Electron Optics Laboratory Co., for the operation in the electron microscopy.

## References

1. R.Maeda and H.Isibe: J.Japan Inst.Metals,29(1965),177.
2. M.Yakushiji,Y.Kondo and K.Kamei: J.Japan Inst.Metals,44(1980), 615.
3. M.Yakushiji,Y.Kondo and K.Kamei: J.Japan Inst.Metals,44(1980), 620.
4. A.E.Gershinskii,A.A.Khoromenko and E.I.Cherepov: Phys.Stat. Sol.,(a)31,(1975),61.

# THE INTERFACE PHASE IN A NEAR-α TITANIUM ALLOY

P. Hallam and C. Hammond

Department of Metallurgy, University of Leeds,
LS2 9JT, U.K.

## Introduction

The existence of a third phase in the (α + β) titanium alloy Ti-6Al-4V was first reported by Rhodes and Williams [1] and designated 'interface phase', because it was found as a layer between the α and β phases. They found that it occurred either as a 'monolithic' single crystal layer with an f.c.c. structure (a ≈ 426 pm) or as a striated layer ∿ 200μm thick consisting of platelets, which gave rise to selected area diffraction patterns with arced reflections characteristisc of 'Type 2' α [2] (hexagonal α phase which is not related to the β by Burgers orientation relationship). Hall [3], working with the same alloy, confirmed the existence of both structures and established an orientation between the primary α and monolithic phase (γ) as: $(0001)_\alpha //(111)_\gamma$; $<11\bar{2}0>_\alpha //<\bar{1}10>_\gamma$. Where the γ structure is either f.c.c. (a = 436 pm) or f.c.t. ($a^\gamma$= 432 pm, $c/a$ = 1.13) according to heat treatment. The hexagonal structure of the striated (p) form in slow cooled material was often, though not always, related to the primary α by the orientation relationship: $(10\bar{1}0)_\alpha //(10\bar{1}0)_p$; $[0001]_\alpha //<1\bar{2}13>_p$. The α- platelets formed on more rapid cooling after β solution treatment were related to the β by Burgers relationship.

Interface phase was also reported in Ti-11.6Mo and Ti-14Mo-6Al alloys [4], but always had the hexagonal α structure (Burgers related) and was often twinned on $\{10\bar{1}1\}_\alpha$ planes. In more recent work Rhodes and Paton [5] found that the thickness of the interfacial layer was a function of cooling rate, and that the striated layer had the same f.c.c. crystal structure as the 'monolithic' form, but contained {111} twins.

The crystal structure of the interface phase requires further investigation in view of the differences between the reports of previous workers. The crystallography of the interface phase in a 'near α' alloy (IMI685) is the subject of the present work.

## Experimental

The alloy used in this work was IMI685 of composition:
Ti-6%Al-5%Zr-0.5%Mo-0.23%Si-0.02%Fe-0.15%$O_2$-20ppm $H_2$.

Material was β solution treated for one hour at 1050°C in an argon atmosphere, and subsequently cooled at various rates, sometimes interrupted by isothermal holds or quenching.

## Results

The microstructure of β solution treated IMI685 consisted of martensite or Widmanstätten α depending on the cooling rate. At the cooling rates used in this work Widmanstätten plates of α related to the β by the same variant of the Burgers orientation relationship formed in colonies. Some plates were separated by thin (∿ 0.2 μm) layers of retained β, representing <3% of the total volume, and it was at the boundary between this β and the α that the interface phase (γ) was found.

Both the striated and monolithic forms were observed but their occurrence depended upon cooling rate. In air cooled (∼400°C/min) material all the interface phase had the monolithic morphology, but the striated form predominated after slow cooling (< 5°C/min). Both forms were present after cooling at intermediate rates. Although precise measurements of interfacial layer thickness were not made, this was much less after air cooling (∼ 0.01μm) than in furnace cooled material (∼ 0.2μm).

Figure 1a is a typical transmission electron micrograph of the structure between two α plates, but it is seen more clearly in a dark field image from an interface phase reflection (figure 1b) which shows a monolithic layer on either side of the inter-plate β. On the evidence of selected area diffraction patterns the crystal structure was probably f.c.c. (a ≃ 435 pm) but reflections often appeared as streaks and an f.c.t. structure could not be discounted. The following orientation relationship was found between the α, β and γ crystal structures, and is illustrated in figures 1c and 1d:

$$[000\bar{1}]_\alpha // [\bar{1}10]_\beta // [\bar{0}01]_\gamma; \quad (1\bar{2}10)_\alpha // (1\bar{1}\bar{1})_\beta // (\bar{1}10)_\gamma.$$

The striated form (figure 2) was polycrystalline and many crystallographic orientations were found in the same vicinity giving rise to complex diffraction patterns. However all interface phase reflections could be indexed to the f.c.c. or f.c.t. structure of the monolithic form, and were related to the β by the aforementioned orientation relationship, but only one variant obeyed the γ to α relationship. This suggests that the significant orientation relationship is between the α and β structures, and that any relationship between γ and α structure occurs solely as a result of the Burgers relationship between α and β. However, the α related variant was present in all the polycrystal γ investigated, and was invariably the predominant orientation.

After interrupting the furnace cool by an isothermal hold at 950°C for half an hour a {111}γ twinned form of the monolithic interface phase was observed (figure 3).

Specimens which had been furnace cooled to a given temperature, and then quenched, were examined to determine the temperature at which the interface phase started to form. It can be stated with certainty that the interface phase was present in specimens quenched from 700°C or lower, but β present at 800°C transformed on quenching to hexagonal α or an f.c.c. martensite, with some retained β. No interfacial layer was detectable at the prior α/β boundaries, but some doubt must be expressed because the f.c.c. martensite (figure 4) has the same crystal structure and orientation relationship to the α and β as the twinned monolithic interface phase. Thus the interface phase probably starts to form between 800 and 700°C.

## Discussion

The f.c.c. or f.c.t. structure of the interface phase is similar to that of the γ titanium hydride which has been found in (α + β) alloys [6, 7] and the orientation relationship between α and the interface phase is the same as that reported by Sanderson and Scully [8] between α and the hydride. However the composition of the alloy in the present study would give a hydrogen content for the interface phase at least ten times lower than the known composition range for the γ hydride of $TiH_{1.5}$ to $TiH_2$. Hydrogen does segregate to the β phase and the α/β interface has been shown to be a preferred site for the precipitation of hydrides [9] and it may be that hydrogen has a role in the nucleation of the interface phase. Hydrogen would be expected to occupy the tetrahedral interstices in the interface phase lattice, as in the hydride, and Hall [3] has identified the interface phase as a nucleation site for strain induced basal hydrides.

The fact that the interface phase is always related to the $\beta$ by the orientation relationship:

$$[001]_\gamma // [\bar{1}10]_\beta; \quad (\bar{1}10)_\gamma // (1\bar{1}\bar{1})_\beta$$

but only one variant is related to the $\alpha$ seems to indicate that it is a product of a $\beta \to \gamma$ transformation. However the $\alpha$ related variant is always the most dominant and in the polycrystalline form the precipitates appear to grow into the $\alpha$ phase (figure 2). The $\gamma$ probably nucleates and grows by transformation of the $\beta$ phase, but that variant which is correctly oriented will also grow by transformation of the $\alpha$.

Mechanisms for $\beta \to \gamma$ and $\alpha \to \gamma$ transformations are proposed, based on a shear of the parent lattice followed by shuffles of the atoms in the unit cell.

The proposed $\beta \to \gamma$ transformation mechanism is illustrated in figure 5. The arrangement of atoms in the (110) planes of the $\beta$ b.c.c. structure is shown on the left, and progressing to the right, it can be seen how the passage of twinning dislocations on alternate $(11\bar{2})_\beta$ planes gives an atomic arrangement which approximates to that on the cube face of the f.c.c. structure. Atom shuffles, and lattice dilatation are required to give the f.c.c. structure (far right), which is in the observed orientation relationship.

Figure 6 shows the proposed $\alpha \to \gamma$ transformation mechanism. Again working from left to right, the arrangement of atoms in the basal plane is changed by the passage of $1/6$ <11$\bar{2}$0> partial dislocations across every $(10\bar{1}0)_\alpha$ plane to a distorted f.c.c. cube face arrangement. The f.c.c. structure is obtained by appropriate atom shuffles and lattice expansions and contractions. The existance of the $1/6$ <11$\bar{2}$0> partial dislocation has not been reported, but Partridge [10] has suggested that the dissociation of a $1/3$ <11$\bar{2}$0> dislocation lying in a {10$\bar{1}$0} plane is geometrically possible in two ways:

$$1/3 <11\bar{2}0> \to 1/18 <42\bar{6}3> + 1/18 <24\bar{6}3>$$

and  $1/3 <11\bar{2}0> \to 1/6 <11\bar{2}1> + 1/6 <11\bar{2}\bar{1}>$

The $1/18$ <42$\bar{6}$3> partial dislocation is considered the most likely to occur because of the corrugated nature of the {10$\bar{1}$0} plane, but either group of partials will produce the shear required by the transformation mechanism. The effects of the additional components in the 'c' direction and normal to the shear plane could be annulled by the passage of different partials on alternate {10$\bar{1}$0} planes.

The lattice expansions required in each transformation are shown in Table 1 and are based on the lattice parameters $a_\alpha$ = 295 pm, $c_\alpha$ = 469 pm, $a_\beta$ = 326 pm, $a_\gamma$ = 435 pm.

Table 1   Lattice Expansions (%)

| Transformation | Direction of Expansion | | |
|---|---|---|---|
| | $[000\bar{1}]_\alpha // [\bar{1}10]_\beta$ $// [\bar{0}01]_\gamma$ | $[\bar{1}2\bar{1}0]_\alpha // [1\bar{1}\bar{1}]_\beta$ $// [\bar{1}10]_\gamma$ | $[\bar{1}010]_\alpha // [\bar{1}\bar{1}2]_\beta$ $// [\bar{1}10]_\gamma$ |
| $\beta \to \gamma$ | -6.7 | + 7.7 | + 13.9 |
| $\alpha \to \gamma$ | -6.3 | + 6.2 | + 21.5 |

The largest expansions are in the $[10\bar{1}0]_\alpha$ or $[\bar{1}\bar{1}2]_\beta$ direction, but a 12% lattice contraction occurs in this direction during the $\beta \to \alpha$ phase transformation. Therefore the formation of the interface phase (particularly the monolithic type) would rely on the volumetric strains resulting from the $\beta \to \alpha$ transformation.

## Conclusions

1.   An f.c.c. (a = 435 pm) or f.c.t. interface phase was observed between the $\alpha$ and $\beta$ phases in IMI 685 after $\beta$ solution treatment and cooling at rates $\leqslant 400^\circ$C/min.

2.   Two morphologies were observed; monolithic (single crystal) which was sometimes twinned, and striated (polycrystalline).

3.   The orientation relationship: $[000\bar{1}]_\alpha // [\bar{1}10]_\beta // [001]_\gamma$; $(1\bar{2}10)_\alpha // (1\bar{1}\bar{1})_\beta // (\bar{1}10$ has been established for the monolithic form, and the polycrystalline form was related to the $\beta$ in the same manner.

4.   Cooling at $400^\circ$C/min produced the monolithic form only, but the polycrystalline form predominated after slow cooling.

## Acknowledgements

The authors wish to thank Rolls Royce Ltd., Derby, for the provision of material, and acknowledge the sponsorship of the Science Research Council.

## References

1.   C.G. Rhodes and J.C. Williams:  Met. Trans. 6A(1975), 1670.
2.   C.G. Rhodes and J.C. Williams:  Met. Trans. 6A(1975), 2103.
3.   I.W. Hall: Scand. J. Met. 8(1979), 17.
4.   H. Margolin, E. Levine, and M. Young:  Met. Trans 8A(1977), 373.
5.   C.G. Rhodes and N.E. Paton:  Met. Trans. 10A(1979), 209.
6.   J.D. Boyd:  Trans. A.S.M. 62(1969), 977.
7.   I.W. Hall:  Met. Trans. 9A(1978), 815.
8.   G. Sanderson and J.C. Scully:  Trans. A.I.M.E. 239(1967), 1883.
9.   G.F. Pittinato and W.D. Hanna:  Met. Trans. 3(1972), 2905.
10.   P.G. Partridge:  Met. Reviews 118(1967), 169).

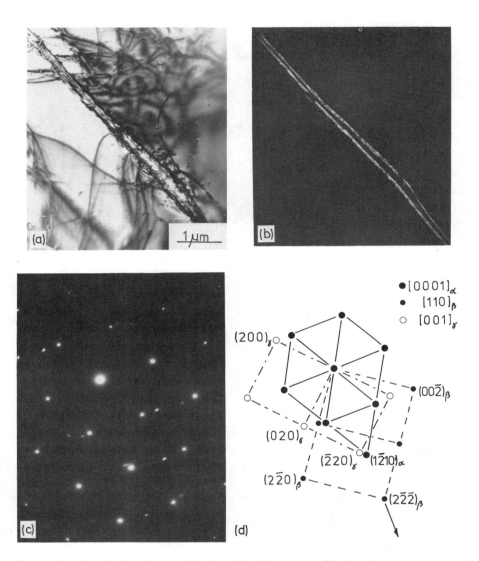

Figure 1.   Monolithic f.c.c. (γ) interface phase at α/β   boundary.

a) Bright field micrograph
b) Dark field micrograph, (020)γ reflection fig. c.
c) Selected area diffraction showing reflections from α, β and γ
d) Indexing of fig. d showing orientation relationships

$$[\overline{0}001]_\alpha // [\overline{1}10]_\beta // [\overline{0}01]_\gamma; \quad (1\overline{2}10)_\alpha // (1\overline{1}\overline{1})_\beta // (\overline{1}10)_\gamma.$$

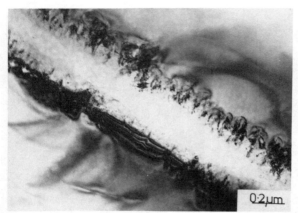

Figure 2.

   Polycrystalline f.c.c.
   interface phase at α/β
   boundary.

Figure 3.

   {111} twinned f.c.c.
   interface phase.

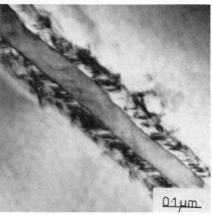

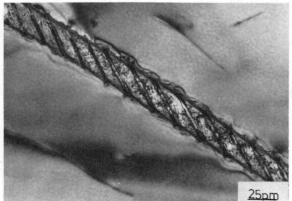

Figure 4.

   {111} twinned f.c.c.
   martensite between α
   plates after 1hr at
   1050°C, furnace cool to
   800 and oil quench.

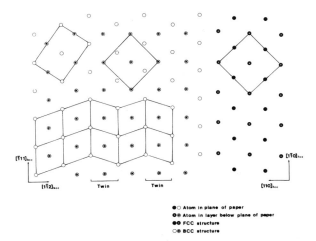

Figure 5.

b.c.c. (β) → f.c.c. (γ) transformation mechanism giving orientation relationship

$$[\bar{1}10]_\beta // [\bar{0}01]_\gamma; \quad (1\bar{1}\bar{1})_\beta // (\bar{1}10)_\gamma.$$

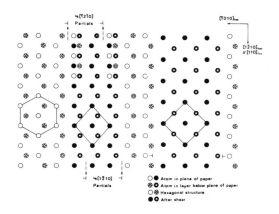

Figure 6.

Hexagonal (α) → f.c.c. (γ) transformation mechanism, giving orientation relationship

$$[\bar{0}001]_\alpha // [\bar{0}01]_\gamma; \quad (1\bar{2}10)_\alpha // (\bar{1}10)_\gamma.$$

THE AGEING CHARACTERISTICS OF Ti-3%Al-8%V-6%Cr-4%Mo-4%Zr
(Ti-38644)

G.C. Morgan and C. Hammond

Department of Metallurgy
University of Leeds
Leeds LS2 9JT

## Introduction

Ti-3%Al-8%V-6%Cr-4%Mo-4%Zr (Ti-38644) is a commercial metastable $\beta$ titanium alloy combining high strength and toughness with deep hardenability. Decomposition of the $\beta$ phase, typical of this alloy group, results in the formation of the hexagonal $\alpha$ phase which, depending upon ageing temperature, may be preceded by the phenomenon variously described as the $\beta \rightarrow \beta + \beta'$ phase separation [1], or the formation of zones [2]. The precipitated $\alpha$ phase is related to the $\beta$ matrix by the Burgers orientation relationship i.e. $(110)_\beta//(0001)_\alpha$ and $<111>_\beta//<11\bar{2}0>_\alpha$. An additional form of $\alpha$ which does not obey this orientation relationship has been observed in Ti-Mo and Ti-Mo-Al alloys [3] (denoted 'Type 2' $\alpha$ or non-Burgers $\alpha$), and also Ti-6Al-4V [4], with a similar effect noted in Hf rich Nb alloys [5]. The presence of 'Type 2' $\alpha$ is characterised by anomalous arced diffraction reflections.

## Experimental Procedure

Ti-38644 was supplied by the Reactive Metals Inc. in the form of sheet approximately 1 mm thick. Chemical composition of the alloy is given in Table 1.

Table 1  Chemical Compositions

| Composition wt% | | | | | | | | | |
|---|---|---|---|---|---|---|---|---|---|
| Al | V | Cr | Mo | Zr | Fe | C | N | O | H(ppm) |
| 3.4 | 8.1 | 5.6 | 4.3 | 3.6 | 0.06 | 0.01 | 0.012 | 0.102 | 72 |

Specimens were solution treated at $800^\circ$C for $\frac{1}{2}$hr in an Argon atmosphere, water quenched, and aged in air at various temperatures between $300^\circ$C and $600^\circ$C for up to 1000 minutes, then air cooled. Hardness measurements were taken to monitor ageing response. Thin foils were prepared using standard electropolishing techniques [6] and observed using a Phillips EM 300 transmission electron microscope with double tilt goniometer stage.

## Results and Discussion

The change in hardness during ageing can be seen in figure 1. No significant hardness response was found after ageing at $300^\circ$C or $350^\circ$C, although an increase was detected after 1000 minutes at $400^\circ$C. Maximum hardness was reached after ageing at $500^\circ$C, VHN $\sim$ 450, and a maximum of $\sim$ 365 at $600^\circ$C. Table two lists the microstructures observed by transmission electron microscopy after ageing at various times and temperatures.

Table 2   Ageing Response:-   Precipitation Observed by TEM

| Temp/$^{o}$C | Time/mins | Observation |
|---|---|---|
| 400 | 30 | Zones, fine |
|  | 250 | Zones, coarse |
| 500 | 5 | Zones, fine |
|  | 30 | Zones, coarse |
|  | 120 | $\alpha$ plates |
|  | 420 | Limited Type 2 $\alpha$ within the $\alpha$ plates. |
| 600 | 5 | Zones, few |
|  | 30 | $\alpha$ plates |
|  | 120 | Type 2 $\alpha$ (tetragonal) within $\alpha$ plate |
|  | 360 | Type 2 $\alpha$ (f.c.c.) within $\alpha$ plates. |

In all cases, the hardness response was caused by the precipitation of Type $\alpha$ (Burgers), which was preceded by the formation of zones, although only a limited number of zones were observed at 600$^{o}$C.

Disc shaped zones, lying on {100}$_\beta$ planes (figure 2) were observed after ageing for 30 minutes at 400$^{o}$C and 5 minutes at 500$^{o}$C when they were approximatel 50 nm in diameter. Coarsening occurred on further ageing until after 250 and 60 minutes at 400$^{o}$C and 500$^{o}$C respectively they were approximately 100 nm in diameter. The few zones observed at 600$^{o}$C were approximately 150 nm in diameter. No extra reflections were obtained in electron diffraction patterns which could be attributed to the zones, and although X-ray diffraction revealed $\beta$ peak broadening, no $\beta$ peak splitting was observed, indicating that the zones were not formed as the result of the reported $\beta \rightarrow \beta + \beta'$ phase separation.

The $\alpha$ precipitated as plates, which when first observed after ageing at 500$^{o}$C for 120 minutes were $\sim$ 500 nm long and $\sim$ 50 nm thick (Fig. 3). The hexagonal crystal structure of the $\alpha$ is in the Burgers orientation relationship with the $\beta$ matrix, and the striations seen in some plates are twins on {10$\bar{1}$1} planes, similar to those found in twinned martensite. Coarsening occurs upon further ageing.

After ageing for 420 minutes at 500$^{o}$C (figure 4) the anomalous arced diffraction reflections, characteristic of Type 2$_\alpha$, appeared on diffraction patterns. The diffraction pattern in Figure 4 (b)$^\alpha$ shows a <100>$_\beta$ zone with <10$\bar{1}$1> zones, which are {10$\bar{1}$1} twin related, and <11$\bar{2}$0>$_\beta$ zones, all of which obey the Burgers orientation relationship. The remaining$^\alpha$ reflections (Type 2 $\alpha$) can be indexed as <110> face centred cubic zones (lattice parameter a $\sim$ 420 pm. which are twin related on the {111} face centred cubic planes.

Ageing at 600$^{o}$C for 120 mins. (figure 5) revealed a similar precipitate, although the diffraction pattern obtained, figure 5(b) can be indexed as a

$<100>_\beta$ zone with $<110>$ face centred tetragonal zones (c/a $\overset{\sim}{} 0.9$) which are twin related on $\{111\}_{f.c.t.}$ The f.c.t. type $2\alpha$ precipitated in the Type $1\alpha$ plates and at the plate boundaries as shown in figures 5(c) and 5(d). The diffraction evidence in figures 4 and 5 would suggest an orientation relationship of the f.c.c. or f.c.t.) phase to the precipitated Burgers $\alpha$ and the $\beta$ matrix of $(111)_{f.c.c.}$ // (0001) // (110)$_\beta$ and $<110>_{f.c.c.}$ // $<11\bar{2}0>_\alpha$ // $<111>_\beta$. Fig. 6(a) shows a selected area diffraction pattern obtained from a single Type $1\alpha$ plate in material aged for 360 minutes at 600°C, with an f.c.c. Type $2\alpha$ $<110>$ zone (which cannot be indexed to the hexagonal $\alpha$). The centred dark field images obtained from reflections in this zone, figure 6(b), all show the same microstructural feature within the $\alpha$ plate.

The formation of an f.c.c. phase in Titanium alloys is not unknown. Rhodes and Williams [7] have found that an f.c.c. phase forms at $\alpha/\beta$ interfaces (interface phase) in Ti-6Al-4V, with a lattice parameter of approximately 420 pm. Recent work by Rhodes and Paton [8] has shown that the interface phase has an orientation relationship of $(110)_\beta$ // $(111)_{f.c.c.}$ // $(0002)_\alpha$ and $[111]_\beta$ // $[1\bar{1}0]_{f.c.c.}$ // $[11\bar{2}0]_\alpha$ and is internally twinned on $\{111\}_{f.c.c.}$. The f.c.c. Type $2_\alpha$ precipitate found in the present work is crystallographically identical to the interface phase found in Ti-6Al-4V.

Significant differences between the two f.c.c. phases are the conditions necessary for their formation: the interface phase was only formed on slow continuous cooling after $\beta$ solution treatment, but the f.c.c. precipitate found in the present work formed on isothermal ageing. Whereas the interface phase is only found at $\alpha/\beta$ interfaces, the f.c.c. Type $2\alpha$ also precipitates within Type $1\alpha$ plates.

## Conclusions

1.    Hardness response during ageing Ti-38644 at temperatures between 300°C and 500°C for up to 1000 minutes is caused solely by the precipitation at hexagonal $\alpha$, which occurs in the Burgers orientation relationship to the $\beta$ matrix.

2.    Zones were found preceding $\alpha$ precipitation, but no evidence was found to indicate that these zones were formed by a phase separation of the metastable $\beta$ $(\beta \rightarrow \beta + \beta')$.

3.    Longer ageing times at 500°C and 600°C resulted in the presence of a face centred cubic phase with a $\overset{\sim}{} 420$ pm (or face centred tetragonal with c/a $\overset{\sim}{} 0.9$) precipitating within the $\alpha$ plates. This f.c.c. phase produced arced diffraction reflections typical of the previously reported non-Burgers 'Type 2'$\alpha$.

4.    Diffraction evidence would suggest an orientation relationship of the f.c.c. Type $2\alpha$ to the Burgers related $\alpha$ and $\beta$ of $(111)_{f.c.c.}$ // $(0001)_\alpha$ // $(110)_\beta$ and $<110>_{f.c.c.}$ // $(11\bar{2}0)_\alpha$ // $<111>_\beta$.

5.    F.c.c. Type $2\alpha$ is crystallographically identical to the f.c.c. interface phase found in Ti-6Al-4V.

## References

1.    C.G. Rhodes and N.E. Paton:  Met. Trans., 1977, vol. 8, pp 1749-1761.
2.    T.J. Meadley and M.J. Rack:  Met Trans., 1979, vol. 10, pp 909-920.
3.    C.G. Rhodes and J.C. Williams:  Met. Trans., 1975 vol. 6, pp 2103-2114.
4.    A Lasalmonie and M. Loubradou:  J. Mat. Sci, 1979, vol. 14, pp 2589-2595.
5.    W.B. Jones, R. Taggart, and D.M. Polonis :  Met. Trans., 1978, vol. 9, pp 723-729.

6.   M.J. Blackburn and J.C. Williams:  Trans. T.M.S. - A.I.M.E., 1967, vol. 239,
     p 287-288.
7.   C.G. Rhodes and J.C. Williams:  Met. Trans., 1975, vol. 6. pp 1670-1671.
8.   C.G. Rhodes and N.E. Paton:  Met. Trans., 1979, vol. 10, pp 209-216.

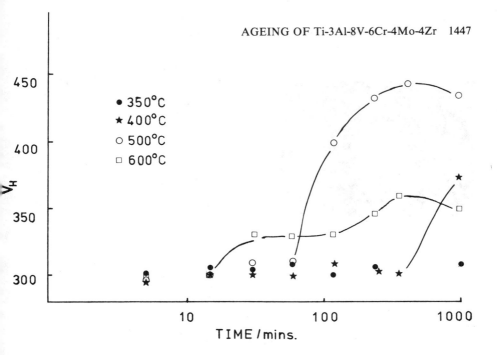

Figure 1.   Hardness response ($V_H$) vs. log time (minutes) on ageing.

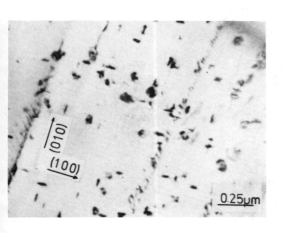

Figure 2.

Zones after ageing for 250
minutes at $400^\circ$C.

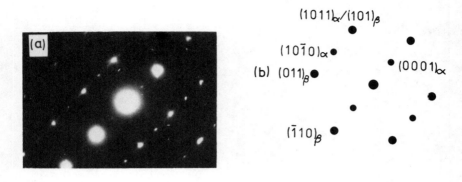

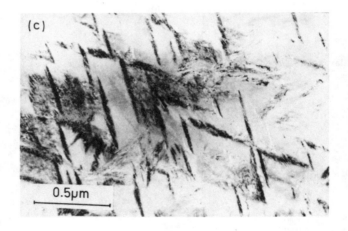

Figure 3.    (a) Selected area diffraction pattern

(b) Indexing

(c) α-plates after ageing for 120 minutes at 500°C.

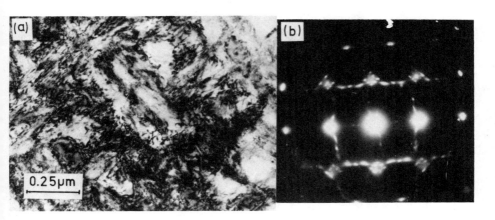

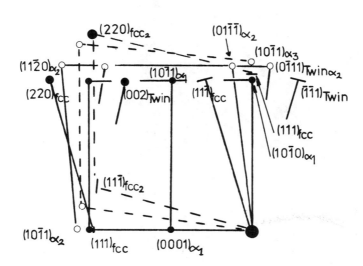

Figure 4.   (a) α-plates after ageing for 420 minutes at 500°C
           (b) Selected area of diffraction pattern
           (c) Indexing.

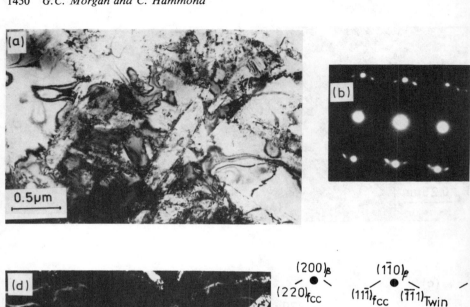

(200)$_\beta$         (1$\bar{1}$0)$_\beta$
(220)$_{fcc}$    (11$\bar{1}$)$_{fcc}$   ($\bar{1}\bar{1}$1)$_{Twin}$

(c)

(110)$_\beta$/(111)$_{fcc}$

(002)$_{fcc}$        ($\bar{1}\bar{1}$1)$_{fcc}$

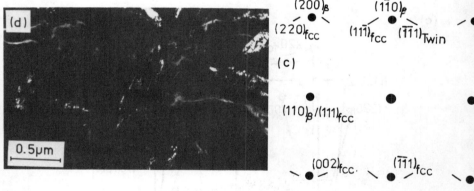

Figure 5.
α-plates (a) after ageing for
120 minutes at 600°C
(b) Selected area diffraction
pattern, (c) indexing, (d) and
(e) dark field images from
(11$\bar{1}$)f.c.t. and ($\bar{1}\bar{1}$1)f.c.t.
twin reflections respectively.

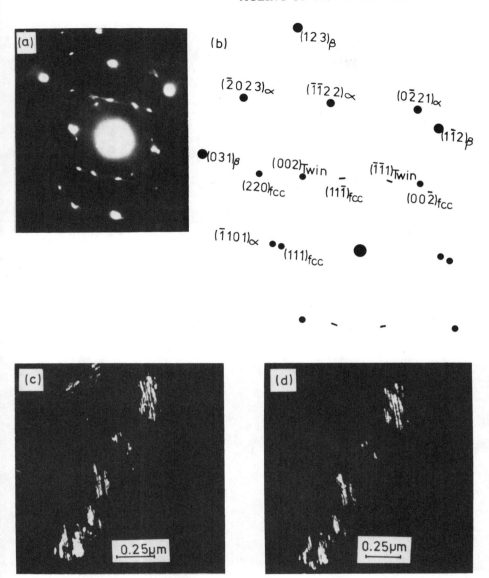

Figure 6.  Ti-38644 aged for 360 minutes at 600°C.
         (a) Selected area diffraction pattern
         (b) indexing
         (c) and (d) dark field images from (11$\bar{1}$)f.c.c. and (220)f.c.c.
         reflections respectively.

# B. Phase Transformation and Heat Treatments

# EFFETS OF THERMAL CYCLES AND SUBSTITUTION ELEMENTS ON THE PHASE TRANSFORMATIONS OF TiNi

Toshio Honma, Minoru Matsumoto, Yoshiro Shugo, Minoru Nishida
and Isao Yamazaki*

Research Institute of Mineral Dressing and Metallurgy,
Tohoku University, Sendai Japan.

## Introduction

TiNi is a typical material which exhibits a unique "shape memory effect" caused by the thermoelastic martensitic transformation. In the Ti-Ni phase diagram, TiNi has a narrow nonstoichiometory near $Ti_{50}Ni_{50}$. The parent phase of TiNi has B2 (CsCl type) crystal structure and transforms thermoelastically to distorted B19 (AuCd type) crystal structure at near room temperature by cooling. Wang et al. [1,2] reported that as for TiNi, the temperature change of the electrical resistivity was affected by heat-treatment and the temperature of the phase transformation varied by the addition of other element. The purpose of this paper is the studies of the effects of thermal cycles and substitution elements on the phase transformations of TiNi.

## Experiment

1. Specimens

The specimens of TiNi were made as follows. Ti, Ni and various substitution elements were balanced by desired proportion and melted by the electron beam or argon arc method. The products were remelted by several times in order to homogenize it. The obtained buttons were shaped to the dimensions for measurements. These specimens were homogenized at about 900°C for several hours in vacuum before measurements.

2. Experimental methods

In the study the temperature change of the electrical resistivity, the magnetic susceptibility and the specific heat accompanied with the phase transformations of TiNi were measured and crystal structures of these phases were investigated by X-ray and neutron diffraction. The electrical resistivity was measured by the conventional method or A.C. method (1kHz). The magnetic balance method was used for the measurement of the magnetic susceptibility. By the use of an adiabatic calorimeter the specific heat was measured.

## Results and Discussions

I. Effect of factors on the phase transformation of TiNi

1. Thermal cycles

Figure 1 shows the effect of incomplete thermal cycle on the temperature

---

*Present address: Sumitomo Metal Industries, Ltd., Kashima, Ibaraki, Japan.

change of the electrical resistivity of TiNi. As shown in Fig.1.(A), the electrical resistivity in the parent phase of TiNi decreased linearly with cooling and at the martensite start temperature ($M_S$ temperature) it decreased abruptly. On heating from low temperature the electrical resistivity increased gradually with the increase of temperature and abrupt temperature change of the electrical resistivity was not observed. In the case of incomplete thermal cycle between the martensite finish temperature ($M_f$ temperature) and the austenite finish temperature ($A_f$ temperature) the electrical resistivity changed with temperature as shown in Fig.1.(B) $\sim$ (D). With the increase of incomplete thermal cycles $M_S$ temperature decreased and the electrical resistivity at $M_S$ temperature increased and the peak of the electrical resistivity appeared clearly in the electrical resistivity vs. temperature curves. In the case of complete thermal cycle between lower temperature than $M_f$ temperature and the higher temperature than $A_f$ temperature, same behaviour of the electrical resistivity were observed. These results are shown in Fig.2. In either case of incomplete and complete thermal cycle, $M_S$ temperature and the electrical resistivity at $M_S$ temperature changed abruptly for initial thermal cycles and these values become gradually constant with the increase of thermal cycles. The figure of the electrical resistivity vs. temperature curve did not almost change with the increase of thermal cycles. Figure 3 shows the difference of the effect of incomplete and complete thermal cycles on the increase of the electrical resistivity. $\Delta\rho/\rho$ is the rate of increase of the electrical resistivity. $\rho$ is the maximum amount of the electrical resistivity at the temperature of peak of the electrical resistivity. $\Delta\rho$ is the difference of $\rho$ and the electrical resistivity value extrapolated from the parent phase region at same temperature. $\Delta\rho/\rho$ of incomplete thermal cycle was higher than that of complete thermal cycle. In incomplete thermal cycle, $\Delta\rho/\rho$ was largely affected by Ni concentration.

Then the effect of aging heat-treatment on the phase transformation of TiNi was studied. Specimens were treated by thermal cycle and successively heated at 400°C for 1 $\sim$ 10 hrs. Figure 4 shows the effect of aging heat-treatment on the electrical resistivity of TiNi. The peak of the electrical resistivity at the austenite start temperature ($A_S$ temperature) was anew observed in the electrical resistivity vs. temperature curves as compared with incomplete thermal cycle in Fig.1.

2.   Substitution of 3d transition metal element

These above unique temperature changes of the electrical resistivity were also observed in the case of substitution of 3d transition metal elements (M) for Ti or Ni in TiNi. On cooling from high temperature, the electrical resistivity of $Ti_{50-x}M_xNi_{50}$ (M = V, Cr and Mn) and $Ti_{50}Ni_{50-x}M_x$ (M = Fe and Co) decreased linearly and it increased at a certain temperature ($M_S^I$ temperature) and then it decreased suddenly at $M_S$ temperature. A typical example is shown in Fig.5. From the present experimental results of the temperature change of the magnetic susceptibility and the specific heat and also X-ray and neutron diffraction of $Ti_{50}Ni_{47}Fe_3$, TiNi which has various substitution elements exhibits two step transformations ( parent phase I (B2) $\rightleftharpoons$ intermediate state II (rhombohedral) $\rightleftharpoons$ martensite phase III (B19) ).

The temperatures ($M_S^I$, $M_S$, $A_S$ and $A_S^I$) of the phase transformations decreased linearly with the increase of the concentration (x) of 3d transition metal elements. The concentration dependence of the temperatures of the phase transformation is shown in Fig.6 in the case of $Ti_{50}Ni_{50-x}Fe_x$. The rates of decrease of these temperatures were different from each other. Thus each phase was actualized as shown in I, II and III. The electrical resistivity vs. temperature curve had not simple figure but a unique one unlike the case of thermal cycle because the decreasing rates of the phase transformation temperatures for x were different from each other.  Figure 7 shows the effect of substitution of 3d transition metal element of the $M_S$ temperature.  The $M_S$ temperatures decreased linearly with the increase of the 3d transition metal element as V, Cr, Mn, Fe and Co.  This rectilinearity of the decrease of the phase transformation temperatures was rearranged by the concept of the average electron concentration per one atom, e/a, which is difined by that total (3d + 4s) electron number divided by total atom number of Ti, Ni and substitution elements.  The results are shown in Fig.8.  $M_S$ temperature was highest at e/a = 7 (TiNi) and changed linearly with e/a in the both case of e/a < 7 and e/a > 7.  It was confirmed that the $M_S$ temperature did not change independently of x of $(TiNi)_{100-x}Mn_x$ for e/a = 7.

### 3.  Atomic size

In order to investigate the effect of atomic size independently of e/a, the temperature change of the electrical resistivity of $Ti_{50-x}Zr_xNi_{50}$ was measured.  As shown in Fig.9 the electrical resistivity showed maximum at $M_S$ temperature for either x = 1 and x = 3.  $M_S$ temperature is lower for x = 3 than x = 1.  The valency of Zr and Ti is +4 and the ion radius of Zr is 1.36 times larger than that of Ti.  Zr has 4d and 5s electrons and so these effect cannot be able to explain by the above discussion of e/a of 3d transition metal element.  Thus it is difficult to compare the effect of substitution of Zr with that of 3d transition metal elements.  Above experimental results show the effect of ion radius on the phase transformation of TiNi.

### Ⅱ.  Effect of Cu on the phase transformation of TiNi

Ni in TiNi is substituted by Cu in order to put this material to practical use for "marmem alloy".  Specimens preparation and experimental methods were as same as above mentioned.  The concentration (x) of Cu in $Ti_{50}Ni_{50-x}Cu_x$ were 5, 10, 20 and 30at%.  From the results of X-ray diffraction, the parent phase of these specimens was single phase with B2 structure and the martensite phase for x = 5 was distorted B19 structure.  Figure 10 shows the temperature change of the electrical resistivity of $Ti_{50}Ni_{50-x}Cu_x$.  In the case of $Ti_{50}Ni_{45}Cu_5$, the electrical resistivity increased at the $M_S$ temperature on cooling and it decreased at the $A_S$ temperature on heating.  These results were reverse as compared to the case of TiNi.  The $M_S$ temperature was lower than the $A_S$ temperature.  These behabiour of the electrical resistivity was different from each other with the Cu concentration.  From these results, the $M_S$ temperatures were determined as shown in Fig.11.  The $M_S$ temperature increased slightly with the increase of Cu concentration.  The temperature hysteresis $(A_S - M_S)$ decreased with the increase of Cu concentration.  The hardness of $Ti_{50}Ni_{50-x}Cu_x$ was measured at room temperature by micro Vickers hardness unit.  The results were shown in Fig.12.  The hardness was mimimum value for near x = 10 at which load was 1kg.  The tensile testing was carried out on x = 5 and 10 at room temperature.  The tensile properties of $Ti_{50}Ni_{45}Cu_5$ was as same as TiNi and the tensile strength of $Ti_{50}Ni_{40}Cu_{10}$ was lower than that of TiNi.

### References

1.  F.E.Wang, B.F.DeSavage, W.J.Buehler and W.R.Hosler: J. Appl. Phys., 39 (1968), 2166.
2.  F.W.Wang: Proceedings of the First International Conf. on Fracture, (1965), BⅡ-103.

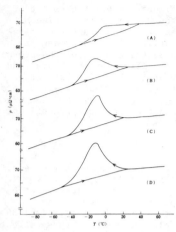

Fig. 1.   The effect of incomplete thermal cycle on the electrical resistivity of $Ti_{49.2}Ni_{50.8}$.  (A) no thermal cycle.   (B) after 1 thermal cycle.   (C) after 3 thermal cycles. (D) after 9 thermal cycles.

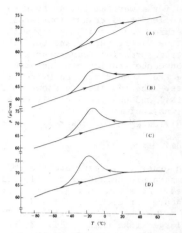

Fig. 2.   The effect of complete thermal cycle on the electrical resistivity of $Ti_{49.2}Ni_{50.8}$.   (A) no thermal cycle.   (B) after 3 thermal cycles. (C) after 9 thermal cycles.   (D) after 18 thermal cycles.

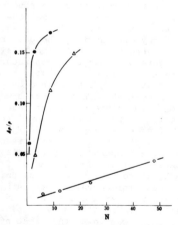

Fig. 3.   The effect of thermal cycles (N) on the rate of increase of the electrical resistivity.

   O : $Ti_{49.8}Ni_{50.2}$
            (incomplete thermal cycle)
   ● : $Ti_{49.2}Ni_{50.8}$
            (incomplete thermal cycle)
   △ : $Ti_{49.2}Ni_{50.8}$
            (complete thermal cycle)

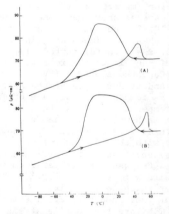

Fig. 4.   The effect of aging heat-treatment (400°C) on the electrical resistivity of $Ti_{49.2}Ni_{50.8}$ (after 9 incomplete thermal cycles).   (A) after heat-treatment for 1 hour.   (B) after heat-treatment for 10 hours.

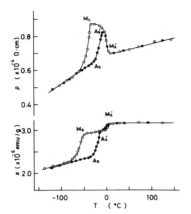

Fig. 5.   The electrical resistivity and the magnetic susceptibility vs. temperature curves of $Ti_{50}Ni_{47}Fe_3$.
O  :  cooling,
●  :  heating.

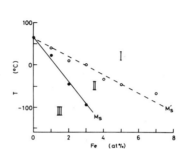

Fig. 6.   The $M_S^I$ and $M_S$ temperature of $Ti_{50}Ni_{50-x}Fe_x$.
I  :  parent phase,
II  :  intermediate state,
III  :  martensite phase.

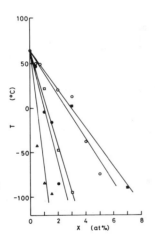

Fig. 7.   The effect of substitution of 3d transition metal element on the $M_S$ temperature ( x:concentration).
O  :  V,      △  :  Cr,      ●  :  Mn,
□  :  Fe,     ■  :  Co.

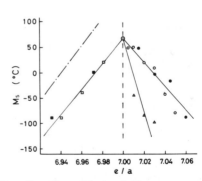

Fig. 8.   The effect of the average electron concentration per one atom, $e/a$, on the $M_S$ temperature.
O  :  V,      △  :  Cr,      ●  :  Mn,
□  :  Fe,     ■  :  Co.
—  ·  —      Wang [2].

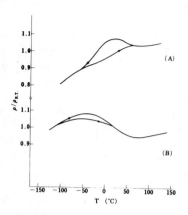

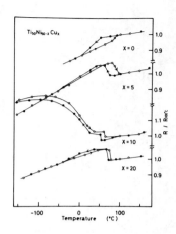

Fig. 9.    The electrical resistivity
vs. temperature curves of $Ti_{50-x}Zr_xNi_{50}$.
(A) x = 1,    (B) x = 3.

Fig. 10.    The electrical resistivity vs. temperature curves of
$Ti_{50}Ni_{50-x}Cu_x$.

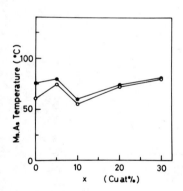

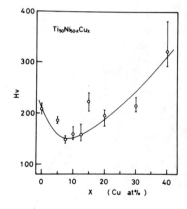

Fig. 11.    The $M_S$ and $A_S$ temperatures
of $Ti_{50}Ni_{50-x}Cu_x$.
    O : $M_S$ temperature,
    ● : $A_S$ temperature.

Fig. 12.    The Vickers hardness of
$Ti_{50}Ni_{50-x}Cu_x$.

DETERMINATION OF PHASE TRANSFORMATION TEMPERATURES OF TiNi
USING DIFFERENTIAL THERMAL ANALYSIS

R. Vincent Milligan

U.S. Army Armament Research and Development Command
Benet Weapons Laboratory, LCWSL
Watervliet, NY 12189

## Introduction

Many different methods have been used to determine phase transformation temperatures in metals. Table 1 lists some of the more popular ones - though not necessarily in their order of importance. References cited in the tables are examples and not meant to be all inclusive. For the most part, the investigators listed studied the TiNi system, but there are some exceptions. Many additional references are listed in references [10] and [11].

Although there have been some efforts using Differential Thermal Analysis [1,4,12] the data is quite sparce and it appears that little effort has been made to correlate the results with other techniques for the TiNi system. The objective of the study was to determine transformation temperatures using Differential Thermal Analysis and then compare these results with those obtained by X-Ray, dilatometry, and electrical resistivity methods.

Table 1  Summary of Methods for Determining Phase Transformations

| Author | Method | Ref. No. |
|---|---|---|
| R.J. Wasilewski et al | X-Ray | 1 |
| I.I. Kornilov et al | Dilatometry | 2 |
| W.B. Cross et al | Elect. Resistivity | 3 |
| H.U. Schuerch | Diff. Thermal Anal. | 4 |
| G.R. Speich et al | Acoustic Emission | 5 |
| R.R. Hasiguti et al | Internal Friction | 6 |
| N.G. Pace et al | Ultrasonic | 7 |
| C.M. Wayman | Metallographic Techs. | 8 |
| K.H. Eckelmeyer | Strain-Temp. Techs. | 9 |

## Experimental

Three different heats of the material were supplied by Titanium Metals Corp. and some additional material by Reactive Metals, Inc. in the form of 1/2 inch diameter rods. These alloys had compositions in the range of 49 to 51 atomic percent titanium. Thin slices 1/16 inch thick were cut from the rods using a low speed metallurgical saw. Small pie-shaped pieces weighing approximately 15 to 25 milligrams were then cut out of the disks so that they would fit into the pan of the Differential Thermal Analyzer.

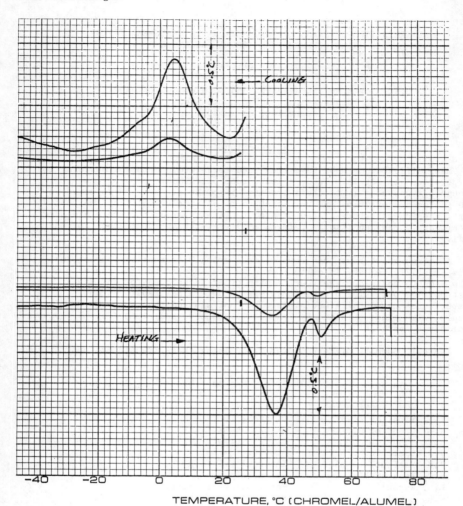

Figure 1.   Change in Differential Temperature vs. Temperature.   Traces were obtained using a Differential Thermal Analyzer.   The top trace shows an exothermic peak on cooling and the bottom trace shows an endothermic peak on subsequent heating.   The smaller peaks are the same data on another pen with a suppressed vertical scale.

Some material was kindly supplied by Goodyear Aerospace Corp. in the form of 100 mil diameter wire. This was cut into thin disks and tested in the DTA without further cutting.

The equipment used was a Dupont 990 Differential Thermal Analyzer. The heating and cooling rates were 10°C per minute. On cooling an exothermal peak was obtained which was indicative of the $M_S$ temperature. The $M_S$ temperature is defined at the temperature at which the parent phase starts to transform to Martensite. On subsequent heating an endothermic peak was obtained and gave the $A_S$ temperature. The $A_S$ temperature is defined as the temperature at which the martensite starts to transform back to the parent phase. For specimens having rather low $M_S$ temperatures, a cold cell filled with liquid $N_2$ was used to provide adequate cooling. Figure 1 is a copy of an actual trace showing these peaks.

### Results and Discussion

Figure 2 is a replot of an $M_S$ transformation curve using data obtained by Kornilov et al [2] by the method of dilatometry. Here I have taken the liberty to plot the transformation temperatures vs. atomic percent titanium rather than atomic percent nickel as done by Kornilov. For brevity, Figure 2 also contains a plot of X-ray data taken from Table 1 of reference [1], as well as a replot of resistivity data by Hanlon et al [13], Figure 7 where I have used °C on the ordinate rather than °K by the authors. In Figure 3, $M_S$ data obtained in the present study by the DTA data superimposed on the line from Figure 3 along with some additional resistivity data by Cross et al [4]. With the exception of the one data point by Wang et al [12], the data agrees quite well with that obtained in the present study. Figure 5 is a summary plot showing a superposition of the results for the $M_S$ temperatures from Figures 2 to 4.

$M_S$ TEMPERATURE vs ATOMIC PERCENT TITANIUM

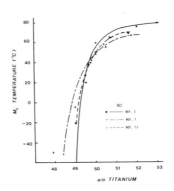

Figure 2

Ms TEMPERATURE vs. ATOMIC PERCENT TITANIUM
Ms TEMPERATURE (°C)
60
40
20
0
-20
-40
48
49
50
51
52
a/o TITANIUM

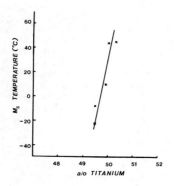

Figure 3

Ms TEMPERATURE vs. ATOMIC PERCENT TITANIUM
Ms TEMPERATURE (°C)
80
60
40
20
0
-20
48
49
50
51
PRESENT STUDY
WASILEWSKI ET ALDTA REF 1
SCHUERCH DTA REF 4
WANG ET AL DTA REF 12
CROSS ET AL RESISTIVITY REF 3
a/o TITANIUM

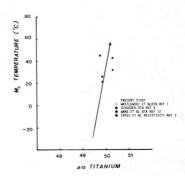

Figure 4

Ms TEMPERATURE vs. ATOMIC PERCENT TITANIUM
Ms TEMPERATURE (°C)
80
60
40
20
0
-20
-40
48
49
50
51
52
53
KEY
M PRESENT STUDY
K REF. 2
W REF. 1
H REF. 13
a/o TITANIUM

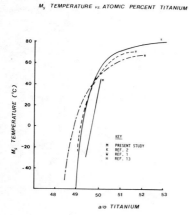

Figure 5

As TEMPERATURE vs ATOMIC PERCENT TITANIUM
As TEMPERATURE (°C)
120
100
80
60
40
20
0
-20
-40
48
49
50
51
52
53
KEY
REF. 9
REF. 7
a/o TITANIUM

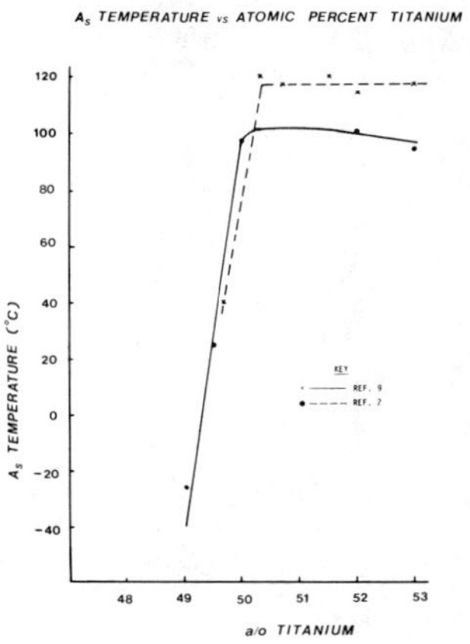

Figure 6

Results from the present study are about as far to the right of the Wasilewski and Hanlon lines as the Kornilov results are to the left. The overall difference in the results in the temperature range from −60° to +60° is less than 1 percent on the composition axis. An average for the four studies would be close to the Wasilewski–Hanlon curves. The very striking thing about these results is how sensitive the transition temperature is to composition.

Figure 6 shows a replot of Kornilov's data for the $A_s$ temperatures. Also included are some data from Figure 3 of Eckelmeyer's report [9]. Eckelmeyer obtained this data by straining the specimens approximately 4 percent, unloading, heating, and then observing the temperature at which the specimen starts to contract (revert back) to its original length. Unfortunately, for this comparison, most of the data is located in the composition range where the transformation temperature is relatively insensitive to temperature. Figure 7 is a plot of data obtained in the present study by the DTA method. Figure 8 is a plot of some miscellaneous DTA data as well as some resistivity data superimposed on the results from Figure 7. In this case the scatter seems to be a little larger than in the plot for the $M_s$ temperatures. Finally, Figure 9 is a superposition of the results from Figures 6 and 7 showing excellent agreement for the three separate studies.

Results from this study indicate that DTA can be a viable method for determining phase transformation temperatures in the TiNi system. If carefully done, it can provide information less expensively than that obtained by using highly sophisticated equipment such as X-ray. It could be a helpful ancillary tool in stress-assisted martensitic, shape memory transformation studies.

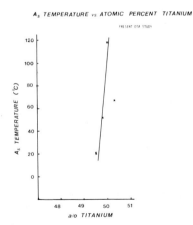

Figure 7

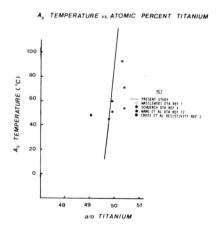

Figure 8

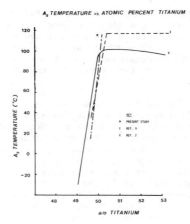

Figure 9

## Conclusions

1.  Based on the excellent agreement of the data from this study with that previously obtained by other methods, it can be concluded that DTA is a credible method for determining phase transformation temperatures in the TiNi system.

2.  The sensitivity of the $M_S$ and $A_S$ temperatures with composition which is of the order of 100°C for 1 atomic percent increase in titanium indicates the potential problems that can arise in seeking repeatability of behavior for TiNi material.

## Acknowledgment

I would like to express appreciation for the help received from Fred Stimler of Goodyear Aerospace and Paul Cote of LCWSL, Watervliet Arsenal in carrying out these experiments.

## References

1.  R. J. Wasilewski, S. R. Butler, and J. E. Hanlon:  "On the Martensitic Transformation in TiNi," Jour. of Met. Sci., Vol., 1967, pp. 104-110.

2.  I. I. Kornilov, Ye V. Kachur, and O. K. Belousov:  "Diffraction Analysis of Transformation in the Compound TiNi," Fiz. Met. Metalloved, Vol. 32, No. 2, 1971, pp. 420-422.

3.  W. B. Cross, A. H. Kariotis, and F. J. Stimler:  "Nitinol Characterization Study," NASA CR 1433, September 1969.

4.  H. U. Schuerch:  "Certain Physical Properties and Applications of Nitinol," NASA CR 1232, November 1969.

5.  G. R. Speich and R. M. Fisher:  "Acoustic Emission During Martensite Formation," ASTM STP 505, May 1972, pp. 140-151.

6.  R. R. Hasiguti and K. Iwasaki:  Correlation Between Plastic Deformation and Phase Changes in the Compound TiNi With Special Reference to Internal Friction," Supp. to Trans. Jap. Inst. of Metals, Vol. 9, 1968, pp. 288-291.

7.  9. G. Pace and G. A. Saunders:  "Ultrasonic Study of the Martensitic Phase Change in TiNi," Phil. Mag., Vol. 22, No. 175, July 1970, pp. 73-82.

8.  C. M. Wayman:  "Deformation, Mechanisms, and Other Characteristics of Shape Memory Alloys," Shape Memory Effects in Alloys, Plenum Press, J. Perkins, Ed., May 1975, pp. 1-27.

9.  K. H. Eckelmeyer:  "The Effect of Alloying on Shape Memory Phenomenon in Nitinol," Scripta Met., Vol. 10, 1976, pp. 667-672.

10. C. M. Jackson, H. J. Wagner, and R. J. Wasilewski:  "55-Nitinol — The Alloy With a Memory:  Its Physical Metallurgy, Properties, and Applications," NASA SP 5110, 1972.

11. Shape Memory Effects in Alloys, Plenum Press, J. Perkins, Ed., May 1975.

12. F. E. Wang, B. F. DeSavage, and W. J. Buehler:  "The Irreversible Critical Range in the TiNi Transition," Jour. of Applied Physics, Vol. 39, No. 5, April 1968, pp. 2166-2175.

13. J. E. Hanlon, S. R. Butler, and R. J. Wasilewski:  "Effect of Martensitic Transformation on the Electrical and Magnetic Properties of TiNi," Trans. AIME, Vol. 239, September 1967, pp. 1323-1327.

# THE THERMODYNAMIC AND STRUCTURAL ASPECTS
## OF THE INVESTIGATION OF TITANIUM NICKELIDE-BASED ALLOYS
## WITH THE SHAPE MEMORY EFFECT

Yu. K. Kovneristii, O. K. Belousov,
S. G. Fedotov, N. M. Matveeva

Baikov Institute of Metallurgy
USSR

The shape memory effect is a rather complex phenomenon caused by thermoelastic martensitic transformations in alloys. Such phenomena require numerical thermodynamic, kinetic, structural and mechanical examination.

A detailed study of the Ti–Ni system phase diagram in the transformation region shows that the alloys lying to the left of the equiatomic composition are insensitive of heat treatment, meanwhile the $980^{\circ}$C quenching of alloys with large nickel content leads to a lowered $A_s$ point in the negative temperature range. A long annealing at $t = 500^{\circ}$C evens the values to approximately uniform values and is 70–80$^{\circ}$C. The phase diagram reveals a horizontal line at $\sim 670^{\circ}$ (heating) and 625$^{\circ}$C (cooling).

The results suggest that below $t = 450^{\circ}$C the homogeneous region is virtually absent, and the Ti–Ni/$Ti_2Ni_3$ boundary has a very gentle incline to the concentration axis compared to the TiNi/$Ti_2Ni$ boundary. This horizontal line was observed earlier and reported in ref. [1], however, up to date there is no universal opinion as to whether it is caused by a new phase, $Ti_2Ni_3$, or results from the solid state transformation in the $Ni_3Ti$ compound.

The experimental evidence generated in testing the electric conductivity vs. composition and heat treatment, Fig. 1, also

indicates that the alloys of the composition with more than 50
at.% Ni may be exposed to a heat treatment, the 980°C quenching
(the dark points), the annealing and 600°C quenching (the open points) , and finally, a long 500°C annealing (marked with crosses).
The 980°C quenching results in a greater ambient temperature stability of the alloys. These alloys display no property variations
after holding at 20°C for 5,500 h compared to the as-quenched
alloys. The 50 at.% Ni alloy has the most stable properties and
is insensitive to heat treatment. The $M_s$ point in this alloy =
63°C, $A_s$ =75°C and $T_o$ defined as 1/2 ($M_s$ +$A_s$) =342°K. The experimental value of the transfer energy in TiNi, i. e. $\Delta H^{\beta \to M}$, varies in the range from 310 cal/g-at to 370 cal/g-at [2] , the most
likely transfer energy value being $\pm$350 cal/g-at, and the corresponding enthropy increment $\Delta S^{\beta \to M}$ = $\pm$1.01 cal-g-at.grad.

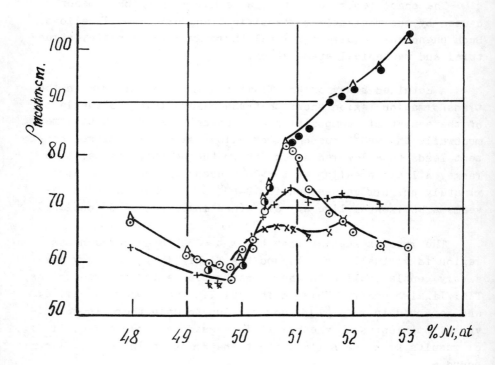

Fig. 1 The influence of heat treatment on the concentration
        dependence of electric resistance in the Ti–Ni sys-
        tem; 1 – the 980°C quenching; 2 – the 600°C quenching;
        3 – after a long 500°C annealing.

From these values and the Gibbs-Helmholz potential

$$\Delta F^{\beta \rightarrow M} = \Delta H^{\beta \rightarrow M} - T\Delta S^{\beta \rightarrow M} \tag{1}$$

the $T_o$ value is $346^{\circ}C$ which is in a very good agreement with the experiment. In [3] was reported that the thermoelastic martensitic transformations, the variation of the free energy could be nearly entirely described by the vibrational component, the other contributions to the free energy being negligible. Therefore, we have:

$$F_{vibr} = E_o + N\kappa T \cdot \left[ 3\ln (1 - e^{-\frac{Q}{T}}) - D\frac{Q}{T}\right] \tag{2}$$

where $E_o$ can be expressed as the ordering energy difference betweet $\beta$ and M, and

$$E_o = \frac{(V - V_o)^2}{2 \kappa_s V_o} + 9/8 \ N\kappa Q_D \tag{3}$$

where $E_o \sim 9/8 \ N\kappa Q_D$ (4); the second term in $E_o$ can be neglected because it is small, i. e.

$$\frac{(V - V_o)^2}{2 \kappa_s V_o},$$

in the temperature range under study. The $\Delta F_{el}^{\beta \rightarrow M}$ value, i. e. the electron component, is small, however, it is desirable to include this for the present type of transformations. For example, it is $\sim 70$ cal/g-at for TiNi. Thus, the calculation shows (Fig. 2) that the $\Delta F^{\beta \rightarrow M}$ curve gives an incline from which $\Delta S^{\beta \rightarrow M} = 0.9$ cal/g-at.grad. and approaches the experimental value, and the motive forces of the transformation, i. e. $\Delta F^{\beta \rightarrow M} = 10$ cal/g-at. This value is the biggest one among the known compounds with the shape memory effect and having an ordered structure. In the Fe $\neq$ 30% Ni alloy (the data are cited from ref. [4] ). $\Delta F_{IM}^{\gamma \rightarrow \alpha'} = 330$ cal/g-at. These transformations, however, do not exhibit the complete retention of shape, and resulting from a large hysteresis, i. e. the $A_S - M_S$ difference, the elastic energy can not be fully realized during the transformation. Alloying of titanium nickelide with various elements of the periodic system leads to variations in the temperature of martensitic transformation, the structure,

physical properties of the alloys, inclusive of the nonelastic
region property complex related to the shape memory effect [5, 6].
X-ray structural and differential thermal (DTA) analyses were
used to determine the temperatures of martensitic transformation,
to study the nature of phase  transitions and the atomic/crystal
structure in the $TiNi_{1-x}Pd_x$ alloys at x = 0 - 0.45. Some alloys
were additionally alloyed with nickel up to 1 at.% for the stabilization to the ambient temperature and below the high-temperature modification of titanium nickelide, β-phase. The X-ray
structural analysis was made on the DRON-0.5 installation in the
filtered $CuK_\alpha$ radiation using the KRN-190 attachment, DTA, in an
automatic thermal weighing system. The specimens cut from ingots
in the form of plates were exposed to a homogenizing annealing
at 900°C for 24 h and quenched. The Table shows the composition
of the alloys.

| Alloy No. | Ti, at.% | Ni,at.% | Pd, at.% |
|-----------|----------|---------|----------|
| 1 | 50 | 50 | – |
| 2 | 50 | 47 | 3 |
| 3 | 49 | 47.5 | 3.5 |
| 4 | 50 | 43 | 7 |
| 5 | 48.9 | 35.4 | 15.7 |
| 6 | 50 | 34 | 16 |

The X-ray structural analysis at the ambient temperature
showed the presence of the β-phase (B2) in all alloys, which suggests the uncompleted martensitic transformation during quenching
of the alloys.

Fig. 3 gives the DTA heating curves for the same alloy mass
depending on the composition. When contained at up to 9 at.%,
palladium reduces the temperature of the martensitic transformation M(B19)→β(B2). A decrease in the thermal effect may be observed with increasing palladium content in the alloys. The external stress (a 5% unaxial compression) leads to an increased
extent of the B2→B19′ transformation, increased martensite content
and accumulated macroscopic strain due to ordered local shears [7].
This will explain an appreciable enhancement of the thermal ef-

fects of the reversed B19→B2 transformation (Fig. 3b). The additional martensite transformation and orientation during straining is among the shape recovery mechanisms during martensitic transformation, and subsequently, the DTA data give information on the behaviour of the shape memory effect.

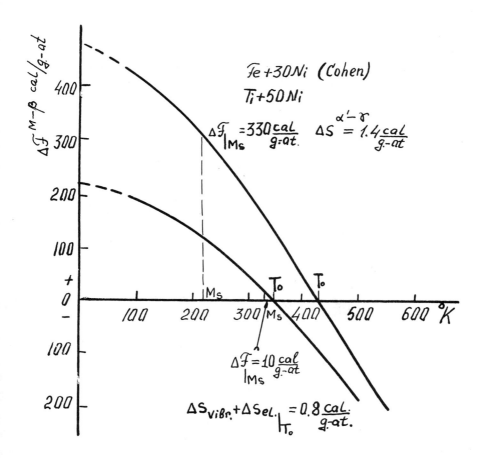

Fig. 2 The temperature dependence of the free energy differences of β (γ)-phase and M martensite in Ti+ 50 at.% Ni alloys (1) and Fe+ 30 at.% Ni (2).

The additional alloying of the alloys with nickel (alloys Nos. 3, 5) resulted in the transition temperatures below the am-

bient. At $-40^\circ$C the X-ray patterns of these alloys show lines of
a new phase. This phase was identified as the rhomb-type B19' ($C_2^5$)
structure distorted toward the monoclinic arrangement. Fig. 4
shows the temperature dependence of the B19' phase fraction during transformations in the alloys with various palladium contents as derived from the X-ray structural data. In the 3.5 at.%
palladium alloy, down to $-180^\circ$C, the B2→B19' trasnformation is
incomplete in a wide temperature range and reaches saturation at
a small B19' phase fraction, about 0.2 (Fig. 4a). In the 15.7 at.%
palladium alloy, a rapid increase in the B19'-phase fraction occurs in a narrow temperature range, and the B2→B19' transformation is more complete (Fig. 4b).

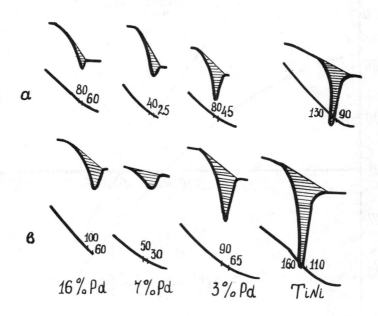

Fig. 3 The differential/thermal heating curves for TiNi$_{1-x}$
Pd$_x$ system alloys: a – as-quenched; b – as-quenched, worked.

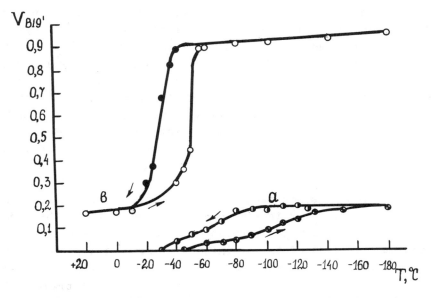

Fig. 4 The temperature dependence of the B19' phase fraction in the alloys with 3.5 (a) and 15.7 (b) at.%
palladium.

The temperature dependence of lattice parameters in the
B19' phase and the monoclinic angle $\beta^o$ display a rapid variation
during cooling down to -190°C and heating to the ambient tempe-
rature (Fig. 5). The temperature dependence of $\beta^o$ shows that the
B19' structure tends to approach the undistorted rhomb structure
B19 ($D_{2h}^5$) on departing from the phase transition.

The temperature dependence of lattice parameters and the
linear expansion coefficient of the B2 phase (Fig. 5) also display a sharp variation in the phase transformation range. The
anomaly behavior of the crystal structure parameters suggests
the existence of blurred interphase transitions in the B2 and
B19' phases of the titanium nickelide alloy, in addition to the
phase transition of the 1st kind [8, 9] .

The titanium nickelide alloy of the stoihiometric composition containing 50 at.% (55 wt.%) Ni retains its shape in the
temperature range of 100-150°C which is not always in agreement
with the required component operating conditions. For example,

the spontaneous – grip fasteners must have the shape retention
effect at temperatures below the ambient temperature. In [5, 10,
11] it is reported that the introduction of small additions (up
to 3.5 wt.%) of V, Cr, Mn, Fe, Co instead of Ni, as well as of
W, Cr, Al, Mn instead of Ti decreases the temperature range of
the martensitic transformation, and vise versa, the introduction
of Zr instead of Ti and Au instead of Ni increases the latter.
The addition of Fe and Co to the titanium nickelide of the stoihiometric composition decreases the shape retention temperature
[12] .

     The present study dealt with the influence of alloying with
Cr, Fe, Co, Nb, Ta, V in a wide composition range on the temperature of the direct and reverse martensitic transformation in
titanium nickelide of the stoihiometric composition, possessing
the most remarkable shape memory effect. The initial and final
temperatures of the direct and reverse martensitic transformations were determined with an accuracy within $\pm 10^{\circ}C$ by the differential thermal analysis (DTA) method. The generalized results
are shown in the Table. Cr, Co and Fe have the greatest tempera-
ture-decreasing effect, which agrees with the data reported in
[10, 12] . The introduction of these as the only alloying additions seems to show no promise for the development of a material
with the shape memory effect in the negative temperature range,
since this envolves processing difficulties related to the accurate duplication of the specified composition. Vanadium and niobium smoothly decrease the temperature range of the transformation, and tantalum does not have any significant effect. These
elements can be added together with the first three in order to
smooth the active effects of the former. It should be noted that
some titanium nickelide alloys with cobalt and iron exhibit a
two-stage martensitic transformation. V. N. Khachin et al. [6]
also pointed to the two-stage transformation mechanism in titanium nickelide-based alloys.

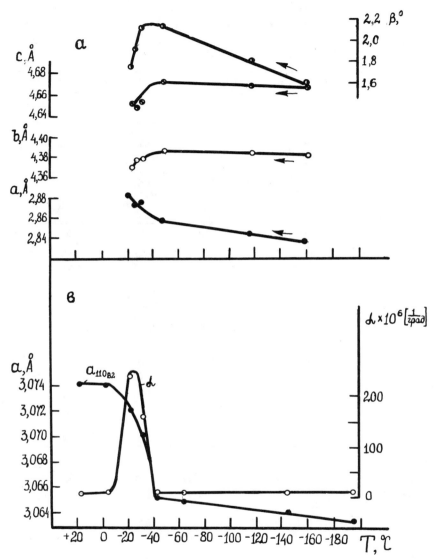

Fig. 5  The temperature dependence of the B19' monoclinic
lattice parameters (a) and the B2-phase lattice
parameter and linear expansion (b) in the alloy
with 15.7 at.% palladium.

Table. The Influence of Alloying of the Temperature Range
of Martensitic Transformation in Titanium Nickelide

| Alloying element, % of the total weight | | $M_s$, °C | $M_f$, °C | $A_s$, °C | $A_f$, °C |
|---|---|---|---|---|---|
| Cr | 1 | +15 | 0 | +23 | +60 |
| | 2 | -120 | -140 | -98 | -53 |
| | 3 | -* | - | - | - |
| | 4 | - | - | - | - |
| Co | 1 | -10 | -25 | +45 | +22 |
| | 2 | -25 | -47 | -80 | -34 |
| | 3 | -84 | -101 | - | - |
| | | - | - | - | - |
| Fe | 1 | -2 | -16 | -8 | +17 |
| | | -27 | -47 | | |
| | 2 | - | - | - | - |
| | 3 | - | - | - | - |
| V | 1 | +60 | +50 | +52 | +110 |
| | 2 | +21 | -12 | +20 | +80 |
| | 3 | +20 | -24 | +22 | +60 |
| | 5 | +13 | -2 | +17 | +35 |
| | 7 | -8 | -25 | 0 | +22 |
| | 8 | 0 | -20 | 0 | +20 |
| Nb | 1 | +60 | +50 | +47 | +123 |
| | 2 | +32 | +26 | +39 | +90 |
| | 3 | +21 | -25 | +19 | +73 |
| | 4 | +33 | +21 | +34 | +75 |
| | 8 | +11 | -13 | +26 | +37 |
| | 9 | -7 | -29 | +16 | +29 |
| | 10 | -31 | -56 | +2 | +17 |
| Ta | 1 | +55 | +50 | +65 | +115 |
| | 2 | +74 | +61 | +61 | +98 |
| | 3 | +60 | +52 | +71 | +116 |
| | 4 | +50 | +42 | +42 | +70 |
| | 5 | +50 | +40 | -50 | +65 |

*) Transformations at temperatures -196°C–300°C were not obser-
ved.

## References

1. R. J. Wasilewski, S. R. Butler, J. E. Hanlon and D. Worden. Metallurg. Trans., 1972, v. 2, No. 1, pp. 229-238.
2. H. A. Berman, E. D. West and A. G. Rosten. J. Appl. Phys., 1967, v. 38, pp. 4473-4476.
3. Y. Murakami. J. of Phys. Soc. of Japan, 1972, v. 33, No. 5, pp. 1350-1360.
4. L. Kaufman, M. Kohen. The Thermodynamics and Kinetics of Martensitic Transformations (in Russ.), FM, 1961, v. 4, p. 192.
5. I. I. Kornilov, O. K. Belousov, E. V. Kachur. Titanium Nickelide and Other Alloys with the Memory Effect (in Russ). M., Nauka Publishers, 1977, pp. 98-105.
6. V. N. Khachin, Yu. I. Paskal, V. E. Gyunter, V. P. Monasevich, V. P. Sivokha. Fizika metallov i metallovedenie (in Russ.), 1978, 46, 3, pp. 511-520.
7. L. A. Solovyev, V. N. Khachin. Fizika metallov i metallovedenie (in Russ.), 1973, 36, 2, pp. 400-401.
8. B. N. Rolov, V. E. Yurkevich. The Theory of Landau Phase Transitions and its Use. Study Aid Material of the Latvian State University. Riga, 1972.
9. Yu. A. Skakov, A. M. Glezer. In: Metallovedenie i termoobrabotka. Itogi nauki i tehniki. (in Russ.), M., 1975, v. 9, VINITI.
10. R. J. Wasilewski. Shape Memory Effect in Alloys. Proc. of Symposium, Toronto, May 19-22, 1975, pp. 245-271.
11. Eckelmeyer. Scr. Met.,      , v. 10, p. 667.
12. D. B. Chernov, D. A. Murzov, O. K. Belousov. Metallovedenie i termicheskaya obrabotka metallov, 1978, No. 2, pp. 72-73.

# PHASE TRANSFORMATIONS IN A Ti-10%V ALLOY

E.S.K. Menon, J.K. Chakravartty, P. Mukhopadhyay and R. Krishnan

Metallurgy Division, Bhabha Atomic Research Centre,
Trombay, Bombay-400 085, India.

## Introduction

Many metastable phases can occur in titanium alloys containing beta stabilizing additions on quenching from the beta phase field [1, 2]. In dilute alloys, beta quenching produces the metastable martensite phase which can be hcp or orthorhombic, depending on the nature and the amount of the alloying additions. Increasing levels of alloying bring down the $M_S$ temperature and a situation can arise where the beta phase can be retained in a metastable manner on quenching to room temperature. Over a limited range of alloy composition close to the beta retention threshold, the athermal omega phase can form during quenching. At higher alloy contents, the omega phase does not form on quenching but can precipitate on ageing at temperatures below about 450°C. In alloys still richer in beta stabilizing solutes, the beta phase can be retained as an equilibrium phase at room temperature.

Vanadium is a very commonly used beta stabilizing alloying addition in titanium. Many of the commercial titanium alloys (e.g. Ti-8Al-1Mo-1V, Ti-6Al-4V, Ti-6Al-6V-2Sn, Ti-13V-11Cr-3Al, Ti-10V-2Fe-3Al etc.) contain vanadium. The titanium-vanadium binary system is beta isomorphous. Vanadium is a relatively weak beta stabiliser compared to elements like manganese, iron, chromium, tungsten, nickel and molybdenum.

The effect of solutionising temperature on the nature and the distribution of phases in the Ti-10%V alloy was studied in this work. The equilibrium phases in this alloy at temperatures lower than about 770°C are the alpha and the beta phases having compositions determined by the temperature in question. However, the formation of metastable phases like the martensite phase, the metastable beta phase and the omega phase could be brought about by rapid cooling from different solutionisation temperatures. Solutionisation was carried out in the beta phase field and subsequently, the samples were isothermally transformed at different temperatures in the alpha + beta phase field. Varying the temperature in the two phase field caused variations in the volume fractions of the alpha and the beta phases. Since vanadium is a relatively weak beta stabilizer, changes in the ageing temperature in the alpha + beta field could bring about quite pronounced changes in the vanadium content of the beta phase.

## Experimental Procedure

The alloy used in this study was prepared by melting weighed amounts of iodide pure titanium and high purity (99.99%) vanadium granules ⊃ in a nonconsumable arc furnace. The finger was initially hot rolled and subsequently cold rolled to a final thickness of about 0.8 mm. A portion of the rolled sheet was rolled down to a thickness of about 0.1 mm for obtaining specimens for transmission electron microscopy. All samples were wrapped in tantalum foils and encapsulated in silica capsules under a protective atmosphere of helium at about 175 mm pressure. The samples were beta solutionised at 1000°C for 15 minutes. In one set of experiments, samples were rapidly quenched into water from the solutionisation temperature while in another, the samples were quickly transferred into another furnace maintained at various temperatures (700, 675, 600 and 550°C), held for one hour, and subsequently water quenched. A tandem arrangement of furnaces was used for this purpose. Some of these isothermally treated samples were subjected to an ageing treatment at 300°C for various periods of time. For transmission electron microscopy, thin foils were prepared by a multistage procedure involving mechanical grinding, chemical thinning and electropolishing. The electropolishing solution contained 35 parts of perchloric acid, 165 parts of n-butanol and 300 parts of methanol. The electrolyte temperature was maintained below -50°C and the voltage used was 20V. Tensile tests were carried out in an instron tester using a strain rate of $7.78 \times 10^{-3}$ min$^{-1}$.

## Results and Discussion

The observations made on the nature and the distribution of different phases and on their effect on the tensile properties of the alloy are described in this section.

### 1. Martensite

Beta quenching from 1000°C produced an acicular martensite structure (Fig.1a). X-ray diffraction did not reveal the presence of any other phase and showed that the martensite had the hcp structure. The quenched alloy was found to be partitioned by large primary plates (which were formed in the initial stages of the transformation), the intervening regions between these plates being occupied by smaller secondary plates belonging to later generations. Many, though not all, of the primary plates were internally twinned, while twinning within secondary plates was infrequent (Fig.1b). The twins were found to be predominantly of the $\{10\bar{1}1\}$ type (Fig.2a & b). Williams et al [3] have suggested that the twins in the primary plates are generated due to post-transformation shear. They have pointed out that if twinning is to be produced by a component of the transformation shear, then twin formation would be more favourable in the secondary plates which are formed at temperature lower than those corresponding to the primary plates. In the present work, however, the occurrence of almost equispaced twins of nearly equal thickness in many plates and the fact that in these cases the measured ratio of the twinned to the untwinned portions of the plates was generally close to the theoretically predicted ratio of 1 : 3 for $\{10\bar{1}1\}$ twinning [4] suggested that these twins were transformation twins. Further, the internal twinning of the primary plates did not, by and large, appear to result from secondary plate impingement. A similar observation has earlier been made in Zr-Nb and Zr-Ti martensites [5, 6] . The absence of twinning within the secondary plates implied that the transformation temperature is not the only factor controlling the fine structure of the martensite plates. However, it should be mentioned in this context that in some primary plates, twins of varying thicknesses and spacings were observed in the present work (Fig.2c). In many instances, the twin bands in the primary plates showed a regular distribution of dislocations aligned in parallel arrays (Fig.3a). Trace analysis indicated that these dislocations were on the $(01\bar{1}1)$ planes. This

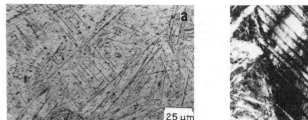

Fig.1 : Beta quenched alloy: (a) Optical microstructure. (b) Internally twinned primary plate and twin free secondary plates.

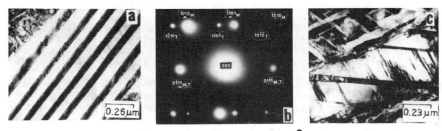

Fig.2 : Beta quenched alloy : (a) Equispaced {10$\bar{1}$1} twins  (b) Corresponding SAD pattern and (c) Unevenly spaced twins.

Fig.3 : Beta quenched alloy : (a) Distribution of dislocations within twin bands (b) faulting in a primary plate.

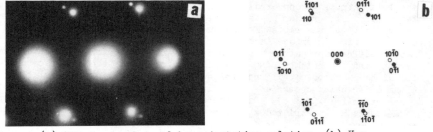

Fig.4 : (a) SAD showing beta-alpha orientation relation. (b) Key.

This observation showing a uniform distribution of dislocations within the twin bands suggested that an inhomogeneous shear occurred by slip along the $(01\bar{1}1)$ plane even within the $(10\bar{1}1)$ internal twin. This implied that the total inhomogeneous shear was made up of two parts, one operating on the $(10\bar{1}1)$ plane and the other on the $(01\bar{1}1)$ plane, through a twin mode and a slip mode respectively. Slip along the $\{01\bar{1}1\}$ planes would occur in either the $\langle 1\bar{2}13 \rangle$ or in the $\langle 2\bar{1}\bar{1}0 \rangle$ directions by the movement of 'c+a' or 'a' dislocations respectively. The visibility of the dislocations under the $(0002)$ and the $\{11\bar{2}0\}$ reflections showed that these were indeed of the 'c+a' type, indicating that the multiple shear was composed of a $(10\bar{1}1)$ twin shear and a $(01\bar{1}1)$ $[1\bar{2}13]$ slip shear. It can be seen that these two components of this multiple shear would be mutually compatible since the direction of the second shear, $[1\bar{2}13]$ is parallel to the plane of the first shear, $(10\bar{1}1)$. The appearance of such dislocation arrays within $\{10\bar{1}1\}$ twin bands has been reported in martensites of Zr-Nb, Zr-Ti and Zr-Ta alloys [5-7].

In addition to the $\{10\bar{1}1\}$ twins, some primary martensite plates contained features (Fig.3b), which appeared to be similar to stacking faults observed in equiatomic titanium-zirconium alloys containing 0.7 at% oxygen [8]. Trace analysis indicated that these features formed along $\{10\bar{1}1\}$ planes.

Though X-ray diffraction failed to reveal the presence of any retained beta phase in the quenched alloy, electron microscopy showed that such retention did occur to a very small extent, pointing to a sub-ambient $M_f$ temperature for the alloy. Collings [9] has reported on the basis of magnetic susceptibility measurements that beta retention should occur in alloys containing more than about 10.5 at%V. The diffraction patterns obtained from the beta-alpha prime regions were found to be consistent with the Burgers orientation relation (Fig.4).

2. Alpha Phase

The isothermal decomposition in the alpha + beta phase field consequent to beta solutionisation led to the emergence of the primary alpha phase. As the reaction temperature was lowered, the volume fraction of this phase increased while its vanadium content decreased. The beta matrix became correspondingly enriched in vanadium. It was seen that even at the highest reaction temperature (700°C) the vanadium enrichment of the beta phase was **sufficient** to prevent the formation of martensite on quenching and to retain the beta phase in a metastable manner. Thus, in samples quenched from the alpha + beta field, the entire alpha phase consisted of primary alpha, which formed as Widmanstatten plates obeying the Burgers orientation relation with respect to the beta matrix. It appeared that these plates nucleated uniformly within the beta grains rather than preferentially only at the grain boundaries (Fig.5a). The majority of the plates formed at 700°C and at 550°C were associated with layers of fine interface features at the precipitate matrix boundaries(Fig.5b & c). Diffraction patterns obtained from areas comprising the interface region and the neighbouring beta region showed that the interface phase was the alpha phase but it did not follow the Burgers orientation relation. The alpha diffraction spots were arced in these patterns, indicating a scatter in orientation (Fig.5d & e). Dark field microscopy revealed that these interface layers had a striated appearance, suggesting that they were not continuous but comprised fine, individual crystallites (Fig.5c). The presence of such interface layers in isothermally transformed titanium alloys has been

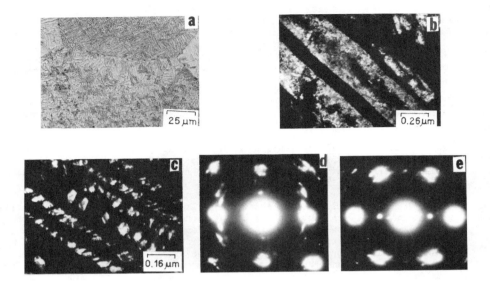

Fig.5 : (a) Alpha plates in samples transformed at 700°C. (b) and (c) Dark field micrographs showing type 2 alpha interface phase in samples transformed at 700° and 550°C respectively. (d) and (e) arcing of alpha spots. Zone axis : (d) ⟨100⟩$_\beta$ and (e) ⟨111⟩$_\beta$

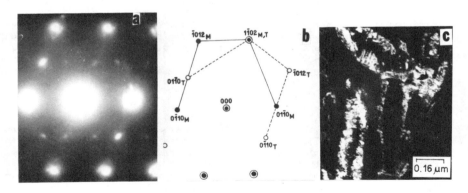

Fig.6(a) SAD pattern demonstrating $\{10\bar{1}2\}$ twin relation between type 1 and type 2 Alpha phases. (b) Key. (c) Dark field micrograph with $(01\bar{1}0)$ twin reflection showing type 2 alpha layers.

observed by many workers [10-12] . This non-Burgers alpha boundary phase
has been termed type 2 alpha to distinguish it from the Burgers oriented
Widmanstatten plate which is designated as type 1 alpha. In the present work,
it was found that the type 2 alpha phase was crystallographically related to
the type 1 alpha phase through {10$\bar{1}$2} twinning (Fig.6). This observation
was similar to that made by Rhodes et al [10] in Ti-Mo and Ti-Mo-Al alloys.

In contrast to the morphology of the alpha plates formed at 700°C and
550°C, those precipitated at 600°C were found to be largely free of the type 2
alpha interface phase (Fig.7a & b). The majority of the plates here showed
sharp interfaces and the usual interface dislocations and fringe contrast.
The Burgers orientation relation was seen to be valid for these plates. The
presence of small angle boundaries (Fig.7c) was frequently noticed within
these, pointing to the possibility that these boundaries separated individually
nucleated alpha crystals of nearly identical orientation. This feature was
also observed, to a limited extent, in the alpha plates obtained at the
other reaction temperatures. It was possible that the Widmanstatten alpha
plates were not monolithic but consisted of low misorientation arrays of contiguous crystals. Such a morphology of the alpha plates was suggestive of the
occurrence of sympathetic nucleation mechanism described by Aaronson [13] .
Since the formation of an alpha precipitate would involve the rejection of
vanadium atoms from the region where the precipitate forms to neighbouring
beta regions, the occurrence of sympathetic nucleation would be difficult from
considerations of chemical free energy alone. However, the sympathetic nucleation of alpha crystals on the surface of pre-existing plates could be possible
if the specific surface energy of the alpha/alpha interface were much smaller
than that of the beta/alpha interface. The alpha/alpha interface energy would
be low, if the boundaries separating contiguous alpha crystals were small
angle boundaries as in the present case.

Rhodes et al [10] have suggested that type 2 alpha is more stable than
type 1 alpha and grows by a nucleation and growth mechanism by consuming the
type 1 alpha phase which forms earlier. It has been suggested that the driving
force for the reaction is the composition difference between the two types of
alpha phases. However, the observation made in the present work that while the
type 2 alpha phase was abundant in samples transformed at 700°C and 550°C,
those transformed at 600°C were relatively free of this phase was not consistent with the type of TTT diagram proposed by these authors for the Ti-Mo and
the Ti-Mo-Al systems.

## 3. Omega Phase

The existence of the athermal omega phase was noticed in samples quenched
from 1000°C, 700°C, 675°C and 600°C. Quenching from 550°C, however, did not
show characteristic streaking associated with the beta-omega diffraction
patterns, suggesting that the vanadium enrichment of the beta phase on isothermal holding at 550°C was sufficient to depress the $\omega_s$ temperature below
the room temperature. Sharp omega diffraction spots could be obtained only
in samples quenched from 700-600°C (Fig.8a). Diffraction patterns obtained
from the beta quenched samples showed very weak streaks (Fig.8b). This was
consistent with the fact that the volume fraction of the retained beta phase
in these samples was very small. In samples isothermally transformed in the
temperature range 700-600°C, the beta matrix surrounding the alpha plates was
found to contain a uniform dispersion of fine omega particles (Fig.9a). The

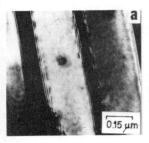
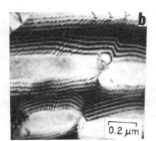
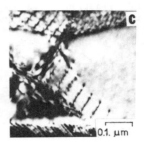

Fig.7 : Ti-10%V alloy, transformed at 600°C; (a) and (b) show alpha plates with interfacial fringes and dislocations. The absence of type 2 alpha may be noticed. (c) small angle boundaries within alpha plate.

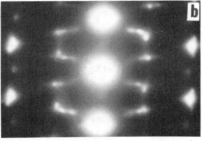

Fig.8 : (a) Well formed omega spots in sample transformed at 700°C and subsequently quenched. (b) Diffuse omega streaks in SAD pattern from beta quenched sample. Zone normal : $\langle 113 \rangle_\beta$ in both patterns.

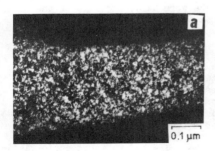
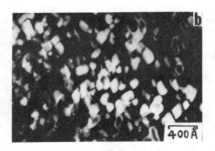

Fig.9(a) : Dark field micrograph showing a fine distribution of omega particles in a region of retained beta in samples isothermally transformed at 600°C. (b) Cuboidal omega precipitates in sample aged at 300°C for 100 hours.

volume fraction of the omega phase in the beta-omega aggregates was found to be quite large. This was consistent with the observation of Hickman [14] that the volume fraction of the omega phase in a beta quenched Ti-15at%V alloy is as high as 0.7.

The growth of omega particles was studies by ageing the samples quenched from 600-700°C for prolonged periods at 300°C. On ageing for 100 hrs., the omega precipitates grew to an average size of 150A° (Fig.9b). These particles exhibited a cuboidal morphology with the cube faces parallel to the {100} planes of the matrix beta phase, as expected in the high misfit Ti-V system [1] .

4. Tensile Properties

The true stress versus true plastic strain plots of tensile specimens isothermally transformed for 1 hour respectively at 700°C, 650°C and 600°C and tested at room temperature are shown in Fig.10, alongwith the flow behaviour of the beta quenched alloy. The corresponding values of the 0.2% proof stress, the ultimate tensile stress and the uniform elongation are presented in Table I.

Table I    Tensile Properties

| Treat-ment | Yield strength ($Kg/mm^2$) | Ultimate tensile strength ($Kg/mm^2$) | Total uniform elongation (%) | Total plastic strain (%) |
|---|---|---|---|---|
| 900 WQ | 74.7 | 94.1 | 7.4 | 3.33 |
| 700 IT | 81.5 | 85.2 | 8.2 | 4.83 |
| 650 IT | 93.2 | 96.6 | 5.2 | 1.94 |
| 600 IT | 83.3 | 89.9 | 7.4 | 2.63 |

WQ  :   Water quenched
IT  :   Isothermally treated

It could be seen that isothermal decomposition at 650°C was associated with the highest strength and the lowest ductility. The work hardening rate of the isothermally treated samples was low and appeared to be insensitive to the reaction temperature. The beta quenched sample, on the other hand, showed a relatively rapid strain hardening behaviour. It is known, in general, that in titanium alloys containing the alpha and the beta phases, the strength of the former is lower than than of the latter, the strain hardening rate of the beta phase is low, and the strength of the two phase alloy is closer to that of the continuous phase [15] . In the present case, for the isothermally decomposed alloys, the beta phase was the continuous phase. The dispersion of the athermal omega phase strengthened this phase and embrittled it. In some of the alpha + beta solutionised samples, the volume fraction and the size of the omega particles was allowed to grow by prolonged ageing at 300°C. While testing, these samples were found to fail at the grips before macroscopic yielding could occur. From the results of the tensile tests, it appeared that the volume fraction of the omega phase formed on quenching the alloy from 650°C was higher than those corresponding to the other reaction temperatures. Fractographic examination of the tested samples showed that even those showing the least uniform elongation were characterised by dimple rupture - a ductile fracture mode (Fig.11).

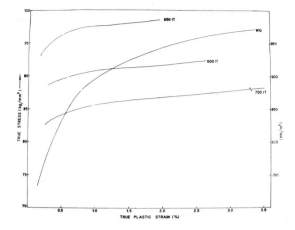

Fig.10 : True stress versus true plastic strain plots.

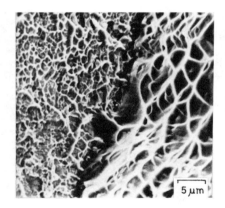

Fig.11:Dimple fracture in sample transformed at 650°C and
tested at room temperature. The bimodal size distribution of dimples may be noted.

It has been shown that dislocations do not move in omega particles which are non-deformable [15, 16] . Consequently, during the deformation of the beta-omega aggregate, the dislocation density in the beta phase matrix increases rapidly. The accumulation of a critical local strain could lead to microvoid nucleation at the beta/omega interfaces, leading finally, to microvoid coalescence and rupture.

The observed high strength associated with the almost fully martensitic structure could be attributed to the solute supersaturation and the fine sub-structure of the martensite. It should be noted, however, that the latter factor was likely to be more important since vanadium is not a strong solid solution hardener in titanium [15] . The rapid work hardening was characteristic of the flow behaviour of the alpha phase in titanium alloys [15] .

## References

1. J.C. Williams: Titanium Science & Technology, 3 (1973) 1433. Plenum Press, N.Y.
2. B.S. Bickman: J. Mat. Sci., 4 (1969), 554.
3. J.C. Williams, R. Taggart and D.H. Polonis: Met. Trans. 1 (1970),2265.
4. G.T. Higgins and E. Banks : Special Report No.93, Iron & Steel Institute, London (1966).
5. S. Banerjee and R. Krishnan : Acta Met., 19 (1971), 1317.
6. S. Banerjee and R. Krishnan : Met. Trans. 4 (1973), 1811.
7. P. Mukhopadhyay, E.S.K. Menon, S. Banerjee and R. Krishnan : Zeit. Metallkde 69 (1978), 725.
8. H.M. Flower and P.R. Swann : Titanium Science & Technology, 3 (1973), 1507, Plenum Press, New York.
9. E.W. Collings : J. Less Comm. Met., 39 (1975), 63.
10. C.G. Rhodes and J.C. Williams : Met. Trans. 6A (1975), 2103.
11. C.G. Rhodes and N.E. Paton : Met. Trans., 8A (1977), 1749.
12. H. Margolin, E. Levine and M. Young : Met. Trans. 8A (1977), 373.
13. H.I. Aaronson : In Decomposition of Austenite by Diffusional Processes, Eds. V.F. Zackay and H.I. Aaronson, (1962), 387. Interscience Publishers, New York.
14. B.S. Hickman : J. Inst. Met., 96 (1968), 330.
15. R.I. Jaffe : Titanium Science & Technology, 3 (1973), 1665 : Plenum Press, New York.
16. J.C. Williams, B.S. Hickman and H.L. Marcus : Met. Trans., 2 (1971), 1913.

# PHASE TRANSFORMATION STUDY OF Ti-10V-2Fe-3Al

Jeffrey R. Toran and Ronald R. Biederman

Worcester Polytechnic Institute
Worcester, Massachusetts, USA

Phase transformations in the near beta titanium alloy Ti-10V-2Fe-3Al have been studied and are presented in the form of an isothermal transformation diagram. Quantitative optical and electron microscopy, x-ray diffraction, and Differential Thermal Analysis (DTA) methods were used to measure the extent of transformation. Grain growth analysis for β titanium in the 860C through 1050C temperature range are presented.

The decomposition of β (beta) to α (alpha) follows the well known Burger orientation relationship $<111>\beta||<11\bar{2}0>\alpha$ and $(110)\beta||(0001)\alpha$ referred to as Type 1α, at temperatures above 555C. At temperatures below 555C, β transforms to Type 1α and Type 2α which follows a $\{10\bar{1}2\}<10\bar{1}1>$ twin relation similar to that found in Ti-Mo alloys.

Direct quenching in water from above the β transus temperature results in ω (omega) phase formation with ellipsoids of ω following the orientation relationship with the β phase proposed by Silcock. The ω crystal structure has been confirmed as hexagonal by electron diffraction analysis. Aging experiments indicate β + ω transforms to β + both Type 1α and Type 2α.

DTA performed on samples quenched from above the β transus temperature of 788C ± 8C indicate the martensite start temperature to be 555C ± 5C. The morphology of the martensite is acicular laths.

## Introduction

A renewed interest in near β titanium alloys for aircraft structural applications has occurred with recent studies (1,2) of Ti-10V-2Fe-3Al. Considerable effort has been spent on this alloy attempting to develop optimum mechanical properties with only limited knowledge of phase transformations and kinetics. While phase transformation studies in other near β titanium alloys (3-6) have revealed a variety of complex transformations, no systematic study of phase transformations in Ti-10V-2Fe-3Al, (hereafter referred to as Ti1023, has been performed. This work presents recent microstructural analysis and kinetics for a substantial portion of the isothermal transformation diagram for a commercial heat of Ti1023.

An understanding of metastable phase formation in Ti1023 is necessary to the development of this alloy. The conversion of β titanium to β + α titanium, β + martensite to ω, and the subsequent low temperature conversion of ω to β + α titanium are of particular interest.

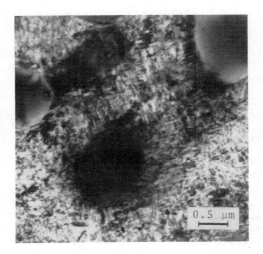

Figure 1A-Brightfield electron micrograph of Ti-10V-2Fe-3Al, as-received.
Islands are primary α with the matrix
a mixture of α and β phase.

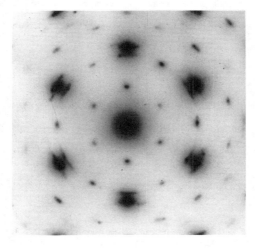

Figure 1B-Electron diffaction pattern
from the matrix region in Figure 1A
oriented to a [111]β zone axis. Three
<112̄0> variants of type 1 α occur. Arced
{101̄0}α reflections indicate Type 2 α.

## Experimental Procedure

A single commercial heat of Ti1023, TMCA heat V-4910, of composition 9.70%V, 2.08%Fe, 2.98%Al, 0.016%C, 0.113%O, 0.016%N, 0.0035%H balance titanium was used in this study. Forged and heat treated transverse samples 15mm x 15mm x 3mm were cut from billets to provide a convenient size and shape for solutionizing, aging, and quenching experiments. A 375C through 1050 ± 10C temperature range was possible using graphite protected lead bath furnaces. Samples prepared for both the grain growth and isothermal transformation studies were sectioned so that microstructural analysis was carried out on an interior region to eliminate effects due to the heat treating or quenching atmospheres.

Standard metallographic polishing and etching techniques were used. For electron transmission microscopy analysis, longitudinal specimens in relation to the original billet were examined. This orientation was obtained by electric discharge machining a 3mm diameter by 15mm long plug which was diamond wafer blade sliced and hand lapped prior to electropolishing using established techniques (7). Samples were examined in a JEOL 100C ASID transmission electron microscope.

Quantitative phase analysis was obtained primarily by quantitative metallography techniques (8) with some independent analysis performed by quantitative x-ray diffraction analysis. Also, the start of the $\beta \to \beta + \alpha$ transformation and the martensite start temperature were independently determined by DTA methods (9).

## Experimental Results and Analysis

Figure 1A shows the microstructure of the starting material for this phase transformation study. Numerous islands of spherical primary $\alpha$ in a fine $\beta + \alpha$ matrix is apparent. A selected area diffraction pattern of the fine matrix structure oriented to a $[111]\beta$ zone axis is presented in Figure 1B. Analysis of this pattern indicates the presence of $\beta$, Type $1\alpha$ and Type $2\alpha$ according to the nomenclature and analysis method used by Rhodes and Williams (4) in studying similar near $\beta$ titanium alloys.

Arcing of $\{10\bar{1}0\}\alpha$ diffraction spots or the absence of arced spots has been suggested by Rhodes and Williams in their study of Ti-Mo alloys as a convenient method of distinguishing Type $1\alpha$ and Type $2\alpha$ transformation products. Both types of $\alpha$ form according to the same habit; however, the substructure within the Type $2\alpha$ appears to be more complex than that found in Type $1\alpha$. This readily is observable in the bright and darkfield images which show a complex scattering image for Type $2\alpha$.

The crystallographic relationship between $\beta$ and $\alpha$ follows the Burger orientation similar to that observed in many other titanium alloys (10): $<111>\beta||<11\bar{2}0>\alpha$ and $\{110\}\beta||\{0001\}\alpha$. Precise parameter x-ray diffraction analysis for the as-received material resulted in the following cell dimensions: $\alpha$ titanium – $a = 2.9392A$ and $c = 4.6847A$; $\beta$ titanium – $a = 3.2125A$.

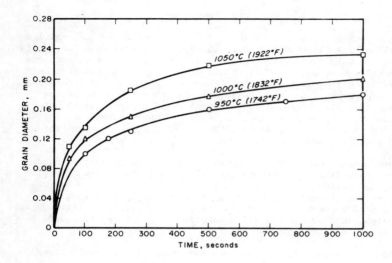

Figure 2-Isothermal Grain Growth for β Ti-10V-2Fe-3Al.

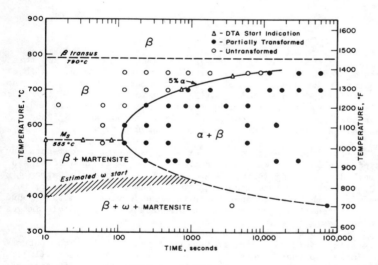

Figure 3-Isothermal Transformation Diagram for Ti-10V-2Fe-3Al
solutionized at 860C for 240 seconds.

Solutionizing at 860C for 240 seconds was sufficient to fully dissolve the primary and secondary α while maintaining a β grain size of ASTM 6-7. Solutionizing at higher temperatures resulted in significantly more rapid β grain growth as shown in Figure 2. Attempts to analytically characterize grain growth in this alloy as a power function of time with an exponential temperature dependence considering both limited and unlimited grain growth as observed in other titanium alloys (11), indicates a more complex β growth kinetics relationship than the usual Arrhenius kinetics. This departure from classical grain growth kinetics is partially due to the significant amount of inclusions present in this commercial heat of material.

## Isothermal Transformations

Isothermal phase transformations in Ti1023 for a β solutionize at 860C for 240 seconds are presented in Figure 3. The following reactions occur with decreasing temperature: 1) β → β + α, 2) β → martensite + β → β + α, and 3) β → martensite + β + ω → β + α. In the unstressed condition isothermal transformation of β titanium is sluggish with both the morphology and size of the transformation product substantially changing with decreasing temperature.

Typical optical photomicrographs for high, intermediate, and low temperature partial isothermal transformations in this system are shown in Figure 4 along with a direct water quenched speciman. At high temperatures, α forms directly from β, preferentially, on β grain boundaries growing into the β matrix. However, as the temperature is lower near the knee on the transformation diagram, but above the martensite start temperature, α precipitates and grows throughout the β matrix as well as in the β grain boundaries. At still lower temperatures, a very fine transformation product involving ω phase results which is resolvable only by electron microscopy thin foil techniques. Direct quenching from the β solutionizing temperature results in formation of acicular martensite as shown in Figure 4D.

## 1. The β → β + α Transformation

At temperatures above 555C, the β phase transforms according to the classical Burger orientation to α titanium. Figure 5A is a brightfield electron photomicrograph typical of early transformation with less than 10% transformation to acicular α. Electron diffraction for a [111]β zone axis is presented in Figure 5B and clearly shows well defined α diffracting conditions with the α phase oriented to a [11$\bar{2}$0]α zone axis. Distinct α spots occur indicating that the α formed directly from β above 555C is characterized as Type 1α. No indication of Type 2α has been observed for transformation of β to α above 555C in this alloy. Samples held for longer times show increasingly sharper α spots due to coarsening and growth of α rather than arcing of {10$\bar{1}$0}α reflections which would indicate conversion of Type 1α to Type 2α. The morphology of Type 1α is predominately acicular. Tilting of Type 1α through 120° shows no significant change in thickness indicating that this type of α is "needlelike". The substructure of Type 1α formed above 555C appears to be more complex than the monolithic characterization by Rhodes and Williams (4) of Type 1α in a Ti-Mo-Al alloy.

## 2. The β → β + martensite → β + α Transformations

The start of the martensite reaction for this heat of Ti1023 as determined by DTA occurs at 555C. Three different cooling rates were performed and the martensite start line was placed at 555C ± 5C, well above room

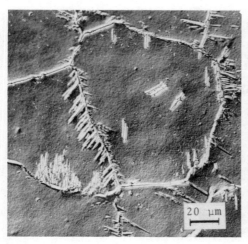

Figure 4A–Sample quenched to 700C, held for 900 seconds, partially transformed to α phase.

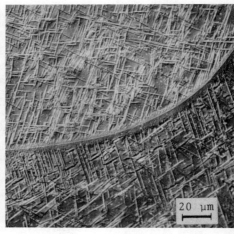

Figure 4B–Sample quenched to 600C, held for 240 seconds showing nucleation of the α phase throughout the β matrix and preferentially at β boundaries.

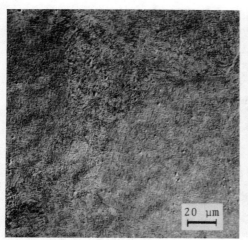

Figure 4C–Sample quenched to 500C, held for 900 seconds showing a fine dispersion of the α phase.

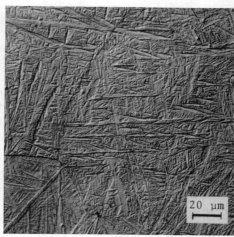

Figure 4D–Sample direct quenched to 0C showing acicular martensite.

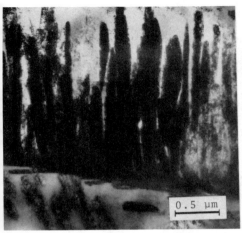

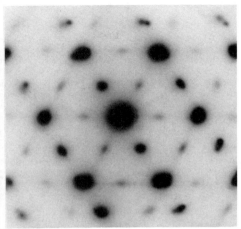

Figure 5A–Electron micrograph of the sample in Figure 4B. Type 1 α growing from β grain boundary.

Figure 5B–Electron diffraction pattern for a [111]β zone axis with three <11$\bar{2}$0>α variants. These variants indicate type 1 α is present. No arced spots indicate Type 2 α is absent.

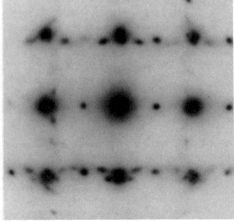

Figure 6A–Electron micrograph of the sample in Figure 4C showing uniform α growth in three directions.

Figure 6B–Electron diffraction pattern for a [100]β zone axis with {0001}α spots indicating Type 1 α and arced {10$\bar{1}$0}α spots inside the {110}β spots indicate Type 2 α.

temperature similar to other Ti-V alloys (5,6,12,13).

Thin foil observation of the martensite laths shows extensive twinning and dislocation tangles similar to the hexagonal martensite reported by Blackburn (12) for a Ti-8Al-1Mo-1V alloy. Subsequent electron diffraction analysis shows the crystal structure of the martensite is BCC with the same lattice parameters as the β phase. Spurling et al (14) reported electro-thinning samples for thin foil electron microscopy observation caused the martensite to change morphology and revert to the β crystal structure in some titanium alloys.

At temperatures below the martensite start line, but above the estimated starting transformation to ω (6,13,16), β + martensite transforms to β + α. The nucleation and growth of α occurs on a fine scale throughout the micro-structure. It appears that nucleation initiates in the martensite at β/mar-tensite interfaces, but is so fine that it is difficult even in darkfield to observe early transformation to α. Figure 6A is a brightfield electron photomicrograph of a sample quenched to 50C below the martensite start line and partially transformed to α phase. Three orientations of α similar to that seen in high temperature transformations are seen. However, the size of the α phase is substantially finer. A [100]β zone electron diffraction pattern for the sample shown in Figure 6A is presented in Figure 6B. Analysis clearly indicates both Type 1α and Type 2α are present. Arcing of the {10$\bar{1}$0}α reflections just inside the {110}β spots indicates Type 2α is present. Distinct {0001}α reflections clearly indicate the presence of Type 1α also. It appears that some conversion of Type 1α to Type 2α occurs in samples near the maximum extent of transformation (∿85%α) which shows predominantly Type 2α to be present.

## 3. The β → β + ω + martensite → β + α Transformation

Direct quenching of Ti1023 results in martensite and ω phase coexisting at room temperature. ω phase formed in direct quenched samples appears as small randomly oriented ellipsoids as shown in Figure 7A, a darkfield electron photomicrograph using a {10$\bar{1}$0}ω reflection. This extremely small phase forms uniformly throughout the sample and is usually observed in two or three ori-entations with [0001]ω||[111]β and (11$\bar{2}$0)ω||(110)β similar to ω formation in other titanium alloys (3,15,16).

Figure 7B is a SAD pattern aligned to a [110]β zone axis. Three orienta-tions of ω and streaking of {10$\bar{1}$0}ω reflections are observed. Streaking of these diffracting conditions is due to the thin dimension for each of the ω orientation variants perpendicular to the growth direction. Figure 8 is a darkfield electron photomicrograph of a sample quenched to 375C and aged for 1 hour, illustrating the preferred growth direction the ω phase exhibits upon aging. Careful examination will show three different crystallographic growth directions. A small change in the size of ω is noted in comparing direct quenched and 375 aged specimens. Electron diffraction analysis confirms the crystal structure of ω phase as hexagonal after Silcock (15) and Blackburn and Williams (3). Lattice parameters obtained from electron diffraction analysis indicate a= 4.79A and c= 2.93A for ω in direct quenched Ti1023. Aging at 375C for times longer than 20 hours results in transformation to both Type 1α and Type 2α phase. The morphology of the α phase is extremely fine acicular α showing distinct growth directions as shown in Figure 9A.

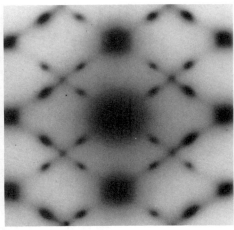

Figure 7A-Darkfield electron micrograph using a {10$\bar{1}$0}ω reflection showing a uniform dispersion of ω phase in
a sample direct quenched from 860C.

Figure 7B-Electron diffraction pattern
for a [110]β zone axis with two <$\bar{1}$2$\bar{1}$0>ω
variants showing streaking due to a
thin dimension in the ω phase.

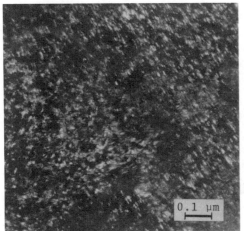

Figure 8-Darkfield electron micrograph
of a sample quenched to 375C and aged
1 hour showing ellipsoidal ω with a
growth direction which develops upon
aging.

Figure 9A–Electron micrograph of a
sample aged 20 hours at 375C showing
α phase in three orientations.

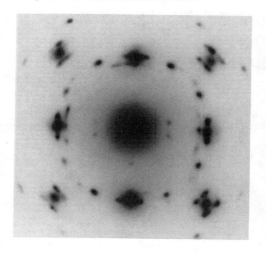

Figure 9B–Electron diffraction pattern
for a [110]β zone axis showing {0001}α
reflections indicating Type 1 α and
arcing of {10$\bar{1}$0}α reflections indicating
Type 2 α.

Analysis of the electron diffraction pattern in Figure 9B, shows {0001}α spots and arcing of the {10$\bar{1}$0}α spots indicating the presence of Type 1α and Type 2α respectively.

## Conclusions

In analyzing the transformations that occur in Ti1023, the following conclusions are drawn:

1) Only Type 1α is formed directly from β at temperatures above 555C.
2) Both Type 1α and Type 2α are formed at temperatures below 555C.
3) Acicular martensite starts to form at 555C.
4) ω phase is formed on direct quenching.
5) ω phase transforms to both Type 1α and Type 2α upon aging.

## Acknowledgements

The authors are grateful to Dr. C. C. Chen and the Wyman-Gordon Company for sponsoring this research program. We would like to thank George Schmidt for preparing the figures for this paper.

## References

1.   C. C. Chen, R. R. Boyer:  Journal of Metals, 31 (1979), July, 33.
2.   E. Bohanek:  *Titanium Science and Technology*, 3 (1973). Plenum Press, NY.
3.   M. J. Blackburn, J. C. Williams:  Trans. TMS-AIME, 242 (1968), 2461.
4.   C. G. Rhodes, J. C. Williams:  Met. Trans. A, 6A (1975), 2103.
5.   F. H. Froes et al:  ONR Technical Report. JWTR-3, (1977), August.
6.   E. L. Harmon, A. R. Troiano:  ASM Trans., 53 (1961), 43.
7.   R. A. Spurling:  Met. Trans. A, 6A (1975), 1660.
8.   R. T. DeHoff, F. N. Rhines:  *Quantitative Microscopy*, (1968). McGraw-Hill Inc., New York.
9.   J. J. Fipphen, R. B. Sparks:  Metal Progress, 115 (1979), April, 56.
10.  A. J. Williams, R. W. Cahn, C. S. Barrett:  Acta Met., 2 (1959), 117.
11.  B. B. Rath et al:  Met. Trans. A, 10A (1979), August, 1013.
12.  M. J. Blackburn:  ASM Trans. Quart., 59 (1966), 876.
13.  R. Klima et al:  CEA-R-4542, 1974, France.
14.  R. A. Spurling, C. G. Rhodes, J. C. Williams:  Met. Trans. A, 5 (1974), December, 2597.
15.  J. M. Silcock:  Acta Met., 6 (1958), 481.
16.  G. H. Narayanan, T. F. Archibald:  Met. Trans. A, 1 (1970), 2281.

# STRESS ASSISTED TRANSFORMATION IN Ti-10V-2Fe-3Al

T. W. Duerig,[*] R. M. Middleton,[**] G. T. Terlinde,[*] and J. C. Williams[*]

[*]Carnegie-Mellon University, Pittsburgh, PA U.S.A.

[**]AMMRC, Watertown, MA, U.S.A.

## Introduction

In $\alpha$ and $\alpha+\beta$ Ti alloys, a martensitic decomposition of the $\beta$-phase occurs during quenching from above the $\beta$-transus temperature, to below the martensite start temperature, designated $M_s$. In $\beta$-Ti alloys, the $M_s$ temperature is depressed, and the $\beta$-phase can be retained in a metastable state during quenching. In many of the "leaner", or less stabilized $\beta$ alloys, martensitic decomposition can be encouraged by externally stressing the $\beta$ solution treated and quenched ($\beta$-ST) material. The temperature below which deformation assisted martensite can form is designated $M_d$. In a number of $\beta$-alloys $M_d$ lies above room temperature. As a result, stress "assisted" or stress "induced" reactions have been reported in the $\beta$-stabilized Ti-Mo[1] and Ti-V [2] systems, as well as in other more complex systems. The structure of the stress assisted martensitic products in $\beta$-Ti alloys has been reported as FCC [3], HCP [4], and most recently as orthorhombic [5,6] or $\alpha''$. This paper will present some characteristics of such a stress assisted reaction, as observed in a commercial $\beta$-alloy, Ti-10V-2Fe-3Al.

## Experimental

The material used in this study was forged and then hot rolled into plate at TIMET. The exact heat chemistry was determined as:

| Element | V | Fe | Al | O | N | C | Ti |
|---|---|---|---|---|---|---|---|
| concentration (weight percent) | 10.3 | 2.2 | 3.2 | 0.15 | .009 | .016 | Bal. |

All specimens were encapsulated 'in vacuo', solution treated, and quenched in agitated water. The resulting microstructures were verified to contain no martensite. Specimens solution treated above the $\beta$-transus ($\beta$-ST) contained only recrystallized $\beta$ grains, with a random distribution of inclusions, rich in Ti, P, Si and S. Specimens solution treated below the $\beta$-transus ($\alpha+\beta$-ST) were found to contain unrecrystallized $\beta$ grains, inclusions, and coarse (> 10 $\mu$m dia.) globular primary $\alpha$ (or $\alpha_p$). All conditions were also found to contain athermal $\omega$ ($\omega_{ath}$).

External stressing, and tensile testing of the ST material was done using an Instron machine at a strain rate of 0.00055 sec$^{-1}$. Strains were measured using a clip-on extensometer.

Optical metallographic specimens were prepared by electropolishing in a 5% $H_2SO_4$ + 1% HF + 94% methanol solution and etching with equal parts of 10% oxalic acid in water and 1% HF in water. Thin foils for Transmission Electron Microscopy (TEM) were prepared in a twin jet electropolishing unit, using a 5% $H_2SO_4$ + 95% methanol solution. A small amount of ion thinning was also done to eliminate any artifacts of preparation. Both JEOL 120CX, and JOEL 100C electron microscopes were used.

Specimens selected for X-ray diffraction study were cut so that the rolling direction was in the plane of the surface. The X-ray equipment used for this study consisted of a Norelco-Schultz type pole figure goniometer, fitted to a G. E. horizontal X-ray diffractometer. Copper $K_\alpha$ radiation was used. A modified pole figure diffraction technique was employed, producing 20 diffraction scans as a function of declination angle. Due to a strong β rolling texture,declination angles from 0° to 60° had to be used to obtain a complete set of peaks.

## Results

An extensive stress assisted reaction was observed upon externally stressing Ti-10V-2Fe-3Al in the β-ST condition. The resulting plates were visible both optically and in TEM (Fig. 1). The X-ray diffraction techniques discussed earlier conclusively identified that the transformed microstructure contained orthorhombic martensite (or α″). Electron diffraction showed that the martensite plates were accompanied by mechanical twinning of the {112} <111> type. Although the relative amounts of these two products is difficult to quantitatively determine, it appears that the amount of twinning is relatively small. The location of the α″ X-ray peaks, and the indexing of these peaks, is illustrated in Table 1. The lattice parameters of the α″ unit cell were determined to be:

$$a = 3.01\text{Å}$$
$$b - 4.83\text{Å}$$
$$c = 4.62\text{Å}$$

Selected Area Electron Diffraction (SAD) demonstrated that the α″ plates are oriented with the β matrix in the following manner (Fig. 2):

$$(110)_\beta || (001)_{\alpha''}$$
$$[1\bar{1}1]_\beta || [110]_{\alpha''}$$

Dark field imaging of α″ spots highlighted the martensitic plates (as shown in Fig. 1c). The plates appeared "mottled" or "spotty". Imaging of $\omega_{ath}$ diffraction spots demonstrated that $\omega_{ath}$ appears to exist uniformly throughout the β matrix and within the deformation products (Fig. 3a). Imaging of the twinned β reflections and $\omega_{ath}$ spots simultaneously illustrates that ω exists within twinned β plates (Fig. 3b). It also appears that the ω distribution is unaltered by the α″. That is, $\omega_{ath}$ is contained within α″ plates.

The tensile stress-strain behavior of ST Ti-10V-2Fe-3Al is shown in Figure 4. The stress needed to initiate the $\alpha''$ reaction at room temperature is shown to be dependent upon the ST temperature. In the $\beta$-ST condition, a stress of only 250 MPa was required. Decreasing the ST temperature, and increasing the $\alpha_p$ content, suppresses the $\alpha''$ reaction. The general shape of these curves is also of interest. Following the initial yield, a nearly flat, low work hardening region was found. Following this flat region, the curve takes a sudden upturn and passes through an inflection point. Finally, a "second yield point" is achieved, after which all conditions exhibited nearly the same behavior.

Before discussing these results, two practical observations of this reaction should be presented, which may be of interest to prospective users of Ti-10V-2Fe-3Al, and $\beta$-Ti alloys, in general. First, when quenching thick sections of $\beta$-ST material (greater than $\sim$ 2cm thickness), quenching strains were found to be sufficient to martensitically transform the $\beta$ to $\alpha''$ at the material's surface. Secondly, all machining and, in fact, simple scribe marking, done while the material was in the $\beta$-ST or near $\beta$-ST condition was found to locally initiate the $\alpha''$ reaction.

## Discussion

The structural transformation involved in the $\beta\rightarrow\alpha''$ reaction can be more easily visualized if the bcc $\beta$ matrix is thought of as its orthorhombic crystallographic equivalent.* This equivalence is demonstrated in Figure 5. Similarly, it is useful to visualize the $\alpha$ and $\alpha'$ hexagonal structures in terms of their orthorhombic representations, for example, the Burger's $\alpha$-$\beta$ orientation relationship:

$$(1\bar{1}0)^{bcc}_{\beta} || (0001)^{hcp}_{\alpha}$$

$$[111]^{bcc}_{\beta} || [11\bar{2}0]^{hcp}_{\alpha}$$

can simply be rewritten as:

$$(001)^{ortho.}_{\beta} || (001)^{ortho.}_{\alpha}$$

$$[110]^{ortho.}_{\beta} || [110]^{ortho.}_{\alpha}$$

Indeed, the $\alpha\rightarrow\beta$ Burger's reaction itself can be visualized as a simple "adjustment" of the $\beta$ orthorhombic structure. Specifically, it involves a distortion of the cell parameters:

$$a_{\beta} \rightarrow a_{\alpha}$$

$$b_{\beta} = \sqrt{2}\, a_{\beta} \rightarrow b_{\alpha} = \sqrt{3}\, a_{\alpha}$$

$$c_{\beta} = \sqrt{2}\, a_{\beta} \rightarrow c_{\alpha} = 1.58 a_{\alpha}$$

---

*This representation is fully equivalent to bcc but is not normally used because it is customary to describe a structure by its highest possible symmetry group, i.e. bcc.

and the following repositioning of the atoms:

$$(0,0,0)_\beta \rightarrow (0,0,0)_\alpha$$

$$(a/2,b/2,0)_\beta \rightarrow (a/2,b/2,0)_\alpha$$

$$(0,b/2,c/2)_\beta \rightarrow (a/6,b/2,c/2)_\alpha$$

$$(a/2,0,c/2)_\beta \rightarrow (a/2,b/6,c/2)_\alpha$$

The above movements provide a viable mechanism for hexagonal martensite formation. The $\alpha''$ structure, however, is best thought of as a compromise structure; that is, an orthorhombic compromise of both the $\beta$ and $\alpha'$ structures. In $\alpha''$ forming systems, the structural shifts required to form hexagonal $\alpha'$ do not go to completion, but instead stop midway, before the $\alpha'$ orthorhombic structure is achieved. It is not clear what controls the progression of this reaction but one generalization can be presented. Systems with highly unstable $\beta$ structures at room temperature ($\alpha$ alloys, for example) generally complete the $\beta\rightarrow\alpha'$ reaction during quenching, while alloys of greater $\beta$ stability prefer to "stop" at an intermediate structure. Further, all carefully characterized stress assisted martensites have been reported as orthorhombic [5]. Two reports of the non-orthorhombic stress assisted martensites [4,7], were not studied in a fashion which would permit resolution of an orthorhombic lattice.

The athermal omega "sub-structure" found in the $\alpha''$ plates is also more fully understood using the above viewpoint. It is known that the hexagonal $\omega$ structure also can be visualized as a bcc structure, with the {111} planes periodically displaced in the <111> direction. If the $\omega$-phase structure is visualized this way, the coherency between the $\omega$ structure and the surrounding $\beta$ matrix is readily understood. Upon quenching, Ti-10V-2Fe-3Al is found to form a uniform distribution of fine $\omega_{ath}$. Upon stressing at room temperature, which is below $M_d$, the residual orthorhombic $\beta$ matrix is distorted to $\alpha''$. The islands of $\omega_{ath}$ may either revert to $\beta$ and then to $\alpha''$, or they may maintain their identity while the surrounding matrix transforms. Our observations indicate the latter occurs, at least in Ti-10-2-3.

Energetically, retention of $\omega_{ath}$ inside $\alpha''$ can be justified using the following argument. In the unstressed $\beta$-ST condition, the $\omega$ structure is energetically more favorable than either $\beta$, $\alpha'$, and $\alpha''$. Since the $\beta\rightarrow\omega$ transformation is known to require little or no thermal activation, $\omega$ forms during the quench. But this reaction is prevented from going to completion, due to the unusual nature of this phonon controlled transformation mechanism. Upon stressing, the $\alpha''$ phase forms because it is more stable than $\beta$, but it is not necessarily more stable than $\omega$. Thus the $\beta$ can be transformed to $\alpha''$, while the $\omega_{ath}$ is left intact. The apparent implication of this finding is that the changes in coherency and strain energy as the orthorhombic $\beta$ lattice is distorted to $\alpha''$ are not large enough to disrupt the $\omega_{ath}$ particles.

The finding of $\omega_{ath}$ within twinned $\beta$ plates is also expected, since $\omega$ particles which are destroyed by twinning should quickly reform after the twin interface has passed.

The orthorhombic distortion associated with $\alpha''$ formation also provides a basis for understanding the unusual stress-strain behavior of ST Ti-10V-2Fe-3Al. Since the lattice undergoes a shape change, $\alpha''$ formation is accompanied by plastic strain. Under an externally applied stress, the $\beta \to \alpha$ reaction distorts the lattice in such a manner as to alleviate the applied stress. More completely stated, the transformation strains from the $\beta \to \alpha$ reaction tend to accommodate elastic strains, and thereby reduce the strain energy of the system.

Referring to the $\beta$-ST condition of Figure 4, it can be seen that the nucleation of $\alpha''$ plates begins at 250 MPa. These first plates grow until a barrier, such as a grain boundary, is encountered. There also is evidence that, upon encountering a grain boundary, a plate may stimulate nucleation of new $\alpha''$ plates in a neighboring grain. Since the stress required to nucleate $\alpha''$ is fairly constant during these early stages, an extended flat portion is found on the stress-strain curve. As more and more $\beta$ is transformed in this manner, the frequency of nucleation, as well as the volume transformed per nucleation event begin to decrease. As this happens, the stress required to produce a given volume of $\alpha''$ (or a given plastic strain increment) begins to increase. Since the stress-strain curves are obtained at constant strain rate, the result is an increasing increment of stress for each corresponding increment of strain leading to an inflection in the curve. Following this inflection, there appears to be a region of rapid work hardening. In reality, this is simply a return to nearly elastic behavior, due to difficulty in producing additional $\alpha''$. Finally, the applied stress becomes sufficiently large to move dislocations in a more classical sense, and a second and true yield point is observed.

In the $\alpha+\beta$-ST conditions the $\beta$ matrix contains $\alpha_p$. Although the $\alpha_p$ is too coarse to directly affect strength, it does tend to increase the chemical stability of the surrounding $\beta$ matrix because of the solute partitioning between $\alpha$ and $\beta$. Thus, as the ST temperature is reduced, and the amount of $\alpha_p$ is increased, the $M_s$ temperature is further suppressed and $M_d$ approaches room temperature. A greater stress is then needed to initiate the $\alpha''$ distortion. This is illustrated in Figure 4. Finally, where the $\beta$ composition is rich enough in Fe and V to depress $M_d$ below room temperature, the material deforms by slip and no $\alpha''$ is formed.

## Conclusion

It has been shown that the $\beta$-phase in Ti-10V-2Fe-3Al is sufficiently stabilized to depress the $M_s$ temperature below room temperature, but that the deformation induced martensite start temperature $(M_d)$ lies above room temperature. It also has been shown that solution treating in the two phase $(\alpha+\beta)$ region enrich the $\beta$-phase composition with respect to Fe and V and thus depresses the $M_d$ temperature. In the fully diluted $\beta$-phases, a stress of 250 MPa was required before the alloy began to transform to an orthorhombic martensite $(\alpha'')$. A small amount of mechanical twinning of the $\beta$ matrix was also observed. Further, the preceding $\omega_{ath}$ dispersion was found to be preserved within the $\beta$ twins, and appears to also be preserved during the $\beta \to \alpha''$ reaction.

The crystallography of the $\beta \to \alpha''$ transformation has been analyzed in a novel way by converting the bcc $\beta$-phase structure to an equivalent orthor-

hombic structure.  Using this analysis, the β→α" can be seen to require
simple atomic shuffles.  The relatively low activation barrier for α" formation can be understood on this basis.  This analysis is described and illustrated.

## Acknowledgments

Three of us (T.W.D., G.T.T.,and J.C.W.) gratefully acknowledge the
support of the Office of Naval Research under contract N00014-76-C-0409.
Facilities for the conduct of this research have been provided by the Center
for the Joining of Materials.  We also acknowledge the experimental assistance of M. Glatz and the secretarial help of Mrs. A. Crelli.

## References

1. Y. C. Liu:  Trans. AIME, vol. 206 (1956), 1036.
2. M. K. Koul and J. F. Breedis:  Acta Met., vol. 18 (1970), 579.
3. J. C. Williams and M. J. Blackburn:  Trans ASM, vol. 60 (1967), 373.
4. F. A. Crossley and R. W. Lindberg:  Proc. of ICSMA-2, vol. 3, (1970),
   ASM, 841.
5. R. M. Middleton:  Proc. of Moscow Ti Conference, in press.
6. J. C. Williams:  Ti Science and Technology, vol. 3 (1973), Plenum
   Press, 1433.
7. M. Oka, C. S. Lee, and K. Shimizu:  Met. Trans., vol. 3 (1972), 37.

Table 1.   X-ray Diffraction Peaks for Ti-10V-2Fe-3Al
β Solution Treated, Quenched, and Compressed
to 900 MPa

| Diffraction Angle – 2θ* Cu K$_\alpha$ radiation; λ = 1.5418Å | Crystal Plane β | α" | d Spacing |
|---|---|---|---|
| 35.35 | | (110) | 2.54 |
| 36.90 | | (020) | 2.44 |
| 39.35** | | (002) | 2.29 |
| 39.50 | (110) | | 2.82 |
| 40.35+ | | (111) | 2.24 |
| 41.60* | | (021) | 2.17 |
| 53.60 | | (112) | 1.71 |
| 54.60** | | (022) | 1.68 |
| 57.05 | (200) | | 1.62 |
| 61.65+ | | (200) | 1.51 |
| 65.10 | | (130) | 1.43 |

*Broad low intensity peak

**Broad very low intensity peak

+Narrow high intensity peak

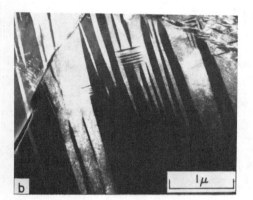

Fig. 1 – Stress induced martensite
shown– a) optically, b)
in TEM bright field, and
c) in TEM dark field.

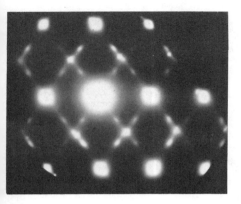 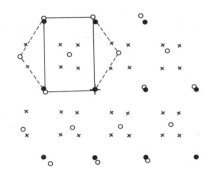

Fig. 2 - a) <110>$_\beta$ zone electron diffraction spot pattern, and b) schematic
representation of the same, with open circles representing the
<001> zone $\alpha''$ reflections, solid circles representing <110> zone
$\beta$ reflections, and crossed lines representing <11$\bar{2}$0> zone $\omega$ reflections (two variants).  The solid lines in a) represent the <001>
orthorhombic cube face symmetry, while the dashed show the distortion
from hexagonal symmetry.

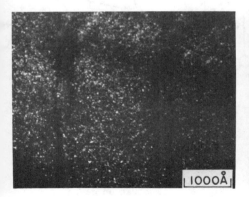 

Fig. 3 - a) Athermal $\omega$, uniformly dispersed throughout the $\beta$ and the stress-
assisted transformation plates, and b) shows dark field imaging of
twinned $\beta$ reflection and $\omega_{ath}$ reflection.

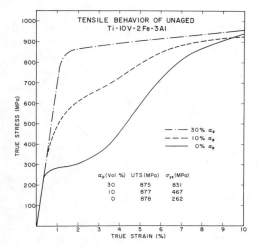

Fig. 4 – True stress-true strain
curves for Ti-10V-2Fe-3Al
solution treated at 720°C,
780°C, and 850°C.

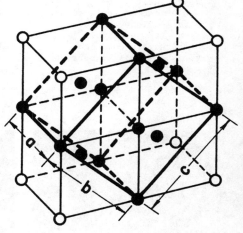

Fig. 5 – Illustration showing the
equivalency of the bcc and
orthorhombic representa-
tions of the β-phase.

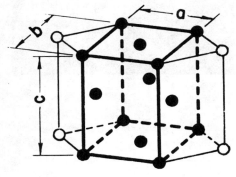

Fig. 6 – Illustration demonstrating the equivalency
between the hcp and the
orthorhombic representa-
tions of the α and α'
phases.

# $M_s$ AND $\beta_s$ TEMPERATURES OF Ti-Fe ALLOYS

Toshimi Yamane and Masao Ito*

Department of Materials Science and Engineering,
Osaka University, Suita, Osaka 565, Japan
* Formerly Graduate Student, now at Shinkokosen
Industry Co., Ltd., Izumisano, Osaka, Japan

## Introduction

It is basic and important to know exact values of $M_s$ temperatures where the martenstic transformation of titanium alloys starts during rapid cooling from the β phase, and those of $\beta_s$ temperatures where the decomposition of the martensite starts during rapid heating to the β phase. The $M_s$ temperatures of titanium-iron alloys have been reported,[1],[2] but their $M_s$ temperatures are different each other, and there is no report on the $\beta_s$ temperatures of this alloy system.

In this research, the $M_s$ and $\beta_s$ temperatures of the titanium-iron alloys were measured exactly and compared with the calculated.

## Experimentals

Pure titanium (H:0.0012, O:0.067, N:0.04, Fe:0.081, Si:0.015 C:0.011wt%) and electrolysis iron (99.9% purity) were melted in an electron beam melting furnace. Ingots were sealed in a vacuum quartz tube and heated at $1000^\circ$C for 48 hours for homogenization, then rolled and drawn to wires of the 1 mm diameter. The chemical compositions of the specimens were Ti-2.16at%Fe, Ti-3.10at%Fe and Ti-3.60at%Fe.

The heating and cooling of the wire specimens were performed in an apparatus shown schematically in Fig.1. The wire specimen

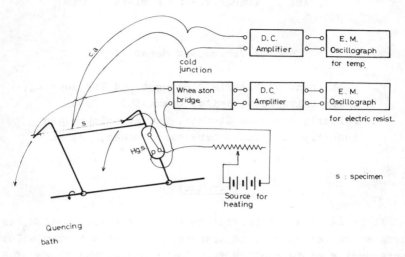

Fig.1    Schematic figure of rapid cooling (quenching)
and rapid heating apparatus

was heated by direct current of electricity from storage batte-
ries. Heating rates were controlled by the direct current density and cooling rates were changed by the exchange of sorts of
quenching bath liquids. The temperatures of the specimen were
measured by the thermocouple of the diameter of 0.1 mm, welded
to the center of the wire specimen. The change in electric resistance of the wire specimen was measured too.


## Experimental Results

Examples of cooling and heating curves are shown in
Fig.2(a) and (b). $M_s$ (starting temperature of martensite formation from β), $\beta_s$ (starting temperature of ß formation from martensite) points were decided as those at the beginning of the bending of temperature-time curves as indicated by $M_s$ and $\beta_s$ in Fig.
2(a) and (b).

The dependencies of the $M_s$ and $\beta_s$ temperatures   on cooling and heating rates are shown in Figs.3 and 4. Transformation

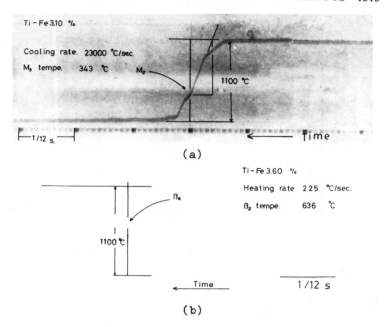

(a)

(b)

Fig.2   Cooling curve (a) and heating curve (b)
(a) Ti-3.10at%Fe    (b) Ti-3.6at%Fe

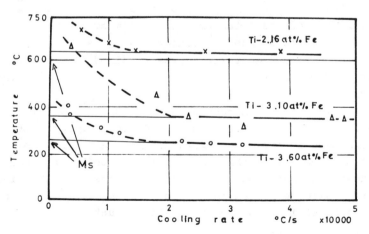

Fig.3   Transformation temperatures as a function
of cooling rates

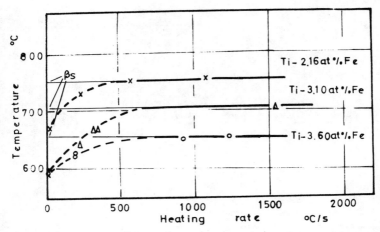

Fig.4 Transformation temperatures as a function
of heating rates

temperatures reach constant values in the high cooling (over
20000°C/s) and heating (over 500°C/s) rates. These constant
temperatures for cooling are defined $M_s$ points and those for
heating are done $\beta_s$ points. The cooling and heating rates
are slops of time- temperature curves before the $M_s$ and $\beta_s$
points.       $M_s$ and $\beta_s$ are plotted in the equlibrium phase
diagram[3] of the titanium-iron system as shown in Fig.5 where
the $M_s$ points of the previous works[1][2] are plotted for
comparison. The $M_s$ temperatures obtained by this investigation
are lower than those obtained by Duwetz,[1] but near those by
Kaneko et al.[2]

Fig.6 shows microstructures of the Ti-3.60at%Fe alloy
quenched in ice water from 1100°C (cooling rate: about 22000
°C/s) and in air (cooling rate: about 200°C/s). Both microstructures have $\alpha'$ which is considered to be martensite. Fig.7
shows the transmission electron micrograph of the Ti-3.10
at%Fe alloy quenched in ice water from 1100°C. This micrograph is similar to the martensite structure observed by
Nishiyama et al.[4]

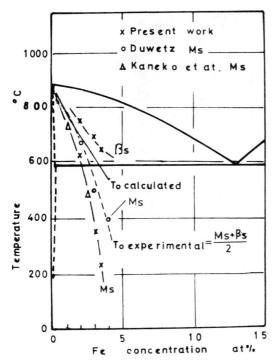

Fig.5   $M_s$, $\beta_s$ and $T_o$
in Ti-Fe system

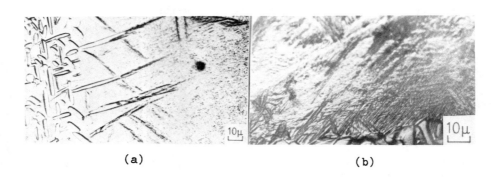

(a)                                    (b)

Fig.6   Optical microstructures of Ti-3.60at%Fe alloy
quenched in ice water (a), and in air (b)
from 1100°C

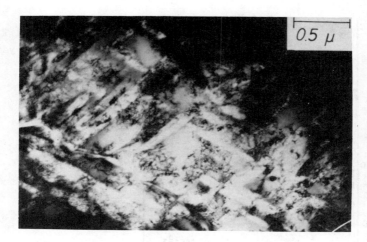

Fig.7    Transmission electron micrograph of
Ti-3.10at%Fe alloy quenched in
ice water from $1100^{0}$C


## Discussion

If the Ti-Fe alloys are the regular solution, the chemical driving force of the martensite formation, $\Delta F_{Ti-Fe}^{\alpha' \to \beta}$ is expressed by the following equation,[5]

$$\Delta F_{Ti-Fe}^{\alpha' \to \beta} = (1 - x)\, \Delta F_{Ti}^{\alpha \to \beta} - x \cdot (RT\, \ln\frac{x_\beta}{x_\alpha} - \frac{(x - 2x_\beta + x_\beta^2)}{x_\beta^2}$$

$$-\frac{(x - 2x_\beta + x_\beta^2)}{x_\beta^2}\, (\,\Delta F_{Ti}^{\alpha \to \beta} + RT\, \ln\frac{1 - x_\beta}{1 - x_\alpha})\,)\ \text{------}(1)$$

where x is an atomic fraction of iron, $\Delta F_{Ti}^{\alpha' \to \beta}$ the difference in free energy between the h.c.p. and b.c.c. modification of titanium tabulated in Table 3 of Reference 5, R the gas constant, T temperature in $^{0}$K, $x_\alpha$, $x_\beta$, atomic fractions of iron at phase boundaries between $\alpha$ and $\alpha + \beta$, between $\alpha + \beta$ and $\beta$ at T.

The chemical driving force of the martensite formation $\Delta F_{Ti-Fe}^{\alpha' \to \beta}$ is calculated by Equation (1) using $F_{Ti}^{\alpha \to \beta}$ tabulated in Table 3 of Reference 5, $x_\alpha$ and $x_\beta$ from Reference 3. The calculated results are shown in Fig.7. In this figure, the temperature where $\Delta F_{Ti-Fe}^{\alpha \to \beta}$ is zero, is defined $T_0$. $T_0$ obtained by the calculation is plotted in Fig.5. Otherwise, experimental $T_0$ is obtained by $T_0 = (M_s - \beta_s)/2$ and is shown in Fig.5 too. In the concentration range lower than about 2 at%Fe, calculated $T_0$ agrees with experimental one, but in the higher concentration than about 2 at%Fe, the experimental $T_0$ is lower than the calculated. This may owe to the deviation of the actual alloys from the regular solution.

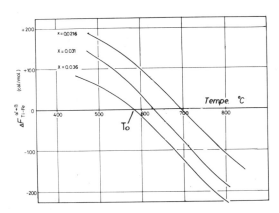

Fig.7  Chemical driving force of martensite formation as a function of temperature and solute concentration

## Summary

1.  The $M_s$ and $\beta_s$ temperatures of the Ti-Fe alloys are constant in the high cooling and heating rates.
2.  The temperatures, $T_0$, at the zero driving force of the martensite formation are obtained by the calculation and the experiment. Both $T_0$ are similar in the lower concent-

ration than about 2 at%Fe, but not in the higher concentration.

## References

1.  P. Duwetz: Trans. Amer. Soc. Metals, 45(1953), 934.
2.  H. Kaneko et al.: J. Jap. Inst. Metals, 27(1963), 393.
3.  M. Hansen: Constitution of Binary Alloys,(1958), 725.
    McGraw-Hill Book Co. Inc.
4.  Z. Nishiyama, M. Oka and Y. Nakagawa: J. Jap. Inst.
    Metals, 30(1966), 16.
5.  L. Kaufman: Act. Met., 7(1959), 575.

# PHASE TRANSFORMATIONS IN A Ti-1.6 a/o N ALLOY

D.Sundararaman, V.Seetharaman and V.S.Raghunathan

Metallurgy Programme, Reactor Research Centre
Kalpakkam-603102, India

## Abstract

The martensitic transformation and the subsequent precipitation of $Ti_2N$ were investigated in a Ti-1.6 a/o N alloy. On quenching, the alloy transforms completely to martensite consisting of several colonies of parallel crystals. While many of the adjacent martensite crystals were found to be twin related, some of them contained $\{10\bar{1}1\}_\alpha$ type internal twins. On ageing, this martensite transforms to a mixture of $\alpha$ and $Ti_2N$. The orientation relationship between the $\alpha$ and the $Ti_2N$ phases has been identified as

$$(10\bar{1}0)_\alpha \parallel (011)_{Ti_2N}$$

$$[\bar{1}2\bar{1}0]_\alpha \parallel [01\bar{1}]_{Ti_2N}$$

## Introduction

It is very well documented that interstitial solutes like C, N and O exert profound influence on the mechanical properties and corrosion behaviour of titanium and zirconium base alloys. Among these, nitrogen is the most effective stabiliser of the $\alpha$ phase in titanium and also imparts maximum strength to the $\alpha$ phase at low temperatures. For example, addition of one atom percent of C, N or O leads to an increase in the tensile strength of titanium at room temperature by 64, 69 or 38 MPa respectively [1] . While several detailed investigations have been carried out on the phase transformations occurring in Ti-O and Ti-C systems, very few studies have been conducted on Ti-N alloys [2,3] This paper reports some results obtained on the martensitic transformation and the subsequent precipitation reaction occurring in a Ti-1.6 a/o N alloy.

The phase diagram for the Ti-N system is shown in Fig.1. The $\gamma$ phase corresponds to a composition $Ti_2N$ and occurs over a very narrow range of compositions. It has a tetragonal (anti C4 type) structure with the lattice parameters, a=4.9414 Å; c=3.0375 Å[4] . The $\delta$ phase corresponding to a TiN stoichiometry is a NaCl type compound with the lattice parameter, a = 4.240 Å.

## Experimental Procedure

Iodide grade titanium foils of about 25 $\mu$m thickness were nitrided in an "ion-nitriding" apparatus described in detail elsewhere[5]. After evacuating the reaction chamber to a pressure of about $10^{-6}$ torr, purified nitrogen gas was admitted

and the system pressure was maintained continuously in the range
` to 5 torr. Pieces of titanium sponge maintained at 500°C were
used as a getter to remove any trace of oxygen present in the
nitrogen stream. Nitriding was carried out at 800°C and the
temperature control of the sample was achieved by adjusting the
ion current and the gas pressure. Typical values of the cathode
voltage and the ion current used were 800V and 30mA respectively.
The nitrogen concentration was estimated by the weight change
technique.

Nitrided samples were homogenised at 1300°C in a vacuum
better than $10^{-5}$torr and then quenched. Subsequently some of the
samples were aged at 200 and 400°C for different durations. Thin
foils for electron microscopy were prepared by chemical polishing
in a solution containing 33% HF, 50% $HNO_3$ and the rest water at
0°C. Since the total time required for polishing was less than
five minutes, the hydrogen pick up as a result of polishing was
expected to be very low. These foils were examined in a Philips
EM400 transmission electron microscope at 120 KV.

Results

**1.** Quenched Microstructures

On quenching from the $\beta$ phase the alloy transforms
completely to the hcp $\alpha$ phase by a martensitic mode. Fig.2
illustrates the typical acicular microstructure consisting of
lenticular plates of martensite. No evidence for the occurrence
of either retained $\beta$ , the fcc martensite or the orthorhombic
martensite was found in this work. Transmission electron
microscopic examination showed that the martensite colonies,
formed as a result of quenching, contained several plates
stacked almost parallel to each other. Some of these were found
to be internally twinned (Fig.3 (a) and (b) ). Diffraction
analysis showed that these twins were all strictly parallel to
$\{10\bar{1}1\}_\alpha$ planes. At the same time it was noticed that many of
the adjacent martensite plates were also twin related; the
twinning plane was identified as $\{10\bar{1}1\}_\alpha$ . The pair of bright
and dark field micrographs shown in Fig.4(a) and (b) provides
an example for this observation.

It was found that the quenching rates achieved in this work
were not sufficient to retain the $\alpha$ phase completely in the
supersaturated solid solution. Evidences for the very early
stages of decomposition were seen within many martensite plates.
Fig.5(a) represents a region which had presumably undergone a
continuous transformation leading to the formation of a fine and
highly coherent product microstructure. The striations were
found to be parallel to $(11\bar{2}1)_\alpha$ and $(1\bar{1}03)_\alpha$ . The selected area
diffraction pattern taken from this region contained many
additional spots which could be indexed as a slightly distorted
(001) zone of $Ti_2N$. The dark field micrograph shown in Fig.5(b)
provides another example for the microstructure obtained after
initial decomposition of the $\alpha$ lattice. The diffraction
pattern taken from this region (Fig.5(c) and (d)) reveals two

variants of the $(0\bar{1}1)$ reciprocal lattice sections of the $Ti_2N$ phase superimposed on the matrix reflections. On analysis of such patterns the orientation relation between $Ti_2N$ and the $\alpha$ matrix has been determined as follows:-

$$(10\bar{1}0)_\alpha \parallel (011)_{Ti_2N}$$

$$[1\bar{2}10]_\alpha \parallel [01\bar{1}]_{Ti_2N}$$

Some of the twins in the martensite crystals were found to undergo a transformation: $\alpha \rightarrow \alpha + Ti_2N$ resulting in the formation of a large number of narrowly spaced parallel interfaces within them. Fig.6 (a) and (b) illustrate the microstructural features so obtained. It can be noticed that the $\alpha$ phase regions sandwiched between adjacent sheets of $Ti_2N$ possess the same orientation as that of the matrix $\alpha$.

## 2. Aged Microstructures

On ageing at 200°C a large number of irregular shaped $Ti_2N$ precipitates of 0.1 - 0.2$\mu$m in length were found to form. Nucleation of such precipitates occurred preferentially on dislocations found within the martensite plates (Fig.7a). In addition to these, thin plate shaped precipitates of $Ti_2N$ were also noticed in many regions of the foils. These plates were about 200 Å thick and about 0.4$\mu$m long. The dark field micrographs shown in Fig.7(c) confirm that these plates are indeed $Ti_2N$. All these $Ti_2N$ plates were found to be partitioned by a large number of closely spaced planar features (Fig.7(b)). Similar internal features have been observed in many precipitates and interpreted as strain relief twins [6] in systems such as Ta-O, Nb-O, $Ni_3V$, V-N etc.[7,8] and as internal faults in some other systems such as Nb-C, Hf-N etc. [9,10].

Ageing for one hour at 400°C resulted in the formation of large spherical precipitates of $Ti_2N$ along the plate boundaries, besides a high density of fine $Ti_2N$ precipitates formed within the plates (Fig.8). The adjacent martensite plates in this micrograph were found to be misoriented by a very small angle only. The pair of dark field micrographs shown in Fig.9(a) and (b) clearly reveal that the central plate of martensite is twin related to the adjacent plates. Since the $\alpha$ and $Ti_2N$ reflections were so close that they could not be isolated, one set of $Ti_2N$ and $\alpha$ reflections was enclosed by the objective aperture for obtaining each of these micrographs. Here again, the twin plane was identified as $(10\bar{1}1)_\alpha \parallel (210)$ $Ti_2N$. The stereographic projection shown in Fig.9(e) reveals that the angle between $(0001)_\alpha$ and $(0\bar{1}0)$ $Ti_2N$ is about 7°. It is seen that this finding is in reasonable agreement with the orientation relations mentioned earlier. When the duration of ageing was extended to three hours, it was noticed that the $Ti_2N$ precipitates assumed a spherical or equiaxed morphology. This aspect is evidenced in the photograph shown in Fig.10.

## Discussion

1.   Morphology and substructure of the α phase

It is well known that addition of substitutional or interstitial solutes causes a change in the morphology and substructure of the titanium martensites. The exact concentrations at which such transitions occur would depend on the type of solute and its influence on the transformation temperatures as well as on the strength and the deformation mode of the parent and the product phases. For example addition of 2.4 w/o Cr, or 6 w/o Cu or 20 w/o Zr to titanium brings about a change in morphology from the massive to the acicular type [11]. However, no such data are available for interstitial solutes in titanium. Though the scanning electron micrograph shown in Fig.2 suggests that the martensite obtained in the present alloy possesses an acicular morphology, it is clear from the transmission electron micrographs that martensite forms in this alloy both as plates and as laths. Thus it appears that the Ti - 1.6 a/o N alloy falls in the range of compositions over which a transition from lathy to plate martensite takes place.

$\{10\bar{1}1\}_{\alpha}$ twins have been identified as the transformation twins in many titanium alloys [12]. However, in the present work only a limited number of the martensite crystals were found to contain such internal twins. Moreover, the spacing between these twins was found to be highly non-uniform. Therefore, it is proper to conclude that these twins could not have been caused by the transformation per se. Indeed it is likely that the post-transformation stresses generated within the martensite crystals could have led to the formation of such twins. In such a situation, the lattice invariant shear associated with the transformation must be accommodated by slip. Of course, the observed density of dislocations within the martensite crystals was quite low; this could be attributed to the fact that the $M_s$ temperature of this alloy was very high.

2. Precipitation of $Ti_2N$

The experimental results obtained on the precipitation of $Ti_2N$ from supersaturated α phase could be summarised as follows:

i) There exists a specific orientation relationship between the α phase and the $Ti_2N$ phase.

$$(10\bar{1}0)_{\alpha} \parallel (011)_{Ti_2N}$$
$$[1\bar{2}10]_{\alpha} \parallel [01\bar{1}]_{Ti_2N}$$

The $Ti_2N$ precipitates always form on definite habit planes.

ii) The transformation is very rapid - it initiates even during quenching.

iii) The volume change associated with the transformation α ⟶ $Ti_2N$ is very small ( ᴗ 5%).

iv) The $Ti_2N$ plates contain many internal twin-like features.

The above mentioned observations suggest that this transformation could be described as 'bainitic' [13] . Similar results have been reported in Cu-Au, Ta-O, V-H etc.[13,14] . Flewitt et al[15] have succeded in applying the phenomenological theories of martensitic transformation in order to predict the crystallographic features of the niobium hydride plates formed spontaneously in a Nb-Zr alloy.

The simplest lattice correspondence between $\alpha$ and $Ti_2N$ is shown in Fig.11. The principal strains along the three mutually perpendicular directions are given below:-

$$\epsilon_{[1\bar{2}10]} = 4.88\%$$

$$\epsilon_{[0001]} = -5.52\%$$

$$\epsilon_{[10\bar{1}0]} = 1.28\%$$

The necessary and sufficient condition for a pair of planes to remain undistorted during a homogeneous deformation is that one of the principal strains be zero and the other two of opposite signs [16] . Since this condition is nearly satisfied for the present case, the amount of slip or twin which needs to be added to the pure lattice strain to produce an invariant plane strain is very small.

At temperatures below 200°C, it appears that the decomposition of $\alpha$ proceeds by a continuous transformation in the initial stages followed by the ordering of the solute rich phase. Fig.5 lends support to this 'spinodal ordering' mechanism. At temperatures above 200°C, the precipitation reaction occurs presumably by nucleation and growth processes. The observation of discrete and large precipitates along the lath boundaries (Fig.8) confirms this view.

## Acknowledgements

The authors wish to thank Shri S.Vaidyanathan for his kind cooperation and assistance in carrying out the experiments. They are also grateful to Dr.P.Rodriguez, Senior Metallurgist for his keen interest and encouragement during the course of this investigation.

## References

1. O.N.Magnitzkii: Titanium Science and Technology, 3 (1973), 1915. Plenum Press, New York.
2. K.Okazaki, M.Momochi and H.Conrad: ibid, p.1649.
3. F.W.Wood and O.G.Paasche: Microstructural Science, 2(1974), 101. American Elsevier Publishing Co., New York.
4. W.B.Pearson: Handbook of Lattice Spacings and Structures of Metals and Alloys: 2(1967),1442. Pergamon Press, Oxford.

5.  D.Sundararaman, S.Vaidyanathan, V.Seetharaman and V.S.Raghunathan: in preparation.

6.  L.E.Tanner and M.F.Ashby: Phys. Stat. Soli. (a), 33(1969),59.

7.  J.Van Landuyt and C.M.Wayman: Acta Met., 16(1968),803.

8.  D.I.Potter: J.Less Common Metals., 31 (1973), 299.

9.  J.B.Mitchell: Metallography, 8(1975),5.

10. R.K.Viswanadham and C.A.Wert: J.Less Common Metals, 48 (1976) 135.

11. S.Banerjee, S.J.Vijayakar and R.Krishnan: Titanium Science and Technology, 3(1973), 1597.  Plenum Press, New York.

12. Z.Nishiyama: Martensitic transformations, (1978), 68. Academic Press, New York.

13. H.Warlimont: Electron Microscopy and Structure of Materials, (1972) 505. University of California Press, Berkeley.

14. M.S.Rashid and T.E.Scott: J.Less Common Metals, 31(1973),377.

15. P.E.J.Flewitt, P.J.Ash and A.G.Crocker: Acta Met., 24 (1976), 669.

16. A.Kelly and G.W.Groves: Crystallography and Crystal Defects, (1970), 321. Addison-Wesley, Massachusetts, USA.

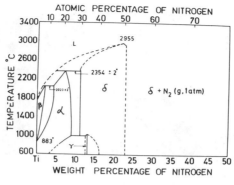

Ti−N PHASE DIAGRAM

Fig.1: Phase diagram for the Ti-N system.

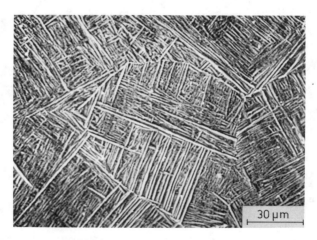

Fig.2: Scanning electron micrograph of the Ti-1.6 a/o N alloy
quenched from the $\beta$ phase. The micrograph reveals an
acicular morphology characterised by a large number of
parallel, lenticular plates of martensite.

Fig.3: (a)Bright field and (b)dark field electron micrographs
demonstrating the presence of narrow {10$\bar{1}$1} internal
twins within martensite plates.

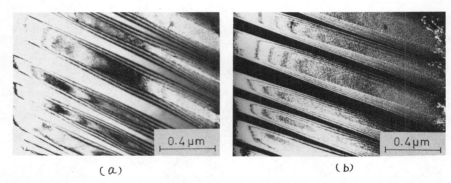

Fig.4: Adjacent plates of martensite found to be mutually twin
related. (a)Bright field and (b)dark field obtained using
a twin reflection.

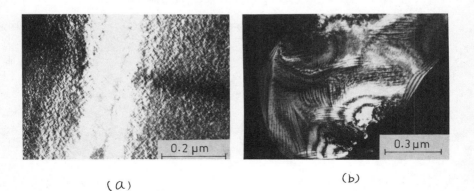

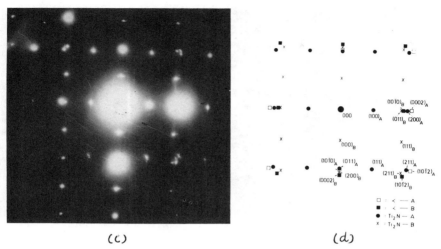

(c)                              (d)

Fig.5: Evidences for the initial stages of decomposition of the
α matrix found in the quenched sample. The diffraction
pattern and the key shown in (c) and (d) respectively
correspond to the micrograph shown in (b).

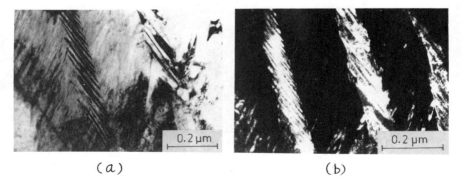

(a)                              (b)

Fig.6: Decomposition of the martensite within the internal twins
resulting in a mixture of α and $Ti_2N$: (a) bright field
(b) dark field. The α region within the transformed
twins and the matrix   have identical orientation.

(a)                    (b)                    (c)

Fig.7: Microstructures observed on ageing at 200°C for one hour.
Ti$_2$N plates containing 'strain relief twins' within them
are visible.  The dark field photograph shown in (c)
confirms that these plates are indeed Ti$_2$N.

Fig.8: Sample quenched and aged at 400°C for one hour.  Besides
fine and uniform precipitates formed within the martensite
plates, large and discrete precipitates of Ti$_2$N are
observed along the plate boundaries.

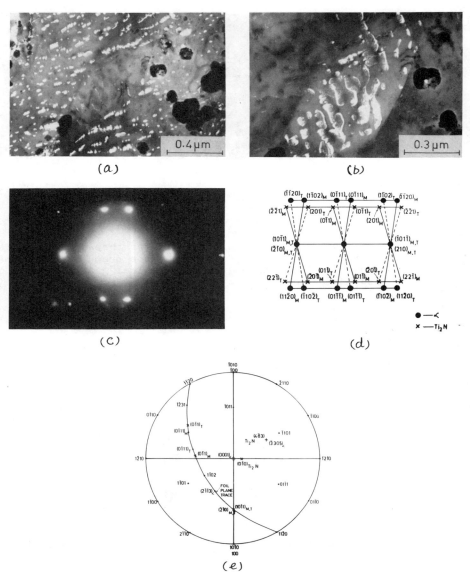

Fig.9: (a) and (b) The dark field micrographs reveal precipitates of Ti$_2$N seen within the adjacent twin related martensite crystals. (c) selected area diffraction pattern (d) key and (e) stereographic projection.

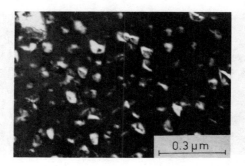

Fig.10: Sample quenched and aged at 400°C for three hours. It is
clear that the Ti₂N precipitates have grown and that their
morphology has changed from ellipsoidal plates to more or
less equiaxed crystals.

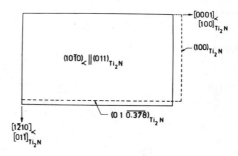

Fig.11: Lattice correspondence between the α matrix and the
Ti₂N precipitates.

# NEW OBSERVATIONS OF THE
# TRANSFORMATIONS IN Ti-6Aℓ-4V

M. Ashraf Imam and Charles M. Gilmore

School of Engineering and Applied Science
The George Washington University
Washington, D.C.   20052

## Introduction

Titanium alloys such as the Ti-6Aℓ-4V are two phase alloys because of the Aℓ and V additions.  The vanadium stabilizes the β (BCC) phase of titanium and the aluminum stabilizes the α (HCP) phase so that at low temperatures it is possible to have in equilibrium a vanadium rich β phase and an aluminum rich α phase.  At temperatures above the β transus titanium alloys are entirely of the β phase.  Rapid cooling of the β phase from above the β transus results in the formation of a hexagonal close packed martensitic phase α´.

The phase transformations of the alloy Ti-6Aℓ-4V have been investigated by several authors.  One of the studies of solution treated Ti-6Aℓ-4V was conducted by Fopiano et al. [1].  It was observed that the β phase when water quenched from heat treatment temperatures of 800°C (1472°F) or higher always transformed to the α´ phase.  Thus, Ti-6Aℓ-4V heat treated at temperatures above 800°C and below the β transus (approximately 1000°C) and water quenched should always form a mixture of primary α that existed before the water quench and α´ formed by martensitic transformation of the high temperature β phase.  Fopiano also observed that as the heat treatment temperature increased the hardness increased, but the ductility did not decrease.  The explanation for this was that the titanium martensite transformation is not a strengthening mechanism.  However, Ti-6Aℓ-4V martensite formed by quenching from above the β transus has very low ductility.  Thus it appeared that the microstructures formed by quenching Ti-6Aℓ-4V from below the β transus may have some special properties that were not revealed in Fopiano's work.  Fopiano utilized X-ray diffraction and surface replication to study the microstructures.  We utilized transmission electron microscopy with selected area diffraction and microprobe analysis.  In this paper only the results from heat treatment temperatures of 900°C and 1065°C are reported.  A more complete analysis contains results from other temperatures and related mechanical property data [2].

## Experimental Procedure

The material utilized in this study was provided by The Naval Ship Research and Development Lab, Annapolis, MD (Heat No. 304171-06) as 3.18 cm diameter extruded rod with a composition in weight percent:

| Aℓ | V | O | Fe | N | C | H |
|------|------|------|------|------|------|-------|
| 6.3 | 4.2 | .188 | .17 | .01 | .02 | .0067 |

The alloy was mill annealed after extrusion for two hours at 704°C.

The thermal treatment was performed in a vertical Centorr air furnace. Once the furnace temperature of 900°C was attained, the specimens, attached to one end of a steel wire, were pulled into the heating zone of the furnace. A chromel-alumel thermocouple was placed in the heating zone. After 10 minutes of thermal treatment at temperature, the suspended specimens were dropped directly from the furnace into a bucket containing room temperature water. The temperature of the furnace was controlled to within ± 2°C. The oxide affected surface layer of 0.05 cm was removed from the specimen surface. After quenching from 900°C, specimens were aged at 500°C for times up to 8 hours. A vacuum tube furnace was used for aging and a pressure of $10^{-5}$ Torr or lower was maintained; the furnace temperature was controlled to within ± 2°C.

Specimens for microhardness measurements were prepared by electropolishing in a solution of 55.4% methanol, 38% butanol and 6.6% of 70% perchloric acid at 14 volts and 0°C. Microhardness measurements were made with a Wilson Tukon Microhardness Tester using a Knoop indenter with a load of 100 grams. The measurements were repeated 8 times for each specimen.

Microstructural analyses were conducted using the methods of optical microscopy, scanning and transmission electron microscopy, and selected area electron diffraction. Specimens for optical microscopic observations were mounted, mechanically polished to a 0.05 μM (alumina powder) surface finish, and etched with Kroll's etchant (3.5% $HNO_3$, 1.5% HF and 95% $H_2O$). The optical microscopy was done with a Carl Zeiss, Ultraphot II microscope or a Leitz Wetzlar Microscope.

Thin foils for transmission electron microscopy and selected area electron diffraction were prepared by electropolishing sections of the material which were cut from the bulk material by spark discharge. The electropolishing solution was 62.5% methanol, 31% butanol and 6.5% of 70% perchloric acid at 13.9 volts and −40 to −50°C. For transmission electron microscopy and selected area electron diffraction a Phillips 200 transmission electron microscope was used with an excitation voltage of 100 Kv. For thick specimens, a JOEL transmission electron microscope was used with an excitation voltage of 200 Kv. Phase structure and lattice parameters of α-Ti, β-Ti, α´-Ti martensite, were determined by selected area electron diffraction.

Electron probe microanalysis was performed to determine the elemental chemistry of the phase. This was done using an X-ray emission and energy dispersive technique [3,4]. Mass fractions of elements were calculated utilizing FRAME developed at the National Bureau of Standards [5,6]. Specimens for electron probe microanalysis were mounted, mechanically polished to a 0.05 μM (alumina powder) surface finish and etched very lightly with Kroll's etchant. The electron probe microanalysis was performed with an Applied Research Laboratory EMX electron probe microanalyzer. An optical microscope, attached with the electron probe microanalyzer, was used to locate the area for analysis. The electron beam diameter used was 0.5 micron and the analytical volume was 1.5 to 2 cubic microns.

## Test Results

Optical and transmission electron micrographs of the alloy heat treated at 900°C and water quenched are shown in Figures 1 and 2. From the optical micrograph the dark α-Ti phase was determined to be approximately 50 volume percent of the total. The transmission electron micrograph in Figure 2 shows that one phase was a mixture of α´-Ti and retained β-Ti; the other phase was the α-Ti grains. The phases were identified by the technique of selected area diffraction. The morphology of the martensite was observed in the dark field image of Figure 3 taken of a martensite reflection. From the standard TEM and the dark field image it appeared that approximately 50 volume percent of the matrix had transformed to martensite and approximately 50 volume percent remained as retained β. Note that on the polished and etched surface in Figure 1 of the as quenched alloy there was no evidence of surface relief.

Specimens that were heat treated at 900°C and water quenched were subsequently aged. The optical and transmission micrographs of these specimens aged for one hour and 8 hours at 500°C are presented in Figures 4 through 7. In the alloy aged for one hour at 500°C (see the TEM of Figure 5) the nuclei that were present after quenching (Figure 2) appeared to grow in the retained β matrix. Also after aging for one hour, surface relief was observed on the polished and etched surface in Figure 4. Because of the lath morphology of the new phase and the appearance of surface relief we conclude that the new phase was martensite. This conclusion was supported by a microhardness measurements in Table 1 from the aged material. After one hour at 500°C the hardness was much higher than the hardness of the as quenched material and it was much higher than the hardness of α´ quenched from 1065°C.

### Table 1
#### Microhardness of Thermally Treated Ti-6Aℓ-4V Alloys

| Thermal Treatment | Knoop Hardness* | Standard Deviation |
|---|---|---|
| 900°C + WQ | 455 | 7.1 |
| 900°C + WQ 1 Hr. at 500°C | 792 | 11.6 |
| 900°C + WQ 8 Hrs. at 500°C | 547 | 5.1 |
| 1065°C + WQ | 511 | 0.0 |

*Based upon 8 measurements

Aging the 900°C as quenched alloy for 8 hours resulted in the optical microstructure shown in Figure 7. The surface relief was also apparent in this figure. Selected area diffraction of the lath microstructure observed in Figure 6 indicated a hexagonal close packed structure, this structure appears to be the growth product of the phase observed after one hour at 500°C. No body centered cubic β phase could be detected. It was concluded that the lath phase observed after 8 hours of aging at 500°C was tempered martensite. This conclusion was supported by microhardness measurements in Table 1, the 8 hour aged specimen hardness was less than that observed after one hour indicating a tempering type of process. The optical microstructure shown in Figure 8 was obtained by heat treating at 1065°C and water quenching;

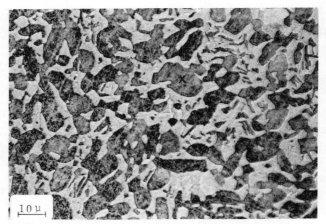

Figure 1.   Optical micrograph of Ti–6Aℓ–4V heat treated
            at 900°C and water quenched showing α-Ti
            (labeled x) and β-Ti.   The β-Ti phase contains
            α´-Ti martensite.

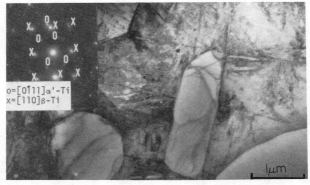

Figure 2.   Transmission electron micrograph of Ti–6Aℓ–4V heat
            treated at 900°C and water quenched.   The diffrac-
            tion pattern was taken from the area of the alloy
            that indexes as a mixture of α´-Ti martensite and
            β-Ti.

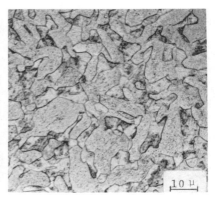

Figure 3. Transmission electron micrograph (dark field) of the same specimen as in Figure 2 showing the morphology of martensite. Dark field reflection is the α´-Ti (011).

Figure 4. Optical micrograph of Ti-6Aℓ-4V heat treated at 900°C and water quenched followed by aging at 500°C for one hour.

Figure 5. Transmission electron micrograph of the same specimen as in Figure 4 showing α-Ti and the phase region where α´-Ti martensite has grown.

Figure 6.   Transmission electron micrograph of Ti-6Aℓ-4V
heat treated at 900°C and water quenched followed
by aging at 500°C for 8 hours.

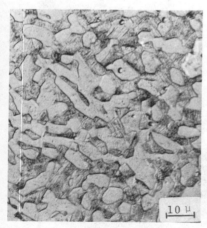

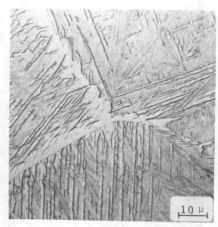

Figure 7.   Optical micrograph of
the same specimen as in Figure 6.

Figure 8.   Optical micrograph of
Ti-6Aℓ-4V solution treated at 1065°C
and water quenched showing α´-Ti martensite.

1065°C was at least 80°C above the β transus. The TEM of the martensite formed by water quenching from 1065°C in Figure 9 showed that this martensite was quite coarse in comparison to that observed in the alloy quenched from 900°C.

In a related experiment, transmission electron micrographs were obtained from the alloy in the as quenched from 900°C condition following failure in tension-compression fatigue (see Ref. 2). Figure 10 shows that after cycling there was a lath like phase that had a HCP structure based upon selected area diffraction. The lath like phase appeared to be martensite formed by a strain induced martensitic transformation from retained β. The strain induced martensite in Figure 10 and the phase formed after a one hour age at 500°C in Figure 2 were of similar morphology.

Electron probe chemical microanalysis was determined in the phases present in alloy water quenched from 900°C and 1065°C. This information was desired for correlation with phase stability. The data presented in Table 2 shows that after heat treating at 900°C the β phase is rich in vanadium and this is why the β phase is retained. In the alloy solution treated at 1065°C the β phase was lean in β stabilizer and when it was quenched the β was unstable and transformed to martensite.

Table 2
Microchemistry of Thermally Treated Ti-6Al-4V Alloys

| Thermal Treatment | Phases & Volume Percent Observed | Phase Composition | |
|---|---|---|---|
| | | Aℓ | V |
| 900°C + WQ | α (50%) | 7.49 ± 0.10 | 2.51 ± 0.12 |
| | α´ (25%) + β (25% | 5.80 ± 0.26 | 6.66 ± 0.65 |
| 1065°C + WQ | α´ (∿100%) | 6.61 ± 0.90 | 4.74 ± 0.07 |

## Discussion

The detailed microstructural analysis utilizing TEM and electron probe microchemical analysis permits an understanding of the ductility observed by Fopiano. The high ductility of the alloy heat treated at 900°C and water quenched was a result of a strain induced transformation of β to α´. After the water quench the most unstable β phase material had transformed to α´ martensite. The retained β phase although more stable than the material that had transformed was in a metastable condition and could transform to α´ if activated. The metastable retained β was activated for transformation by both mechanical and thermal energy. The metastable retained β transformed during fatigue cycling to martensite. The experimental observations indicated that the β retained after quenching from 900°C could also be thermally activated to transform into martensite, thus this martensitic transformation must be isothermal in character. Williams' recent review article emphasized that the martensitic transformation of the β phase Ti and its alloys was athermal, and that no valid isothermal transformations had been observed [7]. Therefore; it is necessary to prove conclusively that the transformation observed was an isothermal martensitic transformation. The aging experiments of the β retained after the water quench conclusively show that the transformation was isothermal. The more difficult part to prove is that it was a martensitic

Figure 9.    Transmission electron micrograph of the same
specimen as in Figure 8.  The electron diffraction
pattern indexes to be α´ (hexagonal) martensite.

Figure 10.    Transmission electron micrograph of the same
specimen as in Figure 2 taken close to the fracture
surface after tension–compression fatigue failure
showing sharp needles of α´-Ti martensite and α-Ti
grains.

transformation.  There is a significant amount of evidence that the result of the transformation was martensite:

a.   The fine lath morphology of the transformation product after aging for one hour at 500°C.

b.   The high microhardness after aging for one hour at 500°C.

c.   The absence of surface relief in the optical micrograph of the specimen as quenched from 900°C and the appearance of surface relief with aging at 500°C for one hour and 8 hours.

d.   The morphological similarity of the strain induced martensite and the isothermal transformation product.

e.   The absence of any detectable β phase resulting from nucleation and growth after aging the 900°C as quenched material for 8 hours at 500°C.

Williams has observed that the retained β phase transforms directly to Widmanstätten α phase that obeys the Burgers relationship [7].  The results of our work show that the β phase that was retained after a water quench from 900°C transformed by an isothermal transformation to martensite.  Continued aging of the martensite resulted in a tempered martensite.  It is expected that the tempered martensite would eventually transform to equilibrium α plus β of the Widmanstätten type structure.

It was interesting to go back to the original paper by Wiskel et al. where the kinetics of the β to α phase transformation was first reported [8]. Wiskel et al. cooled high purity titanium from 1000°C at a rate of 2°C per minute.  They reported that "these structures presumably indicate a shear transformation β → α athermally nucleated, but capable of isothermal progression."  This appears to be what we were observing in this 900°C alloy. In the as quenched condition nuclei can be observed that formed during the quench.  Then the TEM's after one hour and 8-hour aging at 500°C depict the isothermal growth of the nuclei that formed on quenching.  This type of model is that of an isothermal martensite.  However, in the literature Wiskel's work is quoted as evidence that the β to α transformation is athermal [8]. From the work reported by Wiskel et al. it could be concluded that even in pure titanium the β to α transformation was isothermal.

## Acknowledgement

The authors acknowledge the financial support provided by the Naval Air Systems Command.  The authors also thank the staff at The National Bureau of Standards for permitting the use of their microscopy and microanalysis facility, and specially to Dr. Anna Fraker for her assistance.  Discussions with Dr. George Yoder of The Naval Research Lab. were of considerable help during the progress of this work.

References

1. P. J. Fopiano, M. B. Bever, and B. L. Averbach:  *Trans. ASM*, Vol. 62 (1969), 324.
2. M. A. Imam:  *D.Sc. Thesis Microstructure and Fatigue Properties in Ti-6Aℓ-4V*, (May 1978).
3. K. F. J. Heinrich:  *NBS Technical Note 719*, (Appendix II, *NBS Technical Note 521*), (May 1972), 35.
4. R. Castaing, Doctoral Thesis, University of Paris, (1951).
5. H. Yakowitz, R. L. Myklebust:  and K. F. J. Heinrich, *NBS Technical Note 796*, (October 1973).
6. J. Philibert and R. Tixier:  in "*Quantitative Electron Probe Microanalysis*", ed. K. F. Heinrich, NBS Special Publication 298, (1968), 13.
7. J. C. Williams:  *Titanium Science and Technology*, ed. R. I. Jaffee and H. M. Burte, Vol. 3, (1973) 1433.  Plenum Press.
8. S. U. Wiskel, W. V. Youdelis, and J. G. Parr: *Trans. Met. Soc. AIME*, Vol. 215, (1959), 875.

# PHASE TRANSFORMATIONS OF TITANIUM AND SOME TITANIUM ALLOYS.
## KINETICS AND MORPHOLOGY OF PHASE TRANSFORMATION.
## APPLICATION TO HEAT TREATMENT.

E. Etchessahar*, J.P. Auffredic**, J. Debuigne*

* Laboratoire de Métallurgie. Institut National des Sciences Appliquées
Rennes (France)

** Laboratoire de Cristallochimie. Université de Rennes (France).

## ALLOTROPIC TRANSFORMATION OF TITANIUM

The allotropic transformation of titanium which connects the h.c.p. low temperature phase to the c.c. high temperature phase, is very sensitive to the impurities contents of the metal. Titanium has a particularly high reactivity towards oxygen and nitrogen, whose solid solutions cause a very sharp rise of the transition temperature.

### Dilatometric study of metals of different purity levels

The kinetic investigations about the allotropic transition necessitate very pure base materials and high vacuum technique, in order to minimize contamination. The dilatometric experiences were conducted in a direct type apparatus built in our laboratory [1][2]. Its working conditions are : high vacuum ($10^{-6}$ to $10^{-7}$ Pa) at maximum temperatures of 1050 to 1100° C. A high sensibility i.e. a maximal commonly usable amplification coefficient of $25.10^3$. A particularly high stability of the base line allows very long time studies of cumulative annealing type. It is therefore possible to have very fine studies of the samples' evolution in the neighbourhood of the phase transition, enabling precise definitions of transition temperatures.

The present study was carried out using thermal cycling at different speeds of temperature variations vs time and, on the other hand, using the cummulative annealing method. The metals used are R.C.P. Titanium [3] and Van Arkel Titanium [4], their chemical analysis are given in Table 1.

Table 1 : Analysis (in $10^{-6}$ atomic)

| Materials | O | O+N | C | Hf | Al | Fe | Ni | Zr |
|---|---|---|---|---|---|---|---|---|
| R.C.P. Titanium | 900-1800 | - | 280 | - | 10-200 | 150-250 | 50-70 | 10-40 |
| Van Arkel Titanium | 100-150 | - | - | - | - | 70-100 | - | - |

Thermal cycles. The temperatures of the beginning of transformation during heating of the Van Arkel Titanium are 882 -884 - 886° C for heating speeds of 50, 150 and 300° C/h. On the other hand, the β → α transformation temperature is unique, i.e. 881° C for the three cooling speeds. The amplitudes of the dimensional variations vary from cycle to cycle in proportions like 1 to 5. The α → β and β → α transformations show composite phenomenons : contraction, expansion, sudden variation of slope. Such a dilatogram is given in figure 1.

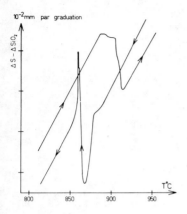

Fig. 1 : Phase transition
of titanium

Thermal cyclings of less pure titanium at speeds of 300° C/h and 150°C/h, give the beginning of a α → β transition at 886° C and of β → α at 681° C. The thermal spreading of the transformation has an approximate amplitude of 25 to 30° C. The dilatations and contractions at transformation vary in great proportions (1 : 20) from cycle to cycle, these phenomenons are always composite.

Cumulative annealings. During heating as well as during cooling, the α ↔ β transformation of Van Arkel titanium (figures 2, 3) is an isothermal transformation at 882 ± 2° C which is composed of a slow dilatation and a slow contraction. The transformation time is 36 h. at heating and 28 h. at cooling. The α → β transformation of R.C.P. titanium begins at 884° C and spreads over approximately 17° C at cooling, the β → α transformation runs in the same limits.

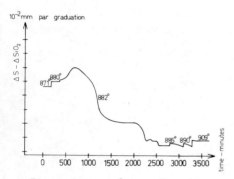

Figure 2 : α → β transformation

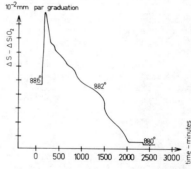

Figure 3 : β → α transformation

As a whole, the influence of impurities is globally α-gen and therefore, the phase transition temperature is raised in the limit of two degrees and an α + β two phase domain appears with a width of 17° C. In this domain, a determined transformation rate corresponds to each temperature, as we showed previously in another study on the phase transition of zirconium (1)(5).

Possible influence of retrograde solubilities of some impurities.

Recently, J. Matyka et al. (6) established experimentally the retrograde solubility of iron in titanium. Our thermodynamic exploitation of the results (7) convinced us of the retrograde solubility of other impurities.

The examination of figure 4 concerning the titanium-rich side of iron-titanium phase diagram with superposition of vertical lines corresponding to the three grades of titanium sutdies (with respect to their iron concentrations) gives evidences for differences in the developments of phase transformation (nature of phase embryos, phase germination, kinetics of phase transformation). The comparison with the dilatometric curves is very instructive. On the other hand, if we consider other solutes with retrograde solubilities, we have a

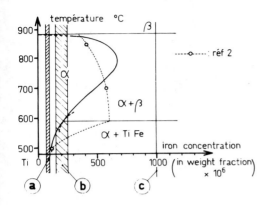

rather complicated problem in accounting for the details of their influence.

In industrial alloys, the kinetics of transformations may also be sufficiently influenced by the impurities to generate some variations in the results of identical heat treatments on different grades of the same type of alloy.

Figure 4 : Iron-titanium phase diagram

ALLOTROPIC TRANSFORMATION OF TITANIUM ZIRCONIUM ALLOYS

The outlines of the titanium-zirconium binary diagram were given as early as 1930 and since, only modifications in the limits of the two phase α + β were given. This diagram is shown on figure 5.

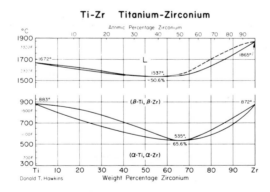

Figure 5 : titanium-zirconium binary diagram

The indifferent point β solid-α solid concerns the equiatomic composition. While the different authors agree with the general features of the diagram, they don't agree with the position of the indifferent point β solid-α solid ; Fast ( 8 ), Hayes et al. ( 9 ) give their point at 545° C, Duvez (10 ) at 485° C ; Farrar and Adler ( 11) at 535° C.

The precise determination of the indifferent point appears as a fundamental requirement before developping other investigations in this binary system.

## Experimental methods

The phase transition of equiatomic Ti-Zr alloys was studied by high vacuum dilatometry in the apparatus previously described and by microcalorimetry in a high temperature Calvet microcalorimeter equiped with special calorimetric cells. The calibration of this calorimeter was obtained by Joule-effect either "in situ" or in special calibration run  when the equipment occupied too great a space to allow "in situ" calibration.

Three  Ti-Zr alloys at different impurity levels, were studied. The first one, Ti-Zr 1, is at industrial purity level, the other two, Ti-Zr 2 and Ti-Zr 3 were synthetized under very pure Argon. The impurities concentrations of starting metals, are given in table 2.

Table 2

| Samples | Materials | Composition at $10^{-6}$ | | | | | | | | | | |
|---------|-----------|---|---|---|----|----|----|----|----|-------|----|----|
|         |           | O | N | C | Cr | Hf | Al | Fe | Ni | H | Zr | Ti |
| Ti Zr 1 | Ti Zr | 3130 | 500 |     |     | 4000 |     |     |    |       |    |    |
| Ti Zr 2 | Ti | 900 1800 |     | 280 |     |     | 10 200 | 150 250 | 50 70 |       | 10 40 |    |
|         | Zr | 2250 | 293 | 1520 | 105 | 102 |     |     |    | 23600 |    | 38 |
| Ti Zr 3 | Ti | 100 |     |     |     |     |     | 70 |    |       |    |    |
|         | Zr | 8 | 8 |     |     | 27 | 0 | 2 |    |       |    |    |

## Dilatometry

Thermal cycles.  At the same speed of thermal cycling, differences appear in the transition temperatures from one alloy to another. Table 3 gives the experimental results. We note that with increasing purity of the alloys an increasing hysteresis appears, between the beginning of the $\alpha \to \beta$ transition and the beginning of the $\beta \to \alpha$ transition.

Table 3

| Alloys | $\alpha \to \beta$ | | $\beta \to \alpha$ | | Hysteresis (°C) |
|--------|--------------------------------|-------------------------|--------------------------------|-------------------------|-----------------|
|        | transition beginning (°C) | transition end (°C) | transition beginning (°C) | transition end (°C) |                 |
| Ti Zr 1 | 616 | 642 | 586 | 567 | 30 |
| Ti Zr 2 | 616 | 641 | 583 | 568 | 33 |
| Ti Zr 3 | 621 | 642 | 581 | 562 | 40 |

The dilatometric amplitudes at the transition are similar for the $\alpha \to \beta$ and for the $\beta \to \alpha$ transition when the maximum temperature of the sample does not exceed by more than a hundred degrees the temperature of transformation. On the contrary, when the maximum temperatures of the sample are higher, systematic variations of the dilatometric amplitudes take place. Above 940° C, the dilatometric amplitudes at the phase transition are maximum for identical annealing times. The variations of these amplitudes at transition were studied using four dilatometric cycles α → β and β → α caracterized by a heating and cooling speed of 300° C/h. Before the experience, the sample was annealed at 940° C during 20 minutes. The dilatograms are given figure 6.

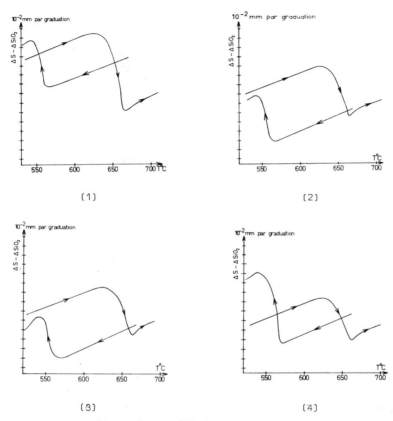

Figure 6 : amplitudes at transition

Cycle 1 : after the α → β transition, the sample is annealed at 700° C during 20 minutes. At cooling, the amplitude of the β → α transition is much lower than that observed at heating.

Cycle 2 : The amplitude of the α → β transition is comparable to that of the β → α transition in the first cycle. After a 20 minutes β-phase annealing at 700° C, the amplitude of the β → α transition is within the same range of values as the amplitude observed at heating for the α → β transition.

Cycle 3 : The transition amplitudes at heating and cooling are in the same range.

Cycle 4 : In this cycle, the β-phase annealing temperature is 900° C for 20 minutes ; a consequence of such an annealing is the increase of amplitude of the β → α transition. This amplitude may be compared with that of dilatations observed during transitions in the previous dilatograms.

This phenomenon makes evident the influence of the β-phase annealing temperature and is observed whatever the purity of the alloy. In experimental conditions similar to those of Cycle 4 in figure 5, the amplitude ratios for α → β and β → α transformations are always in the same range.

    Ti Zr 1 alloy : α → β amplitude ≃ 58 % of β → α amplitude

    Ti Zr 2 alloy : α → β amplitude ≃ 60 % of β → α amplitude

    Ti Zr 3 alloy : α → β amplitude ≃ 57 % of β → α amplitude

    This phenomenon, common to the three alloys of different purity levels, is certainly specific of the type of transformations involved and we suggest a noteable effect of stresses generated in the metals during the phase transition.

### Cumulative annealing

    This study was made for the different alloys.

α → β transition (figure 7

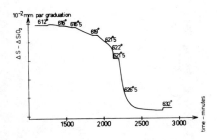

Ti Zr 1 sample                                        Ti Zr 2 sample

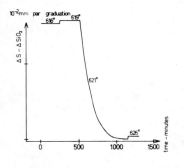

Figure 7 : cumulative annealings
α → β transition

Ti Zr 3 sample

    For the less pure alloy (Ti Zr 1), the phase transition begins at 616° C and extends over 10° C, is slow and for each temperature in this domain, corresponds a determined transformation rate.

    The transformation of Ti Zr 2 begins at 617° C and is comparable to that of alloy 1.

The transformation of Ti Zr 3 alloy, the purer one, is much more rapid and is isotherm at 621° C. The transformation time is approximately 490 minutes.

After α → β transformation, the final temperature is 940° C for the three alloys, which are annealed for 30 minutes at this temperature before cooling.

β → α transition (figure 8)

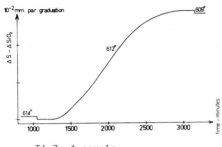

Ti Zr 1 sample

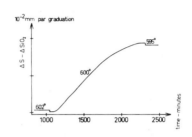

Ti Zr 2 sample

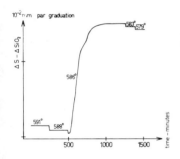

Ti Zr 3 sample

Figure 8 : cumulative annealings, β → α transition

This transformation is isothermal at 612° C for Ti Zr 1 alloy ; it is slow and lasts approximately 28 hours. For alloy 2, it is also isothermal but at 600° C and lasts approximately 20 hours. The purer alloy, Ti Zr 3, shows an iso-thermal transition at 586° C during approximately 7 hours 30 minutes. This kinetics, in spite of its slowness, is the quickest obtained.

Figure 8 (4) shows a metallography of the surface of Ti Zr 3, after iso-thermal β → α transition ; in the acicular structure observed, the needles have not altered and show the relief effect due to β → α transformation. These are the last martensitically transformed zones.

## Calorimetric study

Experimental procedures. To prevent pollution, the sample is at the bottom of a closed zirconium cylinder which is placed in a silica specimen holder moving with smooth friction in the calorimetric cell. This cell is a transparent silica tube closed by a double Wilson's joint metallic head, with a metallic valve and a manometer to measure pressure inside the cell. In a first time, the cell alone is disgassed under vacuum at 900° C ; during this operation, the zirconium cylinder is annealed at 800° C in ultra high vacuum.

After this, the different elements are assembled and introduced in the calorimetric cell, where titanium getters are also placed.

The calorimetric cell is then connected to a high vacuum ionic pump for disgassing at 400° C before introduction of highly purified argon. Finally, the titanium getters are inductively heated at 1000° C during 48 hours.

These operations avoid pollution of the sample during the calorimetric determination which proceeds under static argon.

Thermal cycles. The $\alpha \to \beta$ transition is endothermic and the $\beta \to \alpha$ transition is exothermic for the three alloys studied.

Tables 4, 5, 6 give the experimental results on the three alloys submitted to three successive thermal cycles. The temperatures of the beginning and of the end of the transition are respectively noted $\theta_i$ and $\theta_f$: $\theta_{Mx}$ being the temperature of the maximum of the endo or exothermic calorimetric peak. $\Delta H$ is the heat of transition. The calorimetric curves for each of the three alloys are given on figures 9, 10, 11, the roman numbers I, II and III are for the successive thermal cycles.

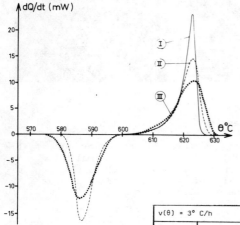

Figure 9 :

Calorimetric curves, sample n° 1

Table 4 : transformation results

Ti Zr, sample 1

| $v(\theta) = 3° C/h$ | Ti Zr 1 alloy | | | | | | | m = 3,40884 g |
|---|---|---|---|---|---|---|---|---|
| | $\alpha \to \beta$ | | | | $\beta \to \alpha$ | | | |
| Cycle n° | $\theta_i$ (°C) | $\theta_{Mx}$ (°C) | $\theta_f$ (°C) | $\Delta H$ (J/mole) | $\theta_i$ (°C) | $\theta_{Mx}$ (°C) | $\theta_f$ (°C) | $\Delta H$ (J/mole) |
| I | 602 | 623 | 627 | 2617 | | | | |
| II | 603 | 624 | 629 | 2530 | 601,5 | 587 | 577 | 2559 |
| III | 603 | 623 | 631 | 2500 | 601 | 585 | 572 | 2530 |

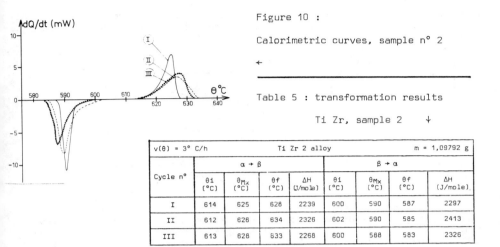

Figure 10 :

Calorimetric curves, sample n° 2

←

Table 5 : transformation results

Ti Zr, sample 2   ↓

| v(θ) = 3° C/h | | Ti Zr 2 alloy | | | | | m = 1,09792 g | |
|---|---|---|---|---|---|---|---|---|
| | α → β | | | | β → α | | | |
| Cycle n° | θi (°C) | θMx (°C) | θf (°C) | ΔH (J/mole) | θi (°C) | θMx (°C) | θf (°C) | ΔH (J/mole) |
| I | 614 | 625 | 628 | 2239 | 600 | 590 | 587 | 2297 |
| II | 612 | 626 | 634 | 2326 | 602 | 590 | 585 | 2413 |
| III | 613 | 628 | 633 | 2268 | 600 | 588 | 583 | 2326 |

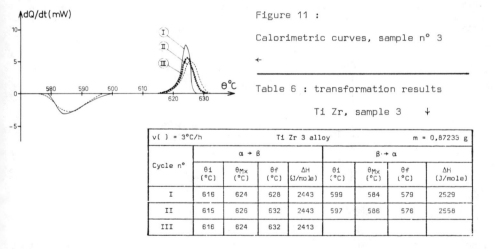

Figure 11 :

Calorimetric curves, sample n° 3

←

Table 6 : transformation results

Ti Zr, sample 3   ↓

| v( ) = 3°C/h | | Ti Zr 3 alloy | | | | | m = 0,87233 g | |
|---|---|---|---|---|---|---|---|---|
| | α → β | | | | β → α | | | |
| Cycle n° | θi (°C) | θMx (°C) | θf (°C) | ΔH (J/mole) | θi (°C) | θMx (°C) | θf (°C) | ΔH (J/mole) |
| I | 616 | 624 | 628 | 2443 | 599 | 584 | 579 | 2529 |
| II | 615 | 626 | 632 | 2443 | 597 | 586 | 576 | 2558 |
| III | 616 | 624 | 632 | 2413 | | | | |

All three samples have the same thermal history before the microcalorimetric study : an annealing at 940° C for 30 minutes in high vacuum ($10^{-6}$ Pa) and then a cooling at a speed of 300° $C.h^{-1}$ to room temperature.

. No significative variations were observed in the values of transformation heat at heating or cooling, the average value of ΔH = 2425 ± 131 J/mole.

. The temperatures of the beginning of the α → β transition lower with the increase of impurities content and the hysteresis between both temperatures at beginning of α → β transition and beginning of β → α transition increase with lowering of impurities content.

. In a general way, it can also be pointed out that the kinetics of either α → β or β → α transformation of the second and third cycle is noticeably different

from that we had in the first cycle. In this case, the curves maximum was noticeably higher. The thermal history of a sample submitted to three successive cycles
is :

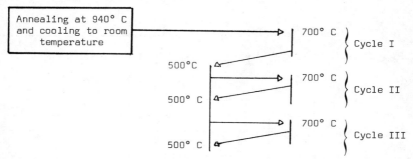

The ultimate temperature of heat treatment influences the shape and the
spreading of the thermal peak. For the same alloy, the spreading of the peak is
minimum for an experience after an annealing at 940° C, on the other hand the
ultimate annealing temperature has no influence on the value of heats of the
$\alpha \rightarrow \beta$ and $\beta \rightarrow \alpha$ transformations.

These features are probably due to stresses generated in the sample during
thermal cyclings. These stresses can be annealed at an appropriate temperature
as shown on figure 12, where the temperature raise of the sixth cycle on Ti Zr 3
after annealing at 940° C in ultra-high vacuum is. represented.

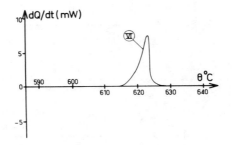

Figure 12 :        Ti Zr 3, sixth cycle

The influence of sample pollution. The goal is to study the influence on
the $\alpha \leftrightarrow \beta$ phase transition of Ti-Zr equiatomic, of pollution due to the contact
of the sample with silica in the apparatus. In this experience, we do not use a
zirconium container and consequently, the sample is put directly in the transparent silica container. The sample used is of the same grade as Ti Zr 1 ; the
five thermal cycles given on figure 13 and table 7 show evidence of a clear
modification of the calorimetric curves.

During heating and from the first cycle a 45° C magnitude is measured for
the lowering of the transformation beginning temperature. The calorimetric curve
shows two non-separated peaks, the first one having the lowest heigth and spreading over 67° C. In the course of the five cycles, the calorimetric curve shifts
steadily towards lower temperatures. The transformation heat average value is
$\Delta H_{\alpha \rightarrow \beta}$ = 2916 ± 33 J/mole, then higher than 2425 ± 131 J/mole.

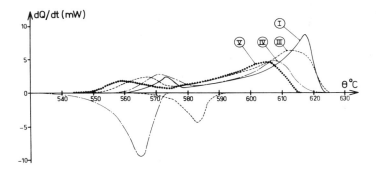

Figure 13 : Ti Zr 1, influence of sample pollution

Table 7

| | Ti Zr 1 Alloy | | | | m = 3,69519 g | | | | |
|---|---|---|---|---|---|---|---|---|---|
| Cycle n° | α → β | | | | | β → α | | | |
| | $v(\theta)$ (°C/h) | $\theta i$ (°C) | $\theta_{Mx}$ (°C) | $\theta f$ (°C) | $\Delta H$ (J/mole) | $v(\theta)$ (°C/h) | $\theta i$ (°C) | $\theta_{Mx}$ (°C) | $\theta f$ (°C) | $\Delta H$ (J/mole) |
| I | 3 | 557 | 617 | 624 | 2878 | | | | | |
| II | 4,5 | 555 | 614 | 625 | 2936 | 1 | 594 | 583 | 563 | 2297 |
| III | 3 | 549 | 607 | 621 | 2936 | 3 | 575 | 565 | 532 | 2268 |
| IV | 3 | 544 | 605 | 618 | | | | | | |

During cooling, the β → α transformation gives only one peak on the calor-imetric curves, with a shift towards lower temperatures as the successive cycles take place. The evidence of the sample evolution appears by comparison between the values of previously defined caracteristical temperatures for the fourth cycle (θi = 575° C - θm = 565° C - θf = 532° C) with those found in good exper-imental conditions on the same alloy (see table 3) (θi = 601° C - θm = 585° C - θf = 572° C).

The differences between the calorimetric curves during heating and during cooling for the polluted sample can be explained in the following manner :
. on heating, the partially polluted sample begins to transform (first peak) and then the whole core transforms (second and highest peak).
. on cooling, the hysteresis may be the origin of the two reverse peaks over-lapping giving apparently a unique reverse peak.

This viewpoint must be discussed further and, above all, controlled pollut-ion experiments are necessary to complement our results. These experiments are being carried at this time in our laboratory and we think they might explain the results of Farrar and Adler (11)

General conclusion

The results obtained in this study of phase transformation of titanium and of an equiatomic titanium-zirconium solid-solution, both with samples of differ-ent grades, show that at slow heating and cooling rates, the transformation is slow and its characteristics depend on the impurities content. Pulse heating (12) and fast quenching (13) experiments give very high transformation rates. The rate of transformation is probably dependent on the heat flux density input or output. The different variables considered here must be kept in mind when examining industrial heat treatment of titanium and titanium alloys.

## References

1.    E. Etchessahar : *Thèse de spécialité*, Nantes (1975)

2.    E. Etchessahar, J. Debuigne : *Brevet ANVAR 9350.*

3.    R.C.P. N° 244 du C.N.R.S. *"Propriétés mécaniques du Titane et de ses Alliages".*

4.    J. Bigot, C.E.C.M. Vitry, France

5.    E. Etchessahar, J. Debuigne *(to be published)*

6.    J. Matyka, F. Faudot, J. Bigot, *Scripta Metallurgica*, vol. 13, p. 645-648 (1979).

7.    B. Jounel, J. Debuigne *(to be published)*

8.    J.D. Fast : *Rec. Trav. Chim.*, 58 (1939) 973-983.

9.    E.T. Hayes, A.H. Roberson, O.G. Paasche : *U.S. Bur. Mines Dept. invest.*, 4826 (nov. 1951)

10.    P. Duwez, *J. Inst. Met.* 80 (1951-52) 525-527.

11.    P.A. Farrar, S. Adler : *Trans. A.I.M.E.*, 326 (1966) 1061-64.

12.    R. Parker : *Met. Trans.* 233 (1965) 1545-1549.

13.    M. Cormier, F. Claisse : *J. Less Common Metals* 34 (1974) 181-189.

# EVOLUTION OF THE EQUIAXED MORPHOLOGY OF PHASES IN Ti-6Al-4V

Harold Margolin, Polytechnic Institute of New York, Brooklyn, NY 11201
Paul Cohen, Con Edison Co., Astoria, NY 11105

## ABSTRACT

The evolution of the equiaxed (E) morphology was studied in Ti-6Al-4V which had been cooled from the β field to a series of temperatures in the α/β field and water quenched. The resulting Widmanstätten plus grain boundary (W+GB)α structures were warm worked and recrystallized at the temperature from which they were quenched. Alpha nuclei were found to form at α/β interfaces and grow across W and GBα . New α grains would cross one or more α platelets. When a single platelet was traversed, surface tension requirements reduced the dihedral angle to a value less than 180° and in the process the new α grains became larger than the original α platelets. Grains crossing two or more α platelets initially grew along the length of the platelets and were longer than they were wide. When nuclei crossed GBα the W sideplates were pinched off from the new grains as β penetrated the α/β boundaries. When β recrystallized prior to α, "scallops" would form along the α/β interfaces, as the α attempted to satisfy surface tension requirements with β/β boundaries. Thus α could penetrate β/β boundaries producing two αβ boundaries, each of which attempted to move toward its center of curvature. Thus the narrow region, where bridging initially occurred, became enlarged, the β was shifted to other locations and the length of α/β interface was reduced. If recrystallization of the bridged α occurred, the α/α boundary formed parallel to the original α/β interface at the position of minimum thickness of the bridge.

## INTRODUCTION

Ti literature on mechanical properties abounds with examples of the dependency of tensile ductility, fracture toughness, fatigue life, fatigue crack propagation, and other properties on microstructure. Yet, little attention has been paid to the processes by which W morphology is converted to an E assembly of phases. Some studies on recrystallization of two phase "massive" structures have been carried out on α/β Cu-Zn brasses (1-4), on α/β Ag-Mg alloys (5), Ag-Ni alloys (6), and Cu-Ni-Zn alloys (7). A study of recrystallization of Ti-6Al-4V has been reported by Majdic and Ziegler (8), who found that recrystallization began 5° to 90°C above the temperatures required for commercially pure Ti and reported that nucleation is not hindered by the β phase.

The results of the preceeding work on two phase alloys indicate that recrystallization behavior will be affected by recrystallization temperature, not only for the normally expected reasons, but also because the composition and volume fraction of the phases depend on temperature, the possibility of α precipitation in the β phase and the probability of non-uniformity of deformation of the two phases.

### Experimental Procedure

In an attempt to avoid the problem of changing volume fractions of phases, specimens of 1.59 Cm dia, swaged bar were solution treated in air in the β field at 982°C for 2.5h, furnace cooled to 760, 788 or 843°C, held 15 min and water quenched. They were subsequently annealed, after rolling, at the quenching temperature. The details of heat treatment, rolling and subsequent annealing are given in Table I.

Table I

HEAT TREATMENTS - ANNEALING OF SPECIMENS

<u>SET I</u>

a)   982°C - 2.5 hrs., furnace cooled to 760°C - held 15 min., water quenched; b) Rolled 34% at 100°C at 0.0076 cm. per pass from initial thickness of 1.10 cm., water quenched after final pass; c) Annealed at 760°C for: 1, 3, 5, 15, 24, 168, 336, 504 hrs., water quenched.

<u>SET II</u>

a)   982°C - 2.5 hrs., furnace cooled to 788°C - held 15 min., water quenched; b) Rolled 48% at 500°C at 0.013 cm. per pass from initial thickness of 1.08 cm., water quenched after final pass; c) Annealed at 788°C for: 1/4, 3, 10, 24, 168 hrs., water quenched.

<u>SET III</u>

a)   982°C - 2.5 hrs., furnace cooled to 843°C - held 15 min., water quenched; b) Rolled 44% at 500°C at 0.013 cm. per pass from initial thickness of 1.03 cm., water quenched after final pass; c) Annealed at 843°C for: 1/4, 1, 3, 24, 50, 168, 336 hrs., water quenched.

Specimens were electropolished and etched in a solution of 4ml KOH (sat'd in $H_2O$), 26 ml $H_2O$, 5 ml $H_2O_2$ (30%) heated to 60 $\pm$ 2°C. The $H_2O_2$ was added after the solution reached 50°C and samples were immersed in the etchant for 1 min. For the anneals subsequent to rolling specimens were encapsuled in quartz tubes which were dropped into water and broken at the end of the annealing time.

RESULTS

A. Worked Structure:  After rolling the colony structure, produced during cooling, was highly distorted, Fig. 1, and frequently showed twinning, which could traverse interlamellar β, Fig. 2

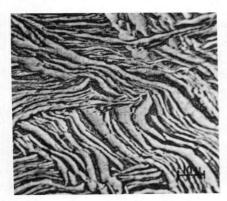   

Fig. 1   Set III specimen, Table I, As-rolled Distorted W+GBα Structure

Fig. 2   Set I specimen, Table I, As-rolled Wα+β+ twinned α

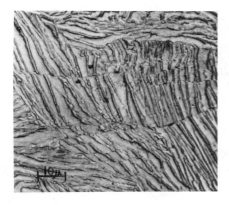

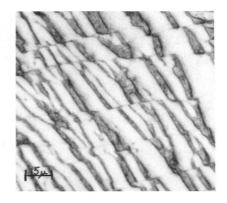

Fig. 3.   Set III Specimen, Table I, As-rolled
As-
Shear bands with voids

Fig. 4   Set III Specimen, Table I,
As-
rolled. Almost continuous α
along shear bands

Occasionally, when α platelets were oriented at a large angle to the rolling plane, shear bands would form, Fig 3, and in some instances voids were seen along them, Fig. 3. The local shear along the bands was frequently so extensive that the α and β phases were shifted into positions where α was opposite β, rather than being separated by β, Fig 4. This shift in effect produced a continuous or nearly continuous region of α, which, after recrystallization, produced a continuous line of recrystallized α grains.

B.   Alpha Recrystallization

1.   Grain Boundary α  - Recrystallized grains were seen to nucleate at αβ interfaces and to traverse the GBα, Fig. 5.  In most instances the new grains did not consume W side plates with

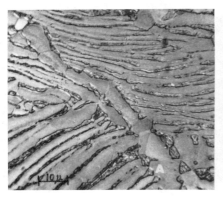

Fig. 5   Set III Specimen, Table I, Annealed
1 hr at 843°C, W.Q. Recrystallized
grains traversing GBα

Fig. 6   Set I Specimen, Table I
Annealed 30 hr. at 760°C,
W.Q. Recrystallized grains
traversing Wα

the same orientation as the original GBα.  Instead, when the boundary of the recrystallized grain crossed the width of the α platelet, surface tension requirements forced the β along the boundary, on both sides of the platelet, and pinched it off from the mother GBα, Fig. 5, at A.  This behavior indicates that it was possible for the GB to remain in position across the α platelet long enough for the β to penetrate the boundary.

2. <u>Wα</u> — At 760°C there is some tendency for the interlamellar β to dissolve into the α, Fig. 6 at A. It may well be that the 15 min hold time at the quenching temperature was insufficient to permit equilibrium amounts of α and β to form, since the companion paper (9) indicates that the amount of β decreased during annealing at 760°C. It is also possible that some local dissolution could occur in the process of reducing interface area. When the β dissolves into the α, a recrystallized grain can traverse this larger α region as at B, Fig. 6, or if β particles remain, growth around the particles can take place as at C. Thus, the tendency is for the recrystallized grains to be larger than the Wα platelets, and this tendency would be present, even if a single platelet were traversed, as will be discussed later.

C.  β <u>Recrystallization</u> - Recrystallization of β is most readily detected at 843°C. When β recrystallizes, the requirements of surface tension cause "scallops" to form at the α/β interfaces, Fig. 7 at A.

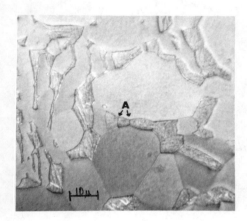

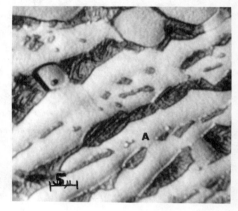

Fig. 7     Set III specimen, Table I, Annealed 168 hr at 843°C, W.Q. Recrystallized β grains

Fig. 8     Set III specimen, Table I Annealed 24 hr at 843°C, W.Q. Bridging of β by Wα at A and B

The rather large dihedral angle at A suggests that the β/β boundaries are low energy boundaries. When higher energy β/β boundaries form, it appears possible that bridging across the β/β boundaries to form a continuous α structure can occur. This can be seen in Fig. 8. The shape of the β particles adjacent to A suggest that an α/α boundary joins the two β particles. Indeed a faint boundary, parallel to the length of the W platelets can be detected. The multiply scalloped appearance of the α in Fig. 9, and the frequent instances of bridging in the same photomicrograph indicates that recrystallization of β is extensive.

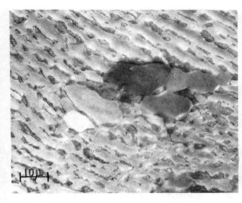

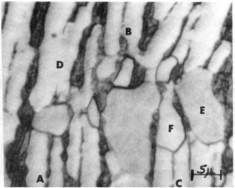

Fig. 9    Set III Specimen, Table I, Annealed
          24 h at 843°C Recrystallized α
          and bridging by α of recrystallized β

Fig. 10  Set III specimen, Table I
          Annealed 3 h at 843°C, W.Q.
          Recrystallized α and Wα
          bridging recrystallized β

Fig. 8 indicates that the β transforms on quenching from 843°C. Thus, apparently the deformed structure of this transformed β which existed at 500°C retained sufficient work in it to permit recrystallization of the β on heating to 843°C. Evidence of recrystallization of β at 788°C was obtained after 1 week at temperature and after two weeks at 760°C.

Examination of Fig. 9 and Fig. 10, in which bridging can also be seen at A, B and C, suggests that bridging occurs when α is not yet recrystallized, i.e. β at 843°C recrystallizes prior to α. Also area D, Fig. 10, suggests that, as bridging takes place, β diffuses to other locations, so that the entire area becomes α of the same orientation. Such areas can recrystallize to form grains with new orientations, areas E and F, Fig. 10.

After elimination of the β/β boundary, recrystallization of α may occur, and this can create a boundary between the contacted α platelets, as at A Fig. 11. In this instance the boundary formed is parallel to the length of the β lamella, and surface tension requirements for the α/α and α/β interfaces emphasize the presence of the α/α boundary at A, as also pointed out earlier in Fig. 9. The β particles below A and above B in Fig. 11 are in contact with two α/α boundaries. In these cases the β becomes larger than the adjacent β to accommodate surface tension requirements. In this way, β tends to become equiaxed. At C in Fig. 11 recrystallization of two α particles, which have joined and eliminated β produces a boundary between the remaining β particles. Several α grains have formed, and their boundaries impinge on the β particles, stabilizing them.

Fig. 12 shows the continuous zone of equiaxed α grains, produced by recrystallization along a shear band.

## Discussion

A.  Wα  —  The sequence of events by which Wα is converted to an equiaxed structure is shown in Fig. 13. Fig. 13a shows the unrecrystallized Wα platelets. When the new grains, $α_R$, grow across a Wα platelet, the α/α boundaries meet β, Fig. 13b. Surface tension requirements do not permit a 180° dihedral angle to exist. The required

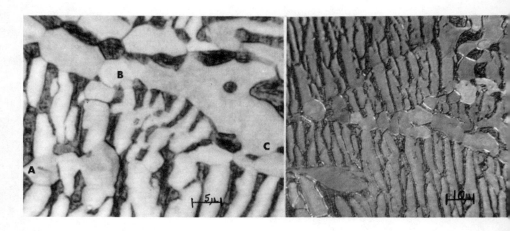

Fig. 11    Set III Specimen, Table I, Annealed
3 h 843°C, W.Q. Bridging of β at
A and formation of α/α grain boun
after α

Fig. 12    Set II specimen, Table I
Annealed 3 h at 843°, W.Q.
Recrystallization of α along a
shear band recrystallization

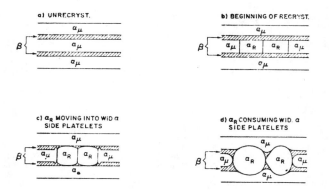

Fig. 13  Schematic Representation of the Conversion W  to E

dihedral angle is obtained by movement of β into the α/β boundary and a rotation of the
α/β boundaries toward one another, Fig. 13c. This rotation in effect enlarges the recrystallized grain so that it is now thicker than the original platelet.  This increase in size
and concommitant shift of β into the α/β boundary permits contact of the recrystallized
grain with the adjacent lamella of unrecrystallized α, Fig. 13c.  Growth of these new
grains continues into the adjacent α, Fig. 13d.  When recrystallization in the same manner
occurs in two adjoining regions, new α grains contact one another, as shown in the
recrystallized region of Fig. 9.  This same process would be expected to operate, if the
recrystallized grain crossed several platelets.

When bridging of interlamellar β takes place prior to recrystallization of α, a driving
force is created for the diffusion of the surrounded β away from its position between the

two connected platelets. This driving force is the reduction of $\alpha/\beta$ surface energy. If bridging can take place at a number of sites simultaneously, then the simultaneous movement of $\alpha/\beta$ interfaces, at the point of bridging, toward their centers of curvature can cause $\beta$ to move to other sites, thus permitting the gradual joining of two adjacent $\alpha$ platelets, Fig. 10.

 B. GBα    The process by which Wα sideplates are separated from recrystallized GBα is shown in Fig. 14. When the recrystallized grain traverses the GBα and reaches the Wα sideplate, there are two $\alpha/\beta$ interfaces which meet at an $\alpha/\alpha$ boundary, Fig. 14a and b. The dihedral angle most probably is not the equilibrium configuration. Two driving forces are in competition: 1) the movement of the $\alpha/\alpha$ boundary into the Wα sideplate and 2) the establishment of surface tension requirements. A small movement of the $\alpha/\alpha$ boundary into the Wα sideplate would make establishment of dihedral angle requirements easier. Initial penetration of the $\alpha/\alpha$ boundary by $\beta$ anchors the $\alpha/\alpha$ boundary, Fig. 14C and subsequent penetrations of the $\alpha/\alpha$ boundary pinches the sideplate off from the GBα, Figure 14c and d.

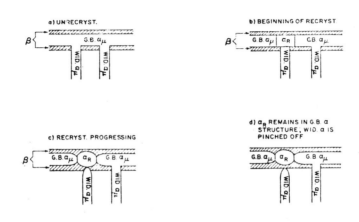

Fig. 14.   Steps in the Conversion of GBα to Equiaxed α

## Acknowledgment

    This work was sponsored in part under Office of Naval Research Sponsorship under Contract No. N-00014-75-C-0793.

## References

1.    R.W.K. Honeycombe and W. Boas:  Nature 159 (1947) p.847
2.    R.W.K. Honeycombe and W. Boas:  Aust. J. Sci. Res. 1(1948 A) p. 70
3.    N.V. Zimin:  Phys. Metals Metallogr (N.Y.) 20 (1965) p. 92
4.    K. Mader and E. Hornbogen:  Scripta Met. 8(1974) p. 979
5.    L.M. Clarebrough:  Aust. J. Sci. Res. 3(1950A) p. 72
6.    A.R. Vasudeven, J.J. Petruvic and J.A. Roberson:  Scripta Met 8 (1974) p. 861
7.    H. Kreye and U. Brenner:  J. Mat Sci 9 (1974) p.1775
8.    M. Majdic and G. Ziegler:  Z. Metallkunde, 65 (1974) p. 173
9.    H. Margolin and P. Cohen:  Proceedings of the Fourth International Conference on Titanium, May 19-22, 1980, Kyoto Japan

# THE EFFECT OF THE COOLING CONDITIONS ON THE RESIDUAL STRESSES
# IN QUENCHED Ti 685 CYLINDERS

S. Denis, J.C. Chevrier and G. BECK

Laboratoire de Métallurgie de l'Ecole des Mines
Institut National Polytechnique de Lorraine - Nancy - France

## Introduction

The TA6ZrD titanium alloy requires a quenching process, after a solution treatment in the beta phase field at 1050°C for one hour, to obtain the structure compatible with good service requirements. This rapid cooling gives rise to thermal gradients which result in a high level of residual stresses. These stresses can have a marked effect on the alloy behaviour especially on its fatigue life. The quenching stresses decrease with lower thermal gradients, i.e. if cooling is slower. Then the structure and the mechanical properties become inadequate. In order to find the quenching conditions leading to the best compromise between good strength properties and low residual stresses, a detailed study of the phenomena governing the appearance of the residual stresses is necessary.

This work was carried out according to a procedure already used for high strength aluminium alloys [1].

A computer program has been developed to predict the stresses and the strains at each stage in the quenching process. Experimental investigations have been carried out to obtain input data : temperature gradient and mechanical behaviour of the material during the quenching process. The results obtained enabled the form of the cooling law which avoids the residual stress formation to be predicted accurately.

## 1. Input Data for the Stress Calculation

### 1.1 Mechanical properties of the TA6ZrD alloy within the temperature range of the quenching process

Tensile tests at elevated temperatures were performed.

Figure 1 illustrates a typical stress strain curve and a matching linear elastic-linear strain hardening function. The considered strain interval is limited to plastic strains less than 1 %.

The quantities $E_p$ : strain hardening modulus and $\sigma_{Eap}$ : apparent yield stress are defined. These quantities and E the Young's modulus have been determined from 1050°C down to room temperature 20°C (figure 2).

### 1.2 Dilatation studies

The quantity of interest here is the thermal expansion coefficient. The sources of internal stresses in quenching are the fluctuations in specific volume due to thermal shrinkage and phase transformation. In the TA6ZrD alloy, the volume changes due to phase transformation are small. Therefore, in this work we consider the thermal strain caused by volume changes due to thermal shrinkage alone. This thermal strain can be easily calculated with the coefficient of linear thermal expansion. The experimental information is given from continuous cooling dilatation curves . Figure 3 shows the temperature dependance of the thermal expansion coefficient. It is described by the same function for the α phase and β phase.

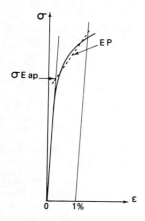

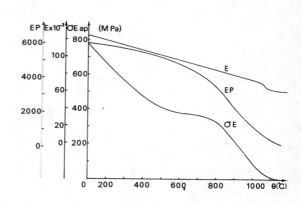

Figure 1 - Typical stress-strain curve
    σEap  : apparent yield stress
    Ep    : strain hardening mo-
          dulus

Figure 2 - Variation as a function of temperature of : E : Young's modulus [2]
        σEap : apparent yield stress
        Ep   : strain hardening mo-
            dulus

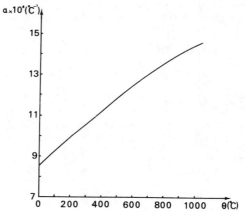

Figure 3

The thermal expansion coefficient versus temperature

## 1.3 - Temperature distribution in TA6ZrD cylinders during the quenching process

Heat conduction in the cylinders during cooling introduces thermal gradients between center and surface. They give rise to permanent strains which
are responsible of the development of residual stresses. Therefore, the knowledge of the temperature distribution in quenching is necessary.

Cylinders with diameters of 40 mm and 60 mm were quenched in oil at 50°C
and in water at 20°C after a solution treatment at 1050°C for one hour. In each
treatment four temperatures for various locations in the mid-plane were measured
simultaneously. These experimental cooling laws enabled the radial temperature
distribution at all stages of the cooling process to be known. The surface temperature is determined by extrapolation. Figure 4 shows the cooling laws and

the temperature distribution obtained in a cylinder with 60 mm in diameter quenched in oil at 50°C.

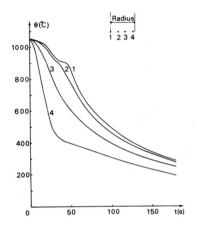

Figure 4 - a  - The cooling curves for various   b  - The radial temperature dis-
                locations in the mid-plane of          tributions at different
                a cylinder with 60 mm in dia-          stages of the cooling in the
                meter                                   same conditions.
                1    r =  0
                2    r =  9,2 mm
                3    r = 18,3 mm   r : radius
                4    r = 27,2 mm.

## 2. Calculation Model

The level of residual stresses depends on the temperature difference between the center and the surface of the solid during cooling. This temperature
difference can vary in two ways during the quenching process. If the stresses
induced have not exceded the yield point of the alloy at any moment during cooling, all the strains are reversible and the stresses are cancelled out when
the solid is at a uniform temperature ; in this case quenching does not introduce residual stresses. If stresses tend to exceed the yield stress, plastic
flow occurs. Thus, it is necessary to calculate the permanent plastic strains
by using a method of elastic-plastic analysis.

### Calculation conditions

In a mathematical model for stress and strain calculation, the material is
assumed to satisfy some defined conditions.
    The model material is :
    - homogeneous
    - isotropic
    - time independent (non viscous and independent of the strain rate )
    - linear elastic-linear strain hardening.

The treatment here is limited to long circular cylinders which have a radial temperature distribution.
    It is supposed that the distribution of temperature is symmetric with
respect to the axis of the cylinder because the specimens are quenched in a
vertical position.
    The experimental bar has a length equal to three times the diameter : it is
supposed that the influence of the ends is negligeable for the stress creation

in the mid-plane.

The model material defined above must satisfy some stress and strain rela-
tions.These are given in basic text books [3][4][5].

The stress state is described by the stress tensor. In cylindrical bars
we suppose that this tensor is reduced to the stress components in the princi-
pal directions: $\sigma_r$ the radial stress $\sigma_\theta$ the tangential and $\sigma_z$ the longitudinal
stresses. In order to answer the question of knowing in what combination of
stress components acting at a point will the material begin to deform plastically, it is necessary to use a yield criterion.

The Von Mises yield criterion has been choosen; it can be written :

$$F = \Sigma\,(\sigma_i - \sigma_j)^2 - 2\,\sigma_E^2$$

where $\sigma_i$ are the stresses in the principal directions and $\sigma_E$ the yield stress in
simple tension. This yield function depends on the stress state, the temperature
and the accumulated plastic strain.

If $F \gtrless 0$ the material deforms plastically
If $F \le 0$ the strains remain elastic.

## 3. Calculation Results

The data previously determined are used in the calculation. Thus, the distribution of strains and stresses in the cylinders will be known at any moment
during quenching.

Different quenching conditions have been investigated : cylinders with
diameters of 40 mm and 60 mm were quenched in oil at 50°C and in water at 20°C.

### 3.1 The development of the internal stresses

Let us consider the cylinder with 60 mm in diameter taken at the solution
temperature of the TA6ZrD alloy (1050°C) and quenched in oil. Figure 5a shows
the cooling laws for the geometrical center and surface of the cylinder ; figure 5b shows the variations of tangential and axial stresses at the center and
the surface of the cylinder as a function of time. It can be seen that cooling
can be divided into several steps.

- at the start of cooling, the surface is cooled more rapidly than the
center. The volume contraction in the surface is prevented by the higher specific volume of the center. This means that compressive stresses will arise in
the center and tensile stresses will arise in the surface.

The stresses increase with increasing temperature difference between
the center and the surface.

- at the point when the temperature difference reaches a maximum, the cooling rate becomes higher at the center than at the surface.

The stresses then decrease inversely, the center being in tension and
the surface in compression.

It can be seen that the temperature difference between the center and
the surface remains nearly constant at the end of the cooling ; this fact explains why stresses vary very slowly. They reach their residual level when
the solid is at a uniform temperature.

At the completion of the quench, the residual stress at the surface is
compressive while the corresponding stress at the center is tensile (figure 6).
The letter p on figure 5b shows the moments when plastic strain occurs. It will
be noted that the surface and the center are plastically deformed at the start
of the cooling.

In fact when the metal passes into the plastic region at high temperature,
it is subjected to permanent plastic strain due to the low yield point at high
temperature. At the end of cooling the mechanical properties are higher and
the accumulated plastic strain can introduce a high level of residual stresses.

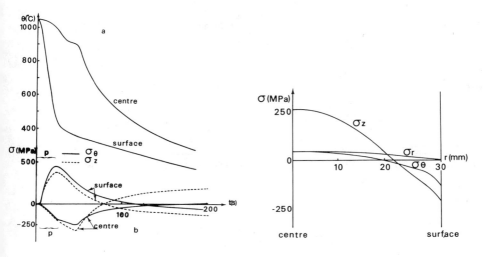

Figure 5 – Variation as a function of time of :
a – temperature at the center and the surface for a cylinder with 60 mm in diameter quenched in oil at 50°C.
b – tangential ($\sigma_\theta$) and axial ($\sigma_z$) stresses at the center and the surface for the same conditions.
The letter p shows the moments when plastic flow occurs.

Figure 6 – Residual stress profiles in a 60 mm diameter cylinder quenched in oil at 50°C.

## 3.2 Effect of the diameter of cylinders and the quenching conditions on the level of residual stresses

If the diameter of the cylinders increases, the temperature difference between the center and the surface at the start of cooling is higher. The permanent plastic strain which occurs is getting more important. The result is a higher level of residual stresses.

The residual stress state of the material can be characterized by the effective stress, a scalar defined as :

$$\sigma_e = \frac{1}{\sqrt{2}} \sqrt{(\sigma_r - \sigma_\theta)^2 + (\sigma_\theta - \sigma_z)^2 + (\sigma_z - \sigma_r)^2}$$

with the von Mises criterion for cylinders.

Figures 7b and 7a show the effective residual stress profile for an increased bar diameter from 40 to 60 mm when the quenching conditions remain constant.

For a constant diameter of the cylinder, an increased cooling rate implies higher temperature gradients between the center and the surface. On figure 7a it can be seen that the cooling in water at 20°C of a cylinder with 60 mm in diameter gives rise to a higher level of residual stresses than the cooling in oil at 50°C. Figure 7b illustrates the same fact in a cylinder of 40 mm in diameter.

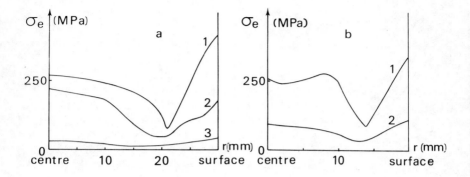

<u>Figure 7</u> - Effective residual stress
profile in cylinders quenched in oil
at 50°C (1), in water at 20°C (2), in
the optimum conditions (3)
a - with 60 mm in diameter
b - with 40 mm in diameter

## 4. Optimum cooling conditions

The results of the calculations have enabled us to show that the permanent
plastic strain is generated already during the initial stage of the quenching
and that this strain governs the level of residual stresses.

Thus, to limit the amount of residual stresses it is very important to limit the transition to the plastic region at the start of the cooling. This result is obtained by retarding cooling. The most rapid cooling law which avoids
the residual stresses formation in a cylinder with 60 mm in diameter has been
determined by a method of successive solutions based on the solution of the
equation of heat flow. Figure 8c shows that this cooling law reduces the difference in temperature between the center and the surface as long as the material is in the β phase field, at high temperature. But with this type of cooling, the structure and the mechanical properties of the TA5ZrD alloy become
inadequate.

In order to obtain the structure compatible with good strength properties
it is necessary to accelerate the cooling within the temperature range 1050°C -
900°C. With this new cooling law (figure 8b), the temperature difference between center and surface is much less than when quenching in oil (figure 8a).
Small plastic strains are generated and the residual stresses are then significantly lower than these obtained when cooling in oil, although the mechanical
properties and structure are not affected.

Figure 7a (curve 3) shows the effective residual stress profile calculated
in this case.

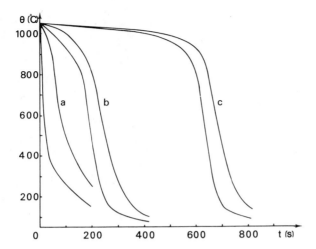

Figure 8 – Cooling curves for the
center and the surface of a cylinder ($\emptyset$ = 60 mm)
a - Cooled in oil at 50°C
b - Cooled in the optimum conditions
c - Cooled to obtain zero residual stresses

## Conclusions

This study, the fundamentals of which are given here, enables the important factors which govern the formation of residual stresses in TA6ZrD alloy
cylinders to be determined more precisely.

It has made it possible to predict that plastic strain occurs at the initial stage of the cooling when the alloy still is in the β phase field.

It has led to the definition of a cooling law by which the passage of
the metal into the plastic region at high temperatures is limited, thereby
giving rise to a low level of residual stresses. This cooling with a low cooling rate at high temperature has been achieved on a cylinder with 60 mm in
diameter : this case is close to the practical size of many parts made with
the TA6ZrD alloy.

It imparts to the alloy the structure compatible with good strength properties associated with the theoretically minimal value of residual stresses.

## References

[1] P. Archambault, J.C. Chevrier, G. Beck, J. Bouvaist - Optimum quenching
    conditions for aluminium alloy castings - Proc. 16 th Int. Heat Treatment
    Conf., Stratford, May 1976, Metals Society, pp. 105-109.
[2] D.F. Neal - New Metals Metallurgical Research Section-Technical letter
    n° 40/75 Imperial Metal Industries Ltd., P.O. Box 216, Kynach Works, Bir-
    minghan B67BA, Gt-Britain
[3] O.C. Zienkiewicz - La méthode des éléments finis, Ediscience Paris (1973)
[4] A. Mendelson - Plasticity Theory and application - Mac Millan New York
    (1968)
[5] S.S. Manson - Les contraintes d'origine thermique, Dunod, Paris (1967).

# THE EFFECT OF HEAT TREATMENT ON MICROSTRUCTURE AND TENSILE PROPERTIES OF Ti-10V-2Fe-3Al

G. T. Terlinde, T. W. Duerig, and J. C. Williams

Carnegie-Mellon University, Pittsburgh, PA U.S.A.

## Introduction

The use of high strength titanium alloys is still dominated by α and α+β alloys such as Ti-5Al-2.5Sn and Ti-6Al-4V. However, there now is an increasing interest in a class of alloys known as metastable β-Ti alloys. These alloys are so-named because they can contain 100% metastable β-phase upon quenching to room temperature. Such alloys also can develop very high strength levels during aging. Beside excellent strength to density and good strength to toughness combinations, the β-Ti alloys exhibit a highly improved deep hardenability for thick section applications. In addition to these advantages, most of these alloys exhibit considerable flexibility in strength and a wide range of microstructures.

Among the more attractive β-alloys is the relatively new alloy Ti-10V-2Fe-3Al (Ti-10-2-3). This alloy has a slightly higher Young's modulus, a lower density and a less sluggish age hardening response than many of the other β-alloys[5,6]. Much work has already been done to develop the manufacturing techniques and forging conditions for achieving attractive tensile properties as well as fracture toughness and fatigue properties of Ti-10-2-3 [7-12]. However, only a limited amount of work has been done to systematically change heat treatments after processing and to relate the different microstructures to the corresponding mechanical properties because the final properties depend both on the processing history and on the subsequent heat treatments.

From earlier studies it is known that the morphology of the α-phase (primary α, secondary α, and grain boundary α) as well as its volume fraction can influence the mechanical properties in this alloy as well as in other β-Ti alloys[1-4,9,11,13-18]. The main goal of this study was to establish a variety of microstructures with a wide range of yield stresses for a fixed forging condition with globular primary α. Using these, a comparison of the other mechanical properties was made. In addition, a limited study has been performed on material with a different forging history which resulted in elongated primary α. Some microstructures from forgings which differ only in primary α morphology have been tested. This allows some conclusions about the influence of the primary α morphology on mechanical properties.

## Experimental Methods

The majority of the Ti-10-2-3 alloy used in this study was supplied by TIMET (Heat #P1452) as hot rolled plate. It had been thermomechanically processed starting in the β-phase field with a final working step in the (α+β) field starting from 730°C. The β-transus was determined as 805°C ± 3°C. The alloy contained (by weight) 10.3%V, 2.2%Fe, 3.2%Al, 0.15%O, 0.009%N and 0.016%C, balance Ti. The other material was in the form of pancake forgings

which were worked in the β-field with a finish upset in the (α+β) field
(774°C).  The β-transus of this alloy was slightly lower than that of the
plate.  For heat treatments above 600°C, specimens were wrapped in Ta foil
and vacuum encapsulated.  Most of the heat treatments below 600°C were performed in a liquid nitrate salt bath.

Specimens for optical microscopy were electropolished in a solution of
5% $H_2SO_4$ in methanol at room temperature at a voltage of 21V.  The etchant
consisted of equal parts of 10% oxalic acid and 1% HF aqueous solutions.

Thin foils for transmission electron microscopy (TEM) were prepared in
a twin jet electropolishing unit with an electrolyte of 59% methanol, 35%
butanol and 6% perchloric acid cooled to -50°C.  A voltage of 12-15V was used.

Tensile tests have been performed on electropolished and etched cylindrical specimens with a diameter of 6.4mm and 32mm gage length using an Instron
testing machine with a clip-on extensometer.  The tests were carried out in
the L-direction of the plate at a strain rate of 5.5 x $10^{-4}sec^{-1}$.  The fracture surfaces of the tensile specimens have been studied by scanning electron
microscopy (SEM).


Results

I.  Microstructures

The microstructures studied here have been chosen using the results of
a recent investigation of the phase transformation behavior of this alloy[19].
Therefore, the microstructural aspects of this work will only be briefly described here.  A schematic phase diagram of a β-alloy (Fig. 1) is useful to
illustrate the heat treatments we performed and the resulting microstructures.
All heat treatments consisted of an elevated temperature solution treatment
(ST) followed by a lower temperature aging treatment.  Solution treating above
the β-transus, e.g. at 850°C, leads to a recrystallized β-phase with equiaxed
grains.  The upper optical micrograph in Figure 1 illustrates this structure.
Numerous inclusions are also visible.  Solution treating in the two phase
α+β field, in this case between 700°C and 780°C (see upper shaded area in
Figure 1), results in a matrix of β-phase containing coarse, almost equiaxed
primary α ($\alpha_p$) particles.  The lower micrograph in Figure 1 shows a typical
(α+β) ST microstructure.  The $\alpha_p$ volume fraction decreases with increasing
solution treatment temperature, here from 30% at 725°C to 0% at 805°C (the
β-transus).  This material also contains a fine subgrain structure which is
stabilized by primary α.  Poorly defined, pancake-like grains can be seen at
low magnifications.  These probably are due to the working history.

The solution treatments were followed by aging at temperatures between
200°C and 500°C (see lower shaded area in Fig. 1).  Between about 200°C and
450°C the metastable ω-phase forms which transformed after sufficiently long
aging times to blocky α-phase precipitates (Fig. 2a).  This α is present as
a uniform distribution of very small plates (length ∿ 1000-1500Å, thickness
150 ∿ 300Å).  Above 400°C a second type of α-phase forms directly in an autocatalytic manner.  This α has a much higher aspect ratio than the aforementioned α, typical dimensions were 3μm-8μm in length and a thickness of
∿ 0.1μm (Fig. 2b).  When aging above 400°C both the aging temperature and the
rate of heating to the aging temperature affect the precipitation reaction.
Heating rapidly to the aging temperature results in the autocatalytic reaction, whereas heating slowly leads first to ω formation during heating and

then to the finer $\alpha$ plates which replace the $\omega$ during holding at temperature. These differences have a marked influence on the mechanical properties as will be shown later.

Especially in the $\alpha$-aged conditions subgrain boundary $\alpha$ and grain boundary $\alpha$-phase has been observed which is illustrated by TEM micrographs in Figures 2c and 2d, respectively (arrows). Figure 2e shows $\omega$-particles in a TEM dark field micrograph.

## II.  Mechanical Properties

These results will be divided into the ($\alpha+\beta$) ST plus $\alpha$-aged and the $\beta$-ST plus $\alpha$-aged conditions to simplify presentation. A limited number of tests were also performed on $\omega$-aged specimens. These results will be described also.

a.  ($\alpha+\beta$) ST plus $\alpha$-aged conditions – It is well known that strengthening during aging is mainly due to the small secondary $\alpha$ precipitates[4,15] and the strength level is controlled by the size and volume fraction of these precipitates. On this basis, we have varied the yield stress in two ways, as described below.

First, we started with microstructures which had varying volume fractions of equiaxed $\alpha_p$ (from 0% to 30%) and which have been aged to the maximum yield strength attainable at a constant aging temperature of 500°C. In this case, the volume fraction of secondary $\alpha$ increases as the amount of primary $\alpha$ decreases. Figure 3 shows the yield stress $\sigma_{0.2}$, the true fracture stress $\sigma_F$ (after Bridgeman correction), and the true fracture strain $\varepsilon_F$ as a function of the solution treatment temperature or volume fraction of primary $\alpha$. The yield stress can be increased from 1030MPa (149 ksi) to about 1350MPa (196 ksi) by decreasing the volume fraction of $\alpha_p$ from 30% (725°C) to 0% (805°C), however this causes the true fracture strain to drop from 0.99 to 0.05. The true fracture stress, however, shows no significant change except a small decrease for specimens solution treated close to and above the $\beta$-transus. The values are between 1500-1550MPa (217-225 ksi).

Second, we studied microstructures with a constant $\alpha_p$ content of 10%, but with changing morphologies and also volume fractions of the secondary $\alpha$-phase. This was achieved by solution treating at 780°C ($\sim$ 10% primary $\alpha$) and isothermally aging. We have chosen an aging temperature of 500°C, which produces yield stress levels around 180-200 ksi, which also are commercially interesting. Figure 4 illustrates the aging response; the yield stress is plotted versus aging time. We have tested conditions where the specimens were heated rapidly (liquid salt) as well as slowly (air furnace), in order to assess the heating rate effect on the secondary $\alpha$-phase. Both conditions show a peak in yield stress, however the slowly heated specimens, with the smaller particles, show significantly higher values for all aging times. For example, after 1 hr the air aged condition has a yield stress of 1445MPa (209 ksi), while the salt bath aged condition shows only 1202MPa (175 ksi) (see also conditions III and IV in Table 1). The yield stress difference becomes smaller after longer aging times. Figure 4 also contains the true fracture strain for the slowly heated and isothermally aged conditions. At peak strength the true fracture strain, $\varepsilon_F$, passes through a minimum of 0.09 (with an elongation to fracture of $e_F$ = 2.6%) and rises to 0.55 ($e_F$ = 10.4%) at 1250MPa (181 ksi) yield stress.

For some microstructures with a constant volume fraction of primary $\alpha$ different aging temperatures have been chosen which resulted in increased yield stresses for the lower aging temperature (see conditions I.and II in Table 1). This step seems to be interesting in order to achieve a most wide variety of primary $\alpha$ and secondary $\alpha$ combinations and the corresponding yield stress and ductility response. Possibly a more interesting way to use different aging temperatures was to age microstructures with different volume fractions of $\alpha_p$ to a comparable yield stress (see conditions II and III in Table 1). A microstructure with 30% $\alpha_p$ aged at 370°C to 1246MPa (180 ksi), for example, shows a true fracture strain of 0.19, while a condition with only 10% $\alpha_p$ aged at 500°C to a comparable yield stress has a much higher fracture strain of 0.63.

b.  β-ST plus α-aged conditions – The β-ST solution treated and α-aged microstructures were characterized by equiaxed β-grains with grain boundary $\alpha$ and secondary $\alpha$ within the grains. The main aging temperature used for a comparison with the (α+β) ST plus α-aged conditions was 500°C. Since aging in air led to macroscopic embrittlement, salt bath aging was performed which led to coarser secondary $\alpha$ particles and lower yield stresses. As expected from the aging response of the microstructures with 10% $\alpha_p$ (see Fig. 4), overaging is observed; the peak yield stress is higher due to the increased volume fraction of secondary $\alpha$. The fracture strain passes through a minimum of 0.03 at the peak yield stress of about 1370MPa (199 ksi). Overaging to 1200MPa increases the fracture strain to about 0.17. The true fracture stress is between 1510MPa and 1440MPa, the slight decrease with increasing aging time is probably within experimental scatter. Aging in salt at lower temperatures such as 400°C has an embrittling effect similar to aging in air at 500°C. However, in this case, the embrittlement probably is due to an increased volume fraction of secondary $\alpha$ instead of a fine particle size.

In addition to the above, we have studied several ω-aged conditions in the (α+β) ST and β-ST material. These conditions have been described to a limited extent elsewhere recently[19], but the results are summarized here because they complement the α-aged results. It is possible to reach the same strength levels as the α-aged conditions aging by ω-phase. Especially in the high strength conditions around 1250MPa (181 ksi) the ω-aged specimens consistently exhibit lower ductilities than the α-aged conditions (compare conditions II, III, V with VI in Table 1). In the low strength region (yield stress values of ∿965MPa (140 ksi),the ω-aged conditions can show ductilities comparable to the α-aged conditions. The uniform elongations, as well as the tensile strengths, however, are always smaller.

Beside strength control, the most important goal of this study was to investigate the tensile ductility or, more generally, the fracture behavior of the different types of microstructures at different strength levels. For this purpose it is very useful to illustrate the results in a diagram where the yield stress $\sigma_{0.2}$ is plotted versus the true fracture strain $\varepsilon_F$ or reduction in area RA, respectively (see Fig. 5). We organized the data in Figure 5 by forming groups of microstructures each having a constant volume fraction of $\alpha_p$ and having been aged to different yield stresses. From such a diagram it is, for example, possible to find out  1) which correlation exists between $\sigma_{0.2}$ and $\varepsilon_F$ for such a group and  2) if the ductilities of the various groups are different at comparable yield stresses. From our diagram we found the following results:  1) At a constant volume fraction of $\alpha_p$, an increase in yield stress leads to a reduction in ductility. This well-known tendency is illustrated by trend lines.  2) For the (α+β) St plus α-aged microstructures an increase in primary $\alpha$ volume fraction leads to a reduction in ductility at comparable yield stresses. This tendency is' observed over the whole yield stress range investigated. At about 1250MPa (∿ 181 ksi) yield stress, for

example, the microstructure with 10% primary $\alpha$ shows a true fracture strain of about 0.5 compared to 0.18 for the condition with 30% primary $\alpha$ and the same yield stress.  3) The $\beta$-ST plus $\alpha$-aged microstructures show lower ductilities than the $\alpha+\beta$ ST plus aged structures for all yield stresses investigated.  The trend line for $\beta$-ST conditions is steeper than those of the $(\alpha+\beta)$ ST plus $\alpha$-aged conditions which means that there is also less ductility gain by reducing the yield stress.  At about 1250MPa (181 ksi) yield stress, however, the ductility is close to that of a microstructure with 30% $\alpha_p$.
4) For the forging (pancake) with elongated $\alpha_p$, the ductilities of equivalent $(\alpha+\beta)$ ST plus $\alpha$-aged conditions compared to the plate forging with globular $\alpha_p$ are lower at comparable yield stresses.  At 1300MPa (188 ksi), for example, the microstructure with $\sim$ 15% $\alpha_p$ (ST at 760°C) has a fracture strain of about 0.33 compared to 0.23 for an equivalent microstructure with elongated $\alpha_p$.  In the $\beta$-ST plus $\alpha$-aged conditions with no primary $\alpha$ the ductilities at comparable yield stresses are the same in both forgings as expected.  It should be noted, however, that identical heat treatments for both forgings resulted in different yield stresses.  The pancake showed a lower yield stress in the $\beta$-ST plus $\alpha$-aged condition (compare conditions V and VII in Table 1) and a higher yield stress (100MPa) after solution treating at 760°C and aging at 500°C for 1 hr.  This opens the question of whether or not this difference is due to a texture effect or, more importantly, can the ductility difference between microstructures with globular and elongated primary $\alpha$ at a constant yield stress be solely attributed to the difference in $\alpha_p$-morphology.  5) $\omega$-aging to a yield stress around 1250MPa ($\sim$181 ksi) results in very low ductilities (1-2% reduction in area) compared to all $\alpha$-aged conditions. At lower yield stresses ($\sim$ 1050MPa) the ductility increases considerably and is comparable to that of a microstructure with 30% primary $\alpha$.

III. Fractography

Fractographic studies of some of the microstructures by scanning electron microscopy revealed some qualitative explanations for the tensile test results.  At least three different types of fracture topography have been observed, as shown in Figure 6.

The $(\alpha+\beta)$ ST plus $\alpha$-aged conditions show a complex dimple rupture fracture mode (Fig. 6a).  The size of the most obvious dimple is comparable to the spacing of the $\alpha_p$, but there are also much smaller dimples (< 1µm) present for several microstructures; these are possibly related to the secondary $\alpha$.

The $\beta$-ST and $\alpha$-aged microstructures especially in the high strength condition ($\sigma_{0.2}$ > 1300 MPa) show mainly intergranular fracture (Fig. 6b).  Very small dimples (< 1µm) have been observed on the grain boundary facets.  With decreasing yield stress the amount of intergranular fracture decreases and, instead, more transgranular dimple type fracture is observed (Fig. 6c).  This dimple fracture is rather irregular with small dimples in the range of 1µm or less and a coarser structure in the range of 10-15µm.

Figure 6d illustrates the fracture of an $\omega$-aged microstructure ($\beta$-ST + $\omega$-aged, RA = 0%).  It shows the line of intersection between the specimen surface and the fracture surface.  It is apparent that very intense slip bands have formed in the $\omega$-aged matrix, and fracture occurred along these slip bands.

Discussion

The results have shown that a variation in heat treatment at a constant forging history can lead to a wide variety of microstructures in Ti-10-2-3. The main microstructural variables were the $\alpha_p$ volume fraction, the secondary $\alpha$ volume fraction and morphology and the grain boundary $\alpha$. Also, on selected microstructures the $\alpha_p$ morphology has been varied (elongated instead of globular) by a different forging history. Also, a limited number of $\omega$-aged conditions have been studied. Based on the mechanical properties for different combinations of these parameters (Fig. 5) we will discuss some more macroscopic aspects of an optimum microstructure. We also will offer some thoughts which may help to understand the microscopic mechanisms of strengthening and fracture.

For all the groups of microstructures (constant volume fraction of $\alpha_p$), it is possible to produce a wide range of yield stresses. Increasing the aging temperature would extend the yield stress range to still lower values. If only an optimum combination of yield stress and ductility is desired, it seems most appropriate to take a microstructure with very little $\alpha_p$ (Fig. 5). However, fatigue properties as well as fracture toughness or corrosion properties usually have to be taken into account in structural design. Thus, other microstructures with comparable yield stresses remain interesting because they may be superior with respect to other properties.

A preliminary explanation of the different ductilities for the various groups of microstructures is possible based on the fracture surface studies (Fig. 6).

The relatively low ductilities of the $\beta$-ST and $\alpha$-aged conditions compared to the ($\alpha+\beta$) ST plus $\alpha$-aged conditions can be attributed to a difference in fracture mode. As has been shown earlier[14], the long, soft grain boundary $\alpha$ layers deform preferentially which leads to a high plastic deformation in a very small volume before the yield stress of the matrix is reached. Because of the long slip length, stress concentrations form at grain boundary triple points where the slip is stopped. This can lead to fracture at very low macroscopic strains, although locally a high plastic deformation occurs. On a Ti-Mo alloy it has been shown that reducing the length of those soft zones by a smaller grain size can considerably increase the ductility by lowering the stress concentrations at the triple points[14]. It may be useful to try this also with Ti-10-2-3.

In the $\omega$-aged microstructures the $\omega$-particles are sheared. Intense slip bands form which create high stress concentrations at grain boundaries or primary $\alpha$-particles and can lead to early fracture. In an $\omega$-aged $\beta$-Ti alloy it has been shown that a reduction in grain size improved the ductility significantly, because it reduced the stress concentrations at the grain boundaries by a smaller pile-up length[16]. Again, for the $\omega$-aged conditions a test with a smaller grain size seems to be useful in order to improve ductility.

The ($\alpha+\beta$) ST plus $\alpha$-aged conditions are of special interest, as they are commercially most interesting. Although a useful discussion of microscopic deformation and fracture is not possible from the present data, we do want to mention some factors which we think are important. For example, varying the volume fraction of $\alpha_p$ changes the matrix composition. This in turn results in different volume fractions and possibly different morphologies of the secondary $\alpha$ for a certain aging treatment. Assuming that both types of $\alpha$ can play a role in the fracture process, it is quite difficult to separate their contributions. In addition, one has to consider that the $\alpha$-phase is softer than the aged matrix and begins to deform plastically at lower stresses. Thus, for an increasing volume fraction of $\alpha_p$ the matrix has to be hardened more strongly in order to achieve the same macroscopic yield stress for such a microstructure compared to one with less primary $\alpha$. As both types of $\alpha$ have been changed, again their contributions to fracture have to be separated. Consequently it appears to be promising to study the micromechanisms of deformation and fracture in more detail, which we already started on the basis of the present results.

## Conclusions

In Ti-10-2-3 a variety of aged microstructures with different volume fractions and morphologies of primary and secondary $\alpha$ and also grain boundary $\alpha$ has been tensile tested. The results can be summarized as follows:

1) For a constant volume fraction of primary $\alpha$ it is possible to vary the yield stress in a wide range; it is also possible to reach comparable yield stress levels for different volume fractions of $\alpha_p$ by an appropriate choice of aging temperatures and times as well as heating up rates to the aging temperatures.

2) Increasing the volume fraction of primary $\alpha$ reduces the ductility at comparable yield stresses for the $\alpha$-aged conditions. Only the $\beta$-ST conditions with no primary $\alpha$ are an exception. They have lower ductilities than the $(\alpha+\beta)$ ST conditions. The $\omega$-aged microstructures have the lowest ductilities at comparable yield stresses.

3) The differences in ductilities between the $(\alpha+\beta)$ ST and $\alpha$-aged, the $\beta$-ST and $\alpha$-aged and the $\omega$-aged microstructures are qualitatively related to different fracture modes.

4) Microstructures from a different forging which instead of the above tested globular primary $\alpha$ have elongated primary show lower ductilities at comparable yield stresses.

## Acknowledgments

This work has been supported by the Office of Naval Research. Experimental work has been conducted using facilities provided by The Center for the Joining of Materials. The authors gratefully acknowledge the experimental assistance of M. Glatz and the secretarial help of Mrs. A. M. Crelli. Helpful discussions with Professor G. Luetjering are also acknowledged.

## References

1. E. Bohanek: "Titanium Science and Technology," Vol. 3, Plenum Press, New York, (1973), 1983.

2. F. H. Froes, J. C. Chesnutt, C. G. Rhodes, and J. C. Williams: ASTM, STP 651, "Toughness and Fracture Behavior of Titanium," (1978), 115.

3. F. H. Froes, R. F. Malone, V. C. Peterson, C. G. Rhodes, J. C. Chesnutt, and J. C. Williams: AFML-TR-75-4, Vols I and II, Air Force Materials Laboratory, Dayton, OH, 1975.

4. J. C. Williams, F. H. Froes, J. C. Chesnutt, C. G. Rhodes, and R. G. Berryman: ASTM, STP 651, "Toughness and Fracture Behavior of Titanium," (1978), 64.

5.  H. W. Rosenberg: Joint Conference: Forging and Properties of Aerospace Materials, University of Leeds, January 1977.
6.  E. Bohanek: Titanium Metals Corporation of America, Technical Report Number 55, May 1972.
7.  C. C. Chen and C. P. Gure: Report RD74-120, Wyman-Gordon Company, North Grafton, MA, November 1974.
8.  C. C. Chen: Report RD-75-118, Wyman-Gordon Company, North Grafton, MA, November 1975.
9.  R. R. Boyer, J. W. Tripp, and J. E. Magnuson: presented at the TMS-AIME Fall Meeting, Milwaukee, WI, September 1978.
10. C. C. Chen and R. R. Boyer: J. of Metals, 31 (1979), 33.
11. G. Lenning: RD 012, TIMET, Marcn 1976.
12. I. A. Matwell: AFML-TR-78-114, Air Force Materials Laboratory, Dayton, OH, 1978.
13. T. Hamajima, G. Luetjering, and S. Weissman: Met. Trans. 4 (1973), 847.
14. M. Peters and G. Luetjering: Zeitschift f. Metallkunde, 67 (1976), 811.
15. M. G. Mendiretta, G. Luetjering, and S. Weissman: Met. Trans., 2 (1971), 2599.
16. A. Gysler, G. Terlinde, and G. Luetjering: Proc. 3rd Int. Conf. on Titanium, Moscow, 1976.
17. J. C. Chesnutt and F. H. Froes: Met. Trans., 8A (1977), 1013.
18. J. C. Williams, B. S. Hickman, and H. L. Marcus: Met. Trans., 2 (1971), 1913.
19. T. W. Duerig, G. T. Terlinde, and J. C. Williams: Phase Transformations and Tensile Properties of Ti-10V-2Fe-3Al, Met. Trans. (in press).

Table 1   Mechanical properties of selected microstructures in Ti-10-2-3

| | Condition | Heat Treatment | Microstructure | $\sigma_{0.2}$ (MPa) | UTS (MPa) | $e_F$ (%) | $\varepsilon_F$ |
|---|---|---|---|---|---|---|---|
| | I | 725°C 20 hrs, WQ + 500°C 1 hr(salt) | 30% prim.α + "large" sec. α | 1063 | 1106 | 17.7 | 0.99 |
| | II | 725°C 100 min. WQ + 370°C $10^3$ min. | 30% prim.α + "small" sec. α | 1246 | 1419 | 7.6 | 0.19 |
| Globular prim.α | III | 780°C 3 hrs WQ + 500°C 1 hr (salt) | 10% prim.α + "large" sec. α | 1202 | 1247 | 10.3 | 0.63 |
| | IV | 780°C 3 hrs WQ + 500°C 1 hr (air) | 10% prim.α + "small" sec. α | 1445 | 1544 | 2.4 | 0.09 |
| | V | 850°C 2 hrs WQ + 500°C 4 hrs (salt) | 0% prim.α + grain bound.α + "large" sec. α | 1250 | 1308 | 3.9 | 0.16 |
| | VI | 700°C 8 hrs WQ + 200°C 6800 min. | 35% prim.α + ω | 1218 | 1266 | 0.5 | 0.02 |
| Elongated prim. α | VII | 850°C 2 hrs WQ + 500°C 4 hrs (salt) | 0% prim.α + "large" sec. α + grain bound. α | 1182 | 1265 | 3.8 | 0.17 |
| | VIII | 760°C 75 min. WQ + 500°C 1 hr (salt) | ∿10% prim.α + "large" sec. α | 1298 | 1381 | 4.6 | 0.23 |
| | IX | 700°C 75 min. WQ + 350°C $10^3$ min. | ∿30% prim.α + "small" sec.α | 1239 | 1395 | 3.9 | 0.11 |

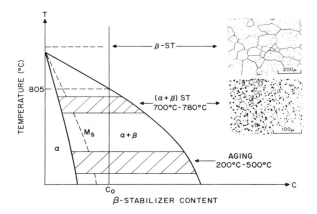

Fig. 1 – Schematic phase diagram with solution treated microstructures and heat treatment temperature ranges.

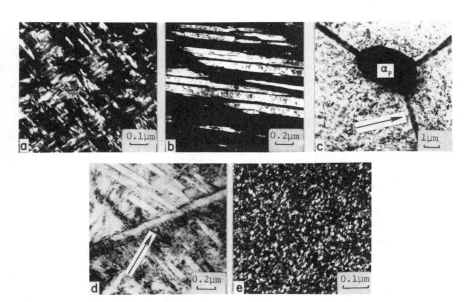

Fig. 2 – Aging products in Ti-10-2-3
a) "Fine" secondary α, which follows metastable ω (TEM darkfield)
b) "Large" secondary α, which forms autocatalytically (TEM darkfield)
c) Subgrain boundary α (TEM brightfield)
d) Grain boundary α (TEM brightfield)
e) ω-phase (TEM darkfield)

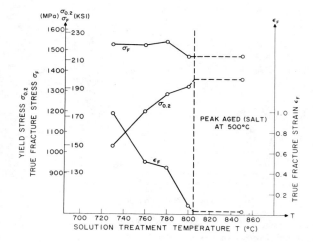

Fig. 3 – Yield stress $\sigma_{0.2}$, true fracture stress $\sigma_F$ and true fracture strain $\varepsilon_F$ as a function of ST temperature for specimens aged to maximum strength at 500°C.

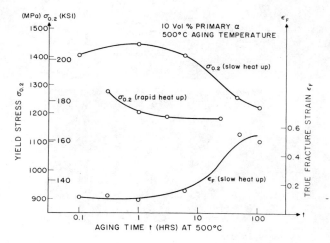

Fig. 4 – Yield stress $\sigma_{0.2}$ and true fracture strain $\varepsilon_F$ for microstructure with 10% $\alpha_\rho$ isothermally aged at 500°C.

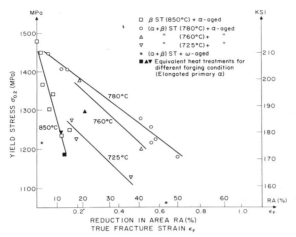

Fig. 5 – Relation between yield stress and true fracture strain $\varepsilon_F$ or reduction in area RA, respectively, for various microstructures.

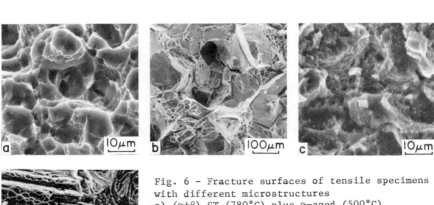

Fig. 6 – Fracture surfaces of tensile specimens with different microstructures
a) (α+β) ST (780°C) plus α-aged (500°C)
b) β-ST (850°C) plus α-aged (500°C), high yield stress (1380MPa)
c) β-ST (850°C) plus α-aged (500°C), lower yield stress (1250MPa)
d) β-ST (850°C) plus ω-aged (370°C), intersection of fracture surface and gage length surface.

# STRESS RELAXATION DURING AGING OF Ti-6Al-6V-2Sn[*]

H. J. Rack

Mechanical Metallurgy Division, Sandia Laboratories[†]
Albuquerque, New Mexico, 87185, USA

## Introduction

Recent investigations [1-3] indicate that the stress relaxation behavior of ferrous alloys is strongly dependent upon their microstructural stability. If the alloy is stable, that is the microstructure does not change during exposure at elevated temperature, the relaxation event involves time-dependent slip-yielding or creep processes [3]. If however the microstructure is unstable, for example if precipitation occurs concurrent with relaxation, the kinetics of the relaxation process are accelerated and actually parallel those of the microstructural alterations [2]. Further, the fractional stress relaxation in the latter case is independent of the initially applied elastic stress. Additionally, if this enhanced relaxation treatment is followed by aging at or above the prior relaxation temperature, a creep-back phenomenon is observed. This phenomenon takes the form of a time/temperature-dependent closure of the unloaded ring sample, that which has previously been used to establish the extent of stress-relaxation.

These observations of enhanced stress relaxation in ferrous alloy systems have prompted Brown et al. [2] to develop a relaxation model which does not depend on generalized slip-yielding or climb processes and which takes into consideration the microstructural variations occurring during the relaxation event. The model postulates that relaxation under these circumstances involves the formation of plastic zones, the zones having been induced by the volume changes associated with the specific microstructural alteration. These plastic zones then respond to--and relieve--the imposed stresses. In principle this model should not only be able to explain the enhanced relaxation and creepback observed during tempering of hardened steels, it should be applicable to any relaxation event which occurs during concurrent aging. The present investigation of stress relaxation concurrent with aging in solution treated and quenched (martensitic) Ti-6Al-6V-2Sn was intended to examine the possible applicability of the plastic zone model to a system where the precipitational processes do not involve carbide formation.

## Experimental Procedure

The chemical composition of the Ti-6Al-6V-2Sn alloy used in this study was (wt. pct.):  5.7 Al, 5.7 V, 2.1 Sn, 0.155 O, 0.77 Cu, 0.75 Fe, 0.011 N, 0.02 C, 45 ppm H, bal. Ti.  It was supplied in the form of a 470 mm outside diameter, 57 mm wall thickness hollow extrusion.  The material had been $\alpha$-$\beta$ blocked at 1173 K prior to $\beta$ extrusion at 1258 K, the final extrusion ratio

[*] This work was supported by the U. S. Department of Energy (DOE) under Contract DE-AC04-76-DP00789.

[†] A U. S. DOE Facility.

being 4.4:1. Microstructural examination [4] showed that this procedure resulted in an essentially equiaxed prior β grain structure with a prior β grain size of 216.3 μm.

Evaluation of the stress relaxation characteristics of Ti-6Al-6V-2Sn initially involved solution treating 32 mm × 57 mm × 153 mm length cord sections of the extrusion for 1h at 1123 K followed by water quenching. Following solution treatment 6.4 mm thick wedge-loaded, eccentric-ring samples were prepared from the heat treated cords. The specimen dimensions, Fig. 1, were selected to assure that the maximum outer fiber stresses utilized in this study did not invalidate the portionality between the gap deflection and the outer fiber stress. Gap preparation itself involved sectioning the rings, with Knoop hardness impressions being placed near the gap faces to serve as fiducial marks for all subsequent gap measurements.

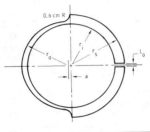

Fig. 1.   Dimensions of eccentric split ring specimens.

| RING SIZE | a (mm) | $r_i$ (mm) | $r_o$ (mm) | $r_s$ (mm) | $L_o$ (mm) |
|-----------|--------|--------|--------|--------|--------|
|  | 0.758 | 22.6 | 24.7 | 27.7 | 1.0 |

Stressing of the ring samples prior to aging was accomplished by inserting a solution treated and water quenched Ti-6Al-6V-2Sn wedge into the gap. Letting the original gap spacing be $L_o$, the wedged gap spacing $L_i$, the initially applied elastic stress due to the insertion of the wedge is [1,5]

$$\sigma_i = KE\delta_i \qquad (1)$$

where K is a constant which depends upon the radial position across the thickness of the ring and ranges from negative (compressive stress) at the outer surface to positive (tensile stress) at the inner surface, E is the Young's modulus and $\delta_i = (L_i - L_o)$ is the gap deflection in the wedged condition.[*] If it is assumed[†] that the lower initial stresses across the thickness of the loaded ring will be relaxed in a fashion similar to that of the higher

---

[*]To insure that the loaded rings were only loaded elastically the wedge was removed and reinserted a number of times prior to any relaxation treatment. Unless the unwedged gap spacing returned to $L_o$ each time, the ring was discarded.

[†]Previous studies [1,2] suggest that this assumption appears to be valid if the decrease in stress due to relaxation ($\sigma_i - \sigma_r$) is proportional to the initial stress ($\sigma_i$), that is the observed fractional stress relief is independent of $\sigma_i$. Fig. 2 supports this assumption since in this study the percent stress relaxation observed in Ti-6Al-6V-2Sn during aging does not depend on the initial outer fiber stress, at least within the range examined.

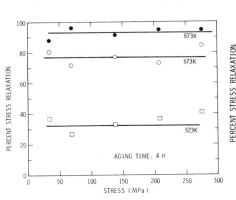

Fig. 2.  Percent stress relaxation as a function of initially applied elastic stress. Relaxation temperature as indicated.

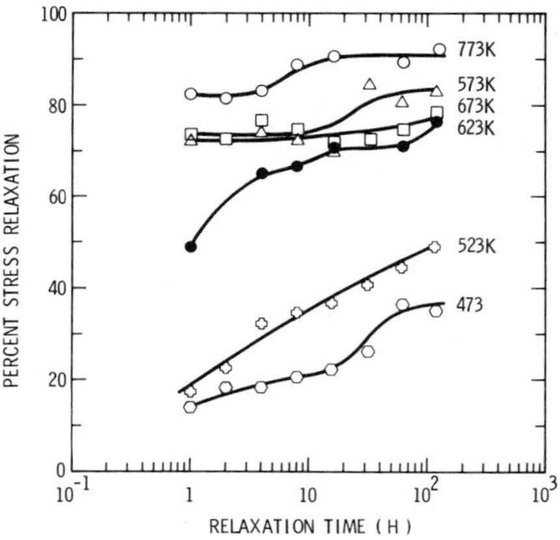

Fig. 3.  Percent stress relaxation as a function of the stress-relieving time and temperature. Initial outer-fiber stress: 138 MPa.

outer fiber stress then it is possible to determine the fractional stress relief by measuring the original gap spacing ($L_o$), the wedged gap spacing ($L_i$), and the gap spacing after a recovery treatment and removal of the wedge ($L_r$). If $\delta_r = (L_r - L_o)$ is now taken to be the gap deflection after some relaxation treatment and wedge-removal then the fractional stress relief will be:

$$(SR) = \frac{\delta_r}{\delta_i} = \frac{L_r - L_o}{L_i - L_o} \qquad (2)$$

This means that, if the gap spacing ($L_r$) after a relaxation treatment and wedge-removal springs back to the original gap spacing ($L_o$), then there has been no stress relief, whereas if $L_r$ remains at its original wedged gap spacing ($L_i$), the stress relief has been complete.

Any subsequent creepback[*] after a relaxation treatment and wedge-removal may also be determined by measuring the gap spacing ($L_{cb}$) after a given creepback treatment. The corresponding gap deflection is then simply $\delta_{cb} = L_{cb} - L_o$ or expressed in the form of a fractional creepback (CB):

$$(CB) = \frac{\delta_r - \delta_{cb}}{\delta_r} = \frac{L_r - L_{cb}}{L_r - L_o} \qquad (3)$$

### Results

The data plotted in Fig. 2 show that the percent stress relaxation occurring concurrent with aging of Ti-6Al-6V-2Sn is not sensitive to the

---

[*] Creepback occurs when the unloaded specimen is given a thermal treatment at or above the previous stress-relieving temperature.

initially applied elastic stress within the range studied.  Fig. 3 indicates
that the percent stress relaxation also does not increase in a simple
fashion with increasing stress relaxation temperature or time.  Indeed it
is possible to obtain more stress relief, at a constant aging time, by stress
relieving at 573 K rather than at 623 K.  Above 623 K the percent stress
relaxation does however again increase with increasing relaxation (aging)
temperature.

Typical measurements of the creepback effect are shown in Fig. 4.  All
the ring specimens were wedge loaded and stress relieved at 773 K ior the
indicated prior relaxation time.  They were then unloaded at room temperature
and reheated in the unwedged condition at the same temperature as the prior
stress relaxation.  The percent creepback, as defined by Eq'n. (3), increases
with the time of heating after each prior stress-relieving treatment.
Furthermore, the percent creepback becomes smaller the larger the amount of
prior percent stress relaxation.  Finally like the percent stress relaxation
the percent creepback is independent of the initially applied stress.

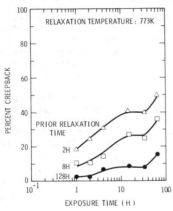

Fig. 4.   Percent creepback as a
function of creepback-time
and temperature.  Initial
outer-fiber stress:  138
MPa.

## Discussion

While the kinetics of stress relaxation appear to be slower in Ti-6Al-
6V-2Sn than those which have previously been reported for hardened steels [2]
the results of this investigation are similar in many other respects to those
for hardened steels.  For example, the percent stress relaxation in both
cases is independent of the initially applied elastic stress.  Furthermore,
the percent stress relaxation does not, in either instance, increase in a
simple fashion with increasing relaxation time or temperature.  Finally both
systems exhibit a creepback phenomena, that is a time/temperature dependent
closure of the ring specimen gap when the partially recovered ring is
further aged in the unwedged state.

These observations suggest that an ordinary creep-type of thermally-
activated slip yielding or a thermally-activated viscoelastic model cannot
account for the stress-, time- or temperature-dependence of the stress relaxation in Ti-6Al-6V-2Sn.  An ordinary creep-type model would require that the
percent stress relaxation be a function of the initially applied stress and
that stress relaxation take place more readily the higher the aging temperature.  Figs. 2 and 3 show that neither of these requirements are met during
stress relief of Ti-6Al-6V-2Sn.  Moreover, progressive changes in the
martensitic dislocation substructure cannot be the operative mechanism

throughout the temperature range investigated since it too would require an essentially continuous increase in percent stress relaxation with aging temperature.

While the thermally-activated viscoelastic model has been successfully used for the time-dependent relaxation of superplastic nickel-base alloys [6] it also does not appear to be applicable to the relaxation of quenched Ti-6Al-6V-2Sn.  This model again requires that the relaxation rate increase monotonically with temperature and that the percent stress relaxation depend upon the initial stress, both requirements being in conflict with the experimental findings reported here.

It is possible to understand the stress relaxation behavior of quenched Ti-6Al-6V-2Sn by considering the plastic zone model alluded to previously.  This model requires that there be a direct relationship between the precipitation processes occurring during aging of Ti-6Al-6V-2Sn and the stress relaxation phenomenon.  Previous studies [4,7-10] of Ti-6Al-6V-2Sn quenched from 1123 K indicate that in the as-quenched condition this alloy will contain primary α, retained β, hexagonal α′ martensite and a transition phase.  Examination of the x-ray diffraction evidence presented by Gueret et al. [7] and Klima et al. [8] suggests that this transition phase is related to the orthorhombic α″ martensite formed in more concentrated α-β titanium alloys [11].  Fig. 5 shows that the decomposition of this alloy seems to occur over two rather broad temperature regions with the elastic modulus going through a minimum after aging for 8h at 623 K.  A similar minimum is observed if the percent stress relaxation at a constant relaxation time of 8h is plotted as a function of aging temperature, Fig. 6.  This similarity strongly suggests that those factors which control the changes in modulus with aging temperature, i.e., the decomposition of the as-quenched structure, also control the stress relaxation behavior of quenched Ti-6Al-6V-2Sn.

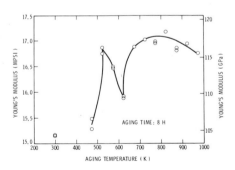

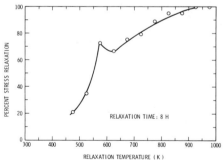

Fig. 5.  Young's modulus of Ti-6Al-6V-2Sn solution treated at 1123, water quenched and aged.

Fig. 6.  Percent stress relaxation as a function of aging temperature for quenched Ti-6Al-6V-2Sn.  Initial outer-fiber stress: 138 MPa.

Adaptation of the plastic zone model not only requires that the precipitation reaction(s) and the stress relaxation response be related but that the local volume change associated with the precipitation reaction be sufficiently large to produce a plastic zone. De Jong and Rathenau [12] have given an approximate criterion which must be satisfied in order for this local volume change to be sufficiently large to produce such a zone:

$$\frac{2E}{3S_o} \left|\frac{\Delta V}{V}\right| = \left(\frac{r_2}{r_1}\right)^3 > 1 \qquad (4)$$

where E is Young's modulus of the matrix, $S_o$ is the yield strength of the matrix, $\Delta V/V$ is the fractional change in volume, $r_1$ is the radius of the transforming (precipitation) center and $r_2$ is the radius of the plastic zone. $(r_2/r_1)^3$ must be greater than unity for a plastic state to exist. Since aging at Ti-6Al-6V-2Sn will involve decomposition of $\alpha'$ martensite, retained $\beta$ and perhaps $\alpha''$ martensite it is possible that each of these reactions may lead to the creation of a plastic zone. Unfortunately our knowledge of the $\alpha''$ martensite decomposition reaction in Ti-6Al-6V-2Sn is fragmentary. However, Table 1 does show representative calculations* which suggest that either of the other reactions, the decomposition of retained $\beta$ by $\alpha$ precipitation or $\beta$ precipitation from $\alpha'$ martensite can result in local plastic straining. Although the yield strengths and moduli shown in Table 1 are room temperature values and should be reduced according to the relaxation temperature being studied, this would only make $(r_2/r_1)^3$ larger and the above inequality would then be satisfied by an even larger margin. Finally some experimental support for the formation of these plastic zones has recently been presented by Young et al. [13]. They showed that $\alpha$ precipitation from $\beta$ is accompanied by the generation of dislocations both within the $\beta$ matrix, at the $\alpha/\beta$ interface and within the $\alpha$ precipitate.

Table 1

Calculation for Plastic-Zone Conditions

Associated with Decomposition of Ti-6Al-6V-2Sn

|  | $\left|\Delta V/V\right|$ | $S_o$ (MPa) | E (GPa) | $(r_2/r_1)^3$ |
|---|---|---|---|---|
| $\alpha$ from $\beta$ | 0.06 | 830 | 96 | 6.8 |
| $\beta$ from $\alpha'$ | 0.03 | 830 | 120 | 2.9 |

(a) $V_\alpha = 17.58 \ \overset{\circ}{A}^3$ per atom, $V_{\alpha'} = 17.53 \ \overset{\circ}{A}^3$ per atom and $V_\beta = 17.01 \ \overset{\circ}{A}^3$ per atom [7,8]

The proposed relationship between the stress relaxation during aging of quenched Ti-6Al-6V-2Sn and production of plastic volumes can also provide an explanation for the creepback phenomenon. When the wedge is removed to measure the extent of relaxation, the plastic zones which are originally

---

*These calculations assume that the mechanical properties of the retained $\beta$ can be characterized by those obtained from an unaged metastable $\beta$ titanium alloy and that the $\alpha'$ martensite can be represented by cold worked $\alpha$ titanium.

relaxed when the wedge is in place are now reloaded with opposite sign by the springback of the elastic regions in the ring. When reheated continued volume changes due to precipitation occur and the local stresses about the original plastic zones are relieved, this local relaxation results in a closure of the gap (creepback).

## Conclusions

The preceding investigation of stress relaxation in quenched Ti-6Al-6V-2Sn during concurrent aging has shown that:

1. The percent stress relaxation is independent of the elastically applied stress over the range of stresses investigated.

2. The percent stress relaxation, while not a simple function of treatment, time or temperature, does appear to follow the microstructural changes which are thought to occur during aging of Ti-6Al-6V-2Sn.

3. Following a stress-relaxation treatment and removal of the applied load subsequent aging at the prior relaxation temperature produces a creepback phenomena.

4. These observations are consistent with a previously proposed model of stress relaxation concurrent with precipitation in which plastic zones, induced by the volume changes associated with the precipitation reaction, respond to--and relieve--the imposed stresses.

## Acknowledgement

The author wishes to acknowledge J. H. Gieske, T. J. Headley and J. C. Smith for their assistance respectively with the elastic moduli, transmission electron microscopy and relaxation measurements.

## References

1. R. L. Brown and M. Cohen:  Metal Progress, 81(1962), 66.
2. R. L. Brown, H. J. Rack and M. Cohen:  Mat'l. Sci. Eng., 21(1975), 25.
3. H. J. Rack and M. Cohen:  Sandia Laboratories Report, SAND-76-0282, June 1976.
4. M. G. Ulitchny, H. J. Rack and D. B. Dawson:  Toughness and Fracture Behavior of Titanium, ASTM STP 651, (1978), 17, ASTM, Philadelphia, PA.
5. I. O. Oding, V. S. Ivanova, V. V. Burduksku and V. N. Gemmov:  Creep and Stress Relaxation in Metals, (1965), 252.  Oliver and Boyd, Edinburgh.
6. H. W. Hayden and J. H. Brophy:  Trans. ASM, 61(1968), 542.
7. G. Gueret, B. Houssin, J. Fries, G. Cizeron and P. LaCombe:  Jn. Less-Cons. Metals, 38(1974), 31.
8. R. Klima, F. Quemper, C. Beauvair, C. Roux, M. Rapin and B. Hockeid: CEA-R-4542, C:E.N. Saclay, B.P. no. 2, 91 190 Gif-sur-Yvette, France, May 1974.
9. M. Majdec, G. Welsch and G. Ziegler:  Proc. Third Int. Conf. Titanium, in press.

10.  P. J. Fopiano and C. F. Hickey, Jr.:  Jn. Testing Eval., 1(1973), 514.
11.  J. C. Williams:  <u>Titanium Science and Technology</u>, 3(1973), 1433.
     Plenum Press, New York.
12.  M. De Jong and G. W. Rathenau:  Acta Met., 9(1961), 714.
13.  M. Young, E. Levine and H. Margohn:  Met. Trans., 10A(1979), 359.

# HEAT TREATMENT AND MICROSTRUCTURE OF THE
## METASTABLE BETA-TITANIUM ALLOY TB2

Chen Hai-Shan

General Research Institute for Non-Ferrous Metals
Beijing, China

TB2 Ti-alloy(Ti-5Mo-5V-8Cr-3Al) is a metastable $\beta$-alloy which began to be developed in 1966(1). Although the composition of the metastable $\beta$-alloys vary widely, strengthened mechanism is roughly the same. These alloys are strengthened by the solution treatment, and followed by aging. Therefore, the effect of varying parameters of heat treatment on aging decomposition of $\beta$-phase must be studied. For $\beta$-alloy, $\beta$-transus temperature and recrystallized temperature are found to be basic temperature parameters. The significance of these temperature parameters is well known. In accordance with our experiment, there are other two temperature parameters for TB2 alloy--the effective quenching temperature and the transition temperature--that are associated with the change of the microstructure morphology. In this paper, the detail of these two temperature parameters have been discussed. The effect of the change in several heat treatment parameters on the precipitation of the second phase has been investigated. Also it is found out how to control the types of aging structure through the heat treatment.

## Experimental Procedure

The alloy studied was received from a plant in form of 3.5mm thick solution treated sheet. This material was cold rolled to 1.8mm thick at our laboratory. The forging used for this investigation was a disk which was forged between 1000-1100°C to 150-250mm in diameter and 20-50mm in thickness. The materials used for this investigation prossessed the following chemical composition(wt pct): 4.9 Mo, 5.0 V, 7.5 Cr, 2.7 Al,

0.12 Fe, 0.14 O, 0.025 N and 0.016 C.

The β-transus temperature for TB2 alloy was determined at 750°C. Metallographic samples were cut from the cold rolled sheet(and from the forging). These samples were solution treated at several temperatures above and below β-transus temperature, followed by aging for 8h at 500°C. Several kinds of pretreatment were adopted in the process of heat treatment for this investigation. For example, the solution treated samples were exposed in the temperature range from 450-630°C for a short time, air cooled, then were also aged 8h at 500°C. We have investigated the variations of the microstructures in the various conditions of heat treatments using an optical microscopy. The microstructure of samples aged at lower temperature (300-400°C) were observed by JEM-1000 transmission electron microscopy.

### Experimental Results

1. The effective quenching temperature

For cold rolled TB2 sheet, the solution treatment in single phase β-field(800°C) and aging 8h at 500°C produced a microstructure as shown in Fig.1a. The precipitate free zone (PFZ) near the grain boundaries are very clearly demonstrated, but single α-phase particle within the grains is impossible to resolve with an optical microscopy. This type of microstructure

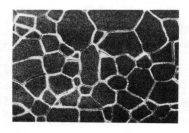
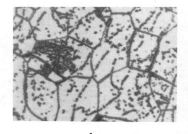

a                                    b

Fig.1 Two types of aging structures(x200)
a. Aging structure with the PFZ
b. "Snowflake-like" aging structure

is the most common for TB2 alloy. For TB2 forging produced at
elevated temperature, a microstructure produced by the same
heat treated was found to be different from above-mentioned
one, as shown in Fig.1b. A large number of α-phase precipitated
at the grain boundaries and an heterogeneous precipitation
within the grains can be seen. We call this microstructure as
the "Snowflake-like" structure.

Further investigations have found that the "Snowflake-like"
structure can be also obtained from sheet of TB2 alloy. For TB2
sheet, solution treatment for 30min at 800°C following by aging
for 8h at 500°C usually produced a microstructure containing
the dispersive precipitated α-phase within the β-grain and the
PFZ near the grain boundary. The "Snowflake-like" structure
generally did not occur(we call the micristructure with PFZ as
type 1 aging structure, call the "Snowflake-like" structure as
type 2 aging structure). If solution treatment was conducted at
the β-transus temperature or slightly above it(about 760°C),
type 2 aging structure was often observed. The solution treatment was conducted between the initial recrystallization and
the β-transus temperature(710-750°C) to produce a microstructure with the parts of recrystallized β-grains. After aging the
resultant structure was also a heterogeneous precipitated α-
phase within the recrystallized β-grains. Therefore, the solution treatments in 710-760°C temperature range--either above
the β-transus or below it-- and subsequent aging produced the
α-phase in the "Snowflake-like" morphology within the recrystallized β-grains. It is found that the β-transus temperature do
not play the role of a "critical point" of altering this heterogeneous precipitation.

In order to ensure that TB2 sheet has a good cold formability, usually the sheet is solution treated in β-phase field.
Numerous heat treatment tests and the metallographic observations have indicated that with the same aging condition the
amount of α-phase precipitates is increased with increasing of
the solution treatment temperature, and the dispersion degree
of α-phase precipitates is also increased. But only if the solution treatment temperature is above a definite temperature,

type 1 structure can be obtained after aging. This temperature
is called the effective quenching temperature. The effective
quenching temperature of TB2 sheet was determined to be about
780°C.

## 2. The transition temperature

A new heat treating process has been presented in this investigation: after solution treatment, material is exporsed for
a period of time in a definite temperature range, and followed
by aging. This exposure prior to aging is called the pretreatment. For example, the heat treatment for five samples were
carried out as follows respectively. Sample 1 was directly aged
for at 500°C after solution treatment. Sample 2-5 were pretreated for 2,4,6,8min at 550°C respectively, and followed by aging
for 8h at 500°C. It was found from Fig.2 that the pretreatment
had a significant effect on the microstructures although the
samples were only pretreated for a very short time. The pretreating for 2min at 550°C and subsequent aging caused the PFZ
to broader markedly(Fig.2b), and a longer time pretreatment
resulted in an heterogeneous precipitation of α-phase(Fig.2c).

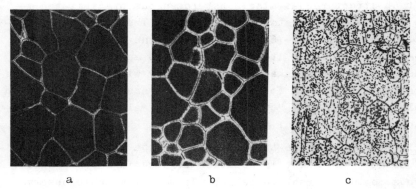

a                         b                         c

Fig.2 Effect of high temperature pretreatment
on microstructure of TB2 alloy (x200)
a. ST + 500°C,8h AC
b. ST + 550°C,2min AC + 500°C,8h AC
c. ST + 550°C,8min AC + 500°C,8h AC

After the solution treatment, the pretreatment prior to aging results in two variations in microstructure. The first one is that a short time pretreating at a lower temperature in the 450-630°C range causes the PFZ to broaden, but this pretreatment does not vary type of aging structure. The second one is that a short time pretreating at a higher temperature(630°C, 2min) or a longer time pretreating at a lower temperature (500°C,30min) produces aging structure in the "Snowflake-like" morphology, that is, the dispersive precipitation of α-phase becomes the heterogeneous one. As compared with the common STA, the higher temperature pretreatment plays the role of reducing nucleation sites and the dispersion degree of the precipitates.

The effects of the pretreating above 450°C on microstructure have been mentioned as above. But when the temperature of this pretreatment continues to be dropped to a temperature below 400°C, the results were contrary to above-mentioned ones. After the sample was solution treated in the same condition, the pretreating for 1-2h between 300-350°C, or for 20min at 400°C prior to 500°C aging did not cause the PFZ to be broaden. On the contrary, it caused the PFZ to be narrow. The higher temperature pretreatment caused the PFZ to be broaden, or resulted in the "Snowflake-like" structure. The lower temperature pretreatment caused the PFZ to be norrow and the precipitates to be fine. There is certainly a "critical point" between the two pretreating temperatures. We call this "critical point" as the transition temperature. The pretreating below the transition temperature plays the role of increasing nucleation sites of α-phase during the aging. The transition temperature of TB2 alloy was found to be about 400°C.

## 3. The practicality of the transition temperature

As noted above, for TB2 alloy, the solution treatment above β-transus prior to aging produced two types of aging structure(the discussion is only limiled to the temperature range of α-phase precipitation). Type 1 structure is a α-phase dispersive precipitated structure with the PFZ and type 2 is a "Snowflake-like" structure. By means of the pretreatment, it is found that the two types of structure could only occur in given condition with a certain regularity. For example, for the sample with type 1 structure, after solution treatment a longer time pretreatment at a higher temperature and subsequent aging could result in type 2 structure. In the following, we shall discuss the way that type 1 structure can occur in the sample with type 2 structure by control of the external condition.

Fig.3a is shown the microstructure of the forging sample which is solution treated at 770°C and aged at 500°C. It clearly demonstrates a lower density and coarser dispersion of α-phase. But the pretreatment for 1h at 300°C prior to 500°C

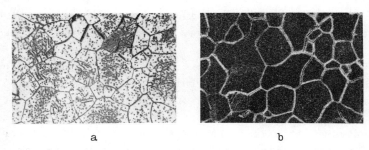

a                          b

Fig.3 Effect of low temperature pretreatment on
microstructure of TB2 forging(x200)
a. 770°C,1h AC + 500°C,8h AC
b. 770°C,1h AC + 300°C,1h AC + 500°C,8h AC

aging led to significant change in the microstructural morphology. The ordinary type 2 structure became type 1 structure, as exemplified by Fig.3b. This experiment confirms that the pretreatment below the transition temperature has a strong promoting effect on dispersive nucleation of second phase.

## Discussion

At present, there have been two mechanisms to explain the PFZ in aluminum alloys: first, the model of "denudation of solute atoms"; secondly, the model of "denudation of vacancies". We thought originally that the PFZ was caused by a heterogene-

ous distribution of alloying elements in the range of a grain
and richment of β-stabilizer near the grain boundary, but this
view is difficult to explainsome exprimental phenomena. For
example, in order to make the solution treatment enough for the
homogeneous distribution of alloying elements, the solution
treatment time was prolonged intentionally. The PFZ should have
been narrowed or eliminated due to the prolongation of solution
treatment time is favourable for the homogeneous distribution
of alloying elements. But the experimentdid not turn out to be
expected: the prolongation of the 800°C solution treatment time
from 0.5h to 5h prior to 500°C aging produced a broadened PFZ.
On the other hand, we exerted an external force upon the solution treated sample to give a small deformation at room temperature, then it was aged for 8h at 500°C. It was observed that
there also occurred the PFZ near the slip bands within the
grain(Fig.4). The slip can not result in redistribution of
alloying elements because the deformation is carried out at room
temperature. Thus, whether the distribution of alloying element
is homogeneous or not is difficult to explain these phenomena.

By means of the theory presented by R.B. Nicholson et al.
for studying Al-alloys(2), the PFZ occurred in TB2 alloy was
explained. Nicholson thought that the vacancy concentration and
its distribution in a quenching alloy play a very important
role of the homogeneous nucleation of second phase. The precipitated nucleus is some form of vacancy/soluted atom cluster,
such clusters could form during the quenching. The grain boundaries are sinks or conductors for vacanies. The zone of the

Fig.4 The PFZ near the slip bands within
the grain (x1000)

denudation of vacancies can form near the grain boundaries as
a result of the annihilation of part of vacancies to the grain
boundaries during the quenching. The effect of the annihilation
vacancies of the grain boundary results in that the concentration of vacancies near the grain boundary is lower than the
critical concentration of excess vacancies. Therefore, second
phase can not be nucleated in these zones, and the PFZ occurred.

The PFZ near slip bands within the grain could be thought
to be the zones of "the denudation of vacancies" in which the
vacancies are annihilated by slip dislocation prior to precipitation of second phase in the early stages of aging. After solution treatment in single phase field, the heat for a short
time(2min) above the transition temperature(550°C) favoures the
deposition of the vacancies near the grain boundary, that is,
the vacancies are annihilated by the grain boundary, which make
the PFZ broaden.

The pretreatment below the transition temperature has a
strong promoting effect on the nucleation of second phase. The
effect of the lower temperature pretreatment on the tensile
properties of $\beta_{\text{III}}$ and Ti-18Mo alloys had been reported by L.A.
Rosales et al.(3). They had conjectured that the effect of the
pretreatment resulted from ω-phase. While F.H. Froes et al.(4)
studied a deep hardened Ti-alloys, they also found that the
lower temperature pretreatment has a considerable effect on
tensile properties of alloy. Authors assumed that the effect
was caused by $\beta'$-phase formed during the phase separation reaction of $\beta$-solid solution. At first, we suspected that the
narrowed PFZ in the aging structure of TB2 alloy after the
pretreatment of lower temperature resulted from the effect of
the phase separation reaction. So we have studied the changes
in the low temperature aging structure for 800°C solution
treated TB2 alloy by using JEM-1000 transmission electron
microscopy. The selected aging temperatures were 400°C, 350°C
and 300°C.

The expreimental results are shown that any trace of precipitates is not found out in the sample treated for 30min at
400°C. The circular shape precipitates(Fig.5a) were observed in

the sample treated for 1-2h at 400°C. The results of the electron diffraction for the selected area have proved that the
precipitates are bcc β'-phase. The prolongation of treatment
time up to 4h caused the circular shape precipitate to become
a cross-shape one(Fig.5b). Prolongation time up to 8h produced
a cross-shape α-phase precipitates along a certain orientation
(Fig.5c). The treatment for 30min at 400°C do not result in any
trace of precipitates. The treatments of 350°C and 300°C, even
if the treatment time was prolonged up to 2h, also do not lead
to any trace of precipitates. But the pretreatment tests showed
that the pretreatment for 20min at 400°C, or for 1h at 350°C,
or for 1h at 300°C had a significant effect on the microstructure of TB2 alloy aged for 8h at 500°C.

The experimental results are indicated that the effect of
the lower temperature pretreatment is unlikely to result from
β'-phase. Before the precipitation of β'-phase, the changes
occurred in the alloy have already had the effect on the aging
results. There is no doubt the fact that the β'-phase precipitated by a longer time treatment at a lower temperature has an

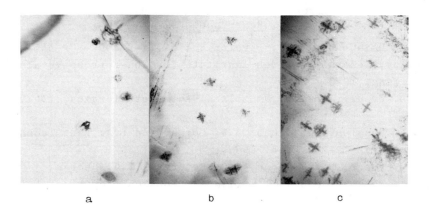

a                     b                     c

Fig.5 Structures of TB2 alloy aged at 400°C(x20000)
    a. ST + 400°C, 1-2h AC
    b. ST + 400°C, 4h AC
    c. ST + 400°C, 8h AC

effect on the final aging results. But even if there is no precipitation of $\beta'$-phase, the lower temperature pretreatment could
affect the final aging results. This problem is very interesting and is worth discussing and studying in further.

## Conclusions

The temperature parameters in TB2 alloy--the effective
quenching temperature and the transition temperature --are presented. A higher temperature pretreatment could cause the PFZ
in TB2 alloy to be broaden, and result in the change in microstructural morphology from type 1 to type 2. A lower temperature pretreatment has a strong promoting effect on the nucleation of second phase, it could allow the PFZ to be narrow.

## Acknowledgements

The author would like to thank his colleagues Mr. Wu Xiu-
Ming and Mr. Zhong Jing-Jun for technical assistance and many
helpful discussions. Thanks are also due to Mr. Zhang Zhu for
comments on the manuscript.

## References

1. Chen Hai-Shan et al.: Development of Ti-5Mo-5V-8Cr-3Al
   Alloy, General Research Institute for Non-Ferrous Metals,
   Beijing, 1973.
2. J.D. Embury and R.B. Nicholson: Acta Metallurgica, 13(1965),
   403.
3. L.A. Rosales, Kanji Ono, A.W. Sommer and L.A. Lee: Technical
   Report NA-72-232, Feb. 1972.
4. F.H. Froes, R.F. Molone, V.C. Petersen, C.G. Rhores,
   J.C. Chesnutt and J.C. Williams: Technical Report AFML-TR-
   75-41, Sept. 1975.